ENCYCLOPEDIA OF PHYSICAL SCIENCE

VOLUME I

ENCYCLOPEDIA OF PHYSICAL SCIENCE

VOLUME I

JOE ROSEN, Ph.D., and
LISA QUINN GOTHARD
Katherine Cullen, Ph.D., Managing Editor

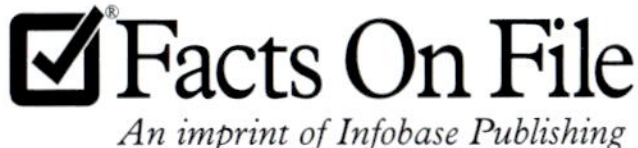

ENCYCLOPEDIA OF PHYSICAL SCIENCE

Facts On File, Inc.
An Imprint of Infobase Publishing
132 West 31st Street
New York NY 10001

Library of Congress Cataloging-in-Publication Data

Rosen, Joe, 1937–
Encyclopedia of physical science / Joe Rosen and Lisa Quinn Gothard.
p. cm.
Includes bibliographical references and index.
ISBN-13: 978-0-8160-7011-4
ISBN-10: 0-8160-7011-3
1. Physical sciences—Encyclopedias. I. Gothard, Lisa Quinn. II. Title.
Q121.R772009
500.203—dc22 2008036444

Facts On File books are available at special discounts when purchased in bulk quantities for businesses, associations, institutions, or sales promotions. Please call our Special Sales Department in New York at (212) 967-8800 or (800) 322-8755.

You can find Facts On File on the World Wide Web at http://www.factsonfile.com

Text design by Annie O'Donnell
Illustrations by Richard Garratt and Melissa Ericksen
Photo research by Suzanne M. Tibor

Printed in China

CP Hermitage 10 9 8 7 6 5 4 3 2 1

This book is printed on acid-free paper.

Contents

Acknowledgments

We would like to express appreciation to Frank K. Darmstadt, executive editor, and to Katherine E. Cullen, managing editor, for their critical review of this manuscript, wise advice, patience, and professionalism. Thank you to the graphics department for creating the line illustrations that accompany the entries in this work, to Alana Braithwaite for expertly assembling all of the manuscript materials, and to Suzie Tibor for performing the expert photo research. The guest essayist, Dr. Amy J. Heston, deserves recognition for generously donating time to share her expert knowledge about forensic chemistry. Appreciation is also extended to the production and copyediting departments and the many others who helped in the production of this project. Thank you all.

INTRODUCTION

Encyclopedia of Physical Science is a two-volume reference intended to complement the material typically taught in high school physics and chemistry and in introductory college physics and chemistry courses. The substance reflects the fundamental concepts and principles that underlie the content standards for science identified by the National Committee on Science Education Standards and Assessment of the National Research Council for grades 9–12. Within the category of physical science, these include structure of atoms, structure and properties of matter, chemical reactions, motions and forces, conservation of energy and increase in disorder, and interactions of energy and matter. The National Science Education Standards (NSES) also place importance on student awareness of the nature of science and the process by which modern scientists gather information. To assist educators in achieving this goal, other subject matter discusses concepts that unify the physical sciences with life science and Earth and space science: science as inquiry, technology and other applications of scientific advances, science in personal and social perspectives including topics such as natural hazards and global challenges, and the history and nature of science. A listing of entry topics organized by the relevant NSES Content Standards and an extensive index will assist educators, students, and other readers in locating information or examples of topics that fulfill a particular aspect of their curriculum.

Encyclopedia of Physical Science provides historical perspectives, portrays science as a human endeavor, and gives insight into the process of scientific inquiry by incorporating biographical profiles of people who have contributed significantly to the development of the sciences. Instruments and methodology-related entries focus on the tools and procedures used by scientists to gather information, conduct experiments, and perform analyses. Other entries summarize the major branches and subdisciplines of physical science or describe selected applications of the information and technology gleaned from physical science research. Pertinent topics in all categories collectively convey the relationship between science and individuals and science and society.

The majority of this encyclopedia comprises more than 200 entries covering NSES concepts and topics, theories, subdisciplines, biographies of people who have made significant contributions to the physical sciences, common methods, and techniques relevant to modern science. Entries average approximately 2,000 words each (some are shorter, some longer), and most include a cross-listing of related entries and a selection of recommended further readings. In addition, 12 essays are included covering a variety of subjects—contemporary topics of particular interest and specific themes common to the physical sciences. More than 150 photographs and more than 150 line art illustrations accompany the text, depicting difficult concepts, clarifying complex processes, and summarizing information for the reader. A chronology outlines important events in the history of the field, and a glossary defines relevant scientific terminology. The back matter of *Encyclopedia of Physical Science* contains a list of additional print and Web resources for readers who would like to explore the discipline further. In the appendixes readers can find a periodic table of the elements, some common conversions, and tables of physical constants, astronomical data, abbreviations and symbols for physical units, the Greek alphabet, and multipliers and dividers for use with SI units.

Lisa Gothard has 10 years' experience teaching high school and college level physical science, biology, and chemistry. She obtained a bachelor of science degree in molecular biology from Vanderbilt University in Nashville, Tennessee, where she received a Howard Hughes Fellowship for undergraduate research. She holds a master of science degree in biochemistry from the University of Kentucky. She has coauthored peer-reviewed journal articles in the field of gene expression and the role of chaperones in protein folding. She is currently teaching at East Canton High School in East Canton, Ohio, where she lives with her husband and four children.

Joe Rosen has been involved in physics research and teaching for more than four decades. After obtaining his doctorate in theoretical physics from the Hebrew University of Jerusalem, he did one year of postdoctoral work in the physics department of Boston University and three at Brown University. He then joined the faculty of the School of Physics and Astronomy of Tel Aviv University, where he spent the largest portion of his academic career. He has taught most of the standard undergraduate physics courses as well as many graduate-level courses. His last full-time job was chair of the Department of Physics and Astronomy of the University of Central Arkansas. Since retiring from Tel Aviv and Central Arkansas, Joe Rosen has been living in the Washington, D.C., area, where he does visiting and adjunct teaching at colleges and universities in the area, currently at The George Washington University. He has written or edited 11 books and continues writing and carrying out research in his current fields of interest: symmetry, space, time, space-time, and the quantum. His career has allowed him to keep abreast of, and even contribute to, the exciting developments in physics that have been taking place since the late 20th century.

Both authors hope that this encyclopedia will serve you as a valuable reference, and that you will learn as much from referring to it as we have from writing it.

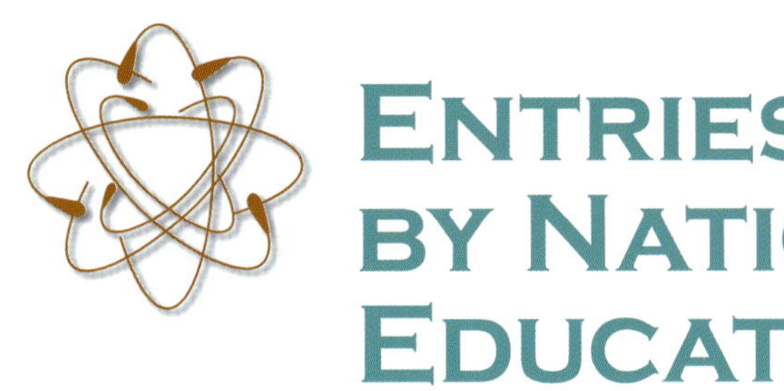

Entries Categorized by National Science Education Standards for Content (Grades 9–12)

When relevant, an entry may be listed under more than one category. For example, Stephen Hawking, who studies cosmology, general relativity, quantum theory, and much more, is listed under all of Physical Science (Content Standard B): Motions and Forces, Conservation of Energy and Increase in Disorder, Interactions of Energy and Matter; Earth and Space Science (Content Standard D); and History and Nature of Science (Content Standard G). Biographical entries, topical entries, and entries that summarize a subdiscipline may all appear under History and Nature of Science (Content Standard G), when a significant portion of the entry describes a historical perspective of the subject. Subdisciplines are listed separately under the category Subdisciplines, which is not a NSES category, but are also listed under the related content standard category.

SCIENCE AS INQUIRY (CONTENT STANDARD A)

analytical chemistry
buffers
calorimetry
centrifugation
chromatography
classical physics
colligative properties
concentration
conservation laws
Dalton, John
DNA fingerprinting
duality of nature
enzymes
Ertl, Gerhard
inorganic chemistry
inorganic nomenclature
invariance principles
laboratory safety
Lauterbur, Paul
Lavoisier, Antoine-Laurent
Mach's principle
Mansfield, Sir Peter
mass spectrometry
measurement
Mullis, Kary
nanotechnology
nuclear chemistry
organic chemistry
pH/pOH
physics and physicists
Ramsay, Sir William
rate laws/reaction rates
ribozymes (catalytic RNA)
scientific method
separating mixtures
simultaneity
solid phase peptide synthesis
solid state chemistry
stoichiometry
surface chemistry
symmetry
theory of everything
toxicology

PHYSICAL SCIENCE (CONTENT STANDARD B): STRUCTURE OF ATOMS

accelerators
acid rain
acids and bases
analytical chemistry
atomic structure
big bang theory
biochemistry
blackbody
black hole
Bohr, Niels
bonding theories
Bose-Einstein statistics
Broglie, Louis de
buffers
chemistry and chemists
citric acid cycle
classical physics
clock
Compton effect
conservation laws
Curie, Marie
duality of nature
Einstein, Albert
electron configurations
energy and work
EPR
Fermi, Enrico
Fermi-Dirac statistics
Feynman, Richard

fission
Franklin, Rosalind
fusion
Gell-Mann, Murray
gravity
Heisenberg, Werner
Hodgkin, Dorothy Crowfoot
isotopes
lasers
Lauterbur, Paul
Lavoisier, Antoine-Laurent
magnetism
Mansfield, Sir Peter
mass
matter and antimatter
Maxwell-Boltzmann statistics
Meitner, Lise
Mendeleyev, Dmitry
metallic compounds
Mössbauer effect
nuclear magnetic resonance (NMR)
nuclear physics
Oppenheimer, J. Robert
oxidation-reduction reactions
particle physics
Pauli, Wolfgang
Pauling, Linus
periodic table of the elements
periodic trends
photoelectric effect
pH/pOH
physics and physicists
Planck, Max
quantum mechanics
radioactivity
Ramsay, Sir William
representing structures/molecular models
Rutherford, Sir Ernest
Schrödinger, Erwin
special relativity
spectrophotometry
states of matter
theory of everything
Thomson, Sir J. J.

PHYSICAL SCIENCE (CONTENT STANDARD B): STRUCTURE AND PROPERTIES OF MATTER

accelerators
acid rain
acids and bases
acoustics
agrochemistry (agricultural chemistry)
air pollution (outdoor/indoor)
alternative energy sources
amino acid metabolism and the urea cycle
analytical chemistry
aspirin
atmospheric and environmental chemistry
biochemistry
blackbody
black hole
bonding theories
Bose-Einstein statistics
Boyle, Robert
buffers
carbohydrates
Cavendish, Henry
centrifugation
chemistry and chemists
chemistry in art and photography
chemistry in wine making
chromatography
cigarette smoke
citric acid cycle
classical physics
colligative properties
computational chemistry
concentration
conservation laws
covalent compounds
Crick, Francis
Crutzen, Paul
Dalton, John
Davy, Sir Humphrey
degree of freedom
diffusion
DNA fingerprinting
DNA replication and repair
Einstein, Albert
elasticity
electrical engineering
electrochemistry
electron configurations
electron emission
enzymes
equilibrium
Ertl, Gerhard
Faraday, Michael
fatty acid metabolism
Fermi-Dirac statistics
fluid mechanics
fossil fuels
Franklin, Rosalind
fullerenes
gas laws
glycogen metabolism
glycolysis
greenhouse effect
harmonic motion
heat and thermodynamics
Hodgkin, Dorothy Crowfoot
industrial chemistry
inorganic chemistry
inorganic nomenclature
ionic compounds
isomers
isotopes
Kornberg, Roger
lasers
Lauterbur, Paul
Lavoisier, Antoine-Laurent
lipids
liquid crystals
magnetism
Mansfield, Sir Peter
mass spectrometry
materials science
Maxwell, James Clerk
Maxwell-Boltzmann statistics
membrane potential
Mendeleyev, Dmitry
metallic compounds
Molina, Mario
Mössbauer effect
Mullis, Kary
nanotechnology
Newton, Sir Isaac
nondestructive evaluation
nuclear chemistry
nucleic acids
Oppenheimer, J. Robert
organic chemistry
organic nomenclature
oxidation-reduction reactions
Pauli, Wolfgang
Pauling, Linus
periodic table of the elements
periodic trends
Perutz, Max
photoelectric effect
pH/pOH
physical chemistry
physics and physicists
Planck, Max
polymers
Priestley, Joseph
proteins
quantum mechanics
reactions in aqueous solution
representing structures/molecular models

ribozymes (catalytic RNA)
Rowland, F. Sherwood
Schrödinger, Erwin
separating mixtures
solid phase peptide synthesis
solid state chemistry
special relativity
spectrophotometry
states of matter
statistical mechanics
superconductivity
surface chemistry
textile chemistry
theory of everything
toxicology
Watson, James
waves
Wilkins, Maurice
X-ray crystallography

PHYSICAL SCIENCE (CONTENT STANDARD B): CHEMICAL REACTIONS

alternative energy sources
aspirin
biochemistry
bonding theories
buffers
carbohydrates
chemical engineering
chemical reactions
chemistry and chemists
chemistry in art and photography
chemistry in wine making
citric acid cycle
computational chemistry
concentration
conservation laws
covalent compounds
Crutzen, Paul
Davy, Sir Humphrey
electrochemistry
electron transport system
enzymes
equilibrium
Ertl, Gerhard
fossil fuels
glycogen metabolism
glycolysis
greenhouse effect
heat and thermodynamics
industrial chemistry
inorganic chemistry
inorganic nomenclature
ionic compounds
isomers
isotopes
Kornberg, Roger
Lavoisier, Antoine-Laurent
liquid crystals
Maxwell, James Clerk
membrane potential
Mendeleyev, Dmitry
metabolism
metallic compounds
Molina, Mario
Nobel, Alfred
nucleic acids
nutrient cycling
organic chemistry
organic nomenclature
oxidation-reduction reactions
periodic table of the elements
periodic trends
pharmaceutical drug development
photosynthesis
pH/pOH
physical chemistry
polymers
radical reactions
rate laws/reaction rates
reactions in aqueous solution
Rowland, F. Sherwood
solid phase peptide synthesis
solid state chemistry
stoichiometry
surface chemistry
toxicology

PHYSICAL SCIENCE (CONTENT STANDARD B): MOTIONS AND FORCES

acceleration
acoustics
black hole
Bose-Einstein statistics
Broglie, Louis de
center of mass
chemistry and chemists
classical physics
clock
Compton effect
conservation laws
coordinate system
cosmology
degree of freedom
diffusion
Doppler effect
duality of nature
Einstein, Albert
elasticity
electrical engineering
electricity
electromagnetic waves
electromagnetism
electron emission
energy and work
engines
equilibrium
Fermi, Enrico
Fermi-Dirac statistics
Feynman, Richard
fission
fluid mechanics
flux
force
fusion
Galilei, Galileo
Gell-Mann, Murray
general relativity
gravity
harmonic motion
Hawking, Stephen
heat and thermodynamics
Heisenberg, Werner
Kepler, Johannes
machines
magnetism
mass
matter and antimatter
Maxwell-Boltzmann statistics
mechanical engineering
Meitner, Lise
momentum and collisions
Mössbauer effect
motion
Newton, Sir Isaac
nuclear physics
optics
particle physics
photoelectric effect
physics and physicists
Planck, Max
power
pressure
quantum mechanics
radioactivity
rotational motion
Schrödinger, Erwin
simultaneity
special relativity
speed and velocity
states of matter
statistical mechanics
superconductivity
telescopes

PHYSICAL SCIENCE (CONTENT STANDARD B): CONSERVATION OF ENERGY AND INCREASE IN DISORDER

PHYSICAL SCIENCE (CONTENT STANDARD B): INTERACTIONS OF ENERGY AND MATTER

LIFE SCIENCE (CONTENT STANDARD C)

buffers
carbohydrates
centrifugation
cigarette smoke
citric acid cycle
concentration
Crick, Francis
Crutzen, Paul
DNA fingerprinting
DNA replication and repair
electrochemistry
electron transport system
enzymes
fatty acid metabolism
Franklin, Rosalind
glycogen metabolism
glycolysis
Hodgkin, Dorothy Crowfoot
industrial chemistry
ionic compounds
isomers
Kornberg, Roger
lipids
materials science
membrane potential
metabolism
Mullis, Kary
nucleic acids
organic chemistry
oxidation-reduction reactions
Pasteur, Louis
Perutz, Max
pharmaceutical drug development
photosynthesis
pH/pOH
proteins
radical reactions
reactions in aqueous solution
ribozymes (catalytic RNA)
solid phase peptide synthesis
toxicology
Watson, James
Wilkins, Maurice

EARTH AND SPACE SCIENCE (CONTENT STANDARD D)

big bang theory
cosmic microwave background
cosmology
Crutzen, Paul
Galilei, Galileo
general relativity
gravity
green chemistry
greenhouse effect
Hawking, Stephen
Kepler, Johannes
Mach's principle
magnetism
matter and antimatter
nutrient cycling
particle physics
physics and physicists
ribozymes (catalytic RNA)
telescopes
theory of everything
time
toxicology

SCIENCE AND TECHNOLOGY (CONTENT STANDARD E)

accelerators
acid rain
acids and bases
acoustics
agrochemistry (agricultural chemistry)
alternative energy sources
aspirin
centrifugation
chemical engineering
chemistry in art and photography
chemistry in wine making
computational chemistry
Crick, Francis
Crutzen, Paul
DNA fingerprinting
electrical engineering
electron emission
engines
enzymes
Ertl, Gerhard
Faraday, Michael
Fermi, Enrico
fission
force
fossil fuels
Franklin, Rosalind
fullerenes
fusion
graphing calculators
greenhouse effect
Heisenberg, Werner
Hodgkin, Dorothy Crowfoot
inorganic chemistry
isotopes
Kornberg, Roger
lasers
Lauterbur, Paul
liquid crystals
machines
magnetism
Mansfield, Sir Peter
mass spectrometry
materials science
Meitner, Lise
Molina, Mario
Mullis, Kary
nanotechnology
Nobel, Alfred
nondestructive evaluation
nuclear chemistry
nuclear magnetic resonance (NMR)
nuclear physics
nucleic acids
Oppenheimer, J. Robert
Pasteur, Louis
pharmaceutical drug development
photoelectric effect
polymers
radioactivity
Rowland, F. Sherwood
solid phase peptide synthesis
solid state chemistry
spectrophotometry
surface chemistry
telescopes
textile chemistry
Wilkins, Maurice

SCIENCE IN PERSONAL AND SOCIAL PERSPECTIVES (CONTENT STANDARD F)

acid rain
acids and bases
acoustics
agrochemistry (agricultural chemistry)
air pollution (outdoor/indoor)
alternative energy sources
analytical chemistry
aspirin
atmospheric and environmental chemistry
buffers
chemical engineering
cigarette smoke
Crick, Francis
Crutzen, Paul
DNA fingerprinting
electrochemistry
fission
fossil fuels
fusion

green chemistry
greenhouse effect
Lauterbur, Paul
Mansfield, Sir Peter
Molina, Mario
nuclear chemistry
nucleic acids
nutrient cycling
Oppenheimer, J. Robert
oxidation-reduction reactions
Pasteur, Louis
pharmaceutical drug development
photosynthesis
pH/pOH
polymers
radical reactions
radioactivity
Rowland, F. Sherwood
textile chemistry
toxicology

HISTORY AND NATURE OF SCIENCE (CONTENT STANDARD G)

alternative energy sources
atomic structure
big bang theory
biophysics
Bohr, Niels
bonding theories
Boyle, Robert
Broglie, Louis de
Cavendish, Henry
classical physics
clock
cosmic microwave background
cosmology
Crick, Francis
Crutzen, Paul
Curie, Marie
Dalton, John
Davy, Sir Humphry
duality of nature
Einstein, Albert
electrical engineering
Ertl, Gerhard
Faraday, Michael
Fermi, Enrico
Feynman, Richard
fission
fossil fuels
Franklin, Rosalind
fusion
Galilei, Galileo
gas laws
Gell-Mann, Murray
general relativity
graphing calculators
green chemistry
greenhouse effect
Hawking, Stephen
Heisenberg, Werner
Hodgkin, Dorothy Crowfoot
invariance principles
Kepler, Johannes
Kornberg, Roger
laboratory safety
lasers
Lauterbur, Paul
Lavoisier, Antoine-Laurent
Mansfield, Sir Peter
materials science
matter and antimatter
Maxwell, James Clerk
mechanical engineering
Meitner, Lise
Mendeleyev, Dmitry
Molina, Mario
motion
Mullis, Kary
Newton, Sir Isaac
Nobel, Alfred
nuclear chemistry
nuclear magnetic resonance (NMR)
Oppenheimer, J. Robert
particle physics
Pasteur, Louis
Pauli, Wolfgang
Pauling, Linus
periodic table of the elements
Perutz, Max
physics and physicists
Planck, Max
polymers
Priestley, Joseph
quantum mechanics
radioactivity
Ramsay, Sir William
representing structures/molecular models
ribozymes (catalytic RNA)
Rowland, F. Sherwood
Rutherford, Sir Ernest
Schrödinger, Erwin
telescopes
textile chemistry
theory of everything
Thomson, Sir J. J.
Watson, James
Wilkins, Maurice

SUBDISCIPLINES

agrochemistry (agricultural chemistry)
analytical chemistry
atmospheric and environmental chemistry
biochemistry
biophysics
chemical engineering
computational chemistry
cosmology
electrical engineering
electrochemistry
green chemistry
industrial chemistry
inorganic chemistry
inorganic nomenclature
materials science
mechanical engineering
nuclear chemistry
nuclear physics
organic chemistry
particle physics
physical chemistry
solid phase peptide synthesis
solid state chemistry
surface chemistry
textile chemistry
toxicology

Entries A–L

acceleration Instantaneous acceleration, the phenomenon usually meant by the word *acceleration,* is the time rate of change of velocity, the change of velocity per unit time. Like velocity, acceleration is a vector, having both magnitude and direction. In a formula

$$\mathbf{a} = \frac{d\mathbf{v}}{dt}$$

where **a** denotes the acceleration, **v** the velocity, and t the time. The SI unit of acceleration is meters per second per second (m/s^2). Velocity is expressed in meters per second (m/s), and time is in seconds (s). The derivative in this formula is the limit of a finite change of velocity divided by a finite time interval as the time interval approaches zero. In other words, the instantaneous acceleration is the limit of the average acceleration for vanishing time interval, where we now define average acceleration.

The average acceleration $\mathbf{a}_{av}$ over the time interval between times t_1 and t_2 equals the change in velocity between the two times divided by the elapsed time interval:

$$\mathbf{a}_{av} = \frac{\Delta\mathbf{v}}{\Delta t} = \frac{\mathbf{v}(t_2) - \mathbf{v}(t_1)}{t_2 - t_1}$$

Here $\mathbf{v}(t_1)$ and $\mathbf{v}(t_2)$ denote the velocities at times t_1 and t_2, respectively. As mentioned, the instantaneous acceleration is obtained from the average acceleration by taking the limit of t_2 approaching t_1.

In a simple situation, the motion of a body might be constrained to one dimension, along a straight line. In that case we may drop the vector notation and use the formula

$$a = \frac{dv}{dt}$$

where a and v denote, respectively, the (positive or negative) acceleration and velocity along the line of motion. Thus, being a derivative, the acceleration at any instant possesses the value of the slope of the graph of velocity as a function of time at that instant. An ascending function, representing increasing velocity, has positive slope, thus positive acceleration. Similarly, the slope of a descending function, describing decreasing velocity, is negative, indicating negative acceleration. At those instants when the slope is zero, the acceleration is momentarily zero.

Since the velocity is the time derivative of the position function, for one-dimensional motion we have

$$v = \frac{dx}{dt}$$

where x denotes the position as a function of time in meters (m). The acceleration is then the second derivative of the position function

$$a = \frac{d^2x}{dx^2}$$

In the case of motion in one dimension the formula for average acceleration becomes

$$a_{av} = \frac{\Delta v}{\Delta t}$$

where Δv is the (positive or negative) change in velocity during time interval Δt. If, moreover, the acceleration a is constant (i.e., the velocity changes [increases

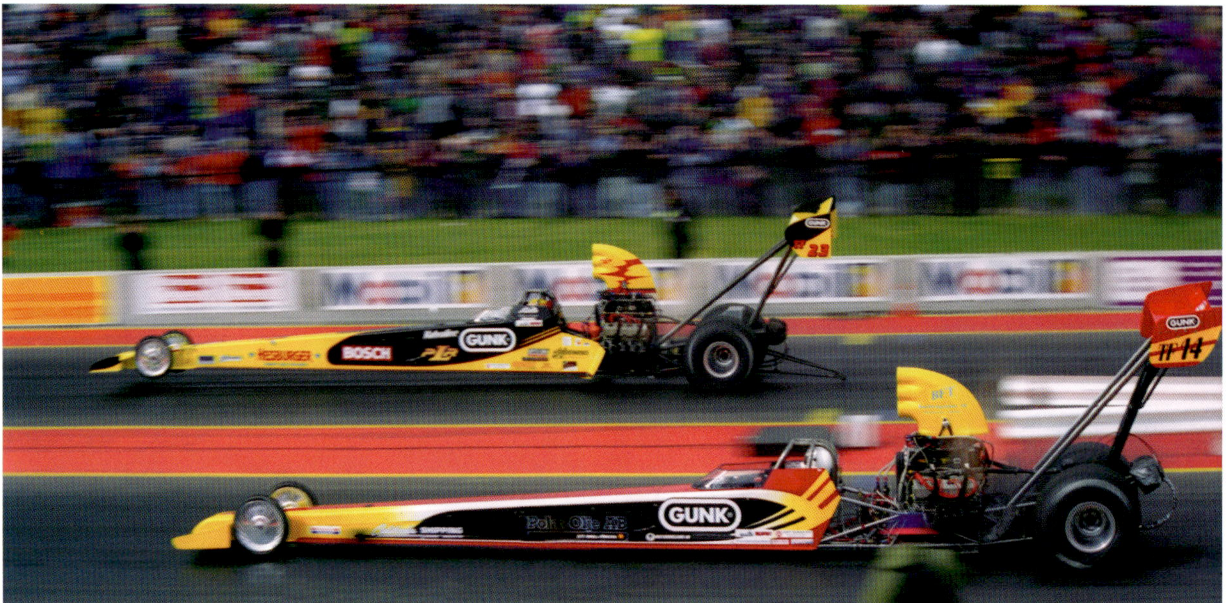

A drag race takes place over a very short distance, usually a quarter of a mile. Thus, it is not the top speed of the racing car that is important, since there is insufficient time for the car to reach its top speed. Rather, it is the car's acceleration, the rate at which it increases its speed, that is all-important in a drag race. *(Tony Watson/Alamy)*

or decreases] by equal increments during equal time intervals), the relation

$$v = v_0 + at$$

will give the velocity at time t, where v_0 denotes the value of the velocity at time $t = 0$. The position at time t will then be given by

$$x = x_0 + v_0t + \tfrac{1}{2}at^2$$

where x_0 is the position at time $t = 0$.

A more restricted use of the term *acceleration* is employed to indicate an increase of speed, as opposed to deceleration, which is a decrease of speed. In this usage deceleration can be thought of as negative acceleration. However, both are included in the general meaning of acceleration that was explained earlier.

Alternatively, the velocity might change in its direction but not in its magnitude, which is motion at constant speed. A simple case of that is constant-speed circular motion. Then the direction of the acceleration is toward the center of the circular path—that is called centripetal acceleration—and the constant magnitude of acceleration, a, is given by

$$a = \frac{v^2}{r}$$

where v denotes the speed in meters per second (m/s) and r is the radius of the circular path in meters (m).

In general, though, acceleration can involve change in both the magnitude and the direction of the velocity.

An important acceleration is the acceleration of a body in free fall near the surface of Earth, also called the acceleration due to gravity, or gravitational acceleration. All bodies dropped from rest at any location near the surface of Earth are found to fall with the same constant acceleration (ignoring the resistance of the air). The magnitude of this acceleration is conventionally denoted by g and has the approximate value 9.8 meters per second per second (m/s^2). Its precise value depends on a number of factors, such as geographical latitude and altitude, as well as the density of Earth's surface in the vicinity of the falling body.

When calculating the acceleration of falling objects one must take into consideration the force of gravity acting on the object. Aristotle first described the motion of falling objects incorrectly by saying that the speed at which an object falls is proportional to the mass of the object. This means that the heavier the object the faster that it would fall, and the lighter the object the slower it would fall. That appeared to be logical, based on what most observers see when an object falls. If a piece of paper and a brick are both dropped at the same time, one can see that the brick falls to the ground more quickly than the paper. Observations such as this led Aristotle and many others to believe that the heavier object falls faster because of its greater mass. The work of the Italian scientist Galileo Galilei in the 1600s showed that the differences in rates of fall were not due to the different masses of the objects, but to the air resistance that acted upon the objects. The paper has a large surface area and is strongly affected by air resistance, so that force of friction slows down the paper. The compact brick is not as affected by the air resistance and therefore falls practically unimpeded and reaches the ground first. If two balls of equal size and shape but different masses are dropped from the same height, they both reach the ground at the same instant. If the paper and the brick from the first example were dropped in a vacuum (in the absence of air), there would be no air resistance and the objects would both fall at the same rate.

The English scientist Sir Isaac Newton's second law of motion states that the magnitude of an object's acceleration a is directly proportional to the magnitude of the net force applied F and inversely proportional to the mass m of the object:

$$a = \frac{F}{m}$$

Here F is in newtons (N) and m in kilograms (kg). When two objects fall, the object with greater mass is acted on by a larger force of gravity F, according to Newton's law of gravitation. Intuitively it would seem that the object with greater mass would have a greater acceleration, but because F is proportional to m, the mass of the object cancels out in the ratio F/m, so that the acceleration remains the same. Thus the brick, the paper, and all other objects on Earth will accelerate at the same rate at the same location, approximately 9.8 m/s^2.

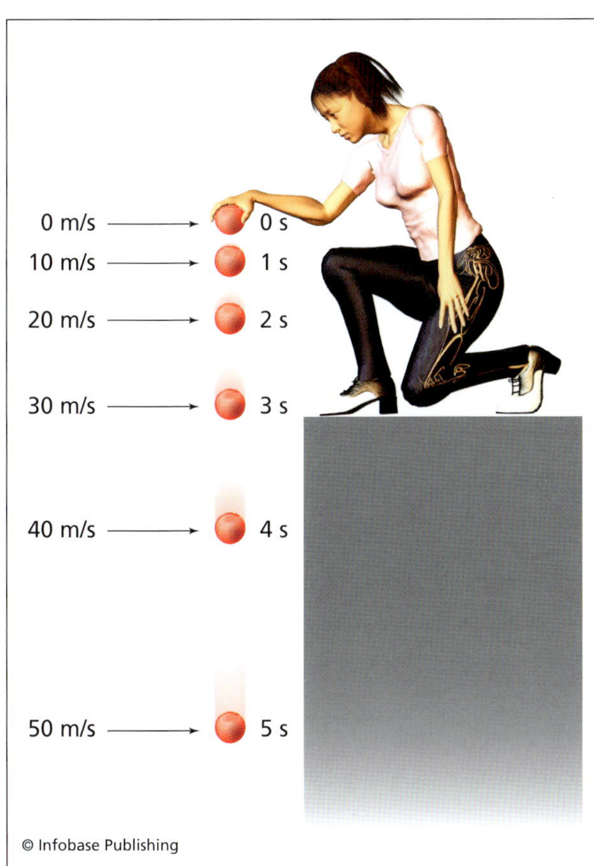

The rate at which an object falls depends on the acceleration due to gravity, 9.8 m/s^2, rounded to 10 m/s^2 in this illustration.

Free fall is a special type of motion that occurs when an object falls under the force of gravity and experiences no air resistance. When observing a freely falling object in slow motion, one will notice that over equal periods the falling object does not travel equal distances. These distances increase the longer that the object falls. Using the formulas presented previously, the speed of a freely falling object at time t after it starts falling from rest is given by

$$v = gt$$

where g denotes the acceleration due to gravity (with the value of approximately 9.8 m/s^2 at the surface of Earth). The distance d that the object falls in time t, again starting from rest, is

$$d = \tfrac{1}{2}gt^2$$

In the real world, however, falling objects do not truly fall freely, since air resistance exerts an opposing force, a force whose magnitude depends on the object's speed and increases with increasing speed. So a falling object will eventually reach a speed at which the magnitude of the upward air resistance force equals the magnitude of the downward force of gravity acting on the object (i.e., the object's weight), giving a net force of zero. According to Newton's second law of motion, the object's acceleration is then zero. Thus, the object will continue falling at a constant speed, called the terminal speed, or terminal velocity, which is therefore the maximal speed a passively falling object can achieve. Skydivers, birds, and other "freely falling" objects experience terminal velocity. In order to accelerate once that velocity is reached, the falling object must change its shape to reduce air resistance. Birds can pull in their wings and skydivers can pull in their arms to "nosedive," creating less air resistance and thus increasing their terminal speed.

See also FORCE; GALILEI, GALILEO; GRAVITY; MOTION; NEWTON, SIR ISAAC; SPEED AND VELOCITY; VECTORS AND SCALARS.

FURTHER READING

Griffith, W. Thomas. *Physics of Everyday Phenomena.* 5th ed. New York: McGraw Hill, 2006.

accelerators A particle accelerator is a device for giving electrically charged particles considerable kinetic energy by accelerating them to high speeds. The accelerated particles can be either subatomic particles or charged atoms (i.e., ions). Acceleration is achieved electromagnetically by means of static and/or time-varying electric and/or magnetic fields. The particles are accelerated in a vacuum, in order to prevent their colliding with air molecules and thus deviating from the desired direction of motion and dissipating their kinetic energy. The uses of particle accelerators are many and varied, including research, industrial, and medical applications. For example, researchers use particle accelerators to study the properties of the elementary particles and their interactions, as well as the properties of nuclei. Industrial applications include, for instance, the production of various kinds of radiation for uses such as sterilization, imaging, and the manufacture of computer chips. Medical uses involve mainly the treatment of cancer.

There are two types of particle accelerators: electrostatic and nonelectrostatic. Electrostatic accelerators accelerate charged particles by means of a force that acts on them that is due to a constant electric field, which is produced by generating and maintaining a constant high voltage. This type of accelerator accelerates particles in a continuous beam. The maximal voltage that can be achieved limits the particle energies it can reach. Van de Graaff accelerators belong to this category.

Alternatively, charged particles can be accelerated by time-varying electric and magnetic fields in various ways, as described in the following. Accelerators of this type are either linear accelerators or circular accelerators. Linear accelerators, or linacs, use only electric fields. With many electrodes arrayed along their length, these accelerators accelerate particles in bunches from one electrode to the next through the accelerator tube by means of precisely timed varying electric fields. The voltages between

A view of some of the superconducting magnets of the Relativistic Heavy Ion Collider particle accelerator at Brookhaven National Laboratory. As ionized gold atoms travel along the collider's 2.3-mile-long tunnel at nearly the speed of light, 1,740 of these magnets guide and focus the particle beams. *(Image courtesy of Brookhaven National Laboratory)*

adjacent electrodes do not have to be especially high, so the limitation of the constant-voltage linear accelerator is avoided, but the cumulative effect of all the electrodes is to boost the particles to high energy. There is no limit in principle to the energy that can be achieved; the longer the accelerator, the higher the energy. However, there are practical limitations to the accelerator's length.

Accelerating the particles in a circular, rather than linear, path avoids the length limitation of linacs, allowing even higher energies to be reached. The use of magnetic fields, which might be constant or might vary in time, maintains the circular path. The acceleration itself is accomplished by time-varying voltages between adjacent pairs of the electrodes that are positioned around the path. Such accelerators go by the names *cyclotron* and *synchrotron*. The maximal achievable magnetic field is a major limiting factor in achieving high energies in such devices. The use of superconducting magnets allows higher magnetic fields and consequently higher energies than are attainable by conventional magnets.

To give an example of what is being accomplished, the Tevatron accelerator at the Fermi National Accelerator Laboratory (Fermilab) in Batavia, Illinois, can accelerate protons to an energy of about 1 tera-electron volt (TeV) = 1×10^{12} electron volts (eV) and has a diameter of approximately one and a quarter miles (2 km). The Large Electron-Positron collider at the European Organization for Nuclear Research (CERN) near Geneva, Switzerland, could achieve around 50 giga-electron volts (GeV) = 50×10^9 eV for electrons and its diameter is about 5.3 miles (8.5 km). CERN will soon start operating its Large Hadron Collider, which will be able to collide beams of protons at an energy of 14 TeV = 14×10^{12} eV and beams of lead nuclei at a collision energy of 1,150 TeV = 1.15×10^{15} eV.

Accelerator engineers have designed many variations and combinations of particle accelerators in order to optimize them for the acceleration of particular particles and adapt them to specific applications. Particles might first be accelerated to some energy in a linear accelerator and then be injected into a circular accelerator for further acceleration. A related device is the storage ring, in which previously accelerated particles are kept in circular orbits. Counter-rotating beams of oppositely charge particles, such as of electrons and positrons, can be maintained in the same storage ring and made to collide with each other again and again.

See also ACCELERATION; ELECTRICITY; ELECTROMAGNETISM; ENERGY AND WORK; MAGNETISM; PARTICLE PHYSICS; SPEED AND VELOCITY; SUPERCONDUCTIVITY.

FURTHER READING

CERN—European Organization for Nuclear Research home page. Available online. URL: http://public.web.cern.ch/Public/Welcome.html. Accessed April 23, 2008.

Fermilab—Fermi National Accelerator Laboratory home page. Available online. URL: http://www.fnal.gov. Accessed April 23, 2008.

acid rain Acid rain occurs when precipitation falls in a form such as rain, snow, or sleet that is more acidic than normal rainfall. The identification of acid rain as an environmental problem began in England in the 1860s. Robert Angus Smith, a Scottish chemist, coined the term *acid rain* and was the first to correlate the production of pollution from factories with an effect on the environment.

CHEMISTRY OF ACID RAIN

The most common way to measure acidity uses the pH scale, a logarithmic scale between 0 and 14 that represents the amount of hydrogen ions (H^+) in solution. The formula for calculating pH is

$$pH = -\log[H^+]$$

The more H^+ the substance releases, the lower the pH value. A pH below 7 is acidic, equal to 7 is neutral, and above 7 is basic. Natural rainwater registers a pH in the acidic range. Carbon dioxide (CO_2) naturally released into the atmosphere reacts with water to create carbonic acid, leading to the lower than neutral pH of normal rainfall at approximately 5.0–5.6.

Sulfur dioxide (SO_2) and nitrogen oxides (NO_x) contribute to the characteristically low pH of acid rain. SO_2 is a covalent molecule formed between sulfur, with six electrons in its outer energy level (valence electrons), and oxygen, which also has six valence electrons. Sulfur and oxygen react according to the following equation:

$$S + O_2 \rightarrow SO_2$$

Nitrogen has a valence number of 5 and reacts with oxygen by forming covalent bonds. Several nitrogen oxides, including nitrogen monoxide (NO) and nitrogen dioxide (NO_2), form according to the following equation:

$$N_2 + O_2 \rightarrow NO_x$$

SO_2 and NO_2 are not damaging molecules on their own. The decrease in pH seen in acid rain occurs when SO_2 and NO_2 react with the water present in the atmosphere, ultimately forming the product sulfuric

Water vapor rises from the two stacks of the emission scrubber system at Tennessee Valley Authority's Cumberland Fossil Plant near Clarksville, Tennessee, November 20, 2002. Plants such as these are sources of pollution. *(AP Images)*

acid (H_2SO_4) or nitric acid (HNO_3). These acids fall to earth in the form of acid precipitation that can dramatically affect the environment.

AFFECTED POPULATIONS

While the acid rain problem was originally characterized in England, it presently affects every industrialized country as well as its neighbors. In response to local pollution issues, many factories increased the height of their smokestacks to reduce the problem in the local area. The increased height of smokestacks caused the emissions to blanket a much larger region. The sulfur dioxides and nitrogen oxides travel with the wind patterns and spread around the world. The combination of the high concentration of factories in the northeastern United States and the wind carrying sulfur dioxides and nitrogen dioxides from the Midwest compounds the problem of acid rain in the northeastern region of the United States. Europe and parts of Asia also feel the effects of acid rain. The regional composition of the soil determines how the area is impacted by the lowered pH. The presence of lime (which is basic) in the soil neutralizes the acid and decreases damage caused by acid rain. Areas with soil that does not have this buffering capacity suffer more dramatic effects on plants, animals, and buildings.

SOURCES OF POLLUTION

Sulfur dioxides are produced during the burning of fossil fuels, primarily coal. Burning the coal releases sulfur, which reacts with atmospheric oxygen to produce SO_2. Oil and gas also release SO_2 during burning, and volcanoes release it naturally. The quantity of sulfur found in coal is dependent on where in the country the coal originates. Coal from the eastern United States contains considerably more sulfur than the coal found in the western United States. For coal users in the northeast regions of the United States, the lower sulfur content of the coal in western states is offset by the increased costs of transporting this coal.

Automobiles are the primary producers of NO_x in the forms of NO and NO_2. More vehicles on the road produce more NO_x. Older style catalytic converters do not efficiently reduce the levels of nitrogen oxides produced by burning gasoline. The continued replacement of older cars with cars that have the more modern catalytic converters contributes to the reduction in the levels of NO_x.

EFFECTS OF ACID RAIN

Acid rain affects vegetation in a variety of ways. The lowered pH of rainwater destroys the waxy cuticle on the leaves of a plant, interfering with the plant's ability to control its water loss via transpiration. Destruction of the leaf structure compromises the plant's control of oxygen and carbon dioxide levels inside the plant. This interferes with the plant's ability to undergo photosynthesis and respiration.

Acid rain also impacts plants by affecting the pH of the soil. The addition of extra hydrogen ions causes many important nutrients, including potassium, nitrogen, and phosphorus, to leach from the soil before the plants have a chance to utilize them, leading to nutrient deficiencies. The acid rain also increases the amount of toxic metals released from the soil. The most common of these metals, mercury and aluminum, are harmful to the plants as well as animals. Some plants depend on an important mutualistic relationship with a fungus that grows on their roots. The fungus-plant associations known as mycorrhizae dramatically increase the surface area of the plant root and enable it to absorb more nutrients. When the pH of the soil decreases, the fungi die, decreasing the plant's ability to take up water and nutrients.

Acid rain damage to *Cleopatra's Needle* in Central Park *(Danger Jacobs, 2008, used under license from Shutterstock, Inc.)*

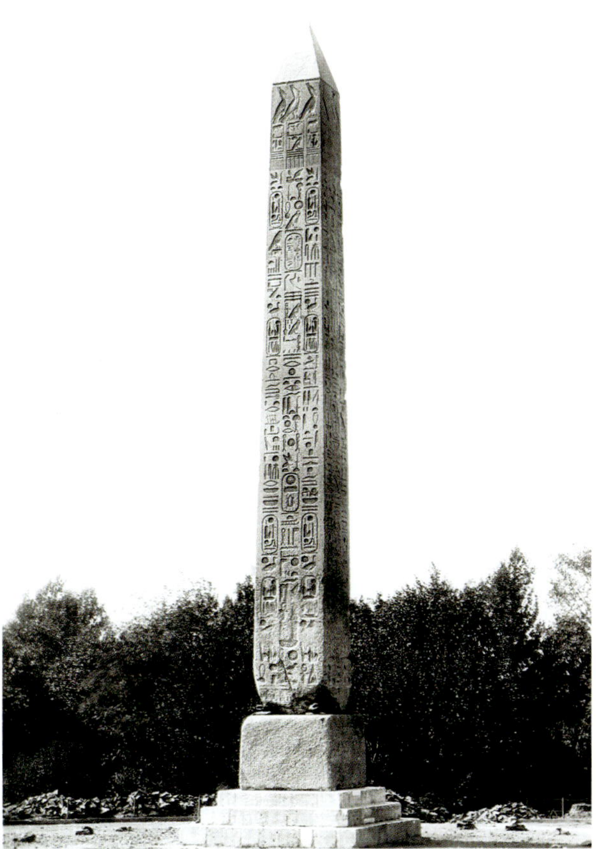

The obelisk *Cleopatra's Needle,* created ca. 1500 B.C.E., is shown here in 1881 in Central Park, New York City. *(© Bettmann/CORBIS)*

Though acid rain affects all animals, aquatic animals feel the effects most directly. As the pH decreases, many aquatic organisms cannot survive. Most fish die at a pH lower than 4.6 as a result of interference with their gill function. In freshwater ecosystems, the microorganisms responsible for decomposing organic material in the water cannot survive in acidic conditions, and without them, the decaying material builds up on the lake bottom.

As in the soil, the levels of metals like mercury and aluminum rise in the more acidic water, allowing toxic substances to build up more quickly than normal. The buildup harms humans and other animals that eat fish from these lakes and streams. By consuming plants and other aquatic animals from an acidic environment, they also suffer the effects of the acid rain. Aluminum exposure has been linked to Alzheimer's disease, and mercury shows a cumulative effect on the nervous system.

From a financial perspective, acid rain devastates the infrastructure of cities and towns in the affected areas. The cost and impact on bridges and buildings in areas suffering from acid rain exceed billions of dollars. Acid rain increases the rate of oxidation of iron, leading to rusting of bridges. H_2SO_4 and HNO_3 react with the carbonates (XCO_3) in limestone and marble according to the following reactions:

$$H_2SO_4 + CaCO_3 \rightarrow CaSO_4 + H_2CO_3$$

$$HNO_3 + CaCO_3 \rightarrow CaNO_3 + H_2CO_3$$

The breakdown of limestone and marble leads to the deterioration of buildings, statues, and other structures. Many historical structures in Europe and the United States are in danger of being destroyed.

WHAT IS BEING DONE TO STOP ACID RAIN?

As countries became aware of the damaging effects of sulfur dioxides and nitrogen oxides, many governments began to legislate the emission levels of these compounds. In 1988 the United Nations started the Long Range Transboundary Air Pollution Agreement, requiring the reduction of NO_2 emissions to 1987 levels. In 1990 the United States amended the Clean Air Act of 1970, requiring the SO_2 from plants to be reduced to 10 million tons by January 1, 2000. The Clean Air Act led to the National Acid Precipitation Assessment Program (NAPAP), which studied the levels of SO_2 and NO_x as well as their effect on the environment. In 1991 the NAPAP found that 5 percent of New England lakes were acidic. President George Bush passed the Clear Skies Legislation in 2003, demanding a reduction in power plant emissions of SO_2, NO_x, and mercury. The law requires that SO_2 and NO_x be cut by 73 percent from 2000 levels. According to the Environmental Protection Agency (EPA), the present level of sulfur allowed in diesel fuel is 500 parts per million (ppm); the limit for gasoline is presently 350 ppm. In 2006 regulations required reductions to only 15 ppm diesel and 30 ppm gasoline.

USEFUL TECHNOLOGY

The need to reduce the levels of gases that cause acid rain dramatically prompted the conception of new technology. Aware of the destructive effect of SO_2 and NO_2 on the environment, scientists have developed new products that prevent the production of the oxides or that eliminate them after production before they are released into the environment. One result is the invention of stack scrubbers or flue gas desulfurization technology. All newly constructed power factories must include a scrubber to be in compliance with emission standards; older factories require retrofitting for their smokestacks in order to be in compliance with these standards. The installation of scrubbers costs companies a significant sum, so not all factories that need them are using them. In 1987 only 20 percent of U.S. coal-powered plants used scrubbers, while in Japan, 85 percent were equipped with scrubbers. Stack scrubbers help to remove sulfuric and nitric acid from the discharged gases prior to release. The majority of scrubbers in use today are wet scrubbers, which often use a calcium carbonate slurry that reacts with SO_2 to form calcium sulfite ($CaSO_3$).

$$CaCO_3 + SO_2 \rightarrow CaSO_3 + CO_2$$

Oxidation converts calcium sulfite into gypsum, a substance useful in the production process for the building industry.

Technology to remove sulfur from the fuels prior to burning, called desulfurization of fuels, is in development. The goal is to prevent the sulfur from being released into the environment, reducing the amount of damage caused by acid rain. Once the pH is lowered, there are limited methods for correcting the problem. Adding lime to lakes and streams that have been affected by acid rain raises the pH by neutralization.

Alternative fuels are becoming increasingly important to our society. Dependency on fossil fuels has serious political and economic as well as environmental effects. The replacement of fossil fuels as an energy source will result in decreased production of sulfur dioxides and nitrogen oxides, leading to a healthier environment.

See also ACIDS AND BASES; ALTERNATIVE ENERGY SOURCES; CHEMICAL REACTIONS; FOSSIL FUELS; PH/POH.

FURTHER READING

"Acid Rain." From Microsoft Encarta Online Encyclopedia. 2008. Available online. URL: http://encarta.msn.com. Accessed July 23, 2008.

Brimblecombe, Peter. *Air Composition and Chemistry.* New York: Cambridge University Press, 1996.

acids and bases Acids are substances that give many of the foods we know their sour taste. Vinegar contains acetic acid, and lemons and other citrus fruits contain citric acid. Regulating the amount of acids present in many solutions, a characteristic known as pH, is critical to the proper functioning of living things. The human body tightly regulates the acidity of the blood, and the health of lakes, rivers, and streams depends on the maintenance of an optimal pH. A pH level that is too low, or too acidic, can kill the organisms living in the water. Decreased pH in rainwater, acid rain, leads to the deterioration and destruction of many roads, bridges, statues and buildings as well as the erosion of geological structures. Solutions that are acidic have pH values below seven. Bases are compounds that are bitter to the taste and slippery to the touch. Solutions of many common household chemicals like baking soda are basic. Bases are found in many cleaning supplies, soaps, and compounds like bleach and ammonia. Solutions that are basic have a pH value above seven and are known as alkaline. Both acids and bases change the color of chemical indicators.

The first definition of an acid and a base was that of the Swedish chemist Svante Arrhenius in the 1880s. The Arrhenius definition of an *acid* is a substance that donates a proton (H^+) in aqueous solution. An aqueous solution is one in which the solute is dissolved in water as the solvent. An Arrhenius base is a substance that donates a hydroxide ion (OH^-) in aqueous solution. This is the most restrictive of all definitions because it requires that the acid or base be dissolved in water. If the acid or base is not dissolved in water, then the definition does not hold. When the hydrogen ion is removed from the acid, it reacts with a molecule of water to form a hydronium ion, H_3O^+. In 1923 two scientists, Johannes Brønsted and Thomas Lowry, independently used different parameters to define acids and bases. According to the Brønsted-Lowry definitions, an acid is any substance that donates a proton, and a base is any substance that accepts a proton. This broadened the definitions to include compounds such as ammonia as a base, since the compound NH_3 is capable of accepting a proton to become NH_4^+. The reaction of hydrochloric acid (HCl) with ammonia yields NH_4^+ and chloride ion (Cl^-).

$$HCl + NH_3 \rightarrow NH_4^+ + Cl^-$$

An American chemist, Gilbert Newton Lewis, further expanded the definitions of acids and bases, when he defined a Lewis acid as a substance that is able to accept electrons and a Lewis base as a substance capable of donating electrons. The Lewis definition does not require an acid to contain a proton, making it the broadest of all the different ways to define an acid.

Many substances can act as either acids or bases, depending on the other reacting substances that are present in the reaction. When an acid reacts, it requires a base to accept its proton, and vice versa. Water can act as an acid in some reactions (with a strong base) and as a base in other reactions (with a strong acid). Substances that can act as both acids and bases are known as amphoteric. The reactions of water with a strong acid and a strong base are shown in the following. Notice that in the first case water is accepting the proton and is acting as a base, while in the second reaction water is donating a proton and acting as an acid.

$$H_2O + HCl \rightarrow H_3O^+ + Cl^-$$

$$CO_3^{2-} + H_2O \rightarrow HCO_3^- + OH^-$$

The names of inorganic acids, acids that do not contain carbon, are derived from the name of the anion that is present in the acid. When the anion ends with *-ic,* the name of the acid contains *hydro-* as a prefix, *-ic* as a suffix, and acid as an ending. For instance, HBr is hydrobromic acid. Anions whose names end in *-ite* form acids that end in the suffix *-ous.* The acid HNO_2 contains the anion nitrite and takes the name *nitrous acid.* When the anion ends in *-ate,* the acid has *-ic* as a suffix. The acid HNO_3 contains nitrate as an anion and is named *nitric acid.* The prefixes in the anion are kept when the acid is named. For instance, perchloric acid, $HClO_4$, contains the perchlorate anion that ends with *-ate.*

The names of Arrhenius bases, those that contain an OH^-, contain the cation name followed by the term *hydroxide.* For example, $Ca(OH)_2$ is named *calcium hydroxide* and NaOH is named *sodium hydroxide.*

When an acid reacts, the product is called a conjugate base. When a base reacts, the substance produced is known as a conjugate acid. The strength of a conjugate acid-base pair depends on the strengths of the original acid and base. When a strong acid dissociates to give a conjugate base, the conjugate base is a weak base. When a weak acid dissociates to give a conjugate base, the conjugate base is a strong base.

The strength of an acid or a base is determined by the degree of dissociation that occurs in solution. How much of the time does the acid spend in the protonated versus the deprotonated state? Strong acids spend their time entirely dissociated into ions. When dissolved in water, the undissociated state is not present, and the reaction lies completely to the right with the reverse reaction not considered.

$$HA \rightarrow H^+ + A^-$$

HA is the protonated form of the acid, and A^- is the conjugate base of the acid. These acids are considered electrolytes because as dissociated ionic compounds they will conduct an electric current. There are many types of strong acids including hydrochloric acid (HCl), sulfuric acid (H_2SO_4), and nitric acid (HNO_3). Remembering that a strong acid or base has a very weak conjugate base or acid (respectively), if HA is a strong acid, then A^- must have very weak basic tendencies. When calculating the pH of solutions of a strong acid, consider the H^+ concentration to be equal to the concentration of the acid itself. Calculating the pH of strong bases assumes the OH^- concentration to be equal to the concentration of the original base itself. Concentrated values in molarity (M, moles per liter) for a few important acids are listed in the following:

Concentrated HCl: 12 M

Concentrated H_2SO_4: 18 M

Concentrated HNO_3: 15.8 M

Weak acids and bases do not completely dissociate; therefore, the reactions of these acids and bases require the use of the equilibrium arrows demonstrating the reversibility of these reactions. A measure of the dissociation, and thereby the strength, of a weak acid is given by the acid-dissociation constant (K_a).

$$K_a = \frac{[H^+][A^-]}{[HA]}$$

A larger value of K_a indicates that more of the acid is dissociated, and, therefore, it indicates a strong acid.

Weak bases work similarly, and the measure of dissociation of a weak base is given by the base-dissociation constant (K_b)

$$K_b = \frac{[OH^-][B^+]}{[BOH]}$$

where B^+ is the conjugate acid of the weak base BOH. The higher the value for K_b, the more the base has dissociated, meaning the base is stronger. K_a and K_b exhibit an inverse relationship. A stronger acid leads to a weaker conjugate base, and vice versa, and their product is equal to K_w, the dissociation constant of water.

$$K_a \times K_b = K_w = 1 \times 10^{-14}\ M^2$$

The pH scale was based on the dissociation constant of water where

$$pH = -\log[H^+]$$

The pH scale has values ranging from 0 to 14. Similarly, K_a and K_b values can be transformed into more usable numbers by calculating the pK_a and pK_b values.

$$pK_a = -\log K_a$$

$$pK_b = -\log K_b$$

$$pK_a + pK_b = 14$$

The lower the pK_a value, the stronger the acid. The higher the pK_a value, the weaker the acid.

BUFFERS

Buffers prevent the pH of a solution from changing even when an acid or base is added to the system. The conjugate acid-base pairs formed from a weak acid and its salt or a weak base and its salt form the basis of a buffer system that absorbs any excess H^+ or OH^- present in the solution and maintains the pH of the system.

$$[HA] \leftrightharpoons [H^+][A^-]$$

$$H_2CO_3 \leftrightharpoons HCO_3^- + H^+$$

If a strong acid is added to the buffer system, then the conjugate base (HCO_3^- in the example) reacts with it and pushes the reaction to the left toward the production of more undissociated acid. If a strong base is added, then the acid (H_2CO_3 in the example) reacts with it and pushes the reaction to the right, favoring the production of more conjugate base. Both of these situations are able to absorb the addition of the acid and base without disturbing the concentration of $[H^+]$ in the reaction, and, therefore, the pH remains constant.

A strong acid or base and the corresponding conjugate base or acid will not be able to function as a buffering system because the dissociation is considered complete and the reverse reaction is not in equilibrium with the forward reaction of the dissociating acid or base. The weak conjugate acid or weak conjugate base of a strong base or strong acid is not strong enough to absorb the additional H^+ or OH^- in order to behave as a buffer.

The concentration of the conjugate acid and base components determines the amount of H^+ or OH^- a system can absorb without changing its pH. The higher the concentration of the buffering components, the more acid or base the buffering system can absorb. All buffer systems function best in a pH range around their pK_a. The Henderson-Hasselbach equation allows the calculation of the pH of a buffer system given the starting concentrations of the acid and base in the system.

$$pH = \frac{pK_a + \log[\text{base}]}{[\text{acid}]}$$

Blood pH must be close to 7.4 in order for the proteins and other cellular components to maintain their proper structure and function. The main buffer system of human blood is the carbonic acid–bicarbonate buffer system shown by

$$H_2O + CO_2 \leftrightharpoons H_2CO_3 \leftrightharpoons HCO_3^- + H^+$$

The carbonic acid (H_2CO_3) dissociates to give the bicarbonate ion (HCO_3^-) and H^+. Carbonic acid can also react to give carbon dioxide (CO_2) and water. In times of high exertion, the carbon dioxide levels in the blood rise because CO_2 is a by-product of cellular respiration. When this happens, the preceding reaction is pushed to the right to take the CO_2 levels back to normal. The human brain also signals an increase in breathing rate that increases the exchange of carbon dioxide in the lungs. As carbon dioxide is removed through the increased breathing rate, the reaction is pushed to the left and excess H^+ participates in the formation of carbonic acid.

See also BUFFERS; EQUILIBRIUM; PH/POH.

FURTHER READING

Brown, Theodore, H. LeMay, and B. Bursten. *Chemistry: The Central Science,* 11th ed. Upper Saddle River, N.J.: Prentice Hall, 2008.

Crowe, Jonathon, Tony Bradshaw, and Paul Monk. *Chemistry for the Biosciences.* Oxford: Oxford University Press, 2006.

acoustics The branch of physics that deals with sound and its production, propagation, and detection is called acoustics. A mechanical wave propagating through a material medium is termed sound. The nature of a sound wave depends on the type of medium. Solids can transmit longitudinal waves (oscillations in the direction of propagation) and transverse waves (oscillations perpendicular to the direction of propagation), as well as other versions, such as torsional waves (rotational oscillations—twisting—in a plane perpendicular to the direction of propagation). Gases and liquids, which do not support shear deformation (i.e., angular deformation), carry only pressure waves (oscillations of pressure), which can also be described as longitudinal waves.

SPEED OF SOUND

The propagation speed, v, of a sound wave is given by an expression of the form

$$v = \sqrt{\frac{\text{modulus of elasticity}}{\text{density}}}$$

where the modulus of elasticity represents the stiffness of the medium, the extent to which it resists deformation, and the density represents the medium's inertia, its resistance to changes in velocity. The appropriate modulus of elasticity is used for each type of wave. In the case of a sound wave in a gas, such as air, the appropriate modulus of elasticity is the bulk modulus, B, which takes the form

$$B = \gamma p$$

Here p denotes the pressure of the gas in pascals (Pa) and γ represents the ratio of the gas's specific heat capacity at constant pressure to that at constant volume. So the speed of sound in a gas is given by

$$v = \sqrt{\frac{\gamma p}{\rho}}$$

with ρ the density of the gas in kilograms per cubic meter (kg/m^3) and v in meters per second (m/s).

For an ideal gas the expression becomes

$$v = \sqrt{\frac{\gamma RT}{M}}$$

where R denotes the gas constant, whose value is 8.314472 joules per mole per kelvin [J/(mol·K)], T is the absolute temperature in kelvins (K), and M is the mass of a single mole of the gas in kilograms (kg). The latter is the number of kilograms that equals one-thousandth of the molecular weight (or effective molecular weight for a mixture such as air) of the gas. A real gas can be sufficiently well represented by an ideal gas, and this formula will be valid for it, as long as the density of the gas is not too high and its temperature is not too low. For a given gas the speed of sound increases with increase of temperature. This is related to the fact that as wind instruments warm up, their pitch tends to go sharp. For the same temperature and value of γ, the speed of sound increases as the molecular weight of the gas decreases. This relation underlies the demonstration in which the instructor inhales helium gas and for a few seconds sounds like Donald Duck. In both cases a rise in sound speed brings about an increase of frequency. Since the frequency of sound correlates with the sound's perceived pitch, this causes the pitch of the wind instrument and of the instructor's voice to rise.

The speed of sound in dry air at comfortable temperatures is around 340 meters per second.

AUDIBLE SOUND

For sound to be audible to humans, the frequency of the sound wave impinging on the eardrum must lie within a certain range, nominally taken as 20 hertz (Hz) to 20 kilohertz (kHz). The human female voice typically involves higher frequencies than does the voice of an adult male and is thus generally perceived as having a higher pitch than that of the adult male. This effect is due to the typically smaller size of the female larynx and vocal cords compared to the male's sound-producing equipment, causing the production of sounds of shorter wavelengths and correspondingly higher frequencies.

The perception of loudness of sound correlates with the physical quantity of sound intensity and is reasonably well described by a logarithmic function of intensity, called intensity level, and most often expressed in decibels (dB). The intensity level is calibrated to zero for an intensity of 10^{-12} watt per square meter (W/m^2), called the threshold of hearing, and is given by the expression

$$\text{Intensity level (in decibels)} = 10 \log \frac{I}{I_0}$$

where I denotes the intensity of the sound in W/m^2 and I_0 is the intensity of the threshold of hearing, 10^{-12} W/m^2. The logarithm is to base 10. As an example, if the intensity of a sound at the eardrum is 1 million times the threshold of hearing, or 10^{-6} W/m^2, the sound's intensity level is

$$10 \log \frac{10^{-6}}{10^{-12}} = 10 \log 10^6 = 10 \times 6 = 60 \text{ dB}$$

A further result of this intensity level formula is that every doubling of intensity increases the intensity level by 3 dB. Thus, a quartet of, say, guitars, whose intensity is four times that of a single instrument, *sounds* louder than a single guitar by 6 dB. The following table lists several examples.

The spectral composition of the sound (i.e., the relative power of each of the various frequency components that make up the sound wave) determines the perception of tone quality, or timbre. That is

VARIOUS SOUNDS AND THEIR REPRESENTATIVE INTENSITIES AND INTENSITY LEVELS

Sound	Intensity (W/m^2)	Intensity Level (dB)
Jet engine	100	140
Threshold of pain	1	120
Rock concert	1	120
Pneumatic hammer	0.1	110
Loud car horn	3×10^{-3}	95
Hair dryer	10^{-4}	80
Noisy store	10^{-6}	60
Office	10^{-7}	50
Whisper	10^{-10}	20
Threshold of hearing	10^{-12}	0

The intensity level of sound at a rock concert usually reaches, and can even exceed, the threshold of pain, which is nominally taken to be 120 decibels. The large speakers play an essential role in amplifying the sound to such an extent. *(David McGough/Time & Life Pictures/Getty Images)*

The design of Boston's Symphony Hall is the result of careful and rigorous measurements and calculations, to ensure that all the listeners hear the performance optimally no matter where they sit in the hall. This hall is considered one of the two or three finest concert halls in the world. *(AP Images)*

related to the difference between, say, the tone of a trumpet and that of an oboe. A pure tone is one consisting of only a single frequency.

The study of sound perception is called psychoacoustics. The difficulty some people have in following normal speech can serve as an example of what might be investigated in this field. Such a situation might be related to specific neural problems that can be identified, possibly allowing the therapist to devise computer-aided acoustic exercises for the patient, to help him or her overcome the handicap.

Both the motion of the source of a sound and the motion of the observer, relative to the propagation medium, affect the observed frequency, and thus pitch, of a sound through the Doppler effect. This comes about since the motion, whether of the source or of the observer, affects the wavelength of the sound wave as observed by the observer, and that in turn affects the sound's frequency for the observer. When the source and observer are approaching each other, effectively "compressing," or shortening, the sound's wavelength, the observer detects a frequency that is higher than that of the source. However, when their separation is increasing, thus "stretching," or lengthening, the wavelength, the frequency detected by the observer is lower than the frequency that the source is producing. A good example of that is the sound of the siren or electronically generated wailing of a police or fire vehicle. While the vehicle is approaching, the listener hears the sound at a higher pitch than she would if the vehicle were at rest. Then the vehicle passes and the pitch abruptly and very noticeably drops and remains at a lower pitch as the vehicle recedes.

APPLICATIONS OF ACOUSTICS

The following are some applications of acoustics:

- Music: understanding how musical instruments generate the sounds they do. This allows the design of improved, and even new, types of instruments.
- Architecture: predicting and controlling the sound in structures such as concert halls. This is done by considering and determining the reflection of sound—which creates echoes and reverberation—and the absorp-

tion of sound, by surfaces and by the audience. Such factors are controlled by the shape of the hall and the shapes of various surfaces in the hall, by the materials coating the ceiling, walls, and floor, and even by the clothes that the audience wears.

- Noise abatement: controlling the levels of unwanted noise. Noise must be identified and measured. It might be controlled by redesigning the noise source to be quieter or by introducing sound-absorbing materials to reduce the sound intensity at the hearers' ears.
- Sonar: the use of underwater sound for the detection and identification of submerged objects. The sonar device emits underwater sound waves, and a computer times and analyzes their echoes, allowing for the detection, identification, and location of underwater objects. Submarines make use of sonar routinely. Whales and dolphins are naturally equipped with sonar, in addition to their good vision.
- Medicine: applying high-frequency sound (ultrasound) for imaging and for treatment. For imaging, an ultrasound source on the skin sends sound waves into the body. They are reflected at every boundary between tissues of different density. A computer detects and analyzes the reflections to create images, such as of a fetus, an organ, or a tumor. Ultrasound treatment is based on the heat-producing property of such waves. They are introduced into the body by a source on the skin as a relatively high-intensity focused beam. In the region where the waves converge, the tissue is heated with therapeutic effect.

See also DOPPLER EFFECT; ELASTICITY; GAS LAWS; HEAT AND THERMODYNAMICS; POWER; PRESSURE; SPEED AND VELOCITY; STATES OF MATTER; WAVES.

FURTHER READING

Fletcher, Neville H., and Thomas D. Rossing. *The Physics of Musical Instruments,* 2nd ed. New York: Springer, 2005.

Hall, Donald E. *Musical Acoustics,* 3rd ed. Belmont, Calif.: Thomson Brooks/Cole, 2002.

Rossing, Thomas D., F. Richard Moore, and Paul A. Wheeler. *The Science of Sound.* San Francisco: Addison Wesley, 2002.

agrochemistry (agricultural chemistry) Agrochemistry or agricultural chemistry is a branch of chemistry concerned with the scientific optimization of agricultural products. Agricultural chemists are responsible for creating a food supply that is safe at a reasonable cost. The production of agricultural fibers also falls in this field. Products such as wool and cotton that are produced by farms need to be analyzed. Agricultural chemists also explore the environmental impact of agricultural pursuits. As the population of the world continues to increase and the area in which crops can be raised continues to decrease, the role of agricultural chemists becomes increasingly important. The diversity of the field is greater than that of nearly all other areas of chemistry. Agricultural chemists can work in multiple areas, including animal research, breeding, and genetics; plant and crop production; manipulation of photosynthesis; or fertilizers, herbicides, and pesticides. In order to control weeds and insects, agricultural chemists study insect populations and their responses to various chemicals. In addition to all the fields listed, an agricultural chemist must also serve in the role of toxicologist and environmental chemist. Understanding the impact that treatments, medications, and chemicals have on animals and the food supply helps keep society healthy and safe.

EDUCATION AND TRAINING

The educational requirements for a career in agricultural chemistry and related agricultural sciences include a degree in chemistry, biochemistry, analytical chemistry, environmental chemistry, biology, or even genetics. After the completion of a bachelor of science degree, students can seek a job in the field or pursue higher levels of education. A master of science or doctoral degree can also be obtained in the field of agricultural chemistry. An agricultural chemist can work in a wide variety of workplaces including any company that produces food products or farms and businesses that deal with crops, animals, or fiber production. Nearly 20 years ago, the field of agricultural chemistry exploded as a result of the recognition of the widespread environmental impact of agricultural chemicals, creating a fast-growing job market. According to the American Chemical Society, job opportunities in the area of agricultural chemistry have since slowed, as larger companies are taking over the field and consolidating the work of these chemists.

TECHNIQUES USED BY AGRICULTURAL CHEMISTS

The techniques used in the field of agricultural chemistry include those associated with analytical chemistry, biochemistry, and environmental chemistry as well as cell culture techniques, genetic engineering, and toxicology. Assays for screening the effectiveness of chemical treatments and analytical chemistry isolation and quantification techniques such as mass spectroscopy are common. Soil testing, air testing, and water testing are all utilized in agricultural research. A wide base of laboratory experience is essential for this field.

EFFECT ON NUTRIENT CYCLING

Agricultural pursuits affect every aspect of environmental interest. One of the simplest ways to evaluate the impact of agricultural chemistry is to measure the effect of agriculture on the natural cycles of the earth. The constant use and regeneration of chemical elements through geological and biological processes are known as nutrient cycling. Examples of nutrient cycles include the water cycle, the nitrogen cycle, and the phosphorus cycle. Agricultural processes affect each of these cycles.

Water makes up more than two-thirds of the human body and three-fourths of the surface of the planet Earth. The water cycle involves the changes of state of the water molecule from a solid to a liquid to a vapor. The cycling of water on the planet is a continual process with no beginning or end. Once the water hits the Earth, it either enters the ground in a process known as infiltration or stays on top of the ground as surface runoff. The water that infiltrates can become part of freshwater storage systems such as aquifers or springs. The total amount of global freshwater represents only 2 to 3 percent of the total amount of all global water. Gravity forces all of the surface runoff water to find the lowest point, which is the oceans; thus, all water ultimately returns to the oceans. Surface runoff collects in natural channels such as creeks, streams, and rivers, which serve a major function in returning the water back to the ocean in order to continue the cycle. As the water runs off, it picks up any contaminants present in the soil. When this water flows through soil where crops are being grown, any fertilizer, herbicide, or other chemical placed on the crops has the chance of flowing into streams and creeks. Water can also become polluted when water flows through areas contaminated by animal waste such as cesspools on large pig farms and poultry farms.

Agricultural chemists are responsible for the development of processes that protect these waterways (discussed later). Examples of innovations in this area are shown in new developments in recycling poultry waste and treatment of cesspool waste to remove the solid contaminants, isolate the nutrients such as nitrogen and phosphorus from it, and return clean water back into the environment, thus preventing contamination of the water supply. Finding creative uses for poultry waste removed from the soil reduces the impact on the environment. For example, agricultural chemists have found that heated poultry waste is excellent for the remediation of toxic, heavy metal contamination in the soil. The charcoal form of poultry waste is able to chelate (remove) these heavy metals, removing them from the soil. Removing the waste from the natural system diminishes the amount of water and soil contamination that occurs. This in itself is a positive development. The additional positive impact is the newfound usefulness of the waste in other types of pollution.

Another nutrient that is cycled through earth is nitrogen. Because nitrogen is a critical component of the biomolecules proteins and nucleic acids, all living organisms require this nutrient. Nitrogen makes up 78 percent of the air. This atmospheric form of nitrogen is diatomic (N_2) and is bonded together by a triple bond between the nitrogen atoms. While this form of nitrogen is readily abundant, most organisms are unable to utilize the nitrogen in this form. In order to be useful for most life-forms, it first must undergo a process known as nitrogen fixation, carried out by certain types of bacteria that live in the soil or are catalyzed by lightning. Bacteria perform the majority of this work, fixing an estimated 140 million metric tons of nitrogen each year. Nitrogen-fixing bacteria often participate in mutualistic relationships (symbiotic relationships in which both partners benefit) with the roots of legumes, such as peas, soybeans, clover, alfalfa, and beans. Human activities significantly impact the nitrogen cycle, especially through fossil fuel emissions, sewage leaching, and agricultural fertilization. The burning of fossil fuels produces nitrous oxide compounds that are important contributors to acid rain and the acidity of soil. The application of high concentrations of nitrogen in the form of fertilizers leads to excess that can wash away. When farmers apply animal waste directly to a field, they increase the possibility that excess nitrogen is added. Scientific reclamation projects by agricultural chemists allow the treatment of these wastes to isolate the useful components and reduce excess in the environment. Increased use of fertilizers by farmers raises the nitrate content in the soil, and, as a result, the amount of runoff in groundwater or streams and lakes increases. The same is true for leaching (percolation through the soil) of water-soluble nitrates from sewage into waterways. When the nitrate content in a lake, river, or slow-moving stream increases, the algal population increases (causing algal blooms). This depletes the supply of oxygen dissolved in the water, a condition that can eventually lead to eutrophication of the lake and death of most of its inhabitants. Eutrophication is a process whereby excess nutrients lead to the overgrowth of plants, usually algae. Responsible farming practices are necessary in order to prevent environmental damage by nitrogen-based fertilizers. Modification of the crops being planted through genetic means can reduce the amount of fertilizer required to grow them, reducing the negative environmental impact of the crop. Scientists, including those at the U.S. Department of Agriculture (USDA), are developing new ways of processing animal waste to form useful products while decreasing the amount of nitrogen being released. For example, agricultural chemists have developed an effective way to clean and purify pig waste, removing the solids and ammonia. The process, tested in North Carolina, removed more than 95 percent of all of the pollutants.

All living things require phosphorus, as it makes up the backbone of deoxyribonucleic acid and ribonucleic acid and is part of the primary energy source of the cell (i.e., adenosine triphosphate [ATP]). Phosphorus is second only to calcium in abundance in human bones. The elemental form of phosphorus is rare; most phosphorus is found in the form of phosphates (PO_4^{3-}), either inorganic or organic (associated with a carbon-based molecule). Animals can use phosphates in the form of either organic or inorganic phosphates, but plants can only use the inorganic phosphate form. Soil and rocks contain much of Earth's phosphorus, and when these rocks break down through erosion, the phosphorus enters the soil in the form of phosphates. Plants can absorb phosphates from the soil, and then primary consumers can eat the plants as their source of phosphorus. Secondary consumers eat the primary consumers, and the phosphates travel up the food chain. As the animals produce wastes, phosphates can be returned to the soil for use by the plants or enter into waterways, where they can be dissolved and taken up by aquatic plants, which are then ingested by aquatic animals. When these organisms decompose, the phosphates become part of the sediment that makes its way into the rocks and soil, thus completing the phosphorus cycle.

Phosphorus is an important nutrient for plants and is a component of most fertilizers. In the United States, in an average year, corn, wheat, and alfalfa crops remove from the soil 1.5 billion pounds of phosphorus that needs to be replenished in order to continue growing crops. However, as for any other chemical, the overproduction and overuse of phosphates by humans in the environment can have a large ecological impact. The overuse of fertilizer can lead to excessive phosphates in surface runoff because phosphates do not bind to the soil well and run off easily. This surface runoff makes its way to the lakes, rivers, and streams, where it can lead to the overgrowth of weedy plant life and algae that can lead to depletion of the dissolved oxygen, killing aquatic life through eutrophication, as happens with excess nitrogen. Human impact on the excess amount of phosphorus in the form of phosphates is due to fertilizer runoff, sewage leachate, as well as the use of phosphate-based detergents in washing of clothes. These phosphate-based detergents have been nearly eliminated today because of their negative environmental impact.

Phosphorus used to create fertilizers is generally mined from underground phosphorus deposits. In February 2008 agricultural chemists at the USDA determined it was possible to purify phosphorus from poultry waste. When adding all of the poultry waste to a field, there is much more phosphorus than the plants can use and the rest will run off, potentially contaminating lakes and rivers and streams. Chemists have also determined methods to turn pig waste into phosphorus sources to use in fertilizers.

IMPACT OF AGRICULTURAL CHEMISTRY ON ANIMALS

Many agricultural chemists work in the field of animal breeding, animal health, and food production. Maximizing the amount of product made from animal sources while keeping costs down and food safe is a difficult task. Genetic engineering techniques as well as hormonal treatments can often create bigger and more productive animals. This leads to increased profits. For example, the addition of growth hormones to cows that dramatically improved milk production is one area of research for agricultural chemists, who were responsible for determining whether there were chemical impacts on the milk supply and the possible ramifications to humans.

Improved breeding techniques to create positive effects on the animal populations can be seen in a chicken population. For years laying hens (chickens used for egg production) were bred for the size and quantity of eggs. Other traits were not necessarily a concern. In many cases, this led to aggressive behavior in the animals, and losses were seen when the chickens established a pecking order, killing many of the farmer's animals. Recent breeding techniques have involved the goal to remove aggression to prevent pecking and killing due to competition among the animals.

Medicinal improvements for animals have created a much safer food supply. Vaccinations of animals, quarantine and testing of sick animals, as well as blockage of sick animals from reaching the food supply are the responsibility of the USDA. The system is not perfect, and mass recalls of potentially contaminated meat sources are seen periodically. Agricultural chemists are also involved in the field of animal competition. The development of improved testing for animals, such as racehorses, that move from one country to another allows for the protection of all animals in this country. Microbiological techniques aim to prevent germs in animals to reduce the amount of medication required and the possible impact of these medications on the food supply.

OTHER AREAS OF INTEREST

The responsibility for the food supply and all agricultural crops has led to several interesting innovations by agricultural chemists. Some of these include nontoxic glues developed from the bacteria found in cows' bellies. Scientists studying the process of cellulose digestion in ruminants found that the bacteria in these animals held on to the cellulose very strongly. The sticky nature of these bacteria and their secretions led to the creation of nontoxic, gentler adhesives to replace many of the harsh glues.

The poultry industry, as well as producing massive amounts of poultry waste discussed earlier, produces pollution in the form of feathers. Discarding of these feathers takes up landfill space, so reusing them has become an area of interest to agricultural chemists. Feathers that are ground up have been used for everything from plastic toy boats to insulations. Reusing these types of waste reduces the negative impact of agriculture on the environment.

THE AGRICULTURAL RESEARCH SERVICE (ARS)

The Agricultural Research Service (ARS) is the research agency for the USDA. This organization is the primary research organization for the governmental protection of our nation's food supply. Their self-described mission is to seek ways to produce more and better food and fiber, to control insect pests and weeds safely, to get food to consumers in better condition, to improve human nutrition and well-being, to improve animal health, to find new uses for agricultural commodities, and to help protect soil, water, and air. The ARS is divided into the following four areas of research:

- nutrition, food safety/quality
- animal production and protection
- natural resources and sustainable agricultural systems
- crop production and protection

Private companies, such as Switzerland-based Syngenta (the world's largest agribusiness), also research these areas and are contributing to the optimization of agricultural pursuits. These companies have become quite large and have the potential for bias, depending on who is funding their research. The ARS is theoretically an unbiased body that can be counted on to keep the best interest of humans, animals, and the environment in mind.

See also ACIDS AND BASES; AIR POLLUTION (OUTDOOR/INDOOR); ANALYTICAL CHEMISTRY; ATMOSPHERIC AND ENVIRONMENTAL CHEMISTRY; BIOCHEMISTRY; GREEN CHEMISTRY; NUTRIENT CYCLING; PHOTOSYNTHESIS.

FURTHER READING

Chesworth, J. M., T. Stuchbury, and J. R. Scaife. *An Introduction to Agricultural Biochemistry*. New York: Kluwer Academic, 2007.

air pollution (outdoor/indoor) Air pollution is the presence of unwanted particles and gases in the air that can adversely affect the environment and the health of the organisms exposed to it. Two types of pollution sources exist: stationary and mobile. Stationary sources include factories with smokestacks and industrial processes. As they burn fossil fuels and complete the production process, they release multiple pollutants. The amount of pollution in the air is a local, regional, and global problem. Air pollution is a local problem in that the geographical area immediately surrounding the plant or polluting source is the first to experience the pollution. To relieve the local effects of pollution, many plants began installing taller smokestacks, reducing the amount of pollution released in the immediate vicinity of the plant. This practice creates a more regional problem by spreading pollutants over a larger geographical area. Weather patterns and winds also affect the amount of pollution that collects in an area.

Mobile pollution sources include cars, trucks, trains, boats, and airplanes. The regulation of mobile pollution is difficult because the pollution sources are able to move from one location to another. Because of this, mobile pollution sources are generally regulated at the production level. Automobile manufacturers must follow emissions requirements for new automobiles, but the upkeep on these vehicles and transportation fleets is then left primarily to the owners. Many states require emission testing for vehicle licensure.

TYPES OF AIR POLLUTION

The age and overall health of an individual help determine how strongly he or she is affected by air pollution. The most seriously impacted are the elderly, the young, and anyone who has a compromised respiratory system, such as someone who is asthmatic, but air pollution can also negatively affect healthy adult individuals. Exposure to airborne chemicals can cause symptoms such as respiratory irritation, nausea, and rashes. The primary goal of regulating air pollution is to preserve human health, followed by the desire to protect property and the environment from damage. The Environmental Protection Agency (EPA) recognizes six primary pollutants (criteria pollutants) of concern to human health and property protection

This view of the Los Angeles skyline shows the smog that is created around large cities. *(Jose Gil, 2008, used under license from Shutterstock, Inc.)*

including ozone (O_3), sulfur dioxides (SO_2), nitrogen oxides (NO_x), carbon monoxide (CO), particulate matter (PM), and lead (Pb).

Three covalently bonded oxygen atoms make up a molecule of ozone. While ozone is essential in the stratosphere for protecting Earth's surface from harmful ultraviolet rays from the Sun, ozone at ground level can be very damaging. Ozone forms by the reaction of oxides, such as nitrogen oxides, and hydrocarbons in the presence of sunlight. Smog is composed of ground-level ozone, nitrogen oxides, particulate matter, and aerosols. When the air near the ground is cooler than the air above it, the pollution particles and gases become trapped at the surface of Earth, leading to the haze that is found over big cities.

Sulfur dioxide is a combustion product of sulfur-containing fuels such as coal and oil. When sulfur dioxides are released into the atmosphere, they react with water to form sulfuric acid, causing acid rain. Normal rainwater is slightly acidic with pH values around 5 to 6. High levels of sulfur dioxide in the air lower the pH of precipitation; pH levels of rainwater have been recorded as low as 3.5. The environmental impact of acid rain is seen in waterways as well as vegetation. Fish and other organisms living in lakes, rivers, and streams are not able to thrive and in some cases cannot survive in water with a lowered pH. Vegetation also suffers the effects of acid rain, leading to damaged plants, crops, and trees. The effects of acid rain caused by sulfur dioxides do not stop there—nonliving matter is also affected. Acid rain causes deterioration of buildings, statues, bridges, and roadways. Sulfur dioxide can also form sulfate particles that can cause respiratory problems for everyone who breathes them but especially the young, the old, and those with asthma or other preexisting respiratory or cardiac conditions.

Nitrogen oxides, NO_x, refer to a class of covalent compounds formed between various numbers of nitrogen and oxygen atoms. As sulfur dioxides do, nitrogen oxides form as a result of the combustion process. Motor vehicles, followed by industrial processes that burn fuels, are the primary sources of nitrogen oxide pollution. When released into the atmosphere, nitrogen oxides react with water to form nitric acid, which contributes to the creation of acidic rain. Nitrogen oxides can also contribute to the formation of ground-level ozone, which causes serious respiratory problems, as mentioned, and they can form nitrate particles that can enter the body and damage lung and heart tissue. The environmental impact of nitrogen dioxides includes an increased level of nitrogen in lakes and rivers, leading to eutrophication, or oxygen depletion, and death of many organisms in the lake. Dinitrogen oxide N_2O is considered a greenhouse gas that contributes to global warming. Nitrogen oxides are the only criteria pollutant tracked by the EPA that has not been reduced since the 1970s.

Carbon monoxide is a covalent compound of one carbon and one oxygen atom, formed from the incomplete combustion of fossil fuels. More than half of the carbon monoxide in air is released from automobile exhaust. This percentage is greater in highly congested traffic areas. Wood-burning fires, including forest fires, also release carbon monoxide into the atmosphere. Carbon monoxide is a colorless, odorless, tasteless gas that can be deadly. By competing with oxygen for binding sites on hemoglobin, the oxygen carrier in red blood cells, carbon monoxide limits the amount of oxygen that a body receives. Symptoms of carbon monoxide exposure include dizziness, nausea, and an inability to concentrate as well, as heart and respiratory damage. The development of the catalytic converter in the 1970s was a major step in reducing the level of carbon monoxide emissions from vehicles. These, along with other controls, have helped reduce the amount of emissions that new cars produce but do not control how much is released as the vehicle ages. The increased number of drivers on the road and the increased amount of time people spend in their cars are coming close to offsetting the improvements in technology in carbon monoxide emissions.

Particle pollution (or PM for particulate matter) is a class of inhalable particles smaller than 10 microns (a micron equals one-millionth of a meter) in diameter, from various sources that are regulated by the EPA. PM is a mixture of solids and liquids such as smoke, dirt, mold, and dust and is divided into fine particulate matter that is less than 2.5 microns in diameter and particles with a diameter of 2.5 to 10 microns, termed inhalable coarse particles. Because of the small size, particle pollution is most detrimental, since the particles are able to pass deeply into lung tissue and create severe respiratory problems.

The element lead is a metal in group IVA of the periodic table of the elements with an atomic number of 82 and an atomic mass of 207.2. In the 1970s the primary source of lead pollution in the air was due to automobiles' using leaded gasoline. The use of leaded gasoline was banned by the EPA in 1995, and the development of automobiles that utilize unleaded gasoline has greatly reduced the amount of automobile lead emissions. The majority of lead pollution presently originates in metal-processing plants.

AIR POLLUTION REGULATION

As pollution became more of a problem, Congress passed the first federal air pollution law, known as the Air Pollution Control Act of 1955. The primary impact of this legislation was to call attention to air pollution as a national problem and to grant research funds for exploring the problem more thoroughly. By 1963 sufficient research had been done that Congress passed the Clean Air Act of 1963, setting emissions standards for stationary pollution sources. Several amendments to this act (in 1965, 1966, 1967, and 1969) addressed issues such as monitoring the air quality in regions of the country and regulating emissions from mobile pollution sources such as automobiles. The Clean Air Act of 1970 contained some of the most stringent air pollution requirements and standards that had been written to date and significantly reduced the allowable limits for stationary sources and mobile pollution sources. After this act, multiple amendments pushed back deadlines for companies to meet these standards as well as raised the allowable levels in order to make it easier for companies to be in compliance.

The Clean Air Act of 1990 revised and strengthened the older Clean Air Acts to account for newly identified problems such as acid rain and ozone depletion. This more recent legislation focused on investigations into controlling air pollution from motor vehicles and research into alternative energy sources to replace fossil fuels. This act also addressed the atmospheric levels of chlorofluorocarbons (CFCs) in order to reduce the effect on the stratospheric ozone layer according to the Montreal Protocol of 1985, which was amended in 1990, 1992, 1995, 1997, and 1999.

INDOOR AIR POLLUTION

Indoor pollution is potentially more damaging than outdoor air pollution as more and more Americans spend their time indoors. Since the air flow inside the building is not fresh and is restricted, the pollutants are "trapped" within the system. People are exposed to more concentrated doses of pollutants and for a longer period, with dangerous effects on those that live and work in the structure. Pollutants such as cigarette smoke, dust, mold, pet hair, toxic chemicals released from building supplies, carbon monoxide, and radon are indoor air pollution concerns. Proper ventilation, appliance installation and maintenance, proper ambient humidity, and restriction of smoking help minimize the harmful effects of indoor air pollution.

Cigarette smoke is a predominant indoor pollutant. Secondhand cigarette smoke is now known to be as dangerous to a nonsmoker as smoking is to the smoker. Working and living in areas where cigarette smoke is present increase one's risk of lung cancer and respiratory problems, including asthma. As with most respiratory pollutants, cigarette smoke has the greatest effect on the very young and the very old. In order to improve indoor air quality, many states, such as New York, California, and Ohio, are adopting antismoking legislation that bans cigarette smoking in public places.

Biological contaminants, such as pet dander, pet hair, mold, and other particulate matter, can pose a risk for respiratory complications as well as allergies. The presence of toxic mold in a home can lead to memory loss, allergies, and respiratory problems in healthy individuals. Young children and those with compromised immune systems can have even more serious reactions, including death.

Chemicals released from indoor building supplies also pollute the air. Building materials that release formaldehyde include particle boards, pressed wood used in furniture, and adhesives often used in carpets and other furniture. Formaldehyde affects those with asthma and other breathing disorders and has also been linked to cancer. The release of formaldehyde decreases over time, so the greatest risk occurs when the furniture is new. Asbestos is another building product that has adverse affects on indoor air. Found primarily in older homes and buildings, asbestos was used in insulation, siding, tiles, and other building uses and poses its greatest risk when it is released as particulate matter into the air. Asbestos particles build up in the lungs and can cause cancer. The EPA and Occupational Safety and Health Administration (OSHA) began regulating asbestos in the 1970s.

As well as being an outdoor pollutant, carbon monoxide is a deadly indoor pollutant. Improper functioning and venting of furnaces and other heaters that burn fossil fuels can produce carbon monoxide. Because it is an odorless, tasteless gas, its detection requires an electronic carbon monoxide detector. Regular maintenance of furnaces and heaters and installation of carbon monoxide detectors are important measures for preventing carbon monoxide deaths.

Lead is also a harmful indoor pollutant. Prior to 1978 many paint products contained lead, and most homes and furniture were painted with lead-based paint. When that paint chipped or released dust, the lead became airborne and could be inhaled. Lead accumulates in bodily tissues and is especially harmful to infants and small children. Low levels of lead in the body can damage many organs including the heart, kidneys, liver, and brain.

Radon is an odorless, tasteless, radioactive noble gas with an atomic number of 86. As radium naturally present in rocks decays and releases radon, the gas seeps into homes through basements and foundations. Radon is the number one cause of lung cancer in nonsmokers. In 2007 the EPA estimated that radon exposure causes 14,000–20,000 deaths per year, primarily from lung cancer. Radon exposure is second only to smoking as a cause of lung cancer. Specific testing for radon is the only way to know whether this gas contaminates a structure. Radon levels above four picocuries per liter (pCi/L) need to be mitigated. Simple measures such as sealing cracks in the floor and walls and ventilating basements can help reduce radon levels. Venting of the gases from below the foundation is known as subslab depressurization. This method removes the radon gas before it can enter the home. These repairs should be performed by a qualified radon contractor.

See also ACID RAIN; ALTERNATIVE ENERGY SOURCES; ATMOSPHERIC AND ENVIRONMENTAL CHEMISTRY; CIGARETTE SMOKE; FOSSIL FUELS; GREENHOUSE EFFECT.

FURTHER READING

U.S. EPA and U.S. Safety Product Commission. "The Inside Story: Indoor Air Pollution." Available online. URL: http://www.epa.gov/iaq/pubs/insidest.html#Intro1. Accessed July 23, 2008.

Wright, Richard, and Bernard Nebel. *Environmental Science: Toward a Sustainable Future,* 10th ed. Upper Saddle River, N.J.: Prentice Hall, 2008.

alternative energy sources While energy cannot be created or destroyed, it can be transformed between types. The major sources of energy on Earth are the fossil fuels, including coal, natural gas, and oil, all of which are nonrenewable resources. Each of these was produced over a period of millions of years under the surface of the Earth, and it is not possible to replenish these supplies. The primary uses for the energy obtained from the fossil fuels include heating our homes and buildings, generating electricity to run everything from computers to alarm clocks, and providing transportation. In addition to being limited in supply, the burning of coal as an energy source releases sulfur dioxide into the atmosphere, where it contributes to the formation of acid rain. As a result of the government's decreasing the allowable sulfur dioxide levels, coal is losing popularity as an energy source. The world's population is growing, and the number of industrialized countries is increasing. Whereas in the past a relatively small predictable group of nations competed for fossil fuels, both competition and demand have significantly increased in recent years. There was a time when electrical systems in homes were only required to power such items as lamps, ovens, and television. Present-day society and technology place a much greater demand on energy production. Scientists and engineers are engaged in ongoing research to develop alternative energy sources, innovative methods for meeting the world's energy demands to reduce the dependence on traditional, nonrenewable resources.

As demand increases, the reserves of fossil fuels are diminishing. Dr. Marion King Hubbert, an American geophysicist and leading expert on energy supplies, predicted that the world's oil production would peak in 1995 and would then decrease according to a bell-shaped curve. This peak of oil production, known as Hubbert's peak, predicts that if the current usage stands, the world's production of fossil fuels will be depleted by the year 2050. If this is the case, then the world's primary energy source may need to be replaced within a generation. A political component also affects the energy situation. The United States is becoming increasingly dependent on foreign sources of oil and gas, placing American consumers at the mercy of supply and demand. The prices can fluctuate significantly, and consumers have no real choice except to pay the asking price.

One short-term solution that would increase the U.S. production of oil but has been controversial because of the environmental impact of the proposal has been suggested. President George W. Bush and members of the House of Representatives attempted several times to gain approval of oil drilling of the Alaskan Arctic National Wildlife Refuge (ANWR). Many people are opposed to this drilling because of the negative impact on the environment and the area's animal populations, such as polar bears and caribou. Estimates project that the ANWR contains approximately 4 billion barrels of oil that could reduce U.S. dependence on foreign oil sources. To date this question has not been resolved.

No really viable contenders have emerged to replace fossil fuels as our energy source. Nuclear fission reactors presently meet approximately 20 percent of our energy needs. These reactors depend on a chain reaction of breaking up large uranium 235 nuclei and producing two smaller nuclei. The dangers inherent with this type of energy are controversial. Radioactive release from the plants as well as storage and disposal of nuclear wastes produced in the plant are problems that demonstrate that without significant improvements, nuclear fission reactors will not completely replace fossil fuels for energy production. Several alternatives have presented themselves as potential candidates as replacements for fossil fuels. Further research and development will be necessary in order to determine how best to implement these methods and develop more cost-effective and operable energy sources. Potential energy sources include hydroelectric power, burning of biomass, wind power, solar power, and geothermal power.

The highest producer of alternative energy to date is hydroelectric power. Hydroelectric power utilizes the potential energy of stored water and turns it into kinetic energy of a moving turbine that powers an electrical power generator. The Tennessee Valley Authority (TVA) is the largest public electrical power company, servicing 8.5 million customers. TVA controls 29 hydroelectric plants and supplies energy to the entire Tennessee Valley, stretching from southwestern Virginia to the Mississippi River. The water is stored behind a dam, creating potential energy, and is released in a controlled manner through a turbine, causing it to turn and power a generator that uses the kinetic energy of the turbine to create electrical energy. TVA operates one such pumped storage plant, the Raccoon Mountain Pumped Storage Plant, near Chattanooga, completed in 1978. The stored water is located at a higher elevation, on the top of Raccoon Mountain, increasing the potential energy available to generate electricity. When experiencing peak usage, TVA can release this water down through the center of the mountain to turn a turbine and generate electricity. The water is collected in the Nickajack Reservoir at the bottom of the mountain. When the electricity demand decreases, as on weekends and evenings, the water is pumped back up the mountain, replenishing the upper reservoir for use during the next peak demand. Pumped storage plants serve as backup energy storage for the power company, allowing it to respond to high demand. Another advantage to this type of power plant is that the community can use the reservoir for recreational activities such as boating, swimming, and fishing.

The second largest U.S. alternative energy source is biomass, the organic matter that makes up living things. Burning firewood is an example of using biomass to produce energy for heating. Biomass burning can also produce electrical power. The largest categories of biomass types used for energy production today are wood, pulping liquor from wood processing plants, and municipal solid waste. The burning of biomass does not technically increase the total concentration of carbon dioxide in the atmosphere because plants created the biomass by photosynthesis, a process that removes carbon dioxide from the environment, but there is a time delay between the incorporation of atmospheric carbon dioxide to produce the biomass and the return of carbon dioxide to the environment by burning the plant product. The carbon dioxide released today increases today's concentration. This is the only alternative energy source that has a polluting component—the production of additional greenhouse gases is a negative effect of burning biomass. An advantage to the use of biomass is that the process can actually remove damaging waste products from the environment through burning. A waste-to-energy plant that utilizes the organic waste from a landfill reduces the amount of trash stored in the landfill and at the same time produces electrical energy.

The U.S. Department of Energy operates the National Wind Technology Center (NWTC) located in Colorado. This center researches and develops technology for electrical power production by wind-powered turbines. The turbines harness the kinetic energy of the moving wind and use it to create electrical power. Wind turbines are very tall in order to take advantage of the upper-level winds, can face into the wind or downwind, and generally have three blades. Wind-powered movement of the turbine causes a generator to turn; through electromagnetic induction, that turning creates electricity, which can then be distributed to places that require the energy. Many wind turbines operate together in locations known as wind farms. Large-scale wind turbines can produce from 50 to 750 kilowatts of power. The NWTC estimates that wind power has the potential to supply 20 percent of the United States's energy supply. As with other alternative energy sources, the use of wind power is nonpolluting and enables ranchers and farmers to utilize their land in a profitable venture. Wind is essentially an unlimited and free energy resource, but the technology required to use the moving wind's energy is expensive, and because of the vast amount of space required to harness the wind's energy, wind power is not yet practical for large-scale energy production.

Solar power involves harnessing the energy of the Sun to perform work on Earth. The amount of energy released from the Sun would be sufficient to power all work on Earth, but the total amount of solar radiation does not reach the Earth's surface. Much of the radiation is either reflected back into space by the atmosphere or absorbed by the atmosphere. The energy that reaches the Earth's surface can be used for many heating purposes as well as electrical production. Solar panels can be utilized in buildings to heat water or to heat a house by collecting sunlight and heating contained water, which circulates and radiates that heat into the room. Solar energy can also be used to produce electricity directly from sunlight using solar cells called photovoltaic cells.

Geothermal energy is energy that is derived from the heat of the rocks and fluid in the Earth's crust. This energy can be used for electrical power production or the simple heating of buildings or homes. The geothermal energy can be used to run heat pumps or employed in electric power plants. The heat pump simply moves the heat from one place to another. In the summertime, the heat pump removes heat from a building and sends it into the heat sink of the Earth. In the winter, the heat pump uses the ground as a heat source to warm the building. Direct-use geothermal heat systems do not use heat pumps or power plants; they directly heat water that circulates to heat homes and buildings. Geothermal heating systems produce no pollution and, as solar and wind power do, utilize energy found in the completely free resource of the Earth's natural heat. According to the Geothermal Resource Council, the present U.S. production of geothermal energy is at the rate of approximately 2,200 megawatts.

Hydroelectric power, biomass power, wind power, solar power, and geothermal power can all be used as alternative energy sources for two of the three main uses of fossil fuels—electricity generation and heating. Energy for transportation requires the production of alternative fuels for vehicles, as well as vehicles that can utilize the alternative fuels and fueling stations that can supply the alternative fuels. The Energy Policy Act (EPAct) of 2002 addressed these issues and mandated that percentages of certain vehicle inventories be powered by alternative fuel choices. According to the U.S. Department of Energy, those vehicles impacted include governmental and alternate fuel providers that have more than 50 vehicles operating in high-concentration metropolitan areas. These and other transportation fleets that contain more than 20 vehicles in local fleets and more than 50 in U.S. fleets were responsible for adding alternatively powered vehicles to their fleets. Those fuels that are considered alternative fuel sources by EPAct are methanol, ethanol, and other alcohols; blends of 85 percent or more of alcohol with gasoline; natural gas; liquefied petroleum gas (propane); coal-derived liquid fuels; hydrogen; electricity; and biodiesel fuels (other than alcohol) derived from biological materials.

The Energy Policy Act of 2005, passed by Congress and signed into law by President George W. Bush, allows for incentives for the purchase of alternative fuel vehicles and appropriates money for the research and development of new technologies that enhance the use of alternative fuel sources for vehicles. In order to run a vehicle on alternative fuel comfortably, the buyer needs access to the sale of that fuel. The United States presently has 5,259 alternative fueling stations. Alternative fuels are taking on a political and mainstream societal edge. According to the U.S. Department of Energy, by December 2008 there were tax incentives for the use of alternative fuels or conservation of fuel and efficiency regulations for fuel in every state except Alaska. Other programs in development include a comprehensive Web site known as Yahoo Autos Green center that compares the advantages and disadvantages of alternative fuel technology and other types of vehicles that are available. Presently there are 146 so-called green cars available for purchase, indicating that the use of alternative fuels is becoming more of a reality.

This hybrid Nissan Qashqai is an off-road compact car for the European market, 2006. Hybrid vehicles combine a gasoline-powered engine with an electric motor, reducing the amount of gasoline required to run the vehicle. *(AP Images)*

Alternative energy sources offer nonpolluting energy choices that utilize virtually limitless supplies of energy from sources such as the Sun and wind. Clearly these two facts make the use of alternative energy sources preferable to use of fossil fuels. Also, alternative energy sources can be produced in the United States without influence from foreign countries, keeping control of the energy supply within the country. The present technology in power plants and vehicles does not allow for these programs to tap into their full potential. As the fossil fuel problem becomes a crisis, more research and development should go into these alternatives in order to ensure a dependable, nonpolluting, and limitless amount of energy for generations to come.

See also ELECTRICITY; ENERGY AND WORK; FISSION; FUSION.

FURTHER READING

U.S. Department of Energy. "Energy Sources." Available online. URL: http://www.energy.gov/energysources/index.htm. Accessed July 23, 2008.

amino acid metabolism and the urea cycle

The metabolism of amino acids, the building blocks of proteins, is complex compared to that of other biomolecules. All amino acids consist of a central carbon atom, bonded to a hydrogen atom (–H), a carboxyl group (–COOH), an amino group ($–NH_3$), and a variable side chain (–R). The catabolism, or breakdown, of amino acids is complex and varies among the 20 naturally occurring amino acids. Most amino acids can feed into the citric acid cycle directly or indirectly and therefore can be used to synthesize glucose when necessary. Under fasting conditions, the body catabolizes muscle protein for energy; the first step in this process is to break the peptide bonds of the polypeptides, releasing individual amino acids. The liver can convert the carbon skeletons from amino acids into glucose and ketone bodies and the nitrogen into urea [$CO(NH_2)_2$], an end-product of protein degradation that is excreted in the urine. The urea cycle is the cyclical biochemical pathway that synthesizes urea from ammonium ions (NH_4^+), carbon dioxide (CO_2), and the amino group from the amino acid aspartate.

AMINO ACID SYNTHESIS

Though 79 percent of the atmosphere consists of nitrogen gas (N_2), this form of nitrogen is unusable to animals. The triple bond has a bond energy of 225 kilocalories per mole (kcal/mol), making it quite unreactive. Animals obtain their nitrogen already fixed by ingesting other animals or plants. Soil bacteria, many living in symbiotic associations with plants, and a type of photosynthetic bacteria called cyanobacteria carry out the energetically expensive process of nitrogen fixation, the assimilation of atmospheric nitrogen into ammonia (NH_3), catalyzed by the enzyme complex nitrogenase. The reductase activity of the enzyme complex transfers electrons (e^-) from a high-potential donor to the nitrogenase component, which then adds those electrons to N_2 to make NH_3, which quickly converts to ammonium (NH_4^+) in solution. The following reaction summarizes the fixation of a single N_2 molecule.

$$N_2 + 8e^- + 16ATP + 16H_2O \rightarrow \\ 2NH_3 + H_2 + 16ADP + 16P_i + 8H^+$$

The amino acids glutamate and glutamine play an important role in the next step of assimilating nitrogen into biomolecules. Most other amino acids obtain their α-amino group from the α-amino group of glutamate by a process called transamination. Glutamate itself is synthesized from α-ketoglutarate, a five-carbon intermediate of the citric acid cycle, in a reaction catalyzed by glutamate dehydrogenase, an enzyme that is especially active in the liver and kidney.

$$NH_4^+ + \alpha\text{–ketoglutarate} + NADPH + H^+ \rightarrow \\ \text{glutamate} + NADP^+ + H_2O$$

The same enzyme catalyzes the reverse reaction, releasing an ammonium ion, depending on the relative concentrations of reactants. In the catabolic reaction, oxidized nicotinamide adenine dinucleotide (NAD^+) is the reducing agent. Glutamine synthetase incorporates another ammonium ion into glutamate to create glutamine.

$$\text{glutamate} + NH_4^+ + ATP \rightarrow \\ \text{glutamine} + ADP + P_i + H^+$$

Depending on the organism, of the 20 amino acids necessary for protein synthesis, some are nonessential, meaning the organism has enzymes capable of biosynthesizing them, and the rest are essential, meaning they must be obtained through diet or absorbed from the external environment. For example, adult humans are capable of synthesizing 11 amino acids, and the other nine are essential; they are shown in the following table. Many different biochemical pathways cooperate to synthesize the nonessential amino acids, but the carbon skeletons are all obtained from the central metabolic pathways of glycolysis, the pentose phosphate pathway, and the citric acid cycle. The synthesis of the essential amino acids is much more complex than that of nonessential amino acids.

NONESSENTIAL AND ESSENTIAL AMINO ACIDS

NONESSENTIAL AMINO ACIDS	ESSENTIAL AMINO ACIDS
alanine	histidine
arginine	isoleucine
asparagine	leucine
aspartate	lysine
cysteine	methionine
glutamate	phenylalanine
glutamine	threonine
glycine	tryptophan
proline	valine
serine	
tyrosine	

Synthesis of the nonessential amino acids is not very complex, and many share common metabolic intermediates as precursors. As already mentioned, the glutamate carbons can be derived from α-ketoglutarate, an intermediate of the citric acid cycle, and glutamate amination results in glutamine. Glutamate also serves as a precursor for proline and arginine. Proline is synthesized by first phosphorylating glutamate, then reducing it to form an intermediate that cyclizes and is further reduced. Alternatively, the reduced intermediate can be transaminated into ornithine, an intermediate of the urea cycle that can be converted into arginine by several additional steps. Transferring an amino group from glutamate to pyruvate, the end product of glycolysis, results in alanine and α-ketoglutarate, and transferring an amino group from glutamate to oxaloacetate, an intermediate of the citric acid cycle, yields aspartate and α-ketoglutarate. The addition of an amide group to aspartate results in asparagine. An intermediate of glycolysis, 3-phosphoglycerate, converts to serine by an oxidation step, a transamination, and hydrolysis of the phosphate. Glycine can be synthesized from serine or glyoxylate, but also by using the essential amino acid threonine as a precursor. Serine provides the carbons, and the essential amino acid methionine provides the sulfur for synthesizing cysteine. Hydroxylation of the essential amino acid phenylalanine produces the 11th nonessential amino acid, tyrosine.

AMINO ACID DEGRADATION AND THE UREA CYCLE

Unlike carbohydrates or lipids, cells cannot store an excess of amino acids. The body breaks them down for fuel by first removing the amino group, which is converted to urea. The carbon backbone feeds into an amphibolic pathway, one that plays a role in both anabolism and catabolism, via acetyl coenzyme A (acetyl CoA), acetoacetyl CoA, pyruvate, or an intermediate of the citric acid cycle.

The liver is the major site for amino acid degradation in mammals. Aminotransferases transfer an α-amino group from amino acids to α-ketoglutarate, forming glutamate in the process. In this manner, glutamate collects the nitrogen from other amino acids. Oxidative deamination of glutamate by glutamate dehydrogenase releases NH_4^+. The high-energy molecules adenosine triphosphate (ATP) and guanosine triphosphate (GTP) allosterically inhibit glutamate dehydrogenase, whereas adenosine diphosphate (ADP) and guanosine diphosphate (GDP) activate it. Because of this, when the energy charge is low, such as during starvation, oxidation of amino acids occurs, creating intermediates for catabolic pathways. The enzymes serine dehydrogenase and threonine dehydrogenase directly deaminate the amino acids serine and threonine to release NH_4^+.

Depending on metabolic needs, some of the NH_4^+ will be used to synthesize other nitrogenous biomolecules, but the majority becomes nitrogenous waste destined for excretion. Most terrestrial vertebrates excrete nitrogenous waste as urea, but reptiles and birds excrete uric acid, and aquatic animals excrete NH_4^+. One molecule of urea contains two nitrogen atoms—one from NH_4^+ and the other removed from the amino acid aspartate. The urea cycle, the first metabolic pathway to be discovered, is a cyclical biochemical pathway responsible for the synthesis of urea. Part of the cycle occurs in the mitochondrial matrix, and part occurs in the cytoplasm of the cell. In the first mitochondrial reaction, ornithine transcarbamoylase transfers a carbamoyl (–C=O) from carbamoyl phosphate to ornithine, generating citrulline. After citrulline is transported into the cytoplasm, argininosuccinate synthetase catalyzes the condensation of citrulline and the amino acid aspartate to create argininosuccinate using energy obtained from ATP. Argininosuccinase breaks argininosuccinate into the amino acid arginine and fumarate, a four-carbon intermediate that links the urea cycle to the citric acid cycle. Arginase hydrolyzes arginine into urea and ornithine, completing the cycle. The net reaction for the synthesis of urea is

$$CO_2 + NH_4^+ + 3ATP + \text{aspartate} + 2H_2O \rightarrow \text{urea} + 2ADP + 2P_i + AMP + PP_i + \text{fumarate}$$

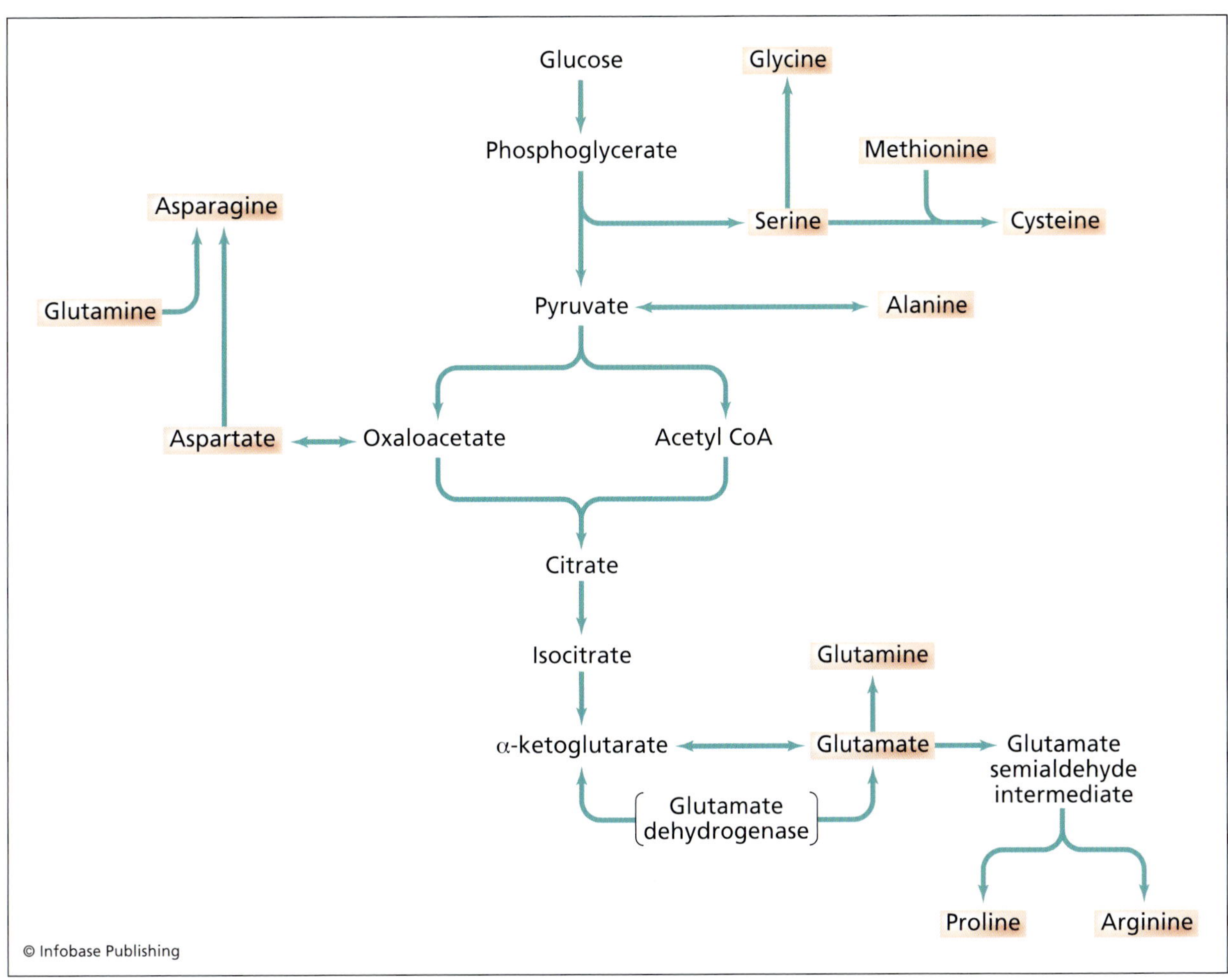

Humans can synthesize 10 nonessential amino acids from intermediates of glycolysis or the citric acid cycle. Hydroxylation of the essential amino acid phenylalanine creates the 11th—tyrosine. (Note: A single arrow does not necessarily denote a single reaction in this diagram.)

The carbon skeletons that remain after removal of the amino group are converted into glucose derivative or feed into the citric acid cycle. Amino acids can be grouped according to the metabolic fate of their carbon skeletons after deamination. Glucogenic amino acids are degraded into pyruvate, α-ketoglutarate, succinyl CoA, fumarate, or oxaloacetate. Ketogenic amino acids are degraded into acetyl CoA or acetoacetyl CoA. Because mammals do not have a mechanism for synthesizing glucose from acetyl CoA or acetoacetyl CoA, these compounds give rise to ketone bodies, any of three intermediates of fatty acid degradation—acetoacetic acid, acetone, or β-hydroxybutyric acid. Categorizing amino acids by this method is not straightforward, as isoleucine, phenylalanine, threonine, tryptophan, and tyrosine give rise to both glucose precursors and ketone bodies.

The three-carbon amino acids alanine, serine, and cysteine are converted into pyruvate, as are glycine, threonine, and tryptophan. The four-carbon amino acids aspartate and asparagine enter the citric acid cycle through oxaloaceate. Arginine, histidine, glutamine, and proline are all converted into glutamate, which bridges the entry of these five-carbon amino acids into the citric acid cycle by deamination into α-ketoglutarate. The nonpolar amino acids valine, isoleucine, and methionine are converted into succinyl CoA. Phenylalanine and tyrosine can be degraded into acetoacetate and fumarate. Lysine and leucine, two strictly ketogenic amino acids, are degraded to either acetyl CoA or aceteoacetate. Degradation of several amino acids can occur by alternate pathways—for example, threonine can be converted to acetyl CoA or succinyl CoA in addition to pyruvate, and aspartate can enter the citric acid cycle through fumarate as well as oxaloacetate.

Enzyme deficiencies of amino acid metabolism cause diseases such as phenylketonuria (PKU). More than 100 different genetic mutations in the gene encoding the enzyme phenylalanine hydroxylase,

Mitochondrion
$CO_2 + NH_4^+$
2 ATP
2 ADP + P_i
Carbamoyl phosphate synthetase 1 (CPS1)
1
Carbamoyl Phosphate
P_i
2 Ornithine transcarbamoylase
Ornithine
Citrulline
Cytosol
Ornithine
Citrulline
Urine
Urea
5 Arginase
H_2O
Arginine
4 Argininosuccinase
Fumarate
To citric acid cycle
Urea Cycle
Argininosuccinate
3 Argininosuccinate synthetase
ATP
AMP + PP_i
Aspartate

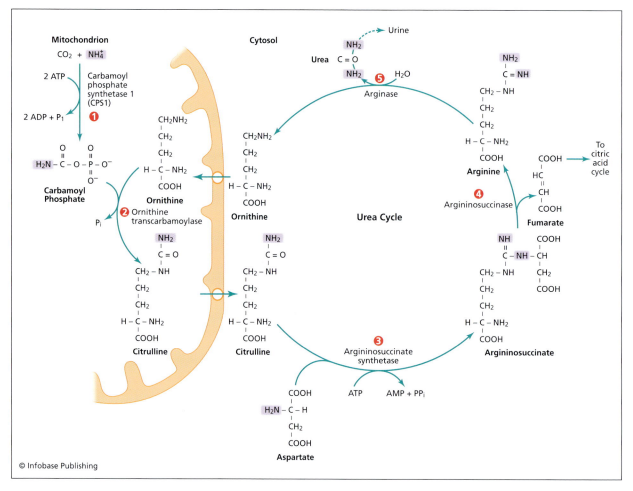

Different stages of the urea cycle, the metabolic pathway that synthesizes urea, occur in the cytosol and the mitochondrial matrix.

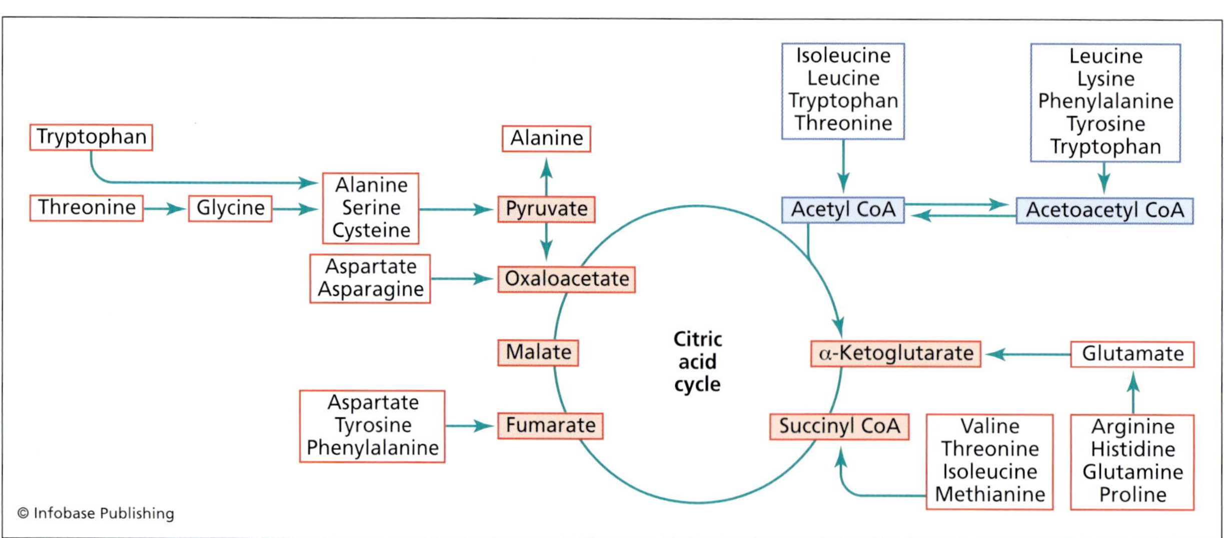

The carbon backbones of some amino acids are converted into pyruvate or citric acid cycle intermediates—these are called glucogenic and are indicated in red. Others, called ketogenic, are converted into ketone bodies and are indicated in blue.

which converts phenylalanine to tyrosine, have been identified. These mutations lead to PKU, a disease characterized by the accumulation of phenylalanine and its degradation products. If it is diagnosed by a simple blood test at birth, the implementation of strict dietary restrictions can prevent brain damage and progressive, irreversible mental retardation. Alcaptonuria results from defective phenylalanine and tyrosine catabolism and is relatively harmless but causes the urine to darken when it is exposed to air. Maple syrup urine disease, distinguished by its symptom of sweet-smelling urine, occurs when the catabolism of valine, leucine, and isoleucine is defective. The myelin in the central nervous system does not form properly, and those who have this disorder are mentally retarded, develop abnormally, and have shortened life spans. Related disorders of amino acid metabolism, such as hypermethioninemia, homocystinuria (or -emia), and cystathioninuria, represent biochemical abnormalities, as opposed to specific disorders, and thus can each result from a variety of causes.

Mutations in the genes involved in the urea cycle or in the synthesis of carbamoyl phosphate (a necessary reactant) lead to hyperammonemia, excess NH_4^+ in the blood. Without the ability to synthesize urea, brain damage and death result, though some forms are milder than others. Treatments include strict control of the amino acid and total protein quantities through dietary restrictions to force other metabolites to carry nitrogen out of the body or to bypass defective pathways by activating other latent biochemical pathways.

See also CITRIC ACID CYCLE; ENZYMES; METABOLISM; PROTEINS.

FURTHER READING

Berg, Jeremy M., John L. Tymoczko, and Lubert Stryer. *Biochemistry,* 6th ed. New York: W. H. Freeman, 2006.

King, Michael. "The Medical Biochemistry Page." Available online. URL: http://themedicalbiochemistrypage.org/. Last modified July 11, 2008.

Smith, Colleen M., Allan Marks, and Michael Lieberman. *Marks' Basic Medical Biochemistry: A Clinical Approach,* 2nd ed. Philadelphia: Lippincott, Williams, & Wilkins, 2004.

analytical chemistry Analytical chemistry is the branch of chemistry responsible for the separation, identification, and quantification of substances present in a sample. Research in the field of analytical chemistry is divided into two types of analysis: qualitative analysis and quantitative analysis. Qualitative analysis concerns the identity and properties of components in a sample, whereas quantitative analysis studies the amount of substances in a sample. The steps to quantitative analysis include either complete analysis or partial analysis. Complete analysis attempts to determine the quantities of every substance of a sample. Partial analysis, the most common technique, determines the amount of one or a small number of substances in a sample. The statistical analysis of the data collected in any of the experimental techniques is also the responsibility of an analytical chemist.

EDUCATION AND TRAINING

Education and training in the field of analytical chemistry require a strong background in general chemistry, physical chemistry, computers, and statistics. Laboratory training is critical, and analytical chemistry education involves extensive study and training in analytical techniques. The type of degree required for employment as an analytical chemist depends upon the type of job desired. Degrees in analytical chemistry range from four-year bachelor of science degrees (in chemistry) to research-based master of science degrees and doctoral degrees. In general, laboratory technician positions require a bachelor's or master's degree. Supervisory roles require either a master's degree or a doctoral (Ph.D.) degree. Full-time academic positions generally require a Ph.D., and many require additional postdoctoral work in the field.

Analytical chemistry positions are as diverse as the areas of study. Fields of research include varied topics such as drug development, water analysis, disease study and treatment, food inspection, and nutrition. Analytical chemists can hold positions in academia and business as well as governmental jobs.

EXPERIMENTAL DESIGN AND DATA ANALYSIS

Analytical chemistry requires solid experimental design and analysis. Sampling should accurately represent the entire set of components and be performed in a random manner in order to prevent any bias. Analysis of results involves statistical hypothesis testing procedures such as analysis of variance (ANOVA) to compare results from multiple populations or a Student's t-test, which compares two populations to each other or one population to a mean value. A t-test allows for the comparison of several samples to make an overall statement about the amount of a particular substance in the source sample. An example of an application of these data analysis techniques is a water-testing experiment. An analytical chemist might measure the chlorine levels in a sample of drinking water. Statistical analysis of the data allows one to conclude whether the levels are within safe limits for drinking through the comparison of the chlorine levels in different water sources.

TECHNIQUES

The field of analytical chemistry draws conclusions from a vast array of techniques, capitalizing on the physical properties of the sample components. Many classes of techniques are used; these include gravimetry, titrimetry, electrochemistry, spectrometric techniques, and chromatography.

Gravimetry

Studies of solution formation and reaction equilibria are common in analytical chemistry. A solution is a homogeneous mixture of a solute, the substance that dissolves, and a solvent, the substance that does the dissolving. The most common type of solution is an aqueous solution, meaning water is the solvent. Gravimetric analysis techniques are based on the weight of a sample. The amount or concentration of a solute can be determined by using gravimetric techniques such as equilibrium determinations or electrolyte analysis.

The determination of electrochemical properties due to conduction of electricity involves the identification of electrolytes versus nonelectrolytes. Electrolytes are substances that conduct an electric current when dissolved in aqueous solution. Most ionic compounds are electrolytes, as they ionize in water to give positive and negative ions that are capable of conducting an electrical current.

The study of equilibrium reactions is essential to analytical chemistry. Equilibrium is the condition at which the net amount of reactants and products does not change over time. The equilibrium constant (K_{eq}) of a particular reaction relates to the concentrations of reactants and products in the following way:

$$A + B \rightarrow C + D$$

$$K_{eq} = \frac{[C][D]}{[A][B]}$$

Analytical chemists study the equilibrium of reactions as well as which factors affect the equilibrium point of a reaction. LeChatelier's principle explains the response when a system at equilibrium is put under stress: when a system at equilibrium experiences a stress, such as change in temperature, concentration, or pressure, the system will respond to the stress by shifting the equilibrium point of the system. For example, if the concentration of product increases dramatically, then the reaction equilibrium will shift toward the right (toward formation of products). Removal of product as the reaction is occurring will also push the equilibrium to the right.

The quantity of acidic and basic compounds in a solution is important to analytical chemistry. Measuring and understanding the amount of H^+ present in a solution determine the pH using the formula

$$pH = -\log[H^+]$$

Titrimetry

Titrimetry techniques allow for the analysis of a sample based on volume. Titrations are performed to determine the concentration of a substance in a sample by comparison to a known volume of a concentration standard. The classic titration experiment involves acids and bases. If a sample contains a strong acid of unknown concentration, one can determine the concentration of the acid by determining how much base is required to neutralize it. Acid-base neutralization occurs when a strong base is added to the strong acid, forming an ionic salt and water as products.

$$NaOH + HCl \rightarrow NaCl + H_2O$$

This example utilizes both a strong acid (HCl) and a strong base (NaOH). Because HCl and NaOH both completely ionize, the reaction is written as irreversible. When the hydrochloric acid is completely neutralized, the pH should be 7.0; this is known as the end point or equivalence point of a titration. At this point, the amount of equivalents of base added is equal to the amount of equivalents of acid present in the original sample. Determining the amount of NaOH required to neutralize the HCl (since the OH^- and H^+ are in the same molar ratio) determines the number of moles of HCl present in the starting sample. Knowing the volume and concentration of base necessary to titrate the sample, and knowing the volume of the sample with the unknown quantity, one can calculate the concentration of the initial acid sample.

In a titration, the addition of an indicator aids in the determination of the end point. Indicators are chemical compounds that have different colors in acids and bases. A broad range of indicators exists with different color changes at different pH values. A typical titration takes place with the sample to be analyzed in a flask and the base or acid of known concentration in a burette, which allows for analytical delivery of the standard solution.

The amount of solute present in a solution (concentration) of a sample can be described using multiple units, depending on the most useful units for the experiment. Common concentration units are shown in the table on page 31.

The concept of titration can also be used for reactions in which a precipitate (a solid that falls out of solution) forms. Using a known concentration of a reactant, one can determine the concentration of

THE CHEMISTRY OF FORENSIC SCIENCE

by Amy J. Heston, Ph.D.
Walsh University

Chemistry, the study of matter, has been a valuable tool in the field of forensic science. The focus of this essay is the chemistry of forensic science, its importance, and its history. Examples of chemicals that assist in crime scene investigation will be discussed as well as techniques used by forensic scientists. Fingerprinting techniques and poisons will be described in detail. First, cyanoacrylate fuming assists in the identification of latent or hidden fingerprints on smooth surfaces. After fuming, the fingerprint will appear white and can be seen by the naked eye. Second, an overview of poisons will be given and their effects on the human body.

Forensic science is science as it is applied to the law. The first reports of forensic science date from as early as 44 B.C.E., when Antisius, a Roman physician, examined a body for the cause of death. Investigations were reported in China in the 1200s, and fingerprinting began in the 1600s. In 1901 Karl Landsteiner, an Austrian pathologist, invented the ABO blood-grouping system to illustrate the three main types of blood. He was awarded the 1930 Nobel Prize in physiology or medicine for his work. Recent advances in the field of forensic science include firearm databases and databases of fingerprints from 65 million people to assist in criminal identifications.

In recent years, techniques have been made to improve the visualization of latent or hidden fingerprints. The Japanese chemist Shin-ichi Morimoto conducted experiments using Super Glue®, also known as cyanoacrylate ester, to facilitate the visualization by the naked eye of fingerprints on nonporous, smooth surfaces such as plastic bags, metal weapons, and electric tape. In this process, Super Glue® is heated, producing a white fume or vapor. According to the British chemist Mark Stokes, the fumes interact with moisture in the perspiration of the fingerprint to make polymers, hardening the fingerprint, and the print turns white. While this facilitates visualization in many cases, the white prints from fuming are difficult to see on white or light-colored surfaces. To improve visibility, Morimoto incorporated two dyes: a red-purple dye and a blue dye. He mixed the dyes with black carbon powder, put the mixture onto a cigarette paper, and rolled it up tightly. In addition, he made incense sticks by mixing the dye with a thick cohesive, moistening them, and rolling the substances into sticks. These two techniques allowed the dye to sublime at a constant rate while it burned. (Sublimation is the process of a solid's changing to a vapor without going through the liquid state.) The cyanoacrylate ester was used in a fuming box and the dyes were sublimed. The main purpose for using the sublimed dyes was to improve the contrast of the print compared to that of Super Glue® fuming alone. This combination of techniques successfully increased the visibility of latent fingerprints found on a cigarette box and on a white plastic bag by coloring them. The results can assist crime scene investigators, leading to criminal identification. For example, some reports indicate that this technique can be used to fume the entire inside of a car associated with a crime.

Poisoning is another challenge that forensic scientists face. Poisons may include chemicals, metals, or plants. Two examples of potentially fatal chemical poisons are arsenic and cyanide. Arsenic affects the stomach and gut, and the victim will have blue-green fingers and toes. Cyanide affects the blood's ability to carry oxygen, and the victim becomes unconscious and then asphyxiates. Antimony, lead, and thallium are examples of heavy metal poisons. Antimony causes heart failure, and the warning signs are sweating, depression, and shallow pulse. Lead affects the brain and liver, and the victim will experience stomach pains and eventually drift into coma. Thallium destroys the nerves and is often mistaken for the flu because of its similarity in symptoms. Plants, fungi, and their products can also serve as poisons. The death cap mushroom affects the gut and causes the liver to fail, and symptoms include delirium and coma. Consumption of the perennial flowering plant called deadly nightshade paralyzes the lungs and heart, and the victim suffers from hallucinations, enlarged pupils, and coma. Strychnine, a white, toxic, odorless powder isolated from the tree *Strychnos nux-vomica,* paralyzes the lungs, causes violent muscle spasms, and eventually tears the muscle tissue.

Chemistry plays a critical role in forensic science and is a very important part of the evaluation of crime scene evidence. Research groups are working to improve the techniques used by crime scene investigators and develop new ways to identify those responsible for committing crimes.

FURTHER READING

Genge, Ngaire E. *The Forensic Casebook: The Science of Crime Scene Investigation.* New York: Ballantine, 2002.

Morimoto, Shin-ichi, Akira Kaminogo, and Toshinori Hirano. "A New Method to Enhance Visualization of Latent Fingermarks by Sublimating Dyes, and Its Practical Use with a Combination of Cyanoacrylate Fuming." *Forensic Science International* 97, no. 2–3 (1998): 101–108.

Platt, Richard. *Crime Scene: The Ultimate Guide to Forensic Science.* New York: DK, 2003.

Siegel, Jay A. *Forensic Science: The Basics.* Boca Raton, Fla.: CRC Press, 2007.

Stokes, Mark, and John Brennan. "A Free-Standing Cabinet for Cyanoacrylate Fuming." *Forensic Science International* 71, no. 3 (1995): 181–190.

Takekoshi, Yuzi, Kiyohito Sato, Susumu Kanno, Shozi Kawase, Tadashi Kiho, and Shigeo Ukai. "Analysis of Wool Fiber by Alkali-Catalyzed Pyrolysis Gas Chromatography." *Forensic Science International* 87, no. 2 (1997): 85–97.

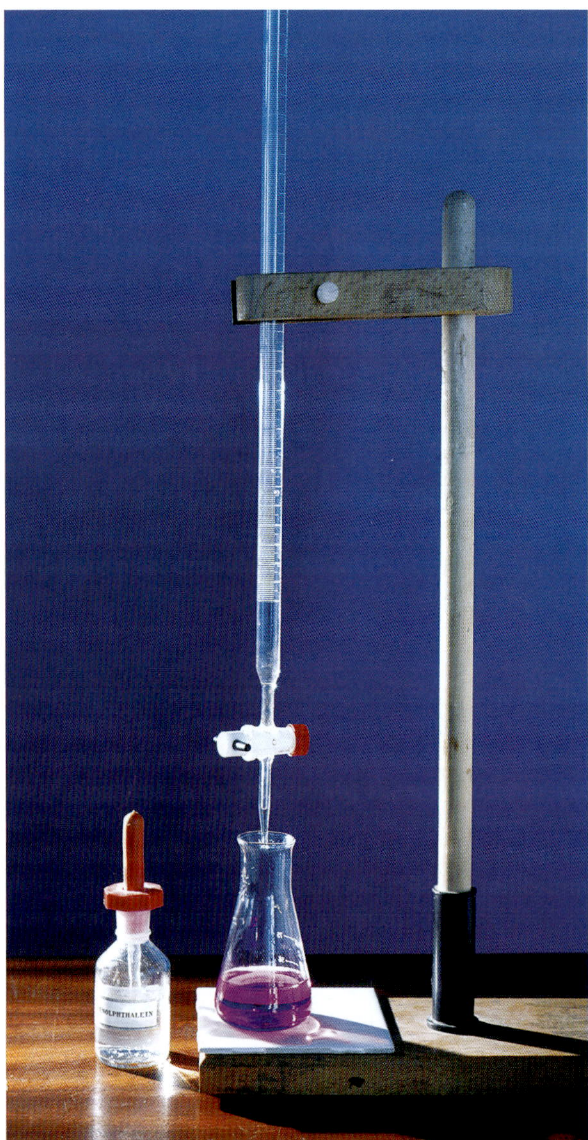

In a titration experiment a known concentration of an acid (in burette) is used to determine the concentration of a given volume of a base (in conical flask). *(Andrew Lambert Photography/Photo Researchers, Inc.)*

another reactant by measuring the amount of precipitated product formed upon the addition of the known quantity. This titration leads to the calculation of an unknown concentration from the volume of a standard solution.

Electrochemistry

Analytical chemistry determinations often require electrochemistry techniques that are based on the potential difference between sample components. Electrochemistry's role in analytical chemistry relates chemical reactions and the electrical properties associated with the reaction. Oxidation-reduction reactions involve the transfer of electrons between reactants. The compound that loses electrons is oxidized and is known as a reducing agent. The compound that gains electrons becomes reduced and is known as the oxidizing agent. The type of electrochemistry is categorized on the basis of the type of measurements made. Potentiometry is the study of quantitative measurements of potential differences in samples. An example of an analytical technique that utilizes potential differences is pH measurement. The concentration of H^+ ions in a solution can be determined on the basis of the electric potential differences measured by the pH meter. Electrolytic studies involve adding an electric current to a sample and measuring the resulting signal from the sample.

Spectrometric techniques

Spectrometric techniques give analytical chemists vast quantities of information about the sample based on the interaction of the sample components with electromagnetic radiation. The frequency (cycles/second) and wavelength (distance from one crest to another) of electromagnetic radiation determine the amount of energy present in the radiation in terms of the equation

$$E = hf$$

where E stands for the energy of the wave, h represents Planck's constant, and f represents the frequency of the radiation. Since the frequency f and wavelength λ of a wave are inversely related, the following formula represents the relationship:

$$f\lambda = c$$

where c represents the speed of light in a vacuum. Electromagnetic radiation is classified on the basis of the frequency of the waves, listed in order from lowest frequency to highest: radio waves, microwaves, infrared radiation, visible light, ultraviolet radiation, and gamma rays.

Techniques that utilize electromagnetic (EM) radiation are based on either the absorption of electromagnetic radiation by the sample being studied or the emission of electromagnetic radiation from the sample. Both mechanisms give information regarding the structure and/or amount of components in a sample.

Absorption of electromagnetic radiation can cause vibrational changes (change in bond length) or electronic changes (change in electrons). Analytical chemists study the absorption of electromagnetic radiation by atoms (atomic absorption) or molecules (molecular absorption). Absorption spectrum can be determined using ultraviolet radiation (UV), visible

CONCENTRATION UNITS

Concentration Unit	Definition	Symbol
molarity	Moles solute/liter solution	M
normality	Molarity × (number of H^+)	N
parts per million	Quantity of substance per million	ppm
part per billion	Quantity of substance per billion	ppb
% volume/volume	(vol. of solute)/(vol. of solution)	% v/v
% mass/volume	(mass of solute)/(vol. of solution)	% m/v

light (Vis), and infrared (IR). IR is commonly used for vibrational absorption, while UV/Vis is utilized for electronic changes. As UV and visible light have similar wavelengths, they are often measured on the same piece of equipment. Infrared readings require a separate machine.

Emission of electromagnetic radiation occurs when a sample becomes excited and its electrons are raised to a higher energy level. When the electrons lose that energy and fall back down to a lower energy state, they release electromagnetic radiation in a recognizable frequency that can be compared between samples. Emission spectra include line spectra (contain only certain wavelengths), band spectra (wavelengths in specific regions), continuous spectra (light of all wavelengths), phosphorescence (absorbing high-energy radiation and releasing it over time), and fluorescence (reradiating energy with a different frequency).

Spectroscopy applies to all of the techniques that utilize absorption or emission of electromagnetic radiation to reveal information about a sample. Molecular spectroscopy utilizes UV/Vis spectroscopy and IR spectroscopy. Atomic spectroscopy utilizes such techniques as scintillation detectors, X-rays, and X-ray diffraction. Scintillation detectors detect and measure emissions given off from radioactive samples. X-ray diffraction is the scattering of X-rays as they pass through a crystallized form of the sample, leaving a diffraction pattern that can be translated into information concerning the molecular structure of the compound. Mass spectroscopy is a separation technique based on the mass-charge ratio of the ions. Analysis of the ions present in a sample can be quantified to determine the exact concentration of a substance in a sample.

Chromatography

Chromatographic techniques separate mixtures on the basis of the differential attraction of the mixture components to a specific matrix. The mobile phase contains the mixture and is transferred along through the stationary phase. Paper chromatography uses porous paper as the stationary phase. Placing a mixture on the paper and using a liquid mobile phase move the components of the mixture up through the paper with the smallest components and those least attracted to the paper moving the farthest, and the largest or those most attracted to the paper moving the least. Column chromatography utilizes a stationary phase that is placed in a column. The mobile phase containing the mixture to be separated runs through the column. The attraction of the individual mixture components to the matrix of the column determines how long they take to come off the column (elute). After the sample has run through the column, substances attached to the column can be removed (eluted) by changing the chemical properties of the solvent.

Gas chromatography requires that the sample be vaporized before traveling through the column, while a carrier gas moves the sample along. Because liquid chromatography does not require that the sample be vaporized, more samples can be subjected to this process.

COMBINATIONS OF ANALYTICAL CHEMISTRY TECHNIQUES

Analytical chemists have developed multiple combinations of techniques that identify mixture components more accurately. Examples of these combination methods are gas chromatography/mass spectroscopy (GC/MS), gas chromatography/Fourier transform infrared spectroscopy (GC/FTIR), gas chromatography/atomic emission spectroscopy (GC/AES), and liquid chromatography/mass spectroscopy (LC/MS). These combinations involve an initial separation of mixture components, followed by identification and quantification of the ions or molecules present in a sample.

See also CHROMATOGRAPHY; CONCENTRATION; ELECTROCHEMISTRY; ELECTRON EMISSION; EQUILIBRIUM; MASS SPECTROMETRY.

FURTHER READING

Rubinson, Judith F., and Kenneth A. Rubinson. *Contemporary Chemical Analysis,* 2nd ed. Upper Saddle River, N.J.: Prentice Hall, 2008.

aspirin Aspirin, acetylsalicylic acid, is a member of a class of pain relievers known as nonsteroidal anti-inflammatory drugs (NSAIDs). The medicinal applications of aspirin have been known since ancient times. Teas and elixirs made of willow bark relieved pain and reduced fevers. The chemical compound responsible for this effect, known as salicin, was first isolated in 1829 by the French chemist Henry Leroux. The Italian chemist Raffaele Piria isolated the active component from salicin, salicylic acid, in 1838. This form of the medication was widely used in the treatment of pain and fever for decades. Although powders of salicylic acid effectively treated pain and fever, they were damaging to the esophagus, stomach, and intestinal tract, causing stomach ulcers as well as internal bleeding in those treated. The extremely low pK_a (measure of acid dissociation) of salicylic acid causes this tissue damage. In 1853 the French chemist Charles Gerhardt first isolated an improved version, known as acetylsalicylic acid (ASA), with the chemical formula $C_9H_8O_4$. This new compound was not widely used at the time, but it prevented some of the adverse effects of salicylic acid treatment. Widespread use and production of ASA began in 1897, with Felix Hoffmann, a German chemist and employee of the German pharmaceutical company Bayer. Hoffmann rediscovered the more stable form previously abandoned by Gerhardt. Comparing salicylic acid to acetylsalicylic acid, the only structural difference between the two is the addition of an acetyl group. Bayer produced the first water-soluble tablet form of acetylsalicylic acid in 1900.

SYNTHESIS OF ASPIRIN

The synthesis of aspirin is a standard experiment conducted in introductory organic chemistry laboratory classes. Commercial industry also uses the laboratory synthesis process for the production of pure aspirin.

The starting materials are salicylic acid and acetic anhydride, which is a dehydration product of the combination of two acetic acid molecules. The reaction takes place using acetic anhydride, which is a liquid, as the solvent. In order for the reaction to proceed at an appreciable rate, the addition of sulfuric acid as a catalyst is necessary.

Acetic anhydride + salicylic acid →
acetylsalicylic acid + acetic acid

After the reaction, the acetylsalicylic product can be isolated by the addition of cold water, which allows any unreacted acetic anhydride and excess acetic acid to stay in solution, while causing the aspirin, which has a low solubility in cold water, to precipitate. By filtering the sample, the aspirin stays in the filter paper and the acetic acid and acetic anhydride pass through in solution.

Testing the purity of the compound is critical if the sample is to be used and sold as a pharmaceutical agent. Given pure reactants, the only possible contaminant at this point would be unreacted salicylic acid. A ferric chloride test for salicylic acid produces a strong violet color if salicylic acid is present. An analysis of the infrared spectrum of the sample also reveals whether the sample is sufficiently pure.

PROSTAGLANDINS

The primary effects of aspirin include fever reduction, pain relief, reduction of inflammation, and prevention of heart attacks in those who have had previous heart attacks. Aspirin completes these tasks by inhibiting the production of a class of biochemical compounds called prostaglandins, a group of local hormones that were originally discovered in secretions from the prostate gland, giving rise to their name. Evidence suggests that almost every cell type in the body produces prostaglandins in response to stress. There are several classes of prostaglandins named prostaglandin A (PGA) through PGI. All of the classes contain 20 carbons and include a five-carbon ring. The amounts of double bonds outside the ring are represented by a subscript after the name. Every class of prostaglandin is created from a precursor prostaglandin H_2 molecule. PGH_2 can be converted to all the other classes of prostaglandins or into thromboxanes (a class of molecules involved in platelet aggregation, vasodilation, and vasoconstriction). Body cells produce PGH_2 through a series of reactions that release the precursor molecule arachadonate from phospholipids and diacylglycerols. The reactions that produce PGH_2 require the enzyme cyclooxygenase, the molecular target for aspirin action. Humans have two forms of the cyclooxygenase (COX-1 and COX-2), and aspirin inhibits them both, thus blocking the production of prostaglandins.

FEVER REDUCTION

Aspirin and the precursors to aspirin have been known for centuries to have antipyretic properties, meaning they help relieve fever. The region of the brain known as the hypothalamus regulates body temperature by sending out signals to the body in response to temperature shifts. For example, the body may respond to a decrease in temperature by shivering, an act that increases body temperature. An increase in body temperature may stimulate the sweat glands to produce and secrete sweat and increase circulation to the extremities, where heat is lost from the body, in an attempt to lower the body temperature. The release of chemicals called pyrogens from macrophages in response to an invading organism or tissue damage causes the body temperature of an organism to rise. If the body temperature rises too high, then the hypothalamus sends signals that cause sweating and vessel dilation in order to lower the body temperature back to normal. Prostaglandin production by damaged tissues inhibits the hypothalamus's ability to regulate and reduce temperature. Since aspirin blocks the production of prostaglandins, it can help reduce fever.

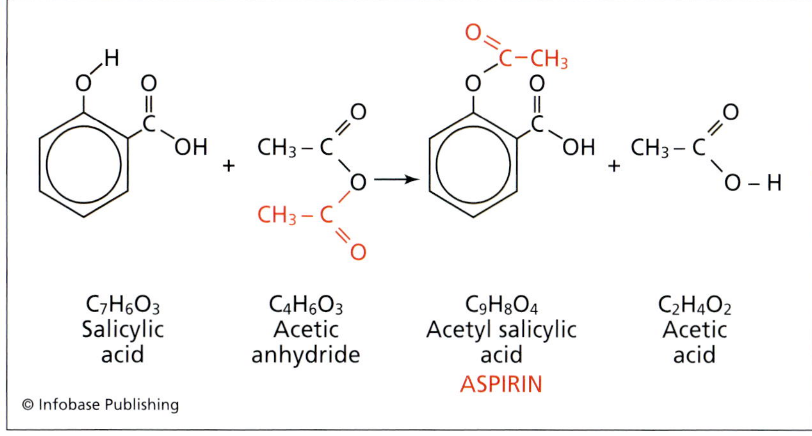

The reaction of salicylic acid and acetic anhydride forms acetic acid and acetylsalicylic acid, or aspirin.

ANTI-INFLAMMATORY AND ANALGESIC EFFECTS OF ASPIRIN

The inhibition of prostaglandins also leads to the inhibition of the inflammatory response in the body. The inflammatory response is part of the body's second line of defense and can be triggered by chemicals, trauma, heat, bacteria, or viruses. The inflammatory response is nonspecific and leads to four characteristic signals: redness, heat, pain, and swelling in response to tissue damage. In response to damage, cells release chemical signals, such as histamine and prostaglandins, which lead to dilated blood vessels, leaky capillaries, and white blood cells moving to the area of damage. All of these responses are protective, in that they increase the amount of blood, clotting proteins, white blood cells, and nutrients to the area in order to repair the damage; however, inflammation also causes discomfort and pain. Aspirin affects the inflammatory response by blocking the production of prostaglandins needed for this response as well as inhibiting the prostaglandins required for blood clotting.

Aspirin reduces the painful effects of injury and tissue damage. One particularly useful application is pain relief from inflammation caused by rheumatoid arthritis. The relief offered by aspirin results from the inhibition of prostaglandin production in damaged tissues. The effective dose range for an adult is between 600 and 1,000 mg repeated every four hours, up to six times per day.

HEART ATTACK AND STROKE PREVENTION

According to the American Heart Association, taking aspirin at low doses over an extended period can also reduce the risk of blood clotting. The mechanism for this anticoagulant function of aspirin lies in its ability to inhibit two enzymes: COX-1 and COX-2. Blockage of these two enzymes inhibits the ability of platelets to adhere to one another in order to form a clot. Because aspirin blocks the synthesis of prostaglandins, it also blocks the production of thromboxane A-2, a chemical derived from prostaglandins. This prevents aggregation of the blood cells and subsequent clot formation, decreasing the risk of a blood clot's forming and blocking a blood vessel, especially vessels that are already at risk due to plaque buildup or fat blockages. If a clot forms and blocks blood flow to the brain, a stroke occurs; if the blockage prevents blood flow to the heart, then a heart attack can result from the lack of blood supply to the heart tissue. Physicians often recommend that patients with such a history take a low-dose or "baby" aspirin (81 mg) every day.

ADVERSE EFFECTS OF ASPIRIN

According to the U.S. Food and Drug Administration (FDA), overdose of aspirin is a dangerous possibility. Aspirin is present in many medications such as Alka-Seltzer and Pepto-Bismol, so one must be aware when taking multiple medications simultaneously to prevent an accidental overdose. The side effects of aspirin overdose include gastrointestinal pain and bleeding, ringing in the ears, nausea, and vomiting. A doctor should be contacted immediately if one experiences dark, bloody vomit or stools that resemble coffee grounds as these may be indicators of stomach or intestinal bleeding.

Reye's syndrome is a rare, potentially fatal disease that occurs in children who have flu, chicken pox, or other viral infection. The liver and brain are the tissues that are most damaged by the disease. The use of aspirin and other NSAIDs during the viral infection increases the incidence of Reye's syndrome; thus advisories warn against allowing children under the age of 12 to take aspirin. Pregnant and nursing mothers also should not take aspirin, as it passes the placental barrier as well as into breast milk and has damaging effects on the fetus or baby.

ALTERNATIVES TO ASPIRIN

According to the Aspirin Foundation, aspirin is one of the least expensive pain relievers in the world and is used in more than 50 over-the-counter medications. As scientific advances increase and organic synthesis techniques improve, replacements and modifications to aspirin have reached the market. Acetaminophen is not an NSAID and does not affect the inflammatory process but affects the brain's perception of pain; thus it is still considered an analgesic.

The inflammatory response results in the release of COX-2, so specific inhibition of COX-2 helps reduce the side effects of aspirin. One area of research includes the selective inhibition of COX-2. The two versions of cyclooxygenase appear to have similar functions, but COX-1 is found in nearly all cell types, while COX-2 is present only at sites of inflammation and infection. Selective inhibition of COX-2 would be productive and would prevent the gastrointestinal side effects found with aspirin. The production of COX-2 inhibitors, such as Vioxx®, was initially thought to be an incredible advance in pain treatment for such cases as rheumatoid arthritis. Doubled risk of heart attacks compared to that of placebo groups caused Vioxx® (known by the generic name *rofecoxib*) to be removed from the market in 2004. Other COX-2 inhibitors, such as Celebrex®, (known by the generic name *celecoxib*), are still on the market and do not appear to have this effect.

See also CHEMICAL REACTIONS; ORGANIC CHEMISTRY.

FURTHER READING

Jeffreys, Diarmuid. *The Remarkable Story of a Wonder Drug.* New York: Bloomsbury, 2004.

atmospheric and environmental chemistry Environmental issues that have developed in industrialized economies have led to an emerging multidisciplinary field of chemistry known as environmental chemistry, beginning in the late 1980s. Primarily concerned with understanding how environmental processes function naturally and how humans impact that natural environmental flow, environmental chemistry includes the study of water, soil, and air chemistry. Atmospheric and environmental chemists compare the natural state of these resources and determine how they are affected by human activities. Clearly one of the most diverse fields of chemistry, environmental chemistry combines the fields of organic chemistry, inorganic chemistry, biochemistry, analytical chemistry, and toxicology as well as biology and ecology. Many industries require the use of environmental chemistry, including the petroleum, mining, pesticide, agricultural, marine and fishing, and food production industries. Environmental chemists must be able to understand the chemical reactions and interactions taking place in the environment as well as the biological impact of those reactions. Environmental chemists spend a large part of their time separating, isolating, and analyzing the components of the soil, water, and air. Environmental chemistry can be divided into several areas of focus, including the impact of human activity on the following: the atmospheric chemistry of the upper and lower atmosphere; the chemistry of water, including the oceans, lakes, rivers, streams, and drinking water; the study of weather and its impact on the planet; air pollution in cities and towns; and food production.

CONCERNS OF ENVIRONMENTAL CHEMISTS

Atmospheric chemists study the changes that take place in the upper and lower (troposphere) atmosphere. These include but are not limited to the study of ozone (O_3) formation and the greenhouse effect and global warming. Ozone, a molecule that exists in two resonance structures, resides in the upper atmosphere (specifically the stratosphere) and acts as a shield to the exposure to ultraviolet radiation from the Sun. Ozone in the lower atmosphere (the troposphere) is one of the main contributors to air pollution and smog. Many large cities, such as Los Angeles, now advertise which days are ozone-action days, days that the elderly or young are advised to remain indoors. The primary source of ozone production in the lower atmosphere is fires. Naturally started (primarily through lightning) and intentionally started fires constitute the major contributors to ozone as air pollution. The burning of the rain forests as well as the boreal forests by humans dramatically increases the amount of ozone produced. Burning estimates include 100,000 square kilometers (10 million hectares) of northern boreal forests, 400,000 square kilometers (40 million hectares) of tropical forests, and 5–10 million square kilometers (500 million–1,000 million hectares) of savannah.

The study of water in lakes, rivers, and streams includes the determination of impact of substances being discharged into these water sources. The environmental effect of the pollutants from business, industry, and citizens is analyzed, and cleanup efforts are coordinated on the basis of the impact and the chemicals involved. Acid rain caused by the release of nitrogen oxides and sulfur dioxides from vehicles and factories has a dramatic impact on the water and structures of an area. When the pH of the rainwater decreases, or becomes more acidic, the aquatic life of the affected bodies of water can be destroyed. Acid rain is capable of destroying bridges, buildings, and other structures as well.

The effects of human activities on weather patterns are most clearly demonstrated by the greenhouse effect. Emissions of carbon dioxide from the burning of fossil fuels and other carbon-based fuels leads to an increase in the amount of atmospheric carbon dioxide that insulates the planet and increases the amount of the Sun's energy that is retained in the atmosphere, leading to an increase in the planet's temperature. Regulations regarding carbon dioxide emissions and the effect of these emissions are at present a hotly debated topic.

The research and development of methods to dispose of nuclear waste from nuclear power plants and nuclear weapons projects are of special importance to environmental chemists. Controlling the release of nuclear waste into the environment is critical in order to protect the air, water, and soil surrounding the nuclear waste. Many types of nuclear waste will not decay for thousands of years; thus its storage or disposal affects many generations to come.

Other areas of interest to environmental chemists include exposure to consumer chemicals such as asbestos, lead paint, mercury in the water, PCB, Freon®, Teflon® in cooking implements, and dry cleaning chemicals, to name a few. Environmental chemists are also studying alternative fuel source development, food quality, air quality, water contamination, safety and cleanup of hazardous chemicals, as well as environmental disasters such as Chernobyl, the Exxon *Valdez* and other oil spills, and wetlands preservation.

CAREERS IN ENVIRONMENTAL CHEMISTRY

Education and training required for environmental chemists include a bachelor's degree in environmental chemistry, environmental science, or chemistry in order to work as a technician. A master's degree with lab experience will create more opportunities in laboratory work and industry, and a Ph.D. is required for most academic research and teaching positions at colleges or universities and laboratory director positions. Coursework involved in environmental chemistry includes much cross-curricular coursework. A well-trained environmental chemist completes coursework in toxicology, organic chemistry, inorganic chemistry, analytical chemistry, atmospheric chemistry, nuclear chemistry, petroleum chemistry, and geochemistry. Proficiency in technical writing and research is also recommended. Business courses are useful because many of those chemists work in industry and need to understand the business side of the industry. Communication skills are especially important for environmental chemists because they often act as liaisons between environmental groups and businesses in an area or among the environmental groups, business interests, and government issues in an area. If one is interested in working in the field, then much work of an environmental chemist can take place outdoors.

According to the U.S. Bureau of Labor Statistics, in 2006 environmental scientists filled approximately 92,000 jobs in business, industry, and academic positions. About 44 percent of environmental scientists were employed in state and local governments; 15 percent in management, scientific, and technical consulting services; 14 percent in architectural, engineering, and related services; and 8 percent in the federal government. About 5 percent were self-employed. Job opportunities for an environmental chemist range from studying solid, water, and air samples in a laboratory; to acting as a liaison for business and industry with the government to ensure that the businesses are complying with the best standards regarding pollution and environmental impact of their product and production process. Environmental chemists are often responsible for maintaining production standards that comply with permits and governmental regulations. Environmental chemists can work as toxicology consultants, risk assessors, research and development chemists, project chemists, and safety specialists.

Several organizations pursue environmental chemistry, including the American Chemical Society (ACS), which has an environmental chemistry branch. Another leading organization in environmental chemistry and environmental science is the Cooperative Institute for Research in Environmental Science (CIRES), which is a joint cooperative between the University of Colorado at Boulder and the National Oceanic and Atmospheric Administration (NOAA). This partnership began in the 1960s and now fosters research in all aspects of environmental chemistry. Climate system variability includes the study of both the short-term and long-term effects of climate changes on the environment and on living organisms. Short-term climate effects include altered weather patterns and the effects of weather events, such as floods and tornadoes. Another organization of environmental chemists is the Society of Environmental Toxicology and Chemistry (SETAC), which combines the interests of educational institutions, business, and governmental agencies. The goals of SETAC in North America include maintaining, regulating, and protecting environmental issues as well as educating the public about the information that is available about environmental issues. SETAC states as their goal to be a communication forum for the interactions of business, government, and academia for the protection and welfare of the general public.

One of the fundamental tasks of all environmental chemists is to separate (speciate), analyze, and quantify the amount of a particular chemical or pollutant found in the environment. Scientists need to be able to gather information about the elemental, chemical, and structural nature of substances in addition to the structural integrity of samples in order to determine whether a sample is being detrimentally affected.

A wide variety of techniques exist for the separation of samples including distillation, chromatography, and centrifugation, as well as some techniques that are particular to this field. Distillation separates a mixture on the basis of differential boiling points of the substances and is especially useful when solid contaminants are present in a liquid solution. Chromatography is based on the attraction of particles in the solution to the chromatography matrix. Substances can be separated by size, charge, or affinity for any substance attached to the matrix. Centrifugation involves spinning samples and separating in terms of differential size or density.

Many analytical techniques are available as well as combinations of techniques that allow for precise determination of sample components. Spectroscopy is an important method of analysis that is based on the absorption, emission, or scattering of light. Fourier transform infrared spectroscopy (FITR) combined with microscopy allows for the identification of unknown samples and the analysis for quality and purity of samples. Auger spectroscopy allows for the contamination analysis of samples and trace analysis of samples. This method is a nondestructive surface level analysis that does not destroy the integrity of the sample.

Electron microscopy, including both transmission electron microscopy (TEM) and scanning electron microscopy (SEM), is a valuable tool for evaluating the composition and structural characteristics of a substance at the surface (SEM) and on the interior (TEM) of the substance. TEM allows for the observation of crystal defects in substances that would not be detectable from the surface.

Environmental chemists often use detection systems such as inductively coupled plasma-atomic emission spectrometer (ICP-AES) and inductively coupled plasma-mass spectrometers (ICP-MS) in order to analyze samples. ICP involves the introduction of a solution into a chamber known as a torch that contains plasma (usually ionized argon). ICP-AES is based on the presence of excited atoms in a substance introduced to plasma; the excited atoms give off characteristic wavelengths according to the elements found in the sample. The stronger the electromagnetic radiation that is given off, the more concentrated the sample must be for that particular element, allowing for quantitative calculations of component concentrations in a mixture. In order to be used in ICP-AES, the sample must first be dissolved in water. If the sample is a liquid sample, it can be used directly. In ICP-AES the sample is then analyzed using an atomic emission spectrometer. ICP-MS allows for the detection of quantities as small as one part per trillion. Chemists use this sensitive technique to analyze samples of many metals and a smaller range of nonmetals. Once the sample is introduced into the plasma torch, analysis using mass spectrometry aids in the identification of the substances in the mixture, which is based on the mass-to-charge ratio. The strength of the signal is based on the concentration of the substance in the sample.

See also ACID RAIN; AIR POLLUTION (OUTDOOR/INDOOR); ALTERNATIVE ENERGY SOURCES; ANALYTICAL CHEMISTRY; CENTRIFUGATION; CHROMATOGRAPHY; GREEN CHEMISTRY; GREENHOUSE EFFECT; MASS SPECTROMETRY; SEPARATING MIXTURES.

FURTHER READING

Andersen, Jesper, L. Schluter, and G. Ertebjerg. "Coastal Eutrophication: Recent Developments in Definitions and Implications for Monitoring Strategies." *Journal of Plankton Research* 28, no. 7 (2006): 621–628.

Cornelis, Rita, Joe Caruso, Helen Crews, and Kalus G. Heumann. *Handbook of Elemental Speciation.* Hoboken, N.J.: John Wiley & Sons, 2005.

atomic structure An atom is the smallest particle that retains the chemical properties of a specific element. These particles are extremely small, with 6.02×10^{23} (602 billion trillion) hydrogen atoms weighing only one gram. Lining up 100,000,000 atoms side by side, the line that they form is only 0.4 inch (1 cm) long. In 1981 scientists first saw atoms through a scanning tunneling microscope, an instrument designed by two Swiss scientists, Gerd Binnig and Heinrich Rohrer, in which atoms look like rows of bright spots on a monitor. No apparatus exists to date that permits scientists to visualize the subatomic structure of an atom. Relying on indirect observations and experimental data, such scientists as John Dalton, Sir J. J. Thomson, Sir Ernest Rutherford, and Niels Bohr made it their lives' work to elucidate a feasible model of atomic structure. Their combined efforts culminated in a subatomic picture of negatively charged particles called electrons surrounding a positively charged core, or nucleus, containing two other kinds of subatomic particles, called protons and neutrons. The electrons are often intuitively pictured as orbiting the nucleus in defined paths, or orbitals, much as the planets in the solar system revolve around the Sun. Such a picture, although it has its uses, is not rigorously allowed by quantum mechanics, which expresses ideas in terms of the probability of finding an electron at any location.

HISTORY OF THE ATOM

More than 2,000 years ago, Democritus of Abdera suggested the existence of small indivisible particles that he called in Greek what we call atoms. (The word *atom* is derived from Democritus's similar word meaning "incapable of being cut.") Yet, no scientists generated experimental data that might support the existence of these particles until the late 18th and early 19th centuries. Antoine-Laurent Lavoisier (1743–1794), a French chemist who is often called the founder of modern chemistry, discovered that the weight of the reactants in a chemical reaction equaled the weight of the products generated, suggesting that chemical reactions do not create or destroy matter. Lavoisier's quantitative studies led to the development of a fundamental principle in modern chemistry, the law of conservation of mass. Another French chemist, Joseph-Louis Proust (1754–1826), concluded from his studies of iron oxides that the ratio of elements to one another in a specific compound always stays the same; thus, he established through his work the law of definite proportions. Until Proust's studies, scientists believed that the types of compounds produced depended on the amount of reactants present in the reaction. For example, if one added two parts hydrogen to one part oxygen, the result would always be H_2O, or water, but adding three parts hydrogen to one part oxygen would result in H_3O, water with an extra hydrogen atom. Another principle, the law of multiple proportions, addressed how elements combine to form different compounds, or molecules, in specific whole number ratios with respect to their masses. If H_2O and H_2O_2 (hydrogen peroxide) are the compounds being compared, each compound has two hydrogen atoms in it, weighing two grams in each mole of the compound. In one H_2O molecule, one oxygen atom exists weighing 16 grams per mole. Hydrogen peroxide contains two oxygen atoms weighing 32 grams per mole. The ratio of oxygen in H_2O_2 and H_2O is 32:16 or 2:1.

Despite these studies suggesting the particulate nature of elements, some scientists still argued that matter was continuous. These scientists believed that you could divide and subdivide a lump of gold endlessly, never reaching a point where the material was no longer gold. Other scientists, including the English chemist and physicist John Dalton (1766–1804), believed that matter was made of atoms: that is, at some point one could no longer divide the gold and have it remain gold. Dalton proposed that his meteorological observations regarding the rise and fall of air pressure combined with his chemical analyses of various atmospheric gases suggested that air, as do all other compounds, consisted of small particles. He believed that tiny, indestructible particles were the basic building blocks of elements as well. Dalton called these particles atoms. According to Dalton's atomic theory, atoms of the same element shared the same physical and chemical properties, but atoms of different elements possessed different characteristics. Dalton's atomic theory also predicted that elements combine in fixed ratios of simple whole numbers following the laws of definite proportions and multiple proportions. In compounds, atoms always combine in whole number ratios with other atoms in exactly the same way each time. A molecule of H_2O always has exactly two hydrogen atoms and one oxygen atom. Water can never have 1.5 or 2.5 hydrogen atoms, because atoms cannot be divided during chemical reactions. Dalton assumed that atoms obey the law of the conservation of matter. Chemical reactions only change the arrangement of atoms with respect to one another; these processes neither create nor destroy the particles themselves. In 1803 John Dalton announced his atomic theory, which gained further support from the experimental studies of others. The scientific community quickly accepted the theory, which grew into a foundational building block of modern chemistry.

STRUCTURE AND SUBATOMIC PARTICLES

Dalton's atomic theory served as a starting point for further investigations of the atom. Over time his idea that atoms were the smallest possible entities proved false with the discovery of the subatomic particles: electrons, protons, and neutrons. Yet, scientists needed another 100 years or so to decipher the interior landscape of the atom and the relationships of these three particles with respect to one another. The resulting model has negatively charged electrons following orbitals around a positively charged nucleus. The nucleus consists of protons and neutrons, and all three kinds of particles impart mass to each atom. Because electrons behave as both discrete subatomic particles and continuous waves, scientists also incorporated concepts from a newer field of physics, quantum mechanics, to explain electron movement from one orbital with less associated energy to an orbital with more associated energy and back again into a model of the atom that is still credible.

In 1897 the British physicist Sir J. J. Thomson performed experiments in which voltage was applied to a vacuum tube (cathode-ray tube) containing a positive electrode (anode) and a negative electrode (cathode). Thomson observed the generation of a glowing beam of particles that traveled from the cathode to the anode in the vacuum tube. Using north and south poles of magnets on the outside of the cathode-ray tube, Thomson found that the beam of particles that he generated was moving away from the cathode toward the anode; that observation led him to the conclusion that the particles were negatively charged. Because these particles originated at the cathode, Thomson first called them cathode rays, but he later renamed them *electrons*. Thomson determined that electrons have a very small mass relative to an atom, and he also demonstrated that the overall electric charge of an atom is neutral, leading him to suspect that positively charged particles exist that offset the electrons within the atom. The "plum pudding" model of the atom that Thomson put forth suggested that a cloud of positive charge covered the volume of the atom with tiny electrons dispersed throughout like plums dotting a pudding or chocolate chips in a chocolate chip cookie.

Sir Ernest Rutherford, a physical chemist from New Zealand, worked under Thomson's supervision at the Cavendish Laboratory while Thomson refined his "plum pudding" model. During his time there, Rutherford became interested in the newly discovered phenomenon of radiation, namely, the energetic emissions (alpha, beta, and gamma rays) that radioactive elements released as they decayed. The characterization of alpha rays, which carry a positive charge, particularly interested Rutherford, and much of his work centered on understanding this particular type of radiation. In 1907 Rutherford combined his interest in alpha rays with the desire to test the validity of Thomson's model of atomic structure. He directed alpha particles, which had been identified as helium atoms missing their two electrons, at targets consisting of gold foil. If an atom consists of a cloud of positive charge dotted with negatively charged electrons, Rutherford predicted that most of the alpha particles in his studies would pass through the gold atoms in the foil. The majority of them did, but a small fraction of the alpha particles did not pass through the gold foil. This fraction of particles deflected at large angles away from the foil and, in some cases, deflected straight back toward the source of the particles. Rutherford concluded that Thomson's model was invalid, and he proposed that an atom has a small dense core, or nucleus, composed of positively charged particles that makes up most of an atom's mass and that negatively charged electrons floated around in the remaining space. If an atomic nucleus were the size of a Ping-Pong ball, the rest of the atom with its encircling electrons linked through electromagnetic forces to the nucleus would be about three miles (4.8 km) across. More precisely, nuclei are approximately 10^{-12} m in diameter, and the total diameter of an atom is about 10^{-8} m, roughly 10,000 times larger.

The Danish physicist Niels Bohr joined Rutherford's laboratory in 1912, and he spent the summer of that year refining the new model of atomic structure. Bohr added energy to hydrogen atoms in a controlled environment, then studied the light released (light emissions) by the atoms using a spectrophotometer, an instrument that detects these particles, or photons, of light. These experiments yielded patterns (line spectra) of light emissions suggesting that atoms lost, or released, energy only in certain discrete amounts rather than in any amount. These results led to the development of the planetary model of the atom. Bohr proposed that electrons traveled around the nucleus in defined paths, or orbitals, much as the planets in our solar system travel around the Sun. Electrostatic attraction between the positively charged nucleus and the negatively charged electrons keeps the electrons inside the atom as gravity keeps the planets circling around the Sun. In addition, Bohr proposed that the farther away from the nucleus the electrons were, the more energy they had. Electrons did not have a random amount of energy—the order of the orbitals around the nucleus dictated the possible energies of an electron in that atom. This order was denoted by the letter n, with the nearest orbital designated $n = 1$. In Bohr's model, when energy is added to an atom, electrons move from an orbital close to the nucleus to orbitals that are farther away. The orbital with $n = 1$ is called the ground state, and the orbital that the electron jumps to is called the excited state. Eventually, the electrons move back to the ground state and in doing so release the added energy in the form of light. The energy of the light being released is equal to the difference in energies between the excited and the ground states. Because many possible excited states exist, many different light emissions occur. Bohr realized that the laws of classical physics could explain neither the nature of the ground and excited states nor the nature of an electron in the transition state between ground and excited states. That led Bohr to use concepts from quantum mechanics to predict accurately the energies of the ground and excited states of hydrogen atoms and the hydrogen line spectrum.

With the big picture of subatomic structure in place, scientists spent the next 20 years identifying the remaining subatomic particles and defin-

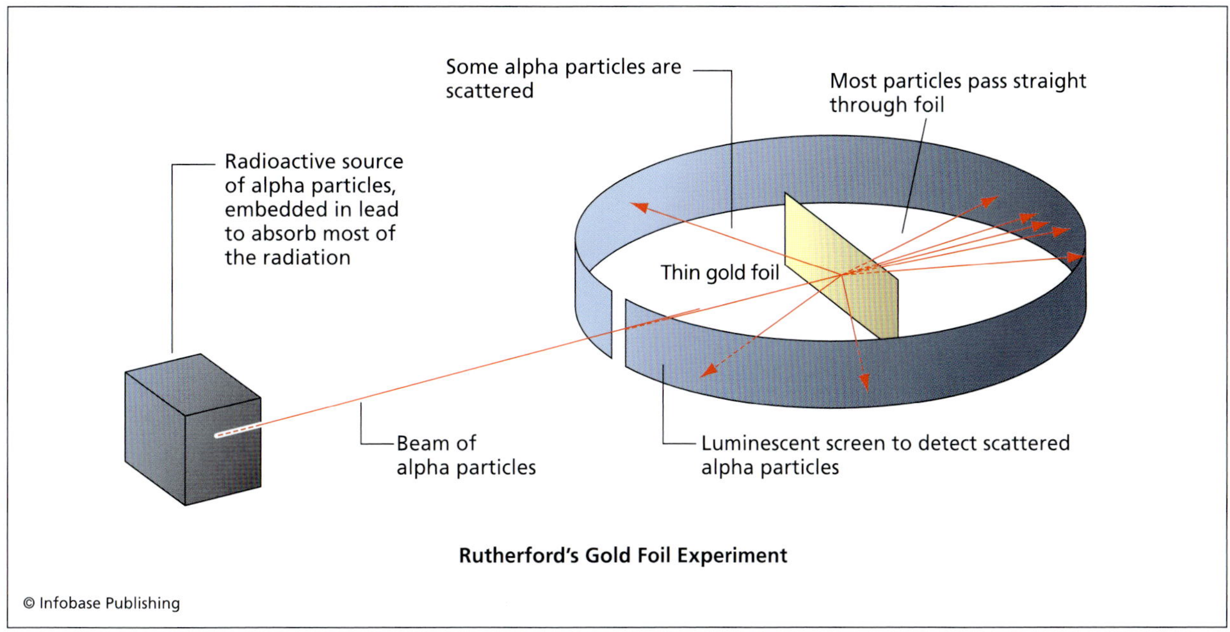

After bombarding a thin piece of gold foil with alpha particles, Rutherford observed that some of these particles were deflected away from the atoms. He concluded that each atom contains a dense core of charge, or nucleus, responsible for the deflections.

ing their relationships with one another. In 1919 Rutherford identified the subatomic particle that imparted mass to an atom and carried a positive electric charge. He called the particle a proton. The number of protons in an atom defines the element. Protons are components of the dense nuclear core, and the positive charge of these particles cancels out the negative charge of the electrons, making atoms with the same number of protons and electrons neutral. James Chadwick, a physicist who had worked with Rutherford, identified the third subatomic particle, the neutron, in 1932. Chemists had known that another particle must exist, because the atomic weights of the elements seemed off if they considered only the mass of protons in their calculations. Neutrons have approximately the same mass as protons but carry no electric charge. Without neutrons, the close proximity of multiple protons in a nucleus would threaten the integrity of the core, as like charges repel each other electrostatically. The short-range "strong nuclear force" that acts among protons and neutrons overcomes the electrostatic repulsion among the protons that would otherwise make the nucleus unstable.

In the second half of the 20th century, scientists modified the Bohr model to predict the orbital energies of other elements. In general, one must use four numbers, called quantum numbers, rather than just one to describe atomic structure: principal quantum number, n; angular momentum quantum number, l; magnetic quantum number, m_l; and spin quantum number, m_s. The principal quantum number is the same n that Bohr used and describes the energy level of an electron. Values of n range from 1 to infinity. The angular quantum number determines the shape and type of the orbital. Possible values for l begin at 0 and end at $n - 1$. For example, if $n = 2$, the possible values for l are 0 and 1. A spherical *s*-orbital is defined as $l = 0$ and a dumbbell-shaped *p*-orbital is defined as $l = 1$. The magnetic quantum

SUBATOMIC PARTICLES AND THEIR CHARACTERISTICS

Particle	Location	Mass (kg)	Charge
Proton	Nucleus	1.673×10^{-27}	+1
Neutron	Nucleus	1.675×10^{-27}	0
Electron	Orbitals	9.109×10^{-31}	-1

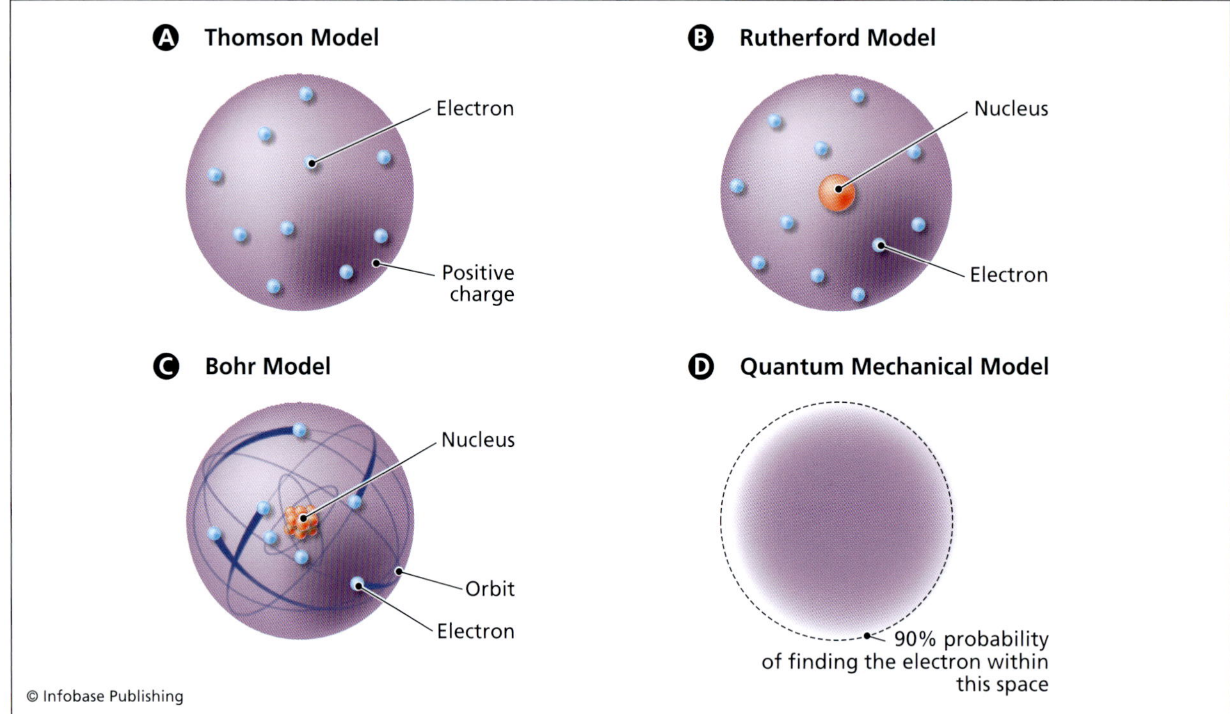

Over time, the atomic model has changed as scientists discovered new properties of the atom. (a) In the 19th century, Sir J. J. Thomson set forth the plum pudding model of the atom, describing the atom as a cloud of positive charge dotted throughout with negatively charged electrons. (b) In the early 20th century, Sir Ernest Rutherford demonstrated that a dense core of positive charge (the nucleus) exists within an atom, and he envisioned that electrons took up the remaining space. (c) Building on Rutherford's model, Niels Bohr proposed that electrons travel in certain defined orbits around the nucleus, much as the planets in the solar system orbit the Sun. Bohr also suggested that atoms gain and lose energy in discrete quantities as their electrons jump among the various allowed orbits. (d) Modern atomic theory describes electronic structure of the atom as the probability of finding electrons within a specific region in space.

number defines the direction that the orbital points in space. Possible values for m_l are integers from $-l$ to l. The spin quantum number, m_s, has a value of +1/2 or −1/2. The spin quantum number distinguishes between two electrons in the same orbital. Whereas previous models of the atom predicted that electrons were particles that moved in predictable paths, the Austrian physicist Erwin Schrödinger developed an equation with these four variables that predicted only possible locations of electrons. A mathematical function called a wave function predicted the probability of finding an electron in a particular spot. As a result, the perception of orbitals evolved from a two-dimensional circular orbit to three-dimensional shapes. This new model, called quantum mechanics, embodies the understanding of atomic structure to the present.

See also Bohr, Niels; Dalton, John; electron configurations; nuclear chemistry; nuclear physics; Planck, Max; quantum mechanics; Rutherford, Sir Ernest; Thomson, Sir J. J.

FURTHER READING

Asimov, Isaac, and D. F. Bach. *Atom: Journey across the Subatomic Cosmos.* New York: Plume, 1992.

Johnson, Rebecca L. *Atomic Structure (Great Ideas of Science).* Breckenridge, Colo.: Twenty-First Century Books, 2007.

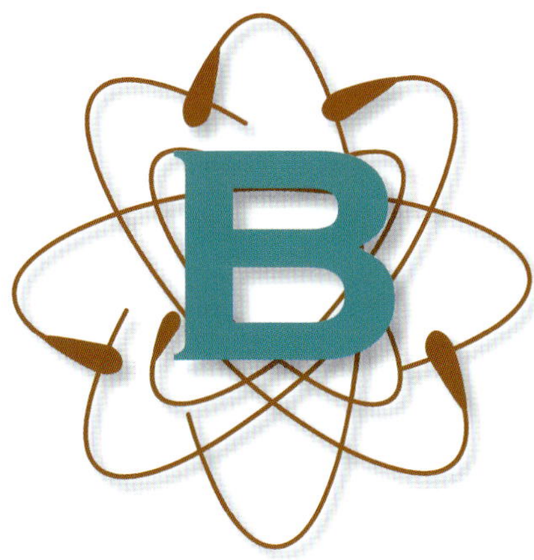

big bang theory The big bang theory is the presently generally accepted model of cosmology that describes the evolution of the universe as starting in an explosion 13.7 billion years ago from an extremely small and dense state. Initially the entire cosmos was compacted into a size smaller than an atom. Since the beginning, the time point of the big bang, the universe has been expanding and cooling. The theory describes the universe from a fraction of a second after the explosion and explains the origin of galaxies and stars. No current theory explains the universe before the big bang.

THE VERY EARLY UNIVERSE

One principle governing the origin of the universe is the inverse relationship between the temperature of the universe and its size and age. Temperature is a measure of heat, and therefore the internal energy associated with a substance. A high temperature is associated with a lot of heat, or a large quantity of energy. At the time point of the big bang, the temperature of the universe was infinitely hot, and since then the universe has been cooling.

The furthest back modern physics can adequately attempt to explain is to about 10^{-43} second after the big bang. At this time, the size of the universe was around 10^{-35} meter and its temperature was some 10^{32} kelvins. Particle physics models predict that at this temperature, the four natural forces (the gravitational, electromagnetic, strong nuclear, and weak nuclear forces) were unified into a single superforce. As the universe expanded and cooled, the gravitational force became distinct from the superforce, marking the beginning of the grand unified era, in which the other three forces were still unified. The universe was still extremely dense, and the particles continually collided with one another.

The universe continued to expand and the temperature continued to decrease until it reached a critical point, when the universe was about 10^{-35} second old and the temperature was around 10^{27} kelvins, a time referred to as the inflationary phase. During inflation, the expansion rate of the universe accelerated rapidly. Inflation ended at 10^{-33} second, when expansion began decelerating and the universe continued cooling. Then the strong nuclear force became distinct from the grand unified force, leaving the electromagnetic and weak nuclear forces still unified as the electroweak force. That era ended when the universe was 10^{-10} second old, and the weak force split from the electromagnetic force when the temperature was 10^{15} kelvins, giving the four distinct fundamental forces that govern the natural world.

The next era was the quark era. Quarks are among the elementary particles that compose all matter and interact with one another via the strong nuclear force that holds atomic nuclei together. At this temperature, however, the strong nuclear force was too weak, and gluons, the exchange particles for the strong force, could not hold quarks together. When the temperature cooled to 10^{12} kelvins, quarks began to combine with other quarks, ending the quark era. By the time the universe was 10^{-4} second old, protons and neutrons formed, but the temperature was still too high to allow the formation of stable nuclei. Further cooling to about 1 billion kelvins around 100 seconds after the big bang allowed the strong force to hold protons and neutrons together to form nuclei, a process called nucleosynthesis. Only very light nuclei formed, mostly hydrogen (consisting of a single proton), deuterium (consisting of one proton and one neutron), and a few isotopes of helium (consisting of two protons and one or two neutrons).

The temperature continued to drop, and the universe continued to expand for about 300,000 years. When the temperature reached about 3,000 kelvins, electromagnetic radiation became sufficiently weak that it allowed the electrons to remain associated with nuclei. Atoms formed, ushering in the era of matter. Electromagnetic radiation interacts only with charged particles, but as the negatively charged electrons became stably associated with positively charged nuclei, forming neutral atoms, the majority of the radiation remained intact from this point on. In other words, light and matter decoupled, and the existing photons could roam the universe more or less unaffected by matter. The detection of this leftover radiation in the universe provided key evidence supporting the big bang theory in the 1960s. Since the formation of atoms, the nature of matter has not changed appreciably, thus marking the end of the scenario of evolution from the big bang.

SUPPORTING EVIDENCE

Two major observations of the big bang theory are that the universe is expanding, with more distant galaxies moving away at a faster rate than nearby ones, and that background cosmic radiation fills space as a remnant of the original explosion. The notion of a big bang that birthed the universe was first suggested in the 1920s, when several physicists and astronomers began collecting both experimental and theoretical evidence that the universe was expanding, probably the most important astronomical discovery since the 16th-century astronomer Nicolaus Copernicus proposed that the Sun, not Earth, was the center of the universe. This led to the inference that at one time the universe must have been exceedingly small, and "in the beginning" a huge explosion sent the dense matter flying into space with tremendous speed. In the late 1940s, physicists realized that if the universe did originate from a huge explosion, then an afterglow might still exist in the form of cosmic background radiation.

The notion of the big bang theory developed initially from work performed in the early 1920s by a Russian cosmologist and mathematician named Alexander Friedmann. Albert Einstein developed his general theory of relativity in 1915, and he found that it described the universe as either expanding or doomed to self-destruction by eventual collapse. He invented a cosmological constant that he added to his equations to compensate for this by counteracting the tendency for all the cosmological bodies to attract one another, leading to his proposal that the universe was constant in size. Friedmann showed that using different values for the cosmological constant led to different models for the universe, including the possibility of a dynamic, changing universe by setting the constant to zero, essentially returning to Einstein's original formula for gravity.

An astronomer from the Lowell Observatory at Flagstaff, Arizona, Vesto Melvin Slipher, had collected evidence that nebulae (as they were called then—they are now known to be galaxies) were moving rapidly away from the Earth at a rate of approximately 373 miles (600 km) per second. In 1924, while working at the Mount Wilson Observatory of the Carnegie Institution of Washington, the American astronomer Edwin P. Hubble had use of the world's largest telescope, the 100-inch (254-cm) Hooker telescope. Hubble discovered that the Andromeda nebula was far outside the Milky Way, our own galaxy—at least three times as far away as the outer limit of the galaxy itself. This proved wrong the assumption that all nebulae were within the Milky Way. A Belgian physicist and clergyman, Georges Lemaître, translated this information into a theoretical model decreeing that the universe was expanding and surmised that as one went backward in time, the universe must have been progressively smaller until reaching the origin of the universe. Lemaître's ideas are the forerunner of the big bang model. Lemaître even predicted that the recession velocity of each galaxy—the speed at which the galaxy is moving away from us—would be proportional to its distance from us. Unaware of Lemaître's work, in 1929 Hubble proposed what is now called Hubble's law, stating that galaxies recede at a faster rate the farther away they are, a claim that has turned into important evidence for an expanding universe.

In 1949 physicists from Johns Hopkins University, George Gamow, Ralph Alpher, and Robert Herman, carried out physical calculations predicting the existence of cosmic background radiation originating from the primordial explosion. In 1964 Arno Penzias, a radio astronomer working at Bell Laboratories, and Robert Wilson, who had just joined Penzias in New Jersey, were attempting to measure radio emissions from the Milky Way. They could not account for a residual background hissing noise in the radio they were using even though it was specially designed to prevent unwanted reception and only detected frequencies near 4,080 megahertz. In December 1964, from a colleague, Penzias learned of a soon-to-be-published paper by James Peebles, a cosmologist at Princeton University working with the physicist Robert Dicke. Peebles predicted that background radio waves resulting from the intense heat of the big bang should fill space uniformly and constantly and have a temperature of about 10 kelvins (i.e., 10 degrees above absolute zero). Two of Peebles's associates were already assembling a radio receiver specifically to search for the background radiation. Penzias called Dicke, who then inspected Penzias's apparatus and data. Dicke confirmed that the background noise was radiation left over from the universe's origination.

The other prevailing theory of the universe was the steady state theory, which claimed that the universe had always existed. Penzias's and Wilson's findings supported predictions made by the big bang theory, making it the more credible explanation. In 1978 Penzias and Wilson received the Nobel Prize in physics "for their discovery of cosmic microwave background radiation." Technology at the time, however, could not overcome interference from Earth's atmosphere to examine this radiation clearly. So, though Penzias and Wilson detected the radiation through their earthbound measurements, they could not confirm its origin or verify that it was indeed background noise, meaning that it was the same in all directions.

In 1989 NASA launched the *Cosmic Background Explorer (COBE)* satellite to study the cosmic background radiation from space, outside Earth's atmosphere, a project led by the physicist John C. Mather. The distribution of the radiation emitted from the glowing body of the big bang explosion would depend solely on its temperature. This type of radiation is called blackbody radiation, and its spectrum displays a characteristic shape related to the distribution of energy (intensity) at different wavelengths. The data collected from COBE after only nine minutes clearly displayed the exact blackbody spectrum predicted from the first light in the universe and allowed calculation of the current temperature of background radiation, finding it to be 2.7 kelvins. Though the radiation itself has cooled significantly over the past 13.7 billion years, the spectrum retains the same overall shape, just shifted to longer wavelengths. Other COBE studies led by George Smoot, a physicist at the University of California at Berkeley, found and measured tiny temperature variations in different directions, called anisotropies, in the background radiation. This information helped scientists figure out how matter began to aggregate, forming seeds that developed into the galaxies and galactic clusters that make up the universe as it is today. This experimental evidence supported another prediction of the big bang theory. In 2006 Mather and Smoot were awarded the Nobel Prize in physics "for their discovery of the blackbody form and anisotropy of the cosmic microwave background radiation." The *Wilkinson Microwave Anisotropy Probe (WMAP)* satellite soon collected even clearer pictures of the background radiation, and studies will continue with the upcoming launch—scheduled for spring 2009—of the European *Planck* satellite.

Other evidence in support of the big bang theory includes the abundances of the elements. Whereas Earth and its atmosphere are composed of heavier elements, the Sun is composed mostly of light elements, 74 percent hydrogen and 25 percent helium. Heavy elements make up only 1 percent of the total of all matter in the universe. The ratio of hydrogen to helium, 3:1, also supports the big bang model, as do the structure and distribution of the galaxies.

See also BLACKBODY; COSMIC MICROWAVE BACKGROUND; EINSTEIN, ALBERT; GENERAL RELATIVITY; THEORY OF EVERYTHING.

FURTHER READING

Dicke, Robert H., P. James E. Peebles, Peter G. Roll, and David T. Wilkinson. "Cosmic Black-Body Radiation." *Astrophysical Journal* 142 (1965): 419–421.

Hubble, Edwin. "A Relation between Distance and Radial Velocity among Extra-Galactic Nebulae." *Proceedings of the National Academy of Sciences USA* 15 (1929): 168–173.

Lawrence Berkeley National Laboratory, Physics Division. "The Universe Adventure." Available online. URL: http://universeadventure.org. Accessed July 23, 2008.

Lidsey, James E. *The Bigger Bang.* Cambridge: Cambridge University Press, 2000.

Penzias, Arno A., and Robert W. Wilson. "A Measurement of Excess Antenna Temperature at 4080 Mc/S." *Astrophysical Journal* 142 (1965): 419–421.

Singh, Simon. *Big Bang: The Origin of the Universe.* New York: Fourth Estate, 2004.

biochemistry Biochemistry is the branch of chemistry concerned with chemical reactions that take place within or by living organisms. Encompassing a broad scope of scientific topics as well as a vast array of techniques, biochemistry forms the link between the chemical understanding of what happens in a reaction and the physical manifestations that reaction causes in an organism. The study of biochemistry includes in-depth study of each class of macromolecules that are present in living organisms: proteins, lipids, carbohydrates, and nucleic acids. Each of these macromolecules is a polymer, a large molecule made of repeated chemical subunits known as monomers. Biochemistry studies the buildup, breakdown, and regulatory mechanisms of each of these macromolecules. One fundamental principle underlying all of biochemistry is that the structure of a compound directly relates to its function. The study of the structures of biologically important molecules allows the elucidation of their function, and vice versa.

EDUCATION AND TRAINING

Education and training in the field of biochemistry require comprehensive study of each of the classes of macromolecules, including the structure and function of proteins, gene structure and regulation, structure and function of cell membranes, regulation and processes of photosynthesis, and metabolism of carbohydrates, lipids, and proteins. A strong background in chemistry, molecular biology, and genetics is also required. Laboratory training is critical in the field of biochemistry, and biochemistry education involves extensive study and training in biochemical techniques. The type of degree required for employment as a biochemist depends upon the type of job desired. Degrees in biochemistry range from four-year bachelor of science degrees to research-based master of science degrees and doctoral degrees. In general, laboratory technician positions require a bachelor's or master's degree. Supervisory roles require either master's degrees or doctoral (Ph.D.) degrees. Full-time academic positions generally require a Ph.D., and many require additional postdoctoral work in the field.

Biochemistry positions are as diverse as the areas of study. Biochemists can work on any aspect of living organisms as well as the interactions among organisms. Careers are in varied fields such as drug development, environmental issues, disease study and treatment, and nutrition. Biochemists can hold positions in academia and business, as well as in government jobs.

PROTEINS

Proteins are polymers of amino acids that are responsible for most of the cellular activity in a living organism. Having roles in movement, structure, catalysis of reactions, defense against invading organisms, cellular signaling, and cell division, proteins are the most diverse of all the macromolecules and are often considered the workhorses of the cell. Protein polymers are made up of monomers known as amino acids. There are 20 different amino acids in the human body that make up all of the known proteins. Each amino acid has the same basic structure, with the difference occurring in the side chain of the amino acid, (R). Each amino acid contains a central carbon that has an amino group, a carboxyl group, hydrogen, and a variable R group.

The amino acids are joined together through a dehydration reaction (a bond formed by the removal of water) between the carboxyl group of the first amino acid and the amino group of the next amino acid, creating a peptide bond. Each peptide and polypeptide has an amino terminus (an end with a free amino group) and a carboxyl terminus (an end with a free carboxylic acid group). The reactivity, polarity, and size of the R groups on the component amino acids of a polypeptide chain help control the way, the protein is folded and the way it functions. Biochemists study the arrangement of amino acids, the folding and function of each of these proteins, and the ways they can impact the function and well-being of the organism in which they are found.

NUCLEIC ACIDS

Nucleic acids form a class of polymers that are responsible for the storage of hereditary information within the cell and the transmittance of that information from the parent cell to any daughter cells it may produce. The forms of nucleic acids are known as deoxyribonucleic acid (DNA) and ribonucleic acid (RNA). The monomers of nucleic acids are known as nucleotides, with each nucleotide containing a five-carbon sugar (ribose or deoxyribose), a phosphate group, and a nitrogenous base. There are two categories of nitrogenous bases, known as purines (including adenine, A, and guanine, G) and pyrimidines (including thymine, T; cytosine, C; and uracil, U). Adenine, guanine, cytosine, and thymine are all found in DNA, while uracil replaces thymine in RNA. Nucleotides are joined together by dehydration synthesis reactions between the hydroxyl located on the 3′ carbon of one sugar and the phosphate bonded to the 5′ carbon of a second sugar, forming a phosphodiester bond. Each nucleic acid polymer has directionality, having a 5′ end (with the free phosphate group) and a 3′ end (with the free hydroxyl group).

DNA within the nucleus of a cell contains all of the information necessary for carrying out every process within the organism. The precise copying of DNA (DNA replication) before a cell divides is essential to maintaining the integrity of the cell and the organism. In order to serve its purpose, DNA acts as a template that leads to the production of proteins through two important biochemical processes known as transcription (copying of DNA into a messenger RNA transcript) and translation (using the messenger RNA to assemble amino acids in the correct order to form a functional protein). Biochemists and biochemical techniques have established the mechanism of each of these processes, determining the steps involved, the enzymes and proteins required for the process to proceed, the ramifications of mistakes in the processes, as well as the control mechanisms of replication, transcription, and translation. Biochemists also work to reveal the processes of gene regulation, or which sections of DNA (genes) actively encode for proteins and which genes are turned on or off within a particular cell type or at a specific time.

Another important application of nucleic acids applies particularly to the nucleotide adenine triphosphate, or ATP. The cell tightly controls and regulates the levels of ATP, which serves as the primary energy currency of the cell. The analysis of the cellular metabolism and the energy required for cellular functions is an important focus of biochemical studies.

CARBOHYDRATES

Carbohydrates are a class of polymers whose main function inside the cell is as an energy source. Other functions of carbohydrates include cellular signaling, cellular attachment, and structural functions, such as composing cell walls in plants. The monomers of carbohydrates are known as monosaccharides, single-sugar molecules that are bonded together to form disaccharides (two sugars) and polysaccharides (many sugars) through glycosidic linkages.

Monosaccharides include such molecules as glucose, fructose, galactose, and ribose. Most carbohydrates that are taken into the body are broken down into glucose, which then enters glycolysis and cellular respiration in aerobic eukaryotes. The study of biochemistry also is necessary for understanding the regulation of glucose levels in the bloodstream.

Disaccharides include such familiar compounds as lactose (milk sugar), sucrose (table sugar), and maltose (breakdown product of starch). Lactose contains glucose and galactose, sucrose contains glucose and fructose, and maltose contains two glucose molecules. The breakdown of these molecules into monosaccharides is performed by protein enzymes that are coded for in the DNA of the cell. This exemplifies the links among multiple aspects of biochemistry. Proper function of the cell requires that all of the macromolecules within the cell be in the correct location and concentration at the right time. For example, individuals who are not able to tolerate milk products do not have appropriate levels of lactase, the enzyme required to break lactose into glucose and galactose.

Polysaccharides include molecules such as starch, glycogen, and cellulose. Starch and glycogen are storage forms of glucose in plants and animals, respectively. Glucose-level regulation in the body is a well-studied biochemical process. When blood sugar levels are too high (hyperglycemia), the pancreas secretes insulin, a hormone that causes glucose to be removed from the bloodstream and incorporated into a polymer of glycogen by the liver, leading to a reduction of blood glucose levels. In hypoglycemic conditions (low blood glucose levels), the pancreas secretes glucagon, a hormone that causes the breakdown of glycogen polymers and the release of glucose into the bloodstream, raising the blood glucose level.

Starch and cellulose are both glucose polymers in plants. The glucose monomers in starch are bonded together with alpha- (α-) linkages, meaning that the monomers are all oriented in the same direction. Cellulose is a plant polymer made of glucose monomers bonded together with beta- (β-) linkages, meaning the each monomer is rotated 180 degrees relative to its neighbors. As a structural polymer in plants, cellulose lends strength and rigidity to cell walls. Starch is found in plants such as potatoes, corn, and wheat. The enzyme that cleaves starch is only able to break alpha-linkages. Humans do not have the gene that encodes for an enzyme that breaks beta-linkages, and therefore humans cannot break down cellulose.

The contributions of biochemistry to the understanding of carbohydrates include the structure and function of each type of carbohydrate as well as the medical implications of defects in metabolism of carbohydrates.

LIPIDS

The study of lipids is one of the oldest areas of biochemical research. Lipids are a class of hydrophobic (water-fearing) covalent compounds that perform multiple functions within cells. Lipids (triglycerides) serve as energy storage molecules, are the main component of cellular membranes (phospholipids), and make up cholesterol and steroid molecules.

Triglycerides (commonly known as fats) provide an excellent means of energy storage as fat contains nine calories per gram, compared with proteins and carbohydrates, which each contain four calories per gram. The structure of triglycerides consists of three fatty acid chains attached to a glycerol molecule as a backbone. The length of the fatty acid chains as well as the saturation level (number of double bonds) in the fatty acid chain determine the structure and function of the triglyceride. Breakdown of triglycerides into glycerol and fatty acid molecules results in molecules that can enter directly into the biochemical pathway of cellular respiration. The glycerol backbone enters glycolysis as dihydroxyacetone phosphate (DHAP) and the fatty acid chains are converted via β-oxidation into molecules of acetyl coenzyme A (acetyl CoA), which travels through the citric acid cycle.

Phospholipids are made up of two hydrophobic fatty acid chains attached to a hydrophilic (water-loving) phosphate group. The overall molecule is amphipathic, meaning that it has both hydrophobic and hydrophilic domains. The structure of this molecule lends itself to the formation of the cell membrane. The phospholipids form layers with the phosphate heads on one side and the fatty acid tails on the other. When two of these layers join to exclude water, the fatty acid tails are on the interior, and the structure is known as a phospholipid bilayer. The phospholipid heads interact with the exterior and interior of the cell, and the fatty acid tails make up the central portion of the double-layered membrane. One of the important aims of biochemistry is the study of molecule movement through the cell membrane.

BIOCHEMISTRY OF METABOLISM

Biochemists have been instrumental in elucidating the metabolic processes of organisms. The resolution of the processes of photosynthesis, cellular respiration, and fermentation has applications across biological sciences, physiology, and medicine. Bioenergetics involves the study of the production and utilization of ATP, a high-energy molecule used by cells or living organisms. Metabolic processes that involve the building up of compounds by using cellular energy are known as anabolism, an example of which is photosynthesis. Processes involving the breakdown of compounds are known as catabolism, an example of which is glycolysis.

Photosynthetic organisms (autotrophs) capture the energy of the Sun and use carbon dioxide and water to produce carbohydrates. These carbohydrates are an energy source for all organisms (autotrophs and heterotrophs). Biochemical research into photosynthesis has involved the determination of all chemical participants in the process, the visualization and structural identification of each of the cellular structures that take part in photosynthesis, as well as the steps that make up the process.

In aerobic organisms (organisms that utilize oxygen), the catabolism of food molecules to generate energy takes place in three stages: glycolysis, the citric acid cycle (Krebs cycle or tricarboxylic acid [TCA] cycle), and oxidative phosphorylation. Glycolysis results in the breakdown of glucose into two molecules of pyruvate, produces a net gain of two ATP per glucose molecule, and transfers some energy via electron carriers to the electron transport chain for oxidative phosphorylation. The pyruvate molecules from glycolysis are converted into acetyl CoA with the transfer of energy to more electron carriers by reducing them. Acetyl CoA enters the citric acid cycle, where the carbon chain is broken down and released as carbon dioxide and ATP guanosine triphosphate (GTP), accompanied by the reduction of electron carriers. Oxidative phosphorylation takes place in the mitochondrial matrix of the cell and utilizes the electrons transported via the electron carrier molecules to generate a proton gradient across the membrane that supplies the energy required for ATP production. The net gain from the oxidation of one glucose molecule is approximately 30 ATP. This deviates from the theoretical value of 36–38 ATP per glucose molecule because of rounding up of the values for the amount of energy stored in reduced electron carriers, the use of reduced nicotinamide adenine dinucleotide (NADH) for other reduction processes in the cell, and the use of the proton gradient to transport other molecules across the mitochondrial membrane.

In anaerobic organisms (organisms that do not utilize oxygen), the main process of energy production is glycolysis. The oxidized forms of electron carriers generated during glycolysis are regenerated via fermentation.

TECHNIQUES IN BIOCHEMISTRY

The field of biochemistry utilizes a wide variety of techniques to study cellular processes both in vivo (within the cell) and in vitro (outside the cell). When studying processes that take place within the cell, a biochemist performs techniques common to cell biology, such as cell culture and cell fractionation using centrifugation. Cell culture is a technique in which cell lines are grown and multiplied under controlled conditions inside culture dishes. The researcher tightly controls the food supply, temperature, pH, and cell concentration in order to maintain the conditions optimal for cell growth. Cell culture produces excellent reaction systems to study the cellular effects of chemical treatments. The process of cell fractionation begins with breaking open the cell membranes to release the internal components in a process known as homogenizing. Spinning this suspension of cell particles in a centrifuge at increasingly faster speeds causes the organelles and cell structures to sediment in different fractions, facilitating further study of the function of certain isolated organelles.

Study of the individual classes of macromolecules can be performed using in vitro techniques such as gel electrophoresis, Southern blot analysis (DNA study), Northern blot analysis (RNA analysis), Western blot analysis (protein study), and far-Western blot analysis (protein-protein interactions).

Gel electrophoresis is a molecular separation technique consisting of a gel matrix through which a sample subjected to an electrical field passes. The separation in gel electrophoresis is based on size of the molecule. The sample is loaded in a well at the top of the gel and subjected to an electric field from negative at the top to positive at the bottom of the gel. The sample moves through the gel matrix containing pores of a consistent size. The smaller molecules have an easier time passing through the matrix and therefore move faster through the gel. The largest molecules do not travel far in the gel because of their inability to pass through the pores. A molecular weight–based separation is achieved with the smallest molecules of the sample found at the bottom of the gel and the largest at the top. Controlling the parameters of the electrophoresis process such as

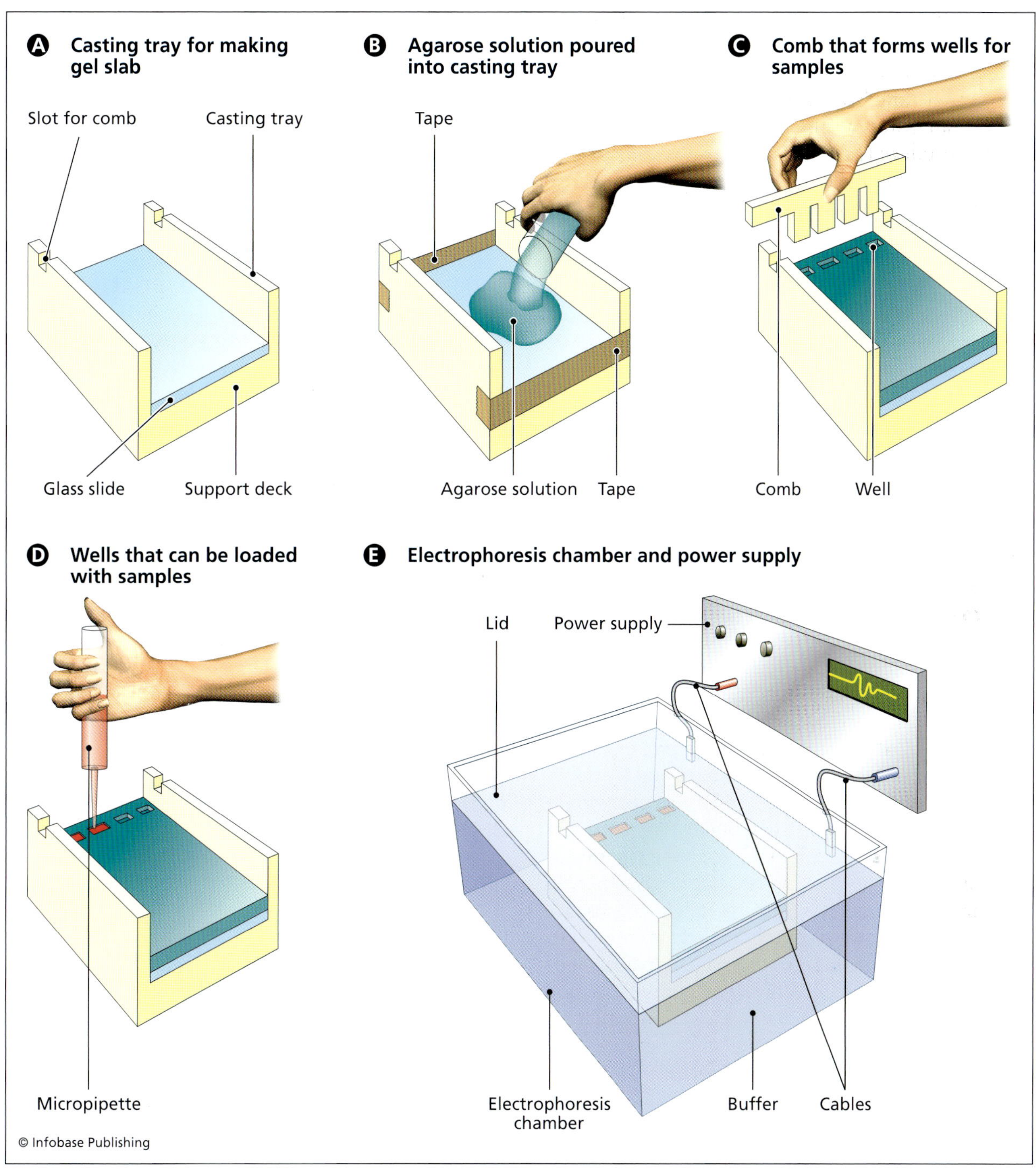

Gel electrophoresis is a separation technique that utilizes the ability of DNA, RNA, and proteins to travel in an electric current. A gel slab is poured into a casting tray and a comb inserted to create wells. The sample is then pipetted into the wells and the electric current is supplied. The sample will move down the gel.

pore size, length of gel, time of run, and voltage of electrophoresis determines the degree of separation of a sample.

The two most common gel matrixes are agarose (a carbohydrate) and acrylamide (a complex polymer). When acrylamide gels are used the process is known as polyacrylamide gel electrophoresis (or PAGE). Separation of DNA fragments is performed most often by using agarose gel electrophoresis unless the fragment size is extremely small (less than 250 nucleotide base pairs), such as in DNA sequencing techniques, for which acrylamide gels are used. Protein separations are most often carried out using PAGE after treatment of the sample with an anionic detergent such as sodium dodecyl sulfate (SDS). This protein separation technique is known as SDS-PAGE.

Separation by gel electrophoresis is a precursor to many analytical biochemistry techniques.

Southern blot analysis allows for the identification of specific DNA sequences. After agarose gel electrophoresis, the DNA fragments are transferred to a sheet of nitrocellulose and cross-linked to prevent movement of the fragments. A radioactively labeled probe that contains a DNA sequence that is complementary to the DNA sequence of interest is incubated with the nitrocellulose membrane. The nitrocellulose is washed to remove any probe that is bound nonspecifically, and the Southern blot is exposed to film. After appropriate exposure times, the film is developed to reveal the pattern of DNA fragments specific to the labeled probe.

Northern blot analysis is carried out in much the same way as Southern blot analysis. Messenger RNA samples are subjected to agarose gel electrophoresis, transferred to nitrocellulose, and probed with a radioactively labeled probe. The importance of Northern blotting lies in the visualization of the messenger RNA that is produced within a cell, giving information about gene regulation and protein production within a particular cell type.

Western blot analysis is used for the detection of specific proteins in a sample, and it begins with the separation of a protein sample by SDS-PAGE. The gel products are transferred to nitrocellulose, and the presence of proteins is determined by application of an unlabeled primary antibody specific for the protein of interest. Visualization of the protein products results from a labeled secondary antibody that is specific for the primary antibody, such as anti-immunoglobulin G (anti-IgG). The secondary antibody can be radioactively labeled and the Western blot exposed to film, or the secondary antibody can be tagged with a fluorescent compound that is detected and measured. Western blot analysis can also be carried out using a tagged primary antibody.

Far-Western blot analysis is used to detect protein-protein interactions. The procedure is carried out in much the same way as Western blotting; however, care is taken to maintain the protein sample in the native conformation as much as possible. Native conformation is required in order for the protein to bind appropriately to other proteins. After electrophoresis and nitrocellulose transfer, the far-Western blot is exposed to a bait protein (one that binds the protein of interest) that is enzymatically labeled or unlabeled. Labeled bait proteins can be visualized directly. If the bait protein is unlabeled, the process requires a labeled antibody that recognizes the bait protein for visualization.

Analytical chemistry techniques are often utilized in biochemical research. One can determine the chemical nature and structure of biological molecules by using X-ray crystallography, mass spectroscopy (MS), and nuclear magnetic resonance (NMR). X-ray diffraction is a technique that reveals the structure of a crystalline solid. Crystals of a particular protein are subjected to X-ray beams, and the diffraction patterns of the X-rays indicate the spacing and location of the atoms within the protein. Analysis of these diffraction patterns leads to a structural model of the protein. MS is a technique that separates samples on the basis of their mass/charge ratio, allowing analysis and detection of the molecules present in a sample. Trace amounts of a compound can be detected using mass spectroscopy. NMR is a technique that gives information about the position of atoms and the structure of compounds by measuring the changes in spin of the nuclei of the atoms after being subjected to a strong magnetic field. Each of these techniques is used in biochemistry to elucidate the structure of molecules, which provides information on how these molecules function.

See also ANALYTICAL CHEMISTRY; CARBOHYDRATES; CENTRIFUGATION; CHROMATOGRAPHY; DNA REPLICATION AND REPAIR; ENZYMES; FATTY ACID METABOLISM; GLYCOGEN METABOLISM; GLYCOLYSIS; LIPIDS; MASS SPECTROMETRY; METABOLISM; NUCLEAR MAGNETIC RESONANCE (NMR); NUCLEIC ACIDS; PROTEINS; X-RAY CRYSTALLOGRAPHY.

FURTHER READING

Berg, Jeremy M., John L. Tymoczko, and Lubert Stryer. *Biochemistry*, 6th ed. New York: W. H. Freeman, 2007.

biophysics Also referred to as biological physics, biophysics is an interdisciplinary science consisting of many diverse studies, whose sole unifying characteristic is the application of the understanding and methods of physics to biology. When an investigator is involved with physics applied to biology with concern for solving practical problems, such as designing artificial vision or a mechanical hand, the field is better called bioengineering.

Biophysicists most often hold physics degrees and are physicists with a special interest in issues of biology. They might be found on the faculty of colleges and universities as well as in research institutions and in industry and government. In academic institutions it is rare for there to exist a department of biophysics. Rather, biophysicists mostly carry out their research and teaching in biophysics specialty groups within or across classic departments, mostly physics, biology, or biochemistry and molecular biology departments, but also in various departments of medicine.

In order to give an idea of what the field of biophysics comprises, the following are some examples of areas in which biophysicists are actively pursuing research and examples of research projects being carried out in those areas:

- *Biomechanics.* The application of the principles of mechanics to animal and cell locomotion, the action of muscles, and molecular motors: How do birds and insects fly? What are the forces acting on human limbs? How do dolphins and whales propel themselves so efficiently?
- *Molecular biophysics.* The study of the structure and physical properties of biological molecules, such as nucleic acids and proteins, and of assemblies of such molecules: How can nanoscale structures for detecting and manipulating single biological molecules be constructed? How does the structure of very large biomolecules relate to their biological function?
- *Cell membrane physics.* The investigation of the mechanical and electrical properties of cell membranes: How does the voltage across a cell membrane affect the membrane's permeability to specific molecules? Can neurons transmit signals by means of density pulses along their membrane?
- *Bioenergetics.* The application of energy and thermodynamic considerations to biological processes at all scales, from the scale of molecules, through the scale of cells, to that of entire organisms: What energy can be derived by an organism from the making and breaking of chemical bonds, such as in carbohydrates? How does energy move through the food chain via predator-prey relationships?
- *Photobiophysics.* The study of the interaction of light with biological systems, such as vision and other light detection systems as well as light-producing systems: How do the firefly and other animals produce light so efficiently? What is the effect of ultraviolet radiation on human skin?
- *Informatics.* The investigation of how biological systems, from single molecules to organs to entire organisms, collect, store, and transmit information: How does the brain store and process information? By what mechanism do cells signal each other? How is genetic information coded?
- *Neural networks.* The study of the function of networks, or circuits, of interconnected real neurons or of artificial neuronlike devices: How do such networks remember and recognize patterns? How do they solve problems? Can this process lead to artificial intelligence?

Biophysicists have organized themselves into the Biophysical Society, founded in 1957, with a worldwide membership of more than 7,000. The society's purpose is to encourage development and dissemination of knowledge in biophysics, as it does through such as meetings, publications (including the monthly *Biophysical Journal*), and outreach activities.

FURTHER READING

Biophysical Society home page. Available online. URL: http://www.biophysics.org. Accessed January 9, 2008.

Cotterill, Rodney. *Biophysics: An Introduction.* Hoboken, N.J.: Wiley, 2002.

blackbody Also written as *black body,* this is the name given to a body that absorbs all electromagnetic radiation impinging on it. Although a blackbody is an idealization, it can be approximated by a small hole in the wall of a cavity, when the cavity and its enclosure are in thermal equilibrium at some temperature. Then practically all electromagnetic radiation falling on the hole from the outside will be absorbed by the cavity. The concept of a blackbody is an important one for the theory of the interaction of electromagnetic radiation with matter.

A blackbody emits electromagnetic radiation of all wavelengths, with a particular wavelength for which the intensity of radiation is maximal and the intensity tapering off for longer and shorter wavelengths. The wavelength of maximal intensity λ_{max} is inversely proportional to the blackbody's absolute temperature T, according to Wien's displacement law:

$$\lambda_{max}T = 2.898 \times 10^{-3}\ \text{m·K}$$

where the wavelength is in meters (m) and the absolute temperature in kelvins (K). This underlies the effect that real bodies, such as a chunk of iron or a star, at higher and higher temperatures appear to glow successively red, orange, yellow, white, blue, and violet. At extremely low or high temperatures, they might not be visible at all, since most of their radiated energy would then be in the infrared or ultraviolet range, respectively, and thus invisible to the human eye.

The total power that a blackbody emits at all wavelengths is proportional to the fourth power of the absolute temperature. That is expressed by the Stefan-Boltzmann law:

$$P = \sigma T^4$$

A lightbulb filament is not strictly a blackbody. When it is heated by an electric current passing though it, the elctromagnetic radiation it emits approximates that emitted by a blackbody at the same temperature. This filament's color can serve as an indication of its temperature. *(Alfred Pasicka/ Photo Researchers, Inc.)*

where P denotes the electromagnetic power (energy per unit time) radiated per unit area in watts per square meter (W/m^2) and σ is the Stefan-Boltzmann constant, whose value is 5.67051×10^{-8} $W/(m^2 \cdot K^4)$.

Nothing about blackbody radiation—neither the distribution of intensity over wavelengths (the black-body spectrum), Wien's law, nor the Stefan-Boltzmann law—is understandable in terms of classical physics. Early in the 20th century the German physicist Max Planck, making the nonclassical assumption that energy is exchanged between the electromagnetic radiation and the matter composing the body only in discrete "bundles," derived the formula correctly describing the blackbody spectrum:

$$\Delta E = \frac{8\pi ch}{\lambda^5} \frac{\Delta\lambda}{e^{\frac{hc}{kT\lambda}} - 1}$$

Here ΔE denotes the energy density (energy per unit volume) of the electromagnetic radiation, in joules per cubic meter (J/m^3), in a small range of wavelengths $\Delta\lambda$ around wavelength λ for a blackbody at absolute temperature T. The symbols h, c, and k represent, respectively, the Planck constant, the speed of light in vacuum, and the Boltzmann constant, with values $h = 6.62606876 \times 10^{-34}$ joule-second (J·s), $c = 2.99792458 \times 10^8$ meters per second (m/s), and $k = 1.3806503 \times 10^{-23}$ joule per kelvin (J/K). Albert Einstein helped to improve understanding of the concept of discrete energy "bundles" underlying this formula through his explanation of the photoelectric effect, and that work led to the development of quantum mechanics. Both Wien's displacement law and the Stefan-Boltzmann law follow from Planck's formula for the black-body spectrum.

See also CLASSICAL PHYSICS; EINSTEIN, ALBERT; ELECTROMAGNETIC WAVES; ELECTROMAGNETISM; EQUILIBRIUM; HEAT AND THERMODYNAMICS; PHOTOELECTRIC EFFECT; PLANCK, MAX; POWER; QUANTUM MECHANICS; WAVES.

black hole A black hole is an object whose gravity is so strong that nothing, not even light or other electromagnetic radiation, can escape from it. According to Albert Einstein's general theory of relativity, bodies of sufficiently high density will exert gravitational forces on other bodies and on electromagnetic radiation, including light, in their vicinity, so that neither the other bodies nor the radiation will ever be able to leave that vicinity. A theoretical surface, called the event horizon, surrounds the black hole and determines the fate of other bodies and of radiation. A body or light ray originating within the event horizon will never pass through the event horizon and descends inexorably toward the center of the black hole. A body or light ray heading toward the event horizon from outside is drawn in, becomes trapped, and suffers the fate just described. However, a body or light ray that is outside the event horizon may escape, if its direction is appropriate. As a rule of thumb, derived from the idealized situation of a nonrotating, spherically symmetric black hole, the event horizon can be thought of as a spherical surface surrounding the black hole and centered on it, with radius of about 1.8 miles (3 km) times the mass of the black hole expressed in terms of the mass of the Sun. Thus a five-solar-mass black hole possesses an event horizon with radius of about nine miles (15 km) and is wholly contained within that radius. The idea of five solar masses compacted within a sphere of radius nine miles (15 km) gives an impression of the tremendous density of a black hole (around 10^{18} pounds per cubic foot, or 10^{19} kg/m^3) and of the gigantic gravitational forces holding it together.

Since they do not emit light, black holes cannot be seen directly. Nevertheless, they can be detected in two ways. One is their gravitational effect on nearby bodies, such as when a black hole is one member of a pair of stars rotating around each other, called a binary-star system. Then the motion of the other member of the pair gives information on the black hole. Another way to detect black holes is through the X-rays emitted by matter falling into them. Black holes have been observed by both means.

Black holes are understood to form as the final stage of the evolution of stars whose masses are greater than about 25 times the mass of the Sun. Such stars, after catastrophically throwing off much of their material, end their lives as black holes of at least some three solar masses. In addition to being end-of-life stellar remnants, black holes lie at the centers of galaxies. The latter are of mammoth size, possess masses of millions of stars, and continually grow by sucking in more stars.

In spite of their not emitting anything from inside their event horizon, black holes can nevertheless "evaporate," as the British physicist Stephen Hawking discovered theoretically. The process commences with the spontaneous production of a virtual particle-antiparticle pair just outside the black hole's event horizon. Normally a virtual particle-antiparticle pair would immediately mutually annihilate and disappear, as if they had never existed. Such goings on characterize the quantum vacuum and are a consequence of the Heisenberg uncertainty principle. This principle allows the "borrowing" of energy for the creation of short-lived, virtual particle-antiparticle pairs and other particles, as long as the particles self-annihilate and the "loan" is repaid within a sufficiently short time interval. However, near the event horizon of a black hole, one member of such a pair might become sucked into the black hole before the pair can mutually annihilate. Then the other member becomes a real particle and leaves the black hole's vicinity, as if it had been emitted. As a result, the black hole loses energy, in the amount carried away by the escaping particle, and thus loses mass. In this manner, black holes eventually "evaporate" and disappear. The rate of evaporation is inversely related to the mass of the black hole. Thus, the effect is negligible for large-mass black holes, such as those that develop from the death throes of stars.

See also EINSTEIN, ALBERT; ELECTROMAGNETIC WAVES; ENERGY AND WORK; GENERAL RELATIVITY; GRAVITY; HAWKING, STEPHEN; HEISENBERG, WERNER; MATTER AND ANTIMATTER; QUANTUM MECHANICS.

FURTHER READING

Melia, Fulvio. *The Black Hole at the Center of Our Galaxy.* Princeton, N.J.: Princeton University Press, 2003.

Scientific American Special Edition: Black Holes 17, no. 1 (2007).

Thorne, Kip S., and Stephen Hawking. *Black Holes and Time Warps: Einstein's Outrageous Legacy.* New York: Norton, 1995.

Tucker, Wallace, Harvey Tananbaum, and Andrew Fabian. "Black Hole Blowback." *Scientific American* 296, no. 3 (March 2007): 42–49.

Bohr, Niels (1885–1962) Danish *Physicist* The work of Niels Bohr in the early part of the 20th century bridged the worlds of classical and quantum physics to create an image of the subatomic world that is still valid today. In the laboratory of Sir Ernest Rutherford, Bohr developed a model for the atom consisting of a nucleus surrounded by electrons that are often pictured much like the planets revolving around the Sun. Bohr incorporated concepts from Max Planck's quantum theory to explain changes that occur as electrons move from one orbital, or energy state, to another, and to predict where electrons would most probably exist within the respective orbitals. Later in his career, he refined the principles of quantum physics through his theoretical studies. During World War II, he played a significant role in the development of the atomic bomb, and he spent his last years endorsing more constructive uses of nuclear energy than nuclear warfare.

Niels Bohr was one of the founding fathers of quantum physics and was awarded the 1922 Nobel Prize in physics. His model of the hydrogen atom was the first to incorporate quantum ideas. *(AP Images)*

Born in Copenhagen, Denmark, on October 7, 1885, Niels Bohr grew up in an environment ideal for inquisitive minds. His father, Christian Bohr, was a professor of physiology at the University of Copenhagen, and his mother, Ellen Adler, was a member of a family of educators. In 1903 Bohr entered the University of Copenhagen, where he worked under the supervision of Professor C. Christiansen, a gifted physicist in his own right. Bohr obtained a master's in physics in 1909 and a doctorate in physics in 1911. During his doctoral studies, Bohr became familiar with many aspects of quantum theory while he addressed the properties of metals as they related to Planck's electron theory.

Bohr's working knowledge of the emerging field of quantum physics proved invaluable when he joined the laboratory of Sir Ernest Rutherford in March 1912. After a brief theoretical examination of alpha ray absorption, he began building on the theoretical model of the atom first proposed by Rutherford, which suggested an atom contains a nucleus with electrons surrounding it. Bohr hypothesized that these electrons orbited the nucleus in defined paths, or orbitals, much as the planets circle the Sun. Because electrons are negatively charged and the nucleus is positively charged, the electrostatic pull of the positive charge from the nucleus serves to keep the electrons nearby, just as gravity keeps the planets close to the Sun. If this subatomic arrangement followed the rules of classical physics, electrons would quickly spiral inward toward the nucleus while radiating away their energy in bursts of photons. This phenomenon could be measured by a spectrophotometer; however, spectroscopic data did not indicate that this situation ever occurred. Applying quantum theory to these experimental results, Bohr proposed that electrons normally orbit the nucleus in a specific orbital in which they do not lose, or release, energy. This initial configuration has a specific energy associated with it and is called the ground state. Bohr showed that other orbitals were possible as well and the farther away an orbital was from the nucleus, the more energy the electrons in that orbital possessed. According to his model, electrons did not have random amounts of energy associated with them; rather, the position (order) of the orbital with respect to the nucleus determined the amount of energy an electron in that specific orbital possessed. The variable n denoted the order, with the nearest orbital, that associated with lowest energy, being $n = 1$. Adding energy to the system causes an electron to move from its original orbital to another orbital farther away from the nucleus; in other words, it enters an excited state. Eventually, electrons spontaneously return to their ground state. When an electron returns to its original orbital, it releases a photon of light that can be measured spectroscopically. The energy of the released photon equals the difference in energies between the excited state orbital and the ground state orbital. Many different excited states are possible; therefore, many different light emissions (wavelengths of light being released) are possible. Spectroscopic data supported this model as a line spectrum from hydrogen showed discrete, rather than continuous, emissions of light, depending on the wavelength (related to the energy) released. Bohr's model of the atom combined aspects of classical physics and quantum mechanics to describe the ground and excited states of electrons in the hydrogen atom and to describe the transition between the ground and excited states. In 1922 Bohr received the Nobel Prize in physics for elucidating this theoretical model of atomic structure.

After four months in Rutherford's laboratory, Bohr held lectureship positions in physics at the University of Copenhagen and later at the University of Manchester before he accepted the position of professor of theoretical physics at the University of Copenhagen in 1916. He became director of the Institute for Theoretical Physics at the University of Copenhagen in 1920 and remained there until his death, except for a brief time during World War II. During the years at the institute, Bohr's research focused on atomic nuclei. He found that nuclear processes, unlike atomic processes, obeyed the rules of classical physics. Bohr attributed this phenomenon to the condensed size of the nucleus relative to the atom, which allows the strength of interactions to be greater than in the atom as a whole. Thus, transition processes that occur in the nucleus, such as neutron capture, follow the rules of classical physics rather than quantum physics. Bohr also developed the liquid droplet theory, which figuratively describes how a nucleus can be split (i.e., nuclear fission), by comparing a nucleus to a droplet of water. During the last 30 years of his life, he continued to clarify concepts in quantum physics through a series of essays found in the two-volume set *Atomic Physics and Human Knowledge* (1958, 1963). In these volumes, Bohr introduced complementarity, a concept that addresses the wave-particle duality of electrons and light, the phenomenon of both entities' simultaneously possessing characteristics of discrete particles and continuous waves. Bohr suggested that a complete description of either entity requires incorporating both its wave properties and its particle properties. For example, one can map the discrete energy and momentum of a specific electron in space as it behaves as a particle and characterize the same electron as a wave with a particular wavelength and frequency. The electron behaves both ways simultaneously according to quantum mechanics; yet, available instruments measure only one aspect while suppressing the other.

During the Nazi occupation of Denmark in World War II, Bohr first escaped to Sweden, then moved to the United States. He joined the Atomic Energy Project, otherwise known as the Manhattan Project, to create an atomic bomb that could be used to end the war. Years earlier, he had proposed the liquid droplet theory that explained the process of nuclear fission, whereby a neutron bombards the nucleus of a heavy atom like uranium, causing the nucleus to split into fragments of approximately equal mass and charge. The created fission products release neutrons, and these neutrons then split other nuclei, culminating in a chain reaction capable of releasing a huge amount of nuclear energy. Scientists confirmed this theoretical event in Chicago, in 1942, by initiating the first self-sustained nuclear reaction, followed by the creation and use of the atomic bomb. Bohr tried to urge the allied leaders to limit the use of atomic weapons although the Manhattan Project succeeded in its goal, but his pleas fell on deaf ears. In later years, he openly supported peaceful, not destructive, uses of nuclear physics. He believed that openness among nations about their knowledge of nuclear physics could prevent other bombs from being used in times of war.

Niels Bohr married Margrethe Nørlund in 1912. They had six sons during the course of their lives together. The four sons who survived to adulthood were all professionals: Hans Henrick Bohr became a medical doctor, Erik studied chemical engineering, Ernest practiced law, and Aage followed in his father's footsteps in theoretical physics, replacing him as director of the Institute for Theoretical Physics at the University of Copenhagen. Bohr's son Aage also won the Nobel Prize in physics in 1975 for his studies of the collective and particle motion of atomic nuclei. Niels Bohr died in Copenhagen on November 18, 1962.

See also ATOMIC STRUCTURE; CLASSICAL PHYSICS; FISSION; NUCLEAR CHEMISTRY; NUCLEAR PHYSICS; OPPENHEIMER, ROBERT; PLANCK, MAX; QUANTUM MECHANICS; RADIOACTIVITY; RUTHERFORD, SIR ERNEST.

FURTHER READING

The Nobel Foundation. "The Nobel Prize in Physics 1922." Available online. URL: http://nobelprize.org/physics/laureates/1922/index.html. Accessed July 24, 2008.

Pais, Abraham. *Niels Bohr's Times: In Physics, Philosophy, and Polity.* New York: Oxford University Press, 1991.

bonding theories Bonding theories explain how atoms interact when they are bonded together. The types of atoms involved in chemical bonds determine the nature, strength, and geometry of the bond formed. Electronegativity is the measure of the ability of an atom in a chemical bond to attract electrons to itself. The higher the electronegativity of an atom, the more it likes electrons and the higher its tendency to take electrons from other atoms. The element with the highest electronegativity is fluorine, making it a very reactive element.

Two main types of chemical bonds are ionic bonds and covalent bonds. Ionic bonds involve the loss of one or more valence electrons by a less electronegative element to a more electronegative element. After the loss of the electron(s), one of the atoms has a net negative charge while the other has a net positive charge. The electrostatic attraction between the positive and negative charges causes an ionic bond.

Covalent bonds, on the other hand, involve sharing of electrons between two atoms. Attempts to understand how the electrons are shared gave rise to multiple bonding theories that explain how electron sharing causes such properties as bond length, molecular geometry, and bond dissociation energies.

The structure of a compound determines its function in chemical reactions. Deviation of the shape or conformation of the compound from the predicted shape can affect the ability of the compound to function as expected during a chemical reaction. Therefore, understanding the three-dimensional geometry of a molecule is essential to legitimate discussion of chemical reactions.

LEWIS THEORY

The oldest and simplest version of bonding theory was put forward by the American scientist Gilbert N. Lewis and is known as the Lewis theory. Lewis electron dot structures depict the valence electrons (those in the outermost energy level) of an atom. They can be used to draw simple representations of the interaction of the atoms in a covalent bond by showing how the valence electrons of different atoms interact.

To write Lewis structures, place the most electronegative atom in the center and the less electronegative atoms around it. Then count the total number of valence electrons contributed by each of the atoms in the molecule. Place the electrons around the atoms, giving every atom (except hydrogen) eight electrons (in four pairs) to satisfy the octet rule. Finish up the Lewis structure by counting the total number of valence electrons in the drawing to ensure the correct number has been used. If fewer electrons are allowed, then remove lone-pair electrons and increase the number of bonds by forming double and triple bonds within the compound.

While Lewis structures explain which atoms are bonded together, they do not account for the differences seen in bond length between the atoms of certain molecules and bond strength within covalent molecules. Drawings of Lewis structures do not demonstrate that atoms share electrons in different ways and do not reveal the three-dimensional structure of the molecule.

VSEPR THEORY

The valence shell electron pair repulsion theory (VSEPR) combines the basic information depicted in Lewis structures with the concept of electron-electron repulsion to give a deeper understanding of the true structure of covalent molecules. One can accurately predict the shape of a molecule by applying this theory.

VSEPR models reveal three-dimensional geometrical approximations based on electron pair repulsions. Predicting the bond angles and bond lengths helps determine the spatial relationships among the atoms in a molecule. According to VSEPR theory, the number of electron pairs present in the central atom influences the geometry of the molecule. The fundamental principle of VSEPR theory maintains that lone-pair electrons (those electrons that are not involved in bonding) take up more space than bonding pairs of electrons. Therefore, the molecular geometry of a molecule will be altered on the basis of the repulsion of the lone-pair electrons and other electrons, either bonding or nonbonding, in the molecule.

VSEPR theory requires calculation of the steric number (SN), the total number of bonding and lone-pair electrons. Calculation of a steric number allows one to predict the shape of a molecule with accuracy. Data from structural experiments of molecules support these predictions. The relationship between the SN number of a molecular and its overall structure is summarized in the following table.

By studying the shapes of the molecules in the table, it becomes apparent that since the lone pairs are larger and therefore cause more steric hindrance between the electrons by way of repulsion, the angle between the atoms is slightly less than the angle with the same number of bonding pairs without the lone pairs. For example, two bonding pairs of electrons in the absence of any lone pairs results in a molecule with a linear shape and a bond angle of 180°. The addition of a lone pair of electrons to the central atom of that molecule changes the shape to bent, with a bond angle of less than 180°. The same is seen for three bonding pairs that without lone pairs would give a trigonal planar shape, whereas with a lone pair of electrons on the central atom pushing the other bonding pairs of electrons away, the bond angles are smaller and a trigonal pyramidal shape is formed.

VALENCE BOND THEORY

Valence bond theory is based on the overlap of electron orbitals, the allowed energy states of electrons in the quantum mechanical model. Chemists developed this type of model to explain further the differences in bond length and bond energy. Bonds form when the potential energy between the two atoms is at a minimum. Orbital types include *s*, *p*, *d*, and *f*. Single bonds called sigma (σ) bonds form by the end-to-end overlap of *s* orbitals. Pi (π) bonds are formed by the side-to-side overlap of *p* orbitals. Double bonds are made of one σ bond and one π bond. A triple bond consists of one σ bond and two π bonds. The overlap of orbitals permits electrons to be shared between the two atoms in the region of space that is shared between the two orbitals. As two atoms approach one another, the electron density between the two atoms increases and the positively charged nuclei of both atoms forming the bonds attract the electrons. As two atoms move close enough for their orbitals to overlap, the potential energy of the atoms is at a minimum, and the bond strength is at its maximum. Moving the atoms any closer together will cause the potential energy to increase dramatically as a result of the repulsion of the two nuclei for each other. Valence bond theory explains why bond length and bond energy will be different for similar molecules. The nuclei assume the positioning that minimizes the potential energy, which determines the distance and degree of attraction between them.

GEOMETRY OF ELECTRON PAIRS

Number of Bonding Pairs	Number of Lone Pairs	Steric Number	Shape	Example
2	0	2	Linear	CO_2
3	0	3	Trigonal planar	BF_3
2	2	4	Bent	H_2O
4	0	4	Tetrahedral	CH_4
3	1	4	Trigonal pyramidal	NH_3

HYBRID ORBITALS				
Number of Atomic Orbitals	**Type of Atomic Orbitals**	**Number of Hybrid Orbitals**	**Type of Hybrid Orbital**	**Molecular Geometry**
2	one ***s*** and one ***p***	2	***sp***	Linear
3	one ***s*** and two ***p***	3	sp^2	Trigonal planar
4	one ***s*** and three ***p***	4	sp^3	Tetrahedral

This description of valence bond theory explains much about bond formation, although it does not explain the formation of molecules that contain more than two atoms (polyatomic molecules). In many polyatomic molecules, the overlap of the orbitals does not fully explain the bond angles of the structures. For example, methane (CH_4) contains four equivalent bonds between a central carbon (C) and four hydrogen (H) atoms. The bond lengths between each C and H are the same, as are the bond energies. Hydrogen only contains one electron with an electron configuration of $1s^1$, and carbon has a valence electron configuration of $2s^22p^2$. If simple orbital overlap occurred, methane would have two different bond types, C–H bonds formed by the overlap between 1*s* and 2*s* orbitals and C–H bonds formed by the overlap of 1*s* and 2*p* orbitals. These bonds would have different lengths and strengths that are not observed in the methane molecule.

The valence bond theory explains the structure of methane and other polyatomic molecules by the hybridization of electron orbitals—the combining of orbitals of the individual atom into orbitals that are representative of the molecule. The number of hybrid orbitals formed is equal to the total number of atomic orbitals that are used to make them. In the case of methane, the carbon atom would have hybrid orbitals by combining the single 2*s* orbital with three 2*p* orbitals. Four equivalent orbitals would result from the hybridization of four atomic orbitals. These four equivalent orbitals could then each overlap with the atomic orbitals of hydrogen to form four identical bonds, as observed in the methane molecule.

Hybridization of one *s* and one *p* orbital results in two orbitals of equivalent energy and is known as *sp* hybridization. The hybrid orbitals have intermediate energy between the low-energy *s* orbital and the higher-energy *p* orbital. When a central atom forms two bonds using two *sp* hybrid orbitals, the molecule has a linear arrangement.

Hybridization of one *s* atomic orbital and two *p* atomic orbitals results in sp^2 hybrid orbitals. The three equivalent orbitals formed from this hybridization can give three bonds of equal length and strength. The sp^2 hybrid orbitals lead to an arrangement of the atoms in a covalent compound that is trigonal planar.

The hybridization of one *s* atomic orbital and three *p* atomic orbitals gives rise to four sp^3 hybrid orbitals. The covalent structure of the atom using sp^3 hybridization is tetrahedral. All of the four bonds have equivalent energy and equivalent bond angle. Methane, CH_4, utilizes this type of hybridization. The valence bond theory fully explains a mechanism that gives equivalent orbitals and equal bond lengths. The following table shows the typical hybrid orbitals.

MOLECULAR ORBITAL THEORY

The 1966 Nobel Prize in chemistry was awarded to Robert Mulliken for developing the molecular orbital theory to explain bonding. The molecular orbital theory is a more sophisticated model of bonding that takes into account the three-dimensional shape of a molecule. The main principle of this theory involves the formation of molecular orbitals from atomic orbitals, similar to the valence bond theory overlap of atomic orbitals. The molecular orbitals belong to the molecule and not individual atoms. Using hydrogen gas (H_2) as an example, one can see the interaction of two 1*s* atomic orbitals. When the two atomic orbitals overlap, two new orbitals are formed. One with lower energy than either of the original 1*s* orbitals is known as a bonding molecular orbital. The other molecular orbital formed has higher energy than both the bonding molecular orbital and the original 1*s* orbital of the hydrogen atoms and is known as an antibonding molecular orbital. Molecular orbitals are designated bonding or antibonding by placing an asterisk after the antibonding molecular orbitals.

If the electron density is centered along the axis of the two bonded nuclei, the orbital type is known as a sigma (σ) molecular orbital, and if the electron density is concentrated on opposite sides of the axis of the nuclei, it is known as a pi (π) molecular orbital. The type of atomic orbital used to create the molecular orbital is written as a subscript.

Applying these rules to the hydrogen gas example, hydrogen would have two sigma molecular orbitals known as σ_{1s} and σ^*_{1s}. The bonding molecular orbital would be σ_{1s}, and the antibonding molecular orbital would be σ^*_{1s}. Electrons are placed in molecular orbitals from the lowest energy (most stable) to the highest energy (least stable). The two electrons in hydrogen are placed into the σ_{1s} bonding molecular orbital. The stability of the covalent bond formed between these molecular orbitals is determined by the bond order.

Bond order = ½ (number of bonding electrons – number of nonbonding electrons)

The bond order explains the stability of the covalent bond formed. If the bond order equals 0, then no bond forms. A bond order of 1 represents a single bond, a bond order of 2 forms a double bond, and a bond order of 3 represents a triple bond. In the case of hydrogen gas, two bonding electrons and zero nonbonding electrons exist. The bond order is calculated as follows.

Bond order = ½ (2 – 0) = 1

Thus, hydrogen gas contains a single bond.

Bond order correctly predicts the unreactivity of noble gases. Two helium atoms would have a total of four electrons: two in the bonding molecular orbital and two in the antibonding molecular orbital. The bond order for He_2 would be

Bond order = ½ (2 – 2) = 0

A bond order of 0 correctly predicts that no bond would form between two helium atoms.

See also COVALENT COMPOUNDS; ELECTRON CONFIGURATIONS; QUANTUM MECHANICS.

FURTHER READING

Curry, John E., and Robert C. Fay. *Chemistry,* 5th ed. Upper Saddle River, N.J.: Prentice Hall, 2008.

Wilbraham, Antony, Dennis Staley, Michael Matta, and Edward Waterman. *Chemistry.* Upper Saddle River, N.J.: Prentice Hall, 2005.

Bose-Einstein statistics The statistical rules governing any collection of identical bosons are called Bose-Einstein statistics, named for the Indian physicist Satyendra Nath Bose and the German-Swiss-American physicist Albert Einstein. A boson is any particle—whether an elementary particle or a composite particle, such as a nucleus or an atom—that has an integer value of spin, that is, 0, 1, 2, ... All the particles that mediate the various interactions among the elementary particles are bosons. In particular, they comprise the gluons, which mediate the strong interaction; the intermediate vector bosons, which mediate the weak interaction; and the photon, which mediates the electromagnetic interaction. These are all spin-1 bosons. The graviton, the mediator of the gravitational interaction, is a boson with spin 2. At the level of atomic nuclei, the mediation of pions, which are spin-0 bosons, holds together the nucleons (i.e., protons and neutrons) constituting nuclei. (More fundamentally, this interaction derives from the strong interaction among quarks, mediated by gluons, since nucleons and pions consist of quarks.)

Bose-Einstein statistics are based on (1) the absolute indistinguishability of the identical particles and (2) any number of the particles being allowed to exist simultaneously in the same quantum state. The latter is a characteristic property of bosons.

One result of Bose-Einstein statistics is that in a system of identical bosons in thermal equilibrium at absolute temperature T, the probability for a particle to possess energy in the small range from E to $E + dE$ is proportional to $f(E)\,dE$, where $f(E)$, the probability distribution function, is

$$f(E) = \frac{1}{Ae^{\frac{E}{kT}} - 1}$$

Here E is in joules (J), T is in kelvins (K), f is dimensionless (i.e., is a pure number and has no unit), and k denotes the Boltzmann constant, whose value is $1.3806503 \times 10^{-23}$ joule per kelvin (J/K). The dimensionless coefficient A is determined by the type of system.

Photons are the quanta of electromagnetic waves. They are the manifestation of the particlelike nature of those waves, according to quantum physics. Similarly, phonons are the quanta of acoustic waves, in particular, vibrational waves in crystals. Both types of particles are bosons. For photons and for phonons, the probability distribution takes the simple form

$$f(E) = \frac{1}{e^{\frac{E}{kT}} - 1}$$

Note that this distribution function goes to infinity as the energy goes to zero and decreases to zero as the energy increases. That indicates a "bunching" of particles in low-energy states, with the distribution becoming sparser for increasing energies. The bunching is a consequence of any number of particles being allowed to exist simultaneously in the same quantum state, as mentioned earlier.

Bose-Einstein statistics predicts that at sufficiently low temperatures, very close to 0 K, a well-isolated collection of identical bosons can become so mutually correlated that they lose their individual identities and form what amounts to a single entity. That state is called a Bose-Einstein condensate. It has been created and investigated in research laboratories with sets of certain kinds of atoms.

See also ACOUSTICS; ELECTROMAGNETIC WAVES; ENERGY AND WORK; FERMI-DIRAC STATISTICS; HEAT AND THERMODYNAMICS; MAXWELL-BOLTZMANN STATISTICS; PARTICLE PHYSICS; QUANTUM MECHANICS; STATISTICAL MECHANICS.

FURTHER READING

Cornell, Eric A., and Carl E. Wieman. "The Bose-Einstein Condensate." *Scientific American* 278, no. 3 (March 1998): 43–49.

Serway, Raymond A., Clement J. Moses, and Curt A. Moyer. *Modern Physics,* 3rd ed. Belmont, Calif.: Thomson Brooks/Cole, 2004.

Boyle, Robert (1627–1691) British *Chemist*

Robert Boyle was one of the premier British scientists of all time. Boyle was interested in chemistry, natural philosophy, and physics as well as alchemy and religion. He made significant contributions to the development of experimental chemistry in order to demonstrate theoretical concepts. He subscribed to the new philosophy that depended on experimentation and observation to understand scientific principles. As one of the founding members of the Royal Society, of which he was a member from 1660 to 1691, Boyle is best known for his experiments utilizing an air pump to create a vacuum and his work on the relationship of gases that ultimately led to what is now known as Boyle's law.

Robert Boyle was a 17th-century English chemist who is best known for his work on the inverse relationship between pressure and volume of a gas that is now known as Boyle's law. *(Stock Montage/ Getty Images)*

EDUCATION AND AFFILIATIONS

Robert Boyle was born at Linsmore Castle in Munster, Ireland, on January 25, 1627. He was the 14th of 15 children born to the first earl of Cork and was the youngest son. His mother was the earl's second wife, Catherine. The earl of Cork was incredibly wealthy, so Boyle lived a privileged life. He attended Eton for four years; however, the majority of his education was provided by private tutors. As was the style of the wealthy at the time, Boyle spent several years studying on what was known as the Continent, including France, Italy, and Switzerland. When Boyle returned home in 1644, he lived in his estate of Stalbridge in Dorset, England, where he began his writing career. The majority of his writing in these early years was about religion rather than science. Boyle set up a laboratory at his estate in 1649 and began his scientific experimentation. Although he was never formally educated at a university, he was in residence at Oxford from 1656 to 1668. While living in Oxford, Boyle was surrounded by a very strong intellectual community.

Because of his family ties, Boyle had many connections, including one to King Charles II. Boyle was a founding member of the Royal Society and maintained membership until his death. King Charles appointed Boyle to the board of the East India Company and made him a member of the Company of Royal Mines. Boyle's religious ties were recognized as he was appointed governor of the Society for the Propagation of the Gospel in New England from 1661 to 1689.

Boyle lived in London with his sister from 1668 until his death on December 31, 1691.

PUBLICATIONS

Boyle was a prolific writer and published more than 40 books in his lifetime. At his estate in Stalbridge, he published several manuscripts discussing religion. When he moved to Oxford, his writing became much more productive, and he began to publish more on the fields of science. His first publication was called *New Experiments Physico-Mechanicall, Touching the*

Spring of the Air and Its Effects (1660). In this work, he published his creation of the first experimental vacuum with an air pump. This publication was followed by a second edition (1662) and included the work that became known as Boyle's law.

In 1661 Boyle published *Sceptical Chymist,* in which he defined the term *element* as "certain primitive and simple, or perfectly unmingled bodies; which not being made of any other bodies, or of one another, are the ingredients of which all those called perfectly mixt bodies are immediately compounded, and into which they are ultimately resolved." In this year he also published *Certain Physiological Essays.* Boyle's other significant scientific works included *Some Considerations Touching the Usefulness of Experimental Natural Philosophy* in 1663 and 1671, *Experiments and Considerations Touching Colours* in 1664, *New Experiments and Observations Touching Cold* in 1665, *Hydrostatical Paradoxes* in 1666, and the *Origin of Forms and Qualities* in 1666. Some of Boyle's later works included continuations and expansions of his experiments with the vacuum pump as well as medical manuscripts including *Of the Reconcileableness of Specifick Medicines to the Corpuscular Philosophy* in 1685. After his death, *The General History of the Air* was published, in 1692. An updated version of the works of Robert Boyle, a compilation of his publications showing their application to present-day scientific understanding, was published in 1999–2000.

BOYLE'S VACUUM EXPERIMENTS

A vacuum is completely devoid of all matter. A true vacuum does not exist on Earth naturally; however, a vacuum pump can be used to remove much of the air and any other matter from the container. Robert Boyle attempted to dispute the long-held Aristotelian belief that matter was continuous, meaning that matter could not be removed in order to create a vacuum. This was also expressed in the saying "Nature abhors a vacuum." Working on the initial design of the German scientist Otto von Guericke, Boyle perfected an air pump that was sealed and was able to be evacuated. Boyle developed his air pump with the help of his assistant, Robert Hooke, an English scientist. Robert Boyle utilized this air pump to determine the effect of air on experiments, flames, and living things. Experiments with this air pump demonstrated that living things including plants and animals required air to breathe and that combustion required air, as a flame would be extinguished when the vacuum was created.

BOYLE'S LAW

The properties of gases were of great interest to Boyle, and he is probably best remembered by students and scientists today for the law that is named after him, Boyle's law. The four significant variables concerning gases are the volume, pressure, amount of the gas, and the absolute temperature. Understanding the relationships among these variables allows for the prediction of the behavior of a gas. Boyle worked on the relationship between volume and pressure of a gas in 1662. Boyle's law demonstrates the relationship between pressure and volume when the amount of the gas and the temperature remain constant. The relationship between volume and pressure is given by

$$PV = \text{constant}$$

where the constant is a proportionality constant. By this formula, the relationship between volume and pressure is an inverse relationship. When the volume of a gas decreases, the pressure increases by the same factor. When the volume of a gas is increased, the pressure of the gas is decreased for the same reason. We now understand that on the basis of kinetic molecular theory, as the volume that the gas occupies decreases, the molecules are forced closer together and will therefore have more collisions. The more collisions that gas particles have with the sides of the container, the higher the pressure of a gas. The particles will spread out when the volume increases, and therefore they will have fewer collisions with each other and the sides of the container, thus reducing the pressure of the gas. When calculating the pressure or volume change in a gas using Boyle's law when the temperature and number of moles of gas remain constant, the formula is

$$V_1P_1 = V_2P_2$$

where V_1 is the initial volume, P_1 is the initial pressure, V_2 is the final volume, and P_2 is the final pressure. To illustrate the utility of the preceding equation, consider the following problem. If the initial pressure of a gas is 101.5 kPa and it takes up a volume of 3.2 L, what is the new pressure of the gas if the volume is expanded to 6.4L?

$$V_1 = 3.2 \text{ L}; P_1 = 101.5 \text{ kPa}; V_2 = 6 \text{ L}; P_2 = ?$$

$$(3.2 \text{ L})(101.5 \text{ kPa}) = (6.4 \text{ L})\ (P_2)$$

$$P_2 = 50.75 \text{ kPa}$$

Without performing the preceding calculation, one can estimate the final pressure on the basis of the Boyle's law relationship. When the volume changed from 3.2 L to 6.4 L, it doubled. One can predict that when the volume doubled, the pressure was halved, showing the inverse relationship between pressure and volume in Boyle's law.

BOYLE AND ALCHEMY

Given Robert Boyle's intense belief in God, one may find his proven interest in the field of alchemy surprising. Many at the time believed that alchemy (the pursuit of transforming elements into gold) had an illicit or immoral side to it. Many self-proclaimed religious individuals steered clear of alchemy, and some even condemned it. Robert Boyle had a strong interest in alchemy, which peaked in the 1670s with the publication of a paper on alchemy in *Philosophical Transactions*. Boyle associated with many alchemists. He firmly believed in the complexity of the universe, and alchemy helped to fit into his beliefs. Boyle was not averse to believing that God could create magical events, allowing him to reconcile his religious and scientific convictions.

Robert Boyle was one of modern chemistry's founding fathers. He contributed significantly to the field of gases with his work on the vacuum pump and the establishment of the relationship between pressure and volume, Boyle's law. Boyle was also influential in the establishment of chemistry as an experimental science. He helped set the standard at the time for experimental technique as well as publication of his work.

See also CHEMISTRY AND CHEMISTS; GAS LAWS; STATES OF MATTER.

FURTHER READING

Anstey, Peter. *The Philosophy of Robert Boyle*. London: Routlege, 2000.

Hunter, Michael, and Edward B. Davis, eds. *The Works of Robert Boyle*. 3 Vols. London: Pickering and Chatto, 1999–2000.

Broglie, Louis de (1892–1987) French *Physicist* A theoretical physicist, Louis de Broglie is famous for his well-confirmed proposal that matter possesses a wave nature. Even a moving automobile behaves as a wave in principle, although its wavelength is too small to be detectable. However, the wave nature of an electron, or other such particle, not only is detectable but forms an essential component of the understanding of nature at the submicroscopic scale.

ARISTOCRATIC HERITAGE

Louis-Victor-Pierre-Raymond de Broglie was born to Victor de Broglie and Pauline d'Armaillé Broglie on August 15, 1892, in Dieppe, France. Louis's father, who held the inherited title of duke, died when Louis was 14 years old, and the eldest son, Maurice, became a duke, a title held only by the male head of the family. All the family members shared the title prince, bestowed upon the de Broglies by the Austrians after the Seven Years' War (1756–63). In the wake of his brother's death in 1960, Louis became both duke and prince de Broglie. The youngest of five siblings, Louis received his primary education at home and his secondary education at the Lyceé Janson de Sailly in Paris, graduating in 1909. At the Sorbonne in the University of Paris, Louis studied literature and history with plans to embark on a civil service career, but he became interested in mathematics and physics. After obtaining a bachelor of arts degree from the Sorbonne in 1910, he decided to pursue a degree in theoretical physics.

After receiving his Licencié ès Sciences from the University of Paris's Faculté des Sciences in 1913, de Broglie served in the French army during World War I, when he learned about wireless telegraphy and served as a radio specialist. After the war, de Broglie continued his studies of physics in his brother's private laboratory. Several breakthroughs and new developments in the quantum theory that was proposed by Max Planck in 1900 revolutionized physics in the 1920s. De Broglie's doctoral dissertation, *Recherches sur la théorie des quanta* (Researches on the quantum theory), published in the *Annales de Physique* in 1925, concluded that matter has

Louis de Broglie proposed that matter particles, such as electrons, possess a wave character, in analogy with the particle nature of light waves, manifested by photons. He was awarded the 1929 Nobel Prize in physics for his theoretical discovery. *(AIP Emilio Segrè Visual Archives, Physics Today Collection)*

properties of both waves and particles and served as an important impetus for this revolution.

RECONCILING WAVES AND MATTER

Throughout the 19th century, physicists gathered data supporting the wavelike nature of light. Then in 1905, Albert Einstein published a new theory of electromagnetic radiation, including light, to explain the photoelectric effect. By applying Planck's concept of quanta to radiation, Einstein proposed that light is transmitted in tiny particles of specific energy, now called photons, simultaneously with its transmission as waves. The American physicist Robert A. Millikan experimentally verified Einstein's theoretical idea in 1913 and 1914. The American physicist Arthur H. Compton provided further support for the photon theory of electromagnetic radiation in 1923, when he explained the X-ray scattering phenomenon called the Compton effect. When an X-ray photon hits an electron, they undergo an elastic particle collision, in which the photon loses energy, causing it to have a larger wavelength (since a photon's wavelength is inversely proportional to its energy). Thus, Compton showed that photons carry energy and momentum, both properties of particles. Despite the new experimentally sound evidence for light's being particulate in nature, light clearly also possessed well-documented wavelike qualities, and physicists hesitantly began to accept the fact that light was of a dual nature, sometimes exhibiting wave properties and sometimes particle properties.

This revelation shattered the previously accepted partitioned view of a physical world that consisted of two realms—particulate matter and waves. Because all matter is composed of particles (atoms, or more fundamentally, electrons, protons, and neutrons), the physics of matter was concerned with atoms and particles that were assumed to obey the classical laws of mechanics. Radiation, on the other hand, was described as the propagation of waves through some medium, obeying the laws of wave physics. The form of matter and the form of radiation had been considered to be of incompatible natures until they were linked through the discovery that electromagnetic radiation, including light, was dual-natured.

De Broglie went one step further. He wondered if light is dual-natured, then what about matter? Though observably particulate in nature, might matter also possess wavelike characteristics? De Broglie's physicist's compulsion for unity and simplicity motivated him to consider that possibility seriously and to propose that particles indeed had a wave nature. This was the topic de Broglie chose for his doctoral thesis, a thesis that was awarded the 1929 Nobel Prize in physics.

Without any supporting facts, the bold graduate student de Broglie purported that "matter waves" were associated with everything material; though not apparent at a macroscopic level, wavelike behavior could be detected at the atomic level. In principle all objects have wave behavior, but their wavelengths are so small compared to the size of the objects themselves that they are undetectable in practice. At the atomic level, however, their size can be relatively large in comparison, and they are detectable. De Broglie claimed that the simultaneous possession of both wave and particle properties was not impossible, although it seemed so from the point of view of classical physics, but required different perspectives. The seemingly conflicting characteristics were not mutually exclusive; however, one could not perceive both aspects at the same time. He backed up his complicated proposition with a mathematical analysis using Einstein's equation that already defined a relationship among photon energy, light frequency, and Planck's constant h, whose value is $6.62606876 \times 10^{-34}$ joule-second (J·s). De Broglie assigned a frequency and a wavelength to a particle and determined that the particle's wavelength λ and the magnitude of its momentum p (the product of particle's mass and its speed) are related through the equation

$$\lambda = \frac{h}{p}$$

De Broglie's wave theory explained why electrons can exist in atoms only in discrete orbits, as proposed by Niels Bohr. According to de Broglie, the orbits were not smooth circular particle trajectories, but waves circulating around the atomic nucleus. A whole number of wavelengths was required for constructive interference around the orbit and the existence of a standing wave. So orbits were dependent on electron wavelengths, which de Broglie discovered to be dependent on momentum and thus on angular momentum, which Bohr had predicted must be an integer multiple of Planck's constant.

The faculty members on de Broglie's dissertation committee did not feel qualified to evaluate the profound thesis, so they recommended that Einstein read it. Einstein did read it and declared that de Broglie had discovered one of the secrets of the universe, forcing others to take the dual nature of matter seriously. De Broglie's hypothesis made Einstein's contention, in his special theory of relativity, that energy and mass were equivalent more understandable. The Austrian physicist Erwin Schrödinger also grasped the significance of the wave concept and used it to develop a wave equation to describe the quantum behavior of such as an electron in an atom, earning him a share of the Nobel Prize in physics in 1933.

PROOF OF MATTER WAVES

When de Broglie proposed the particle-wave relation, he had no supporting facts, but the proof for the wave nature of electrons did follow. In 1927 Clinton Davisson and Lester Germer at the Bell Laboratories in New York serendipitously discovered that crystals diffracted an electron beam. Since diffraction was a property of waves, that finding firmly supported de Broglie's proposal. Moreover, the mathematical calculations from their experiments agreed with de Broglie's formula. Later that same year, the English physicist George Paget Thomson (son of the famous Cambridge physicist Sir J. J. Thomson) observed diffraction by passing electrons through metallic foil. Davisson and Thomson shared the 1937 Nobel Prize in physics for their experimental discovery of the diffraction of electrons. Experimental evidence later produced similar results for protons, atoms, and molecules.

After receiving a doctorate from the Sorbonne, de Broglie taught there for two years. In 1927 he attended the seventh Solvay Conference, where the most prominent physicists of the time debated quantum mechanics, including particle-wave duality, and its implications. Some, such as the German physicist Werner Heisenberg, the Danish physicist Niels Bohr, and the English physicist Max Born, believed that matter waves represented the probability of the position of a particle (rather than the particle's possessing an exact position). Others, such as Schrödinger, Einstein, and de Broglie, disagreed. De Broglie proposed what he called the pilot wave theory in response, but the theory was flawed. Aware of this, he temporarily abandoned it, but he returned to it decades later, seeking an alternative to the probabilistic interpretation. Philosophically, de Broglie could not accept that the world acted in a random manner.

THE FOUNDATION OF QUANTUM MECHANICS

In 1928 de Broglie became a professor of theoretical physics at the Henri Poincaré Institute, and in 1932, a professor of theoretical physics at the Faculté des Sciences at the Sorbonne, where he remained until his retirement in 1962. At the Sorbonne he extended his research on quantum mechanics. He authored dozens of books related to quantum mechanics and the philosophical implications of quantum physics, including *Matter and Light: The New Physics* in 1939 and *The Revolution in Physics* in 1953. Though these are less technical than many of his works, they are still quite complicated because of the nature of their content.

He was elected to the Académie des Sciences in 1933 and served as permanent secretary for the mathematical sciences from 1942 until his death. The academy awarded de Broglie their Poincaré Medal in 1929 and the Albert I of Monaco Prize in 1932. In 1943 he established a center for applied mechanics at the Henri Poincaré Institute. The United Nations Educational, Scientific, and Cultural Organization (UNESCO) awarded de Broglie the Kalinga Prize in 1952 for his efforts to popularize physics and to make it understandable to the general public. The French National Center for Scientific Research presented him with a gold medal in 1955. He also was bestowed many honorary degrees and belonged to several international academic organizations, including the National Academy of Sciences of the United States and the Royal Society of London. Prince Louis de Broglie died at age 94, on March 19, 1987, in Paris.

De Broglie interpreted physical phenomena in an entirely new manner and, in doing so, helped physicists better understand the behavior of matter, especially at the atomic level. His proposal that every moving particle has a wave associated with it helped to fuse the concepts of particles and waves, cementing the duality of nature and laying a foundation for modern theoretical physics. As proponent of one of the most significant advances in the field, he is appropriately considered one of the founders of quantum mechanics.

See also ATOMIC STRUCTURE; BOHR, NIELS; COMPTON EFFECT; DUALITY OF NATURE; EINSTEIN, ALBERT; ELECTROMAGNETIC WAVES; HEISENBERG, WERNER; MASS; MOMENTUM AND COLLISIONS; MOTION; PHOTOELECTRIC EFFECT; PLANCK, MAX; QUANTUM MECHANICS; ROTATIONAL MOTION; SCHRÖDINGER, ERWIN; SPECIAL RELATIVITY; THOMSON, SIR J. J.; WAVES.

FURTHER READING

Boorse, Henry A., Lloyd Motz, and Jefferson Hane Weaver. *The Atomic Scientists: A Biographical History*. New York: John Wiley & Sons, 1989.

Broglie, Louis de. *Matter and Light: The New Physics*. Mineola, N.Y.: Dover, 1939.

———. *The Revolution in Physics: A Non-Mathematical Survey of Quanta*. New York: Noonday, 1953.

Cropper, William H. *Great Physicists: The Life and Times of Leading Physicists from Galileo to Hawking*. New York: Oxford University Press, 2001.

Heathcote, Niels Hugh de Vaudrey. *Nobel Prize Winners in Physics 1901–1950*. Freeport, N.Y.: Books for Libraries Press, 1953.

The Nobel Foundation. "The Nobel Prize in Physics 1929." Available online. URL: http://nobelprize.org/physics/laureates/1929/. Accessed July 24, 2008.

buffers A buffer is a solution that resists a change in pH when additional amounts of acids or bases are added. Several definitions of acids and bases exist, the broadest of which are the Lewis acid and base, which names acids as electron pair acceptors and bases as electron pair donors. According to the Brønsted-Lowry definition, an acid is a proton donor and a base is a proton acceptor. The most restrictive definition of acid and base is given by the Arrhenius acid, in which an acid is a substance that adds a proton to water to produce H_3O^+ and a base is a substance that produces OH^- in solution. Regardless of the definition for acids and bases, the function of the buffer remains the same. If acids are added to the system, the buffer helps take up the extra protons (H^+) so that the overall pH of the solution does not decrease. If a base is added to the buffered solution, the buffer will take up the extra hydroxide ions (OH^-), preventing an increase in pH.

ACID AND BASE EQUILIBRIUM CONSTANTS

All buffer solutions are made up of a weak conjugate acid-base pair, demonstrated by

$$HA \leftrightharpoons H^+ + A^-$$

where A^- is the conjugate base and HA is the undissociated acid. Weak acids do not completely dissociate in water, so some of the undissociated acid will still be present in the solution at the same time as the conjugate base and the proton. The equilibrium constant for the dissociation of a weak acid is given by the acid-dissociation constant (K_a). The equilibrium constant for this reaction is given by

$$K_a = [H^+][A^-]/[HA]$$

The more likely the acid is to lose its proton, the higher its dissociation constant, so large values of K_a represent strong acids. Low values of K_a represent the weak acids. The dissociation constants for weak bases are calculated the same way.

$$K_b = [\text{conjugate acid}][OH^-]/[\text{base}]$$

K_b is a measure of the equilibrium constant of the base dissociation. The higher the K_b, the more dissociated base is present when reacted with water. Weak bases are molecules with one or more lone pairs of electrons. Many weak bases are classified as proton acceptors and include molecules like ammonia (NH_3) and amines. The other class of weak bases includes the conjugate bases of weak acids such as carbonate (CO_3^{2-}), which reacts by the following reaction in equilibrium:

$$CO_3^{2-} + 2H_2O \rightarrow H_2CO_3 + 2OH^-$$

The base CO_3^{2-} reacts with water to produce the conjugate acid H_2CO_3 and a hydroxide ion. The relationship between the K_a of an acid and the K_b of its conjugate base in an equilibrium reaction is equal to the K_w of water, given by the following equation:

$$K_a \times K_b = K_w$$

If the K_a of the acid is high, the K_b of the base must be low, and vice versa. This relationship also allows one to calculate the K_a or K_b of an acid-base pair if one of the values is known, given that $K_w = 1 \times 10^{-14}$ M^2. For example, if $K_a = 4.5 \times 10^{-4}$, then K_b can be calculated as follows:

$$K_a \times K_b = K_w$$

$$(4.5 \times 10^{-4})\ K_b = 1 \times 10^{-14}$$

$$K_b = 2.2 \times 10^{-11}$$

The K_a can be used to calculate pH, and vice versa. When the K_a and the initial concentration of an acid are known, then the amount of H^+ in the solution can be determined. This allows the pH of a solution to be calculated.

For example, what is the pH of a 0.5 M solution of nitrous acid? Nitrous acid dissociates according to the following equilibrium formula:

$$HNO_2 \leftrightharpoons H^+ + NO_2^-$$

The K_a for nitrous acid is known to be 4.5×10^{-4} and is given by the equation

$$4.5 \times 10^{-4}\ M = [H^+][NO_2^-]/[HNO_2]$$

In order to determine the pH of this solution, one must first determine how much of the acid has dissociated to give H^+. Given the initial concentration of HNO_2 = 0.5 M and knowing this decreases by an unknown quantity (x) as it dissociates, then the concentration of H^+ and NO_2^- must increase by the same unknown quantity (x) on the basis of mathematical relationship between the reactants and products (stoichiometry) of the equilibrium reaction. This equation then becomes

$$4.5 \times 10^{-4} = (x)(x)/(0.5 - x)$$

which by algebraic manipulation becomes

$$(0.5 - x)(4.5 \times 10^{-4}) = x^2$$

$$(-4.5 \times 10^{-4}x) + (2.25 \times 10^{-4}) = x^2$$

$$0 = x^2 + (4.5 \times 10^{-4}x) - (2.25 \times 10^{-4})$$

Using the quadratic equation to solve for x, the concentration of dissociated H^+ in a 0.5 M solution of HNO_2 can be determined.

$$x = 0.015 = [H^+]$$

From the $[H^+]$, the pH can be calculated as follows:

$$pH = -\log[H^+]$$
$$= -\log[0.015]$$
$$pH = 1.82$$

Thus for a 0.5 M solution of HNO_2, the pH would be 1.82.

The equilibrium constant equation can also be used to calculate the K_a of a solution when the pH and initial concentration are known. To demonstrate this, using the same nitrous acid equilibrium reaction with a given pH of 1.82 and known concentration of acid, 0.5 M, calculate the K_a. First, use the pH equation to calculate the concentration of H^+ given a pH of 1.82.

$$pH = -\log[H^+]$$
$$1.82 = -\log[H^+]$$
$$[H^+] = 0.015\text{ M}$$

So, in terms of the stoichiometry of the reaction,

$$[NO_2^-] = 0.015\text{ M}$$

Inserting the given value of 0.5 M for the concentration of HNO_2, next determine the concentration of HNO_2 that is not dissociated.

$$[HNO_2] = 0.5\text{ M} - 0.015\text{ M} = 0.485\text{ M}$$

Plugging these values into the equilibrium constant equation gives

$$K_a = [H^+][NO_2^-]/[HNO_2]$$
$$K_a = (0.015\text{ M})(0.015\text{ M})/(0.485\text{ M})$$
$$K_a = 4.6 \times 10^{-4}$$

Notice that this is the same value for the K_a of nitrous acid that was given in the first example, after rounding is taken into account. This value can then be compared to the known K_a value for nitrous acid, as the acid-dissociation constants are known for most common acids.

Because the pH relates to the concentration of $[H^+]$ according to the equation

$$pH = -\log[H^+]$$

solving the equilibrium equation for $[H^+]$ demonstrates the effect of changing the concentration of H^+ or OH^- when added to the buffered solution. If the ratio of undissociated acid and conjugate base remains relatively constant, the $[H^+]$ and therefore the pH will remain constant or relatively constant. This is the case when the concentrations of acid and conjugate base are large relative to the amount of added acid or base. When more OH^- is added to the buffered solution, it will react with the acid HA in order to produce H_2O and more A^-. If more H^+ is added, it will react with the conjugate base $[A^-]$ in order to produce more HA. Once again, as long as the ratio between HA and A^- does not dramatically change, then the $[H^+]$ does not change appreciably, and therefore the pH remains relatively constant.

Buffers must use weak acid and base pairs, because a strong acid would push the equilibrium reaction completely to the right (toward more complete dissociation), and the reverse reaction would rarely occur. The most common method of preparing a buffer solution consists of combining a weak acid and its salt, which dissociates to give the conjugate base. An example would be adding acetic acid (CH_3COOH) to sodium acetate ($NaC_2H_3O_2$), which dissociates into Na^+, and $C_2H_3O_2^-$, the conjugate base.

HENDERSON-HASSELBACH EQUATION

Buffering capacity refers to the amount of acid or base that a buffer can absorb before a significant change in pH occurs. The higher the concentration of the acid-base pair in the solution, the more that solution is able to resist pH change. The concentration of acid-base pair that is present in the buffer solution determines buffering capacity according to the Henderson-Hasselbach equation:

$$pH = pK_a + \log([\text{base}]/[\text{acid}])$$

K_a and K_b are often referred to by their pK_a and pK_b, which are simply the $-\log K_a$ and $-\log K_b$, respectively. The $pK_a + pK_b$ must equal 14. Buffers are most effective when they contain an acid with a pK_a close to the desired pH of the solution. One can determine the pH of a buffer system if the concentrations of buffer components are known. For example, calculate the pH of a buffer solution containing 0.24 M carbonic acid (H_2CO_3) and 0.12 M carbonate ions using the following equilibrium equation:

$$H_2CO_3 \rightleftharpoons 2H^+ + CO_3^{2-}$$

If the K_a of carbonic acid is 4.3×10^{-7}, then

$$pK_a = -\log 4.3 \times 10^{-7}$$
$$pK_a = 6.4$$
$$pH = pK_a + \log[\text{base}]/[\text{acid}]$$
$$pH = 6.4 + \log(0.12\text{ M})/(0.24\text{ M})$$
$$pH = 6.1$$

BUFFER PREPARATION

Thus, on the basis of the Henderson-Hasselbach equation, one should choose an acid with a pK_a near the desired pH range when choosing a conjugate acid-base pair to prepare a solution. Every buffer has a pH range within which it works optimally, and it can lose its effectiveness if too much acid or too much base is added to the solution; therefore, it is necessary to know the desired pH range in order to create a buffer that will work best in that situation.

The following example demonstrates how one would prepare a solution of a desired pH from NH_4Cl and aqueous NH_3. When these reactants dissociate, the components of the solution would include NH_4^+, NH_3, OH^-, H_2O, and Cl^-. Since H_2O and Cl^- do not have strong acidic or basic properties, they can be ignored for the purpose of this discussion. Using the base-dissociation constant formula, one can determine the quantity of NH_4Cl to add to 1.0 L of 0.2 M NH_3 to form a buffer with a pH of 8.3.

$$K_b = [NH_4^+][OH^-]/[NH_3]$$

The K_b is known to be 1.8×10^{-5} and the concentration of NH_3 is given, but in order to use the preceding equation, the concentration of OH^- must be calculated. This can be done by first determining the pOH.

$$pOH = 14 - pH$$

$$pOH = 14 - 8.3$$

$$= 5.7$$

Then calculate the concentration of OH^-.

$$pOH = -\log [OH^-]$$

$$5.7 = -\log [OH^-]$$

$$[OH^-] = 1.995 \times 10^{-6} \text{ M}$$

Now, using the equilibrium constant formula, the concentration of NH_4^+ that is necessary to achieve the desired pH can be calculated by plugging in all the known values.

$$K_b = [NH_4^+][OH^-]/[NH_3]$$

$$1.8 \times 10^{-5} = [NH_4^+](1.995 \times 10^{-6} \text{ M})/(0.2 \text{ M})$$

$$[NH_4^+] = 1.8 \text{ M}$$

So, in order for the solution to have a pH of 8.3, the concentration of NH_4Cl in the solution must be 1.8 M. This means 1.8 moles of NH_4Cl should be present in 1 L of solution.

BIOLOGICAL BUFFER SYSTEMS

Buffered solutions play an important role in the function of the human body. The pH of human blood requires tight regulation. The normal blood pH is approximately 7.4. If the pH rises or falls much from this value, then harmful physiological conditions such as acidosis can result. The majority of protein enzymes in the human body only function within a specific pH range. When the pH is altered significantly, these enzymes are broken down or become nonfunctional, causing a breakdown in reactions in the body. The buffer systems that regulate the blood pH include the carbonic acid–bicarbonate buffer system and the phosphate–hydrogen phosphate buffer system. The carbonic acid–bicarbonate buffer system is the fundamental regulator of blood pH and works according to the following two equilibrium formulas:

$$H^+_{(aq)} + HCO_3^-{}_{(aq)} \rightleftharpoons H_2CO_{3(aq)} \rightleftharpoons H_2O_{(l)} + CO_{2(g)}$$

When the body takes in oxygen and utilizes it to break down glucose during cellular respiration, one by-product is carbon dioxide (CO_2), an important component of this buffer system. The acid involved in this buffer system is carbonic acid (H_2CO_3), which dissociates to form bicarbonate (HCO_3^-), its conjugate base, and a hydrogen ion (H^+), which is another by-product of cellular respiration. As the concentration of H^+ increases in the blood, the pH of the blood decreases. Without the buffer system, the pH would drop to dangerous levels. The pH of the buffer system can be determined using the following formula:

$$pH = pK_a(-\log [CO_2]/[HCO_3^-])$$

This indicates that the pH of the carbonic acid–bicarbonate system is dependent upon the amount of CO_2 and the amount of conjugate base HCO_3^- (bicarbonate) that is present in the solution. Under normal circumstances, the amounts of CO_2 and HCO_3^- are quite large relative to the amount of H^+ being produced in the body; thus, the ratio changes very little. As stated, this buffering capacity works best when the buffer is near the pK_a of the acid or within one pH unit of the pK_a. In this case, the pK_a of carbonic acid is 6.1, well below the pH of the blood, meaning that if the buffer had to work alone, the pH of the blood would change rapidly, especially during strenuous exercise. When the rate of cellular respiration increases, then the amount of H^+ increases, and if exercise occurs in the absence of sufficient oxygen, then the additional production of lactic acid via fermentation would also contribute to the H^+ concentration.

Two organs are involved in helping the buffer system maintain blood pH; the lungs and the kidneys.

When the breathing rate of an individual increases during exercise, the carbon dioxide can be released through the lungs. This pulls the equilibrium reactions of this buffer system toward the right, favoring the creation of carbonic acid and the reduction of H^+, maintaining a stable pH. The kidneys work in conjunction with the lungs to maintain blood pH by removing excess HCO_3^- from the body in urine. This removal pulls the reaction toward the left, and the increased production of H^+ from carbonic acid thus lowers the pH. These two organs work in concert to prevent the pH from becoming too high or too low.

A buffer system with a smaller role in the maintenance of blood pH is the phosphate buffer system. This system is made up of dihydrogen phosphate ($H_2PO_4^-$) and hydrogen phosphate (HPO_4^{2-}) according to the following formula:

$$H_2PO_4^- \rightleftharpoons H^+ + HPO_4^{2-}$$

The pK_a for dihydrogen phosphate is approximately 6.8, making it an effective buffer system at normal blood pH levels. This system plays a minor role because of the low physiological concentrations of both $H_2PO_4^-$ and HPO_4^{2-}.

See also ACIDS AND BASES; BIOCHEMISTRY; EQUILIBRIUM; PH/POH.

FURTHER READING

Brown, Theodore, H. Eugene Lemay, and Bruce E. Bursten. *Chemistry: The Central Science.* Upper Saddle River, N.J.: Prentice Hall, 2006.

Zumdahl, Stephen. *Chemistry,* 7th ed. Boston: Houghton Mifflin, 2006.

calorimetry Calorimetry is an experimental procedure to determine the amount of heat gained or lost during a chemical or physical change. Calorimetry is a valuable tool in the field of thermochemistry, the branch of chemistry involving the study of energy and energy transformations that occur in chemical reactions. Energy is the ability to do work, and in the case of most chemical reactions, the work done is in the form of changed heat. Two types of energy are kinetic (energy of motion) and potential energy (energy of position or stored energy). Potential energy in a chemical reaction is stored in chemical bonds and is known as chemical potential energy. As matter increases its kinetic energy, it increases its temperature (measure of the average kinetic energy of matter).

Energy of a reaction involves the system (the part of the reaction that one is studying) and the surroundings (everything else). In an open system, the system and surroundings are able to interact with one another freely, in contrast to a closed system. Chemical reactions that release energy from the system to the surroundings are exothermic reactions. When energy from the surroundings needs to be put into the system in order for the reaction to occur, the reaction is endothermic. Chemical reactions that are exothermic generally occur spontaneously, and endothermic reactions require a deliberate energy input.

UNITS OF ENERGY

Energy of a reaction can be measured in a variety of units. The SI unit for energy is the joule (J), named after the English physicist James Prescott Joule. The joule is defined as the amount of energy it takes to move a one–newton force one meter. Another common unit of energy is the calorie (cal), defined as the amount of energy required to raise one gram of water one degree Celsius. The conversion between joules and calories is

$$1 \text{ calorie} = 4.184 \text{ joules}$$

The dietary calorie (C) found on food labels is equivalent to 1,000 calories, or a kilocalorie. When comparing the amount of energy released or taken up by a chemical reaction, one must consider the composition of the reactants and products. Heat capacity is a measure of the amount of energy required to raise the temperature of a substance 1°C. A more useful version of heat capacity is the specific heat capacity (C) of a substance. When the specific heat capacity is measured at constant pressure, the abbreviation is C_p. The specific heat capacity (or specific heat) is the amount of energy needed to raise one gram of a substance one degree Celsius. Specific heat capacity values generally have the units J/(g × °C) or kJ/(mol × °C). A higher specific heat capacity of a substance indicates that more energy is required to raise the temperature of that substance. Water has a specific heat of 4.184 J/(g × °C); that means it takes 4.184 joules (or one calorie) to raise the temperature of one gram of water by 1°C. This corresponds to the preceding definitions of both joules and calories.

A swimming pool on a hot summer day illustrates the concept of specific heat. The water in the pool and the concrete surrounding the pool are both exposed to the same amount of radiant energy from the sun. While the pool feels warm and comfortable, the concrete burns one's feet when walking across it. The specific heat of water is 4.184 J/(g × °C), but the specific heat of concrete is 0.880 J/(g × °C), much lower than that for water. This means that while the concrete absorbs the sun's energy and readily increases temperature, the water resists changes in temperature as a result of its high specific heat. This also explains why large bodies of water help moderate climates in coastal areas.

ENTHALPY

The first law of thermodynamics dominates the study of energy transformations. This law states that for all isolated systems the total energy of the system remains constant in value over time. Simply put, energy can be neither created nor destroyed. This means that the amount of energy lost from an exothermic reaction in the system must be transformed into another form of energy or lost to the surroundings. The system can lose energy to the surroundings in the form of heat (q) or by performing some sort of work. When a chemical reaction occurs under constant atmospheric pressure in an open system, such as a beaker, flask, or test tube, much of the energy lost is in the form of heat. Enthalpy (H) is the measure of heat lost or gained in a chemical reaction. When the change in enthalpy (ΔH) is positive, the reaction gives off heat (i.e, is exothermic), and if the ΔH is negative, the reaction absorbs heat (i.e., is endothermic).

TYPES OF CALORIMETRY

Calorimetry determines the values for ΔH experimentally for a given system. The apparatus used for calorimetry is known as a calorimeter. Two main types of calorimeters exist: a constant pressure calorimeter and a constant volume calorimeter or bomb calorimeter. Both types of calorimeters allow the amount of energy being given off by the chemical or physical change in the system to be measured as a change in temperature.

Constant pressure calorimetry is performed in a vessel that is not sealed. The calorimeter monitors the temperature of the sample inside while physical or chemical changes occur. As the sample is under constant air pressure during the process, the change in temperature of the sample is considered to be due to the change in state or reaction in the calorimeter. A constant pressure calorimeter can be as simple as a covered Styrofoam™ coffee cup that contains the reactants and a thermometer to measure

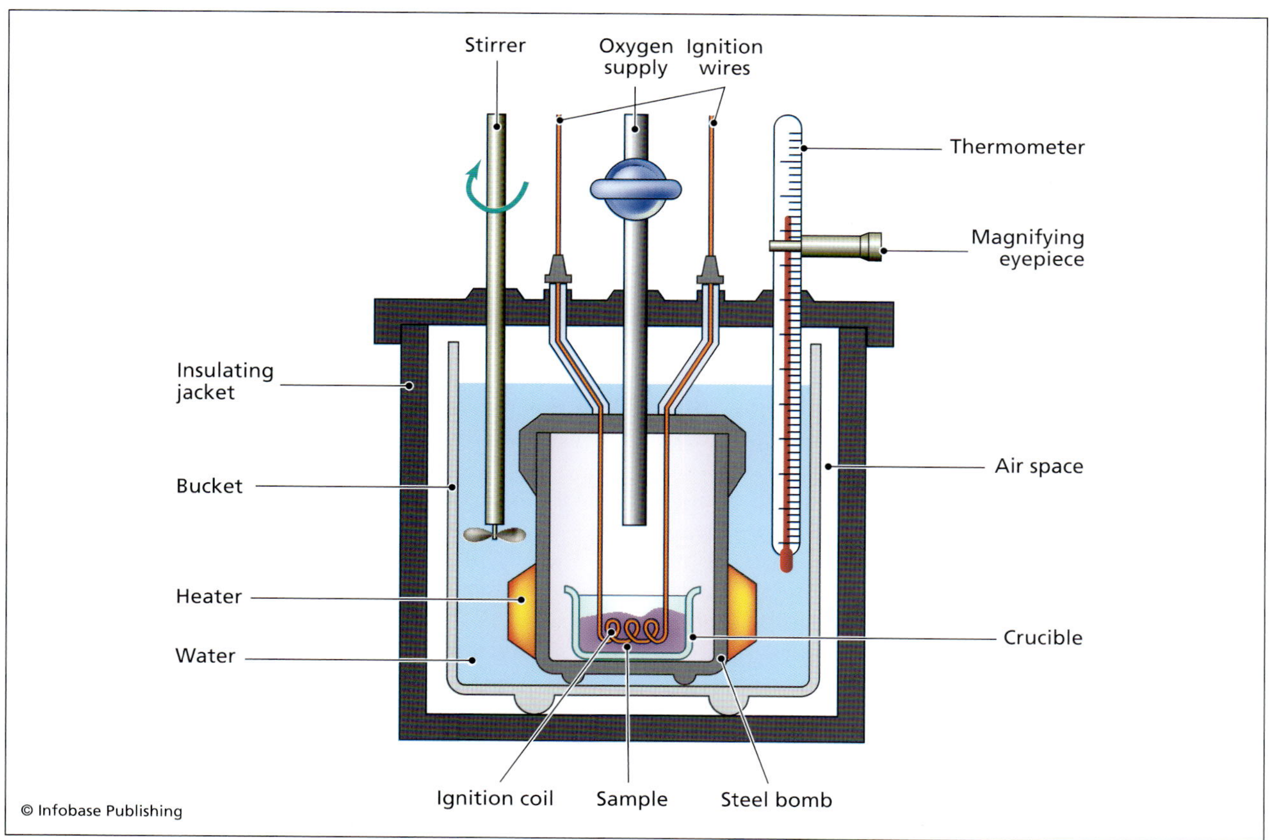

In a bomb calorimeter, the sample to be heated is placed inside a sealed vessel with an oxygen supply surrounded by an outside container holding water. The sample is ignited and combusted inside the sealed vessel. The change in temperature of the water is measured to determine the amount of energy given off by the combusting material.

the temperature change. When there is a temperature change in the system, the amount of heat being transferred to and from the system can be calculated according to the following formula:

$$q = mC\Delta T$$

where q is the amount of heat gained or lost, m is the mass of the substance in grams, C is the specific heat of the substance, and ΔT stands for the change in temperature (final temperature minus the initial temperature).

A typical calorimetry experiment for melting ice is described in the following.

1. Place a known volume of water in the coffee cup and record its temperature (e.g., 25°C).
2. Add an ice cube of known mass and specific heat into the calorimeter (e.g., m = 3 g, C_p = 2.03 J/(g × °C).
3. Record the final temperature of the water as the ice cube melts (e.g., 21°C).
4. The change in temperature ΔT of the water is given by the $T_{\text{final}} - T_{\text{initial}}$.

$$21°C - 25°C = -4°C$$

5. Calculate the amount of energy transferred in this phase change.

$$Q = mC\Delta T$$

$$Q = (3 \text{ g})\ (2.03 \text{ J/g} \times °\text{C})\ (-4°\text{C})$$

$$Q = -24.36 \text{ J}$$

This means that the system (the ice cube) lost 24.36 J of energy to the surroundings (the water).

Bomb calorimetry operates on the same principle as constant pressure calorimetry but holds the test sample in a sealed container at a constant volume rather than a constant pressure. Energy changes that take place during combustion reactions of hydrocarbons are readily studied in bomb calorimeters. Heat of reactions (including heat of combustion) can be determined. A sample to be combusted is placed in the bomb (a sealed reaction vessel) and then placed inside a water-filled calorimeter of a known temperature. A controlled amount of oxygen is piped into the bomb, and to initiate the reaction, an electric current is sent through a wire into the bomb in order to ignite the reactants. As the reaction vessel (the system) has a fixed volume, the reaction is not able to change the volume of the container so all of the energy released from the combustion reaction inside the bomb is transferred as heat that is absorbed by the water in the calorimeter (the surroundings). Insulation around the bomb calorimeter minimizes the loss of heat outside the calorimeter. The amount of heat generated by the reaction is given by

$$q_{\text{reaction}} = -C_{\text{cal}} \times \Delta T$$

where q_{reaction} is the amount of energy given off or absorbed by the reaction inside the bomb, C_{cal} is the heat capacity of the calorimeter that needs to be calculated or known, and ΔT is the change in temperature of the water in the calorimeter before and after the reaction. Calculating the amount of heat given off from the combustion of 3.5 grams of propane in a bomb calorimeter with a heat capacity (C_{cal}) of 5320 J/°C if the temperature of the water in the calorimeter increases by 3°C,

$$q_{\text{reaction}} = -C_{\text{cal}} \times \Delta T$$

$$q_{\text{reaction}} = (-5320 \text{ J/°C})(3°\text{C})$$

$$q_{\text{reaction}} = -15{,}960 \text{ J}$$

The value of −15,960 J represents the amount of energy given off (as denoted by the negative value) by the combustion reaction in the calorimeter. Bomb calorimeters can be used to quantify the amount of Calories in a type of food. If a known quantity of food is placed in the calorimeter and burned, one can measure the amount of joules produced during that reaction. Using the conversions between joules, calories, and Calories, one can calculate the dietary Calorie value of the food being burned. The q_{reaction} of the combustion of a 28-gram brownie (approximately 1/20 of a prepared mix brownie recipe) is equal to 502,080 J. Converting this value into calories is done as follows:

$$502{,}080 \text{ J} \times (1 \text{ cal}/4.184 \text{ J}) = 120{,}000 \text{ cal}$$

Then convert to dietary calories,

$$120{,}000 \text{ cal} \times (1 \text{ Cal}/1{,}000 \text{ cal}) = 120 \text{ Cal}$$

HESS'S LAW

Hess's law states that the amount of energy required to carry out a reaction that occurs in multiple steps is the sum of the required energy for each of the successive steps. Thus, if one calculates the amount of energy required for each of the individual steps of a reaction, adding these values will give the amount of energy required to carry out the reaction to completion. Hess's law can be easily demonstrated using the phase changes of water. When a given mass of ice transforms into steam, multiple steps occur as heat is added to the system. When the addi-

A bomb calorimeter determines the amount of energy transferred during a chemical reaction. Bomb calorimeters utilize a sealed container with a fixed volume. A rotating bomb combustion calorimeter is shown here. *(Oak Ridge National Laboratory)*

tion of heat causes an increase in temperature, the preceding formula can be used ($q = mC\Delta T$). When the result of the addition of heat is not a change in temperature but a change in state, then the ΔT factor in the preceding equation would be zero, reducing the energy (q) to zero. To calculate the amount of energy in these situations, the following formulas are used:

$$q = m\Delta H_{fus}$$

$$q = m\Delta H_{vap}$$

where q is the amount of heat gained or lost, m is the mass of the substance in grams, and ΔH_{fus} and ΔH_{vap} are constants, called enthalpy of fusion and enthalpy of vaporization, respectively. These constants represent the amount of energy per gram required to perform the corresponding change of state and can be looked up in reference tables.

ΔH_{fus}: energy required to change between liquid and solid

ΔH_{vap}: energy required to change between a liquid and a vapor

Knowing that the melting point of ice is 0°C and the boiling point of water is 100°C, the amount of required energy added to the system when 50 grams of ice at -20°C is turned into steam at 110°C would be calculated with the following five steps:

1. (-20°C to 0°C)

$$q = mC\Delta T$$

$$q = (50 \text{ g})(1.7 \text{ J/g} \times°\text{C})(20°\text{C})$$

$$q = 1{,}700 \text{ J}$$

2. (melting of 50 g ice)

$$q = m\Delta H_{fus}$$

$$q = (50 \text{ g})\ (334 \text{ J/g})$$

$$q = 16{,}700 \text{ J}$$

3. (0°C to 100°C)

$$q = mC\Delta T$$

$$q = (50 \text{ g})(4.184 \text{ J/g} \times°\text{C})(100°\text{C})$$

$$q = 69{,}873 \text{ J}$$

4. (vaporizing the water)

$$q = m\Delta H_{vap}$$

$$q = (50 \text{ g})(2260 \text{ J/g})$$

$$q = 113{,}000 \text{ J}$$

5. (100°C to 110°C)

$$q = mC\Delta T$$

$$q = (50 \text{ g})(1.7 \text{ J/g} \times°\text{C})(10°\text{C})$$

$$q = 850 \text{ J}$$

According to Hess's law, the amount of heat required to take the sample from ice to steam is the sum of the heat values for each of the steps. For this example,

$$q_{rxn} = q_{step1} + q_{step2} + q_{step3} + q_{step4} + q_{step5}$$

$$q_{rxn} = 1{,}700 \text{ J} + 16{,}700 \text{ J} + 69{,}873 \text{ J} + 113{,}000 \text{ J} + 850 \text{ J}$$

$$q_{rxn} = 202{,}123 \text{ J}$$

See also CHEMICAL REACTIONS; CONSERVATION LAWS; HEAT AND THERMODYNAMICS.

FURTHER READING

Petrucci, Ralph H., William S. Harwood, Geoff E. Herring, and Jeffry Madura. *General Chemistry: Principles and Modern Applications*, 9th ed. Upper Saddle River, N.J.: Prentice Hall, 2007.

carbohydrates Carbohydrates, or sugars, are the most prevalent type of organic matter found on the earth because they are used by all forms of life as energy stores, fuel, and metabolic intermediates. These biomolecules exist as basic subunits called monosaccharides and as polymers called polysaccharides, sugars linked to form linear and branched chains. Glycogen in animals and starch in plants are examples of polysaccharides that are used as sources of glucose, a sugar molecule frequently used for energy in living organisms. Two specific carbohydrates, ribose and deoxyribose, serve as the skeletal backbones of ribonucleic acid and deoxyribonucleic acid, respectively. Polysaccharides, such as cellulose, make up the cell walls of plants and the exoskeletons of arthropods. Carbohydrates linked to proteins and lipids also play critical roles in cell-cell recognition and cell signaling.

MONOSACCHARIDES

The basic subunits of carbohydrates are simple sugar molecules called monosaccharides that contain three or more carbons serving as a backbone, an aldehyde or ketone functional group, and two or more hydroxyl groups. The general formula $(CH_2O)_n$ describes all monosaccharides. Two common three-carbon monosaccharides, or trioses, are glyceraldehyde, which has an aldehyde group, and dihydroxyacetone, which has a ketone group. Sugars with four- (tetroses), five- (pentoses), six- (hexoses), and seven-carbon (heptoses) backbones also exist. One important pentose is ribose, a sugar containing an aldehyde group that serves as a backbone for the chemically reactive nitrogenous bases of ribonucleic acid. Two common hexoses are glucose, an aldehyde-containing sugar, and fructose, a ketone-containing sugar. These two sugars are simple carbohydrates found in fruits, honey, and juices.

In aqueous solution, a dynamic equilibrium exists between straight-chain and cyclic forms of pentoses and hexoses. The cyclic form is more prevalent. If a hexose has an aldehyde group as glucose has, the aldehyde group at the first carbon reacts with the hydroxyl group at the fifth carbon, forming a six-sided ring with a reduced acetyl group. If a hexose has a ketone group, the ketone group at the second carbon reacts with the hydroxyl group at the fifth carbon to form a five-sided ring structure with a reduced ketone. Fructose forms this type of cyclic structure.

Phosphorylated derivatives of these monosaccharides serve several functions within the cell. Phosphorylated glucose, or glucose-6-phosphate, is the first molecule of the glycolytic pathway. Several intermediates along this metabolic pathway, such as dihydroxyacetone phosphate and glyceraldehyde-3-phosphate, are phosphorylated sugars involved in the transfer of phosphate groups to adenosine diphosphate (ADP) molecules to yield adenosine triphosphate (ATP), the cell's main energy molecule. Another function of phosphorylation is to make the sugars anionic, or negatively charged. These charged sugar molecules cannot pass through the cell membrane, a characteristic that helps the cell retain these biomolecules for use in metabolism.

DISACCHARIDES AND POLYSACCHARIDES

In their cyclic form sugar molecules join together between the first carbon of one sugar molecule and the fourth or sixth carbon of a second sugar molecule. The condensation reaction releases a water molecule, and a glycosidic bond results. Disaccharides consist of two sugar subunits linked via a single glycosidic bond. Polysaccharides form when glycosidic bonds link multiple sugar molecules to form straight or branched chains.

Three common disaccharides are lactose, maltose, and sucrose. Lactose, a disaccharide found in milk, consists of a galactose linked to a glucose molecule. Maltose, the product resulting from the hydrolysis of starch by the enzyme lactase, contains two glucose subunits. Further hydrolysis by the enzyme maltase yields the two constituent glucose molecules. Sucrose, commonly known as table sugar, is a disaccharide with glucose and fructose linked through a glycosidic bond. The sugar industry extracts sucrose from the juice of sugar beets and sugarcane. The enzyme sucrase, sometimes called invertase, catalyzes the hydrolysis of sucrose into glucose and fructose.

Most children are able to digest lactose, but many adults are lactose intolerant because they are deficient in lactase. Without enough enzyme to hydrolyze the lactose, the disaccharide accumulates in the lumen of the small intestine since no alternative mechanism exists for its metabolism. The unabsorbed lactose leads to an influx of fluid into the intestine. The clinical symptoms for lactose intolerance include nausea, abdominal distension, cramping, and watery diarrhea. Lactose intolerance typically manifests itself during adolescence or early adulthood. Lactase deficiency appears to be an autosomal recessive trait that is prevalent in Asian ethnicities.

Polysaccharides are polymers of sugar residues linked by multiple glycosidic bonds. In some cases, these polysaccharides act as nutritional reservoirs.

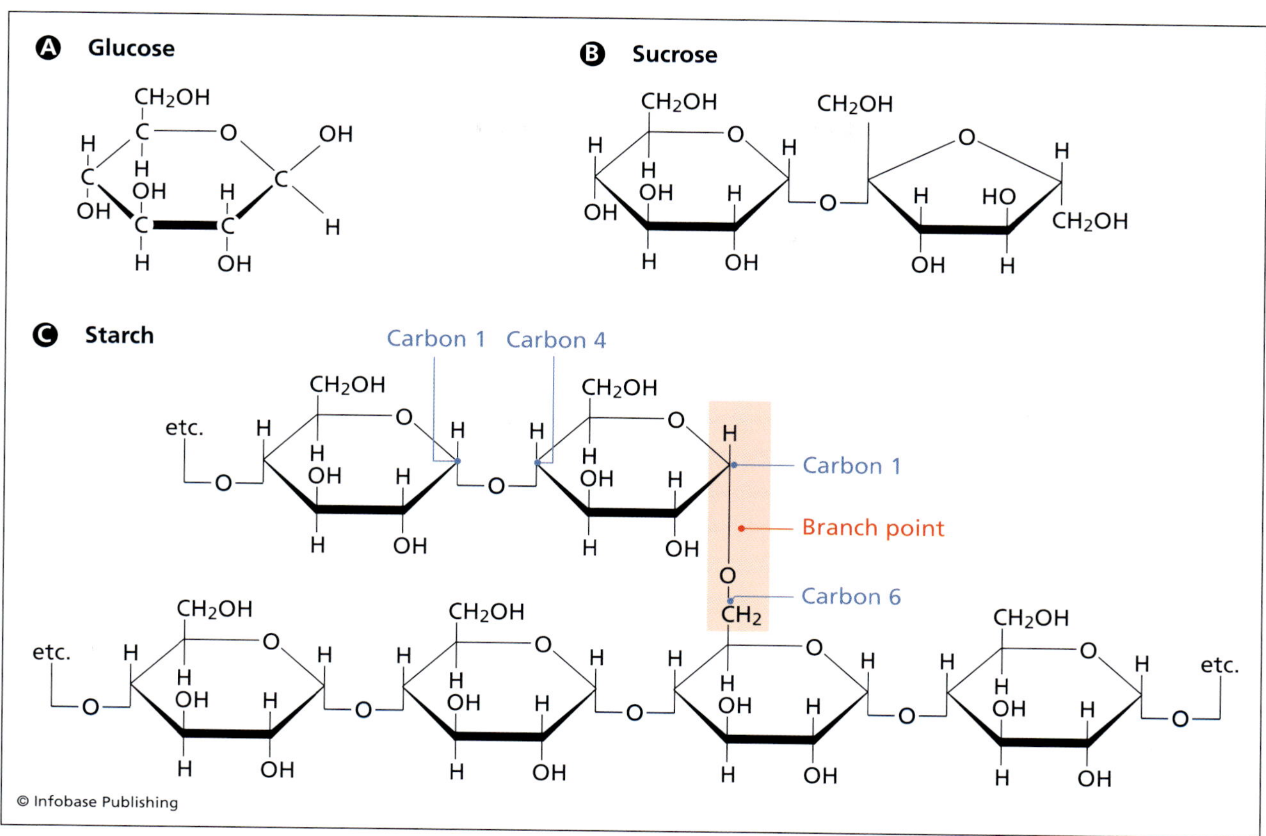

(a) Glucose, a six-carbon sugar (hexose), is an example of a monosaccharide. (b) Sucrose is a disaccharide with glucose and fructose linked through a glycosidic bond. (c) Amylopectin, a type of polysaccharide found in plants, is a branched form of starch consisting of long chains of glucose molecules linked at carbon number 1 and carbon number 4 of adjacent sugars and a second set of glycosidic bonds at carbon number 1 and carbon number 6 of glucose molecules approximately every 30 sugars along the linear chain.

Animals store glucose in the form of glycogen, a large, branched polysaccharide in liver and muscles. Glycosidic bonds link residues of glucose between carbon number 1 of one glucose molecule and carbon number 4 of a second glucose molecule. These bonds create long chains of glucose molecules. About every 10 residues, a second glycosidic bond exists between carbon number 1 and carbon number 6 of two neighboring residues. This second glycosidic bond results in formation of branches within the glycogen molecule. Branches in the polysaccharide increase its solubility in aqueous solution and make releasing glucose quickly from this polymer easier. Plants store glucose in the form of starch, which occurs as amylose or amylopectin. Amylose is the unbranched type of starch, consisting of long chains of glucose molecules linked at carbon number 1 and carbon number 4 of adjacent sugars. Amylopectin is the branched form of starch, which contains a second set of glycosidic bonds at carbon number 1 and carbon number 6 of glucose molecules approximately every 30 sugars along the linear chain. Amylopectin resembles glycogen except that it has fewer branches. More than half of the carbohydrates digested by humans are starch. Alpha-amylase, an enzyme secreted by the salivary glands and the pancreas, hydrolyzes both amylose and amylopectin. Dextran is a storage polysaccharide in yeast and bacteria that also consists of glucose molecules, but the glycosidic bonds exist between carbon number 1 and carbon number 6 of adjacent sugars almost exclusively.

Polysaccharides also play structural roles in biological organisms. Cellulose, the other major polysaccharide of plants, gives plants form and rigidity. This polysaccharide also consists of glucose molecules that are linked but in a slightly different way. As with starch, glucose residues form glycosidic bonds between the carbon number 1 and carbon number 4 of neighboring sugars to form long straight chains; however, each glucose residue rotates 180 degrees with respect to adjoining sugars. The different orientation results in long chains of glucose molecules that can form fibers with a high tensile strength. Mammals do not have cellulases, so they cannot digest plant and wood fibers. Bacteria and fungi do secrete cellulases. Some ruminants harbor cellulose-producing bacteria in their digestive tracts, permitting the extraction and absorption of nutrients from these fibers. Termites depend on a symbiotic relationship with cellulose-producing protozoa in their guts to digest wood. The exoskeletons of insects and arthropods contain chitin, a long-chain polysaccharide consisting of *N*-acetylglucosamine residues that form fibers such as cellulose.

Another role that carbohydrates play in biological organisms involves cell-cell recognition and signaling. Integral membrane proteins often have heterogeneous complexes of sugars, called oligosaccharides, attached covalently to the portion of the protein that faces the outside of the cell membrane. Secreted proteins, such as antibodies and clotting factors, also associate with oligosaccharides. When one or more sugar residues complex with a protein, a glycoprotein forms. Sugar residues bond with proteins at the oxygen in the side chains of serine and threonine to form oxygen (O)-glycosidic linkages or at the nitrogen in the side chain of asparagine to form nitrogen (N)-glycosidic linkages. Oligosaccharides that form N-linked glycosidic bonds usually contain a common pentasaccharide core consisting of three mannose and two N-acetylglucosamine residues. Additional sugar residues bind with this core to produce a variety of potential oligosaccharides that associate with proteins to form glycoproteins. In many cases, five or more different kinds of sugars associate through multiple types of glycosidic linkages to form these oligosaccharides. These highly complex oligosaccharide patterns suggest that the information extracted by interacting molecules from these different combinations of sugars is functionally important.

Evidence suggests that oligosaccharides provide a marker for cells to determine how long a glycoprotein has been present. A carbohydrate-binding surface receptor that exists on liver cells called the asialglycoprotein receptor often removes older glycoproteins from the bloodstream. Newly synthesized glycoproteins frequently have a sialic acid residue at the terminal end of the oligosaccharide. Over a matter of hours or days, depending on the glycoprotein, the enzyme sialase removes the sialic acid residue, exposing a galactose sugar residue in the remaining oligosaccharide. The asialglycoprotein receptor binds to the galactose residue and internalizes the entire glycoprotein, a process called endocytosis, removing these free-floating glycoproteins permanently from the bloodstream.

See also BIOCHEMISTRY; CITRIC ACID CYCLE; ENZYMES; GLYCOLYSIS; LIPIDS; NUCLEIC ACIDS; ORGANIC CHEMISTRY; PROTEINS.

FURTHER READING

Chen, T. Y., W. Smith, J. L. Rosenstock, and K. D. Lessnau. "A Life-Threatening Complication of Atkins Diet." *Lancet* 368 (2006): 23.

Hurtley, Stella, Robert Service, and Phil Szuromi. "Cinderella's Coach Is Ready (Chemistry and Biology of Carbohydrates)." *Science* 291 (2001): 2337.

Tapley, David W. "Carbohydrates." In *Biology*. Vol. 1, edited by Richard Robinson. New York: Macmillan Reference USA, 2002.

Cavendish, Henry (1731–1810) British *Physicist, Chemist*

Henry Cavendish was a British scientist whose primary interest was in the fields of chemistry and physics. Cavendish is best known for his meticulous experimentation and work on the topics of gases, electricity, and the mass of the planet Earth.

EARLY YEARS AND EDUCATION

Cavendish was born in Nice, France, on October 10, 1731, to Lady Anne Gray and Lord Charles Cavendish. He was of aristocratic descent as his maternal and paternal grandfathers were the duke of Kent and the duke of Devonshire, respectively. Cavendish attended school at Dr. Newcome's School in Hackney, London, and in 1749 entered St. Peter's College at Cambridge. Although Cavendish studied there for several years, he left in 1753 without having obtained a degree. Cavendish returned to London, where he lived for the remainder of his life. He became a member of the Royal Society of London in 1760.

Cavendish was extremely introverted, and it was often quoted that he even left notes for his servants rather than discuss matters with them personally. Cavendish's personality affected the amount of his scientific work that he shared with others. He would only publish material that had been thoroughly tested and in which he was completely confident. Much of his work was unpublished and was not truly appreciated until generations later, when it became apparent that Cavendish had already determined some of the concepts that were thought to be newly discovered. Many believe that if Cavendish had published all of his work, the pace at which physical science discoveries proceeded would have increased dramatically.

Cavendish's noteworthy publications included the paper "On Factitious Airs" on the discovery of hydrogen in 1766, the paper in 1784 titled "Experiments on Air," and in 1798 the paper "Experiments to Determine the Density of the Earth" on the measurement of the gravitational constant.

STUDY OF GASES

Cavendish's 1766 publication "On Factitious Airs," reporting the discovery of hydrogen, described a compilation of experiments that he had completed on the study of what were then called "airs" and are now called gases. His paper earned him the Royal Society's Copley Medal. One commonly accepted theory at the time was the phlogiston theory. The scientific community believed that matter underwent a chemical reaction by gaining or losing phlogiston, a mysterious substance. The phlogiston theory stated that all materials released this substance during burning, and the process of burning and releasing phlogiston left behind what they called a "dephlogisticated" substance. The more phlogiston a type of matter had, the more it would burn. Phlogiston was considered to have mass in some cases and negative mass in others (as suited their purposes). When a substance lost mass upon combustion, it was said to lose phlogiston; when a substance gained mass, such as in the combustion of magnesium (forming magnesium oxide), alchemists claimed the cause to be the loss of phlogiston that had negative mass.

Cavendish's work on factitious airs (gases that are contained in other bodies and can be released upon reaction with another substance) proposed that all types of air were not the same, a conclusion in contrast to the accepted thinking of his day that all air was the same. He also demonstrated that the substance given off when metals were reacted with acids was a different type of air than ordinary air. Cavendish determined that the substance known as "inflammable air" was hydrogen. He scientifically demonstrated that the substance formed upon the reaction of metallic compounds with acids was indeed a previously unidentified gas.

Cavendish presented a convincing argument because he actually measured the specific gravity of the gases being produced to show that they were distinct types. Specific gravity is the density of a substance in comparison to the density of water. If a substance is denser than water, it will have specific gravity greater than 1 and those substances that are less dense than water will have a specific gravity less than 1.

$$SG = \rho/\rho_{water}$$

ρ is the density of the substance and ρ_{water} is the density of water at a given temperature. As the specific gravity is a ratio of two densities, it is a dimensionless quantity. The specific gravity of hydrogen is 0.0696. Cavendish was able to use the distinct difference in specific gravities between hydrogen and carbon dioxide (SG = 1.51), the gas produced by fermentation and decay, to show experimentally that they were different gases (airs). Cavendish was the first to demonstrate that the gases (airs) produced under different experimental conditions were truly different substances.

Henry Cavendish was born in 1731 and trained in England as a physicist and chemist. Cavendish is best remembered for his diverse work in science. He worked extensively with gases and discovered hydrogen. Cavendish calculated the universal gravitational constant and worked in electricity. Most of Cavendish's work remained unknown and unpublished until the late 1800s. *(SPL/Photo Researchers, Inc.)*

In 1784 Cavendish wrote the paper "Experiments on Air," which demonstrated that water was not an element, the prevailing belief at the time, but rather a compound formed from the combination of different elements. When hydrogen gas was burned in air, the combination of hydrogen and oxygen present in the air produced water, which Cavendish established as a compound.

STUDY OF ELECTRICITY

Cavendish's work in the mid-1700s on electricity was virtually unknown until long after his death. In the 1870s the Scottish physicist John Clerk Maxwell collected Cavendish's works of nearly a century earlier and made them public knowledge. Cavendish performed thorough studies of the electrical conductivities of different substances.

One of Cavendish's most important discoveries was the experimental quantification of electrical potential. Upon investigation of Cavendish's early work, scientists discovered that he had experimentally demonstrated the relationship between resistance and current, a finding that has been credited to the German physicist Georg Simon Ohm, who performed his research decades later than Cavendish. Cavendish determined that current is inversely proportional to the resistance and directly proportional to the voltage between the two points, as indicated by the following equation:

$$I = V/R$$

where V is the voltage (potential difference) between two points, I is the current, and R represents the resistance between them.

This relationship (now known as Ohm's law) was clearly calculated by Cavendish well before the time of equivalent discoveries. The experimental techniques utilized by Cavendish to make this discovery were somewhat archaic and not quantitative. As there was no true way to measure current, Cavendish used himself as part of the experimental setup. He held on to both electrodes and rated the amount of pain produced. His experimental techniques paid off when he demonstrated the relationship seen in Ohm's law.

Cavendish's work also included the determination of the mathematical relationship representing that the force between a pair of electrical charges is inversely related to the square of the distance between them.

$$F = k(q_a q_b/r^2)$$

where F is the force between the charges; q_a and q_b are the charges of the particles a and b, respectively; r is the distance between the particles; and k is 8.99×10^9 N·m²/C². This relationship is now known as Coulomb's law, named for the French physicist C. A. Coulomb, who was a contemporary of Cavendish's. Once again, Cavendish's calculations were not known until long after his death; thus someone else received the credit for the finding.

STUDY OF EARTH'S DENSITY

Cavendish is perhaps best known for his experimental calculation of Earth's gravitational constant. Newton's law of universal gravitation is given as

$$F_g = G(m_1 m_2/r^2)$$

where F_g is the gravitational force between two objects, G is the universal gravitational constant; m_1 and m_2 are the masses of the individual objects, respectively; and r is the distance between the two objects.

The force by which Earth attracted any object was dependent on the gravitational constant and the mass of Earth. Both of these were unknown at the time, and the discovery of one would allow for the calculation of the other. Cavendish worked diligently on his calculation of the density of the Earth and achieved a value that was not improved upon for hundreds of years. Once the value of G was known, the mass of Earth could be calculated rather than estimated, as it had been since Newton determined the law.

The method that Cavendish utilized to determine the gravitational constant was developed from a torsion balance concept originally proposed by the English scientist John Michell, who died before being able to put it to use. Cavendish used Michell's concept to develop a torsion balance made out of a wooden rod suspended by a wire. At each end of the rod was suspended a small lead sphere that was attracted to two large lead spheres placed in close proximity. As the small spheres were attracted to the larger spheres, the rod would be turned proportionally to the attraction between the spheres.

Using this method, Cavendish was able to determine the density of the Earth as 5.48 times that of water. This calculation was not improved upon for more than a hundred years. The present-day value for the density of the Earth is 5.52 times the density of water.

Henry Cavendish remained single his entire life and died alone on February 24, 1810. Although a wealthy man, he did not aspire for greater riches or fame. He lived on relatively modest funds and intentionally did not share through publication most of his greatest works and discoveries, which easily would have given him fame. Cavendish's contributions to physical science were diverse. He is well known for his work on gases, especially for the discovery of hydrogen. His calculation of the universal gravitational constant contributed to the calculation of the Earth's density. He also demonstrated some fundamental principles of electricity, including Ohm's law and Coulomb's law. In reality, most of Cavendish's contributions to physical science may not have been known had it not been for James Clerk Maxwell, who published in 1879 a compilation of Cavendish's work.

See also ELECTRICITY; LAVOISIER, ANTOINE-LAURENT; MAXWELL, JAMES CLERK.

FURTHER READING

McCormmach, Russell. *Speculative Truth: Henry Cavendish, Natural Philosophy, and the Rise of Modern Theoretical Science.* New York: Oxford University Press, 2004.

Sherman, Josepha. *Henry Cavendish and the Discovery of Hydrogen.* Hockessin, Del.: Mitchell Lane, 2005.

center of mass The center of mass of an extended body is a point at which the mass of the body can be considered to be located, for the purpose of Newton's second law of motion. For this purpose, the body behaves as would a point particle of the same mass located at the body's center of mass. If a force acts on a free body along a line through the body's center of mass, the force will not affect the body's rotation but will cause only linear acceleration in accord with Newton's second law, as if the body were a point particle at the position of the center of mass. A torque acting on a free body in the absence of a net force causes angular acceleration of the body about its center of mass but no linear acceleration. A force on such a body whose line of action does not pass through the body's center of mass has a torque with respect to the center of mass and thus causes both linear acceleration of the body's center of mass and angular acceleration of the body about its center of mass.

If a system is composed of point particles of masses m_i, situated at locations with coordinates (x_i, y_i, z_i), the coordinates of the system's center of mass (x_{cm}, y_{cm}, z_{cm}) are given by

$$x_{cm} = \frac{1}{M}\sum_i m_i x_i$$

$$y_{cm} = \frac{1}{M}\sum_i m_i y_i$$

$$z_{cm} = \frac{1}{M}\sum_i m_i z_i$$

where

$$M = \sum_i m_i$$

is the total mass of the system. For a system made of continuous material, the summations are replaced by appropriate integrations.

For a body of uniform density, the center of mass is located at the center of symmetry, if the body possesses a center of symmetry. The center of mass of a uniform body of each of these forms, as examples, is at the geometric center, which is the center of symmetry: sphere, cube, and straight rod. The center of mass does not have to be located within the volume of the body's material; for example, the center of mass of a uniform hollow spherical shell is at the center of the sphere, where there is no material.

As a simple example of a center-of-mass calculation, consider a body comprising three point masses held together by rods of negligible mass. This might perhaps represent a scaled-up triatomic molecule. The point masses are located at the vertices of a right triangle in the x-y plane: (1) 1.5 kg at the origin (0, 0, 0); (2) 1.0 kg at the point (3.5 m, 0, 0); and (3) 1.2 kg at the point (0, -2.5 m, 0). The total mass of the system is

$$\begin{aligned} M &= \sum_i m_i \\ &= 1.5 \text{ kg} + 1.0 \text{ kg} + 1.2 \text{ kg} \\ &= 3.7 \text{ kg} \end{aligned}$$

The coordinates of the system's center of mass are

$$\begin{aligned} x_{cm} &= \frac{1}{M}\sum_i m_i x_i \\ &= \frac{1}{3.7 \text{ kg}}[(1.5 \text{ kg})\times 0 + (1.0 \text{ kg})\times(3.5 \text{ m}) + (1.2 \text{ kg})\times 0] \\ &= 0.95 \text{ m} \end{aligned}$$

$$\begin{aligned} y_{cm} &= \frac{1}{M}\sum_i m_i y_i \\ &= \frac{1}{3.7 \text{ kg}}[(1.5 \text{ kg})\times 0 + (1.0 \text{ kg})\times 0 + (1.2 \text{ kg})\times(-2.5 \text{ m})] \\ &= -0.81 \text{ m} \end{aligned}$$

$$\begin{aligned} z_{cm} &= \frac{1}{M}\sum_i m_i z_i \\ &= \frac{1}{3.7 \text{ kg}}[(1.5 \text{ kg})\times 0 + (1.0 \text{ kg})\times 0 + (1.2 \text{ kg})\times 0] \\ &= 0 \end{aligned}$$

So the center of mass in this example is located at the point (0.95 m, -0.81 m, 0), which is in the x-y plane.

When calculating the location of the center of mass of a complex system, it might be convenient first to break down the system conceptually into simpler parts. Then one finds the center of mass of each part; conceptually replaces each part with a point particle of the part's total mass, located at the part's center of mass; and calculates the center of mass of that collection of fictitious point particles.

With respect to an inertial reference frame (i.e., a reference frame with respect to which Newton's first law of motion is valid), the center of mass of an isolated system moves with constant velocity (i.e., moves with constant speed in a straight line).

CENTER OF GRAVITY

The center of gravity of a body is the point at which the total weight of the body, the sum of the forces of gravity on all the components of the body, can be thought to be acting. Thus, since the line of action of a body's weight force passes through the center of gravity, gravity exerts no net torque on a body with respect to the body's center of gravity. So if a body is

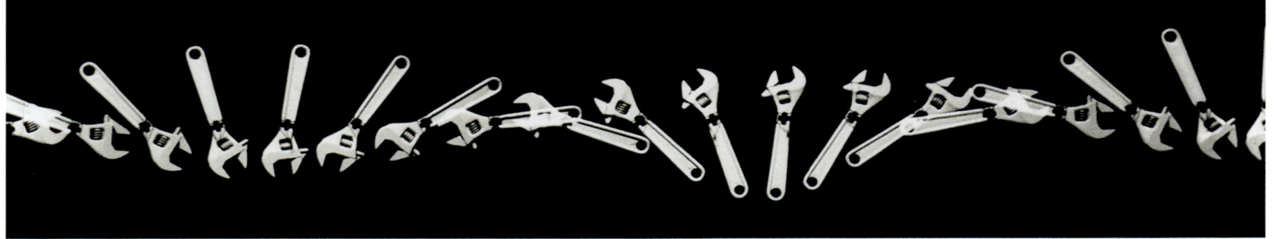

A black mark indicates the center of mass of this wrench, which is moving with only small external forces and torques. The strobe photograph shows the wrench's center of mass moving in a straight line at almost constant speed, while the wrench rotates about its center of mass at almost constant angular speed. *(Berenice Abbott/ Photo Researchers, Inc.)*

suspended at its center of gravity, it will undergo no angular acceleration due to gravity. A body suspended at any other point has a position of stable equilibrium and one of unstable equilibrium, in which its center of gravity is located precisely beneath or above the suspension point, respectively. With its center of gravity at any other location, gravity will exert a torque on the body with respect to the suspension point. If the body is otherwise unimpeded, that torque causes an angular acceleration in the direction of lowering the center of gravity. Consider, for instance, the pendulum of a grandfather clock. Its center of gravity is located somewhere between its pivot, which is the suspension point, and its lower end. When the pendulum is hanging straight down, its center of gravity is then directly beneath the suspension point and at its lowest possible position. Gravity then does not cause the pendulum to swing, if it is at rest there. It is then in stable equilibrium. If the pendulum is displaced from that position, it is no longer in equilibrium, gravity exerts a torque on it, and it swings so as to lower its center of gravity to the stable equilibrium position.

Since the weight of a point particle is proportional to both its mass and the acceleration due to gravity, as long as the acceleration due to gravity is the same (both in magnitude and in direction) for all parts of a body, the center of gravity will coincide with the center of mass. Then the center-of-mass formulas presented earlier can be used for finding the center of gravity. That is the case for sufficiently small bodies, including everyday objects, for which the gravitational field of Earth can be considered uniform in magnitude and direction. But for a body that is larger, whose size is a nonnegligible fraction of the size of Earth (or of whatever other astronomical body is under consideration), the acceleration due to gravity is not the same for all parts of the body. Since this acceleration points toward the center of Earth (or of the astronomical body), it will not have the same direction for all parts of a horizontally extended body. And since the acceleration due to gravity decreases with altitude, it will have different magnitudes for different parts of a vertically extended body. In such cases, the center of gravity of a body does not coincide with the body's center of mass.

See also ACCELERATION; EQUILIBRIUM; FORCE; GRAVITY; MASS; MOTION; NEWTON, SIR ISAAC; ROTATIONAL MOTION; SPEED AND VELOCITY; SYMMETRY.

FURTHER READING

Young, Hugh D., and Roger A. Freedman. *University Physics,* 12th ed. San Francisco: Addison Wesley, 2007.

centrifugation The word *centrifugation* is derived from the Latin *centrum,* meaning "center," and *fugere,* meaning "to flee." The process of centrifugation separates components of a mixture on the basis of their differential molecular weights or their densities by spinning the samples rapidly. The phenomenon known as centrifugal force is not really a force but rather is based on Newton's first law of motion, which states that objects in motion will continue in straight line motion unless acted on by an outside force. When people are riding on a roller coaster that takes a sharp turn to the left, they will feel as if they are being forced to the right as they slide across the seat. Inertia makes them feel as if they are being pushed outward. The riders moving in a straight line would continue moving in a straight line. The seat belt and the door of the roller coaster provide the centripetal force (center-seeking force), pushing back toward the center, that keeps them in the roller coaster cab. In centrifugation, when a solution spins in a circular path, the centrifugal force seems to be pushing the objects toward the outside of the circle away from the center, causing the heaviest or largest objects to move toward the outer path of the circle. Centrifugation serves many purposes in today's world—enriching uranium for use in power plants, separating cream in milk factories, and isolating cell types from blood samples. All of these processes and more are accomplished using a machine called a centrifuge that utilizes centrifugal force.

In 1851 George Stokes, an Irish-born mathematician, developed the first centrifuge and described the forces involved in its function. The principle of separation by centrifugation is based on Stokes's law,

$$v = \left(2r^2 \frac{(\rho - \rho_f)g}{9\eta} \right)$$

where v_t is the settling velocity, r is the radius of the particle being separated, g is the acceleration of gravity (9.8 m/s^2), η represents the viscosity of the fluid, ρ is the density of the particle, and ρ_f is the density of the fluid. Applying Stokes's law shows that a bigger or denser object will move to the bottom of the tube being centrifuged much faster than a smaller, less dense object. When spun fast enough, the larger objects concentrate as a solid in the bottom of the tube, forming a pellet. The liquid portion above the pellet is referred to as the supernatant and can be removed from the pellet by decanting or pipetting.

Theodor Svedberg, a Swedish chemist, developed the first ultracentrifuge in 1925. Ultracentrifuges rotate in a refrigerated vacuum and can spin as fast as 80,000 rpm or 500,000 × *g*. Svedberg received the Nobel Prize in chemistry in 1926 for this invention and the quantitative calculation relating sedimentation rate and molecular weight of a substance. The advent of commercially available ultracentrifuges in the 1940s expanded the research capabilities of all scientists.

TYPES OF CENTRIFUGATION

Centrifugation techniques generally fall into two categories: preparative and analytical. In preparative centrifugation, samples are separated and then isolated, often in large quantities, for further use or study. Preparative centrifugation techniques include two types, known as differential centrifugation and density gradient centrifugation.

Differential centrifugation separates components of a mixture proportionally to the square root of their molecular weight according to the Svedberg equation. Shape plays a role in the separation of components in this type of centrifugation. Differential centrifugation results in the formation of a solid pellet at the bottom of the tube and a liquid supernatant.

Density gradient centrifugation separates the samples on the basis of their densities relative to a viscous medium. Many density gradients utilize sucrose of different concentrations because sucrose does not interact with the samples being separated. One type of density gradient centrifugation is sedimentation velocity centrifugation (also called rate zonal centrifugation or zone centrifugation), which uses a gradient of the medium with the sample placed on top. The sample is then spun for the specified period, and the samples move through the medium on the basis of their velocity from Stokes's law. Density determines the time required to separate the samples. The densest particles travel to the bottom of the tube and the least dense to the top.

A second type of density gradient centrifugation, sedimentation equilibrium centrifugation (also called isopycnic centrifugation), separates samples on the basis of their densities independently of time. A medium is placed in the centrifuge tube with the sample layered on top. As the sample spins, it separates into bands of differing densities. Unlike rate zonal centrifugation, in which all of the sample components can travel to the bottom of the tube given enough time, in sedimentation equilibrium centrifugation, the sample components stay at their specified density band regardless of how long the sample spins.

Analytical centrifugation allows for the separation and real-time analysis of samples, and the components in the sample maintain their native shape in these experiments. The development of the Beckman Coulter XL-A analytical ultracentrifuge in 1992 significantly improved the methods for analytical ultracentrifugation. As in preparative centrifugation, analytical ultracentrifugation utilizes either sedimentation velocity centrifugation or sedimentation equilibrium centrifugation. The machine displays optical measurements of the samples during separation. Analytical ultracentrifugation allows one to study macromolecules in their native conformations, accurately determine molecular mass, observe complex and subunit assembly, and calculate ratios of subunits in complex proteins.

CENTRIFUGE EQUIPMENT

All centrifuges have a fixed base with rotating parts called rotors that hold the samples. Originally hand cranking spun the samples; however, in the early 1900s centrifuges became electric powered. Rotor choice is very important in sample separation and depends on the sample size and centrifuge type. A fixed-angle rotor has a set angle for the sample and separates samples more quickly than the swinging-bucket rotor. The fixed-angle rotor does not have any breakable moving parts. When using a swinging-bucket rotor (horizontal rotor), the tube containing the sample pivots when spun and travels horizontally. This type of rotor minimizes the stress put on the sample by contact with the rotor wall. The slower speed and easier breakage of the swinging bucket are drawbacks to using this type of rotor. Sedimentation velocity centrifugation requires a swinging bucket rotor, while sedimentation equilibrium centrifugation can use either a swinging bucket or fixed-angle rotor. The metal composition of the rotor determines the maximal speed at which the samples can be

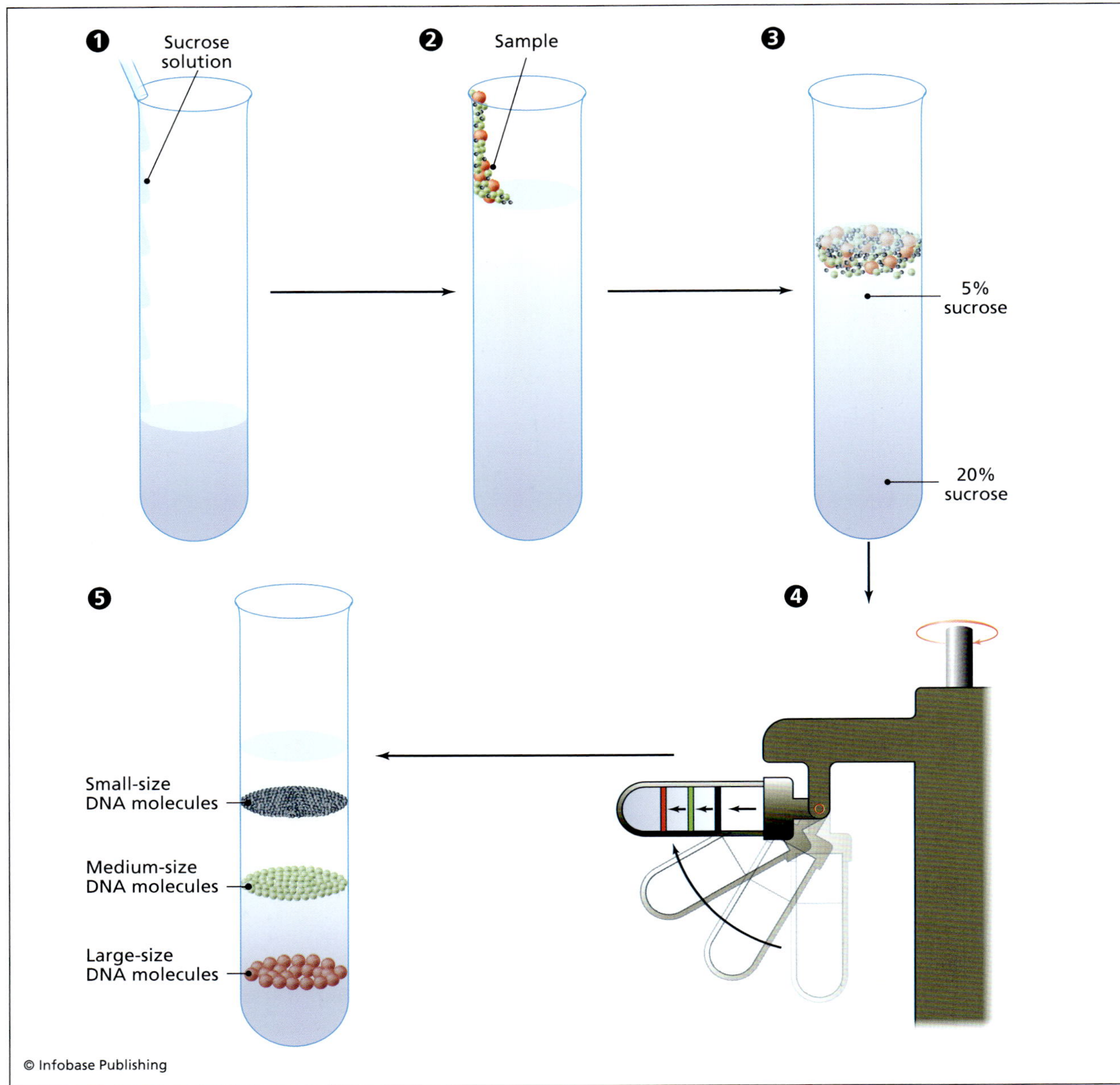

In density gradient centrifugation, a sample is spun on a layered sucrose gradient, separating particles on the basis of size.

centrifuged; the rotor must be able to withstand the g-forces being generated by the centrifuge.

All centrifuges allow the user to control the speed and time that the sample will spin. Revolutions per minute (rpm) are not generally used when calculating centrifugation specifications because rpm varies with rotor size and centrifuge conditions. The common measure of centrifugation speed is based on relative centrifugal force (RCF), defined by the following formula:

$$RCF = 0.00001118rN^2$$

where r is the radius of rotation in centimeters and N is speed (rpm). Using this equation allows centrifugation experiments to be equivalent regardless of the particular centrifuge or rotor being used.

USES OF CENTRIFUGATION

The uses for centrifugation are varied. Hospitals depend on centrifugation for the separation of blood components for testing as well as isolation of specific blood components. Plasma, the lightest part of blood, separates to the top of a density gradient, and red blood cells, the heaviest, fall to the bottom.

White blood cells migrate to the middle of the gradient. The percentage of packed cells versus total blood volume is the hematocrit and indicates the amount of red blood cells in the sample.

The food industry uses centrifugation in the purification and production of goods. Batch centrifugation is used when absolute purity of the sample is the primary concern. Individual samples are centrifuged and separated. Continuous centrifugation, developed in the 1950s, allows a larger batch size and quicker processing time. The dairy industry has separated cream from milk by centrifugation since the 1800s and still uses centrifugation in a variety of processes today. Clarification of milk removes impurities including leukocytes and bacteria. Hot milk separation separates milk fat from skim milk. Altering the separation techniques allows for the production of skim, 1 percent, and all of the available milk fat levels found in the store today. Both of these processes require the use of centrifugation. The production of sugar also uses centrifugation to purify the samples. Sugar factories use either batch centrifugation processes or continuous centrifugation processes to separate impurities from the sugar crystals. The use of continuous centrifugation tends to damage the crystals more than batch centrifugation processes.

Cell biologists depend on centrifuges to study cellular structures and organelles. Cellular fractionation involves homogenization of cells or a tissue sample followed by differential centrifugation at varying speeds and times in order to isolate organelles of specific sizes in large quantities. The largest organelles form the first pellet, and, after decanting, the supernatant is spun again at a faster speed, causing the next largest organelles to pellet. The pellets can be dried and studied or resuspended (dissolved in a buffer) for further purification. Biologists can use these methods to determine which cell fractions are responsible for specific cellular functions in order to isolate the specific organelles responsible for the function further for in-depth analysis.

All centrifuges utilize rotors to hold the samples. Rotors are able to rotate and are manufactured as several types such as fixed-angle rotors and swinging-bucket rotors. Closeup of centrifuge rotor *(Radu-Ion Huculeci, 2008, used under license from Shutterstock, Inc.)*

Cesium chloride (CsCl) gradient and sedimentation equilibrium ultracentrifugation separate linear and circular deoxyribonucleic acid (DNA) molecules or DNA molecules that differ in weight on the basis of nucleotide content. After layering the DNA sample on top of the CsCl, centrifugation causes the Cs^+ to move to the bottom of the tube, creating a density gradient. The DNA samples migrate to the position in the gradient that is equal to their density. The samples can be removed from the bottom of the tube and further characterized.

Astronaut training facilities use centrifuges to simulate the g-forces the astronauts will encounter in space. The centrifuge found at the NASA Goddard Space and Flight center in Greenbelt, Maryland, is 120 feet (36.6 m) long and can reach speeds up to 38 rpm or 30 *g*.

In 1934 Jess Beams at the University of Virginia developed high-speed gas centrifugation, a method that allows for the separation of isotopes. Hundreds of gas centrifuges are linked to provide the primary method of uranium enrichment for use in nuclear power plants as well as making bomb-grade uranium for nuclear weapons. When centrifuged, uranium 235 and uranium 238 separate. The heavier uranium 238 goes to the outside of the tube, and uranium 235 stays on the inside of the tube. The lighter uranium 235 then flows to the next centrifuge, and the enrichment continues. Previous methods for enriching uranium required a tremendous amount of electrical power and were therefore conspicuous when countries were trying to develop nuclear power programs. High-speed gas centrifugation does not require significant power and is easier to perform furtively. The first gas-diffusion centrifuge plant was built in 1945 in Oak Ridge, Tennessee. This type of plant was the primary method for enriching uranium through the 1980s. In 2006 the United States Enrichment Corporation, a supplier of enriched uranium, successfully removed old gas-diffusion centrifuges from the plant in Piketon, Ohio, to prepare for the opening of a demonstration plant showcasing new technology, known as the American Centrifuge Project. The new American Centrifuge technology utilizes less energy and produces less waste than previous gas-diffusion centrifugation methods. The final plant is expected to produce reactor grade uranium by 2010.

See also ACCELERATION; ALTERNATIVE ENERGY SOURCES; ISOTOPES; NUCLEAR PHYSICS.

FURTHER READING

"Centrifuge." In *Encyclopaedia Britannica*. 2006. From Encyclopeadia Britannica Online. Available online. URL: http://search.eb.com/eb/article-1237. Accessed July 24, 2008.

United States Enrichment Corporation. "The American Centrifuge." Available online. URL: http://www.usec.com/v2001_02/HTML/Aboutusec_Centrifuge.asp. Accessed July 24, 2008.

chemical engineering Chemical engineering is a branch of engineering that involves the application of chemical and physical properties of matter to the manufacture of a product. The field of chemical engineering combines knowledge of chemistry with expertise in physics and mathematics to work toward a particular goal and create a compound or product that fills a need of society. Chemical engineers consider all aspects of everyday life including travel, medicine, plastics, rubber, detergents, cement, space, fabrics, and paper.

Chemical engineers are involved in the solution of many societal problems. The development of pharmaceuticals, bioengineered materials, high-strength composites, specialty polymers, semiconductors and microelectronic devices, and safe food supplies are all work of chemical engineers. Many consider chemical engineering the most diverse of all types of engineering because it has applications to so many areas.

The U.S. Bureau of Labor Statistics states that chemical engineers in 2008 held 28,780 jobs with a mean annual salary of $84,240. These jobs are distributed among the following industries:

- architectural, engineering, and related services
- scientific research and development
- basic chemical manufacturing
- resin, synthetic rubber, artificial synthetic fibers and filaments manufacturing
- petroleum and coal products manufacturing

Chemical engineers approach complicated production and development situations as a series of smaller individual steps known as unit operations. An example of an engineering system broken down into unit operations is shown in the figure below. The chemical engineer looks at the individual components of the entire production process and determines the most efficient method to complete that task.

EDUCATION AND TRAINING

Education and training in the field of chemical engineering require a strong background in mathematics, chemistry, and physics. Chemical engineering degrees require coursework in chemistry, organic chemistry, physics, calculus, differential equations, thermodynamics, separations, and chemical kinetics. Laboratory training is also critical, and a chemical engineering education also involves extensive study and training in engineering techniques. Many chemical engineering degree programs focus on laboratory skills and laboratory writing skills.

The type of degree required for employment as a chemical engineer depends upon the type of job desired. Degrees in chemical engineering range from four-year bachelor of science degrees to research-based master of science degrees and doctoral degrees. In general, laboratory technician positions require a bachelor's or master's degree. Supervisory roles require either a master's degree or a doctorate (Ph.D.). Full-time academic positions generally require a Ph.D. Many chemical engineering students minor in a complementary field, such as environmental engineering, materials science, or technology management. In order to become a

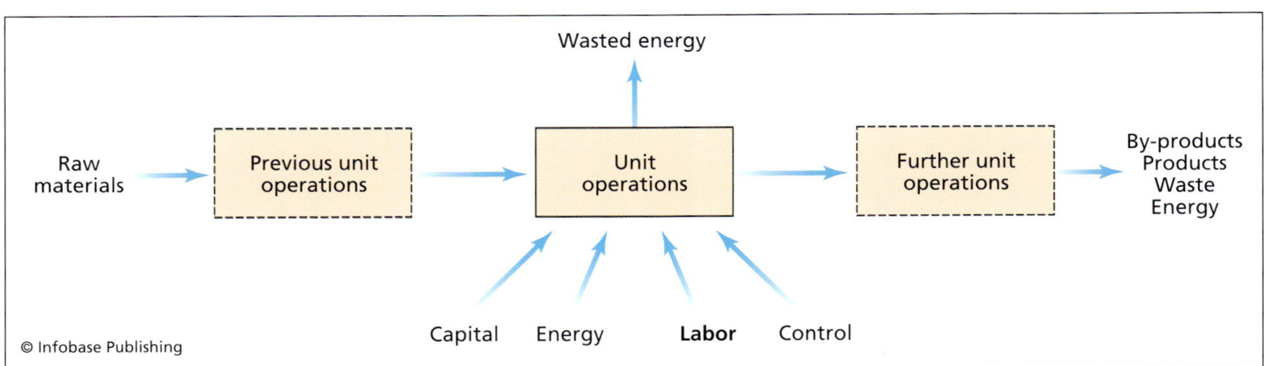

The chemical engineering process is divided into small individual processes known as unit operations. Raw materials enter the unit operation in which money (capital), energy, labor, and control are added. The product from this part of the engineering process is sent to further unit operations leading to final products, wastes, and energy.

professional engineer, all chemical engineering students must pass the Fundamentals of Engineering Exam (FE).

Most chemical engineers find employment in industry. However, chemical engineering jobs are also available in government labs, in academic institutions, or as private consultants.

GREATEST ACHIEVEMENTS OF CHEMICAL ENGINEERING

The American Institute of Chemical Engineers (AIChE) has acknowledged the top 10 greatest achievements in chemical engineering:

- isotope isolation and splitting of the atom
- creation of mass produced plastics
- application of the engineering process to the human body
- mass production of pharmaceuticals
- creation of synthetic fibers
- liquefaction of air
- techniques to clean up the environment
- increased crop yield with fertilizers and biotechnology and food production
- processes to utilize fossil fuels
- mass production of synthetic rubber

Chemical engineers played a large role in the isolation of isotopes and splitting of the atom (nuclear fission). Splitting atoms and isolating isotopes can be utilized to generate power through nuclear power plants and to determine the age of archaeological artifacts through radioisotope dating. Medical applications of isotopes include the use of radioactive tracers to diagnose medical conditions.

Chemical engineers of the 20th century discovered how to produce polymers in an economically feasible manner. In 1908 the first mass produced plastic, known as Bakelite, revolutionized the products available. Plastics and other polymers, such as nylon, polyester, polyethylene, and polystyrene, now impact nearly every aspect of daily life, in clothing, automobiles, medical supplies, and plastic food containers.

Chemical engineers apply engineering principles such as "unit operations" and the evaluation of complex chemical reactions to the human body. This has led to a greater understanding of biological processes and increased the capabilities to treat disease. This field of engineering has also contributed to the creation of therapeutic devices and artificial organs that can replace damaged human organs.

The scale-up principles of engineering have been applied to the production of pharmaceuticals. Beginning with the mass production of penicillin, which had been discovered by Sir Alexander Fleming in 1929, the achievements of chemical engineering have made chemical compounds accessible to more people at a reasonable price. Another example of chemical engineering in the field of pharmaceuticals was the mass production of human insulin, replacing porcine or bovine insulin to treat diabetes. The production of a reliable, safe, and relatively inexpensive alternative has helped to save many lives.

Chemical engineers have developed production methods for synthetic fibers such as nylon, polyester, and spandex that revolutionized the textile world's dependence on natural fibers, such as wool and cotton. Kevlar®, which is used in bullet-proof vests, is an example of a protective synthetic product. Synthetic products are easier to work with and more adaptable than natural fibers. Such fibers are another example of chemical engineers' creating a product that is mass produced such that it is affordable and available to the average citizen. Chemical engineers are continually working on the creation of new textiles including such things as intelligent textiles.

Chemical engineers developed methods to separate different components from air by cooling the air to -320°F (-196°C). When the air reaches this temperature, it turns into a liquid in a process known as liquefaction. In the liquid stage, the different components can be separated. Nitrogen and oxygen separated in this manner can be used for many medical and industrial applications.

The impact of chemical engineering on protecting the environment is tremendous. Cleaning up old problems and helping to prevent new ones with new technology are driving forces behind current research in the field. Such items as catalytic converters in automobiles and smokestack scrubbers in factories help to reduce the amount of pollution created daily.

Chemical engineers have developed biotechnology to increase the yields of crops in poorer areas of the world. They have also made great strides in the development of fertilizers and herbicides that help improve crop yields. Advances have played a large role in the production and distribution systems of society's food supply.

Other successes in chemical engineering have led to the development of processes for synthesizing industrial chemicals and consumer products from fossil fuels. Modifications of petroleum processing and the development of methods to utilize the petroleum products are also chemical engineering feats.

Chemical engineers played a major role in the development of synthetic rubber. Tires, gaskets, hoses, belts, and all of the other rubber components that help keep the world moving were developed by chemical engineers.

COMMON PRINCIPLES OF CHEMICAL ENGINEERING

The use of several common scientific principles in chemical engineering processes maximizes production efficiency and yield and enhances research and design methodology. These include the law of conservation of mass, the law of conservation of momentum, and the law of conservation of energy. These three laws are fundamental principles of physics. The law of conservation of mass states that mass can be neither created nor destroyed. The law of conservation of energy states that energy can be neither created nor destroyed; energy can be transferred from one form to another, but the total amount of energy cannot change. Momentum is the mass of the object multiplied by the velocity of the object. The law of conservation of momentum states that within the same system, the amount of momentum is constant. Chemical engineers utilize these three laws to optimize each individual part of a production process or unit operation.

The field of thermodynamics involves the study of energy transfers and work done within a system. Chemical engineers study the heat flow and energy transfers within a system. Chemical kinetics is the study of the rate at which chemical reactions occur. Understanding which factors affect reaction rates enables engineers to perfect their technology and production processes.

Chemical engineers utilize techniques from many fields, but the most common chemical techniques involved include such processes as distillation, extraction, gas absorption, and crystallization, as well as separation techniques such as filtration, centrifugation, sedimentation, and sieving.

FUTURE APPLICATIONS OF CHEMICAL ENGINEERING

Many fields currently benefit from the work of chemical engineers, as will many in the future. Areas that the field of chemical engineering is focusing on for the future include energy applications, protecting the environment, advanced materials and fibers, electronics, the food and beverage industry, and biotechnology and pharmaceuticals.

Energy applications for chemical engineers include oil and gas production, clean fuel technology, portable energy sources, fuel cell production (solid oxide and polymer electrolyte membrane), and batteries. The practical development of alternative fuels is becoming increasingly important as a result of the decrease in petroleum supplies. Predictions made in 2005 place the amount of fossil fuels consumed to date at approximately 1.1 trillion barrels of oil (nearly half of the total predicted oil reserves). At a rate of 83 million barrels per year, this means that an estimated number of years left at the present consumption rate is approximately 31 years. Given that fossil fuel consumption has been increasing steadily over the last 50 years, these predictions could be overestimates.

The most credible predictions of fossil fuels expenditures and lifetime were given by the Shell Oil geophysicist Marion King Hubbert in 1957. He developed what is known as Hubbert's curve. Hubbard's curve correctly predicted the peak of oil production from 1965 to 1970 in 1956. The true peak occurred in 1970. This added credibility to Hubbert's peak and to predictions that less than 50 years of oil remains at present consumption rates. Chemical engineers study the feasibility of such alternative energy sources as biodiesel, hydrogen fuel cells, and hybrid car technology.

Environmental issues that chemical engineers address include such phenomena as remediation of preexisting environmental concerns such as global warming and acid rain, clean air and water supplies, and construction and factory operation using environmentally sound technology.

The greenhouse effect is a natural phenomenon that traps heat from the Sun in the atmosphere and makes life possible on Earth. Many natural gases in the atmosphere absorb energy that has been reradiated from the surface of the Earth into the atmosphere. After absorption, the energy is not released back into space, resulting in a higher global temperature. The greenhouse gases include carbon dioxide (CO_2), nitrous oxide (N_2O), methane (CH_4), perfluorocarbons (PFCs), hydrofluorocarbons (HFCs), and sulfur hexafluoride (SF_6). One present and future goal of chemical engineering research is finding economically feasible solutions to the problem of greenhouse gases.

Acid rain is an environmental condition in which the pH of rainwater in an area decreases as a result of pollutants. One cause of acid rain is the release of sulfur dioxides and nitrogen oxides from combustion processes. When they are released into the atmosphere, they react with water to form sulfuric acid and nitric acid, causing acid rain. Normal rainwater is slightly acidic, with pH values around 5 to 6. High levels of sulfur dioxide in the air lower the pH of precipitation; pH levels of rainwater have been recorded as low as 3.5. The environmental impact of acid rain on animal life and vegetation of waterways is apparent. The effects of acid rain do not stop there—nonliving matter is also affected. Acid rain causes deterioration of buildings, statues, bridges, and roadways.

The development of advanced materials is another important area for the future of chemical engineering. One of the fastest growing areas of textiles today is the production of products known as intelligent textiles. These are synthetic textiles that are designed and engineered to fit a certain need or solve a particular problem. Some estimates put the intelligent textiles at as high as 38 percent of the textile market. Many of these technologies were based on knowledge gained from the research and development of special clothing for astronauts. Textile chemists and chemical engineers utilized the technologies of space to solve everyday apparel problems. Antibacterial materials that can be used in the manufacture of upholstery materials and temperature-sensitive fabrics that can alert the wearer to overheating or to medical conditions are examples of new technologies. Allergy-resistant materials help those with allergies. The now common Under Armour® technology allows one's clothing to breathe while its wearer sweats, yet still has the insulating power necessary to keep athletes warm and dry outdoors.

Chemical engineers are constantly evaluating the types of electronics and technology available to the consumer. Development of smaller, faster, and more efficient types of electronic devices will continue to be a focus of chemical engineers, as will the creation of new and better microelectronics including cell phones and other digital devices.

The fields of biotechnology and pharmaceuticals are ever-growing areas of research. The production through biochemical engineering of chemicals and medication to treat and prevent diseases has the potential to improve the quality of life for all people. Chemical engineers are working on new and creative drug delivery systems and on the engineering of tissues that could change modern medicine.

See also ACID RAIN; AIR POLLUTION (OUTDOOR/INDOOR); ALTERNATIVE ENERGY SOURCES; ATMOSPHERIC AND ENVIRONMENTAL CHEMISTRY; FOSSIL FUELS; GREEN CHEMISTRY; GREENHOUSE EFFECT; HEAT AND THERMODYNAMICS; PHARMACEUTICAL DRUG DEVELOPMENT; POLYMERS; RATE LAWS/REACTION RATES; TEXTILE CHEMISTRY.

FURTHER READING

Towler, Gavin, and Ray K. Sinnott. *Chemical Engineering Design: Principles, Practice and Economics of Plant and Process Design.* Woburn, Mass.: Butterworth Heinemann, 2007.

chemical reactions Chemical reactions or chemical changes take place when one or more substances transform into one or more new substances by the formation or breaking of bonds. Distinguishing between a chemical and a physical change is often difficult. Unlike a chemical change, a physical change does not involve the breaking or making of bonds, and new compounds are not created. In a physical change, only the state or the arrangement of the compound has been altered. Bubbling, color change, heat, or the formation of a precipitate (a solid that falls to the bottom of a reaction vessel) indicates that a chemical reaction has taken place, although none of these factors guarantees that the process involved a chemical change. The best method for determining whether or not a chemical reaction has occurred is by measuring the disappearance of reactants (compounds going into the reaction) and the formation of products (compounds being formed in the reaction).

Chemical reactions are written using chemical formulas for the compounds involved, separating the reactants from the products by an arrow. Reactions generally appear left to right with the arrow pointing in the direction of the most thermodynamically favorable reaction. Most reactions are reversible, at least to some extent, and are written with arrowheads facing both directions, $\longleftrightarrow$. Information written above and below the arrow represents reaction conditions, such as temperature and presence of a catalyst, that would be required for that reaction to take place, $\xleftrightarrow{37°}$. The physical state of the reactants and products is sometimes indicated beside the symbol of the compound in parentheses, O_2 (*g*), with (*g*) representing gas, (*l*) liquid, (*s*) solid, and (*aq*) aqueous.

CLASSIFICATION OF CHEMICAL REACTIONS

Chemical reactions fall into five main categories: synthesis (combination), decomposition, single-replacement (single-displacement), double-replacement (double-displacement), and combustion. Identifying the reaction type for a set of reactants allows one to makes predictions about the products.

Synthesis, or combination, reactions combine two or more separate reactants to form one product. Synthesis reactions generally are endothermic, meaning they involve an input of energy. Examples of synthesis reactions are

$$A + B \rightarrow C$$

$$H_2 + \frac{1}{2}O_2 \rightarrow H_2O$$

Synthesis and decomposition reactions are often reverse processes. Decomposition reactions involve only one reactant that breaks down into two or more products, one of which is usually a gas, and generally require the addition of heat. Decomposition reactions can be modeled as follows:

$$A \rightarrow B + C$$

$$MgCl_2 \rightarrow Mg + Cl_2$$

TYPES OF DECOMPOSITION REACTIONS

Reactants	Products	Example
Binary compounds	The elements that make up the compound	$2NaCl \rightarrow 2Na + Cl_2$ (*g*)
Carbonates	Form oxides + carbon dioxide	$Na_2CO_3 \rightarrow Na_2O + CO_2$ (*g*)
Chlorates	Binary compound + molecular oxygen	$2KClO_3 \rightarrow 2KCl + 3O_2$ (*g*)
Acids/bases	Forms oxide of acid or base + water	$2LiOH \rightarrow Li_2O + H_2O$ (*l*)

Products of decomposition reactions vary, depending on the type of reactant involved. The following table shows the outcomes of four types of decomposition reactions.

Single-replacement or single-displacement reactions involve the exchange of one positive ion, or cation, for another in an ionic compound as shown:

$$A + BY \rightarrow AY + B$$

$$6Li + Al_2(SO_4)_3 \rightarrow 3Li_2SO_4 + 2Al$$

The cations replace each other on the basis of the activity series of metals, a list of metal reactivity. Each metal has a slightly different reactivity in single-replacement reactions. Several common metals, listed from the most to the least reactive, are lithium, potassium, strontium, calcium, sodium, magnesium, iron, tin, hydrogen, and copper. A cation can only replace another if it is more reactive than the cation it is replacing. Lithium can replace any other metal on the list because it is the most reactive. Copper, on the other hand, would not cause a reaction because of its low reactivity.

Double-replacement reactions or double-displacement reactions occur when two ionic compounds switch their cations or anions as shown:

$$AX + BY \rightarrow AY + BX$$

$$MgCl_2 + Na_2SO_4 \rightarrow MgSO_4 + 2NaCl$$

Most acid-base reactions are double-replacement reactions with the ionic acid and base exchanging partners in order to form a salt and water. Sodium hydroxide and hydrochloric acid react via an acid-base double-replacement reaction.

$$NaOH + HCl \rightarrow NaCl + H_2O$$

Combustion reactions occur when a substance reacts with molecular oxygen (O_2). The combustion of hydrocarbons, organic compounds that contain hydrogen and carbon, always forms carbon dioxide (CO_2), water (H_2O), and heat.

$$C_xH_y + O_2 \rightarrow CO_2 + H_2O + \text{heat}$$

$$C_6H_{12}O_6 + 6O_2 \rightarrow 6CO_2 + 6H_2O + \text{heat}$$

$$C_3H_8 + 5O_2 \rightarrow 3CO_2 + 4H_2O + \text{heat}$$

Not all combustion reactions involve hydrocarbons, and, in these cases, CO_2 and H_2O are not formed.

$$Mg + O_2 \rightarrow MgO_2$$

Isomerization reactions and oxidation-reduction reactions are two other categories of common reactions. The reactant in an isomerization reaction does not combine with other reactants to form a new product or decompose into multiple products—the same substance simply rearranges to become a new compound. This is common in biochemical reactions involved in cellular respiration.

In oxidation-reduction reactions, also called redox reactions, one atom gives up one or more electrons, and another atom gains one or more electrons. Oxidation involves an increase in oxidation number due to the loss of electrons, while reduction involves a decrease in oxidation number due to the gain in electrons. The atom that gives up the electron(s) is called the reducing agent, and the atom that gains the electron(s) is called the oxidizing agent. A useful acronym for redox reactions is *OIL RIG* (Oxidation Is Loss; Reduction Is Gain).

Many synthesis, single-replacement, decomposition, and combustion reactions can also be classified as redox reactions. Double-replacement reactions generally do not involve redox reactions. The following single-replacement reaction helps to visualize an oxidation-reduction reaction:

$$2Fe^{3+} + SnCl_2 \rightarrow 2FeCl_2 + Sn^{4+}$$

The iron begins as Fe^{3+} and ends the reaction as Fe^{2+}. The tin begins as Sn^{2+} and ends as Sn^{4+}. The oxidation number of iron decreased because the iron gained electrons from the tin; therefore, the iron was reduced (think RIG). The oxidation number of the tin increased—the tin was oxidized because it gave the electrons to the iron (think OIL). The iron was the oxidizing agent in this reaction, and tin was the reducing agent.

PREDICTING PRODUCTS OF CHEMICAL REACTIONS

Knowing the types of possible chemical reactions that the reactants can undergo, one can predict the products. If only one reactant is present, the reaction must be a decomposition reaction. A single metal with an ionic compound undergoes a single-replacement reaction. If both reactants are ionic compounds, a double-replacement reaction occurs, and a compound reacting with molecular oxygen will always be a combustion reaction. One can then predict the products by following the rules for each reaction type.

Prediction of the state of ionic compounds follows the solubility rules of ionic compounds. Precipitates are designated (*s*) when assigning states. The ionic rules appear in the following table.

Soluble compounds dissolve in an aqueous solution (*aq*), and insoluble compounds precipitate or remain solid (*s*).

BALANCING CHEMICAL REACTIONS

The law of conservation of matter states that matter can be neither created nor destroyed. This means that when chemical reactions occur, all of the atoms in the reactants must be accounted for in the products. The atoms may have different partners or exist in a different physical state, but each atom must be present. The accounting method used is called balancing equations. First make sure the formulas for the reactants and products are written in the correct form. Then count the atoms on each side in order to see whether the reaction is balanced. The subscripts on chemical formulas of reactants or products cannot be changed in order to balance an equation; only coefficients in front of the reactants or products can be altered. In the following double-replacement reaction, the reactants and products do not contain the same number of hydrogen or oxygen atoms:

$$H_2SO_4 + Mg(OH)_2 \rightarrow H_2O + MgSO_4$$

A coefficient of 2 must be added in front of the water product in order to balance the equation.

$$H_2SO_4 + Mg(OH)_2 \rightarrow 2H_2O + MgSO_4$$

Balancing equations often requires trial and error. Changing the coefficient in front of one reactant or product can disrupt the balance of another. This is to be expected, and the following are several tricks that make balancing easier.

1. Ensure that all formulas are written correctly.
2. Balance monatomic compounds (such as Cl_2) last because the addition of a coefficient in front of them will not affect the amounts of other atoms. Balancing water molecules later in the problem will also be useful.
3. Keep polyatomic compounds that appear on both sides of the equation together rather than balancing the individual atoms.
4. All coefficients must be whole numbers and are distributed throughout the entire compound they precede. For example, $3H_2O$ means there are six hydrogen atoms and three oxygen atoms.
5. After balancing the equation, recount all atoms.

ENERGY OF CHEMICAL REACTIONS

The reactants either take in energy from or give up energy to their surroundings during the formation of products. Endothermic reactions do not occur spontaneously. They require an input from the surroundings in order to take place and generally feel cold to

SOLUBILITY RULES FOR IONIC COMPOUNDS

Compound Type	Soluble/Insoluble	Exceptions
All group IA and ammonium compounds	Soluble	None
All nitrates, chlorates, and acetates	Soluble	None
All halogen compounds (other than F)	Soluble	Ag, Hg, Pb
All sulfates	Soluble	Ag, Hg, Pb, Ba, Sr, Ca
Chromates, carbonates, hydroxides, phosphates, sulfides	Insoluble	Group IA and ammonium compounds

(a) Sulfuric acid (in graduated cylinder) is added to ordinary table sugar, sucrose (in the beaker), to cause a chemical reaction. (b) The sulfuric acid reacts strongly with the sucrose in an exothermic (releases heat) chemical reaction. *(Charles D. Winters/Photo Researchers, Inc.)*

the touch. In endothermic reactions, the products have a higher energy level than the reactants. Imagine a skier at the foot of a mountain. A large amount of energy is required to reach the top. Exothermic reactions occur spontaneously and do not require any input of energy in order to occur. In an exothermic reaction, the products have less energy than the reactants. Imagine a skier at the top of a mountain; traveling down the mountain is easy and spontaneous for the skier.

Many metabolic reactions require help in order to occur. Enzymes supply the help needed to make an endothermic (endergonic in biological terms) reaction take place. An enzyme, usually a protein, has the ability to interact with the reactants and allow the reaction to take place more easily. In the skier analogy, the chair lift could serve the role of enzyme. Rather than the skier's walking up the mountain, which would require a lot of energy input, the chair lift (enzyme) supplies an alternate route up the mountain that requires a lower input of energy. This allows the skier to arrive at the top of the mountain much more quickly. Enzymes change the rate of a reaction. They speed up reactions that would already take place naturally. Enzymes cannot cause a reaction to happen that would not happen on its own or change the equilibrium point of a reaction. An enzyme works by holding both of the substrates in a specific position relative to one another, allowing them to interact in such a way as to facilitate the reaction.

See also COVALENT COMPOUNDS; ENZYMES; GLYCOLYSIS; HEAT AND THERMODYNAMICS; IONIC COMPOUNDS; ORGANIC CHEMISTRY; RATE LAWS/REACTION RATES.

FURTHER READING

Beall, Brower, and Robblee LeMay. *Chemistry: Connections to Our Changing World.* New York: Prentice Hall, 2002.

Wilbraham, Antony B., Dennis D. Staley, Michael S. Matta, and Edward L. Waterman. *Chemistry.* New York: Prentice Hall, 2005.

chemistry and chemists Chemistry is the study of matter, anything that has mass and takes up space. This is a very broad subject; chemistry can be described in more detail as the study of atoms, molecules, and subatomic particles and the reactions that they undergo, the states of matter that they take up, and the energy transformations among them. The field of chemistry has applications in every aspect of daily life including medicine, automobiles, energy resources, building materials, clothing, and aerospace research.

SUBATOMIC PARTICLES AND ATOMS

The fundamental building blocks of chemicals are the subatomic particles that make up atoms. These include protons, neutrons, and electrons. The protons are positively charged particles with approximately 1,840 times the mass of an electron and a mass approximately equal to that of a neutron. Protons are located within the nucleus of an atom and together with the neutron constitute the mass of the atom. Neutrons are neutral particles that are also found in the nucleus. Electrons are negatively charged particles that are significantly smaller than protons and neutrons. They are responsible for the bonding capabilities of different atoms but do not contribute to the mass of the atom.

PERIODIC TABLE OF THE ELEMENTS

All of the known elements are arranged in an organized form known as the periodic table of the elements or simply the periodic table. The original version of the periodic table was developed by the Russian chemist Dmitry Mendeleyev in 1869. This version of the table was based on increasing atomic masses of the elements. The table was later revised by the English physicist Henry Moseley in 1913 to an arrangement based on increasing atomic number (number of protons). There were few differences in the order of the elements, but some could be seen with such elements as cobalt and nickel. Cobalt has an atomic number of 27 and nickel has an atomic number of 28, putting cobalt before nickel in the present periodic table. The atomic masses of these two elements put them in the opposite order in Mendeleyev's periodic table. The location of an element on the periodic table of the elements (as seen in the appendixes) gives chemists much information about the physical properties as well as the chemical reactivities of the individual elements.

CHEMICAL BONDING

Once chemists understood the nature of the atoms involved in compounds, they began to explore the types of bonds in which the chemical elements participated to create new compounds. Three common types of bonds are ionic bonds, covalent bonds, and metallic bonds. An ionic bond is an electrostatic interaction between a positively charged ion (cation) and a negatively charged ion (anion). This type of bond forms between a metal that forms the cation and a nonmetal that forms an anion. Covalent bonds involve sharing electrons between two nonmetals. The driving principle behind the type of bond that forms is the octet rule, which states that every atom wants a full valence shell of electrons. This includes eight valence electrons for all atoms except hydrogen and helium, whose first energy level is full with only two electrons. Metallic bonds form between metal atoms when adjacent atoms loosely hold their valence electrons, creating a "sea of electrons" that is shared with all of the metal atoms. This creates freedom of motion of electrons in metals and contributes to their electrical conductivity.

CHEMICAL REACTIONS

Chemists study the chemical properties of these elements and compounds. Chemical reactions can be categorized into five different types of reactions: combination (synthesis), decomposition, single-replacement, double-replacement, and combustion reactions. Combination reactions involve two reactants that join to create a single product. Decomposition reactions begin with only one reactant and generally involve the addition of heat. These types of reactions result in more than one product. Single-replacement and double-replacement reactions involve ionic compounds and the exchange of one ion for another. In a single-replacement reaction, a lone metal atom replaces the positive ion of a compound. Double-replacement reactions begin with two ionic compounds, and the positive and negative ions switch partners. Understanding how two compounds will react with one another and which products will be formed allows chemists to create desirable products in whatever field they are working.

STOICHIOMETRY

The application of mathematical principles to chemical reactions is known as stoichiometry, the quantitative relationship between reactants and products in a chemical reaction. A balanced chemical equation is like a recipe for the relationship of the components in a reaction. The coefficients in a balanced chemical equation represent the relative quantities of each of the reactants and products involved in the chemical reaction. The coefficients simultaneously represent the ratios of number of atoms in the reaction, the number of molecules in the reaction, or the number of moles (amount of substance) of substances in the reaction. Taking into account the law of conservation of matter, which states that matter can be neither created nor destroyed, one can calculate the quantity of reactants and products going into a reaction and the products being formed from the reaction.

THERMODYNAMICS

Thermodynamics of chemical processes is an important branch of chemistry. It is the branch that involves the study of energy and energy transformations that

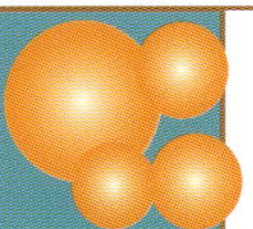

PHYSICS AND CHEMISTRY IN THE HIERARCHICAL STRUCTURE OF SCIENCE

by Joe Rosen, Ph.D.

Science is the human attempt to understand nature. As science developed over the past few centuries, it became apparent from very early on that nature does not at all present a uniform face. Aspects of nature exist that seem largely independent of other aspects and accordingly deserve and possess a particular branch of science for each. That is why one often refers to the sciences, in the plural.

One aspect is that of living organisms of all kinds. To study it, there are the life sciences, which one refers to collectively as *biology.* Another apparently independent aspect of nature is that of matter, including its properties, composition, and structure. The chemical sciences study this aspect of nature, and one groups them together under the heading *chemistry.* Then is the aspect of nature that presents itself in the form of elementary entities, principles, and laws. The various fields of *physics* are concerned with this. Physics, chemistry, and biology make up what are called the *natural sciences.*

Beyond those three aspects, nature reveals the apparently independent aspect of mind. The science that seeks to understand mind is *psychology.* This science is conventionally not grouped with the natural sciences. Some organisms live social lives, and that aspect of nature seems largely independent of others. One lumps the social sciences together as *sociology,* also not considered a natural science.

Now, what is meant by "independent" with regard to the five aspects of nature collected here? The idea is that it is possible to achieve a reasonable understanding of each aspect in terms of concepts that are particular to that aspect, with no, or little, reference to other aspects. This means that the corresponding science is, or can be, practically self-contained. As an example, biology, the science that studies living organisms, can deal with its subject almost wholly in terms of organisms and components of organisms. A better understanding is gained when concepts from chemistry and physics are introduced. However, that is not necessary for biology to operate as a science. In fact, historically the sciences developed very independently, with little reference to each other, if any. Until the end of the 19th century, chemistry was doing very well with no knowledge of the atomic nature of matter. And even fields within a science were considerably isolated from each other. Well into the 20th century, biology was largely separated into botany—the study of plants—and zoology—the science of animals. It has become increasingly recognized that there is much unity within each science and that the various sciences can benefit from each other.

This benefit turns out to be mainly a one-way street, and strictly so in the natural sciences. Take chemistry, as an example. Modern chemists do not make do with merely describing chemical processes and properties and searching for empirical laws to help them organize their observations. They need to know the reasons for their findings, and the reasons invariably are found in physics. Physics underlies chemistry. In principle, all of chemistry can be derived from physics (not that it would be possible in practice to do so). Much in chemistry depends on the forces between atoms. These forces are electromagnetic in nature, and at the atomic level, the quantum character of nature is dominant. Both electromagnetism and quantum mechanics, as well as their unification to quantum electrodynamics, are fields of physics. Take the periodic table of

occur in chemical reactions. Energy is the ability to do work, and, in the case of most chemical reactions, the work done is in the form of heat transfer. Two types of energy are kinetic (energy of motion) and potential energy (energy of position or stored energy). Potential energy in a chemical reaction is stored in chemical bonds and is known as chemical potential energy. As matter increases its kinetic energy, its temperature (measure of the average kinetic energy of matter) increases. Energy of a reaction involves the system (the part of the reaction that one is studying) and the surroundings (everything else). In contrast to a closed system, in an open system the system and surroundings can freely interact with one another. Chemical reactions that release energy from the system to the surroundings are exothermic reactions. When energy from the surroundings needs to be put into the system in order for the reaction to proceed, the reaction is said to be endothermic. Chemical reactions that are exothermic generally occur spontaneously, whereas endothermic reactions require a deliberate energy input.

PHYSICAL PROPERTIES

Physical properties studied by chemists include states of matter, boiling points, melting points, solubility, and concentration of solutions. Matter exists in four states of matter based on the kinetic energy of the particles and the attraction of the particles to one another. These states are solid, liquid, gas/vapor, and plasma. The latter of these states does not exist on the planet Earth but is found on the surface of

the chemical elements, a cornerstone of chemistry. It too is fully explained by physics, through the internal structure of atoms. Physics, on the other hand, has no need of chemistry, biology, psychology, or sociology to explain anything within its purview. It is truly an independent science and thus the most fundamental science.

Return to biology. Where do biologists turn for extrabiological reasons for biological phenomena that do not find an explanation within biology? They seek their explanations from chemistry—largely from biochemistry—and from physics. In principle, all of biology should be reducible to chemistry and physics, and thus ultimately to physics, at least according to many people. There are, however, also many dissenters who hold that biology is, in some way, more than chemistry and physics. It remains true that extrabiological explanations in biology are found in chemistry and physics, never vice versa. It is not only physics that has no need of biology, but chemistry as well.

Among the natural sciences a clear hierarchy, based on the directions of explanations among them, manifests itself. Physics is the most fundamental science, so we might picture it at the bottom of the hierarchy, in foundation position. All explanations from one science to another eventually reach physics. Next is chemistry, one level above physics. Its extrachemical explanations are derived from physics alone. Above chemistry in the hierarchy lies biology. Extrabiological explanations are derived solely from chemistry and from physics.

Actually, more sciences than those three are natural sciences. Earth and space science can serve as an example, as can marine science. Such sciences are modern interdisciplinary fields, which never have, and cannot have, an autonomous existence, independent of physics, chemistry, and biology. Earth and space science, for instance, depends crucially on physics and chemistry as well as on geology, astronomy, and other fields. Thus, such sciences do not hold a well-defined position in the hierarchy of science.

Psychology, the science of mind, has long operated with little or no regard for the underlying mechanism of mind, which seems to be the operation of the brain, a biological organ. There is increasing interest among psychologists in the biological basis of mind. Whether mind is, at least in principle, completely explainable in biological terms is an open question at present. There are those who adamantly insist it is, those who equally adamantly insist it cannot be, and those who are waiting to see what develops. But the position of psychology in the hierarchy of science seems clear: right above biology. The most likely source of any extrapsychological explanation of mind appears to be biology. On the other hand, biology needs nothing from psychology with regard to explanation.

Sociology, the conglomeration of sciences that study society, should then find its hierarchical position above psychology. Its extrasociological explanations must surely derive from the nature of the members of a society and the interactions among them. That leads to the minds and bodies of the members, and thus to psychology and biology. Biology has no explanatory need for sociology. Psychology, however, just might, since real minds do not operate in a social vacuum. So the positions of psychology and sociology in the hierarchy of science might possess some degree of parallelism.

Clearly, then, science is structured in a rather clear-cut hierarchy, a hierarchy that is based on the directions of explanations among the various sciences. The natural sciences are more fundamental than the others. And among the natural sciences, chemistry underlies biology, while physics holds the position of the most fundamental science of them all.

FURTHER READING

Hatton, John, and Paul B. Plouffe. *Science and Its Ways of Knowing.* Upper Saddle River, N.J.: Prentice Hall, 1997.

Wilson, Edward O. *Consilience: The Unity of Knowledge.* New York: Vintage, 1998.

the Sun and within stars. Solid matter has a definite fixed shape and a definite volume. The motion of the particles in a solid is very limited and usually resembles vibrations more than movement in the sense of assuming a new location. Liquids have definite volumes, but their shape can change on the basis of the shape of the container holding them. The molecules in a liquid can flow and have more movement than those in an equivalent solid. The term *gas* is used to describe substances that are gaseous at room temperature, while the term *vapor* is used to describe the gaseous state of a substance that is generally a liquid at room temperature. The change from solid to liquid is known as melting, and the change from liquid to vapor is known as vaporization. The temperature at which melting occurs is known as the melting point, and the temperature at which vaporization occurs is known as the boiling point. Chemists use these physical properties to help identify substances.

Chemical reactions often occur in solution. A solution is formed when there is a homogeneous physical association between two substances. A solution contains a solute, the substance being dissolved, and a solvent, the substance doing the dissolving. The amount of solute dissolved in a solution is critical to the action of that solution. Solvent molecules dissolve solutes through a process known as solvation, during which solvent molecules surround solute molecules and prevent them from interacting with other molecules of solute. Solutions form between the solute and solvent on the basis of the energetics of the interactions. Not all solvents can dissolve all solutes. The attraction of the solvent to the solute has to be strong enough to compete with the solute-solute attraction. The saying "Like dissolves like" describes the likelihood of a solute's being dissolved in a certain solvent. Polar molecules (molecules with a separation of charge) dissolve other polar molecules, and nonpolar molecules dissolve other nonpolar molecules. Polar molecules (such as water) are not able to dissolve nonpolar molecules (such as oil). Determining which reactants and products will be soluble in water is an important aspect of chemistry.

The amount of solute dissolved in the solvent is known as the concentration of the solution. Concentration can be measured in many ways including molarity, molality, and normality. *Molarity* is the number of moles of solute divided by the volume of the entire solution as represented by the following formula:

$$\text{molarity} = \text{moles solute/liter solution}$$

In order to calculate molarity, one must know the number of moles of solute and the final volume of the solution.

Molality is the ratio of the number of moles of solute to the number of kilograms of solvent. The total volume of solution is not computed. Calculation of molality can be performed using the following formula.

$$\text{molality} = \text{moles solute/kilogram of solvent}$$

Normality is a useful concentration unit for the study of acids. The active unit of an acid is the proton or hydrogen ion (H^+). The number of H^+ donated by an acid is known as the number of equivalents. The formula for normality is

$$N = (\text{Molarity} \times \text{number of equivalents})$$

CHEMISTRY EDUCATION

The training to become a chemist involves mastery of both classroom subjects as well as laboratory experiences. General chemistry courses that give an introduction to all areas of chemistry are followed by courses in organic chemistry, physical chemistry, and analytical chemistry. Chemistry degrees also require advanced mathematics courses including calculus and physics courses. Laboratory training is a vital part of chemistry training. Nearly every college or university offers a major in chemistry with some having more specific majors in other aspects of chemistry such as biochemistry or chemical engineering. Most major universities offer graduate training including master of science degrees as well as doctoral programs in chemistry.

BRANCHES OF CHEMISTRY

The field of chemistry is divided into specific branches, depending on the type of chemical reactions studied. Several examples of branches of chemistry are shown in the following table.

AMERICAN CHEMICAL SOCIETY

The American Chemical Society (ACS) is the premier chemical organization and has been in existence for 132 years. The ACS was established on April 6, 1876, in the state of New York and now claims more than 163,000 members. ACS has a large focus on chemistry education. They set guidelines for collegiate chemistry programs to follow in order to ensure consistent levels of education.

The ACS is also responsible for the publication of multiple peer-reviewed journals. Examples of such journals include *Chemical and Engineering News, Analytical Chemistry, Biochemistry, Chemical Reviews,* the *Journal of Organic Chemistry,* the *Journal of Physical Chemistry (A, B, C), Industrial and Engineering Chemistry Research,* and the *Journal of Chemical Education.*

ACS holds national meetings twice a year, in the spring and fall. Local branches of ACS hold regional meetings that generally are based on more specific research interests. In order to facilitate communication among chemists, local sections of the ACS hold monthly meetings that also serve as an excellent opportunity for professional development.

INTERNATIONAL UNION OF PURE AND APPLIED CHEMISTRY (IUPAC)

In 1911 a group of scientists attempting to increase the collaboration among chemists formed a group known as the International Association of Chemical Societies (IACS). This group set forth some goals for the field of chemistry in order to facilitate understanding among different countries and groups of chemists. They considered it important to standardize such chemical concepts as nomenclature (naming of compounds), standard units for atomic weights and physical constants, and a consistent method for the publication of data. The International Union of Pure and Applied Chemistry (IUPAC), formed in 1919, works to "advance the worldwide aspects of the chemical sciences and to contribute to the application of chemistry in the service of Mankind." One of the major contributions of IUPAC has been in the coordination of a systematic method of naming compounds (nomenclature) and weights, as well as

BRANCHES OF CHEMISTRY

Branch	Is the Study of . . .
agrochemistry (agricultural chemistry)	the scientific optimization of agricultural products
analytical chemistry	the composition of substances
atmospheric and environmental chemistry	the impact of chemical substances on the environment
biochemistry	the chemistry of living things
chemical engineering	the application of chemical and physical properties of matter to the manufacture of a product
computational chemistry	theoretical description of how chemical reactions take place by using computers and mathematical modeling
electrochemistry	energy transfers that take place during chemical reactions
green chemistry	environmentally friendly chemical processes
industrial chemistry	application of chemical processes and understanding to all aspects of industrial processes
inorganic chemistry	chemical reactions in compounds lacking carbon
nuclear chemistry	radioactivity
organic chemistry	carbon-containing compounds
physical chemistry	physical laws governing the motion and energy of the atoms and molecules involved in chemical reactions
solid state chemistry	physical and chemical properties of substances in the solid state
surface chemistry	atomic and molecular interactions at the interface between two surfaces
textile chemistry	design and manufacture of fibers and materials

methods of measurement. IUPAC has the following eight divisions:

- Physical and Biophysical Chemistry
- Inorganic Chemistry
- Organic and Biomolecular Chemistry
- Polymers
- Analytical Chemistry
- Chemistry and the Environment
- Chemistry and Human Health
- Chemical Nomenclature and Structure Representation

IUPAC publishes a series of books on chemical nomenclature that set the standard for all areas of chemical work. They are referred to by color and cover the following information:

- Gold Book: Chemical Terminology
- Green Book: Quantities, Units, and Symbols in Physical Chemistry
- Red Book: Nomenclature of Inorganic Chemistry
- Blue Book: Nomenclature of Organic Compounds
- Purple Book: Macromolecular Nomenclature
- Orange Book: Analytical Nomenclature
- Silver Book: Nomenclature and Symbols in Clinical Chemistry

The IUPAC continues to be a standard for cooperation among chemists for the good of the field.

See also ATOMIC STRUCTURE; CHEMICAL REACTIONS; CONCENTRATION; HEAT AND THERMODYNAMICS; PERIODIC TABLE OF THE ELEMENTS; STATES OF MATTER; STOICHIOMETRY.

FURTHER READING

American Chemical Society home page. Available online. URL: http://portal.acs.org/portal/acs/corg/content. Accessed February 14, 2008.

Brown, Theodore L., H. Eugene Lemay, Bruce Bursten, and Catherine J. Murphy (contributor). *Chemistry: The Central Science.* Upper Saddle River, N.J.: Prentice Hall, 2006.

International Union of Pure and Applied Chemistry home page. Available online. URL: http://www.iupac.org/dhtml_home.html. Accessed February 16, 2008.

Pinhole cameras are portable devices that use a light-sealed box that contains a hole the size of a pinpoint and photographic paper in the back of the box. The amount of exposure time determines the intensity of the image formed. *(Steven Collins, 2008, used under license from Shutterstock, Inc.)*

chemistry in art and photography Art and science are often considered radically different territories because they utilize different types of skills and portions of the brain. Science encompasses analytical thought, whereas art utilizes creativity. The chemistry involved in the creation of art is an underappreciated link between these two fields. Creation of artistic works in every field of art involves a vast array of chemical reactions. Artists do not have to understand fully the chemistry behind their art form in order to create, but understanding how the processes work at a molecular level makes tasks such as choosing the right solvent, the right pigment, or the right surface logical rather than trial-and-error. Photography, painting, drawing, ceramics, glass, and sculpting are some areas of art that involve a large amount of chemistry.

SCIENCE OF PHOTOGRAPHY

Original photography has its roots in ancient times. The first version of a camera was the *camera obscura,* which translates from Latin to "dark room." The principles of optics explain that when light passes through a small opening, rather than scattering, the rays transform the image on the outside into an inverted picture. Portable versions of the camera obscura known as pinhole cameras were developed and optimized in the 17th century. The design of the pinhole camera includes a light-sealed box with a small hole the size of a pin or smaller. The back of the camera (box) is covered with photographic paper. When the shutter is removed from the front of the pinhole, all of the light from the area is forced to travel through the same hole, concentrating the light rays and forming an inverted image on the photographic screen. The shutter is manually opened and closed, and the optimal exposure time must be determined for the best resolution of the picture. The size of the hole and the amount of time that the hole is exposed to light determine the brightness of the picture produced. The ratio of hole size to focal length (distance from hole to photographic paper) is approximately 1 to 100.

The problem with the camera obscura was that viewing was necessary at the time of image formation. Creation of a permanent photographic image was accomplished through a series of improvements by several individuals. In 1802 the English physicist Thomas Wedgwood experimentally determined that combinations of nitric acid and silver created a black image when exposed to light. The images created by Wedgwood did not last long, but he is considered by many to be a pioneer in the science of photography. Sir John Herschel, an English chemist, improved on Wedgwood's work, being the first to fix an image by treating the silver with sodium thiosulfate. This reaction dissolves the silver compounds used in creating the image of a photograph, stopping the exposure and fixing the image. Herschel is credited with the concept of negative images.

The French inventor Joseph Nicéphore Niépce combined the pinhole camera with photosensitive compounds to create the first permanent image in 1826. Further improvements to photography occurred in 1834 when Henry Fox Talbot soaked paper in silver iodide and then gallic acid to make a negative. Talbot next created a positive image by placing another sheet of photographic paper below the negative and subjecting it to light again. In 1837 Louis Daguerre used silver-plated copper, silver iodide, and warmed mercury to produce a daguerreotype image. This process became a standard method of photography. George Eastman was the first to use this chemical combination on a photographic film that could be placed inside a camera. In 1888 the first Kodak camera was created when a gelatin-based silver halide was spread on a cellulose nitrate. This became the basis of modern-day film photography.

Present-day photographic film is still produced using these fundamental chemistry principles. The silver halide salt (e.g., AgI, AgBr) is dispersed over a gelatin matrix, which is melted and applied to a cellulose base. Exposure of the silver bromide to light produces Ag^+ and Br and a free electron. If the Ag^+ ion becomes reduced by combining with the electron, Ag can be produced. The silver leads to the dark portions of an image on a negative when it has contact with a developer. Only the areas of the film that have been exposed to light will have reduced silver that forms the dark images on the negative. This explains why the negative is the reverse of the true image. The lightest areas of the original image lead to the darkest areas on the negative, and vice versa. The grain size of silver halide on the film determines the light sensitivity or speed of the film. The larger the grain size, the more light-sensitive the film is.

The developer functions by giving up electrons, or becoming oxidized, while reducing the silver ions. Selectivity is essential for a developer because it must not affect the unexposed grains of silver halide left on the film. The gelatin base of photographic film is uniquely suited to its job of holding the silver halide. Gelatins are polymers made from animal bone. Hydrophilic in nature, gelatins readily absorb water yet dehydrate into a stable structure. Gelatins melt easily, enabling even distribution of the silver halide particles over the film.

As a result of the development of digital technology, the use of film photography by the general public is declining. For example, in 2004 Kodak ended its production of film cameras. The type of camera used is often a matter of preference as some photographers strongly believe that film creates better photographs than computer technology.

The development of digital cameras has radically changed the field of photography. *(Ljupco Smokovski, 2008, used under license from Shutterstock, Inc.)*

CHEMISTRY OF PAINTING

All types of painting depend on the chemistry of the pigments involved as well as the chemistry of the surfaces used to paint. Some paints are specifically designed for wood, cloth, canvas, and paper. The molecular interactions between the paint and the surface determine the strength of the bond. The main components of paint include the pigment that gives the paint color, the binder that causes the paint to adhere to the surface as well as to form a layer on the surface, and the solvent to dissolve the pigment and the binders.

The pigments have a range of colors and can be blended to create colors to suit any project. Different types of binders make up the different types of paints including watercolors, oil paints, tempera paint, and acrylic paint. Latex-based paints use water as the solvent, and oil-based paints (alkyds) use oils such as linseed oil. The evaporation of the liquid from an oil-based paint leaves behind the pigment and binder, which then oxidizes by interaction with molecular oxygen in the air to form a tough coating.

Paint chemists specialize in the science of paints. One responsibility of a paint chemist is determining the correct ratio of pigment to binder to create the paint. Understanding the formulations, production, dry times, viscosity, hardness, and principles behind paint allows the artist to create a work of art and a chemist to produce the paints needed to create it.

CHEMISTRY OF OTHER ARTS

The creation of ceramic art involves heating the shaped or molded material, such as clay, to very high temperatures, 1,832°F (1,000°C), in a process known as firing. The chemical composition of clay primarily includes silica and alumina (i.e., silicon dioxide and aluminum oxide). Different proportions of these compounds make different types of ceramic art, such as porcelain and terra-cotta. Firing evaporates the water and creates a strong, durable product.

Most metal sculptures are made in bronze, which is an alloy, a metallic material that is a mixture of two or more metals. Bronze is a mixture of tin and copper. The physical properties of bronze are an improvement over those of either of its components as well as many other metals. Bronze will oxidize on the surface to develop a characteristic patina, but it will not oxidize past the surface; thus, it will not become weakened over time. The most common method of bronze casting is known as the lost-wax method, in which the sculpture is first made out of wax, surrounded by clay, and then fired. The wax would melt, leaving a hollow mold that could then be cast in bronze. With a lower melting point than steel, bronze casting was used for sculpture as well as weapon production.

Colorful ceramic pots are produced by using different compositions of silica and alumina and evaporating the water by firing the pots. *(Rafal Dubiel, 2008, used under license from Shutterstock, Inc.)*

The production of fabrics and tapestries involves a vast amount of chemistry. Textile chemistry is a broad field that involves the formulation of dyes, the creation of fibers, and the cloth that is produced from them. Different types of dyes are available, ranging from commercially prepared chemical dyes to natural dyes from plants and other organic sources. The dye molecules bind to the molecules of the fabric and create a colored fabric. The fabric can then be used to create works of art.

Drawings are another field of art that has a basis in chemistry. The use of chalks made of calcium carbonate ($CaCO_3$) and pastels that have had dyes added to the calcium carbonate for color creates beautiful images. Pencil drawings utilize pencil lead, which is actually not made of lead but of an allotrope of carbon known as graphite. An allotrope is a structural form of a compound with an identical chemical formula but a different crystal structure. Carbon has multiple allotropes with widely varied properties. One familiar allotrope of carbon is diamond, an incredibly hard, transparent crystal. Graphite, on the other hand, is black and soft enough for writing. Graphite has a natural form that is isolated from metamorphic rock and is greasy, is black, and stains. The type of graphite determines the type of line that is achieved in a pencil drawing. Harder graphite makes darker, sharper lines, while softer graphite produces lighter lines but is smudged more easily.

CHEMISTRY OF GLASSWORK

The main component of glass is silica or silicon dioxide (SiO_2). Silicon dioxide melts when heated to temperatures nearing 3,632°F (2,000°C). Most glass contains additional components that help reduce the melting point, for example, calcium carbonate and sodium carbonate. The popular and valuable type of glass art known as leaded glass utilizes lead oxide instead of calcium carbonate, increasing the glass's strength, weight, and refractive index, making it sparkle more than regular glass. The addition of boric oxide makes the glass less likely to shatter when heated because it expands less with the addition of heat. This is the glass used in such products as Pyrex bake ware. Silicon dioxide does not absorb light; thus glass is transparent. Originally glass contained colors that were due to impurities in the silicon dioxide that was used to create it. Addition of pigments or metals to the silica can lead to color variation of the glass.

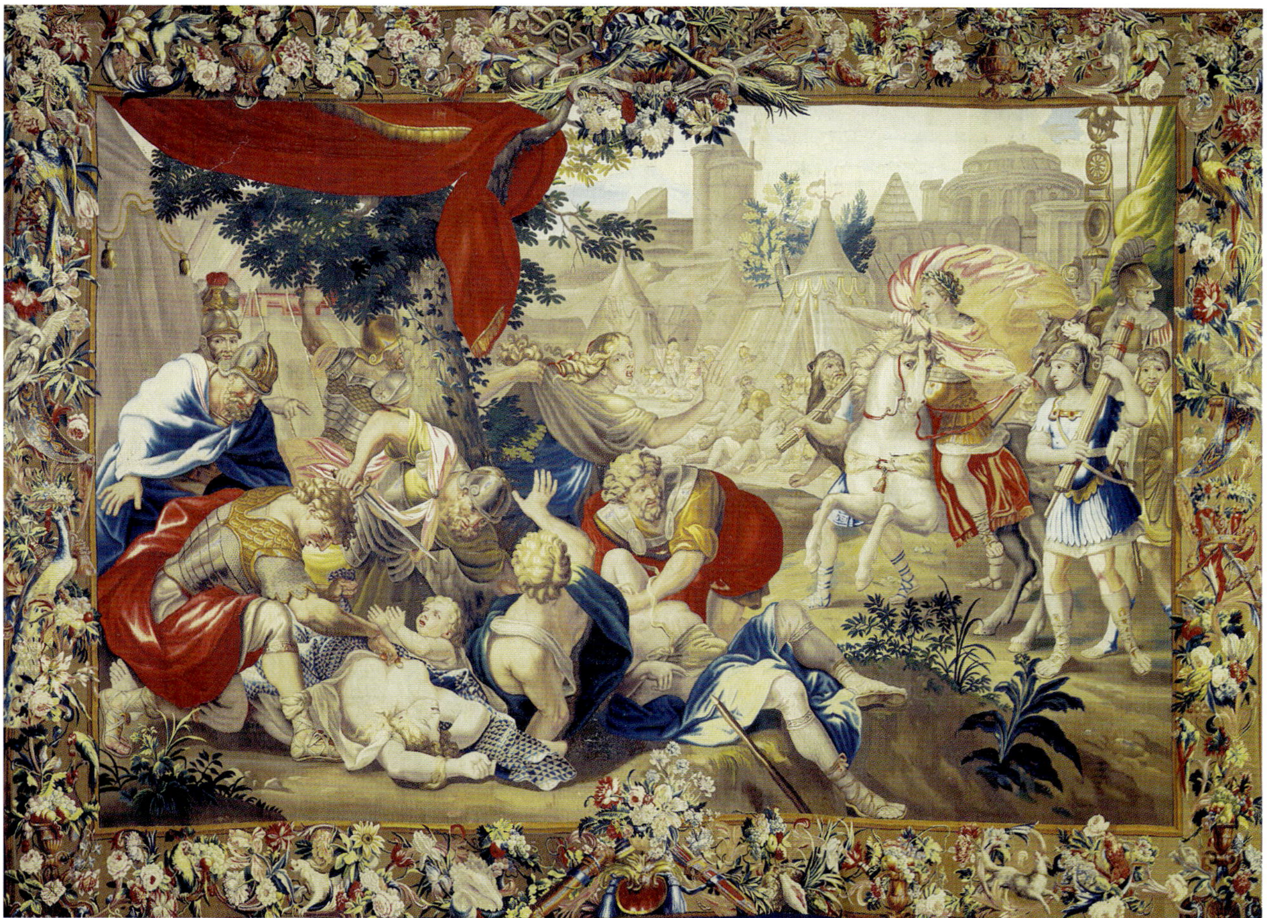

Tapestry from Brussels from the end of the 17th century, showing wars of the Roman emperor Vespasian and his son, Titus, at Jerusalem, ca. 70 C.E. Works of art such as this involve dyes that create the colored fabric. *(Art Resource, NY)*

CONSERVATION SCIENCE

A new field known as conservation science utilizes chemistry knowledge to rejuvenate works of art as well as authenticate works on the basis of analysis of the chemicals present. Restoration of works of art involves chemistry in the development of new techniques and cleaners that remove years of dirt from the painting without harming the original work.

Chemistry is the basis of identification of age and origin for a given work of art. Analytical chemistry techniques are useful in identifying the makeup of artwork. The changes in pigment types over time can identify the age of the pigment. X-ray fluorescence can be used to identify the type of pigments in a painting. X-ray diffraction can determine the chemical makeup of the pigment to identify the pigments in the paint definitively. Mass spectrometry and infrared spectroscopy yield analytical data about the composition of artifacts. Developments in more noninvasive techniques, such as nuclear magnetic resonance (NMR), have led to improvements in conservation science.

See also ANALYTICAL CHEMISTRY; OPTICS; TEXTILE CHEMISTRY.

FURTHER READING

Fujita, Shinsaku. *The Organic Chemistry of Photography.* Berlin: Springer-Verlag, 2004.

Rogers, David N. *The Chemistry of Photography: From Classical to Digital Technologies.* London: RSC, 2006.

chemistry in wine making Wine is an alcoholic beverage made out of fermented grapes. The history of wine production and consumption spans nearly 4,500 years, and drinking wine plays a major role in many social, religious, and cultural rituals and activities. The type of grape, the harvesting methods, and the processing contribute to the wide array of wines available. Chemistry is vital to each stage of wine production: determining the correct harvesting time, analyzing ripeness of the fruit, treatment of the grapes to prevent microbial growth, and the fermentation processes itself.

TYPES OF WINE

The study of wine making is known as oenology, and the process of making wine is known as vinification. Different classes of wine include table wines, fortified wines, and sparkling wines, distinguished by the processes used to make them.

Table wines constitute the largest percentage of the world's wine production. Produced from the juice of pressed grapes, table wine is allowed to ferment naturally or is fermented upon the addition of a yeast culture. Alcohol content can vary from 7 to 15 percent, and the colors vary from white to red. Wine producers divide wine into two main branches based on this color difference. The predominant difference between these two types of wine lies in the inclusion of the skins in the production process. White wines do not include the skins, while red wines do.

Fortified wines have more alcohol added to them after fermentation. These types of wine have a higher alcohol content than table wines, ranging from 14 percent up to approximately 23 percent. Their color ranges from white to dark red. Products such as sherry, port, madeira, and vermouth are all fortified wines.

The most common sparkling wine is champagne. Additional carbon dioxide causes the effervescent characteristic of champagne. This carbon dioxide is produced in a special double-fermentation process that takes place when additional yeast and sugar are added in the bottle. Champagne was first produced in France by Dom Perignon, after whom a well-known type of champagne is named.

More than 5,000 varieties of grapes exist and are utilized in the production of wine. Each of these varieties used belongs to the same species of grapes—*Vitis vinifera*. The different varieties of grapes have variations in their shape, size, juice composition, resistance to disease, and so forth. The predominant forms of grapes that are used for wines are Riesling, chardonnay, cabernet sauvignon, muscat, pinot noir, and sauvignon blanc. The majority of U.S. wine grape production is in California.

CHEMICAL MAKEUP OF GRAPES AT HARVEST

Determining when the grapes are ready to harvest is not an exact science. Although scientific tests are available and used to determine whether grapes are ready, much depends on the soil conditions and temperatures of growth, and the grower still relies heavily on visual and taste techniques to determine when the grapes are ready. Calendar and forecasting methods of harvest scheduling are often subject to changes.

Refractometry is the primary technique for determining ripeness. A refractometer measures the amount of light that is refracted when it passes from air into a liquid. One can use this information to determine the concentration of a solute or the purity of a sample compared to a known curve. The refractometer measures the light refraction in the unit known as refractive index. Vintners can compare the refractive index of the grapes with known refractive indexes of other harvested vintages.

Harvesting of grapes to make good-tasting wine requires the grapes to have the appropriate chemical composition. As grapes ripen, the acidity in them decreases, while the sugar concentration increases. The absolute composition found in a harvest of grapes is not calculable. The growing season temperature, amount of precipitation received that year, and soil conditions alter the composition of the grapes. The average juice from grapes contains approximately 80 percent water and 20 percent carbohydrates, as well as trace amounts of organic acids and phenolics. The grapes chosen for production of grape juice include more than 15 percent soluble solids or sugars. More money is paid to grape producers for higher percentages of soluble solids. If the soluble solid level is too low, the acidity will not be high enough to give the juice its appropriate flavor. The average juice from a grape consists of chemicals that contribute to the taste, color, aroma, and acidity of the juice. Organic acids such as tartaric acids, malic acid, and citric acid and phenolics such as tannins help to give the grapes their flavor. The vitamins and minerals present in the grape juice contribute to the healthful benefits of wine and are required for the proper fermentation process of the grapes. Wines have lower sugar levels than grape juice. The alcohol concentration in wine is between 8 and 15 percent.

CHEMISTRY OF GRAPE PROCESSING

Harvesting of the grapes can be performed by hand or with mechanical harvesters. Each of these methods has potential positive and negative effects. While mechanical processing can be much more efficient and less labor intensive in harvesting, a time delay still exists between picking the grapes and delivery to the processing plant. During the wait, many chemical processes can occur within and to the grapes that can decrease the quality of the final product. Oxidation of the grape components can lead to browning. The addition of sulfur dioxide (SO_2) prevents this oxidation. The maximal legal limit of sulfur dioxide in wine is 200 ppm. SO_2 is the active form that prevents oxidation, but equilibrium exists between it and the inactive form of HSO_3. The minimal sulfur dioxide levels required to prevent oxidation effectively is 80 ppm.

Spontaneous fermentation of the grapes upon harvest is caused by wild-type yeasts that are present on the grapes themselves. Uncontrolled microbial activity is detrimental to the production of both wine and grape juice. Prevention of such microbial activity is carried out by the addition of sulfur dioxide to the grapes after crushing and destemming. Enzymes known as macerating enzymes are added to grapes to help break down the skins in the production of grape juices that use the skins as well as red wines. Research has been conducted showing that rather than adding commercial preparation of these enzymes, adding yeast cultures that already express these enzymes or adding transgenic yeast that overexpress these enzymes causes a comparable breakdown of the skins in red wine, releasing the color compounds, and aroma and flavor compounds such as phenolics and tannins.

Common scientific mixture separation techniques are utilized to separate the grape juice from the unwanted materials such as dead microorganisms, unwanted salts, and particulates from the grapes. The skins, seeds, and pulp from the grapes are collectively known as pomace. Filtration helps remove larger particles including seeds from the juice. This process is often followed by centrifugation, a separation technique in which the sample is spun in a circular motion at a high rate of speed inside a centrifuge. The centrifugal force of the movement causes the sample to be pushed outward, toward the bottom of the centrifuge tube. This causes the heaviest components of a mixture to travel to the bottom and the lighter components to stay on top. If spun under the right conditions, the pomace will end up as a pellet (collection of solids at the bottom of the centrifuge tube) and the juice will all remain in the supernatant.

CHEMISTRY OF THE FERMENTATION PROCESS

The grape juice that is collected is then inoculated (injected with a living culture) with the appropriate strain of yeast. *Saccaromyces cerevisiae* (*S. cerevisiae*), the predominant yeast used in wine fermentation, naturally exists on the skins of many fruits. This common species of yeast belongs to the fungal phylum Ascomycota and reproduces through a process known as budding. *S. cerevisiae* is a valuable genetic tool as well as being useful in fermentation (brewers' yeast). This type of yeast serves as a model eukaryotic system to study genetics because of its fast generation time and the fact that it contains all of the typical eukaryotic organelles. The genome of *S. cerevisiae* was one of the first complete genomes to be sequenced. The wild-type yeast are removed before processing the wine in order to ensure that a pure strain of *S. cerevisiae* is used, in other words, that no additional contaminating microorganisms are present.

Oenococcus oeni is a lactic acid–fermenting bacterium that is used to reduce the acidity of the wine after fermentation. This strain of bacterium can tolerate the high alcohol content that follows alcohol fermentation with *S. cerevisiase* and breaks down malic acid and lactic acid to give the wine a more mellow taste. *O. oeni* is a very slow grower, so manipulation of its genome or growing time would be helpful in order to ensure the timing of the acid reduction relative to the speed of fermentation. This species of bacterium was sequenced in 2002 by the U.S. Department of Energy's Joint Genome Institute.

Fermentation is a process that results in the breakdown of carbohydrates under anaerobic conditions. Yeast, as well as most other organisms, breakdown glucose via the biochemical pathway known as glycolysis. This is a 10-reaction series of steps that begins with the six-carbon molecule glucose and ends with two three-carbon pyruvic acid molecules. If the process occurs in the presence of oxygen, then the pyruvic acid molecules enter the citric acid cycle for complete oxidation to carbon dioxide and water. Under anaerobic conditions, the alternate pathway of fermentation occurs. During alcohol fermentation, yeast convert the pyruvic acid into ethanol, a process that results in the regeneration of oxidized nicotinamide adenine dinucleotide (NAD^+), a necessary component for the continuation of glycolysis.

To produce wine, yeast ferment the sugars in the grapes into alcohol. This alcohol fermentation takes place on a large scale that is strictly controlled to produce the characteristic flavors expected in wines. Fermentation is stopped in the wine after the alcohol content reaches the desired level. Wine fermentation can take 10–30 days to complete, and the optimal temperature is around 75–77°F (24–25°C). Though the biochemical process of fermentation strictly refers to the anaerobic breakdown of energy-rich organic compounds such as pyruvate, the term *fermentation* also refers to the large-scale use of microorganisms during the commercial production of various substances. In this sense, the initial fermentation (called the primary fermentation) of wine making occurs in the presence of oxygen to allow the yeast to reproduce to an acceptable level so that when secondary fermentation takes place, a sufficient concentration of yeast is present to perform the job. This secondary fermentation can take weeks. All of the components involved in the fermentation process aside from the juice and the alcohol need to be separated from the wine solution in a process known as racking, during which the sediment is removed before moving on to the bottling step.

The wine barrels shown are from a storage cave at a winery in Napa Valley, California. The wine is left in barrels such as these to continue the fermentation process, during which the alcohol content is measured. *(Jeff Banke, 2008, used under license from Shutterstock, Inc.)*

The alcohol content of the wine during fermentation is monitored by measuring a property known as specific gravity, which compares the density of a substance to that of water. The denser the substance, the higher its specific gravity. Addition of sugar increases the specific gravity of the sample. When the yeast turns the sugar into alcohol, the specific gravity decreases since alcohol is less dense than water. The device used to measure specific gravity is known as a hydrometer. Specific gravity readings between 990 and 1,050 are utilized for wine-making with the lower values for drier wines and the higher values giving rise to sweeter wines. Fermentation must take place in the presence of carbon dioxide to prevent oxygen from the air from oxidizing the sugars and alcohols in the wine. Oxidation occurs according to the following reaction:

$$CH_3CH_2OH \text{ (ethanol)} + O_2 \text{ (oxygen)} \rightarrow CO_2 \text{ (carbon dioxide)} + H_2O \text{ (water)}$$

During the aging process, the acidity decreases, settling occurs, and the full flavor develops.

Analytical testing of the wine prior to bottling and distribution is essential to ensure proper taste, pleasant aroma, appropriate alcohol concentration, and lack of contaminants.

ENVIRONMENTAL ISSUES IN WINE PRODUCTION

Disposal of the waste products generated during wine production is often a problem. The waste left behind develop a high biological oxygen demand (BOD) and can be damaging to the environment by polluting bodies of water. High BODs can cause algal blooms and destroy the productivity of waterways. Most present-day wineries recycle, mulch, and compost their waste products. Skins and other biological wastes can be spread across a large area and used as fertilizer, but some of the products must go to the landfills.

See also BIOCHEMISTRY; CARBOHYDRATES; GLYCOLYSIS; SEPARATING MIXTURES.

FURTHER READING

Goode, Jamie. *The Science of Wine from Vine to Glass.* Berkeley: University of California Press, 2006.

chromatography The term *chromatography* is formed from the Greek words *chroma,* meaning "color," and *grapho,* meaning "to write." Chromatography is a class of processes that separate mixtures of chemicals into their component parts. A mixture is a physical combination of one or more substances. Chromatography also allows identification of the number and type of components in the mixture. M. S. Tswett of Warsaw University first performed the process of chromatography in 1903 to isolate and identify the pigments found in plants. He ground up leaves using ether as a solvent and then passed the plant mixture over a calcium carbonate column. Using this early method of chromatography, Tswett successfully identified different versions of chlorophyll. Important in the pharmaceutical and biomedical fields, chromatography is used in both analytical and preparative methods. Examples of its utility include drug testing of athletes and purification of medications prior to their distribution.

BASIS OF CHROMATOGRAPHY

Chromatography involves the passing of a mobile phase (the eluant) that contains the mixture to be separated over a stationary phase.

If the molecules being separated are solids, they must first be dissolved in an appropriate solvent. The term *solvent* refers to a substance in which a solute is dissolved, and *solute* refers to the dissolved substance. The mobile phase then flows over the stationary phase, consisting of either a solid or a liquid in the form of a sheet or column. Several factors determine how quickly the mobile phase passes over the stationary phase, including the size of the solute particles, the affinity of the solute particles for the column, and the charge of the solute particles. The eluted material is collected in samples called fractions. A detector, often connected to a strip chart recorder, indicates the amount of solute molecules present in each fraction.

Efficient separation of the mixture components often entails slight alterations in the methodology. Optimizing the procedure involves changing the stationary phase composition, the mobile phase composition, the flow rate of the column, and/or the solvent used to remove the sample from the column. When the method is optimized, the fractions collected at the end represent the individual components in the mixture.

Retention time, or retention factor (R_f), the amount of time required for the mobile phase to move through the stationary phase, varies with individual components. Differential attractions for the stationary phase affect the rate of travel through the paper or the column. This allows analysis of the number, size, and charge of components of a mixture and isolation of the various components into different fractions. One can then study the fractions to identify unknown components or characterize a known component.

TYPES OF CHROMATOGRAPHY

Three common variations of chromatography allow separation based on the types of components and the degree of desired separation: capillary action chromatography, column chromatography, and gas-liquid chromatography. Capillary action chromatography is most useful for characterizing the number and type of components in a mixture. One disadvantage of this type of chromatography is that further characterization of the separated samples is difficult. An example of capillary action chromatography is paper chromatography, in which filter paper or other absorbent paper serves as the stationary phase. The mixture to be separated is spotted on the chromatography paper, which is placed in the solvent. The solvent moves up the paper via capillary action, the attraction of the liquid for the solid paper that causes the liquid to travel up the paper. When the solvent

Streaks on chromatography filter paper, created when spots of dye are placed on the paper and the paper is placed in a shallow pool of solvent. As the paper absorbs the solvent, the solvent travels up the paper, carrying with it the dye's different components, which travel at different speeds, producing the multicolored streaks. *(Charles D. Winters/Photo Researchers, Inc.)*

reaches the solute spot, it dissolves the solute and carries the solute molecules with it as it continues to move up the paper. The solute particles move at a rate relative to their attraction to the stationary phase. The retention time for different solute particles can be calculated to determine the relative attraction of each of the molecules for the stationary phase. Several factors influence the outcome of paper chromatography. The solute molecules being separated must be soluble in the chosen solvent. If the solute is not soluble, then it will not travel in the solvent, and no separation will occur. The length of the paper will also determine the resolution of the different components of the mixture. Longer paper will allow for more separation, facilitating the distinction of closely related solute particles.

Another type of capillary action chromatography is thin-layer chromatography (TLC), a technique that is similar in principle to paper chromatography. The main difference is that in TLC the plate has been coated with a silica bead layer that allows for greater separation of the molecules. After the samples are spotted on the TLC plate and inserted in the solvent, the solvent moves up the TLC plate via capillary action. When two substances are closely related, further separation by TLC is possible by turning the plate 90 degrees and running it again with the solvent.

Column chromatography differs from capillary action chromatography in that the stationary phase is three-dimensional. The stationary phase involves a resin in a column through which the liquid mobile phase travels. Types of column chromatography include high-performance liquid chromatography (HPLC), fast-protein liquid chromatography (FPLC), size-exclusion chromatography (SEC), ion-exchange chromatography, and affinity chromatography.

HPLC is useful for the identification, purification, and often quantification of the components in a mixture. Widely used in biomedical and pharmaceutical applications, HPLC results in excellent sample purity because of its ability to resolve compounds that are chemically very similar. In HPLC, the mobile phase runs over a column under very high pressure, and the samples are separated on the basis of differences in polarity. After the mobile phase has passed through the column, detectors measure the amount of each component in the fractions creating a chromatograph. Many different types of HPLC detectors allow for measurement of different factors including absorption of ultraviolet radiation, light scattering, fluorescence, mass-to-charge ratio via mass spectroscopy, and molecular structure via nuclear magnetic resonance (NMR) spectroscopy.

The long columns used for HPLC consume too much time for testing a large number of samples. This hindrance led to the development of FPLC, a specialized type of chromatography that utilizes a shorter column, reducing the required volume of solvent as well as the run times of the samples and increasing the number of samples that can be processed per day.

SEC, also called gel-filtration chromatography, separates components of a mixture on the basis of their size. Not only does SEC provide useful information regarding molecular weight, but the technique preserves the structure and therefore activity of proteins during the separation. SEC separates components of a mixture on the basis of the size of the particles. The column is packed with polystyrene beads that contain pores of various sizes and then filled with an inert stationary phase, usually tetra-hydrofuran. Larger particles pass more quickly through the column because smaller particles penetrate into the pores of the polystyrene beads of the column, whereas the larger particles cannot. Since their pathway includes more detours, the smaller particles travel a greater total distance while flowing through the column—so they take more time to pass through it. Because of their size, the larger particles are excluded from most of the column volume, causing them to travel a straighter path from the top to the bottom of the column and therefore exit the column more quickly. The time required for the molecules to come off the column, or to elute, can be used to calculate the molecular weight of the compound.

Ion-exchange chromatography involves a column with either a negatively charged (anionic) stationary phase or a positively charged (cationic) stationary phase. When the mobile phase passes over the column, the oppositely charged particles in the mixture adhere to the column, and those that are like-charged pass through unaffected. This results in separation of the charged particles from the uncharged and oppositely charged particles. The desired charged solute particles can then be released by changing the concentration of the buffer running through the column. The molecules least attracted to the column elute first, and the molecules with the greatest attraction for the column elute last. Water softeners utilize ion-exchange chromatography to remove cations such as Mg^{2+}, Ca^{2+}, and Fe^{2+} from hard water.

Affinity chromatography depends on specific interactions of the solute molecules with the column molecules. The column resin is usually made of agarose, a carbohydrate derived from seaweed. The most commonly used form of agarose is Sepharose. A specific compound attached to the Sepharose beads directly interacts with the molecule being isolated. Molecules of interest interact and stick to the column as the mobile phase flows through the stationary phase. The molecules that do not bind the specific attached compound pass right through the column. When the entire mobile phase batch has passed through the column, the column can be washed to remove any unattached molecules. Breaking the interaction between the molecule and the column by using a competing molecule or changing the pH releases the isolated molecules of interest. Affinity chromatography is useful for isolating antibodies

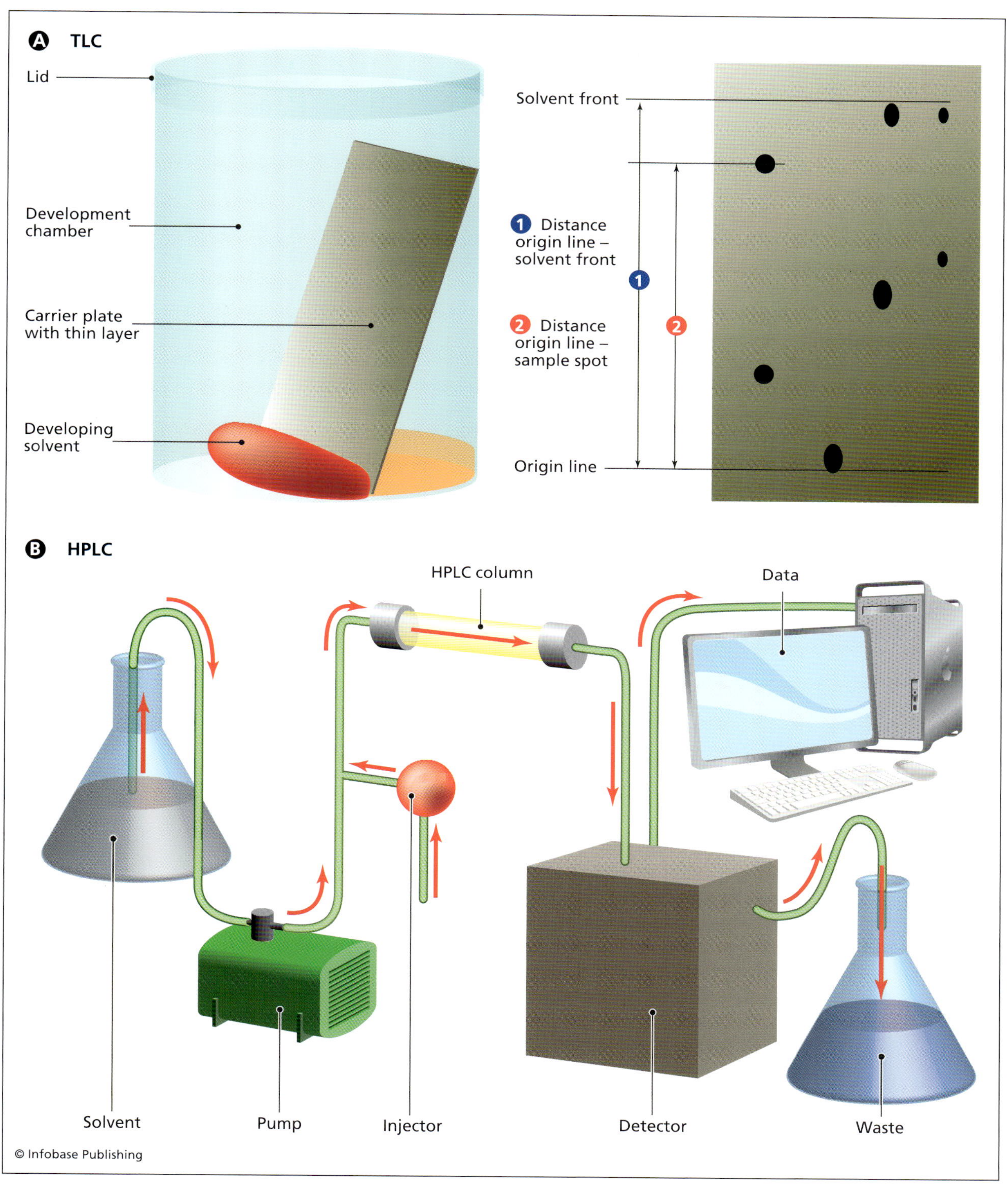

(a) Thin layer chromatography and (b) high-performance liquid chromatography are two types of chromatography used to separate mixtures.

using an antigen on the column, isolating enzymes by using their substrate on the column, or isolating recombinant proteins that have been given a tag that binds them to the column.

In gas-liquid chromatography (GC) the sample is vaporized when it is injected into the port and then mixed with an inert carrier gas, such as nitrogen or helium. The sample travels through the packed column, which contains a liquid-coated, solid stationary phase. A detector records the amount of each specific substance that comes off the column. In gas chromatography, the boiling point of the compounds being separated is critical. Instant vaporization of the sample upon injection into the column requires a temperature several degrees above the boiling point of the compounds in the mixture. Sample loading by rapid injection using a fine syringe allows for a tight peak in the chromatograph. Forensic investigations depend on a form of gas chromatography called mass spectroscopy for the detection of trace amounts of various chemicals.

See also CENTRIFUGATION; MASS SPECTROMETRY; STATES OF MATTER.

FURTHER READING

Miller, James M. *Chromatography: Concepts and Contrasts,* 2nd ed. New York: John Wiley & Sons, 2004.

cigarette smoke

Cigarette smoke is a major component of both indoor and outdoor air pollution. For centuries tobacco has had both recreational and ceremonial use. Corporate cigarette manufacturing began in the United States in 1847 with the establishment of the Philip Morris Company and the J. E. Liggett Company. The R. J. Reynolds Tobacco Company followed in 1875. The use of tobacco by soldiers in World War I increased cigarette consumption in the United States. In the early 1920s, tobacco companies began targeting women as a new class of customers, marketing certain brands of cigarettes as ladylike. World War II cemented the market for cigarette use as soldiers who became addicted while overseas returned home.

During the 1960s, the negative health effects of cigarettes became known, including their addictive power and their link to respiratory disorders such as emphysema and lung cancer. The U.S. surgeon general placed warnings on packaging and began regulating cigarettes for the first time. Despite the health warnings, addicted cigarette smokers were not willing or able to stop. Cigarettes contain nicotine, an addictive substance. Recent studies have shown that teenagers can become addicted to nicotine after a single use. Surveys reveal that 90 percent of smokers become addicted before turning 18 years old.

Beginning in the 1980s, smokers and families of smokers filed lawsuits against the tobacco companies for the harmful effects of smoking. These lawsuits have continued to the present day, and determinations are being made about how much the tobacco companies knew about the addictive and dangerous impact of their product. In 1982 a new attack on cigarette smoking began when the surgeon general issued warnings against the harmful health effects of secondhand smoke, the substance that a smoker exhales while smoking. Secondhand smoke endangers the health of everyone around the smoker, even those who are not smokers themselves.

HEALTH EFFECTS OF CIGARETTE SMOKE

According to the American Heart Association, in 2008 cigarette smoking was the leading cause of preventable death in the United States, causing approximately 440,000 deaths per year. Engaging in this habit increases the risk factors for coronary heart disease, including high blood pressure. Cigarette smoking also reduces one's ability to perform regular physical activity by diminishing lung capacity. Therefore, the cardiovascular health of a smoker is already compromised. Smoking also increases one's risk of coronary heart disease by decreasing the amount of good cholesterol (high-density lipoprotein, or HDL). Cigarette smoking and exposure to secondhand smoke are the leading causes of lung cancer in smokers and nonsmokers. Children who are subjected to cigarette smoke at home have a significantly increased risk of developing asthma, and those who already have asthma experience aggravated symptoms. From an

Cigarette smoke contains toxic and carcinogenic substances that have an effect on the environment as well as those subjected to the smoke. Cigarette smoking is the leading cause of preventable death in the United States. *(Anette Linnea Rasmussen, 2008, used under license from Shutterstock, Inc.)*

aesthetic standpoint, cigarette smoke stains the skin, teeth, and fingernails of the smoker.

CARCINOGENIC COMPONENTS OF CIGARETTE SMOKE

In order to understand the impact of cigarette smoking on public health fully, one must analyze the components of cigarette smoke. The chemicals found in cigarette smoke include toxic chemicals, irritants, and carcinogens (cancer-causing agents).

The carcinogenic compounds that are found in cigarette smoke include benzene, formaldehyde, toluene, chromium, and vinyl chloride. Benzene is a colorless, sweet-smelling, flammable liquid with the formula C_6H_6 that has been classified as carcinogenic to humans. Cigarette smoke is the leading cause of indoor exposure to benzene. Drowsiness, dizziness, rapid heart rate, and confusion are short-term side effects of benzene exposure.

The strong-smelling gas formaldehyde is a leading cause of indoor air pollution. Formaldehyde has been proven to cause cancer in laboratory animals and is considered to be a carcinogen in humans. The health effects of formaldehyde exposure include irritation to eyes, nose, and throat; asthma attacks; respiratory damage; wheezing; and watery eyes. Formaldehyde in pressed furniture and cigarette smoke are the leading causes of formaldehyde pollution indoors.

Toluene is a sweet-smelling liquid with the formula $C_6H_5CH_3$ that is used in the production of benzene. Health effects of toluene exposure include nervous system effects such as trembling, dizziness, confusion, and tremors. Toluene also causes liver and kidney damage.

Chromium, a metallic element with atomic number 24 and atomic mass of 52, is a human carcinogen that also causes eye and skin irritation. Vinyl chloride, a colorless gas that can cause dizziness upon exposure, has also been established as a human carcinogen.

COMPOUNDS AFFECTING THE RESPIRATORY TRACT

Compounds in cigarette smoke that affect the respiratory tract include ammonia, carbon monoxide, carbon dioxide, chloroform, hydrogen cyanide, nitric oxide, phenol, and tar. Ammonia is a strong-smelling gas, has the chemical formula NH_3, and is a severe respiratory tract irritant.

Carbon monoxide (CO) is a toxic, odorless, tasteless, colorless gas that is a by-product of incomplete combustion of organic fuel sources. Exposure to carbon monoxide causes disorientation, dizziness, fatigue, and nausea. When carbon monoxide exposure is high, death can result. Carbon monoxide competes with molecular oxygen (O_2) for binding sites on hemoglobin, causing decreased blood oxygen content. Burning cigarettes also release carbon dioxide (CO_2), a normal component of air. High local concentrations of carbon dioxide decrease the local concentration of oxygen available for respiration.

Chloroform is a colorless gas with a sweet smell. Low exposure to chloroform causes dizziness, fatigue, and headaches. High exposure to chloroform leads to liver and kidney damage. Experimental evidence in mice has shown negative effects on reproduction, causing miscarriage and birth defects.

Three other potentially dangerous components of cigarette smoke are hydrogen cyanide, nitric oxide, and phenol. Hydrogen cyanide (HCN), a compound used as a toxic gas during World War II, forms when cigarettes burn. In low doses, hydrogen cyanide causes headaches, weakness, nausea, and dizziness. Higher dosages of hydrogen cyanide can cause seizures, irregular heartbeat, fainting, and even death. Nitric oxide, NO, is a gas that is an important signaling molecule in the body, but high levels can be detrimental to anyone who has heart disease, asthma, or other respiratory diseases. Exposure to phenol, C_6H_5OH, leads to respiratory irritation and liver damage.

Tar, the sticky black substance that is left over as a cigarette burns, is responsible for the smoke-related staining of fingernails, skin, and teeth. When tar reaches the lungs, it forms a coating on the respiratory surfaces that inhibits their function. This can lead to respiratory disorders such as emphysema, bronchitis, chronic obstructive pulmonary disease (COPD), and lung cancer.

SECONDHAND SMOKE

The risks of cigarette smoke exposure extend beyond the smokers themselves. Anyone who is in the vicinity of a smoker within a vehicle, a restaurant, a home, or a workplace is exposed to the toxic chemicals from the secondhand smoke, the smoke exhaled by the smoker, as well as from the end of the lit cigarette. In 2006 the U.S. surgeon general stated that any level of secondhand smoke was dangerous to health. According to a 2004 study (National Survey on Environmental Management of Asthma and Children's Exposure to Environmental Tobacco Smoke) by the U.S. Environmental Protection Agency, almost 3 million children (11 percent of the children in the United States) age six and younger live with secondhand smoke in their home more than four days per week.

Restaurant smoking and nonsmoking sections began as a method of protecting nonsmokers from the irritating smoke of cigarettes. In 2008 more than 35 states had some level of public smoking ban in place. These bans are based on the scientific findings that secondhand smoke is as dangerous to bystanders' health as if they smoked. Secondhand smoke is also known as environmental toxin smoke. More than 40,000 deaths are attributed to nonsmokers' exposure to secondhand smoke. Many smokers believe that this series of bans robs them of their rights over their own body; however, these laws are based on the belief that in a public place the employees and other patrons of the business have the right to breathe clean, toxin-free air.

See also AIR POLLUTION (OUTDOOR/INDOOR).

FURTHER READING

Harris, Jeffrey. "Cigarette Smoke Components and Disease: Cigarette Smoke Is More Than a Triad of Tar, Nicotine, and Carbon Monoxide." In Monograph 7 of the Tobacco Control Monograph Series published by the National Cancer Institute of the U.S. National Institutes of Health. Available online. URL: http://cancercontrol.cancer.gov/tcrb/monographs/7/m7_5.pdf. Last updated on April 7, 2008.

citric acid cycle

citric acid cycle Cells can extract energy from organic molecules by oxidation in order to convert the energy to a form more readily usable by the cell. The citric acid cycle, also named the Krebs cycle after its discoverer, Sir Hans Adolf Krebs, or the tricarboxylic acid cycle, is a cyclical biochemical pathway for the oxidative metabolism of carbohydrates, amino acids, and fatty acids. These molecules act as fuel for the cell because they have relatively strong reduction potentials, meaning they are able to give up electrons to other molecules, such as the electron carriers oxidized nicotinamide adenine dinucleotide (NAD^+) and flavin adenine dinucleotide (FAD). After accepting electrons from other molecules, the reduced forms of the carriers, $NAD^+ + H^+$ (often simply noted as NADH) and $FADH_2$, take the electrons to the respiratory chain, also called the electron transport chain, where adenine triphosphate (ATP), the cell's principal free energy donor, is made. Molecules in the electron transport chain transfer the electrons between each other, moving them in succession from molecules with higher (more negative) reduction potentials to molecules with slightly lower (meaning less negative or more positive) reduction potentials. In aerobic respiration, oxygen, a strong oxidizing agent (meaning it has a positive reduction potential), serves as the final electron acceptor.

REACTIONS OF THE CITRIC ACID CYCLE

Aerobic respiration consists of three main steps: the glycolytic pathway, the citric acid cycle, and movement down the electron transport chain, which ends in ATP synthesis. Glycolysis can proceed in the presence or absence of oxygen, but the citric acid cycle only occurs when oxygen is available. Oxygen does not directly participate, but the regeneration of oxidized NAD^+ and FAD, a process that does require molecular oxygen, is necessary for the citric acid cycle to operate. The process of glycolysis partially oxidizes a molecule of glucose and converts the sugar to two molecules of pyruvate. Before entering the citric acid cycle, the two molecules of pyruvate must be converted to acetyl coenzyme A (acetyl CoA) by a process called oxidative decarboxylation. In eukaryotic cells, this occurs in the mitochondrial matrix, and in prokaryotic cells, it occurs in the cytoplasm. A large multienzyme complex, the pyruvate dehydrogenase complex, contains a component called pyruvate dehydrogenase that removes a carbon from pyruvate, a three-carbon molecule, with the assistance of the catalytic cofactor thiamine pyrophosphate (TPP), and converts it to a two-carbon compound that is temporarily attached first to the catalytic cofactor TPP, and then to another cofactor, lipoamide. A second component of the multienzyme complex, dihydrolipoyl transacetylase, transfers the two-carbon acetyl group to coenzyme A (CoA). During the last step carried out by the pyruvate dehydrogenase complex, dihydrolipoyl dehydrogenase regenerates the oxidized form of the cofactor lipoamide by transferring two electrons temporarily to an FAD group, and finally to NAD^+. The net reaction for this pyruvate decarboxylation is as follows:

$$\text{Pyruvate} + \text{CoA} + \text{NAD}^+ \rightarrow \text{acetyl CoA} + \text{CO}_2 + \text{NADH} + \text{H}^+$$

After the preliminary step, the activated acetyl component of acetyl-CoA can enter into the citric acid cycle. In summary, the two-carbon acetyl group combines with a four-carbon molecule of oxaloacetate to form the six-carbon molecule called citric acid, hence the name of the cycle. A series of seven more reactions follows, including decarboxylation steps in which carbon atoms are removed and coupled with oxidation-reduction reactions, during which electrons are removed. Along the way a high-energy phosphate bond is formed, generating a molecule of guanosine triphosphate (GTP). The regeneration of oxaloacetate allows the cycle to continue.

During the first step of the cycle, the enzyme citrate synthase transfers the acetyl group of acetyl-CoA to oxaloacetate, using one molecule of water and forming citrate.

$$\text{Oxaloacetate} + \text{acetyl CoA} + \text{H}_2\text{O} \rightarrow \text{citrate} + \text{CoASH} + \text{H}^+$$

In order to undergo oxidative decarboxylation, the goal of the citric acid cycle, isomerization must occur next. Isomers are chemical compounds that contain the same number of atoms and the same elements but possess different properties; isomerization is the process of rearranging the structure, and therefore properties, of a chemical compound.

Rearrangement of a hydrogen atom and a hydroxyl group occurs by the removal of a water molecule from citrate and the subsequent return of the atoms

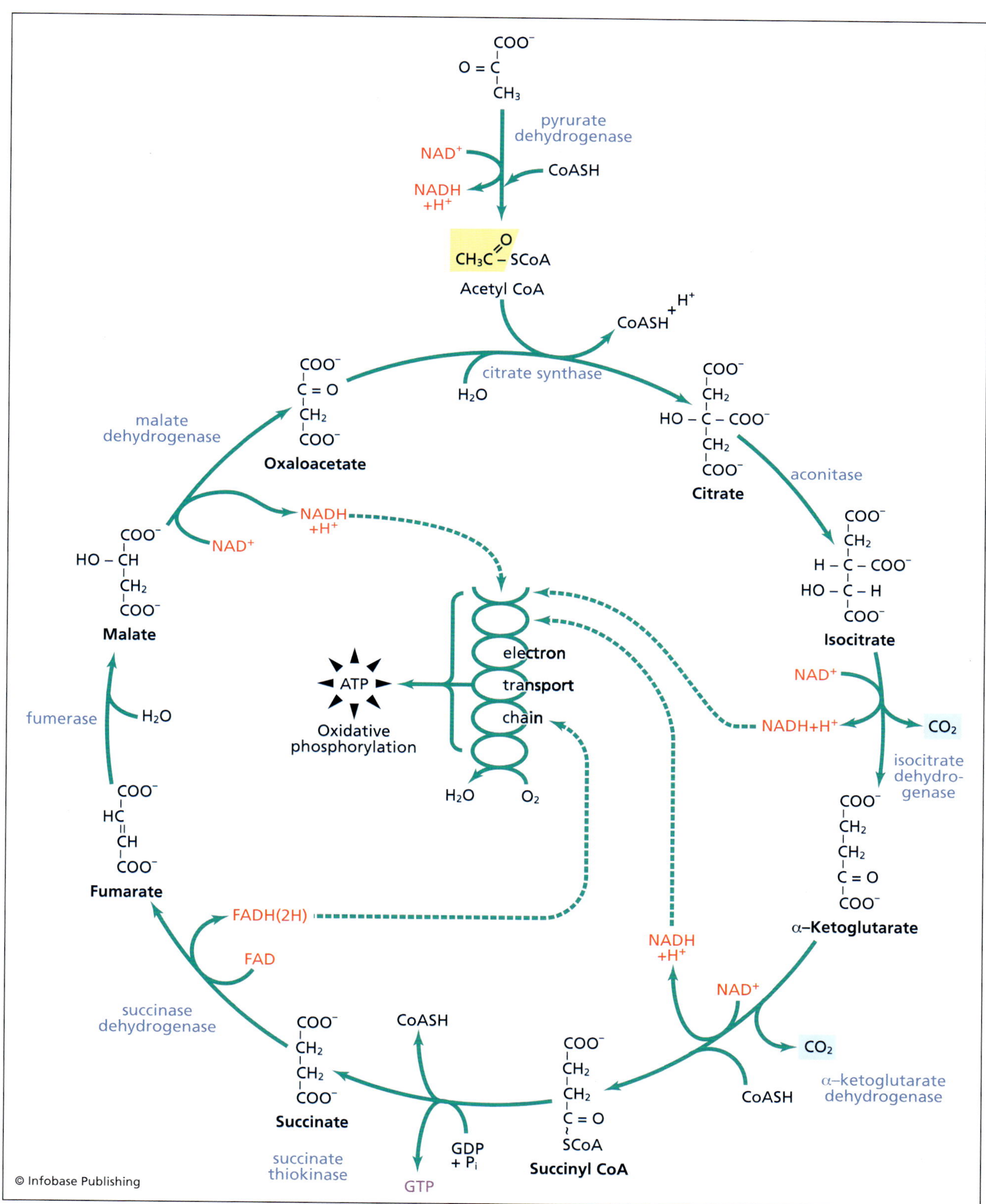

The citric acid cycle is a cyclical biochemical pathway that carries out the final common steps of oxidative metabolism.

to different positions, forming isocitrate. The enzyme that performs this isomerization is called aconitase, because the compound cis-aconitate is an intermediate of the process.

$$\text{citrate} \rightarrow \text{isocitrate}$$

The next step is a reversible oxidation-reduction reaction. Isocitrate dehydrogenase removes a carbon and two oxygen atoms, forming carbon dioxide (CO_2), while transferring two electrons from isocitrate to NAD^+, forming NADH.

$$\text{Isocitrate} + NAD^+ \rightarrow \alpha\text{–ketoglutarate} + CO_2 + NADH + H^+$$

Another oxidative decarboxylation occurs, changing α-ketoglutarate into succinyl-CoA. Again, NAD^+ is reduced to NADH, and CO_2 is released in a reaction catalyzed by the α-ketoglutarate dehydrogenase complex, a multienzyme complex that works in a manner similar to the pyruvate dehydrogenase complex.

$$\alpha\text{–ketoglutarate} + NAD^+ + \text{CoASH} \rightarrow \text{succinyl CoA} + CO_2 + NADH + H^+$$

The next reaction is reversible and is the only one in the citric acid cycle that directly yields a high-energy bond. Succinyl CoA synthetase breaks the bond holding coenzyme A and uses the energy released to form a high-energy bond between an inorganic phosphate and guanosine diphosphate (GDP). This creates GTP, which can be used in protein synthesis, in signal transduction, or in donating a phosphate to ADP to make ATP. Succinate and coenzyme A are released.

$$\text{succinyl CoA} + P_i + \text{GDP} \rightarrow \text{succinate} + \text{GTP} + \text{CoASH}$$

The goal of the last few steps of the citric acid cycle is to regenerate oxaloacetate by oxidizing succinate. Succinate dehydrogenase oxidizes succinate to fumarate, adding the removed electrons to FAD to create $FADH_2$. Though NAD^+ is used as the electron carrier in the other three oxidation-reduction steps of the citric acid cycle, the change in free energy at this step is not great enough to reduce NAD^+. Succinate dehydrogenase is the only enzyme of the citric acid cycle that is embedded into the inner membrane of the mitochondria (only in eukaryotes) and is directly linked to the electron transport chain.

$$\text{Succinate} + \text{FAD} \rightarrow \text{fumarate} + FADH_2$$

Fumarase catalyzes the hydration of fumarate to create malate.

$$\text{fumarate} + H_2O \rightarrow \text{malate}$$

Finally, oxaloacetate is regenerated in a reaction catalyzed by malate dehydrogenase. Another NAD^+ is reduced to $NADH + H^+$.

$$\text{malate} + NAD^+ \rightarrow \text{oxaloacetate} + NADH + H^+$$

The net reaction of the citric acid cycle can be summarized as follows:

$$\text{Acetyl CoA} + 3NAD^+ + FAD + GDP + P_i + 2H_2O \rightarrow 2CO_2 + 3NADH + FADH_2 + GTP + 3H^+ + \text{CoASH}$$

When performing stoichiometric calculations for the oxidation of one molecule of glucose, one must remember that one molecule of glucose yields two pyruvates, both of which are decarboxylated to form acetyl CoA. So two acetyl CoA molecules can enter the citric acid cycle for each molecule of glucose that is oxidized as part of aerobic respiration.

While the citric acid cycle is a major step in the oxidation of glucose to obtain energy for ATP synthesis, the citric acid cycle contains many key metabolic intermediates that can feed into other biochemical pathways depending on the cell's needs. The cell can synthesize many amino acids using α-ketoglutarate or oxaloacetate as precursors. If the cellular levels of oxaloacetate decrease because it is used to synthesize other organic molecules, the citric acid cycle will stop. In mammals, minimal levels of oxaloacetate are maintained by adding a carbon (from CO_2) to pyruvate, a process called carboxylation. This reaction requires the input of energy and water and is catalyzed by pyruvate carboxylase.

$$\text{pyruvate} + CO_2 + ATP + H_2O \rightarrow \text{oxaloacetate} + ADP + P_i + 2H^+$$

REGULATION OF THE CITRIC ACID CYCLE

Animals are unable to convert acetyl CoA into glucose; thus, the preliminary decarboxylation of pyruvate to form acetyl CoA is irreversible. In animal cells, the acetyl CoA can either feed into the citric acid cycle or be utilized in fatty acid synthesis. Because this step is so important, its activity is under strict control. The activity of the enzyme pyruvate dehydrogenase is regulated by phosphorylation, the reversible addition of a phosphate group. When a kinase adds a phosphate group, the enzyme is deactivated, and when a phosphatase removes the phosphate group, the enzyme is activated. A high-energy charge in the cell—meaning the ratios of NADH/NAD$^+$, acetyl CoA/CoA, and ATP/ADP are high—suggests that the energy needs of the cell are being met, and it would be wasteful to continue burning the fuel stores. Shutting off the cycle conserves energy if NADH, acetyl CoA, and ATP are abundant; thus presence of sufficient quantities of these molecules promotes the phosphorylation of the pyruvate dehydrogenase enzyme, deactivating it until the levels decrease enough, allowing the phosphate to be removed, and reactivating the enzyme.

A major regulatory step within the citric acid cycle is the addition of acetyl CoA to oxaloacetate to generate citrate. ATP inhibits the enzyme citrate synthase; thus, when it is in excess, it prevents the entry of more acetyl CoA into the cycle, halting the generation of more ATP. The activity of isocitrate dehydrogenase, the enzyme that catalyzes the decarboxylation of isocitrate to make α-ketoglutarate, is also regulated by ATP levels in addition to NADH levels. When the energy charge of the cell is high, as when plenty of NADH and ATP is available, the activity of isocitrate dehydrogenase is inhibited. Succinyl CoA and NADH, two end products of the reaction catalyzed by α-ketoglutarate dehydrogenase, feed back to inhibit this enzyme's activity. In summary, when the cell has sufficient energy available in the form of ATP and NADH and an abundance of biosynthetic intermediates, several mechanisms cooperate to slow the rate of the citric acid cycle.

Plants and bacteria have similar yet alternative mechanisms for controlling the citric acid cycle and related biochemical pathways.

See also BIOCHEMISTRY; CARBOHYDRATES; ELECTRON TRANSPORT SYSTEM; ENERGY AND WORK; ENZYMES; GLYCOLYSIS; METABOLISM; OXIDATION-REDUCTION REACTIONS.

FURTHER READING

Berg, Jeremy M., John L. Tymoczko, and Lubert Stryer. *Biochemistry,* 6th ed. New York: W. H Freeman, 2006.

Krebs, H. A., and W. A. Johnson. "The Role of the Citric Acid Cycle in Intermediate Metabolism in Animal Tissues." *Enzymologia* 4 (1937): 148–156.

classical physics The term *classical,* as it refers to physics, is used in two different ways. In common discourse, speaking historically, it stands in opposition to *modern* and means the physics of the 19th century. More specifically, this refers to physics that does not involve relativity or quantum considerations. So classical physics is the physics of situations in which speeds are not too high, masses and energies neither too great nor too little, and distances and durations neither too small nor too large.

Among physicists, however, classical physics stands in contrast only to quantum physics, while relativistic effects are included in it. Quantum physics is characterized by the Planck constant, $h = 6.62606876 \times 10^{-34}$ joule·second (J·s), which is nature's elementary unit of a quantity called action. While nature is fundamentally quantum in character and quantum physics seems to give an excellent understanding of it, classical physics can be valid to high accuracy for situations involving actions that are large compared to h, what is referred to as the classical domain. This includes everyday phenomena, of course, and furthermore ranges over lengths, durations, and masses that are not too small, that are in general larger, say, than the atomic and molecular scales.

CHARACTERISTICS OF CLASSICAL PHYSICS

Some of the characteristics of classical physics, which, to a good approximation, reflect properties of nature in the classical domain, are definiteness, determinism, continuity, locality, wave-particle distinction, and particle distinguishability.

Definiteness. Every physical quantity that is relevant to a physical system possesses a definite value at every time. Even if we do not know the value of such a quantity, it nevertheless *has* a definite value. For instance, speed is a physical quantity that is relevant to every body. So at any instant every body possesses a certain value for its speed. And that is true whether one has measured and knows its speed's value or not.

Determinism. The state of a physical system at any instant uniquely determines its state at any time in the future and, accordingly, is uniquely determined by its state at any time in the past. It follows that a full understanding of the working of nature and a complete knowledge of a system's state at any time allow, in principle, prediction of the system's state at any time in the future and retrodiction of its state at any time in the past. Limitations on such predictability and retrodictability, even when the laws of nature are known, are the result of incomplete knowledge of the system's state. In practice, such incompleteness is unavoidable, since all measurements possess only finite precision. (And even if infinite precision were hypothetically possible, inherent quantum uncertainties exist in the values of physical variables.) For example, determinism allows astronomers, using their present-day data on the positions and velocities of the heavenly bodies, to tell us what the sky will look like at any time in the future and how it appeared at any time in the past.

Continuity. Processes occur in a continuous manner, so that physical quantities that vary in time do so continuously—no sudden jumps. Also, the ranges of allowed values of physical quantities are continuous: a physical quantity can possess a certain value, then it can also take values as close to that value as one might like. For instance, if a system can possess 3.14159 joules (J) of energy, then in general the same system is also capable of possessing 3.14158 J of energy.

Locality. What happens at any location is independent of what happens at any other location, unless some influence propagates from one location to the other, and the propagation of influences occurs only in a local manner. What this means is that influences do not instantaneously span finite distances. (Such a hypothetical effect is known as action at a distance.) Rather, the situation at any location directly affects only the situations at immediately adjacent locations, which in turn directly affect the situations at *their* immediately adjacent locations, and so on. In this way an influence propagates at finite speed from its cause to its effect. For example, if one wiggles an electric charge in Washington, D.C., that action will in principle affect all electric charges everywhere in the universe. However, the effect is not instantaneous: rather, it propagates at the speed of light in vacuum, which is approximately 3.00×10^8 meters per second (m/s).

Wave-particle distinction. A wave is a propagating disturbance, possibly characterized, for instance, by frequency and wavelength. It possesses spatial extent, so is not a localized entity. A particle, on the other hand, is localized. It is characterized by mass, velocity, position, energy, and so forth. Waves and particles are distinct from each other and bear no relation to each other.

Particle distinguishability. All particles are distinguishable from each other, at least in principle. For example, this hydrogen atom and that one can always be distinguished one from the other.

In fact, though, nature does not possess any of these characteristics. Quantum theory recognizes that and takes into account nature's uncertainty, indeterminism, discontinuity and discreteness, nonlocality, wave-particle duality, and particle indistinguishability. For example, physical quantities in general possess intrinsic uncertainty in their value over and beyond any uncertainty in knowledge of the value due to inherently imprecise measurement. And as another example, there is no property of a hydrogen atom by means of which two hydrogen atoms in the same state can be distinguished. Nevertheless, classical physics can and does deal with nature to a high degree of accuracy, as long as it is not pushed beyond nature's classical domain.

HISTORY OF CLASSICAL PHYSICS

Although technologies and understanding of nature developed in many parts of the world throughout history, the approach to the comprehension of nature that led to science as we know it today originated in ancient Greece. Thus, the ancient Greek scientists and philosophers are the forefathers of classical physics. One name that stands out is that of Aristotle, who lived in the 300s B.C.E. He observed nature and developed ways of thinking logically about it. With the decline of ancient Greek civilization, physics entered a period of stagnation that extended through the Dark Ages and into the 15th century, when physics picked up again and commenced developing at an ever-increasing rate. The end of the 19th century brought this development to a close, to be followed by the era of modern physics, which started in the early 20th century and continues today.

Some of the events and people in the development of classical physics are these:

- The Italian physicist and astronomer Galileo Galilei, commonly referred to as Galileo, whose life straddled the 16th and 17th centuries, investigated motion and demonstrated the validity of the heliocentric model of the solar system. He is attributed with dropping objects from the Leaning Tower of Pisa in order to study free fall.
- The German astronomer and mathematician Johannes Kepler, a contemporary of Galileo's, found the three laws of planetary motion named for him.
- Isaac Newton, the English physicist and mathematician who lived in the 17th and 18th centuries, discovered the three laws of motion and law of gravitation named for him, among his other accomplishments, which included inventing calculus.
- The English chemist and physicist Michael Faraday, most of whose life was in the 19th century, made important discoveries in electromagnetism.
- James Clerk Maxwell, the 19th-century Scottish physicist, formulated the laws of classical electromagnetism.

By the end of the 19th century, physicists were viewing their field as essentially complete, clear, and beautiful, but with two "small" exceptions, referred to as two clouds in the clear sky. One problem, called the blackbody spectrum problem, was that the energy distribution among the frequencies of the electromagnetic radiation emitted by a radiating body was not explainable by the physics of the time. The other "cloud" was the negative result of the Michelson-Morley experiment, which seemed to indicate that the "ether" that scientists assumed formed the material medium carrying electromagnetic waves did not, in fact, exist. As it happened, both "clouds" served as seeds from which the two revolutions of modern physics sprouted in the early 20th century. The solution of the blackbody spectrum problem by the German physicist Max Planck led to the development of quantum theory. And the negative result of the Michelson-Morley experiment was explained by the German-Swiss-American physicist Albert Einstein's special theory of relativity.

See also DUALITY OF NATURE; EINSTEIN, ALBERT; ELECTROMAGNETIC WAVES; ELECTROMAGNETISM; ENERGY AND WORK; FARADAY, MICHAEL; GALILEI, GALILEO; GENERAL RELATIVITY; GRAVITY; KEPLER, JOHANNES; MASS; MAXWELL, JAMES CLERK; MOTION; NEWTON, SIR ISAAC; PLANCK, MAX; QUANTUM MECHANICS; SPECIAL RELATIVITY; SPEED AND VELOCITY; WAVES.

FURTHER READING

Bernal, J. D. *A History of Classical Physics: From Antiquity to the Quantum.* New York: Barnes & Noble Books, 1997.

clock A clock is any device for determining the time, in the sense of "What time is it?" or for measuring time intervals. It does so by producing and counting (or allowing an observer to count) precisely equal time units. A pendulum swings with cycles of equal time durations. The most modern precision clocks, which are called atomic clocks, count time units that are based on the frequency of the electromagnetic radiation emitted by atoms, when the latter undergo transitions between certain states of different energy.

The fundamental problem of timekeeping is assuring that the time units generated by a clock are indeed equal. How does one know that a pendulum's swings are of equal duration or that the oscillations of an electromagnetic wave emitted by an atom have constant period? Clearly the use of a clock to assure the equality of another clock's time units involves circular reasoning and does not resolve the matter.

In order to resolve the matter, one is forced to make the assumption that a clock's time units are equal. One does not do that for just any device, but only for those for which there is good reason to think it is true. In the example of the pendulum, at the end of each swing cycle, the bob returns to momentary rest at its initial position, whereupon it starts a new cycle. As far as one can tell, it always starts each cycle from precisely the same state. Why, then, should it not continue through each cycle in precisely the same manner and take precisely the same time interval to complete each cycle?

If one assumes that is indeed the case and the pendulum's time units are equal, one obtains a simple and reasonable description of nature based on this equality. Convinced that nature should be describable in simple terms, one then feels justified in this assumption. For more sophisticated clocks, such as atomic clocks, the reasoning is similar, but correspondingly more sophisticated. One ends up defining the cycles of electromagnetic radiation as possessing equal duration. Thus, the International System of Units (SI) defines the time interval of one second as the duration of 9,192,631,770 periods of microwave electromagnetic radiation corresponding to the transition between two hyperfine levels in the ground state of the atom of the isotope cesium 133.

MODERN TYPES OF CLOCKS

In ordinary conversation one uses the term *watch* for a timekeeping device that is carried on one's person, while *clock* indicates any other device for keeping time. However, the present discussion will use the term *clock* for all devices that allow one to tell time or measure time intervals.

Modern clocks are of two general types, analog and digital. Analog clocks indicate the time by means of some physical quantity, almost always by an angle. The most common such clocks possess a circular face, divided into 12 equal central angles that are usually marked from 1 to 12, and at least two rotating arms, called "hands." The 12 angles are subdivided into five equal subangles each, giving 60 equal subangles. The hour hand makes a full rotation every 12 hours and indicates hours by pointing to or near the numbers. The minute hand completes a rotation every hour and indicates minutes past the whole hour by pointing to or near the subangle marks. Such a clock might also be equipped with a second hand, which rotates once every minute and uses the 60 subangle marks to show seconds. Analog clocks operate by means of a mechanical process.

For analog clocks the fundamental unit of time is supplied by a mechanical device designed to allow the clock's mechanism to advance in steps at equal time intervals. All analog watches and some other clocks make use for this purpose of an "escapement" mechanism, which oscillates at a fixed frequency and keeps the clocks showing a reading reasonably close to what it should be. Other clocks use a pendulum for this purpose, since for small amplitudes of swing, a pendulum's period is independent of its amplitude. The most accurate analog clocks are pendulum clocks. The power needed for the operation of analog clocks might derive from a wound spring, batteries, the electric wall socket, or even slowly falling weights, as in certain pendulum clocks.

Digital clocks indicate time in a digital format using a liquid crystal display (LCD) or light emitting diodes (LEDs). Such clocks operate by means of electronic circuits. Their source of fundamental time interval is commonly a quartz crystal oscillator, usually operating at 2^{15} hertz (Hz) = 32.768 kilohertz (kHz). The clocks' circuits contain frequency dividing and counting circuits. The necessary electric power is supplied by batteries or the wall socket, and in some cases batteries are used for backup in case the external electric power fails. Some digital clocks receive radio timing signals from broadcast stations operated by the National Institute of Standards and Technology (NIST) and maintain accuracy in that way.

HISTORY OF CLOCKS

The earliest clocks were sundials and water clocks, with the former probably predating the latter, although their antiquity precludes knowing when they were first used. Sundials make use of the position of a shadow cast by the sun to indicate the time of day. They are obviously limited by not operating when the sun is covered by clouds and at night. Water clocks are based on the controlled flow of water. In the simplest cases, the time required for a bowl of water to drain serves as the time unit. But more complex devices were developed over time, some involving intricate mechanisms.

Well-made pendulum clocks can be very accurate and have long served as time standards. A long-case pendulum clock, such as this, is called a grandfather clock. This one was made for Marie Antoinette in 1787 and is located in the Petit Trianon, Versailles, France. *(Réunion des Musées Nationaux/Art Resource, NY)*

An ancient time-measuring device, the hourglass, is designed so the sand will fall through the narrow neck from the upper chamber to the lower in one hour. *(Scott Rothstein, 2008, used under license from Shutterstock, Inc.)*

Other, similar early types of clocks included the hourglass and burning candles. The hourglass is based on the flow of fine sand from the device's upper chamber to its lower through a tiny orifice. When the sand runs out, one unit of time, such as an hour, has passed, and the clock can be inverted for measuring the next time unit. A uniformly made candle can be marked in equal distances along its length to make it into a clock. As the top edge of a burning candle sinks from one mark to the next, an equal unit of time passes.

Over time, more accurate and more convenient clocks were developed. The 15th century saw the development of clocks powered by wound springs, based on various escapement mechanisms. The invention of the pendulum clock in the 17th century introduced a great improvement in accuracy, which continued to improve as electricity, and then electronics, were introduced into clocks.

ATOMIC CLOCKS

The most accurate clocks in use today are based on the stable and reproducible properties of atoms. Atoms of some isotope are made to deexcite from a certain energy level to a certain lower one, thereby emitting electromagnetic waves of a certain frequency. Electronic circuits count the cycles of the radiation and in this way determine a time interval of a second as a certain number of cycles. The most accurate of atomic clocks today uses the isotope cesium 133. As mentioned earlier, SI defines the time interval of one second as the duration of 9,192,631,770 periods of microwave electromagnetic radiation corresponding to the transition between two hyperfine levels in the ground state of the atom of the isotope cesium 133. NIST operates a cesium clock and with it supplies the time standard for the United States. The uncertainty of this clock is rated at 5×10^{-16}, which means it would neither gain nor lose a second in more than 60 million years.

This cesium 133 clock at the National Institute of Standards and Technology (NIST) at Boulder, Colorado, provides the standard unit of time. The second is defined as the duration of 9,192,631,770 periods of electromagnetic radiation emitted during the transition between the two hyperfine levels of the ground state of the atom of the isotope cesium 133. *(Copyright 2005 Geoffrey Wheeler Photography/ National Institute of Standards and Technology)*

See also ATOMIC STRUCTURE; ELECTROMAGNETIC WAVES; ENERGY AND WORK; ISOTOPES; MEASUREMENT; TIME.

FURTHER READING

Bruton, Eric. *The History of Clocks and Watches.* Secaucus, N.J.: Chartwell Books, 2002.

North, John. *God's Clockmaker: Richard of Wallingford and the Invention of Time.* London: Hambledon and London, 2007.

"The Official U.S. Time." Available online. URL: http://www.time.gov. Accessed July 10, 2008.

"Time and Frequency Division, Physics Laboratory, National Institute of Standards and Technology." Available online. URL: http://tf.nist.gov. Accessed July 25, 2008.

Yoder, Joella Gerstmeyer. *Unrolling Time: Christiaan Huygens and the Mathematization of Nature.* New York: Cambridge University Press, 2004.

colligative properties A colligative property is a property of a solution that is dependent on the total number of solute particles in the solution and is independent of the nature of the solute. A solution, also called a homogeneous mixture, is a type of matter that is made of different substances physically mixed together. Solutions have four types of colligative properties: vapor pressure, boiling point elevation, freezing point depression, and osmotic pressure.

A solution consists of a solute and a solvent. The solute is the component that is being dissolved, and the solvent is the component that is doing the dissolving. Depending on the amount of solute that is dissolved in the solution, the solution will have a high or low concentration. A concentrated solution is one that has a high ratio of solute molecules to solvent molecules. A dilute solution has a low ratio of solute-to-solvent molecules. As an illustration, consider orange juice. Grocery stores often sell orange juice as a concentrated frozen solution that the consumer dilutes by adding water during preparation. The terms *concentrated* and *dilute* describe the solute-to-solvent ratios qualitatively, relating the solution properties before and after preparation. The frozen concentrate contains a large amount of solute (orange juice) particles compared to solvent (water), and the addition of water before drinking dilutes the solution. The total number of solute particles in the pitcher has not changed; only the volume of the solvent changed. As the total volume of solvent increased and the solute molecules stayed constant, the concentrated solution became more dilute. The higher the concentration of solute, the more effect the solute has on the colligative properties of the solution.

Vapor pressure is one colligative property. Pressure is a measure of the force exerted per unit area. Atmospheric pressure is the force exerted by the atmosphere pushing down on the surface of matter. A barometer is a piece of equipment that measures atmospheric pressure by its action on liquid mercury contained within the instrument. Units of atmospheric pressure include the SI unit pascal and the more convenient unit kilopascal (kPa). Based on the construction of standard barometers, the column of mercury (Hg) that the atmospheric pressure can support is reported in millimeters of mercury (mm Hg). Standard atmospheric pressure at sea level is 760 mm Hg, and the unit of atmosphere was developed to simplify these measurements. Standard atmospheric pressure is defined as one atmosphere (atm). The relationship among these three atmospheric pressure units is demonstrated by the following equation:

$$760 \text{ mm Hg} = 1 \text{ atm} = 101.3 \text{ kPa}$$

Vapor pressure results when the molecules of a liquid have enough kinetic energy to escape the surface of the liquid, pushing up against the atmospheric pressure that is pushing down onto the surface of the liquid. The pressure exerted by a volatile solvent (one that vaporizes) is given by Raoult's law, formulated by the French chemist François M. Raoult. To summarize, the law states that a more concentrated solution will have a lower vapor pressure than a dilute solution, as demonstrated by the following equation:

$$P_{soln} = \chi_{solvent} \cdot P^0_{solvent}$$

Where P_{soln} stands for the vapor pressure of the solution, $\chi_{solvent}$ stands for the mole fraction of the solvent, and $P^0_{solvent}$ stands for the vapor pressure of the pure solvent. The mole fraction of the solvent is given by dividing the number of moles of solvent by the total number of moles in the solution.

$$\chi_{solvent} = \frac{\text{moles solvent}}{\text{moles solution}}$$

In order to be released from the surface, the solvent molecules need to overcome the attractive forces between the other solvent molecules, such as the intermolecular forces holding polar molecules together—hydrogen bonds. Because of hydrogen bonding, a molecule of a polar solvent needs to have more energy to break the hydrogen bonds with its neighbors and push upward from the surface of the liquid. Thus, a polar solvent will have a lower vapor pressure than a nonpolar solvent. When a nonvolatile solute (one that does not vaporize) is added to a pure liquid solvent, the solute molecules interact with the molecules of the solvent and take up space in the solution surface. If the surface of the liquid has a lower number of solvent molecules, then fewer solvent molecules can escape the surface of the mixture from a given surface area, and therefore the amount of solvent molecules that are pushing up against the atmosphere is decreased, leading to a lowered vapor pressure. The only requirement is that the solute particles need to be nonvolatile, meaning that they cannot evaporate themselves, or else they would contribute to an increased vapor pressure of the substance.

The fact that a solution has a lower vapor pressure than the solvent it contains means that the amount of pressure pushing up from the surface of the solution is automatically less than the pressure pushing up from the pure solvent. The boiling point of a liquid is the temperature at which the vapor pressure pushing up is equal to the atmospheric pressure pushing down on the surface of the liquid, a condition exhibited by the formation of bubbles rising up from the solution. Lowering the vapor pressure by adding solute to a mixture requires the input of more energy to reach the boiling point; thus, a higher temperature must be reached in order to have enough solvent molecules escape the surface of the liquid to create a vapor pressure high enough to equal the atmospheric pressure—to boil. The lower vapor pressure of polar substances compared to nonpolar substances makes the polar substances boil at higher temperatures even in the absence of any solute, and when solute is added to a polar solvent, the boiling point of the solution is even higher.

This colligative property of boiling point elevation can be demonstrated by the boiling of water to cook pasta. When making pasta, the directions instruct the cook to add salt to the water before heating. This creates a salt solution that will, by definition of boiling point elevation, have a higher boiling point than pure water. Most people do not think about why the salt is added to the boiling water. The cook generally does not take the temperature of the boiling water; one simply looks for the bubbles of a rolling boil. Measuring the temperature at which pure water or salt water begins to a boil would show that the boiling saltwater solution is hotter than the boiling pure water. Cooking the pasta at a higher temperature improves its texture and decreases cooking time.

Another colligative property is freezing point depression. The freezing point of a substance is the temperature at which the molecules of a substance cool off enough to slow down and interact with neighboring molecules to create a solid. The attractive forces of the substance allow the molecules to interact when they lose enough kinetic energy. When a solute is added to a pure solvent, the solute molecules interfere with the interactions among the neighboring molecules, preventing them from changing the physical state from a liquid to a solid. Freezing will occur at a lower temperature—in other words, the solution has a lower freezing point than the pure solvent. The depression of the freezing point of solutions is useful in controlling ice on road surfaces in the wintertime. Safe driving requires that the water on the surface of the road remain a liquid. When the temperature decreases, the water molecules lose kinetic energy and are attracted to one another by hydrogen bonding to create a solid sheet of ice over the surface of the road. Spreading a solute over the road surface will keep the water on the roads in the liquid state, even at temperatures below 32°F (0°C). The type of solute particle determines the temperature at which the solution freezes. Typical road salt decreases the temperature at which ice will form on the road. Road crews also use liquid deicers that are sprayed onto the road to remove the possibility that the salt will be pushed off the road surface. Freezing point depression is also valuable in making homemade ice cream. The salt solution that is placed in the outside container of an ice cream maker makes the solution colder than regular ice or water. This allows for the ice cream to freeze in the container. Freezing point depression also is used when deicing aircraft prior to takeoff in the winter.

The freezing point of a solution is lower than that of pure water. This freezing point depression allows solutions to be used to remove ice from roads, sidewalks, and aircraft. In this photo, type IV deicing fluid is being applied to an aircraft in Cleveland prior to takeoff. *(NASA)*

Colligative properties of solutions explain how the same solution can exhibit two different properties in the car radiator. Ethylene glycol is the solute present in antifreeze/coolant used in automobile radiators. A coolant solution of ethylene glycol mixed with water prevents the coolant from boiling (overheating) in the summertime because the solution has an elevated boiling point compared to pure water. The presence of the ethylene glycol in the solution also prevents the water from freezing in the winter because the solution has a lowered freezing point compared to pure water. The optimal ratio of ethylene glycol to water depends on the season.

Osmotic pressure also depends on the number of solute molecules in solution; thus, it is a colligative property. Osmosis is the diffusion of water across a semipermeable membrane. When water travels across the membrane, it will move from the area of higher water concentration to the area of lower water concentration. In other words, water will move from the area of low solute concentration to an area of higher solute concentration. The more solute particles that are present on one side of the semipermeable membrane, the more water molecules will move across the membrane into the area of high solute concentration, causing the water volume on that side of the membrane to increase. The diffusion of water across a membrane is critical to the survival of plant and animal cells. The osmotic pressure (the force exerted by two solutions separated by a membrane on the basis of their different solute concentration) must not be too high or too low, or the cell will shrink or swell. If an animal cell is placed into a solution that has a higher solute concentration than inside the cell (a hypertonic solution), then the cell will lose water by osmosis, causing the cell to dehydrate. If an animal cell is placed into a solution that has a lower solute concentration than the inside of the cell (a hypotonic solution), then the water on the outside the cell will move into the cell, causing it to swell and possibly burst. An isotonic solution is a solution that has the same total solute concentration as the inside of the cell. Solutions that are used on and inside the body usually contain the same solute concentration as that inside the cell so that osmotic pressure differences between the cells and the solution do not cause cell damage by shrinking or swelling—for example, eye drops or saline solution used to irrigate an open wound. Osmotic pressure is dependent on the concentration of the solute and is given by the following formula:

$$\pi = MRT$$

where π is the osmotic pressure in atmospheres, M is the molarity of the solute, R is the ideal gas constant (8.31 L × kPa/K × mol), and T is the temperature in kelvins.

Calculation of the effect of a colligative property such as boiling point elevation and freezing point depression requires one to know the concentration of solute, which can be determined by calculating the molality (m) of the solution. The concentration unit of molality is given by the following formula:

$$m = \frac{\text{moles solute}}{\text{kilogram solvent}}$$

The formula for the effect of the solute on the freezing point depression is

$$\Delta T_f = iK_f m$$

where the change in the freezing point temperature is ΔT_f, and i, the van't Hoff factor, is the number of solute particles in solution after the substance dissociates. K_f is the molal freezing point depression constant (1.86°C/m), and m is molality. The formula for the change in boiling point of a solution is given by

$$\Delta T_b = iK_b m$$

where ΔT_b is the change in boiling point, i is the van't Hoff factor, K_b is the molal boiling point elevation constant (0.52°C/m), and m is the molality. The van't Hoff factor (the number of solute particles) of the solute is dependent on whether the solute is an ionic compound that when dissolved in water will separate into the individual ions. For example, NaCl will dissociate into Na^+ and Cl^-, giving two particles for every NaCl particle added to a solution. $MgCl_2$ will dissociate to give one Mg^{2+} ions and two Cl^- ions. Sand or sucrose, on the other hand, will not dissociate in solution and will give one mole of particles for every mole of sucrose or sand. The change in each of the four types of colligative properties—boiling point elevation, freezing point depression, osmotic pressure, and vapor pressure—is twice as much for NaCl and three times as much for $MgCl_2$ as for either sand or sucrose. Since the number of particles increases (because of dissociation) when an ionic compound is placed in solution, and since colligative properties are those affected by the number of particles in a solution, ionic compounds have a greater effect on colligative properties than nonionic compounds.

See also CONCENTRATION; IONIC COMPOUNDS; PRESSURE; STATES OF MATTER.

FURTHER READING

Zumdahl, Steven S., and Susan A. Zumdahl. *Chemistry*, 7th ed. Boston: Houghton-Mifflin, 2007.

Compton effect This effect, named for the 20th-century American physicist Arthur Compton, is the increase in wavelength of electromagnetic radiation, specifically X-rays or gamma rays, when it is scattered by the more loosely bound electrons of matter. It is also called Compton scattering. This is the most common mode of interaction of high-energy X-rays with atomic nuclei in organisms and so is important for radiation therapy, such as in the treatment of cancer. In the early 20th century, the Compton effect served as evidence, in addition to the photoelectric effect, of the particle character of light, as it was successfully explained by Compton as a collision between a photon and an electron at rest (compared to the energy of the photon, the electron can be considered to be at rest). Classical electromagnetism cannot explain the effect and would have the scattered radiation possess the same wavelength as the incident radiation. The photon's initial energy and linear momentum are shared with the electron, which recoils and is ejected from the atom. Thus the incident photon loses energy, resulting in a decrease of frequency of the electromagnetic radiation and a concomitant increase of its wavelength.

The analysis of the scattering involves a straightforward application of conservation of energy and momentum. The incident radiation can be scattered at any angle from its initial direction. That angle is called the scattering angle. In terms of the scattering angle θ, the wavelength decrease, called the Compton shift, is given by

$$\lambda_2 - \lambda_1 = \lambda_C(1 - \cos\theta)$$

where λ_2 and λ_1 are the wavelengths, in meters (m), of the scattered and incident radiation, respectively. Thus, the Compton shift is zero for unscattered radiation ($\theta = 0°$, $\cos\theta = 1$), increases with the scattering angle, and is greatest for back-scattered radiation ($\theta = 180°$, $\cos\theta = -1$), when $\lambda_2 - \lambda_1 = 2\lambda_C$. The following quantity

$$\lambda_C = \frac{h}{m_{ec}}$$

is known as the Compton wavelength of the electron, where h, m_e, and c denote, respectively, the Planck constant, the mass of the electron, and the speed of light in vacuum. The values of these constants are $h = 6.62606876 \times 10^{-34}$ joule·second (J·s), $m_e = 9.1093897 \times 10^{-31}$ kilogram (kg), and $c = 2.99792458 \times 10^8$ meters per second (m/s). Thus $\lambda_C = 2.4263 \times 10^{-12}$ m. The angle at which the electron is ejected ϕ, as measured from the direction of the incident radiation, is given by

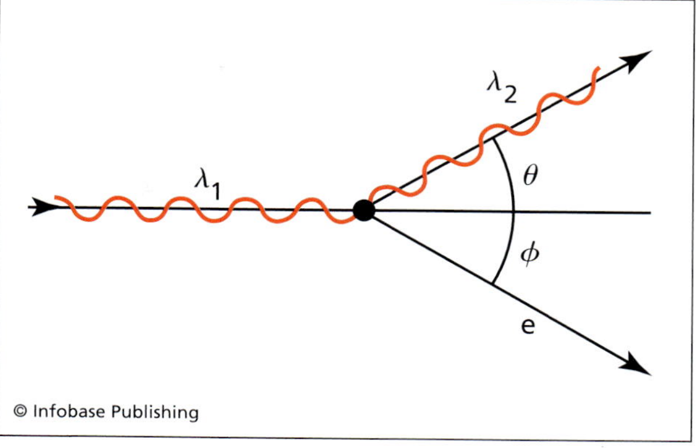

When X-rays or gamma rays scatter from atomic electrons (considered to be at rest), the wavelength of the scattered radiation is greater than that of the incident radiation. The incident and scattered radiation are labeled by their wavelengths, λ_1 and λ_2, respectively. The initially resting electron is represented by the black circle. The directions of the scattered radiation and ejected electron e make angles θ and ϕ, respectively, with the direction of incident radiation. For any scattering angle θ the increase in wavelength, the Compton shift, is given by $\lambda_2 - \lambda_1 = \lambda_C(1 - \cos\theta)$ and the electron's direction by $\cot\phi = (1 + \lambda_C/\lambda_1)\tan(\theta/2)$. λ_C denotes the Compton wavelength of the electron, given by $\lambda_C = h/(m_e c)$, where h, m_e, and c denote, respectively, the Planck constant, the electron mass, and the speed of light in vacuum.

$$\cot\phi = \left(1 + \frac{\lambda_c}{\lambda_1}\right)\tan\left(\frac{\theta}{2}\right)$$

See also ATOMIC STRUCTURE; CONSERVATION LAWS; DUALITY OF NATURE; ELECTROMAGNETIC WAVES; ENERGY AND WORK; MOMENTUM AND COLLISIONS; PHOTOELECTRIC EFFECT; QUANTUM MECHANICS.

FURTHER READING

Serway, Raymond A., Clement J. Moses, and Curt A. Moyer. *Modern Physics*, 3rd ed. Belmont, Calif.: Thomson Brooks/Cole, 2004.

computational chemistry Computational chemistry is a theoretical branch of chemistry that utilizes mathematical and computer modeling techniques to describe the structure of atoms, molecules, and the ways chemical reactions take place. A fundamental principle in chemistry and biology is that the structure of a molecule correlates to its function. Determining molecular structures gives insight into how chemical reactions proceed as well as other molecular interactions. Computational chemists develop three-dimensional molecular models utilizing electron locations to determine the most stable conformation of a compound, such characteristics as active sites of an enzyme, and interactions between molecules to describe reaction mechanisms and binding sites. Calculations of the energy levels of transition states that often do not exist long enough to be measured experimentally can be carried out theoretically using methods common to computational chemistry. Computer programs predict structures of intermediates formed when a reaction takes place, the structures of molecules, and their charge distributions once they are modeled.

Computational chemistry is a theoretical science, meaning it complements and describes that which is observed in the experimental world. Much of the field of computational chemistry lies in understanding the field of quantum mechanics. Scientists discovered in the early 1900s that understanding quantum mechanical laws would theoretically describe everything in the field of chemistry including atomic structure, chemical bonding, and the physical properties of elements and compounds. Calculations of atomic structure and determinations of electron locations using quantum mechanical laws are incredibly complicated and difficult to use in a practical manner. Implementing computer programs that approximate these quantum mechanical equations supplies excellent structural information for atoms and molecules that is more convenient to use in an experimental setting. Computational chemistry is generally thought of as the development of theoretical models for the direct application to another field or area of study.

The primary experimental tool of a computational chemist is a computer. Since the early 1960s, the development of reliable and quick computing capabilities and reliable programming for computers has allowed computational chemistry to become a viable part of the field of chemistry.

Education in the field of computational chemistry includes a bachelor's of science degree in chemistry that includes all of the required courses as a regular chemistry major with the addition of advanced math and computer science courses. A master's of science degree or a Ph.D. can also be granted in computational chemistry. Students who undertake an advanced degree in computational chemistry must have a strong background in chemistry, including such areas as quantum mechanics and atomic structures, and molecular modeling, as well as computer programming, and mathematics, including statistics. Scientists who receive their degree in this field will have the ability to supplement nearly every area of experimentation in chemistry and biochemistry. Computational chemists manage databases and computer systems in academic, research, and industrial settings. Jobs as computational chemists are found in large computing facilities and academic research laboratories. On the job, computational chemists perform molecular and macromolecular simulations, carry out quantum mechanical calculations, and utilize bioinformatics and database programming. Employers hiring computational chemists expect applicants to be familiar with tools for graphic display of data, molecular graphics, database programming, and applications graphics.

Computational chemists provide molecular modeling support to drug discovery programs in pharmaceutical research, protein structure predictions, and information on reaction mechanisms for organic chemists. Active sites of enzymes can be modeled to determine the function, substrate interactions, as well as mechanisms to block enzyme action.

Classical quantum mechanics utilizes the Schrödinger equation to determine the structure of individual atoms as well as atoms within molecules. These types of calculations become increasingly difficult as the molecule gets larger. As scientists began studying the function of larger molecules, the development of simplified calculation methods became increasingly important. Pioneering work in the development of computational chemistry methods led to the awarding of the 1998 Nobel Prize in chemistry to the Austrian-born chemist Walter Kohn and the English chemist John Pople. Their work formed the basis of modern computational chemistry. Kohn, who worked at the University of California, Santa Barbara, was responsible for developing the density functional theory, which simplified the theoretical modeling of all molecules by showing that it was not necessary to model every single electron, but rather the electron densities that allowed for reliable and predictable computational models. This assumption radically simplified the calculations required in order to model large molecules. Pople, who worked at Northwestern University in Evanston, Illinois, focused on the development of a computer molecular modeling program that was reliable and functional. Pople's goal was to develop methods to calculate quantum mechanical values correctly. His program, GAUSSIAN, used a series of increasingly specific calculations to approximate the quantum mechanical values. Today, chemists all around the world use versions of the GAUSSIAN program to study a multitude of chemical problems.

TECHNIQUES

Computational chemistry uses numerous techniques to analyze molecular structures. In order to calculate these complex quantum mechanical values theoretically, some assumptions must be made. Each type of computer programming has a different focus and different assumptions that go into the calculations. Types of computational methods include molecular mechanics, molecular dynamics, Hartree-Fock, post–Hartree-Fock, semiempirical, density functional theory, and valence bond theory.

Molecular mechanics involves the study and prediction of energy associated with molecular conformations. In order to determine which conformation of a molecule has the lowest energy and is therefore the most stable, one must investigate several different types of atomic interactions. The types of energy included in a molecular conformation include bond-stretching energy, bond-bending energy, torsion or strain energy, and energy of atoms not involved in bonds (van der Waals interactions). Molecular mechanics is more accessible than quantum mechanics, and it simplifies the conformational calculations by making several assumptions. Rather than studying the location of every electron, the nucleus and electrons are grouped. These atomlike particles are spherical and have theoretical charges. The calculations used to develop these conformational models preassign specific types of atoms to specific types of interactions that in turn determine the location of the atom and the energy of the molecule.

Molecular dynamics is a computer method that simulates the conformations and interactions of molecules by integrating the equation of motion of each of the atoms. The laws of classical mechanics, including Newton's laws, are followed when determining structures, energies, and interactions of atoms.

Hartree-Fock theory is the fundamental principle of molecular orbital theory, in which each electron is located in its own orbital and is not dependent on interactions with other electrons in the atom. Hartree-Fock theory serves as a reference point for more complicated theoretical methods that are used to approximate the values found using the quantum mechanical Schrödinger equation. Hartee-Fock method averages the repulsions between electrons and then uses this information to determine the conformation of the atoms and molecules.

Post–Hartree-Fock method is an improvement of the Hartree-Fock method. This computational method includes the true repulsion values caused by the electrons' interactions rather than averaging the repulsions. This process makes the post–Hartree-Fock useful for studying excited states and molecular dissociation reactions. Post–Hartree-Fock calculations require more complicated and expensive programs than Hartree-Fock calculations, but they are much more accurate calculations.

Semiempirical methods utilize experimental data to correct their calculations of simplified versions of Hartree-Fock methods. The semiempirical method incorporates many types of calculations, each of which has specific assumptions about atoms built into its calculations. Two examples of calculations including assumptions are the neglect of differential diatomic overlap (NDDO) and zero differential overlap (ZDO), in which integrals involving two-electron charge distributions are not included. Semiempirical methods are grouped in several ways. Original methods such as NDDO, introduced by John Pople, tried to satisfy theoretical results rather than experimental results. These methods are not directly used today but provided a starting point for development of later methods.

The Indian-born theoretical chemist Michael Dewar did much of his work at the University of Texas, where he developed methods to fit experimental values for geometry, heat of formation, dipole moments, and ionization potentials. The programs he developed that utilize these methods, MOPAC and AMPAC, allow for calculations of quantum mechanical structures of large molecules. As computer speed has increased, it is now possible to use these programs to determine the structure of molecules containing nearly 10,000 atoms.

When calculating the geometry of lanthanide complexes, methods such as Sparkle/AM1 are able to predict conformations of coordination compounds. Other programs known as ZINDO and SINDO were developed to predict atomic spectra.

Density functional theory, developed by Kohn, dramatically simplified calculations in electronic structure. Hartree-Fock methods and other original methods revolved around the interactions of many electrons. The density functional theory replaces the interactions of each individual electron with an electron density, producing a conceptually and practically simpler calculation.

Valence bond theory is the application of computer-based methods to valence bond theory rather than molecular orbital theory, on which all of the methods described are based. Computer programs that model modern valence bond theory include such programs as the General Atomic and Molecular Electronic Structure System (GAMESS) and GAUSSIAN, developed by Pople.

See also ATOMIC STRUCTURE; BONDING THEORIES; CHEMICAL REACTIONS; REPRESENTING STRUCTURES/MOLECULAR MODELS; SCHRÖDINGER, ERWIN.

FURTHER READING

Jensen, Frank. *Introduction to Computational Chemistry.* West Sussex, England: John Wiley & Sons, 1999.

concentration *Concentration* describes the quantity of solute dissolved in a given amount of solvent to form a solution. All matter can be divided into two categories, substances or mixtures, based on the ability to separate the components of the matter from one another by physical means. Matter that cannot be physically separated is called a substance, a category that includes elements and compounds. The matter is a mixture if it can be separated by physical means. Mixtures can be subdivided further on the basis of whether or not the mixture is the same throughout (homogeneous mixtures or solutions) or has variable composition (heterogeneous mixtures).

A solution is formed when there is a homogeneous physical association between two substances. A solution contains a solute, the substance being dissolved, and a solvent, the substance in which the solute is dissolved. The amount of solute dissolved in a solution is critical to the action of that solution. Solvent molecules dissolve solute through a process known as solvation, in which solvent molecules surround solute molecules and prevent them from interacting with other solute molecules. Solutions form between solutes and solvents on the basis of the energetics of the interactions. In order for a solvent to dissolve a solute, the attraction of the solvent molecules for the solute molecules must be stronger than both the solvent-solvent attraction and the solute-solute attraction. The saying "Like dissolves like" describes the likelihood of a solute's being dissolved in a certain solvent. Polar molecules are able to dissolve other polar molecules, and nonpolar molecules are able to dissolve other nonpolar molecules, but polar molecules (such as water) are not able to dissolve nonpolar molecules (such as oil). The attraction of the water molecules for the oil molecules is not strong enough to overcome the water-water attractions and the hydrophobic interactions between oil molecules.

Solutions can be made up of components in any phase. Although one generally thinks of a solution as a solid dissolved in a liquid solvent, such as salt in water, a gas or a liquid solute can also dissolve in a solvent. Carbonated beverages are an example of a gas dissolved in a liquid. Carbon dioxide is injected into the can under pressure, and when the pressure is released by opening the lid, the gas escapes, making a hissing noise, and the beverage eventually goes "flat." The component that is considered the solute is the one that is present in the lower amount, and the solvent is present in a higher amount. This is very clear when the solute is a solid and the solvent is a liquid, such as sodium chloride and water, but when the solute and solvent are both liquids, it becomes more difficult to distinguish solute from solvent. For instance, when alcohol is dissolved in water, if there is more water than alcohol, then the alcohol is the solute and the water is the solvent. If there is more alcohol than water in the solution, then the water is the solute and the alcohol is the solvent. Alcoholic beverages such as beer and wine contain less than 10 percent alcohol; thus, the alcohol is the solute, and water is the solvent. Rubbing alcohol, however, is generally 70 percent alcohol and 30 percent water; that means that, in this case, the water is the solute, and alcohol is the solvent.

The quantity of solute that can dissolve in a given amount of solvent at a given temperature is

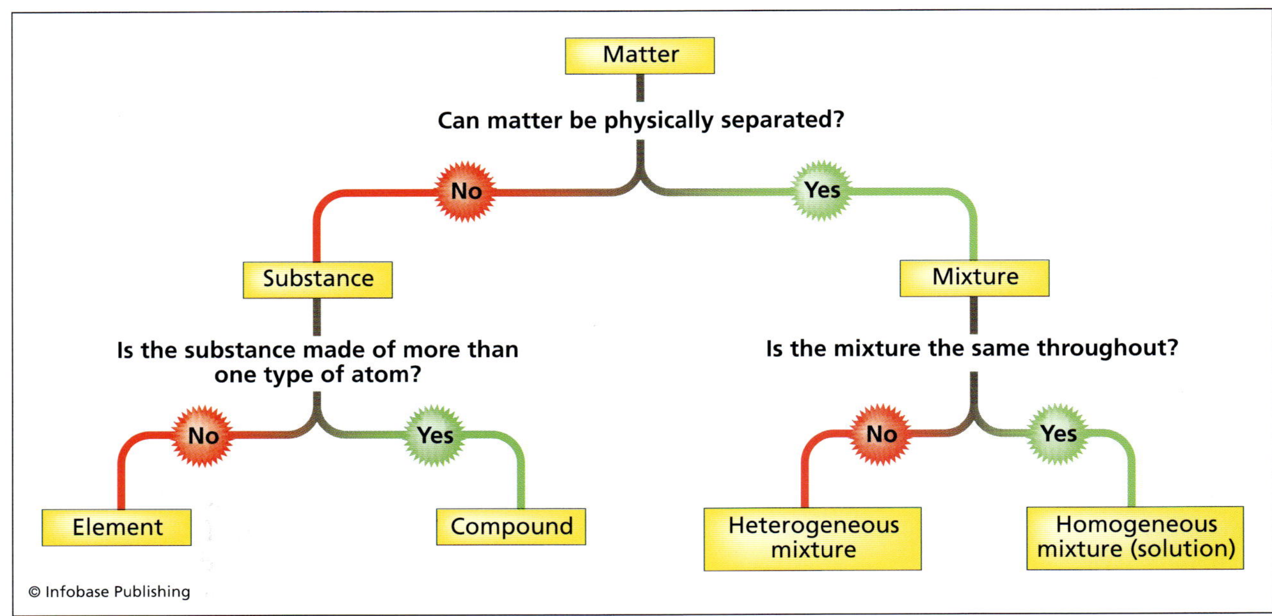

Matter can be classified as physically associated (mixtures) and chemically bonded (substances) matter.

known as its solubility. Solubility curves, graphs representing the quantity of a particular type of solute that can be dissolved in a given solvent over a range of temperatures, allow for the prediction of solution formation. When studying a liquid solute or solvent, solubility curves demonstrate the effect of temperature in determining how much solute will dissolve. The condensing and expanding of water at different temperatures affect the amount of solute that will fit in a given volume of solvent.

REPRESENTING CONCENTRATION

Concentration of a solution can be described qualitatively and quantitatively. Words such as *concentrated* and *dilute* qualitatively describe a solution. *Concentrated* implies that the solution contains "a lot" of the solute relative to a dilute solution that contains a smaller amount of solute. These terms are common in everyday household supplies. Frozen orange juice, laundry detergent, and cleaning supplies are examples of substances that can be purchased in a concentrated form. In order to use these products at the appropriate concentration, they need to be diluted by adding more water (solvent). The dilution process does not change the overall number of moles of component in the solution; it only changes the ratio of solute to solvent in the solution and increases the volume of the solution.

Although not quantitative, classifying solutions as saturated, unsaturated, or supersaturated gives more information than simply calling them dilute or concentrated. A saturated solution contains the maximal quantity of solute that a given amount of solvent can hold at a given temperature. If more solute is added to the solution, it will simply fall to the bottom of the container and not be dissolved.

An unsaturated solution is a solution that is capable of dissolving more solute. The act of dissolving sugar in iced tea demonstrates the saturated/unsaturated distinction. When one adds sugar to the tea, it will dissolve, because the solution is unsaturated, but if one continues to add sugar beyond the solubility of the sugar at that temperature, it will not dissolve and will simply sit at the bottom of the cup. Since a solution must be a homogeneous mixture, the sugar at the bottom is not considered part of the solution. The tea in the lower portion of the glass will contain much more sugar (most of it not dissolved) than the tea at the top of the glass, making the final sips very concentrated relative to the rest of the glass.

A special situation exists when one is able to dissolve more solute in a given amount of solvent at a given temperature than should theoretically be possible. This type of solution is known as a supersaturated solution and only occurs under very specific conditions. Heating the solution will allow dissolution of the solute, and upon cooling, the solute stays in solution as long as there is no surface upon which crystals can form. Once one solute particle falls out of solution, then the rest will quickly follow, leading to a saturated solution with many solute particles at the bottom of the vessel.

As useful as the qualitative descriptions are, some situations require a quantitative description of the amount of solute dissolved in a given solution, in other words, the concentration. Many methods exist to represent concentration, including percentage by mass, percentage by volume, molarity, molality, mole fraction, and normality. The method of representing the concentration does not change the overall amount of solute dissolved in that solution. The selection of a method often depends on the simplicity of measuring different factors. For example, measuring the mass of the solvent might be easier than measuring the volume of the solution in a particular experiment, leading to the use of *molality* as a concentration term rather than *molarity*.

Percentage by mass simply relates the mass of the solute to the volume of the solution in percentage form. The formula for percentage by mass is

$$\text{Percentage by mass} = [(\text{mass of solute})/(\text{volume of solution})] \times 100$$

A 10 percent salt solution contains 10 g of salt for every 100 mL of solution. Percentage by mass is represented by (% m/v).

Percentage by volume is a useful concentration unit when both the solute and solvent are liquids. The formula is

$$\text{Percentage by volume} = [(\text{volume of solute})/(\text{volume of solution})] \times 100$$

Percentage by volume is represented by % v/v.

Molarity is another common method for indicating concentration. Defined as the number of moles of solute divided by the total volume of the entire solution, the following formula represents molarity:

$$\text{Molarity} = (\text{moles solute})/(\text{liter solution})$$

In order to calculate molarity, one must first know the number of moles of solute and the final volume of the solution—not the volume of solvent. The difference between these two factors is the amount of the total solution volume that is due to solute. In order to calculate the number of moles of solute if it is not provided, one must first calculate the given mass of solute into moles using dimensional analysis (unit conversion). This can be done by dividing the given number of grams of solute by the molar mass of that solute. The last step involves converting the volume of solution into liters if it is not already in that form. The following example represents a typical molarity calculation:

1. 35 g of NaCl is dissolved in a final solution volume of 350 mL. First calculate the molar mass of NaCl. Na is present at 23 g/mol, and Cl at 35.5 g/mol. This gives a total molar mass for NaCl of 58.5 g/mol.
2. Convert 35 g of NaCl into moles by dividing by 58.5 g/mol. Moles of NaCl = 0.60 moles.
3. Convert the volume into liters. 350 mL = 0.350 L.
4. Plug these values for the moles of NaCl and the volume of the solution into the molarity formula.

$$\text{Molarity} = (0.60 \text{ moles})/(0.350 \text{ L}) = 1.7 \text{ M NaCl solution}$$

The molarity of different solutions can then be compared. Regardless of the volume, it is possible to determine the strength of various solutions on the basis of their molarity. Preparing dilutions of stock solutions is also possible by converting the molarities. The conversions between volumes of solutions with different concentrations can easily be calculated by the following formula:

$$M_1V_1 = M_2V_2$$

where M_1 is the initial molarity, V_1 is the initial volume, M_2 represents the final molarity, and V_2 represents the final volume. Pharmacies, hospitals, and even veterinarian offices dilute stock solutions in order to prepare the correct dose of medication in an appropriate volume for administration to the patient.

Molality involves the number of moles of solute divided by the number of kilograms of solvent. The total volume of solution is not computed. Calculation of molality (abbreviated with a lowercase, italicized m) requires the following formula:

$$\text{molality} = (\text{moles solute})/(\text{kilogram of solvent})$$

Mole fraction represents the amount of moles of one component in the mixture relative to the total number of moles of all components in the mixture. The symbol for mole fraction is × with the component being studied written as a subscript. The symbol for the mole fraction of NaCl in aqueous solution would be represented X_{NaCl}. The rest of the solution is made up of the moles of water shown as X_{H2O}. The mole fractions of all the components in the mixture must equal 100 percent. In a mixture that contains three components, the mole fraction of component A is given by

$$X_A = (\text{moles of A})/(\text{moles of A} + \text{moles of B} + \text{moles of C})$$

When studying certain substances, specifically acids and bases, it is necessary to represent the amount of equivalents of the component in the solution. An equivalent is the amount of acid or base that will give one mole of H^+ or OH^- in solution. Normality represents the number of equivalents per liter.

$$\text{N} = \text{equivalents/liter}$$

Comparing 6 M HCl, 6 M H_2SO_4 and 6 M H_3PO_4, their respective number of equivalents is 1, 2, and 3. Since molarity equals the moles solute/liter solution, the normality of each of these acids can be calculated by multiplying the number of equivalents by the number of moles. The normality of each of these solutions would be

$$6 \text{ M HCl} = 6 \text{ N HCl}$$

$$6 \text{ M } H_2SO_4 = 12 \text{ N } H_2SO_4$$

$$6 \text{ M } H_3PO_4 = 18 \text{ N } H_3PO_4$$

When solute particles are present at a low concentration in the solution, it is not useful to talk about parts out of 100 as with percentage by mass or volume. The molarity and molality numbers would likewise be too low to be useful. Scientists have developed a system relating parts of solute molecules/10^6 solvent molecules. This system is known as parts per million. Often the concentration of solute is even lower, and units such as parts per billion (ppb) will be used. Air quality and water quality studies often use these two methods of reporting concentration of pollutants or toxic chemicals that must be maintained within acceptable levels.

See also ACIDS AND BASES; PH/POH.

FURTHER READING:

Silerberg, Martin. *Chemistry: The Molecular Nature of Matter and Change*, 4th ed. New York: McGraw Hill, 2006.

conservation laws When a physical quantity maintains a constant value in an isolated system as the system evolves in time, the physical quantity is said to be conserved. When that is the case for all systems, or even for only some classes of systems, a conservation law is said to apply to the physical quantity under

According to the law of conservation of energy, the energy that one uses cannot simply be plucked from thin air but must be the result of conversion of energy from some source. The photograph shows an array of solar panels that convert radiant energy from the Sun to electricity. *(Tobias Machhaus, 2008, used under license from Shutterstock, Inc.)*

consideration. The best known of all conservation laws is the law of conservation of energy, stating that for all isolated systems, the total energy of the system remains constant in value over time (although it might change its form). This law is indeed considered to be valid for all systems, at least as far as is presently known. Note that for conservation of energy to hold, the system must be isolated from its surroundings, so that no energy flows into or out of the system and no work is done on the system by its surroundings or on the surroundings by the system.

As an example of conservation of energy, consider an ideal, frictionless pendulum. With no friction in the pivot or with the air, once set into motion, such a pendulum would continue swinging forever. Setting the pendulum into motion involves performing work on it, which endows the pendulum with a certain amount of energy. As the pendulum freely swings, it retains its original energy as its total energy, but its energy continuously changes form between kinetic and potential energies. When the pendulum is at the end of its swing and is momentarily at rest, all of its energy exists in the form of potential energy, whose quantity equals the original energy. When the pendulum is at its lowest point, its equilibrium point, it is moving at its fastest and all its energy has the form of kinetic energy, whose amount equals the original energy. At any intermediate position, the pendulum has both kinetic and potential energies, and their sum equals the original energy.

For another example of energy conservation, consider calorimetry. To be specific, let a number of objects at different temperatures be put into thermal contact while they are all well insulated from their surroundings. Eventually the collection of objects reaches thermal equilibrium, when they all possess the same final temperature. The analysis of the process is based on conservation of energy, whereby the total amount of thermal energy of the system does not change during the process. The amount of heat that is lost by the hotter objects as they cool down to their final temperature exactly equals the amount of heat that is gained by the colder objects as they warm up to their final temperature.

Additional conservation laws are understood to be valid for all isolated systems. One of them is the law of conservation of linear momentum: the total linear momentum of a system (i.e., the vector sum of the linear momenta of all bodies composing the system) remains constant over time. For conservation of linear momentum to hold, the system must be isolated to the extent that no net external force acts on it. Since linear momentum is a vector, its conservation implies the separate conservation of the x-component of the total linear momentum, of its y-component, and of its z-component. As an example of linear momentum conservation, consider a bomb at rest that explodes in a vacuum (to eliminate the force of air friction). The initial momentum of the bomb is zero. Immediately after the explosion, the bomb fragments all possess momenta of various magnitudes and in many different directions. Yet by conservation of linear momentum, the vector sum of all those momenta equals the bomb's initial momentum, which is zero. The analysis of collisions typically uses conservation of linear momentum.

Angular momentum is another quantity for which a conservation law holds in all isolated systems. As long as no net external torque acts on a system, the total angular momentum of the system, which is the vector sum of the angular momenta of all the bodies composing the system, does not change over time. As is the case for linear momentum, since angular momentum is also a vector, its conservation implies the separate conservation of each of its x-, y-, and z-components. For an example of this conservation, consider an exhibition skater spinning on the tip of her skate with her arms extended. Ignoring any frictional torques, the skater's angular momentum is conserved, and she could ideally continue spinning forever. Now the skater pulls in her arms, thus decreasing her moment of inertia. Note that this action of the skater is internal to her system and does not involve external torques. So angular momentum continues to be conserved. In order for that to happen, the decrease in moment of inertia must be compensated for by an increase in angular speed, since the magnitude of the skater's angular momentum equals the product of her moment of inertia and angular speed. As a result, the skater should spin faster. Indeed, that is what happens. This is one way in which spinning skaters, ballerinas, and athletes control their rate of spin.

The last conservation law to be discussed here is conservation of electric charge. In all isolated systems, the total electric charge, which is the algebraic sum of the electric charges of all the constituents of the system (with negative charge canceling positive charge), remains constant over time. Neutron decay can serve as an example. A free neutron decays into a proton, an electron, and an antineutrino, a process known as beta decay. The system initially consists solely of a neutron, which has no electric charge. The decay products are one proton, which carries a single elementary unit of positive charge $+e$; one electron, which is endowed with one elementary unit of negative charge $-e$; and one neutrino, which is electrically neutral. Here e denotes the elementary unit of electric charge, whose value is approximately 1.60×10^{-19} coulomb (C). The total electric charge of the system after the decay is, therefore,

$$(+e) + (-e) + 0 = 0$$

and indeed equals the total charge before the decay. Conservation of electric charge underlies Kirchhoff's junction rule, one of the theoretical tools for the analysis of electric circuits.

In addition to the conservation laws for energy, linear momentum, angular momentum, and electric charge, conservation laws exist also for a number of quantities that characterize elementary particles. Among these are quantities that are conserved in certain systems but not in others.

All the conservation laws mentioned so far are exact: under the appropriate conditions the quantities are conserved exactly. A well-known *approximate* conservation law is the conservation of mass. If it were an exact conservation law, the total mass of all the components of every isolated system would remain constant as the system evolves. However, as a result of Einstein's famous mass-energy relation $E = mc^2$, changes of internal energy of, say, an atom affect the atom's mass. In this formula E denotes energy, in joules (J); m is the mass, in kilograms (kg), that is equivalent to the energy E; and c denotes the speed of light in vacuum, whose value is approximately 3.00×10^8 meters per second (m/s). So as long as the energy changes are small compared to the energy equivalents of the masses involved, mass is approximately conserved. Such is the case for chemical and biochemical reactions. But in nuclear reactions, notably in fission and fusion reactions, such is not the case, and mass is not conserved in them even approximately.

Note that the meaning of the term *conservation* in physics is very different from its everyday meaning. Ordinarily *energy conservation* means saving energy, by using it sparingly and avoiding waste. But in physics the meaning is as explained here: for all isolated systems, the total energy of the system remains constant in value over time (although it might change its form). A connection *does* exist, however. Because of the law of conservation of energy, energy cannot be obtained from nothing. Its use is always at the cost of depleting resources. Thus it makes sense to use energy sparingly in order to allow energy resources to last as long as possible.

A relation exists between every exact conservation law and a symmetry of the laws of nature, also called an invariance principle. Conservation of energy relates to the invariance of the laws of nature in time, to the fact that the laws of nature do not change over time. Conservation of linear momentum relates to the invariance of the laws of nature in space, to the fact that the laws of nature are the same everywhere. Conservation of angular momentum relates to the invariance of the laws of nature in direction, to the fact that the laws of nature are the same in all directions. The invariance principle that is related to the conservation of electric charge possesses a more abstract character.

See also CALORIMETRY; CHEMICAL REACTIONS; EINSTEIN, ALBERT; ELECTRICITY; ENERGY AND WORK; FISSION; FORCE; FUSION; MASS; MOMENTUM AND COLLISIONS; NUCLEAR PHYSICS; PARTICLE PHYSICS; ROTATIONAL MOTION; SPECIAL RELATIVITY; SYMMETRY; VECTORS AND SCALARS.

FURTHER READING

Young, Hugh D., and Roger A. Freedman. *University Physics,* 12th ed. San Francisco: Addison Wesley, 2007.

coordinate system A grid in space that serves as a reference for the specification of points, or locations, by means of sets of numbers, called coordinates, is a coordinate system. The most commonly known among coordinate systems is the Cartesian coordinate system of three-dimensional space, which consists of three mutually perpendicular straight axes bearing equal length scales. Cartesian coordinate axes are usually denoted x, y, and z, and the coordinates of a point (x, y, z). The coordinates of a point are the readings on the respective axes of the perpendicular projections from the point to the axes. Other coordinate systems are often convenient. They include curvilinear systems and nonorthogonal systems. For three-dimensional space, for example, one often uses a spherical coordinate system or a cylindrical coordinate system. The former is related to the latitude-longitude coordinate system used for the surface of Earth.

The coordinates of a point in a spherical coordinate system are (r, θ, ϕ). They are related to the (x, y, z) coordinates by

$$x = r \sin \phi \cos \theta$$
$$y = r \sin \phi \sin \theta$$
$$z = r \cos \phi$$

and

$$r = \sqrt{x^2 + y^2 + z^2}$$
$$\theta = \tan^{-1} \frac{y}{x}$$
$$\phi = \cos^{-1} \frac{z}{\sqrt{x^2 + y^2 + z^2}}$$

In a cylindrical coordinate system, the coordinates of a point are (r, θ, z), which are related to (x, y, z) by

$$x = r \cos \theta$$
$$y = r \sin \theta$$
$$z = z$$

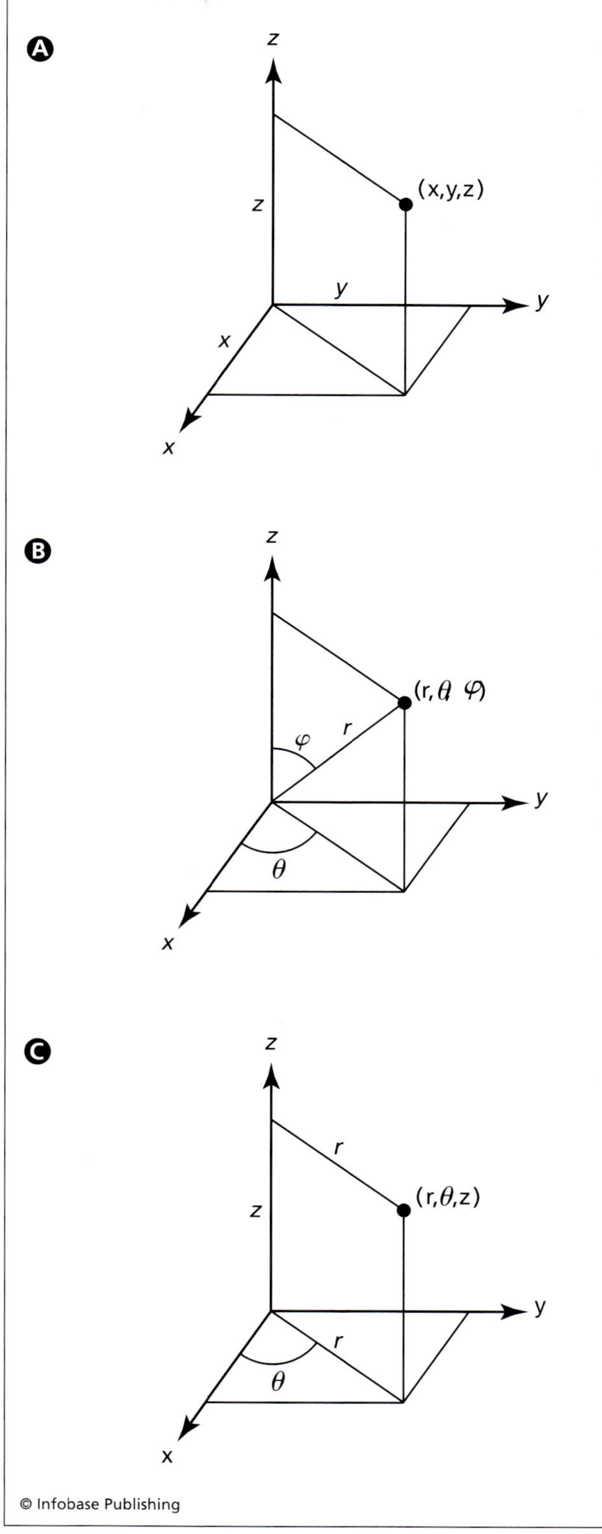

The coordinates of the same point in three-dimensional space are given in three coordinate systems. (a) Cartesian coordinate system, consisting of three mutually perpendicular axes: (x, y, z). (b) Spherical coordinate system: (r, θ, ϕ). (c) Cylindrical coordinate system: (r, θ, z).

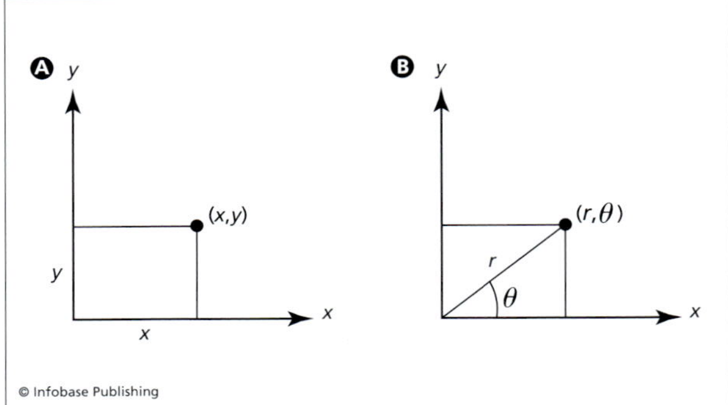

The coordinates of the same point in a plane, which is a two-dimensional space, are given in two coordinate systems. (a) Cartesian coordinate system, consisting of two mutually perpendicular axes: (x, y). (b) Polar coordinate system: (r, θ).

and

$$r = \sqrt{x^2 + y^2}$$

$$\theta = \tan^{-1} \frac{y}{x}$$

$$z = z$$

For a plane, which is a two-dimensional space, in addition to the common Cartesian coordinate system of x- and y-axes, one often uses a polar coordinate system. The coordinates of a point in a Cartesian coordinate system are (x, y) and in a polar system (r, θ). They are related by

$$x = r \cos \theta$$

$$y = r \sin \theta$$

and

$$r = \sqrt{x^2 + y^2}$$

$$\theta = \tan^{-1} \frac{y}{x}$$

Abstract spaces make use of coordinate systems as well. One such use is for four-dimensional space-time. In space-time coordinate systems, one of the four axes indicates time, while the other three serve for space. When all axes are straight and mutually perpendicular, the coordinate system is a Minkowskian coordinate system, especially useful for Albert Einstein's special theory of relativity. Einstein's general theory of relativity makes use of curvilinear coordinate systems called Riemannian coordinate systems. Abstract spaces might have no relation to space or time. States of a gas might be represented by points in an abstract two-dimensional volume-pressure space, specified with reference to an orthogonal coordinate system consisting of a volume axis and a pressure axis, each marked with an appropriate scale.

See also EINSTEIN, ALBERT; GENERAL RELATIVITY; PRESSURE; SPECIAL RELATIVITY; STATES OF MATTER.

cosmic microwave background Observations with radio telescopes reveal that Earth is immersed in electromagnetic radiation, called the cosmic microwave background, the cosmic background radiation, or the cosmic microwave background radiation. This radiation has no apparent source. Its spectrum is characteristic of that emitted by a blackbody at the absolute temperature of 2.73 kelvins (K). It is assumed that this radiation permeates all of space (and is not merely a fluke of some special situation of Earth). The cosmic microwave background is generally understood to be a remnant from an early stage in the evolution of the universe, when the universe was much hotter and denser, after a cosmic explosion from an extremely dense, hot state, called the big bang. As the universe expanded, the radiation cooled to its present temperature. The cosmic microwave background is commonly taken as evidence for the big bang. By this view, in observing the cosmic background radiation, one is "seeing" the actual glow of the big bang. Over some 14 billion years since the big bang, the wavelengths of that glow have stretched, as a result of the expansion of the universe, from the extremely short wavelengths characteristic of the big bang's very high temperature

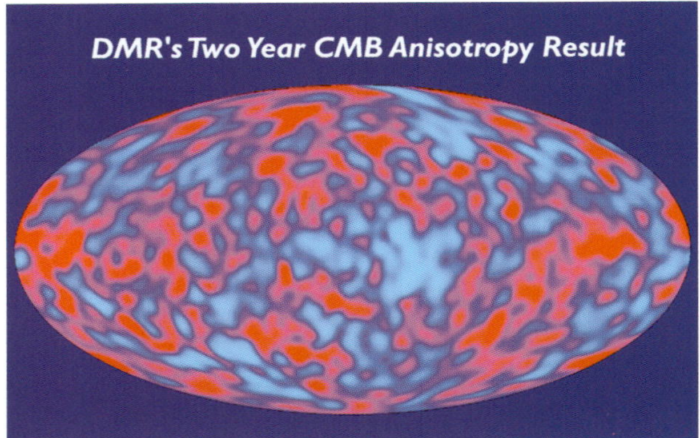

The *Cosmic Background Explorer (COBE)* satellite gathered data to generate this false-color microwave radiation map of the whole sky showing variations in the microwave background radiation that are consistent with the big bang theory of the universe. The blue represents cooler temperatures of the background radiation, and the pink areas are warmer. *[NASA Goddard Space Flight Center (NASA-GSFC)]*

Robert Wilson (right) and Arno Penzias shared the 1978 Nobel Prize in physics for their discovery of the faint cosmic microwave background radiation remaining from the big bang that gave birth to the universe some 14 billion years ago. The antenna they used appears behind them. *(AP Images)*

to the present-day wavelengths, which correspond to a very cold 2.73 K.

The cosmic microwave background is observed to be extremely uniform in all directions. Tiny fluctuations have been detected in it; they are generally assumed to reflect small inhomogeneities in the cosmic distribution of matter soon after the big bang. Such inhomogeneities are thought to have been involved in the formation of the earliest galaxies.

The cosmic microwave background was first predicted in 1948 by the American theoretical physicists George Gamow, Ralph Alpher, and Robert Herman. During 1964–65 the American experimental physicists Arno Penzias and Robert W. Wilson discovered the radiation, which they found as initially unexplained noise in a detector that they had intended for radio astronomy, and determined its temperature. For their discovery, Penzias and Wilson were awarded the 1978 Nobel Prize in physics. In 1992 satellite and spacecraft observations first revealed the small fluctuations in the cosmic microwave background. For their work on the *COBE* (*Cosmic Background Explorer*) satellite the Nobel Prize in physics was awarded to John Mather and George Smoot in 2006.

See also BIG BANG THEORY; BLACKBODY; ELECTROMAGNETIC WAVES.

FURTHER READING

Chown, Marcus. *Afterglow of Creation: From the Fireball to the Discovery of Cosmic Ripples.* New York: University Science Books, 1996.

Close, Frank. *The New Cosmic Onion: Quarks and the Nature of the Universe.* London: Taylor & Francis, 2006.

cosmology The study of the universe, the cosmos, as a whole is called cosmology. In particular, cosmology deals with the large-scale structure and evolution of the universe. One of the most important theoretical tools of modern cosmology is the cosmological principle, which states that at sufficiently large scales the universe is homogeneous (i.e., possesses the same properties everywhere). Another way of stating this is that an observer at any location in the universe would look into the sky and see very much the same sky people on Earth see. In other words, Earth has no privileged position in the universe, and what is observed from Earth can be taken to be representative of the cosmos in general. The cosmological principle is an assumption, but a necessary one, for cosmologists to make any progress in understanding the universe. Another important theoretical tool of modern cosmology is Albert Einstein's general theory of relativity.

The present picture of the large-scale structure of the universe is one of great complexity. Many galaxies are grouped into clusters, and clusters into superclusters. There seem to exist tremendous sheets of galaxies, many millions of light-years in size. Between the sheets appear to reign gigantic voids, which are empty of galaxies. Galaxy clusters are found where the sheets intersect. As far as is presently known, galaxies and clusters affect each other only through gravitational attraction, which is thus considered responsible for molding the universe's structure.

Cosmology also deals with the evolution of the universe, from its beginning—if, indeed, it had a beginning—to its eventual fate. (Cosmogony is the study of the origin of the universe in particular.) One of the most widely accepted determinations in cosmology is that the universe is expanding: all galaxies are, on the average, moving apart from each other. This would seem to offer two possibilities. Either the universe will expand forever, or it will eventually reverse its expansion and collapse into itself. On the basis that gravitational attraction among all the galaxies acts as a retarding force, it would be expected that the rate of expansion is decreasing. Then, depending on the amount of mass in the universe, which determines the gravitational attraction, the cosmic expansion might eventually halt and reverse itself, leading to what is often called the big crunch. However, recent observations seem to indicate that not only is the rate of expansion not decreasing, but it is actually increasing (i.e., the cosmic expansion is accelerating). Much theoretical investigation is taking place into possible causes.

The general expansion of the universe strongly suggests that in the past all matter was closer together and the universe was denser. Evolution scenarios along these lines have the universe starting its existence in an explosion from an extremely dense, hot state, called the big bang, and expanding and cooling ever since. It is generally, but not universally, accepted among cosmologists that the evolution of the universe is following some version of a big bang model.

A possible alternative to the big bang idea is based on what is known as the perfect cosmological principle, stating that not only would an observer at any location in the universe see very much the same as is seen from Earth, but so also would an observer at any time during the universe's evolution. In other words, the universe should remain the same over time (at sufficiently large scales). Such a scenario is called a steady state model. The constant density required thereby and the observed expansion of the universe are reconciled through the assumption that new matter is continuously being created. However, the big bang model of universe evolution is most commonly accepted at present, and the perfect cosmological principle is in disrepute.

Present times are exciting ones for those interested in cosmology. Advancing technologies are allowing observations at larger and ever-larger distances, which, because of the finite speed of light, are revealing the past of the universe from earlier and even earlier eras. New evidence is continually casting doubts on old notions, while active cosmologists are coming up with new ideas. Some of those ideas sound quite weird, such as dark matter (matter that does not radiate), noninteracting matter (matter that does not affect other matter except via gravitation), dark energy (a form of energy that pervades the universe and causes its expansion to accelerate), cosmic strings ("faults" in space-time), and extra dimensions (in addition to the three dimensions of space and one of time). However, these ideas might not yet be weird enough to comprehend fully what the universe is up to. The universe might be much weirder than people think. It might even be weirder than humans are capable of thinking.

Cosmologists are theoretical physicists who specialize in cosmic considerations. Their major theoretical tool is the general theory of relativity, which they must therefore study thoroughly and understand well. That requires very advanced mathematics, including partial differential equations and differential geometry. Their major material tool is the high-powered computer, which they use for studying models of the universe and its evolution and for solving equations numerically. Astronomers supply cosmologists with experimental data, which the cosmologists use for testing their theories and for formulating new theories. Cosmologists are mostly found in universities on the faculty of physics departments.

See also BIG BANG THEORY; COSMIC MICROWAVE BACKGROUND; EINSTEIN, ALBERT; GENERAL RELATIVITY; GRAVITY.

FURTHER READING

Coles, Peter. *Cosmology: A Very Short Introduction.* Oxford: Oxford University Press, 2001.

Ferreira, Pedro G. *The State of the Universe: A Primer in Modern Cosmology.* London: Phoenix, 2007.

Levin, Frank. *Calibrating the Cosmos: How Cosmology Explains Our Big Bang Universe.* New York: Springer, 2007.

covalent compounds A covalent compound is a compound that consists of elements held together by sharing electrons in a covalent bond, a type of chemical bond that forms between two nonmetals. Ionic compounds form when one element steals an electron from another, and the negatively and positively charged ions that result attract one another, forming an ionic bond. In a covalent compound, neither element is able to steal an electron from the other, so the elements share their valence electrons (electrons in the outermost energy level of an atom).

THE NATURE OF COVALENT LINKAGES

The electron configuration of an atom determines its number of valence electrons, the outermost electrons in an atom that have the highest energy and are available for bonding. All of the less energetic sublevels below the valence shell are filled with electrons. The number of valence electrons involved in bonding is determined by the octet rule, which states that in order to achieve maximal stability, each atom strives to obtain the electron configuration of the closest noble gas. In order to achieve this, in every case except hydrogen (H) and helium (He), the octet rule means that an atom needs eight electrons in the outermost shell. Hydrogen and helium only need two electrons to have a full outer shell because they only possess one energy level that contains a single *s* sublevel capable of holding two electrons. Other atoms attempt to fill the *s* and *p* sublevels of their outer energy level. Whereas an *s* sublevel holds a maximum of two electrons in its single orbital, a *p* sublevel has three degenerate (of equal energy) orbitals that can hold a total of six electrons. Atoms want to fill this outer energy level with eight total electrons and will share or steal electrons in order to do so, forming covalent or ionic bonds, respectively, as a result.

Metals generally have low ionization energy, energy required to remove an electron from a nonionized state of an atom. Having a low ionization energy means that another atom can take an electron from a metal atom relatively easily; therefore, metal atoms are likely to donate electrons and become positively charged ions, or cations. Nonmetals, on the other hand, have high ionization energies, meaning that they hold onto their electrons tightly, and they also have a high electron affinity, meaning that they are attracted to the electrons of other atoms and tend to accept additional electrons. This is why when a metal and a nonmetal form a bond, the nonmetal usually steals the electron from the metal, which compared to the nonmetal has a low ionization energy and a low electron affinity.

When two nonmetals interact to form a compound, a clear difference between the ionization energies and electron affinities of the two component atoms does not exist. Because they both have a similar affinity for electrons, neither atom can steal electrons from the other, so an ionic bond cannot form. Rather, the two nonmetallic elements share their valence electrons so that both of the atoms have access to a full octet of electrons without fully gaining or losing any electrons. This type of interaction results in a strong covalent linkage between the two atoms.

The difference in the electronegativities between the two elements in a compound determines whether that compound is covalent or ionic. Electronegativity is the attraction of an atom for the electrons in a bond. When an ionic compound forms between a metal and a nonmetal, the nonmetal has a higher electronegativity, so it has a greater attraction for the valence electron(s) and it literally "steals" it from the metal, which has a low electronegativity. In 1932 Linus Pauling developed a measurement system for electronegativities in which the most electronegative element on the periodic table, fluorine, has a value of 4 and lithium, which is to the far left in the same period on the table, is given a Pauling value of 1. The least electronegative element on the periodic table, francium, has a Pauling value of 0.7. The electronegativities on the periodic table increase from the left to the right and decrease from top to bottom. This makes the metals on the periodic table some of the least electronegative elements and the nonmetals some of the most electronegative elements on the periodic table. The difference between the electronegativities of two elements is known as the ΔEN. If the difference in Pauling value ΔEN between the two elements is greater than or equal to 1.7, the bond formed between the two elements will be an ionic bond. If the difference in electronegativity is between 1.4 and 0.4, then the bond formed between the two elements will be a polar covalent bond. If the electronegativity difference between the two elements is less than 0.2, the bond is considered a nonpolar covalent bond.

POLAR COVALENT BONDS

Polar covalent bonds form when two elements share valence electrons unequally. When one of the elements is more electronegative than the other, it tends to be an electron "hog." The more electronegative element is not strong enough to pull the electron completely away from the less electronegative element. The electron does, however, spend more time in the vicinity of the more electronegative element, giving that element a partial negative charge. The less electronegative element has a partial positive charge because the shared electron(s) spends less time in its vicinity. Because of these partial charges, polar covalent compounds have strong intermolecular forces

Electronegativities (in covalent compounds)

Period \ Group	1	2	3	4	5	6	7	8	9	10	11	12	13	14	15	16	17	18
1	H 2.20																	He
2	Li 0.98	Be 1.57											B 2.04	C 2.55	N 3.04	O 3.44	F 3.98	Ne
3	Na 0.93	Mg 1.31											Al 1.61	Si 1.90	P 2.19	S 2.58	Cl 3.16	Ar
4	K 0.82	Ca 1.00	Sc 1.36	Ti 1.54	V 1.63	Cr 1.66	Mn 1.55	Fe 1.83	Co 1.88	Ni 1.91	Cu 1.90	Zn 1.65	Ga 1.81	Ge 2.01	As 2.18	Se 2.55	Br 2.96	Kr 3.00
5	Rb 0.82	Sr 0.95	Y 1.22	Zr 1.33	Nb 1.6	Mo 2.16	Tc 1.9	Ru 2.2	Rh 2.28	Pd 2.20	Ag 1.93	Cd 1.69	In 1.78	Sn 1.96	Sb 2.05	Te 2.1	I 2.66	Xe 2.60
6	Cs 0.79	Ba 0.89	*	Hf 1.3	Ta 1.5	W 2.36	Re 1.9	Os 2.2	Ir 2.20	Pt 2.28	Au 2.54	Hg 2.00	Tl 1.62	Pb 2.33	Bi 2.02	Po 2.0	At 2.2	Rn
7	Fr 0.7	Ra 0.9	**	Rf	Db	Sg	Bh	Hs	Mt	Ds	Rg	Uub		Uuq		Uuh		Uuo

Lanthanoids *	La 1.1	Ce 1.12	Pr 1.13	Nd 1.14	Pm 1.13	Sm 1.17	Eu 1.2	Gd 1.2	Tb 1.1	Dy 1.22	Ho 1.23	Er 1.24	Tm 1.25	Yb 1.1	Lu 1.27
Actinoids **	Ac 1.1	Th 1.3	Pa 1.5	U 1.38	Np 1.36	Pu 1.28	Am 1.13	Cm 1.28	Bk 1.3	Cf 1.3	Es 1.3	Fm 1.3	Md 1.3	No 1.3	Lr

The Pauling scale for electronegativities is used to calculate the relative attraction of each atom for electrons in a bond. Elements with small differences in electronegativity will form covalent bonds.

(referring to the interactions between molecules of the compound, not between atoms within a molecule). Polar covalent solvents such as water can dissolve polar molecules such as ionic compounds, whereas nonpolar solvents cannot.

The polar nature of the covalent bonds of a water molecule contributes to the unique qualities of water that make life on Earth possible. Water contains one very electronegative element, oxygen (the electron "hog"), and two atoms of hydrogen, an element with a low electronegativity. The shared electron spends more time with the water molecule's oxygen atom, conferring on it a partial negative charge, and a slightly positive charge on each hydrogen atom. The hydrogen and oxygen atoms of adjacent water molecules can participate in hydrogen bonds. While a hydrogen bond is not a true bond like an ionic or covalent bond, it is an important intermolecular force. The capability of water to form extensive hydrogen bonds is responsible for many properties of water such as surface tension, a high boiling point, a high specific heat, lower density of ice than of water, and the ability to act as a versatile solvent. Surface tension is the force that is responsible for a water strider's walking on water and skipped rocks' bouncing along the surface of the water. When the water molecules form hydrogen bonds with the neighboring molecules on the surface of the liquid, a skinlike layer is formed that is able to withstand a certain amount of force. If the force is greater than the hydrogen-bonded surface can withstand, then the rock or the bug will sink. Water has a higher boiling point than most molecules of the same size and shape that cannot form hydrogen bonds. The intermolecular hydrogen bonds hold together the water molecules more tightly than those substances that cannot hydrogen bond, and therefore it takes more energy to excite the water molecules enough to break away from the surface of the water and evaporate. Ice is less dense than water because of hydrogen bonds. When one molecule of water forms four hydrogen bonds with its neighbors, the result is a rigid structure that allows for much open space in the center of the molecules. The less dense ice cube can float on the surface of water, and the ocean's surface can freeze without the ice's sinking to the bottom. This creates an insulation barrier on the surface of the frozen water that allows for life underneath. Water's polarity also allows it to be an excellent solvent for polar solutes. Water can surround positively charged solutes with its oxygen end and negatively charged solutes with its hydrogen end. Water is known as the universal solvent because it can effectively dissolve most polar solutes. All of these valuable properties of water can be attributed to its polar covalent bonds.

PROPERTIES OF NONPOLAR COVALENT COMPOUNDS

Covalent compounds, also called molecules, have properties distinct from those of ionic compounds, including differences in melting and boiling points, hardness, solubility, and electrical conductivity. Covalent compounds have lower melting points than ionic compounds. (The melting point is the temperature at which a substance changes from a solid to a liquid.) Individual molecules of covalent compounds are not strongly attracted to each other and easily separate enough to flow past one another in the form of a liquid. In contrast, ionic compounds have very strong interactions between their formula units and form strong crystal structures that require more kinetic energy, or higher temperatures, to melt.

Molecules of covalent compounds only weakly interact with other molecules of the compound. The amount of energy required to break them is 10–100 times less than the amount needed to break the strong ionic interactions between the cations and anions in an ionic compound. Because of this, covalent compounds such as hydrocarbons have relatively low boiling points. Nonpolar covalent compounds, made from elements with very similar electronegativities, have the lowest boiling points. While the total of intermolecular forces can create a strong effect, they are not nearly as strong as the attractive forces that form ionic bonds.

Covalent compounds are generally softer and more flexible than ionic compounds. By being able to slide and move past each other because they have less attractive forces than ionic compounds, the covalent molecules are not found in the rigid conformations characteristic of ionic compounds. Most ionic compounds are solids in a crystalline pattern, whereas covalent compounds exist as solids, liquids, and gases. Since there are no bonds between the individual molecules, the covalent compounds can be isolated from each other more easily than ionic compounds.

Covalent compounds are not usually very soluble in water, in contrast to ionic compounds. When electrons are shared in a covalent linkage, especially when they are shared equally, there is nothing to promote interaction with the water molecules. Polar covalent compounds, however, are more soluble in water than ordinary covalent compounds. The unequal sharing of electrons and partial separation of charge allow for electrostatic interactions with the polar water molecules. Since ionic compounds easily separate into their positive and negative ions in water, they dissolve easily. The saying "Like dissolves like" means that compounds will dissolve in substances that are similar to them. Nonpolar substances dissolve in nonpolar solvents, and polar solutes dissolve in polar solutes. Covalent compounds are often nonpolar and, therefore, not soluble in water.

Electrical conductivity measures the ability of a substance to conduct an electric current. Covalent compounds do not dissociate in solution, and therefore they are not good conductors of electricity. Sugar dissolved in distilled water has a low electrical conductivity because sugar is a covalent compound. Table salt (NaCl) dissolved in distilled water has a much higher electrical conductivity because salt is an ionic compound that dissociates into Na^+ and Cl^- in water.

LEWIS DOT STRUCTURES OF COVALENT MOLECULES

In order to represent the elements that participate in covalent bonds, Lewis dot structures can be written for each element. The Lewis structure uses a dot for each valence electron available in that element. The filled energy levels in the atom are not depicted. The symbol for the element represents all of the electrons with the exception of the valence electrons. In order to draw a Lewis structure of an element, the total number of valence electrons is written in pairs on each of the four sides of the element symbol. For example: Li· represents the single valence electron of lithium.

The following steps serve as a guide for writing a Lewis structure of a molecule—for example, CO_2:

1. Count the total number of valence electrons from each element present in the compound. In the case of CO_2, the total valence electrons available to the compound would be four from carbon and six each from the oxygen atoms (4 + 6 + 6) for a total of 16 for the whole molecule.
2. Write the symbols of the elements, usually with the unique element written in the center. For CO_2 it would be

   O C O

3. Arrange the dots to give each element a complete octet of electrons. With the exception of hydrogen and helium, which each get two, all the elements need eight.

   :Ö:C̈:Ö: (with pairs of dots below each symbol)

4. Count again to ensure that the total number of electrons used equals the total number of electrons available in step number 1. In the carbon dioxide example, if each element receives eight electrons, then the total number of electrons will be 20, but the total number of valence electrons contributed from all of the participating atoms is only 16.
5. If the number of electrons exceeds the total number available, double and triple bonds are inserted until the number of electrons used equals the number of valence electrons available. The Lewis structure of carbon dioxide is

$$:\ddot{O}::C::\ddot{O}: \quad \text{or} \quad :\ddot{O}=C=\ddot{O}:$$

When more than one Lewis structure can be written for a molecule, the different structures are said to be resonance structures and are separated from each other by a double-headed arrow (↔). Resonance structures are two or more different forms of the same compound, and the distinct forms are indistinguishable in the actual molecule.

See also BONDING THEORIES; ELECTRON CONFIGURATIONS; IONIC COMPOUNDS; PERIODIC TABLE OF THE ELEMENTS; PERIODIC TRENDS.

FURTHER READING

Chang, Raymond. *Chemistry,* 9th ed. New York: McGraw Hill, 2006.

Crick, Francis (1916–2004) British *Biochemist*

Francis Crick and his American colleague, James Watson, elucidated the three-dimensional structure of deoxyribonucleic acid (DNA) in 1953, and their model served as the cornerstone for the developing fields of biochemistry and molecular biology in the 21st century. Both Crick and Watson wanted to determine which macromolecule was responsible for transmitting genetic information from one generation to the next. Because chromosomes carried genes, and chromosomes consisted of protein and DNA, they surmised that either proteins or DNA was the macromolecule responsible for heredity. In 1944 the American microbiologist Oswald Avery determined that DNA isolated from a virulent form of bacteria transformed a harmless bacteria into a deadly one, and these results led many, including Crick and Watson, to believe that DNA and not protein was the molecular carrier of the information for inherited traits. Crick and Watson assimilated these results as well as data from X-ray crystallography, compositional analyses of the nitrogenous bases, and parallels found in the alpha helices of proteins to construct a structural model of a DNA molecule. The structure proposed by Crick and Watson contains two strands of DNA that run in opposite directions (antiparallel) to one another and that are individually composed of a deoxyribose sugar-phosphate backbone with one of four nitrogenous bases protruding from each monomer, or nucleotide. The nitrogenous bases from one strand form hydrogen bonds with the bases in the other strand in specific ways to form a double helix with the bases resembling rungs of a ladder in the interior of the molecule. The entire structure then twists around itself, resembling a spiral staircase. The structure of DNA is stable, and genetic information is stored in the sequence of these bases. Crick and Watson also rightly suggested that the double helix of DNA lends itself to faithful replication from one generation to the next by using the two strands as templates for new strands. Crick, Watson, and a New Zealander from King's College in London named Maurice Wilkins received the Nobel Prize in physiology or medicine in 1962 for determining the structure of DNA. Crick later proposed that information flowed from DNA through an intermediate ribonucleic acid (RNA) to yield a protein. He called the flow of information the central dogma, a term now well known by all molecular biologists and biochemists. Crick also helped to crack the code relating nucleic acid sequence information to amino acid sequences, suggesting the existence of adaptor molecules that read three nitrogenous bases at a time to translate them into one of the 20 amino acids. After more than 30 years acting as a pioneer in the field, Crick spent his last years studying problems in neurobiology.

Born to Harry Crick and Annie Elizabeth Wilkins on June 8, 1916, Francis Harry Compton Crick grew up in Northampton, England. Fascinated by science at early age, he attended the local grammar school, Northampton Grammar School, and subsequently Mill Hill School in London, where he concentrated on mathematics, physics, and chemistry. He enrolled at University College in London and majored in physics, obtaining a B.S. in physics in 1937. Crick wanted to continue his studies at Cambridge University, but he was denied admission because he had not taken any Latin courses. Crick remained at University College, beginning his graduate studies in the laboratory of the physicist Edward Neville da Costa Andrade. The outbreak of World War II interrupted his studies in 1939, when he became a scientist for the British Admiralty and designed magnetic and acoustic mines for the British war effort.

When Francis Crick left the British Admiralty in 1947, he joined a migration of physical scientists into the biological sciences, a movement instigated by physicists such as Sir John Randall, who invented radar during the war. The abrupt change of fields required Crick to teach himself biology, organic chemistry, and crystallography. With the help of family and a scholarship, Crick began work in the Strangeways Laboratory at Cambridge studying the physical properties of cytoplasm. In 1949 Crick joined Max Perutz at the Cavendish Laboratory at a critical point in the field when determining the physical structure of DNA became a heated race among the Perutz group, members of the King's College London laboratory led by Sir John Randall, and the American chemist Linus Pauling to determine the physical structure of DNA.

Francis Crick believed that the secret of the transition of molecules from nonliving to living combined aspects of Charles Darwin's theory of evolution by natural selection with Gregor Mendel's classical genetic rules of heredity and the molecular basis of genetics, the chromosome. Pauling had already made inroads to determining different aspects of protein structure. Crick believed that his training as a physicist qualified him to determine the physical structure of the other component of chromosomes, DNA. Influenced by the work of Pauling and Erwin Schrödinger, Crick believed that covalent bonds in biological molecules were structurally stable enough that they could hold genetic information. During the war, studies by Avery suggested that DNA, not protein, was the genetic molecule, as he changed the phenotype (physical characteristics) of a harmless bacterial strain to a virulent one by injecting it with DNA isolated from a virulent strain. Although Crick believed that DNA was the genetic molecule, he studied proteins and their secondary structure using X-ray crystallography as the other members of the Perutz laboratory did during this time, and he helped to develop a mathematical theory that could be used to analyze the X-ray diffraction patterns generated by helical molecules that would prove invaluable in his studies of DNA.

In 1951, while working on his dissertation, Crick met a 23-year-old graduate student from the United States, James D. Watson. Both were interested in figuring out how genetic information was stored in molecular form, and they began assimilating information gleaned from other studies to guide them in their model building. In the 1930s X-ray crystallography studies by William Astbury indicated that the four nitrogenous bases (adenine, A; cytosine, C; guanine, G; and thymine, T) were stacked one on top of the other with 10 bases per turn of the DNA molecule. Adenine and guanine are two-ringed carbon and nitrogen molecules known as purines, and cytosine and thymine are single-ringed carbon and nitrogen molecules known as pyrimidines. The Austrian-American biochemist Erwin Chargaff experimentally determined that the amount of guanine always equaled the amount of cytosine and the amount of adenine always equaled the amount of thymine in a DNA molecule, a concept that came to be known as Chargaff's ratios. Crick surmised that base pairs formed between the A and the T and between G and C, and these base pairs were the key to faithful replication of genetic information. As a friend of Maurice Wilkins, Crick often engaged in informal discussions with Wilkins and his colleagues, Raymond Gosling and Rosalind Franklin. With a background in chemistry, Franklin pointed out to Crick and Watson that the phosphates and deoxyribose sugars were hydrophilic in nature; thus, these groups would be positioned on the outside of the DNA molecule. The nitrogenous bases, conversely, were hydrophobic in nature, meaning that they would tend to be in the interior of the molecule. Because of his friendship with Wilkins, Crick also controversially became privy to X-ray crystallography results obtained by Rosalind Franklin that indicated DNA has a helical structure. The importance of this unpublished data to the elucidation of the physical structure continues to be a thorn in the side of Crick and Watson's discovery to this day.

Francis Crick was a British biochemist who is best known for his collaborative work with the American biochemist James Watson on the determination of the structure of DNA in 1953. *(AP Images)*

During the model building, Crick and Watson quickly found that the base pairs fit together better if the two strands of DNA ran antiparallel to one another with respect to the way the deoxyribose sugars faced. The sugar and phosphate groups in each nucleotide formed covalent bonds with adjacent nucleotides, creating a backbone or scaffold that supported these base pairs. Held together through hydrogen bonds, the base pairs of A:T and G:C structurally resembled one another, with the overall lengths of the base pairs identical. After determining the relationship of the phosphates, the sugars, and the bases, Crick and Watson constructed a double-helix model of DNA with two sugar-phosphate backbones linked through hydrogen bonds between nitrogenous bases on opposite strands. This structure wrapped around itself to form a right-handed helix that resembled a spiral staircase. Upon elucidating the structure of DNA, Crick and Watson postulated that faithful transmission of genetic information occurs by unzipping the two strands of DNA and using each strand as a template to make new strands of DNA. Their successful interpretation of experimental results that had been obtained by others yielded them, and their friend Maurice Wilkins, the Nobel Prize in physiology or medicine in 1962.

Francis Crick continued to lay the foundations of the fields of biochemistry and molecular biology through his work on the nature of the genetic code and the mechanisms by which proteins are made. After the creation of the double-helix model of DNA, Crick's interests immediately turned to the biological properties of this molecule. Crick hypothesized that adaptor molecules existed that could hydrogen bond to short sequences of nucleic acids as well as to amino acids. These adaptor molecules are transfer ribonucleic acid (tRNA), and they bridge the language barrier between nucleic acids and amino acids, translating triplets of nucleotides into one of 20 amino acids. Crick also suggested the existence of ribosomal RNA–protein complexes (ribosomes) that catalyze the assembly of proteins by acting as an assembly site where RNA, tRNA, and amino acids join to generate new proteins. The discovery of a third type of RNA, messenger RNA (mRNA), led Crick to theorize the sequence of events required for protein synthesis: the DNA sequence stores genetics information, enzymes synthesize mRNA to serve as an intermediary between DNA in nucleus and protein synthesizing machinery in cytoplasm, and protein synthesis occurs in the cytoplasm at the ribosomes. Crick then combined these steps into a visual flowchart that he called the central dogma, an idea summarizing that the information flow between macromolecules was one way.

$$\text{DNA} \rightarrow \text{RNA} \rightarrow \text{Proteins}$$

In 1940 Francis Crick married Ruth Doreen Dodd, and they had one son, Michael F. C. Crick, who became a scientist. During the transition from physics to biochemistry, Crick divorced Dodd. In 1949 he married Odile Speed, and they raised two daughters, Gabrielle and Jacqueline, in a house aptly called the Golden Helix. After an illustrious 30-year career at Cambridge, Crick changed his field of interest a second time, devoting the last years of his life to neuroscience. He joined the Salk Institute for Biological Studies in La Jolla, California, in 1977; taught himself neuroanatomy; and focused his energies on the study of consciousness. He also wrote an autobiography, *What Mad Pursuit* (1988), which presented his personal perspective on the events, discoveries, and choices he made during his life. Vivacious with a great sense of humor, Crick continued to be a larger-than-life character in his later years. He spoke quickly and loudly, and his laugh was said to be infectious. Crick died of colon cancer on July 28, 2004, in San Diego, California.

See also BIOCHEMISTRY; DNA REPLICATION AND REPAIR; FRANKLIN, ROSALIND; NUCLEIC ACIDS; PAULING, LINUS; WATSON, JAMES; WILKINS, MAURICE.

FURTHER READING

Crick, Francis. *What Mad Pursuit: A Personal View of Scientific Discovery.* New York: Basic Books, 1988.

Edelson, Edward. *Francis Crick and James Watson and the Building Blocks of Life (Oxford Portraits in Science).* New York: Oxford University Press, 2000.

Crutzen, Paul (1933–) Dutch *Chemist* Paul Crutzen is a chemist interested in atmospheric chemistry and the consequence of human activities on climate and the ozone layer. He was the first to demonstrate that nitrogen oxides react with ozone, leading to the destruction of the ozone molecule. For his work on atmospheric chemistry and the effect of chemical compounds on ozone, Crutzen shared the 1995 Nobel Prize in chemistry with the American chemists Mario Molina and F. Sherwood Rowland.

EARLY EDUCATION AND POSITIONS

Paul J. Crutzen was born on December 3, 1933, in Amsterdam, the Netherlands, to Josef Crutzen, who was a waiter, and Anna Gurk, who worked in a hospital. Although Crutzen was an excellent student, World War II significantly impacted his education. His elementary years corresponded to the time of the war, and German occupation caused his school to be shuffled from one place to another. Despite this disrupted start, Crutzen was very successful in school, and his experiences made for a rather nontraditional educational background for a Nobel laureate. His interest in school revolved mostly around natural sciences and foreign languages. Crutzen learned several languages, as his father spoke French and his mother spoke German. When he later married and moved to Sweden, he

Paul Crutzen is a Dutch chemist who received the Nobel Prize for chemistry for his discovery of the reaction of nitrogen oxides with ozone. Paul Crutzen is shown in his lab at Tel Aviv University, December 14, 2006. *(Gil Cohen Magen/Reuters/Landov)*

added Swedish to his list of languages. Crutzen proceeded to what was called middle school in the Netherlands, which is comparable to the six or seven years of both middle school and high school in America.

Because he had not received a scholarship to attend the university, Crutzen compromised and attended a technical school after his high school graduation in 1951. Crutzen held positions as a civil engineer from 1951 to 1954 in Amsterdam, the first with the Bridge Construction Bureau for the city of Amsterdam. In 1958 Crutzen met his wife, Tertu Soininen from Finland, married, and moved to Sweden, where he worked for the House Construction Bureau in Stockholm. While there, Crutzen applied for and received a job as a computer programmer for the department of meteorology at the University of Stockholm. He had no background in programming, but he had a strong mathematics background. While at the University of Stockholm he was able to take courses and received enough credits to receive the equivalent of a master's degree in 1963. This opportunity roused Crutzen's interest in atmospheric chemistry and meteorology. He learned about this field by doing the programming for an American scientist who was researching ozone. In 1968 Crutzen completed the requirements for a Ph.D. in meteorology. The topic of his research was the science of ozone in the stratosphere.

Crutzen accepted a postdoctoral fellowship at Oxford at the European Space Research Organization. He received the doctor of science (D.Sc.) degree in 1973. His research focused on the photochemistry of ozone in the stratosphere and the pollution caused by aircraft in the stratosphere. In 1974 Crutzen moved to the United States and expanded his research to include a study of the effects of chlorofluorocarbons on the depletion of the ozone layer. From 1977 to 1980 he served as director of the Air Quality Division at the National Center for Atmospheric Research, in Boulder, Colorado.

OZONE LAYER

Ozone (O_3) is a natural component of the atmosphere as well as being a major component in smog when it exists at ground level. The atmosphere of the Earth depends on the chemistry of ozone, which serves to protect the Earth's surface from the majority of the Sun's ultraviolet radiation. Ozone is formed by the covalent bonding of three oxygen atoms. The structure of ozone involves the resonance structures shown in the figure.

Ozone is formed in the lower portion of the stratosphere when molecular oxygen (O_2) is struck with solar energy according to the following reactions:

$$O_2 + \text{sunlight} \rightarrow O + O$$

$$O + O_2 \rightarrow O_3$$

LOSS OF THE OZONE LAYER

While working at the University of Stockholm, Crutzen was not convinced that the reaction rates being used for the study of ozone decomposition were accurate. He realized that they did not explain the entire problem of atmospheric ozone depletion—nitrous oxides sped the decomposition of ozone. In 1970 Crutzen demonstrated that nitrous oxides released into the stratosphere could react with ozone in the following manner:

$$NO + O_3 \rightarrow NO_2 + O_2$$

$$O_3 + \text{light} \rightarrow O_2 + O$$

$$NO_2 + O \rightarrow NO + O_2$$

The first reaction demonstrates the reaction of nitrogen monoxide with ozone to produce nitrogen dioxide and molecular oxygen. The second reaction shows what occurs when ozone reacts with ultraviolet light to produce molecular oxygen and atomic oxygen. The nitrogen dioxide formed in the first reaction combines with the atomic oxygen formed by the second reaction to recreate the nitrous oxide and form molecular oxygen. The overall process converts two ozone molecules to three molecular oxygen molecules without altering the amount of nitrous oxide present.

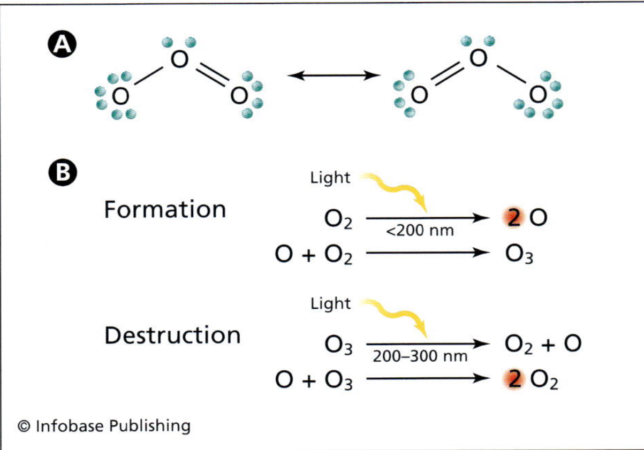

(a) Ozone is a covalent compound formed by three oxygen atoms. The Lewis structure of ozone demonstrates the two resonance forms of ozone, showing the double bond on one side of the ozone molecule or the other. (b) The formation of ozone (O_3) requires two steps: the breaking down of molecular oxygen (O_2) into atomic oxygen (O) and the reaction of this atomic oxygen (O) with molecular oxygen (O_2) to form ozone (O_3). The ozone compounds can be destroyed by the addition of light to form molecular oxygen and atomic oxygen. This atomic oxygen reacts with ozone to give molecular oxygen.

CFCS AND THE OZONE LAYER

Chlorofluorocarbons (CFCs) are synthetic compounds originally developed in the 1920s and 1930s as a safer alternative to refrigerants and coolants of the time. The market for CFCs expanded rapidly up until the 1980s as they were nontoxic, nonreactive, and nonflammable. CFCs were utilized in refrigeration (both home and commercial) under the common name Freon, as aerosol propellants, in insulation (foam-type), and in Styrofoam. After Crutzen moved to the United States, he began researching the effects of CFCs on the ozone layer. Most of the work at this time was being done by Molina and Rowland, the scientists with whom Crutzen shared the Nobel Prize. Molina and Rowland showed that the reaction between chlorofluorocarbons and ozone caused destruction of the ozone molecule. The method of destruction of the ozone layer by CFCs does not involve the entire CFC molecule. Upon being subjected to intense solar energy at the higher levels of the atmosphere, the chlorofluorocarbons break down, leaving primarily reactive chlorine (Cl). The chlorine reacts with a molecule of ozone and atomic oxygen (O) to produce two molecules of molecular oxygen according to the following reactions:

$$Cl + O_3 \rightarrow ClO + O_2$$

$$ClO + O \rightarrow Cl + O_2$$

As is shown in these reactions, chlorine reacts with ozone to create one molecule of molecular oxygen and one molecule of chlorine monoxide (ClO), which then reacts with an atom of oxygen to re-form the chlorine atom with an additional oxygen molecule. The net result of these reactions is that O_3 and O are converted into two molecules of O_2. Chlorine is not consumed in the reaction and thus can continue to react with more ozone, which is consumed.

SUPERSONIC AIRCRAFT

One of the applications of Crutzen's work on nitrogen oxides was the push in the 1970s for fleets of supersonic jets being planned in many countries. Crutzen recognized the threat of supersonic aircraft to the ozone layer. As the airplanes flew, their engines released nitrogen oxides directly into the ozone layer. The American chemist Harold Johnston at the University of California at Berkeley was studying this phenomenon. When it was determined that the nitrous oxides released by these aircraft were a danger to the ozone layer, major studies ensued. The environmental restrictions put on supersonic planes dramatically limited the number of planes produced at the time from fleets to fewer than 10. This success was due in part to the work by Crutzen.

OZONE IN THE TROPOSPHERE

Another interesting finding from Crutzen's work was that while nitrogen oxides contributed to the destruction of ozone in the stratosphere (the second layer of the Earth's atmosphere), in the troposphere (the lowest layer of the Earth's atmosphere), the nitrogen oxides actually added to the ozone that is a component of smog. The reason for this difference is the burning of fossil fuels at ground level. The hydrocarbon oxidation together with the catalytic activity of nitrogen oxides leads to the creation of smog, which can be damaging to the environment and to human health. Crutzen demonstrated that while smog is most evident over large cities, these reactions take place throughout the entire troposphere.

OZONE HOLE OVER ANTARCTICA

The work of Mario Molina, F. Sherwood Rowland, and Paul Crutzen in the 1970s elucidated the destruction of the ozone layer by CFCs. Serious environmental implications of the reactions among nitrogen oxides, CFCs, and ozone surfaced in 1985 when the English scientist Joseph Farman identified a large hole in the ozone layer over Antarctica. The hole encompassed an area larger than that of the United States. This dramatic finding coupled with the solid understanding that CFCs were directly linked to the production of this hole led to immediate action. In

1985 an international treaty known as the Montreal Protocol was passed to reduce the number of CFCs produced from 1985 with plans to phase out their production by 1996. The Montreal Protocol in 1987 totally banned the most harmful gases. CFCs have been taken off the market, and the release of CFCs from older appliances into the atmosphere is unlawful.

EFFECT OF A DAMAGED OZONE LAYER

The loss of the ozone layer or even the depletion of the ozone layer dramatically affects the human population. The ozone layer acts as a protective blanket that absorbs the majority of the Sun's ultraviolet rays. Without it, life on Earth would cease to exist. Destruction of the ozone layer could lead to increases in skin cancer and damage to crops, wildlife, and habitats. Crutzen demonstrated that nitrogen oxides can form ground-level ozone that also impacts the environment by contributing to the greenhouse effect, trapping solar energy in Earth's atmosphere. Greenhouse gases (including ozone) in the atmosphere prevent solar energy reflected from Earth's surface from escaping, leading to a rise in global temperature.

According to a 2006 report by the U.S. State Department on the condition of the ozone layer, the destruction appears to be slowing. The CFCs that have already been released have an incredibly long life span in the stratosphere and can continue to damage the ozone layer for nearly 100 years after they have been released. Despite this, the elimination of CFCs and the understanding of the chemistry of the stratosphere will eventually have a significant effect on preventing further loss of the ozone layer.

Paul Crutzen's contributions to the understanding of atmospheric science and the impact of human endeavors on the ozone layer have implications to all of humankind. Crutzen continued his work on atmospheric chemistry at the Max Planck Institute for Chemistry in the department of atmospheric chemistry in Mainz, Germany, until he retired as professor emeritus in 2000.

See also AIR POLLUTION (OUTDOOR/INDOOR); ATMOSPHERIC AND ENVIRONMENTAL CHEMISTRY; GREEN CHEMISTRY; GREENHOUSE EFFECT; MOLINA, MARIO; ROWLAND, F. SHERWOOD.

FURTHER READING

Graedel, Thomas E., and Paul J. Crutzen. *Atmosphere, Climate, and Change.* New York: Scientific American, 1997.

The Nobel Foundation. "The Nobel Prize in Chemistry 1995." Available online. URL: http://nobelprize.org/nobel_prizes/chemistry/laureates/1995/. Accessed April 8, 2008.

Curie, Marie (1867–1934) Polish *Physical Chemist* Marie Curie significantly changed the fields of chemistry and physics when she identified and studied two new elements, polonium and radium, with the ability to generate energy seemingly from nothing. Her research paved the way for the emerging field of nuclear chemistry and the current understanding of radioactivity. She is also one of only four individuals to receive two Nobel Prizes, one in physics and a second in chemistry.

Born in Warsaw in 1867, Marie Sklodowska was the youngest of four children. Both of her parents were teachers, who provided an environment conducive to learning. Marie was an avid reader, who enjoyed all her studies. Her father taught physics and chemistry at a boys' school, and she was fascinated by science from a young age. Although Marie was extremely bright, she could not pursue an advanced degree in Poland because women were not allowed this opportunity in Russian-controlled Warsaw. Marie decided that she would attain a degree at the Sorbonne in Paris; however, she did not have the financial means to accomplish her goal. She made a deal with her older sister, Bronya, who wanted to go to medical school. Curie would become a governess and use the money she earned to send her sister

Marie Curie received two Nobel Prizes, one in physics and one in chemistry, for her work on radioactivity. She is shown here in her lab. *(AP Images)*

to school. In return, Bronya would send Marie to school after graduation from medical school. Marie's older sister kept her promise and told her to come to Paris when she was 24 years old.

In 1891 Marie began her studies of physics and mathematics at the Sorbonne. She graduated the first of her class in physics in 1893 and the second of her class in mathematics the following year. Marie planned to obtain a teaching certificate, then return to Poland. These plans changed when she met Pierre Curie, a physicist eight years her senior. Pierre Curie studied the magnetic properties of crystals with no center of symmetry and invented many measuring devices important to the early work on radioactivity. Their friendship evolved, resulting in marriage in 1895. Marie Curie was grudgingly allowed to work in a leaky shed at the School of Industrial Physics and Chemistry, where Pierre was head of a laboratory.

The same year that Marie and Pierre Curie married, Wilhelm Röntgen discovered a new type of short electromagnetic wave, or radiation. Röntgen studied cathode rays, streams of high-speed electrons created inside a vacuum tube. He noticed that some of these rays penetrated the glass of the cathode-ray tube along the sides of the tube as well as where the beam of rays was pointed. As does light, these rays traveled in straight lines and affected photographic film. *X-rays,* the name Röntgen gave this new form of radiation, could also penetrate through human skin and provide photographic images of the skeleton.

Because X-rays cause the glass of the cathode-ray tube to fluoresce, the French physicist Antoine-Henri Becquerel hypothesized that fluorescent objects, which glow in the dark, produce X-rays without the need of a cathode-ray tube. Becquerel had access to many different types of fluorescent samples at the French Museum of Natural History, where his laboratory was located. Becquerel designed a straightforward experiment. He exposed fluorescent salts to light, and then he placed a dish containing these "charged" salts on top of photographic film covered with black paper to prevent its exposure to light. Because the weather was cloudy when he first attempted this experiment, Becquerel set aside the film and the dish of unexposed fluorescent salts in a drawer with other samples. Several days later, Becquerel developed the film and found that it had been exposed to some sort of radiation that did not require an external energy source. He repeated the experiment using each of the samples in his desk until he eliminated all but uranium as the source of this new type of radiation.

When Marie Curie heard about these new uranium rays, she chose to perform a detailed analysis of this new form of radiation as her doctoral dissertation project. Becquerel had observed that gases exposed to these rays were able to conduct electricity; therefore, Curie used an electrometer to measure the presence or absence of these rays in her experimental samples. She quickly discovered that a second element, thorium, emits the same type of rays as uranium. Changes in the amount of rays depended only on how much uranium and thorium were present in the compounds. Curie concluded that the generation of the rays originated within the atoms of uranium and thorium, not the arrangement of these atoms within a particular compound. This conclusion became a watershed event as nuclear chemistry and physics was born.

Next Curie decided to examine the natural ores that contain uranium and thorium. When she measured the conductivity of dark and heavy pitchblende, an ore containing uranium, she found that the ore itself was four or five times more active than uranium. She hypothesized that the ore contained other elements that could also spontaneously emit this new type of radiation. With the help of her husband, Pierre, she tried to isolate these new elements. Curie quickly found one such element, which had properties similar to those of bismuth and appeared to be several hundred–fold more active in generating the new form of radiation than uranium. In July 1898, the Curies reported their discovery of the new metal element, named *polonium* after Marie's homeland. Still, Curie's measurements of pitchblende activity suggested a second element contributed to the high radiation levels. Preliminary results indicated that this new element emitted high levels of radiation and behaved chemically similary to barium. In December 1898 the Curies suggested the name *radium* for this second element.

Marie Curie began the monumental task of isolating these new elements. She obtained several tons of pitchblende and processed 44 pounds (20 kg) of the ore at a time. In a decrepit outbuilding, Marie carried out the chemical separations, and Pierre measured the activity of the isolated fractions. Thousands of crystallizations yielded $^{1}/_{10}$ of a gram of almost pure luminescent radium chloride, enough for chemical analysis to determine radium's place on the periodic table and its atomic weight of 225 atomic mass units. Pierre Curie noted an astounding amount of energy from the radium: a gram of water could repeatedly be heated from freezing to boiling with no apparent depletion of energy. This bottomless pit of energy defied the law of conservation of energy, calling into question a fundamental tenet of classical physics. In 1903 Marie Curie defended her doctoral thesis. Later that year Becquerel, Pierre Curie, and Marie Curie received the Nobel Prize in physics for their discovery of radiation.

By the time they were awarded the Nobel Prize, Marie and Pierre Curie showed signs of radiation sickness. The Curies were exposed to large amounts of radiation emanating from the ore and radon gases released during the separation processes. They performed their experiments bare-handed. Both scientists suffered from radiation burns to their hands and from constant fatigue. Unaware of the consequences to their health, Marie and Pierre Curie often carried the luminescent radium around on their persons. To this day, their notebooks are still so radioactive that they are kept in a lead-lined safe, and visitors must sign a disclaimer to see them. These were happy times for Marie Curie nonetheless. The Curies had two daughters, Irène and Eve, during these fruitful years of research and collaboration. Then tragedy struck in 1906, when Pierre Curie was killed by a horse-drawn carriage, leaving Marie Curie a widow at 38.

In the years that followed Pierre's death, Marie Curie continued to make significant contributions to the scientific community. She assumed Pierre's position at the Sorbonne in 1908, becoming the first woman to be appointed a professor at that institution. In collaboration with André Debierne, Curie isolated radium in its metallic form. She became one of only four individuals to receive a second Nobel Prize in 1911, this time in chemistry for the isolation of polonium and radium. During World War I, she helped to equip ambulances and hospitals with X-ray machines and to train young women to use these machines. Curie even served as an ambulance driver to help with the war effort. After the war was over, she concentrated on the development of the Radium Institute. Her daughter Irène discovered artificial radioactivity with her husband, Frédéric Joliot, while performing research at the institute, and the pair received the Nobel Prize in chemistry for their discovery in 1935. Marie Curie did not live to see her daughter awarded the prestigious prize. She died of leukemia, a direct result of her long-term exposure to radioactivity, on July 4, 1934.

See also ATOMIC STRUCTURE; NUCLEAR CHEMISTRY; NUCLEAR PHYSICS; QUANTUM MECHANICS; RADIOACTIVITY; RUTHERFORD, SIR ERNEST.

FURTHER READING

Curie, Eve. *Madame Curie: A Biography.* Translated by Vincent Sheean. New York: Doubleday, 1937.

Fröman, Nanny. "Marie and Pierre Curie and the Discovery of Polonium and Radium." From The Nobel Foundation. Originally delivered as a lecture at the Royal Swedish Academy of Sciences in Stockholm, Sweden, on February 28, 1996. Available online. URL: http://nobelprize.org/nobel_prizes/physics/articles/curie/index.html. Accessed July 26, 2008.

Quinn, Susan. *Marie Curie: A Life.* New York: Simon & Schuster, 1995.

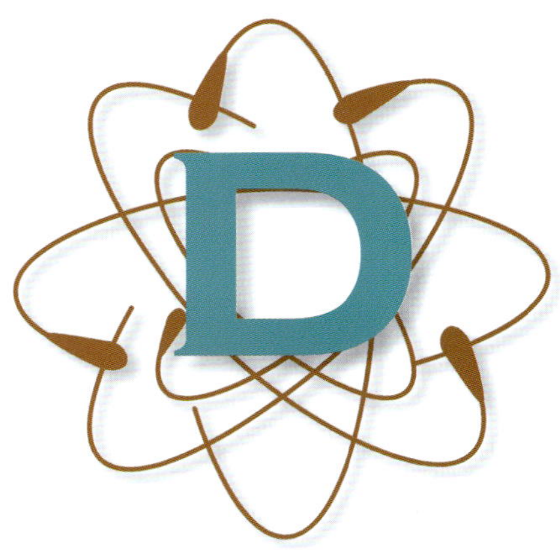

Dalton, John (1766–1844) English *Chemist*
John Dalton was an English scientist best known for his atomic theory, developed in the early 19th century. Dalton shed light on many of the previously unexplained chemical phenomena and laid the foundation of modern chemistry. While he did not invent the idea of the atom, he was the first to define it in terms of weight and to provide chemical evidence for its existence. He is also credited with proposing the law of partial pressures and the law of multiple proportions.

EARLY YEARS

John Dalton was born on September 5 or 6, 1766 (records disagree), to Joseph and Deborah Greenup Dalton in Eaglesfield, near Cockermouth, in England. His father was a cloth weaver and a member of the Society of Friends, also known as Quakers. Quaker religious beliefs impacted John's upbringing and adult lifestyle. John attended the local Quaker school, where they focused on English, arithmetic, the Bible, and a little geography and history. John was a good student and received additional tutoring in science from a wealthy neighbor named Elihu Robinson, who allowed John to use his library. Because there was a shortage of teachers, John took over the local school at the youthful age of 12, when the former schoolmaster retired. However, John was unable to control some of the rowdy older students, who were perhaps resistant to taking direction from a younger boy. Thus, after one year John simply worked around town doing odd jobs.

At the age of 15, John began to tutor at a boarding school that one of his cousins owned. The school contained a small but respectable collection of scientific literature and apparatuses, of which John took advantage. He taught there for a total of 12 years on a variety of subjects. During these years, Dalton spent his free time furthering his own knowledge in mathematics and science under the tutelage of a philosopher named John Gough. Dalton attributed much of his inspiration and early successes to his relationship with this mentor. Gough encouraged Dalton to keep a journal of daily weather observations, including temperature, air pressure, humidity, and rainfall. Dalton embarked on this undertaking in 1787 and kept the journal for the remainder of his life. In 1793 he published his first book, *Meteorological Observations and Essays*. It included much of the organized data he had collected from his journal and described scientific equipment used in meteorology. With Gough's recommendation, in 1793 Dalton accepted a position in Manchester, teaching college mathematics and chemistry at New College.

He continued recording meteorological observations and found himself becoming very interested in weather and atmospheric phenomena. He sought out colleagues who shared his intellectual interests and became a member of the Manchester Literary and Philosophical Society in 1794. The members met regularly to discuss literature and recent scientific advancements. Dalton became very involved with this society, serving as its secretary (1800), vice president (1808), and president (1817 until the time of his death). Over his lifetime, Dalton presented 117 papers to the society, and 52 were published in the *Memoirs*. Dalton never married but made many friends and was active in the community and contributed his time and abilities to the Society of Friends.

Though he performed well at New College, Dalton began his own prosperous private tutoring business while in Manchester. Eight or nine pupils visited each day to be instructed in the subjects of mathematics, natural philosophy, and grammar. When he was dissatisfied with the selection of textbooks for teaching grammar, he wrote his own successful text, *Elements of English Grammar,* published in 1801. In 1800 he resigned his position at New College, which was having financial difficulties anyhow, and supported himself for the remainder of his life by privately tutoring students. The Literary and Philosophical Society allowed Dalton the use of some rooms at their headquarters for teaching and research. He researched scientific topics that fascinated him.

STUDIES IN VISION AND METEOROLOGY

In 1794 Dalton presented his first scientific paper titled, *Extraordinary Facts Relating to the Vision of Colours, with Observations,* to the Literary and Philosophical Society. Dalton suffered from red-green color-blindness. He had observed that he saw colors differently than other people while studying botany a few years earlier. He collected information from other people with a similar disability and offered a hypothesis explaining the affliction. He thought the aqueous medium in the eyeballs of afflicted individuals was bluish in color; thus, it would absorb red light rays and prevent the person from seeing the color red. He made arrangements for his eyeballs to be dissected upon his death in order to prove his hypothesis. Unfortunately, he was incorrect, but his original paper on the matter was pioneering, as it was the first scientific accounting of the disorder.

In *Meteorological Observations and Essays* Dalton had begun a discussion on the behavior of gases. Over the next few years, he expanded his interests and studied rain, evaporation, water vapor, and heat. He did not subscribe to the popular belief that air was a compound; rather, he believed air was a mixture of several different gases. He published this idea in 1801 in *The Journal of Natural Philosophy, Chemistry and the Arts.* After this work, Dalton presented a new concept to the Manchester Literary and Philosophical Society, proposing the total pressure of a gaseous mixture was the sum of the independent pressures of each individual gas in the mixture. In other words, if a mixture contained two gases named A and B, the total pressure exerted by the mixture was the sum of the pressure exerted by gas A plus the pressure exerted by the gas B. Today this is referred to as Dalton's law, or the law of partial pressures.

By late 1802, Dalton had read a series of papers to the Literary and Philosophical Society in which he discussed the composition of the atmosphere. He also emphasized that gases such as CO_2 were absorbed into liquids (as into air) by mechanical forces, not chemical means. He defined the law of multiple proportions, which states that when the same two elements form a series of compounds, the ratio of masses of each element in the compounds for a given mass of any other element is a small whole number. For example, methane (CH_4) and ethylene (C_2H_4) are both composed of carbon and hydrogen, but experimental evidence showed that a constant amount of carbon combined with twice as much hydrogen in methane as in ethylene. The ratios of carbon to hydrogen are 1:4 versus 1:2.

John Dalton was an early 19th-century British chemist who is best known for the development of his atomic theory. Dalton is shown here in an 1834 engraving. *(National Library of Medicine)*

ATOMIC MODELS

The Greek philosopher Democritus was the first to use the word *atom,* meaning "indivisible," to describe the smallest unit of matter. An *atom* today can be defined as the smallest component of an element that retains the chemical properties of that element. Atoms combine to form compounds.

The concept of atoms as indivisible particles of matter was not revolutionary per se. However, Dalton thought to define atoms by their weights. Originally he thought that particles of all gases were the same size, but then why should some particles specifically repel only other similar particles? He decided that all the atoms of a specific element must have equal weights, and atoms of different elements have different weights. For example, all carbon atoms had the same size and atomic weight as each other but different sizes and weights than nitrogen atoms.

Because atoms are so minuscule (a single hydrogen atom weighs approximately 1.67×10^{-24} gram), it was impossible to determine the weight of individual atoms, so Dalton invented a relative system. He assigned hydrogen, the lightest known element, an atomic weight of 1. Then atomic weights of all other elements were reported in relation to the weight of hydrogen. These could be determined experimentally. For example, since hydrogen combined with eight times its own weight in oxygen to form water, he assigned oxygen an atomic weight of 8. He incorrectly assumed the combining ratio of hydrogen to oxygen was 1:1. Today chemists know that two atoms of hydrogen combine with a single atom of oxygen to form water, H_2O. Dalton had no way of knowing the ratios by which atoms combined, so he assumed the simplest possible ratio. Also today, since weight is technically the measure of heaviness of an object, and mass is the total amount of matter in an object, it is more correct to talk of *atomic mass*. In a paper presented in late 1803, Dalton included the first table of relative atomic weights.

Dalton suggested compounds were formed by the combination of atoms of different elements. Later, the Italian chemist Amadeo Avogadro said that atoms of the same element could combine to form compounds, just as atoms of different elements could. Dalton stated that atoms always combined to form simple ratios, such as 1:1 or 1:2. These ratios were predictable and constant for all molecules of the same compound.

After Dalton's death, another 100 years passed before the structure and nature of the atom could be further explained. Of course, today we know that "indivisible" atoms are in fact divisible; they are composed of protons, neutrons, and electrons. And some subatomic particles can be further broken down into quarks. Yet Dalton's bold chemical atomic theory laid the framework upon which future chemists built an entire science.

A NEW SYSTEM OF CHEMICAL PHILOSOPHY

In 1807 Dalton gave a series of lectures in Edinburgh and Glasgow, Scotland, to present the ideas described in a coherent chemical atomic theory. The content of these lectures was really a preview of his book, *A New System of Chemical Philosophy,* the first part of which was published in 1808. The second part was published in 1810. Much of this seminal book was devoted to heat, but the book also delineated his theory on the particulate nature of matter, the chemical atomic theory.

The atomic theory pronounced that central particles or atoms of homogenous bodies all had equal weights, which differed from those of different elements. The book included a discussion of the constitution of bodies as sets of rules for combining elements into compounds. Dalton also stated that central particles could be neither created nor destroyed any more easily than an entire planet could be created or destroyed. When new materials are created, old combinations of particles are simply remixed into new combinations. The idea that atoms combine chemically in simple ratios to create molecules (which Dalton confusingly referred to as "compound atoms") was also articulated. The book reviewed Dalton's law of multiple proportions and the law of constant composition. The law of constant composition was proposed by the French chemist Joseph Louis Proust in 1799 and stated that the relative masses of elements are fixed in a pure chemical compound.

The major contribution of *New System* was the proposal of a method of establishing relative atomic weights from chemical data of composition percentages. Most famous chemists quickly adopted Dalton's *New System* philosophies. The second volume did not appear until 1827 and was much less popular than the first. This may be because the material was outdated by the time it was finally published.

DALTON'S ACHIEVEMENTS

On the basis of the enormous success of the *New System,* English chemist Humphry Davy, future president of the Royal Society, encouraged Dalton to petition for election to membership. The Royal Society of London was the only scientific organization in Britain more influential than the Manchester Literary and Philosophical Society. Dalton never did. He may have been confident about his scientific achievements, but he was too humble and grounded in his Quaker beliefs to seek recognition for them.

John Dalton did receive many unsolicited honors during his lifetime. In 1816 he was nominated to be a correspondent in chemistry for the French Académie des Sciences, and in 1822 he was elected to the Royal Society of London. Royal Society colleagues who were embarrassed that the French Académie des Sciences elected Dalton for membership before he was a member of his own country's analogous organization proposed him for membership without his knowledge. In 1826 Dalton was one of the first two recipients of the Royal Medal from the Royal Society for "promoting objects and progress of science by awakening honourable competition among philosophers." There was a little controversy over this, as the recipients were supposed to have made significant contributions to science in the previous year. In 1830 Dalton was elected one of only eight foreign associates of the French Académie des Sciences. The following year, Dalton attended the foundation meeting for the British Association for the Advancement of Science, an organization in which he actively participated for several years. He was awarded honorary degrees from the University of Oxford and the University of Edinburgh. In addition, in 1833 his government granted him a pension. In 1834 Dalton was elected a member of the Royal Society of Edinburgh.

Dalton continued studying chemistry and was sought after as a public lecturer and as a consulting chemist for many of Manchester's local industries. In 1837 he suffered two strokes, which left him partially paralyzed. His mental powers were declining by this date as well. He presented a few papers to the Literary and Philosophical Society during his last decade of life. In April 1844 Dalton read his last paper to the society, discussing the rainfall in Manchester over a 50-year period and summarizing the data he had collected in his personal meteorological journal. John Dalton died at his home in Manchester on July 27, 1844, at the age of 77.

While some portions of Dalton's chemical atomic theory were incorrect, the main idea remains a tenet of modern chemical theory. Pure substances are composed of tiny particles called atoms. Atoms of each element are defined by their atomic masses all relative to the atomic mass of one element, which today is carbon. These atoms combine in a predictable manner to form chemical compounds in simple ratios of whole numbers. Two centuries ago, only John Dalton was creative and courageous enough to put forth a theory that provided a powerful explanation for previous observations such as the law of the conservation of matter and the laws of constant composition and multiple proportions.

See also ATOMIC STRUCTURE; BONDING THEORIES; CHEMICAL REACTIONS.

FURTHER READING

Brock, William. *The Chemical Tree: A History of Chemistry.* New York: Norton, 2000.

Chemical Heritage Foundation. "Chemical Achievers: The Human Face of the Chemical Sciences. 'John Dalton.'" Available online. URL: http://www.chemheritage.org/classroom/chemach/periodic/dalton.html. Accessed April 12, 2008.

Holmyard, E. J. *Chemistry to the Time of Dalton.* London: Oxford University Press, 1925.

Patterson, Elizabeth C. *John Dalton and the Atomic Theory: The Biography of a Natural Philosopher.* Garden City, N.Y.: Doubleday, 1982.

Davy, Sir Humphry (1778–1829) English *Chemist* Sir Humphry Davy's greatest contributions to chemistry were in the area of electrolysis, and for them he is known as the father of electrolysis. Davy is credited with the isolation of potassium, sodium, magnesium, calcium, and barium. His most practical invention was the miner's lamp that safely kept the flames of the light from the gases in the mine. This lamp became known as the Davy lamp. Many also recognize Davy for the role he held in the scientific development of his young assistant Michael Faraday, who became a notable physicist.

EARLY YEARS

Sir Humphry Davy was born on December 17, 1778, in Cornwall, England, to a poor family, and his father died while the child was very young. As financial considerations were paramount, Davy became an apprentice to an apothecary in Penzance to help the family pay the bills. This helped foster his natural interest in science. Davy made a good name for himself as an apprentice, and from 1799 to 1801 Davy served as the director of the Pneumatic Institute in Bristol, England. Here Davy worked on the science of inhaled gases, including nitrous oxide, or laughing gas. The overexposure to inhaled gases was thought to be a possible cause of Davy's death in 1829. In 1801 Davy took a teaching position at the Royal Institution of London. In 1803 he became a member of the Royal Society of London, and in 1820 he was made president of the Royal Society.

ELECTROLYSIS

The chemical decomposition of a compound by sending an electric current through it is known as electrolysis. When an electric current travels through an aqueous solution or a molten substance, the electrons travel from the negative electrode (the cathode) to the positive electrode (the anode). Electrolysis leads to the breaking of chemical bonds. The simplest system for demonstrating electrolysis is water. When an electric current is sent through water, the H_2O molecule is broken down into hydrogen gas (H_2) and oxygen gas (O_2). Humphry Davy utilized the principles of electrolysis to isolate elements that had not previously been isolated in pure form. The use of electrolysis to isolate these atoms and the work of Davy helped develop the new field of electrochemistry, the branch of chemistry involved with the movement of electrons and the relationship between electrical energy and chemical reactions, which can be utilized to create an electrical current. Areas of study in electrochemistry include oxidation-reduction reactions, galvanic cells and batteries, and electrolytic cells. When spontaneous chemical reactions occur, they can be used to create electrical current, and that electrical current can be used to drive chemical reactions. Electrochemical principles lead to the production of batteries and explain the chemical processes of metal corrosion and rust.

ELEMENTAL DISCOVERIES

The elements Davy isolated by electrolysis fall into the groups now known as alkali metals and alkaline earth metals. Potassium (K) and sodium (Na) were isolated by Davy in 1807 and belong to the alkali metals. Calcium (Ca), strontium (Sr), barium (Ba), and magnesium (Mg) belong to the alkaline earth metals and were isolated by Davy in 1808.

Sodium has an atomic number of 11, with the mass of the most abundant isotope 23. Sodium is a silvery, soft metal that is easily cut with a knife. The electron configuration of sodium has two electrons in the first energy level, eight electrons in the full second energy level, and a lone valence electron in the third energy level. This single valence electron is what leads to the high reactivity of the alkali metals. They readily lose their valence electron to create a +1 cations. Davy recognized both sodium's and potassium's reactivity with water by commenting that it appeared that they danced across the water.

Potassium has an atomic number of 19 with the mass of its most abundant isotope being 39. Potassium is also a silvery, soft metal. The electron configuration of potassium leads to its high reactivity. There are two electrons in its first energy level, eight in the second, 18 in the third energy level, and a lone electron in the fourth energy level. Potassium plays an important role in biological signaling pathways.

Magnesium has an atomic number of 12, with the mass of its most abundant isotope being 24. Magnesium is a grayish metal that is relatively soft, but not as soft as the alkali metals. The electron configuration of magnesium leads to the creation of +2 cations from magnesium atoms. Full first and second energy levels are followed by two valence electrons in the third energy level.

Calcium has an atomic number of 20, and its most abundant isotope has a mass of 40. Calcium is a whitish gray, soft metal. Creation of +2 cations of calcium occurs when the two valence electrons in the fourth energy level are lost.

Strontium is an alkaline earth metal that has an atomic number of 38 and an atomic mass of 88. Unlike most of the other metals discovered by Davy, strontium is yellow. It too forms a +2 cation upon the loss of its valence electrons in the fifth energy level.

The final element isolated by Davy is barium, with an atomic number of 56, and the mass of the most common isotope of 137. The valence electrons of barium are located in the sixth energy level.

In addition to the isolation of the alkali and alkaline earth metals, Davy demonstrated that the elements chlorine and iodine were individual elements. Scientists at the time, including the French chemist Antoine Lavoisier, believed that all acids contained oxygen, but this was not true for acids such as hydrochloric acid (HCl), which consists of hydrogen and chlorine. In 1810 Davy performed experiments that determined there was indeed a new element involved in this acid; he named it *chlorine.* In 1813 Davy and his wife, along with his assistant Michael Faraday, went on a trip to France at the invitation of Napoleon. During the extended trip, Davy discovered that a substance isolated from seaweed the French were working on was a new element. Davy named this element *iodine.*

THE DAVY LAMP

Countless lives of miners have been saved by Davy's invention of the safety lamp. The ignition of gases trapped in mines including methane gas made illumination within coal mines quite dangerous. In 1816 Davy developed a method for encasing the flame of the lamp with metal gauze, making it possible for air to reach the flame without the flames' being able to escape to ignite the gases. This type of safety lamp has been improved upon, but the basic principles

Sir Humphry Davy contributed greatly to the field of chemistry in the early 1800s. He was also responsible for the development of a lifesaving lamp for use in coal mines that prevented the explosion of methane gas by a lit torch. Davy is shown here demonstrating his safety lamp to coal miners. *(Science Source/ Photo Researchers, Inc.)*

of the Davy safety lamp have withstood the test of time.

DAVY AND MICHAEL FARADAY

Despite all of Davy's scientific accomplishments, he may be best known for his decision to hire an assistant named Michael Faraday in 1811. At the time Davy was a professor of chemistry at the Royal Institution (RI) of Great Britain, an organization that supports scientific research and teaching. He was well known for discovering that electricity could be used to liberate metals from compounds and for discovering the gas nitrous oxide. Davy had long been considered a popular, entertaining lecturer. Faraday attended Davy's lectures and recopied all of his lecture notes neatly onto fresh paper, illustrated them, and bound them. By a strange twist of fate, Davy was temporarily blinded in a lab explosion in 1812. (This was not an unusual occurrence in Davy's lab.) Faraday served as his secretary for a few days. Shortly thereafter, Faraday sent Davy the nearly-400-page bound volume of his lecture notes with a request for a position. Davy was impressed but disappointed Faraday as no position was available.

Several months later, a laboratory assistant at the RI was fired for fighting. Davy recommended Faraday for the spot, and in March 1813, at the age of 21, Michael Faraday started working for the RI as a chemical assistant. His duties included assisting Davy in his laboratory research, maintaining the equipment, and helping professors in their lecture preparation. Michael remained at the RI until his retirement. Many believe that Davy was competitive with Faraday and intentionally tried to hold him back. When Faraday attempted to become a member of the Royal Academy of Science, Davy actually prevented his acceptance.

Davy was knighted in 1812 and became a baronet in 1818. Sir Humphry Davy made great contributions to the field of chemistry. His isolation of alkali metals and alkaline earth metals, as well as the nonmetals chlorine and iodine, expanded the understanding of chemical reactivity in the early 1800s. The development of the Davy lamp was able to save the lives of many miners by preventing the explosion of methane gas from their lighting torches. He is also well known for the development of and sometime hindrance to the career of his assistant Michael Faraday. Davy died in Geneva, Switzerland, on May 29, 1829.

See also ELECTRICITY; ELECTROCHEMISTRY; ELECTROMAGNETISM; FARADAY, MICHAEL; PERIODIC TABLE OF THE ELEMENTS; PERIODIC TRENDS.

FURTHER READING

Knight, David. *Humphry Davy: Science and Power*. New York: Cambridge University Press, 1996.

degree of freedom

degree of freedom Each possible independent mode of motion of a system is called a degree of freedom. A point particle, for example, has three degrees of freedom, which correspond to motions in each of any three independent directions in three-dimensional space, say, the mutually perpendicular x, y, and z directions. A system of N point particles then has three degrees of freedom for each particle, giving a total of $3N$ degrees of freedom for the whole system.

A rigid body of finite size possesses—in addition to the three possible independent translational displacement motions of its center of mass that a point particle has also—the possibility of rotation about its center of mass. The possible rotational motion about an axis in each of any three independent directions—again, say, the x, y, and z directions—through the body's center of mass adds another three degrees of freedom, for a total of six. Thus, a system of N bodies has $6N$ degrees of freedom. The system consisting of the Sun together with the eight planets of the solar system possesses $6 \times 9 = 54$ degrees of freedom, six for each of the nine bodies.

The notion of degree of freedom is used somewhat differently when deriving properties of matter from its atomic makeup, such as when calculating the specific heat or molar heat capacity of a gas according to the kinetic theory. Then a degree of freedom is an independent mode for an atom or molecule to possess a significant amount of energy. The number of degrees of freedom plays a major role in determining the specific heat or molar heat capacity of a substance. For instance, the atoms of a monatomic gas, such as helium or neon, can both move in space and rotate about their centers of mass. But they are each considered to possess only the three translational degrees of freedom and no rotational ones. That is because the amounts of rotational kinetic energy they normally acquire are not significant, as a result of their relatively small moments of inertia.

On the other hand, each molecule of a diatomic gas, such as nitrogen (N_2) or carbon monoxide (CO), does possess rotational degrees of freedom (in addition to the three translational ones), but only two. They are possible rotations about any two axes through the center of mass and perpendicular to the molecule's own axis (which is the line connecting the centers of the two atoms), since the molecule's moment of inertia with respect to such axes allows it to possess significant rotational kinetic energy thereby. Rotations about the molecule's axis, however, do not normally endow the molecule with a significant amount of rotational kinetic energy, since its moment of inertia with respect to its axis is too small. Two additional degrees of freedom exist as a result of the molecule's possible vibrational motion along its axis. (Imagine the two atoms connected by a spring and oscillating lengthwise.) Vibration along a single direction involves two degrees of freedom, since such motion possesses two modes for the system to contain energy: kinetic energy and potential energy. Thus a diatomic molecule possesses three translational degrees of freedom, along with two rotational degrees of freedom and two degrees connected to its vibrational motion, for a total of seven degrees of freedom. However, depending on the temperature, not all the degrees of freedom might be effective.

The molar heat capacity of a substance at constant volume, which is the amount of heat required to raise the temperature of one mole of a substance by one kelvin (equivalently, one Celsius degree), is $R/2$ for each degree of freedom. Here R denotes the gas constant, whose value is 8.314 J/(mol·K). Thus, the constant-volume molar heat capacity of any monatomic gas, such as helium (He) or argon (Ar), which possesses only three (translational) degrees of freedom, is

$$3R/2 = (3/2) \times 8.314 \text{ J/(mol·K)} = 12.47 \text{ J/(mol·K)}$$

For a diatomic gas, such as hydrogen (H_2) or carbon monoxide (CO), at temperatures around 500 K only the translational and rotational degrees of freedom are effective, making five effective degrees of freedom. The molar heat capacity at constant volume for such gases is accordingly close to

$$5R/2 = (5/2) \times 8.314 \text{ J/(mol·K)} = 20.8 \text{ J/(mol·K)}$$

At sufficiently high temperatures, more than around 6,000 K, the two vibrational degrees of freedom become effective, giving each molecule of a diatomic gas seven degrees of freedom with a constant-volume molar heat capacity of around

$$7R/2 = (7/2) \times 8.314 \text{ J/(mol·K)} = 29.1 \text{ J/(mol·K)}$$

A crystalline solid can be modeled as a three-dimensional array of atoms connected by springs. These atoms cannot move about, as can atoms of a gas, so they possess no translational degrees of freedom. Neither do they possess rotational degrees of freedom, as their rotational kinetic energy is insignificant. However, each atom can oscillate in three independent directions. As mentioned earlier, each direction of oscillation contributes two degrees of freedom to the substance. So the total number of degrees of freedom for such a solid is six. The constant-volume molar heat capacity of such a solid is, accordingly, approximately

$$6R/2 = 3R = 3 \times 8.314 \text{ J/(mol·K)} = 24.9 \text{ J/(mol·K)}$$

This model explains the empirical rule of Dulong and Petit, which states that the molar heat of such a solid is around 25 J/(mol·K). This value becomes a better approximation to the actual value for such solids as the temperature increases.

A nonrigid macroscopic body, which can bend, twist, expand, contract, and so forth, is, for the purpose of dealing with its mechanics, considered to possess an infinite number of degrees of freedom. A stretched string, such as on a guitar or a cello, is a simple example. A whole building can serve as another, more complex example. A field, too, such as the electromagnetic field or the gravitational field, is treated as a system with an infinite number of degrees of freedom. Such systems can, among other behavior, oscillate in various ways and carry waves.

See also CENTER OF MASS; ENERGY AND WORK; GAS LAWS; HEAT AND THERMODYNAMICS; MOTION; ROTATIONAL MOTION; STATISTICAL MECHANICS; WAVES.

FURTHER READING

Young, Hugh D., and Roger A. Freedman. *University Physics,* 12th ed. San Francisco: Addison Wesley, 2007.

diffusion In one sense of the term, *diffusion* is the spreading-out process that results from the random motion of particles. After the release of perfume, as an example, the odor can soon be detected some distance away. That is due to the diffusion of the perfume molecules through the air molecules by their random wandering and their being buffeted about in all directions. Similarly, a drop of milk in a glass of water is soon spread out and diluted. The molecules that compose the milk travel randomly among the water molecules, continually colliding with them and being redirected at each collision. In a solid, too, diffusion can occur, even though the elementary constituents of a solid—its atoms, molecules, or ions—are relatively fixed in position. Given sufficient time, it is found that foreign particles—say, radioactive tracer atoms—introduced into a solid at a location in it diffuse throughout the material. Since heat is the random motion of particles, its conduction through matter is by diffusion.

The rate of net material transfer by diffusion between two points in a diffusive medium is proportional to the difference between the densities of the material at the two points. In reality, matter is diffusing from each point to the other. But there is more diffusion from a higher concentration to a lower than vice versa, so the net flow is from higher to lower. In this manner, heat flows by conduction from a region of higher temperature to one of lower temperature, with the rate of flow proportional to the temperature difference.

An isolated system in which diffusion is taking place will eventually reach a uniform state, with no further net flow occurring. This is a state of equilibrium, however, and not a static situation. Diffusion is actually continuing between any pair of locations as before, but the rates of diffusion in both directions are equal, so there is no net diffusion.

Quantitatively, steady state diffusion obeys a relation called Fick's law:

$$\text{Diffusion rate} = DA\left(\frac{C_2 - C_1}{L}\right)$$

where the diffusion rate is the net transfer of mass per unit time, in kilograms per second (kg/s), over distance L, in meters (m), through cross-sectional area A, in square meters (m^2), from a region where the concentration is C_2 to one with lower concentration C_1. With the diffusion rate expressed in kilograms per second, the concentrations are expressed in kilograms per cubic meter (kg/m^3). However, other units may be used for the quantity of the diffusing material, such as moles per second (mol/s) for the diffusion rate and moles per cubic meter (mol/m^3) for the concentration. The expression $(C_2 - C_1)/L$ is called the concentration gradient. The quantity D, in square meters per second (m^2/s), is the diffusion coefficient for the particular combination of diffusing material and substrate material and is generally very dependent on temperature. According to Fick's law, the diffusion rate increases with larger diffusion coefficient, larger cross-sectional area, and larger concentration gradient. The latter increases with larger concentration difference and shorter length. Some diffusion coefficients are shown in the following table.

Another sense of the term *diffusion* relates to the scattering of light as it travels through a translucent medium or is reflected from a rough surface. Frosted lightbulbs, for instance, produce diffuse light compared to that from clear lightbulbs, in which the glowing filament can be seen. In this sense, light is diffused when its direction of propagation is spread out into a range of directions. Diffuse light is sometimes preferred because it softens the edges of shadows, when sharp shadows are to be avoided.

The result of diffusion of a drop of dye in water a few seconds after the drop fell into the water. *(Martin Shields/Photo Researchers, Inc.)*

DIFFUSION COEFFICIENTS

Substances and Temperature	Diffusion Coefficient D (m^2/s)
oxygen through air, 0°C	1.8×10^{-5}
oxygen through air, 20°C	6.4×10^{-5}
water vapor through air, 0°C	2.4×10^{-5}
sugar through water, 0°C	5×10^{-10}
sugar through water, 12°C	28×10^{-10}
oxygen through water, 20°C	1×10^{-9}
oxygen through animal tissue, 20°C	1×10^{-11}

See also EQUILIBRIUM; HEAT AND THERMODYNAMICS; MOTION; OPTICS.

FURTHER READING

Serway, Raymond A., Jerry S. Faughn, Chris Vuille, and Charles A. Bennet. *College Physics*, 7th ed. Belmont, Calif.: Thomson Brooks/Cole, 2006.

DNA fingerprinting DNA (deoxyribonucleic acid) fingerprinting is a technique used to distinguish among individuals of the same species by identifying distinctive patterns in their DNA composition. In traditional fingerprinting, fingerprints of an individual are lifted from crime scenes using pigmented powder and cellophane tape; scientists obtain DNA fingerprints by comparing sequences of DNA nucleotides—

the order of adenine (A), cytosine (C), guanine (G), and thymine (T) nucleotides along a strand of DNA—looking for differences in the number of times certain short segments of DNA repeat themselves and for differences in the nucleotide makeup of these regions of DNA. This phenomenon, termed polymorphism, broadly means that a single gene or sequence of DNA may have multiple possible states without affecting the organism. Sir Alec Jeffries, a British geneticist at the University of Leicester, first observed that certain short segments of DNA sequences (minisatellites) that do not contribute to the function of genes encoding proteins exist in all organisms. These minisatellites, with varying nucleotide compositions, may be repeated 20 to thousands of times within the entire genome of an individual, and every individual (except identical twins who develop from a single zygote) has a unique pattern of these minisatellites. Jeffries developed the technique of DNA fingerprinting in 1984, making it possible to discern between two individuals, in a statistically significant way, by the number of minisatellites detected at one of several loci (places) in the genome. With the development of polymerase chain reaction, a technique that increases the amount of DNA sample being studied, scientists needed far less sample DNA to perform DNA fingerprinting analyses; in addition, sequences within the minisatellites could be examined for nucleotide differences between two samples. Forensic scientists find incriminating evidence about individuals by matching DNA fingerprints of suspects with samples of blood, hair, saliva, and semen; conversely, DNA fingerprinting has exonerated previously convicted individuals. Other applications of DNA fingerprinting include the identification of human remains, the determination of paternity, and matching of organ donors with recipients. Certain problems continue to plague the technique, and they center mainly around questions regarding possible sample contamination and biased data interpretation.

The British geneticist Sir Alec Jeffries is shown here at his Leicester University laboratory. Jeffries is credited with the development of restriction fragment length polymorphisms (RFLPs) analysis in the early 1980s used today in DNA fingerprinting. *(TravelStock Collection-Homer Sykes/Alamy)*

Restriction fragment length polymorphism (RFLP) analysis, the method of DNA fingerprinting that Alec Jeffries developed in the early 1980s, compares the number of minisatellites at a particular locus in the genome for different DNA samples. The first step in RFLP analysis involves purifying DNA from cells in a fluid sample (blood, saliva, semen, etc.). After extracting the DNA, restriction enzymes (bacterial enzymes that cleave DNA at sequence-specific sites) cut DNA into fragments of varying sizes, and gel electrophoresis separates the DNA fragments according to differences in size. Because DNA has a negative charge, pieces of DNA move toward a positive charge when exposed to an electric current. The DNA sample is run through a gelatinlike solid substance called an agarose gel. In these horizontal gels, smaller fragments move faster toward the positive charge, leaving larger fragments closer to the place where the sample was originally loaded. To identify the minisatellite regions, Jeffries used a multistep procedure called Southern blot hybridization to visualize the DNA fingerprints. After transferring the separated fragments from the agarose gel to a nylon membrane via capillary action, he denatured (split) the DNA into single strands, then exposed the membrane to a radioactive probe corresponding to regions of the minisatellite sequences. The probe bound to single-stranded DNA that contained those sequences. Once he washed off excess probe, Jeffries placed the Southern blot beside a piece of X-ray film. A dark mark formed on the X-ray film at any point where the radioactive probe attached to a complementary piece of DNA, showing the location of these fragments on the gel. The same Southern blot might be exposed to multiple radioactive probes during RFLP analyses, yielding information about the relative number of minisatellites at multiple loci within the sample DNA and giving rise to the DNA fingerprint of the individual under scrutiny. Disadvantages to the RFLP method include the need for good-quality DNA samples (DNA that is neither too dilute nor too degraded), the qualitative nature of this type of analysis (one can estimate the size and number of the generated restricted fragments, but not the quantity), and the amount of time required to probe multiple loci using a single Southern blot. Yet, despite its shortcomings, the potential applications of DNA fingerprinting, including paternity testing, forensics, and population genetics, were not lost on the scientific community and the public at large. The first practical use of DNA fingerprinting occurred in

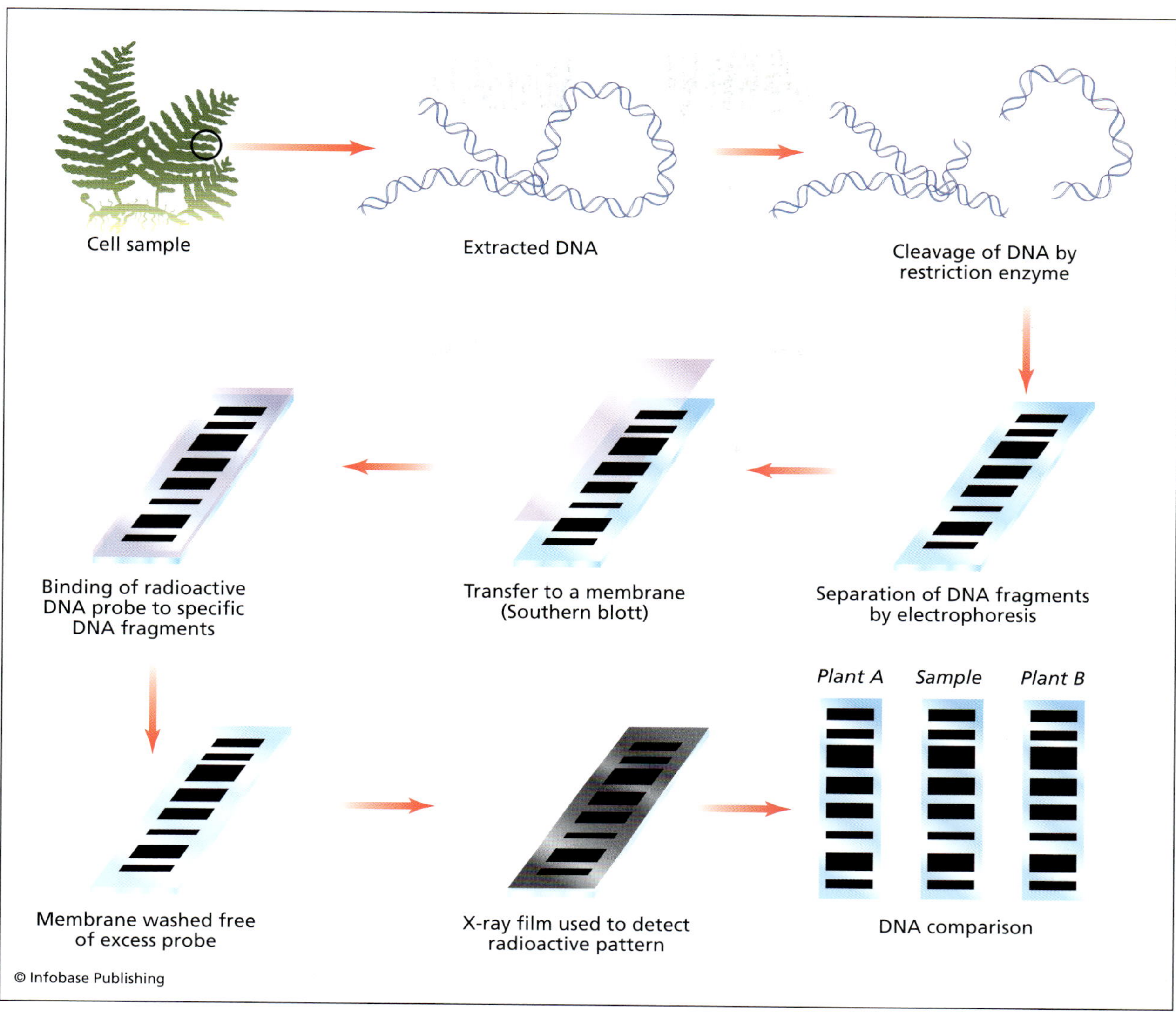

In DNA fingerprinting a DNA sample is extracted from a cell source and cleaved into fragments by restriction enzymes. Southern blotting using a specific probe reveals a pattern of fragments called a DNA fingerprint that one can compare with that of other samples.

1985 as Jeffries determined the paternity of a small boy whom British authorities planned to deport until data showed him to be a citizen's son. The demand for these analyses quickly outstripped the resources of a university laboratory; consequently, Jeffries sold the license to the technology to Imperial Chemical Industries, which formed a subsidiary called Cellmark Diagnostics, with facilities in Great Britain and the United States.

In 1993 Kary B. Mullis received the Nobel Prize in chemistry for devising a technique called polymerase chain reaction (PCR), a process that can make trillions of copies of a single fragment of DNA in only a few hours. To amplify the number of DNA copies, PCR utilizes special synthetic primers (short pieces of DNA) tailored to bind to the DNA fragment of interest and a thermostable (heat-resistant) DNA polymerase to assemble new DNA from free nucleotides using the DNA fragment as a template, or guide. This process can be repeated multiple times, leading to an exponential increase in the amount of DNA sample. With PCR, DNA fingerprinting took huge strides forward in its power to discriminate among different samples in a more timely manner and its ability to glean information from small, less-than-pristine starting samples. Three of the most prevalent methods of DNA fingerprinting incorporate PCR methodologies into their protocols—amplified fragment length polymorphism (AmpFLP), single-nucleotide polymorphisms (SNPs), and short tandem repeats (STR) analyses. Because SNPs look only for nucleotide differences, research and development firms manufacture commercial kits readily available for use by the general public. With these kits, PCR amplifies specific regions of DNA with known variations, and the amplified DNA hybridizes with probes anchored on cards. Different-colored spots form on these cards, corresponding to different sequence variations.

The use of AmpFLP analyses began in the early 1990s because they were faster than RFLP analyses. The AmpFLP technique relies on variable number tandem repeat (VNTR) polymorphisms to distinguish among alleles or versions of a particular gene. After PCR amplification of the DNA sample, the generated fragments are separated on a polyacrylamide rather than an agarose gel to achieve higher resolution between the fragments. To analyze the results, one includes a comparison sample containing fragments of known size called an allelic ladder, and visualization of the fragments requires silver staining of the gel. A disadvantage to this method, like other methods utilizing PCR, surfaces when the DNA sample is very small or highly degraded. Sometimes allelic dropout occurs, leading scientists to think that an individual is a homozygote (both alleles are identical for a particular gene) rather than a heterozygote (two different alleles at the same gene) because the PCR amplifies the DNA of one allele more readily than the other. Another phenomenon that is problematic in analysis presents itself when a high number of repeats exists within a particular sample, giving rise to difficulties at the gel electrophoresis step as the sample stays caught up in the loading well and does not migrate through the agarose gel. In its favor, the AmpFLP analyses lend themselves to automation, so the cost of the procedure is relatively low with its ease of setup and operation costs, which, in turn, make it a popular choice for DNA fingerprinting in less affluent countries.

The most prevalent method used for DNA fingerprinting today is STR analysis. This method distinguishes individuals from each other by looking at highly polymorphic regions that have short repeated sequences of DNA that are three to five base pairs in length. Because different individuals have different numbers of these repeated regions, individuals can be discriminated from one another. From country to country, different STRs are used to create a DNA profile; in North America, the Federal Bureau of Investigation (FBI) created the Combined DNA Index System (CODIS), a system that uses a 13-allele profile for DNA fingerprinting of millions of known individuals and DNA samples from crimes that remain unsolved. Scientists in the United States target these 13 STR loci via sequence-specific primers and PCR. Traditionally, the resulting DNA fragments were analyzed using gel electrophoresis, but investigators are moving toward capillary electrophoresis (CE), as the procedure is faster, allows one to look at all 13 loci simultaneously, and is less prone to human error because it is automated. In CE, a technician injects DNA into a thin glass tube filled with polymer, and the DNA moves through this polymer when an electric current is applied to the tube. As in gel electrophoresis, smaller DNA fragments move faster. The CE is run beside a fluorescent detector that picks up the presence and number of multiple STRs simultaneously, a process referred to as multiplexing. The STR fragments have been marked by using fluorescent dyes attached to the primers in the PCR portion of the procedure. The true power of this method over alternatives lies in its statistical power of discrimination. Because all 13 loci examined sort independently—meaning that having a certain number of repeats at one of the 13 loci does not change the number of repeats at another locus—the power rule of statistics can be applied with a one in a quintillion (10^{18}) probability that two unrelated individuals will match.

In 1994 the notorious O. J. Simpson trial demonstrated to the world some of the intrinsic problems associated with DNA fingerprinting in criminal cases. A former NFL running back and Heisman Trophy winner, Simpson was accused of killing his ex-wife, Nicole Brown Simpson, and Ronald Goldman. His high-profile trial brought to light the possibility that a match between an individual and a crime scene sample could be an accidental random match or that evidence had been planted against the individual. DNA fingerprinting is no less immune to sample contamination or faulty preparation than other forensic tools. Instances of fake DNA samples have been given more than once by a suspect. In 1987 a jury convicted the first criminal caught using DNA evidence, when DNA fingerprinting linked the British baker Colin Pitchfork to the rape and murder of two teenage girls with samples from the crimes. Pitchfork almost avoided conviction by convincing a friend to provide a sample of blood and saliva on his behalf. Had the friend not bragged about their cunning plan, Pitchfork might never have been caught. In 1992, Canadian authorities accused the physician Dr. John Schneeberger of sedating then raping a patient, leaving semen in her underwear as incriminating evidence. Taking fake DNA evidence to new levels, police drew blood from Schneeberger three times before they found a surgically implanted Penrose drain in the physician's arm filled with someone else's blood, a cleverly devised plan by the doctor to avoid conviction. An equally valid concern among some in the criminal justice community lies in the intrinsic possibility of erroneous data interpretation by forensic laboratories biased toward confirming a suspect's guilt. Bench scientists like Sir Alec Jeffries advocate a blinded experimental design in which the technician does not know the source of the DNA samples under study, believing that this approach is crucial in analyzing qualitative data objectively; yet crime laboratories continue to be reluctant to adopt this approach to alleviate doubts concerning their results. Recent technological advances may ultimately render these concerns moot, as chip-based DNA profiling tests that take DNA fingerprinting from the lab to the crime scene move closer to a reality. In the near future, a police officer might swab a piece of evidence, place the sample on a microchip linked to a computer, and download a complete criminal record on the individual to whom the sample belongs in a matter of minutes.

See also MULLIS, KARY; NUCLEIC ACIDS.

FURTHER READING

Dove, Alan. "Molecular Cops: Forensic Genomics Harnessed For the Law." *Genomics and Proteomics,* June 1, 2004, 23.

Fridell, Ron. *DNA Fingerprinting: The Ultimate Identity.* London: Franklin Watts, 2001.

Vuylsteke, Marnik, Johan Peleman, and Michiel J. T. van Eijk. "AFLP Technology for DNA Fingerprinting." *Nature Protocols,* June 1, 2007.

DNA replication and repair Deoxyribonucleic acid, or DNA, exists in all prokaryotes (i.e. bacteria), all eukaryotes (i.e., plants, fungi, protists, animals including humans), and many viruses. DNA contains genetic information, embedded in its structure, that organisms use to transmit inherited traits from one generation to the next. DNA was first identified in 1869; in the 1930s chemists at the Rockefeller Institute for Medical Research, Oswald Avery, Colin MacLeod, and Maclyn McCarty, demonstrated that DNA carried the molecular information to transform a harmless bacterium into a virulent form. The American microbiologists Alfred Hershey and Martha Chase confirmed that DNA is the genetic material in 1952 by labeling viral proteins with radioactive sulfur or viral DNA with radioactive phosphorus. Hershey and Chase infected bacteria with viruses containing one of these two labeled biomolecules and showed that DNA is the hereditary material because they detected the labeled DNA only in the infected bacteria.

In 1953 James Watson and Francis Crick, two young scientists at the Cavendish Laboratory at Cambridge University, elucidated the structure of DNA as two polymers twisted around one another, forming a double helix that resembles a spiral staircase. The two chains of the double helix run in opposite orientation, or antiparallel, to each other, and each chain consists of monomers called nucleotides containing a deoxyribose sugar, a phosphate group, and one of four nitrogenous bases: two purines, adenine (A) and guanine (G), and two pyrimidines, cytosine (C) and thymine (T). Covalent bonds between the phosphate group of one nucleotide and the sugar portion of the adjacent nucleotide form a sugar-phosphate backbone with the nitrogenous bases protruding out from the polymer. Hydrogen bonds link two individual strands together via specific base pairing: adenine always pairs with thymine via two hydrogen bonds, and guanine always pairs with cytosine via three hydrogen bonds. DNA strands that have these specific bonding patterns between nitrogenous base pairs are called complementary. The configuration assumed by DNA with its specific bonding patterns makes for an ideal information storage and retrieval mechanism. Taking advantage of these intrinsic characteristics, the process of replication makes a second DNA molecule by separating the original two strands and using each single strand as a template, or mold, to generate a strand that will base pair correctly to the original. In the event that damage occurs to the DNA, repair mechanisms fix the problem before replication proceeds.

REPLICATION

As soon as Watson and Crick deciphered the structure of DNA, its mechanism for replication became obvious. Watson and Crick envisioned a double helix untwisting, with each half serving as a template for the assembly of a new half. Using free nucleotides within the cell, a new strand grows as these nucleotides form base pairs with nucleotides on the existing strand of DNA through hydrogen bonding. For example, when a double helix is unwound at an adenine-thymine (A-T) base pair, one unwound strand carries the A and the other strand carries the T. During replication, the A in the template DNA pairs with a free T nucleotide in the newly made DNA strand, giving rise to an A-T base pair. In the other template, the T pairs with a free A nucleotide, creating another A-T base pair. Thus, one A-T base pair in the original double helix results in two A-T base pairs in two double helices, and this process repeats itself at every base pair for the entire length of the double-stranded DNA molecule. Covalent bonds between the sugar and phosphates of the free nucleotides cement them together to form the new DNA chain, or strand. This mechanism is termed semiconservative replication because one-half of each new double helix of DNA is taken from a preexisting double helix.

The molecular process of replication involves unwinding the DNA, breaking apart existing base pairs, copying the information, and then gluing the

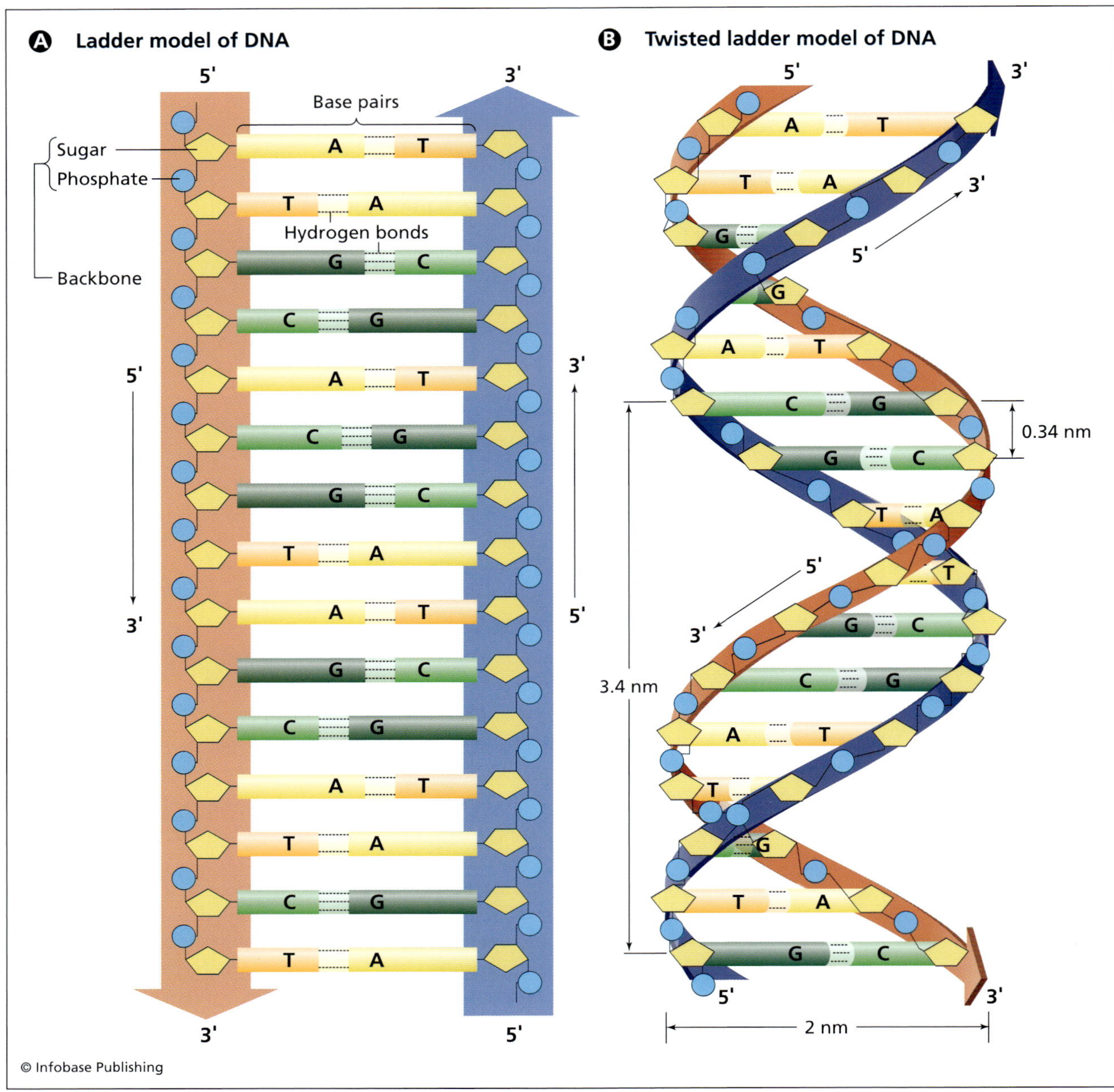

(a) In its untwisted form DNA looks like a ladder with the sugar-phosphate backbones running in opposite directions (antiparallel) and serving as the rails of the ladder and complementary nitrogenous base pairs forming the rungs of the ladder. (b) The ladder twists upon itself to form a right-handed double helix that resembles a spiral staircase.

new strand together before the DNA re-forms a double helix. As in virtually all metabolic events, enzymes control DNA replication. Enzymes called helicases unwind and hold apart DNA into two separate strands in the region about to be replicated. A human genome with 23 pairs of chromosomes begins the process at thousands of these opened portions of the DNA double helix, or replication forks. After helicases create the replication forks by breaking hydrogen bonds between base pairs at an initiation site (site where the process begins), a second enzyme primase (a ribonucleic acid, RNA, polymerase) attracts complementary RNA nucleotides to build a short piece of RNA called an RNA primer at each initiation site. The process requires the synthesis of an RNA primer because DNA polymerase (the enzyme that synthesizes new DNA) cannot initiate a nucleic acid chain on its own. The RNA primer attracts DNA polymerase to the replication fork, and this enzyme draws in DNA nucleotides complementary to the template strand. As the DNA moves past the enzyme, DNA polymerase self-checks (proofreads) for mistakes, taking out wrong nucleotides and replacing them with the correct ones. After removal of the RNA primer, the enzyme ligase covalently binds the nucleotides to each other on the new strand, attaching the phosphate group from one monomer to the deoxyribose from the second nucleotide. Replication proceeds along the length of DNA until reaching completion of a new strand of DNA.

In the simplest model of DNA replication, DNA polymerase adds DNA nucleotides according to the rules of complementarity, simultaneously on both new strands as the existing double helix opens up at the replication fork. A huge kink to this scenario exists because the two strands of DNA run antiparallel to each other. Following in one direction, one strand is oriented $5' \rightarrow 3'$, whereas the other strand is $3' \rightarrow 5'$ (these directions refer to the numbering of carbon atoms on deoxyribose). Since DNA replication involves the generation of two new antiparallel double helices using the old strands as templates, one new strand would have to be replicated in the $5' \rightarrow 3'$ direction, and the second strand would have to be replicated in the $3' \rightarrow 5'$ direction. As do other enzymes, DNA polymerase works directionally, adding new nucleotides via a condensation reaction between the 5′-phosphate group and the exposed 3′-hydroxyl end of the sugar in the growing strand. Thus replication proceeds in a $5' \rightarrow 3'$ direction. Experimental evidence indicates the process occurs simultaneously along both strands in vivo. From the $3' \rightarrow 5'$ strand, DNA polymerase generates one new continuous strand of $5' \rightarrow 3'$ DNA, called the leading strand. DNA polymerase is a highly processive enzyme on the leading strand, meaning that DNA polymerase remains attached to the template until the entire strand is replicated, once it initially associates with the template. A discontinuous form of replication occurs on the complementary strand, as DNA polymerase synthesizes small pieces of DNA from the inner part of the replication fork outward in a pattern similar to backstitching in sewing. These short segments, called Okazaki fragments, average about 1,500 nucleotides in prokaryotes and 150 in eukaryotes. Discontinuous replication requires countless repetitions of four steps: primer synthesis, elongation (lengthening of new DNA strand), primer removal, and ligation of fragments to one another. Because repeating these steps over and over requires more time to complete synthesis, the strand synthesized in this discontinuous fashion is called the lagging strand.

Because DNA consists of two single strands of nucleic acids wrapped around each other, a topological problem also surfaces as the replication machinery opens up a particular section of DNA. The two strands of DNA in front of the fork wind more tightly around themselves, a phenomenon called positive supercoiling, because the DNA overcoils around itself in the same direction that the helix twists upon itself (right-handed). Negative supercoiling occurs when DNA winds about itself in the direction opposite to the direction is which the double helix twists (left-handed). Positive supercoiling increases the number of turns of one helix around the other, referred to as the linkage number, whereas negative supercoiling reduces the linkage number. Enzymes that affect the degree of supercoiling by increasing or decreasing the linkage number are called topoisomerases. Type I topoisomerases address supercoiling problems by breaking one strand of the DNA double helix, passing the other strand between the break, then sealing off the break. Type II topoisomerases break both strands of DNA, pass an intact section of double helix through the gap, and ligate the breaks. Topoisomerases eliminate positive supercoiling in front of the replication fork by either creating negative supercoils ahead of the fork or alleviating positive supercoiling after its creation.

Numerous differences exist between prokaryotic and eukaryotic DNA replication. Often a prokaryotic genome, such as *Escherichia coli,* consists of a single circular chromosome with approximately 4.2×10^6 base pairs of DNA; in contrast, humans, who are higher eukaryotes, have 46 linear chromosomes with a thousand times as much DNA. To get six feet (1.8 m) of human DNA into a single cell, proteins called histones associate with the double helix to compact the nucleic acids into the protein-DNA complex called chromatin. Histones slow the eukaryotic replication machinery because they must be moved by the replication apparatus before synthesis of new DNA strands can occur. As a result of these factors, replication proceeds at a much faster rate in prokaryotes than eukaryotes. The *E. coli* replication fork moves about 25,000 base pairs per minute, whereas the replication fork in eukaryotes moves only about 2,000 base pairs per minute.

The increased complexity of the replication process in eukaryotes versus prokaryotes also manifests itself in the number of enzymes involved in the process itself. Within prokaryotes, three enzymes synthesize DNA: DNA polymerase I, II, and III. DNA polymerase I fills in small DNA segments during replication and repair processes. DNA polymerase III serves as the primary polymerase during normal DNA replication. DNA polymerase II acts as a backup enzyme for I and III. Eukaryotes generally have five DNA polymerases, called DNA polymerase α, β, γ, δ, and ε. DNA polymerase δ seems to be the major replicating enzyme in eukaryotes. In eukaryotes, the

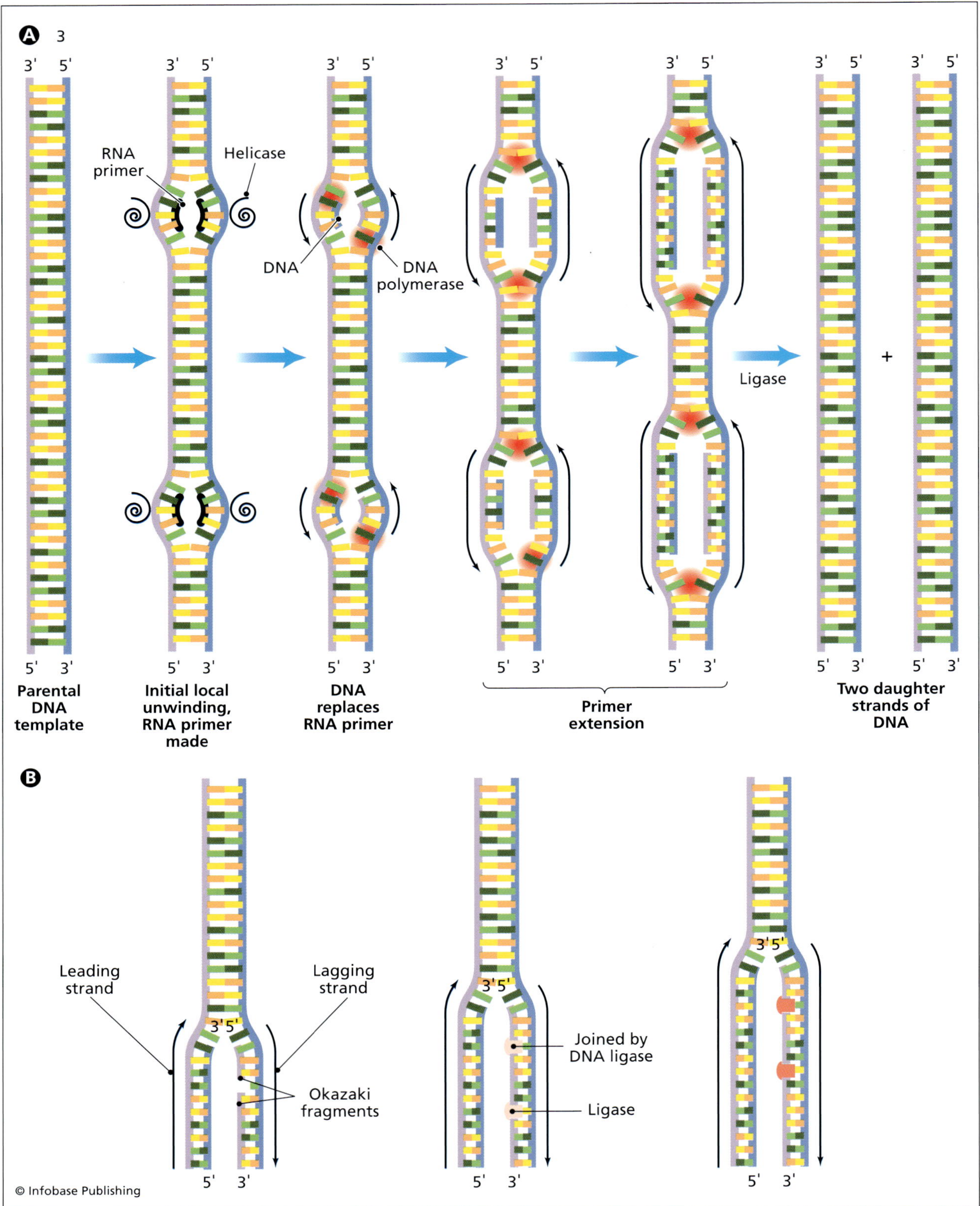

DNA replication is a multistep process. (a) Replication begins with the DNA helix unwinding at many different sites along the parental template and the synthesis of RNA primers corresponding to these regions. DNA polymerase synthesizes new complementary strands of DNA following the rules of base pairing, replaces the RNA primers with DNA, and checks for mistakes. Ligase joins the fragments of DNA. The entire process yields two identical daughter strands. (b) Because DNA polymerase only synthesizes DNA strands in the $5' \rightarrow 3'$ direction, one strand, called the lagging strand, must be replicated in short pieces, from the inner part of the fork outward. Ligase joins the pieces. The leading strand is the DNA strand synthesized in one continuous section from a single RNA primer.

Okazaki fragment primers are added by a primase, but then polymerase α adds a short segment of DNA nucleotides before polymerase δ begins its work. Polymerase β is probably the major repair polymerase, similar to polymerase I in prokaryotes; polymerase ε appears to be involved in repair as well. DNA polymerase γ replicates mitochondrial DNA. In eukaryotes ribonuclease (RNase) (an enzyme that digests RNA) removes RNA primers along the lagging strand rather than DNA polymerase; as in bacteria. Telomerase, another eukaryotic-specific enzyme, adds 250 to 1,000 repeats of a five to eight-base-pair DNA sequence to the ends of newly replicated linear chromosomes. The ends, or telomeres, created by the telomerase mark the termination point of each linear chromosome, and they also serve to protect the rest of the DNA from exonucleases, enzymes that degrade DNA from the ends of DNA strands.

DNA REPAIR

The replication machinery synthesizes new DNA strands with very few mistakes, averaging only about one in 1,000,000 bases incorrectly incorporated. Yet ultraviolet radiation, chemical mutagens, heat, enzymatic errors, and spontaneous decay damage DNA over the course of an organism's life. Repair enzymes exist that fix damaged or mutated DNA so that the cell's ability to carry out faithful replication of the double helix remains uncompromised. These repair enzymes oversee the fidelity of the DNA through four main mechanisms to repair damaged or incorrectly replicated DNA: photoreactivation, excision repair, apurinic/apyrimidinic (AP) repair, and mismatch repair.

Ultraviolet radiation causes an extra covalent bond to form between adjacent pyrmidines, commonly two thymines, creating a kink in the double helix. The linked thymines are called thymine dimers, and the extra bond disrupts faithful replication, making it possible for noncomplementary base pairs to be incorporated during DNA synthesis. One mechanism of DNA repair, photoreactivation, breaks these dimers apart. Unique to bacteria and fungi, the enzyme photolyase binds to dimers in the dark, absorbs energy from the visible light, breaks the covalent bond linking the dimers with the absorbed energy, and then falls free of the repaired DNA.

In the early 1960s researchers discovered an alternative repair mechanism for damage by ultraviolet radiation while studying mutant *E. coli* that were sensitive to light. This second repair mechanism, named excision repair, reversed the damage through another multistep process. Enzymes that carry out excision repair cut out the dimerized nucleotides and those nucleotides immediately adjacent to them. DNA polymerase fills in the gap by incorporating the correct nucleotides using the exposed template as a guide.

AP repair is another type of excision repair, in which nucleotides that are apurinic or apyrimidinic (missing the nitrogenous bases) are excised. Both ultraviolet radiation and DNA glycosylases (enzymes that sense damaged bases) remove bases from nucleotides. Two classes of AP endonucleases have the ability to correct this problem. Class I AP endonucleases nick DNA at the 3′ side of the AP site, and Class II AP endonucleases nick DNA at the 5′ side of the site. An exonuclease removes a short section of the DNA including the AP site. DNA polymerase and ligase repair the gap.

The fourth mechanism of DNA repair is mismatch repair, and this particular mechanism accounts for 99 percent of all DNA repair. In this type of repair, enzymes scan newly replicated DNA for small loops protruding from the double helix. Loops indicate an area where two DNA strands are not aligned properly, as they would be if complementary base pairing occurred. For example, when a guanine in the parent strand incorrectly pairs with a thymine rather than a cytosine in the new strand, a bulge in the double helix results. Such mismatching occurs frequently in chromosomes where very short DNA sequences, or microsatellites, repeat over and over. The mismatch repair enzymes follow behind the replication fork, and they remove incorrect nucleotides by nicking the new DNA strand on both sides of the problem. To determine which strand of DNA needs to be fixed, mismatch repair enzymes recognize methylated and unmethylated states of the DNA. The parent strand is methylated, the new strand is not. The enzymes remove the nucleotides from the unmethylated DNA, DNA polymerase and ligase repair the gap, and the enzyme DNA methylase places methyl groups along the new strand.

See also BIOCHEMISTRY; CARBOHYDRATES; CRICK, FRANCIS; DNA FINGERPRINTING; FRANKLIN, ROSALIND; NUCLEIC ACIDS; WATSON, JAMES; WILKINS, MAURICE.

FURTHER READING

DePamphilis, Melvin L., ed. *DNA Replication and Human Disease.* Cold Spring Harbor Monograph Series 47. Woodbury, N.Y.: Cold Spring Harbor Laboratory Press, 2006.

Friedberg, Errol C., Graham C. Walker, Wolfram Siede, Richard D. Wood, and Roger Schultz. *DNA Repair and Mutagenesis,* 2nd ed. Herndon, Va.: American Society of Microbiology Press, 2005.

Holmes, Frederic Lawrence. *Meselson, Stahl, and the Replication of DNA: A History of the Most Beautiful Experiment in Biology.* New Haven, Conn.: Yale University Press, 2001.

Doppler effect The dependence of the observed frequency of a periodic wave on the motion of the wave source and of the observer toward or away from each other is the Doppler effect, named for the 19th-century Austrian physicist Christian Johann Doppler. Motions that do not alter the distance between the source and the observer do not create a Doppler effect. Any change in frequency due to the Doppler effect is called a Doppler shift.

The effect can be heard whenever a honking car, whistling train locomotive, or siren-sounding emergency vehicle passes by. The pitch one hears is higher as the source approaches, then quickly shifts to lower as the source recedes. That is the Doppler effect for sound waves. One use of the Doppler effect is to detect motion and measure its speed. For that, the observer emits radio or light waves, which are reflected back by the object under observation. Any difference between the frequency of the returning wave and that of the emitted wave indicates that the object is in motion. If the former is higher or lower than the latter, the object is moving, respectively, toward or away from the observer. One can calculate the speed of the object's approach or recession from the magnitude of frequency difference. This is the principle of operation of the "radar guns" used by police to detect speeding cars. The devices emit pulses of radio waves and detect their reflection from the moving vehicle. Similarly, meteorologists use radar to measure the speed of storms.

Let f_s denote the frequency of the wave source in hertz (Hz). For illustrative purposes, imagine that the source is emitting a sequence of pulses at this frequency (i.e., f_s pulses per unit time). (What the source is really emitting at this frequency are cycles of oscillation.) As each pulse is emitted, it leaves the source at the wave propagation speed determined by the medium. The distance between adjacent pulses propagating through the medium represents the wavelength of the wave.

As a result of the Doppler effect, when a fire truck is approaching, its siren is heard at a higher pitch than when it is at rest. As the truck passes, the pitch drops suddenly, and, while the truck recedes, the pitch remains lower than when the truck is at rest. *(blphoto/Alamy)*

MOVING SOURCE

If the observer is at rest with respect to the medium of wave propagation and the source is moving away from the observer, the source is emitting pulses at increasingly greater distances from the observer. That increases the distance between adjacent pulses propagating toward the observer and thus "stretches" the wavelength of the wave reaching the observer. A fundamental relation is valid for all periodic waves

$$f\lambda = v$$

where f, λ, v, denote, respectively, the frequency, the wavelength in meters (m), and the propagation speed in meters per second (m/s) of the wave. According to this relation, for a fixed propagation speed, the frequency and wavelength are inversely proportional to each other. Thus, an increase in wavelength causes a lowering of the frequency, which results in the observer's detecting a frequency that is lower than that of the source, f_s. On the other hand, if the source is moving toward the observer, the wavelength is correspondingly "squeezed," resulting in an observed frequency higher than f_s.

Note that if the source approaches the observer at the same speed as the wave propagates in the medium, the wavelength is squeezed down to zero, since the source is then emitting pulses on top of each other. All the pulses reach the observer at the same time. This phenomenon is known as a shock wave.

MOVING OBSERVER AND COMBINED EFFECT

Now let the source remain at rest with respect to the medium, while the observer moves toward it. While at rest, the observer receives pulses at the frequency of the source, f_s. Moving toward the source, the observer "intercepts" the next pulse before it would otherwise have reached him or her, thus decreasing the time interval between receipt of adjacent pulses compared to when the observer is at rest. That causes the observer to receive more pulses per unit time than the source is emitting, and that is an observed frequency higher than f_s. If the observer is moving away from the source, the pulses have to "chase" him. He then receives pulses at greater time intervals than if he were at rest, meaning fewer pulses per unit time than the source is emitting. That translates into an observed frequency lower than f_s.

Note that if an observer is moving away from the source at the speed of wave propagation or greater, a wave emitted while this motion is occurring will never overtake her. She will then observe no wave at all.

When both the source and the observer are moving with respect to the medium, the Doppler effects are combined. That might enhance the effect or diminish it. Assume that the source and observer are moving, if they are moving at all, along the same straight line. Assume also that the source is moving *away from* the observer at speed v_s in meters per second (m/s) with respect to the medium and that the observer is moving *toward* the source, that is, chasing the source, at speed v_o in meters per second (m/s) with respect to the medium. Let f_s denote the frequency of the source, as previously, and f_o the observed frequency. Then the observed frequency relates to the emitted frequency as follows:

$$f_o = f_s \frac{1 + v_o / v}{1 + v_s / v}$$

The speeds v_s and v_o are taken to be positive when the motions are in the directions specified. If either motion is in the opposite direction, the corresponding speed is given a negative value. This formula is valid only if the observer is actually receiving waves from the source. Then all the effects mentioned follow from the formula. When the source and observer are moving in such a way that the distance between them remains constant, that is, when $v_o = v_s$, the effects cancel each other and there is no net Doppler effect; the observer observes the frequency of the source, f_s.

If v_o and v_s are both small compared to the speed of wave propagation v, the formula for the Doppler effect takes the form

$$f_o \approx f_s \left(1 + \frac{u}{v}\right)$$

Here u denotes the relative speed of the source and observer in meters per second (m/s), which is the speed of one with respect to the other. In the formula u takes positive values when the source and observer are approaching each other and negative values when they are moving apart.

RED AND BLUE SHIFTS

In the case of light or any other electromagnetic radiation, there is no material medium of wave propagation. That requires a modification of the preceding considerations. The qualitative conclusions, however, are similar. When the source is receding from the observer, the observed frequency is lower than that of the source. This is called a redshift, since lowering the frequency (or equivalently, increasing the wavelength) of visible light shifts it toward the red end of the spectrum. The light, or other electromagnetic radiation, is then said to be redshifted. On the other hand, a blue shift results when the source is approaching the observer. The formula that relates the observed frequency to the source frequency for electromagnetic waves is

$$f_o = f_s \sqrt{\frac{1 + u / c}{1 - u / c}}$$

As before, u denotes the relative speed of the source and observer, with the same sign convention, and c is the speed of light, whose value is 2.99792458×10^8 meters per second (m/s), but to a reasonable approximation can be taken as 3.00×10^8 m/s. For u small compared to c, this formula reduces to a form similar to the low-speed formula presented earlier:

$$f_o \approx f_s \left(1 + \frac{u}{c}\right)$$

Astronomers can determine whether stars or galaxies are moving away from or toward Earth by observing their red or blue shifts, respectively, and can also thus measure their speed of recession or approach. In the case of a police radar gun, described briefly earlier, the Doppler effect for electromagnetic waves occurs twice. Assuming that the car is approaching the gun, the car, as observer, "sees" waves at a higher frequency than the frequency emitted by the gun, as source. That is the first Doppler shift, a blue shift. The car then reflects these waves back to the gun and thus serves as a source, with the gun acting as observer. The gun "sees" a higher frequency than that reflected from the car. This is the second Doppler shift, again a blue shift. The radar gun automatically compares the received and emitted frequencies and from the amount of total blue shift calculates and displays the speed of the car.

According to the preceding considerations, motion of the source perpendicular to the line connecting the observer and the source should not produce a Doppler effect, since the distance between the two is not changing. Nevertheless, as a result of the relativistic effect of time dilation, the source's motion in any direction whatsoever causes a decrease in its frequency, as observed by the observer. This effect, termed the relativistic Doppler effect, or relativistic redshift, becomes significant only at speeds that are a sufficiently large fraction of the speed of light. If the source is moving at speed u in a direction perpendicular to the line connecting it to the observer, then the observed frequency and source frequency are related through the relativistic Doppler effect by

$$f_o = f_s\sqrt{1-\frac{u^2}{c^2}}$$

The low-speed approximation for this formula is

$$f_o \approx f_s\left(1-\frac{u^2}{2c^2}\right)$$

See also ACOUSTICS; ELECTROMAGNETIC WAVES; HARMONIC MOTION; SPECIAL RELATIVITY; SPEED AND VELOCITY; WAVES.

FURTHER READING

Young, Hugh D., and Roger A. Freedman. *University Physics*. 12th ed. San Francisco: Addison Wesley, 2007.

duality of nature The concept of duality of nature refers to the phenomenon whereby waves and particles each possess characteristics of the other. This is a quantum effect, also called wave-particle duality, or complementarity. The Danish physicist Niels Bohr pioneered the notion of the duality of nature in the early 20th century.

A particle is a discrete, localized entity, described by its position at any time and by its mass, energy, and momentum. A wave is a continuous, nonlocalized effect, spread out over space, for which such quantities as position, momentum, mass, and energy have no meaning. On the other hand, a periodic wave is characterized by its frequency and wavelength, quantities that are irrelevant to a particle. In spite of the apparent incompatibility of wavelike and particlelike properties, they do coexist in wave-particle duality. An electromagnetic wave such as light, for example, in addition to its wavelike character—exhibited by diffraction, for instance—can also behave as a flow of particles, called photons—demonstrated by the localized, grainy effect on a photographic film. In addition, electrons—particles, which can be counted individually by detectors—can exhibit wavelike behavior—which is the foundation of electron microscopy. The particles associated with sound waves are called phonons.

Whether a wave-particle phenomenon exhibits its wave or particle aspect depends on the manner of its investigation. Measuring a wavelike characteristic (e.g., diffraction) will reveal the wavelike nature of the phenomenon. On the other hand, if one investigates the interaction with a localized entity, such as an atom, particle behavior will become manifest. In each instance, the other aspect is "suppressed." In the first, free particles do not undergo diffraction and travel in straight lines. In the second, waves are not localized and are spread out over space.

Quantum mechanics gives relations between the frequency and wavelength of waves, on the one hand, and the energy and momentum of the particles associated with the waves, on the other. For a wave of frequency f in hertz (Hz), the particles associated with it individually possess energy E in joules (J), such that

$$E = hf$$

where h is the Planck constant and has the value $6.62606876 \times 10^{-34}$ joule-second (J·s), rounded to 6.63×10^{-34} J·s. Thus, the photons of ultraviolet light, whose frequency is higher than that of infrared light, individually possess higher energy than do infrared photons. So, for example, ultraviolet photons are capable of inducing biochemical reactions that infrared photons do not possess sufficient energy to affect. Ultraviolet photons thereby damage the skin and alter the DNA, a process that can lead to cancer-causing mutations, while infrared photons do no more than merely heat the skin.

In addition, the wavelength of a wave λ, in meters (m), and the magnitude of the momentum of each of its associated particles p, in kilogram-meters per second (kg·m/s), are related by

$$p = \frac{h}{\lambda}$$

This relation was proposed by the French physicist Louis de Broglie to explain better the wave nature of matter. The wavelength that corresponds to a particle according to this relation is known as the particle's de Broglie wavelength. According to this formula, the wavelength of electrons in an electron microscope is inversely proportional to the individual momentum of the electrons. To shorten the wavelength and increase the microscope's resolving power, the momentum is increased, by raising the voltage that is used to accelerate the electrons. A possible downside of this, however, is that increasing the momentum also increases the energy, and higher-energy electrons tend to heat the sample more and might damage it.

In the spirit of this discussion, let us consider the double-slit experiment first performed by, and named for, the 18–19th-century English scientist Thomas Young. In the experiment, light from a single source is directed through a pair of parallel slits to a screen or a photographic plate. As a result, an interference pattern appears. The pattern is completely understood in terms of coherent waves—waves that possess definite phase relations among themselves—emanating from the slits and undergoing interference at the photographic plate. That is the wave picture of the experiment. The many silver grains that result from the interaction of light with silver halide molecules in the photographic plate form the interference pattern on the plate. This interaction is completely understood in terms of a photon interacting with each molecule, leading to the particle picture, in which two streams of photons flow from the slits to the plate. But a photon, as a particle, can pass through only one slit or the other, not through both at the same time. Imagine reducing the light intensity to, say, one photon per minute and use a photon detector to find out which slit each photon passes through. One discovers that each photon passes through this slit or that on its way to the plate, but the interference pattern never forms. Instead, one finds a random scatter of silver grains. On the other hand, if one does not use the detector and waits long enough—at one photon per minute—the interference patterns does gradually take form.

Young's experiment is basically asking a wave question: How does the light from the two slits interfere (interference is a typical wave phenomenon), and how does the interference pattern depend on the light's wavelength (wavelength is also a typical wave phenomenon)? As long as that question is in force, there is no meaning to which slit the light passes through; it passes through both. That remains true even at such a low intensity that only one photon would pass through the system per minute if the particle picture were valid. But a single photon does not interfere with itself, and there is no other photon around to interact with it. Still, the interference pattern forms. So the particle picture is not valid. By asking a wave question, one makes the wave picture valid and invalidates the particle picture (i.e., one demonstrates the wave aspect and suppresses the particle aspect).

When one places a photon detector at the slits to discover which slit the photon passes through, one is changing the question to a particle question: Which path does the photon take? When that question is answered, the particle picture is valid; a photon passes through this or that slit. Then the interference pattern disappears, since particles do not undergo interference. In this manner, one makes the particle picture valid and invalidates the wave picture; one demonstrates the particle aspect and suppresses the wave aspect.

MATTER WAVES

The waves corresponding to the wave aspect of matter are called matter waves. As was shown earlier, for a beam of particles of individual energy E, the frequency f of the associated wave is given by

$$f = \frac{E}{h}$$

and if the magnitude of the linear momentum of each particle is p, the wavelength λ of the matter wave is

$$\lambda = \frac{h}{p}$$

The propagation speed, or phase speed, v_p, in meters per second (m/s), of the matter wave associated with a beam of particles of individual energy E and magnitude of momentum p is obtained from the relation, valid for all periodic waves, that it equals the product of frequency and wavelength,

$$v_p = f\lambda = \frac{E}{p}$$

The special theory of relativity relates E and p to the speed of the beam particles v as follows:

$$E = \frac{m_0 c^2}{\sqrt{1 - v^2 / c^2}}$$

$$p = \frac{m_0 v}{\sqrt{1 - v^2 / c^2}}$$

where m_0 denotes the rest mass of a particle in kilograms (kg) and c is the speed of light in a vacuum, whose value is 2.99792458×10^8 m/s, rounded to 3.00×10^8 m/s. From these relations, it follows that the propagation speed of the matter wave and the speed of the particles are related by

$$v_p = \frac{c^2}{v}$$

Since a particle's speed is always less than the speed of light, the propagation speed of matter waves is accordingly always greater than c. That is not in violation of the special theory of relativity, since neither matter nor information is moving faster than light.

It can be shown that

$$v_p = c\sqrt{1 + \left(\frac{m_0 c \lambda}{h}\right)^2}$$

Note that the propagation speed depends on the wavelength. (This effect is called dispersion.) A localized wave effect, as a representation of an individual particle, is described by a wave packet, which is an interference pattern of limited spatial extent produced by overlapping waves of different wavelengths. The propagation speed of a wave packet is called its group speed, denoted v_g and given in meters per second (m/s) by

$$v_g = v_p - \lambda \frac{dv_p(\lambda)}{d\lambda}$$

Taking the indicated derivative of the preceding expression for v_p as a function of λ, one obtains

$$v_g = \frac{c}{\sqrt{1+(m_0 c\lambda / h)^2}}$$

$$= \frac{c^2}{v_p} = v$$

Thus, the speed of the matter wave packet v_g indeed equals the particle speed v, as it should, since the localized wave packet is representing the particle.

The first decisive experimental demonstration of matter waves was carried out in 1927 by the American physicists Clinton Davisson and Lester Germer in what is known as the Davisson-Germer experiment. They directed an electron beam at a nickel crystal and detected electrons scattered in different directions. The scattering pattern corresponded exactly to the interference pattern expected from waves of the appropriate wavelength reflected from the regularly arranged atoms of the crystal lattice.

See also Bohr, Niels; Broglie, Louis de; electromagnetic waves; energy and work; mass; momentum and collisions; optics; Planck, Max; quantum mechanics; special relativity; speed and velocity.

FURTHER READING

Ball, David E. "Light: Particle or Wave?" *Spectroscopy* 21, no. 6 (June 2006): 30–33.

Serway, Raymond A., and John W. Jewett. *Physics for Scientists and Engineers,* 7th ed. Belmont, Calif.: Thomson Brooks/Cole, 2008.

Serway, Raymond A., Clement J. Moses, and Curt A. Moyer. *Modern Physics,* 3rd ed. Belmont, Calif.: Thomson Brooks/Cole, 2004.

Young, Hugh D., and Roger A. Freedman. *University Physics,* 12th ed. San Francisco: Addison Wesley, 2007.

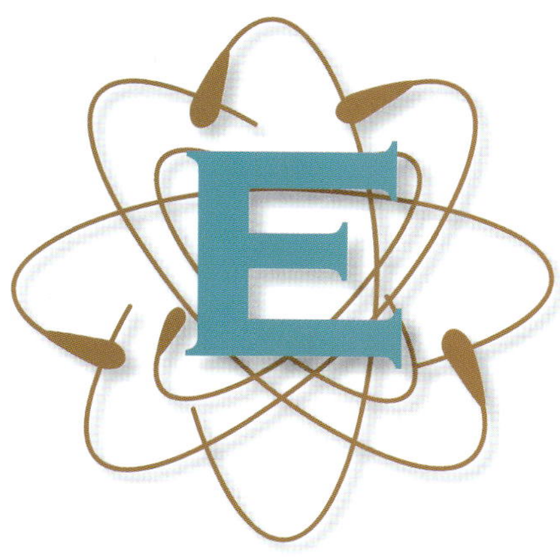

Einstein, Albert (1879–1955) German-born American *Physicist* During the early 20th century, the German-born theoretical physicist Albert Einstein completely transformed the way physicists, and even laypersons, viewed the universe. He is most famous for developing the theories of relativity, including the renowned equation $E = mc^2$, but was awarded the Nobel Prize in physics in 1921 for his explanation of the photoelectric effect. He has been called the greatest scientific mind since Sir Isaac Newton, and perhaps of all time.

A SLOW START

Albert Einstein was born on March 14, 1879, to Hermann and Pauline Koch Einstein in Ulm, Germany. The family moved in 1880 to Munich, where his father opened an electrical equipment business. They lived in a large house shared with Hermann's brother, Jakob, who often took home science and mathematics textbooks for young Albert. Although Albert did not speak until age three, and then took six years to become fluent in his own language, he was not a slow learner. The Einstein family was Jewish but not religious, and Albert was enrolled at a Catholic school when he turned five. When he was 10 years old, he started attending another school, called the Luitpold Gymnasium. He disliked the strictness and rigidity of the school systems but performed satisfactorily in school because of his independent reading. A medical student who often dined with the family gave Albert books and discussed scientific ideas with him. Albert also learned to play the violin and for the rest of his life enjoyed music as a refuge.

When the business in Munich failed in 1894, the family moved to Milan, Italy, and then to Pavia. Albert remained in Munich to finish school, but he eventually quit. Reports claim that he either had or faked a nervous breakdown, and when informing the principal that he was leaving, Albert was told he was expelled for being disruptive. Whatever the

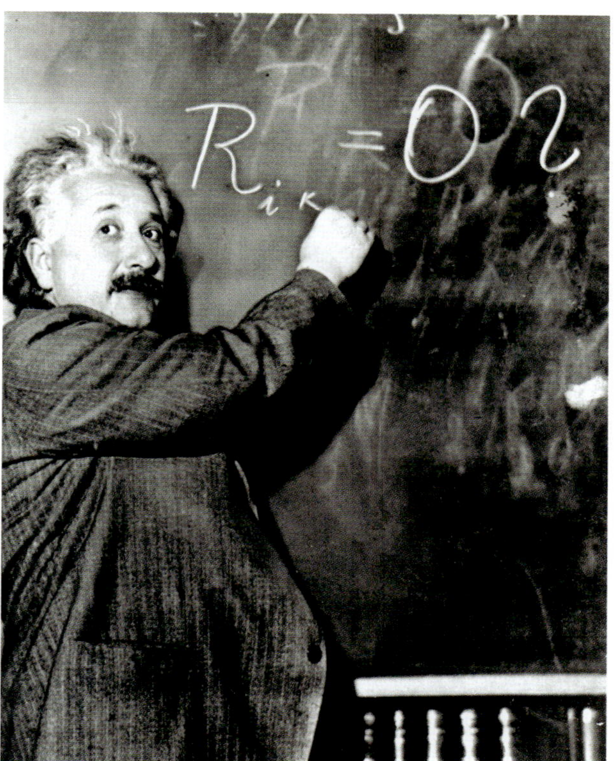

Albert Einstein, the best known of all physicists, is famous for his theories of relativity. Among physicists he is known also for his theoretical work in quantum physics and in statistical mechanics. Einstein was awarded the 1921 Nobel Prize in physics. *(AP Images)*

reason, he left without a diploma. After leaving, he renounced his citizenship so he could later return to Germany without being arrested for dodging the draft. He joined his family and spent the next year writing a paper about the relationship between electricity and magnetism and the ether. The ether was believed to be a medium through which electromagnetic waves were transmitted.

Einstein began studying for the entrance examination to the Swiss Federal Polytechnic Institute in Zurich, perhaps the finest technical school in Europe at the time. His first effort was unsuccessful overall, but his mathematics and physics scores were impressive. Einstein enrolled at the nearby Aarau cantonal school and entered the polytechnic the following year. His father wanted him to enroll in the engineering program so he would be useful to the family business; however, Einstein decided to enter the general scientific section. Einstein enjoyed tinkering in the physics laboratory but was not favored by his instructors. He often skipped lectures, and when he did attend, he was bored and disruptive. He made friends among his classmates, who happily shared their lecture notes with him, so he managed to pass, but without receiving any recommendations for postgraduate assistantships.

When he graduated in 1900, he was certified as a secondary school teacher and independently performed theoretical research for a doctorate in physics. He became a Swiss citizen, worked as a private tutor, and held a series of temporary teaching positions. He desired a permanent position, since he wanted to marry a Serbian physics student, Mileva Marić, whom he met while at the polytechnic. In 1902 she gave birth to his daughter, Lieserl, whom her parents forced her to give up for adoption. That same year, Einstein was offered a position at the Swiss federal patent office in Bern. His responsibility was to review patent applications for newly invented electrical machines. With a secure job, Einstein married Mileva in 1903 and later had two more children, Hans Albert in 1904 and Eduard in 1910.

His job in the patent office provided Einstein with access to the scientific library and left him plenty of time to think. During his employ, he obtained a doctorate degree in physics from the University of Zurich. His dissertation, entitled "A New Determination of the Size of Molecules" and published in the *Annalen der Physik* in 1905, demonstrated a method for determining the size of a sugar molecule. In 1906 he was promoted to the position of technical expert at the patent office.

THE DUAL NATURE OF LIGHT

His doctoral dissertation was actually the second of several articles published by Einstein in 1905. This was the year he transformed physics by publishing a series of landmark articles in the *Annalen der Physik* that suggested an alternative structure to the foundations of physics. The first paper described the nature of light as both wavelike and particlelike.

In the 17th century, the Dutch physicist Christiaan Huygens first suggested that light was made up of waves and that the waves traveled through a massless atmospheric medium that he called "ether." In 1704 the English physicist and mathematician Sir Isaac Newton published *Opticks,* in which he proposed a particle (as opposed to wave) theory of light. Then, research done by the Scottish physicist James Clerk Maxwell in the mid-1800s and the German physicist Heinrich Hertz in the late 1800s further advanced the wave theory of light. Yet no evidence of ether had ever been found, although the wave theory depended upon it, since it was thought that waves required a medium through which they could propagate. In 1887 the Americans Albert Michelson and Edward Morley performed experiments that disproved the existence of ether, and then evidence contradicting the wave theory of light was published. Confused physicists could not explain what light actually was. The German physicist Max Planck discovered that light was emitted in packets of specific amounts, called *quanta,* which led to his quantum theory in 1900. He related the energy of each quantum to the frequency of the light.

Einstein used Planck's quantum theory to explain the photoelectric effect, the release of electrons by a metal when irradiated with light. Brighter lights brought about the emission of more electrons, but the emitted electrons were no more energetic, although classical physics predicted they should be. If light with a higher frequency (and therefore shorter wavelength) was shined on the metal, higher-energy electrons were released. Classical physics did not predict that at all. Einstein explained that if light itself were packaged into discrete quantities (later called *photons*), as suggested by Planck, then the effect could be explained. Einstein's paper, titled "On a Heuristic Viewpoint Concerning the Production and Transformation of Light," demonstrated that light could be treated as both a wave and a particle. A certain color of light had both a specific wavelength and definite bundles of energy. Since light was particlelike, it did not require ether to travel. Many scientists had trouble appreciating Einstein's theory until Planck's quantum theory became more widely accepted. For his work explaining the photoelectric effect, Einstein was awarded the Nobel Prize in physics for 1921.

BROWNIAN MOTION

The third paper Einstein published in 1905, "On the Motion of Small Particles Suspended in a Stationary Liquid, According to the Molecular Kinetic Theory of Heat," also appeared in the *Annalen der Physik*. In 1828 the Scottish botanist Robert Brown had described an erratic motion of pollen grains suspended in water. At first he thought this jerky activity, later called Brownian motion, was due to a life force inside the pollen grains, but he observed the same motion in nonliving materials such as dye particles. For many decades, no scientist could explain this seemingly perpetual movement, and then Einstein came along and not only explained the movement, but used this discovery to prove the existence of molecules. He suggested that an object suspended in a liquid would constantly be bombarded from all directions by molecules of the liquid. Though larger objects would not be moved, smaller objects, such as pollen grains, would be slightly displaced by the force of the impact if hit by more molecules on one side than the opposite. This would make the suspended object appear to be jerking back and forth in a random fashion, just as described by Brown. Since increasingly larger molecules could cause even bigger particles to move by bumping into them, Einstein found a way to estimate the average size of the molecules based on the observable effects. Einstein also used mathematical calculations to approximate the amount of molecules in a certain volume of a liquid or gas. A few years after Einstein's publication, the French physicist Jean Perrin performed experiments validating Einstein's theoretical calculations and experimentally proved the existence of atoms.

SPECIAL RELATIVITY

As if describing the nature of light and proving the existence of atoms were not enough, Einstein next combined these ideas to form a mind-boggling theory in his fourth paper of 1905, "On the Electrodynamics of Moving Bodies." He began to doubt that one could know anything to be certain. The idea of possibly not knowing anything plunged Einstein into a virtual cerebral panic. On the verge of mental collapse one night, his special theory of relativity came to him. He proposed that concepts of space and time were only valid in relation to our experiences. In other words, people could know things compared to, or relative to, other things.

In the 17th century, Newton had assumed space and time to be absolute, immovable reference frames. A reference frame enables an observer to specify his position and velocity while observing an event. For example, a person sitting inside an airplane, watching the flight attendant roll the service cart down the aisle, would consider herself to be sitting still and be able to measure how fast the cart was moving down the aisle toward her seat. However, a person on the ground might be watching the plane fly overhead, and he would consider the plane itself and everything in it to be moving (including the passenger who is sitting "still" in her seat). The reference frame to determine the plane's speed is attached to Earth, and the earthbound observer is at rest in it. Yet from outer space, the earthbound observer himself appears to be moving, as Earth is rotating on its axis and also orbiting the Sun. As far as Newton was concerned, space was the immovable, absolute reference frame to which all motion ultimately could be referred. This manner of thinking remains suitable for most calculations and viewpoints, but it causes problems when measuring the speed of light.

In 1879 American Albert Michelson had determined the speed of light to be 186,350 miles (299,792 km) per second. With Edward Morley, he next set out to determine the speed of Earth's ether wind, the wind created by Earth's movement through the putative motionless ether (similar to the breeze created by a speeding car). Prevailing theories held that the ether was a motionless absolute reference frame from which the motion of celestial bodies could be measured. To an observer on the rotating Earth, the ether would appear to be moving. By measuring the relative speed at which Earth passed through the ether, the existence of ether could be verified.

In 1887 Michelson and Morley set out to do this by measuring the speed of light traveling upwind (parallel to the motion of Earth around the Sun) and comparing it to the speed of light traveling in a different direction (perpendicular to the motion of Earth around the Sun); the difference between the two would allow calculation of the apparent speed of the ether. This would be similar to measuring the speed of a rowboat traveling with the current and across the current, and using the two measurements to calculate the speed of the current. They expected the speed of light to be greater when traveling parallel to the direction of the motion of Earth than when traveling perpendicular to it, just as traveling with the current would increase the speed of a rowboat. Michelson and Morley sent out light in the direction of Earth's rotation and at right angles to it. However, when they measured the speed of the light, it was the same in all directions. Thus, the long-assumed ether was only imaginary.

Because of the Michelson-Morley experiments, Einstein proposed the velocity of light in a vacuum to be constant, independent of the motion of its source and of the observer. Also, believing that light could travel in a particlelike fashion, and therefore not require ether, he dismissed the ether as superfluous. Without ether, there was no absolute space; there was nothing to serve as an immovable, absolute reference frame, and thus, all motion could be described only relative to some arbitrary frame of reference. At this stage, Einstein only considered uniform, nonaccelerated motion, which is why this theory is called the "special" theory of relativity.

This assumption that all motion was relative led to the conclusion that time is relative too. Einstein determined that as one approached the speed of light, time would slow. This thought occurred to him as he was riding in a streetcar looking back at a clock tower. If the streetcar were traveling at the speed of light, the clock tower would be traveling at the speed of light relative to him and the clock hands would appear to be standing still, but the watch in his pocket would continue to advance. For objects traveling at very high speeds, time advances more slowly. As did space, time differed, depending on the observer's ref-

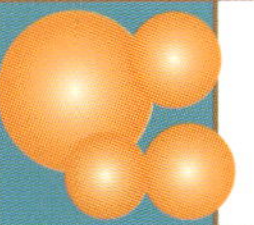

UNIFICATION: FINDING ONENESS IN DIVERSITY

by Joe Rosen, Ph.D.

In the operation of physics, physicists search for pattern and order among the most fundamental phenomena of nature; attempt to discover laws for those phenomena, based on their pattern and order, which allow the prediction of new phenomena; and try to explain the laws by means of theories. (Note that in the language of science, *theory* does not mean "speculation" or "hypothesis," as it does in everyday language, but rather explanation.) Consistently throughout this procedure, what appear to be different phenomena are found to be only different aspects of the same phenomenon, different laws are revealed as particular manifestations of the same law, and different theories turn out to be subsumed under a single, broader, more fundamental one. The coming together and merging of diverse facts or concepts to form a more general, encompassing framework is termed unification. This is the recognition and demonstration that an apparent diversity is, in fact, a unity, whose various elements are but different aspects of a common core.

Unification continually takes place throughout physics as an essential ingredient of physicists' striving to gain understanding—when apparently different things are recognized as manifestations of the same thing, they become better understood. Some noteworthy examples of unification that have occurred at various times and at various levels in the operation of physics follow.

Start with a unification of phenomena, and take the phenomena to be the motions of the Sun and planets in the sky, as observed from Earth. Judged by appearance alone, those motions seem to have almost nothing to do with each other. That was the situation until the 16–17th centuries, when the only known planets were those that can be seen with the naked eye: Mercury, Venus, Mars, Jupiter, and Saturn. Johannes Kepler studied Tycho Brahe's earlier observations of the motions of the Sun and those planets and gained the recognition that the diversity of motion is only apparent. He showed that all the seemingly unrelated motions fit into the solar system model that is taken for granted today, together with three laws of planetary motion, which are obeyed by all the known planets, including Earth. When additional planets—Uranus, Neptune, and Pluto (since reduced in status to a plutoid)—were discovered, they, too, were found to obey Kepler's laws.

That was indeed unification on an astronomical scale! Rather than planets' moving in any which way along just any orbits, the picture became one of nine particular cases of a single set of laws of motion. Perhaps, one might even say, nine manifestations of a single generic planetary orbit around the Sun.

Consider another example of unification of phenomena. Let these phenomena comprise the various and diverse chemical elements: hydrogen, helium, lithium, . . . , etc., etc. In the 19th century, Dmitry Mendeleyev discovered a pattern among the elements, which led to his development of the first periodic table of the elements. In the 20th century, quantum mechanics explained Mendeleyev's periodic table. What became clear was that the fundamental unit of a chemical element is the atom, which consists of a positively charged nucleus and a number of (negatively charged) electrons. Each element is characterized by the number of electrons in a neutral atom of the element, its atomic number, which serves to order the elements in the periodic table. The chemical and physical properties of an element depend on the states that are allowed to its atomic system of nucleus and electrons, in particular its ground state, its state of lowest energy. Quantum mechanics determines those states.

Here was unification on the atomic scale. The chemical elements, as widely different as they are in so many ways, became understood as manifestations of the same basic phenomenon—a number of electrons in the electric field of a positively charged nucleus.

Turn now to an example of unification of laws. Until the 19th century, the understanding of electric and magnetic effects and of relations between the two had been gradually developing. Various patterns that had been discovered were expressed as a number of laws, including Coulomb's law, Ampère's law, the Biot-Savart law, and Gauss's law. In the 19th century, James Clerk Maxwell proposed a theory of electromagnetism in the form of a set of equations, called Maxwell's equations. These equations unified all the previously known laws, in the sense that all the latter can be derived from Maxwell's equations for particular cases. All the various laws became understood as different aspects of the laws of electromagnetism, as expressed by Maxwell's equations. Furthermore, Maxwell's the-

erence frame; time was relative, and time was fused with space. One could not refer to "now" without also asking "where?" and "relative to an observer at what velocity?" Both time and space were necessary to establish a frame of reference.

Special relativity changed many previously accepted aspects of physics. One sensational realization was that the mass of an object increases as its speed and kinetic energy increase. Einstein defined this newly recognized relationship between mass and energy in the world's most famous equation,

$$E = mc^2$$

ory showed that electricity and magnetism themselves must be viewed as two aspects of the more general phenomenon of electromagnetism. In this way, electricity and magnetism were unified within electromagnetism.

Unification takes place also at the level of theories. For example, the modern, quantum treatment of the electromagnetic force (electromagnetism), the weak force, and the strong force is carried out in the framework of quantum field theory, where the three forces are described by means of what are known as gauge theories. During the 20th century, it became apparent that the electromagnetic and weak forces have much in common, although they appear to be quite different. For instance, the electromagnetic force is mediated by the spin-1, massless, neutral photon, while the weak force has three kinds of mediating particle, all with spin 1, all possessing mass, and two carrying electric charge.

Yet, at sufficiently high energy, meaning high temperature, when the mass energy of the weak intermediaries becomes negligible, the neutral one becomes very similar to the photon and the two forces merge into one, the electroweak force. This force, then, forms a unification of the electromagnetic and weak forces that is valid at high energy. The force is mediated by four particles. Each of the electromagnetic and weak forces, with its corresponding intermediating particle(s), is an aspect of the electroweak force. Presumably, that was the situation during an early stage in the evolution of the universe, when temperatures, and thus energies, were very much higher than they are now. As the universe expanded and cooled, the two forces eventually "parted ways," so to speak, each developing its own individual identity. In this way, the theory of the electroweak force unifies the theories of the electromagnetic and weak forces.

Even beyond the electroweak unification, there are indications that a further unification should exist. This one would unify the electroweak force with the strong force, giving what is called a grand unified theory (GUT). If true, this unification should be valid only at even higher energies and temperatures than are required for electroweak unification. And in the life of the universe, the GUT should have held sway for a shorter duration than did the electroweak theory.

The ultimate goal of unifying *all* the interactions dazzles physicists. It would unify the GUT with gravitation and encompass all the known interactions within a single theory, grandly labeled a "theory of everything" (TOE). Its reign in the evolution of the universe would have to have been only for the briefest of times, when the universe was extremely hot. As the expansion of the universe cooled it, the gravitational and GUT forces would become distinct from each other. Later, at a lower temperature, the strong and electroweak forces would separate from each other and gain their identities. Then the weak and electromagnetic forces would enter into separate existence. The unification of electricity and magnetism within electromagnetism remains valid still today.

Physicists are actively working on GUTs, but so far with no definitive result. Work on a TOE must first overcome the hurdle of making Albert Einstein's general theory of relativity, which is the current theory of gravitation, compatible with quantum physics. The result, if and when achieved, is termed a theory of quantum gravity. This theory would—it is hoped—then be unified with the GUT—assuming one is found—to produce a TOE. It is difficult to imagine what, if anything, might lie beyond a theory of everything. The unification scenario just described—leading to a TOE—is merely conjecture, only an extrapolation from the present situation into the future. Probably, as has so often happened in physics, new discoveries will change the picture dramatically, and the path of unification will lead in very unexpected directions.

Whatever might occur in that regard, unification has so far served physics very well, in the ways described as well as in many others. Unification makes for simplicity. Rather than two or more disparate entities needing to be understood (think of electricity and magnetism, or of all the chemical elements), each requiring its own laws and theory, only a single, broader, and deeper entity takes the investigational stage (e.g., electromagnetism, or electrons in the electric field of a nucleus). The many laws and theories are replaced by fewer laws and by a single, encompassing, more fundamental theory (Maxwell's equations, or quantum mechanics, in the examples), and physics takes a further step toward a deeper understanding of nature.

FURTHER READING

Greene, Brian. *The Fabric of the Cosmos: Space, Time, and the Texture of Reality.* New York: Vintage, 2004.

Hatton, John, and Paul B. Plouffe. *Science and Its Ways of Knowing.* Upper Saddle River, N.J.: Prentice Hall, 1997.

Pagels, Heinz R. *Perfect Symmetry: The Search for the Beginning of Time.* New York: Bantam, 1985.

Smolin, Lee. *The Life of the Cosmos.* New York: Oxford University Press, 1997.

Wilson, Edward O. *Consilience: The Unity of Knowledge.* New York: Vintage, 1998.

where E represents energy in joules (J), m represents mass in kilograms (kg), and c is the speed of light with the value 2.99792458×10^8 meters per second (m/s), or 3.00×10^8 m/s to three significant figures. The major implication of this equation is that matter possesses an energy that is inherent to it. Mass can be converted to energy, and a minuscule amount of mass possesses a huge amount of energy. The efficiency of nuclear power and the destructiveness of nuclear weapons are based on this principle. Alternatively, energy has a mass associated with it and can be converted to mass, such as when a sufficiently energetic photon (called a gamma ray) passes by a nucleus and converts to an electron-positron pair.

Another conclusion of special relativity is that nothing can travel faster than light and no massive body can be accelerated to the speed of light. For example, in particle accelerators, electrons can be accelerated to more than 99 percent of the speed of light, but never to 100 percent of c.

People did not immediately understand Einstein's theory. In 1908 one of his former teachers at the institute, Hermann Minkowski, presented Einstein's theory of relativity in the context of the new concept of fused space-time. Though the faculty at the University of Bern initially had rejected Einstein for a job after he submitted a copy of his theory of relativity in 1907, Minkowski's explanation of the space-time continuum helped the world understand and appreciate Einstein's brilliance.

GENERAL RELATIVITY

The next several years were characterized by a series of academic positions of increasing prestige. In 1908 Einstein was named a *privatdozent,* an unpaid lecturer, at the University of Bern. The following year, he was appointed associate professor at the University of Zurich. In 1911 he became a full professor at the German University of Prague. During the period 1914–33, Einstein headed the Kaiser Wilhelm Physical Institute in Berlin, became a professor at the University of Berlin, and became a member of the Berlin Academy of Sciences.

Since proposing his special theory of relativity, Einstein had been contemplating the situation of moving bodies in relation to other moving bodies. He published his general theory of relativity in the *Annalen der Physik* in 1916 in "The Foundation of the General Theory of Relativity." Newton's universal law of gravitation explained many physical phenomena, but certain discrepancies appeared. Einstein's general theory of relativity resolved these discrepancies and lifted the restrictions to uniformly moving reference frames from his special theory of relativity.

A significant factor in the general theory of relativity was the equivalence principle, which stated that no one can determine by experiment whether he or she is accelerating or is in a gravitational field. This principle can be illustrated by one of Einstein's famous thought experiments, situations that he contemplated in order to provide support for or to refute a theory. For example, if a woman in a free-falling elevator were holding a ball and she let go of it, the ball would not fall to the floor but would appear to float in the air at the same position where her hand let go of it. This is true because the elevator would be pulled toward the Earth at the same acceleration that the ball would. If the woman were in a stationary elevator in outer space, away from any gravitational field, and she let go of her ball, it would float, just as if she were in the free-falling elevator. The passenger would not be able to distinguish between the two situations. If the earthbound elevator were stationary, the ball would drop and accelerate toward the floor when she let go. And if the elevator were accelerating upward in outer space, the ball would also appear to accelerate toward the floor. The passenger would not be able to distinguish whether the ball fell because the elevator was ascending in the absence of gravity, or whether gravity pulled the ball downward in the absence of ascension. In other words, there is no physical difference between an accelerating frame of reference (the ball falls because the elevator is ascending) and one in a gravitational field (the elevator is at rest, and the ball is pulled downward by gravity).

Whereas Newton's universal law of gravitation describes gravity as a force that acts between bodies, Einstein described gravity as a curvature of space-time surrounding a body, while the other body moves in a force-free manner along a path that is the analog in curved space-time of a straight line in ordinary space. This motion appears in ordinary space as motion under the effect of gravity. Einstein's general theory of relativity predicted also that a ray of light would be affected similarly and would bend—following a "straight line" in curved space-time—as it passes near a massive body such as the Sun.

To illustrate the idea of motion in curved space-time, in particular of light, consider a rubber sheet stretched over the edges of a bowl. A marble-sized ball, representing a photon, can roll across the rubber sheet in a straight line. If a weight representing the Sun is placed in the middle of the rubber sheet, it distorts the rubber sheet, now representing curved space-time. As the marble rolls by the weight, it follows a curved path. Similarly, as light passes by the Sun, the beam is bent.

Einstein's theory explained also the unusual orbit of Mercury, the planet closet to the Sun, and therefore the most affected by the gravitational field surrounding the Sun, but additional practical proof in support of the general relativity theory was not provided until 1919. By then, Einstein's body was worn out. He spent more time theorizing than caring for his own physical needs. He moved in with his cousin Elsa, who cared for him. After he divorced his first wife, he married Elsa.

In November 1919, the astrophysicist Arthur Eddington presented proof for general relativity to a joint meeting of the Royal Society and the Royal Astronomical Society in London. Eddington had taken photographs of the solar eclipse from the island of Príncipe in the Gulf of Guinea, off the west coast of Africa. Stars that had not been visible in sunlight were visible during the total eclipse of the Sun. The positions of the stars during the eclipse were compared with their positions in photographs taken from exactly the same location six months earlier, when the Sun was on the other side of Earth. It appeared as if the stars had moved. What actually happened was that the light from the stars was bent as it passed by the Sun—however, following a "straight line" in curved space-time—and, moreover, bent by the amount that the general theory of relativity predicted. Einstein was thrilled that his predictions proved true, and the eccentric, scruffy-looking professor suddenly enjoyed fame among the general public as well as his colleagues.

UNIFICATION FAILURE BUT FURTHER ACCOMPLISHMENTS

For the next 30 years, Einstein struggled unsuccessfully to uncover a unified field theory that would embrace all of nature's forces, which at that time were thought to comprise electromagnetism and gravity. He wanted to find a law that would describe the behavior of everything in the universe, from the elementary particles of an atom to the celestial bodies in the cosmos. He published the first version of his efforts toward this goal in 1929. Some of his colleagues thought he was wasting his time. In spite of that, during the last half of his life, Einstein accomplished quite a lot, both in physics and in the social and political arenas. He used his celebrity status to speak out against anti-Semitism, and the Nazis offered a reward for his assassination. In 1933 he left Europe for Princeton, New Jersey, where he accepted a position at the newly founded Institute for Advanced Study. He became a U.S. citizen in 1940.

In 1939 the Danish physicist Niels Bohr informed Einstein that the German scientists Otto Hahn and Lise Meitner had accomplished nuclear fission and described the enormous power released in the process. Ironically, this phenomenon was predicted by the relationship $E = mc^2$, discovered decades before by Einstein himself. Worried about the possible consequences if the Nazis had possession of nuclear weapons, Einstein wrote a letter to President Franklin D. Roosevelt warning him of the possibilities and destructive power that might soon be in the hands of the Germans. As a result, Roosevelt initiated the top secret Manhattan Project to build the atomic bombs that were ultimately used to end World War II. Einstein later admitted regret for his involvement and spent years after the war campaigning for nuclear disarmament.

During his years at Princeton, Einstein worked in turn with a number of young collaborators, who later became well-known theoretical physicists in their own right. One of these was the American, later Israeli, Nathan Rosen, with whom Einstein collaborated closely during 1934–35. Among their investigations, they developed the following three major ideas:

(1) There are problems with the description of reality offered by quantum mechanics. This was expressed in their paper, with American Boris Podolsky, "Can Quantum-Mechanical Description of Physical Reality Be Considered Complete?" The paper was published in *Physical Review* in 1935, is commonly referred to as the EPR paper, and is the most widely cited paper in physics. The subject of the paper is intimately related to contemporary cutting-edge research areas such as quantum cryptography and quantum cloning. The experiment proposed there has since been carried out, and the results confirm the predictions of quantum mechanics, with which Einstein was very uncomfortable.

(2) Another idea put forth by Einstein and Rosen concerned the problematics of an elementary particle in the general theory of relativity, discussed in another 1935 *Physical Review* article, "The Particle Problem in the General Theory of Relativity." Here they introduced the concept of what became known as the "Einstein-Rosen bridge." This is, so to speak, a "direct" connection between events in space-time, allowing time travel and "warp speed" space travel. These days the more common term is *wormhole*. Carl Sagan explicitly refers to it in his science-fiction book *Contact* and in the film based on the book.

(3) The third idea resulted in the seminal paper "On Gravitational Waves," published in the *Journal of the Franklin Institute* in 1937, in which the existence of waves in the gravitational field, or waves in the "fabric of space-time," was proposed. Although indirect evidence of the existence of gravitational waves now exists, their direct detection is the goal of a number of Earth-based projects and the planned space-based LIGO project.

In 1952 Einstein was offered the presidency of the new state of Israel. He declined and continued working on his fruitless calculations for a unified field theory. He died at the age of 76 in Princeton, New Jersey, of a ruptured aorta. His body was cremated and his ashes scattered in an undisclosed location.

The results of Einstein's research caused an upheaval in physics. While his work explained some phenomena and results that had puzzled physicists previously, it also forced scientists to abandon what had seemed to be commonsense knowledge. The photoelectric effect has been applied in technology such as motion and light sensors and exposure meters on cameras. The theories of relativity shelved the separate notions of space and time and fused them into a single space-time. The interconversion of mass and energy has facilitated atomic studies, allowed the harnessing of nuclear power, and shed light on the big bang theory of the origin of the universe. Though Einstein never succeeded in discovering a unifying theory and was even ridiculed for trying to do so, today physicists have returned to this problem, even picking up on ideas that Einstein originated. In 1952 a new element, atomic number 99, was discovered. It was named *einsteinium,* after the man who changed the way humanity looked at matter, time, and space.

See also ACCELERATION; BIG BANG THEORY; ELECTRICITY; ELECTROMAGNETIC WAVES; ELECTROMAGNETISM; ENERGY AND WORK; EPR; FISSION; GENERAL RELATIVITY; MAGNETISM; MASS; NEWTON, SIR ISAAC; NUCLEAR PHYSICS; PHOTOELECTRIC EFFECT; PLANCK, MAX; SIMULTANEITY; SPECIAL RELATIVITY; TIME.

FURTHER READING

Bodanis, David. *$E = mc^2$: A Biography of the World's Most Famous Equation.* New York: Walker, 2000.

Cropper, William H. *Great Physicists: The Life and Times of Leading Physicists from Galileo to Hawking.* New York: Oxford University Press, 2001.

Goldsmith, Donald. *The Ultimate Einstein.* New York: Byron Preiss Multimedia: Pocket Books, 1997.

Heathcote, Niels Hugh de Vaudrey. *Nobel Prize Winners in Physics 1901–1950.* Freeport, N.Y.: Books for Libraries Press, 1953.

The Nobel Foundation. "The Nobel Prize in Physics 1921." Available online. URL: http://nobelprize.org/physics/laureates/1921/. Accessed November 16, 2007.

Nova Online, WGBH Educational Foundation, 1996. "Einstein's Big Idea." Available online. URL: http://www.pbs.org/wgbh/nova/einstein/. Accessed January 11, 2008.

Sagan, Carl. *Contact.* New York: Pocket Books, 1997.

Strathern, Paul. *The Big Idea: Einstein and Relativity.* New York: Anchor Books, 1999.

elasticity The tendency of materials to resist deformation and, after being deformed, to regain their original size and shape after the cause of deformation is removed is termed *elasticity.* This definition refers to solid materials, although liquids and gases possess elastic properties as well, but only with regard to volume change. The cause of deformation is called *stress,* while the response of the material to stress, the deformation itself, is *strain.* Many materials do indeed return to their initial configuration, at least if the stress is not too large. These are termed elastic materials. Examples are steel, wood, some plastic, bone, and rubber. Plastic materials, including chewing gum, caulk, putty, and ice cream, do not return to their original dimensions.

The behavior of elastic materials under increasing stress is typically as follows. For a range of stress from zero up to some limit, called the proportionality limit, the strain is proportional to the stress. The extension of a coil spring, for example, is proportional to the stretching force, up to some maximal force. The proportionality of strain to stress is the content of Hooke's law, named for the 17th-century English scientist Robert Hooke. So materials with a significantly high proportionality limit can be said to obey Hooke's law (with the understanding that above the proportionality limit, Hooke's law becomes invalid for them). Although many elastic materials do obey Hooke's law, some, such as rubber, do not, or do so only approximately.

Materials that obey Hooke's law can be characterized by a number of quantities called moduli of elasticity, all of which are defined as the ratio of the stress to the resulting strain:

$$\text{Modulus} = \frac{\text{stress}}{\text{strain}}$$

In all cases the stress is force per unit area, and its unit is the pascal (Pa), equivalent to newton per square meter (N/m^2). The strain is always a dimensionless ratio of a change in a quantity to the value of the quantity. So the unit of the modulus of elasticity is the same as that of stress, the pascal.

For linear deformation caused by tension or compression, Young's modulus is the ratio of the tensile stress, the longitudinal force per unit cross-sectional area, to the tensile strain, the relative change in length. If a longitudinal force of magnitude F causes a length change ΔL in a length L, then Young's modulus Y is defined as

$$Y = \frac{F/A}{\Delta L/L}$$

where A denotes the cross-sectional area. Here F is in newtons (N), A in square meters (m^2), and ΔL and L in the same units of length. F and ΔL are

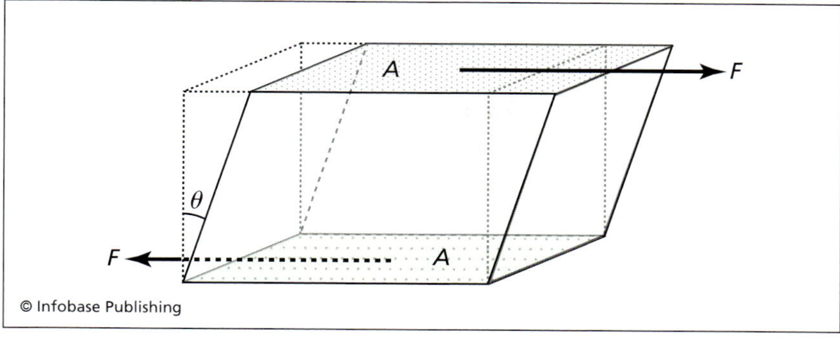

Shear is a deformation of material caused by a pair of equal-magnitude, oppositely directed forces, whose lines of action are displaced from each other. Each force, of magnitude *F,* acts parallel to an area of magnitude *A.* The shear stress is *F/A,* while the shear strain equals the angle of deformation, θ (expressed in radians).

taken to have the same algebraic sign, so a positive force causes extension, while a negative force compresses.

For volume change under pressure, the bulk modulus B is defined as the ratio of the pressure, which is the bulk stress, to the relative change of volume, the bulk strain:

$$B = -\frac{p}{\Delta V/V}$$

Here p denotes the pressure in pascals (Pa), V the volume, and ΔV the volume change, where ΔV and V are in the same units of volume. A minus sign is introduced to make B positive, since a positive pressure, a squeeze, causes a decrease of volume (i.e., a negative value for ΔV).

The situation in which a pair of equal-magnitude, oppositely directed (i.e., antiparallel) forces act on material in such a way that their lines of action are displaced from each other is called shear. The result of shear can be an angular deformation of the material, possibly including twisting, or even separation of the material. We produce shear when we tear paper or use a paper punch, for instance. Scissors (also called shears) and electric shavers operate by means of shear. As examples of noncatastrophic shear deformation, the joining together of the bare ends of a pair of electric wires by twisting them is a shear effect, as is the realignment of the hook of a wire coat hanger by twisting it.

The shear of the force pair is described by the shear stress, whose value is the ratio of the magnitude of one of the forces to the area of the surface parallel to the forces upon which the forces are acting. Its SI unit is the pascal (Pa), equivalent to newton per square meter (N/m^2). But note that although shear stress is force per unit area, it is not pressure. The latter is force per unit area perpendicular to the force, while in the present case the force is parallel to the area upon which the force pair is acting. Shear strain, which equals the angle through which the material is deformed, describes the result of noncatastrophic shear, the deformation of the material. For elastic deformation, the material's resistance to shear is measured by its shear modulus S, also called rigidity modulus, whose value is the ratio of shear stress on the material to the resulting shear strain,

$$S = \frac{F/A}{\theta}$$

REPRESENTATIVE ELASTIC MODULI FOR VARIOUS MATERIALS

Material	Young's Modulus Y (10^{10} Pa)	Bulk Modulus B (10^{10} Pa)	Shear Modulus S (10^{10} Pa)
Aluminum	7.0	7.5	2.5
Brass	9.0	6.0	3.5
Copper	11	14	4.4
Crown glass	6.0	5.0	2.5
Iron	21	16	7.7
Lead	1.6	4.1	0.6
Nickel	21	17	7.8
Steel	20	16	7.5
Titanium	12	11	4.6

Here F denotes the shear force in newtons (N), A the area parallel to the force over which the force is acting in square meters (m^2), and θ the deformation angle in radians (rad).

For a range of stress above the proportionality limit, although the strain is not proportional to the stress, the material still behaves elastically: if the stress is removed, the material will still regain its original configuration. That holds for stresses up to the elastic limit. Beyond the elastic limit, the atoms or molecules that constitute the material rearrange themselves and the deformation becomes permanent. Then the material behaves in a plastic manner. As an example, think of a paper clip. In normal use, it stays within its elastic limit and successfully holds sheets of paper together. If used to clip too many sheets, it becomes bent open and no longer does its job.

Above the elastic limit is the yield point, the stress at which the material starts to flow, its strain increasing over time with no further increase of stress. That leads to the breaking point, the stress at which the material ruptures.

A potential energy can be defined for elastic deformation, since the deforming force is then a conservative force. That is especially simple as long as Hooke's law is obeyed. For an ideal spring, for instance, the applied force, of magnitude F, in newtons (N), and the spring's resulting displacement x, in meters (m), are related by

$$F = kx$$

for any force below the proportionality limit, where k is the spring constant, or force constant, in newtons per meter (N/m). The corresponding potential energy

This machine is used to investigate the tensile properties of metals by stretching uniform rods made of those metals. The sample in the machine has clearly passed its elastic limit. *(Maximilian Stock Ltd/Photo Researchers, Inc.)*

E_p, in joules (J), of a spring that has been stretched or compressed through displacement x is

$$E_p = \tfrac{1}{2} kx^2$$

These formulas are valid as well for any elastic device or object that behaves as an ideal spring.

In order to help clarify the use of these formulas, consider this problem. A bungee jumper drops from the top of a cliff with her ankle attached to the end of a bungee cord. The other end of the cord is fixed firmly to the top of the cliff. The jumper's mass is m = 50.0 kg, the cord's force constant is k = 300 N/m, and its unstretched length is L_0 = 20.0 m. Ignoring air resistance and assuming that the bungee cord obeys Hooke's law and the preceding formulas apply, what is the farthest the jumper will fall? What is the bungee cord's force on the jumper at that instant? For simplicity, ignore also the size of the jumper.

Immediately after leaving the top of the cliff, the jumper is in free fall, affected solely by the downward pull of gravity and accelerating at g = 9.80 m/s^2, until she falls the length of the unstretched bungee cord (which is 20.0 m). From that instant, the cord starts to stretch and exerts an upward force on the jumper, a force that increases in magnitude as the jumper continues falling. During this segment of the fall, the constant downward force of gravity, whose magnitude is the jumper's weight, is opposed by the increasing upward force of the cord. But since the jumper's weight is greater than the magnitude of the cord's force, the jumper continues to accelerate downward, albeit with continually decreasing acceleration, according to Newton's second law of motion. At some amount of stretch, the magnitude of the bungee cord's force equals the weight, and at that instant the net force on the jumper is zero; and, accordingly, she has zero acceleration, according to Newton's first law of motion, but still has downward velocity. From that instant on, the magnitude of the cord's upward force is greater than the weight and continually increases, so an increasing net upward force acts on the jumper, who, following the second law, decelerates at an increasing rate. At some instant, the deceleration has reduced the jumper's speed to zero and she starts to rebound. The total distance the jumper has fallen until this instant answers the first question asked.

This problem involves a varying force and is most easily treated using energy methods, if possible. Indeed that *is* possible: one can use conservation of mechanical energy for the jumper, Earth, and the bungee cord, since no external work is being performed on that system. Take the top of the cliff as the reference level for gravitational potential energy. When the jumper is on the top of the cliff, she is at rest, so her kinetic energy is zero. Her gravitational potential energy is also zero, and the bungee cord's elastic potential energy is zero as well, since the cord is unstretched. So the system's total initial mechanical energy is zero. Denote the greatest distance the jumper falls by L. At that distance, her speed is instantaneously zero, so her kinetic energy is zero. Her gravitational potential energy is then $-mgL$. The bungee cord is then stretched by $x = L - L_0$, so using the formula, its elastic potential energy equals $\frac{1}{2}k(L - L_0)^2$. The total mechanical energy of the system in this state is thus

$$\tfrac{1}{2}k(L - L_0)^2 - mgL = 0$$

and equals zero because of conservation of mechanical energy from the initial state. The answer to the first question of the problem is found by solving this quadratic equation for L, all the other quantities in the equation being given. Two solutions emerge, 29.9 m and 13.4 m. The second solution is meaningless in the context of the problem, giving the result that the bungee jumper falls 29.9 meters before coming to momentary rest and rebounding. In answer to the second question, when the jumper is 29.9 m below the top of the cliff, the bungee cord exerts an upward force on her. The magnitude of this force equals the magnitude of the force that she exerts on the cord, according to Newton's third law of motion. That quantity, by the formula presented earlier, equals $k(L - L_0)$, whose numerical value is 2.97×10^3 N.

See also ENERGY AND WORK; FORCE; MOTION; PRESSURE.

FURTHER READING

Young, Hugh D., and Roger A. Freedman. *University Physics,* 12th ed. San Francisco: Addison Wesley, 2007.

electrical engineering This is the field that studies electricity and electromagnetism and their application for practical purposes. It comprises various subfields, which include electronic engineering, computer engineering, power engineering, control engineering, instrumentation engineering, signal processing engineering, and telecommunications engineering.

Electrical engineers typically study four or five years at the college level toward a bachelor's degree. Their curriculum includes foundational courses in physics and mathematics as well as various courses specific to electrical engineering, such as circuit theory and computer science, with many labs. Toward the end of their studies, they specialize in a subfield and concentrate on courses that are relevant to that subfield. The occupations of electrical engineers vary widely. They might design and develop electric power systems, from the generation of electricity, through its long-distance transmission, to its delivery to the consumer. Or they were and are involved in the design and maintenance of the Global Positioning System (GPS) and in the development of the various devices that make use of that system to help people locate themselves and obtain travel directions to their destinations. Electrical engineers might design mobile phones, laptop computers, DVD players and recorders, or high-definition television sets. They might be called upon to plan the electric wiring of new construction.

The field of electrical engineering emerged as a separate field, distinct from physics, in the late 19th century. It was then that investigations into electric phenomena intensified. Two names that are associated with this work are Thomas Edison and Nikola Tesla. The field developed rapidly in the early 20th century with the invention and application of radio, television, radar, and other technologies, leading to such applications as electronic computers in the mid-20th century. Research in the 20th century led from the vacuum tube to the transistor and on to the integrated circuit, which forms the foundation of today's electronics.

The largest professional organization in the world for electrical engineers, in the broadest sense of the word, is the Institute of Electrical and Electronics Engineers, known as IEEE, or I-triple-E. This organization serves as a source of technical and professional information, determines standards, and fosters an interest in the profession of electrical engineering. As of the end of 2007, IEEE had 375,000 members in more than 160 countries and included 38 member societies. The organization published 144 transactions, journals, and magazines and sponsored or cosponsored more than 850 conferences annually throughout the world. IEEE had 900 active electrical engineering standards and more than 400 in development. Standards are standardized conventions, that is, rules, for the purpose of maintaining consistency and interconnectibility. For instance, the standard for USB plugs and sockets on cables and computers assures that every USB plug will fit into every USB socket. Similarly, the standard for CDs makes it possible to buy a CD and know that any CD player will play it.

SUBFIELDS

Electrical engineering encompasses numerous subfields such as electronic engineering, computer engineering, electric power engineering, control engineering, instrumentation engineering, signal processing engineering, and telecommunications engineering.

Electronic engineering involves the design, testing, development, and manufacture of electronic circuits and devices. The circuits might be composed of discrete components, such as resistors, inductors, capacitors, and transistors. More modern circuits, called integrated circuits, combine microscopic versions of those components, sometimes millions of them, on small plates of silicon called chips. These chips might have a size on the scale of about half an inch (about 1 cm). Devices that make use of electronic circuits include television sets, audio systems, computers, GPS devices, miniature voice recorders, and even various controllers in automobiles.

Computer engineering deals with computers, of course, as the name implies, but also with systems based on computers. Computer engineers design, develop, and test the hardware for ordinary (laptop and desktop) computers, for supercomputers, and for computer-based systems such as consoles of video games, automobile computers, and computers integrated into mobile phones.

Electric power engineering involves the generation, transmission, distribution, and utilization of electric power. Among their many occupations, electric power engineers design electric generators and transmission lines, both conventional and superconducting. They develop transformers and other devices essential for the operation and protection of the electric grid, which is the system of interconnected electric generators and transmission lines throughout the country. And they design electric motors.

Control engineering is related to understanding the behavior of systems and designing methods and devices to control them, that is, to make the systems operate as desired. Control engineers study systems mathematically, by finding equations that describe their behavior and solving the equations, often numerically with the help of computers. Such procedures are called modeling. The systems studied and controlled are diverse: for example, industrial processes, military and civilian aircraft, spacecraft, rockets, and missiles,

Instrumentation engineering deals with designing, developing, and testing devices that measure various physical quantities. These might include flow speed, pressure, acceleration, temperature, electric current, and magnetic field. Such devices are often used in conjunction with control systems. For exam-

This is a distribution center for electric power that is generated at a nearby electric power plant. The transformers raise the voltage to an appropriate value for transmission along the long-distance power lines. *(Anson Hung, 2008, used under license from Shutterstock, Inc.)*

ple, a home thermostat measures the temperature of the air in which it is immersed and uses the measurement result to turn on and off the furnace or air conditioner as needed to maintain a constant temperature.

Signal processing engineering focuses on creating, analyzing, and processing electric signals, such as telecommunications signals. Signal processing engineers might design processes and devices that detect and correct errors that enter into the data that are transmitted by communication satellites, for example. Or they might devise methods for "cleaning up" noisy intercepted voice messages obtained from electronic surveillance of enemy radio communications. On the other hand, signal processing engineers might investigate methods for the secure transmission of secret data.

Telecommunications engineering specializes in the transmission of data, which might represent speech, numerical data, or other information. The modes of transmission include electromagnetic waves through air and empty space (such as radio and television), light waves through optical fibers, and electric signals through conducting wires or coaxial cables. Telecommunication engineers design and develop the transmitters, receivers, and other devices needed for such purposes. They might send data digitally, in the form of pulses, or in analog form. The latter requires the transmission of a carrier wave that encodes the information by what is known as modulation. The most common types of modulation are frequency modulation (FM) and amplitude modulation (AM). In the former, the transmitter varies, or "modulates," the frequency of the carrier wave in a manner that represents the signal that is being transmitted. Television and some radio broadcasts are transmitted in this way. In amplitude modulation, the transmitter encodes the signal in the carrier wave by affecting its amplitude. Amplitude modulation is the transmission mode for AM radio.

See also ELECTRICITY; ELECTROMAGNETIC WAVES; ELECTROMAGNETISM; HEAT AND THERMODYNAMICS; MAGNETISM; MOTION; OPTICS; PRESSURE; SUPERCONDUCTIVITY; WAVES.

FURTHER READING

Baine, Celeste. *Is There an Engineer Inside You? A Comprehensive Guide to Career Decisions in Engineering,* 3rd ed. Belmont, Calif.: Professional Publications, 2004.

Billington, David P. *The Innovators: The Engineering Pioneers Who Transformed America.* Hoboken, N.J.: Wiley, 1996.

Institute of Electrical and Electronics Engineers (IEEE) home page. Available online. URL: http://www.ieee.org. Accessed January 6, 2008.

electricity The phenomenon of electricity is based on the existence in nature of electric charges, which exert nongravitational forces on each other in accord with Coulomb's law (see later discussion). All matter is formed of elementary particles. At the level of elementary particles, one finds, for example, that an electron and a proton attract each other; electrons repel each other, as do protons; while neutrons are immune to the effect. Coulomb's law is valid for those forces when electrons are assigned one fundamental unit of negative electric charge, protons one fundamental unit of positive electric charge, and neutrons are declared neutral (i.e., free of electric charge). The magnitude of this fundamental unit of charge is conventionally denoted e and has the value $1.602176463 \times 10^{-19}$ coulomb (C), rounded to 1.60×10^{-19} C.

Electricity and magnetism make up two aspects of electromagnetism, which subsumes the two and, moreover, in which they affect each other.

COULOMB'S LAW

Named for the French physicist Charles-Augustin de Coulomb, Coulomb's law describes the electric forces that two point electric charges exert on each other. The forces on the two charges are oppositely directed along the line joining the charges and are equal in magnitude. The force on either charge is toward the other charge, if the charges are of opposite sign. In other words, unlike charges attract each other. If the charges are of the same sign, the force on either charge is directed away from the other charge: like charges repel each other. The magnitude of the forces is proportional to the product of the absolute values of the two charges and inversely proportional to the square of the distance between them. In a formula, the magnitude of each force is as follows

$$F = k\frac{|q_1||q_2|}{r^2}$$

where F denotes the magnitude of the force, in newtons (N), on either charge; $|q_1|$ and $|q_2|$ are the absolute values of the two charges q_1 and q_2, in coulombs (C); and r is the distance between them, in meters (m). The value of the proportionality constant k depends on the medium in which the charges are immersed. In vacuum, $k = k_0 = 8.98755179 \times 10^9$ N·m^2/C^2, rounded to 8.99×10^9 N·m^2/C^2. Otherwise

$$k = \frac{k_0}{\kappa}$$

where κ (Greek lowercase kappa) denotes the dielectric constant of the medium and measures the extent to which the medium, by means of its electric polarization, reduces the magnitude of the forces between the charges (see later discussion).

Note that F is the magnitude of the forces and is thus a positive number (or zero). A negative value for F is meaningless. That is why only the absolute values of the charges—ignoring their signs—are used in the formula. The directions of the forces are not obtained from the formula for F, but rather from the rule that like charges repel and unlike charges attract. Nevertheless, the formula for F is sometimes used with the signed charges instead of absolute values. In that case, according to the repulsion-attraction rule, a positive or negative F indicates repulsion or attraction, respectively.

The constant k is often written in the form

$$k = \frac{1}{4\pi\varepsilon} = \frac{1}{4\pi\kappa\varepsilon_0}$$

where $\varepsilon = \kappa\varepsilon_0$ is called the permittivity of the medium and ε_0 the permittivity of the vacuum, whose value is $8.85418782 \times 10^{-12}$ $C^2/(N\cdot m^2)$, rounded to 8.85×10^{-12} $C^2/(N\cdot m^2)$. Permittivity is just the inverse of the proportionality constant k, with a factor of $1/(4\pi)$ thrown in for convenience in other formulas.

$$\varepsilon = \frac{1}{4\pi k}$$

The origin of the term *permittivity* is historical, and it does not indicate that anything is permitted. Accordingly, the formula for the magnitude of the electric forces that two point charges exert on each other, according to Coulomb's law, is often written in the form

$$F = \frac{1}{4\pi\kappa\varepsilon_0}\frac{|q_1||q_2|}{r^2}$$

FORCE AND FIELD

Electric forces are described by means of the electric field, which is a vector quantity that possesses a value at every location in space and can vary over time. The electric field is considered to mediate the electric force, in the sense that any charge, as a source charge, contributes to the electric field, while any charge, as a test charge, is affected by the field in a manner that causes a force to act on the charge. The value of the electric field at any location is defined as the magnitude and direction of the force on a positive unit test charge (a charge of one coulomb) at that location. Its SI unit is newton per coulomb (N/C), or equivalently, volt per meter (V/m).

The vector force **F**, in newtons (N), that acts on an electric charge q, in coulombs (C), as a test charge, at a location where the electric field has the vectorial value **E**, is given by

$$\mathbf{F} = q\mathbf{E}$$

So the electric force acting on a positive test charge is in the direction of the electric field at the location of the test charge, while the force on a negative test charge is in the direction opposite to that of the field.

The contribution that a charge q, as a source charge, makes to the electric field at any point, called the field point, is as follows. The magnitude of the contribution, E, is given by

$$E = \frac{1}{4\pi\kappa\varepsilon_0}\frac{|q|}{r^2}$$

where $|q|$ denotes the magnitude of the source charge and r is the distance from the location of the charge to the field point. The direction of the contribution is away from the source charge, if the latter is positive, and toward the source charge, if it is a negative charge. The total electric field is the vector sum of the contributions from all the source charges. Note that in calculating the electric field and the force on a test charge, the test charge cannot at the same time serve as a source charge too. In other words, a charge cannot affect itself. The relation between the electric field and the source charges that produce it is also expressed by Gauss's law (see the following section).

An electric field line is a directed line in space whose direction at every point on it is the direction of the electric field at that point. Electric field lines make a very useful device for describing the spatial configurations of electric fields. The field lines for the electric field produced by a single positive point charge, for example, are straight lines emanating from the charge and directed away from it, like spines sticking out from a spherical porcupine or hedgehog. Only one electric field line can pass through any point in space. In other words, field lines do not intersect each other.

GAUSS'S LAW

Named after Karl Friedrich Gauss, the 18–19th-century German mathematician, astronomer, and physicist, Gauss's law exhibits a relation between the electric field and the electric charges that are its sources. More specifically, consider an arbitrary abstract closed surface in space, called, for the present purpose, a Gaussian surface. Gauss's law relates the values of the electric field at all points of this surface to the net electric charge that is enclosed within the surface.

A closed surface, such as an inflated balloon of any shape, possesses an inside and an outside. Divide the surface into infinitesimal surface elements. Let $d\mathbf{A}$ denote a vector that represents such an element at any point on the surface: the direction of $d\mathbf{A}$ at a point is perpendicular to the surface at the point and directed outward from the surface, while the magnitude of $d\mathbf{A}$, denoted dA, is the infinitesimal area of the surface element, in square meters (m^2). Let $\mathbf{E}$ represent the electric field at any point, with E denoting its magnitude. Form the scalar product between $d\mathbf{A}$ at any point on the surface and the electric field at the same point:

$$\mathbf{E} \cdot d\mathbf{A} = E\, dA \cos\theta$$

where θ is the smaller angle (less than 180°) between vectors $d\mathbf{A}$ and $\mathbf{E}$. The scalar product $\mathbf{E}\cdot d\mathbf{A}$ equals the product of the area of the surface element, dA, and the component of the electric field that is perpendicular to the element, $E \cos\theta$. Now sum (i.e., integrate) the values of $\mathbf{E} \cdot d\mathbf{A}$ over the whole surface. The result is the net outward electric flux Φ_e through the closed surface under consideration:

$$\Phi_e = \oint_{\text{surface}} \mathbf{E} \cdot d\mathbf{A}$$

The integral sign with the little circle on it and a closed surface designated for it is a special notation for integration over the designated closed surface.

Gauss's law states that the net outward electric flux through an arbitrary closed surface equals the algebraic sum (taking signs into account) of electric charges enclosed within the surface, divided by the permittivity of the medium in which it is all embedded. In symbols this is

$$\oint_{\text{surface}} \mathbf{E} \cdot d\mathbf{A} = \frac{1}{\varepsilon}\sum q$$

where Σq denotes the algebraic sum of electric charges enclosed by the surface and ε is the permittivity of the medium. Note that while all the charges in the universe affect the electric field and determine its value at every point, only the charges enclosed by

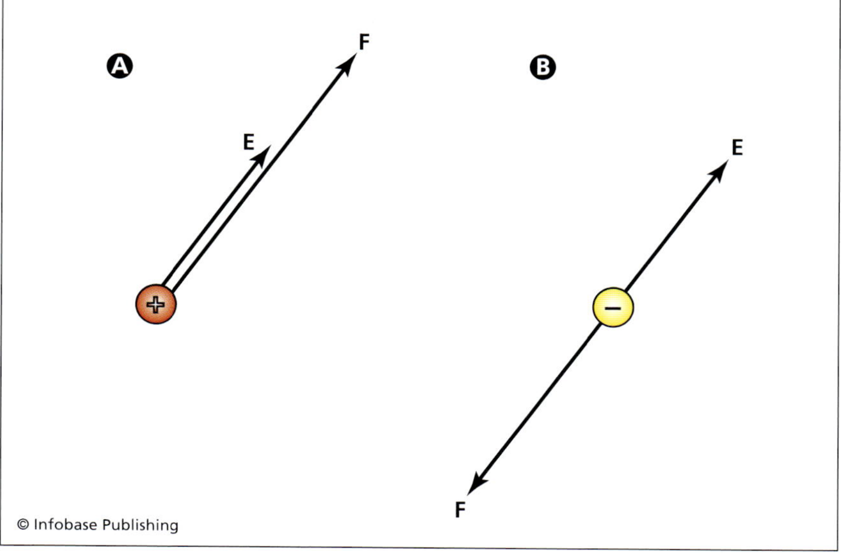

The direction of the force **F** acting on a point test charge in an electric field **E.** (a) When the test charge is positive, the force is in the direction of the field. (b) The force is directed opposite to the field in the case of a negative test charge.

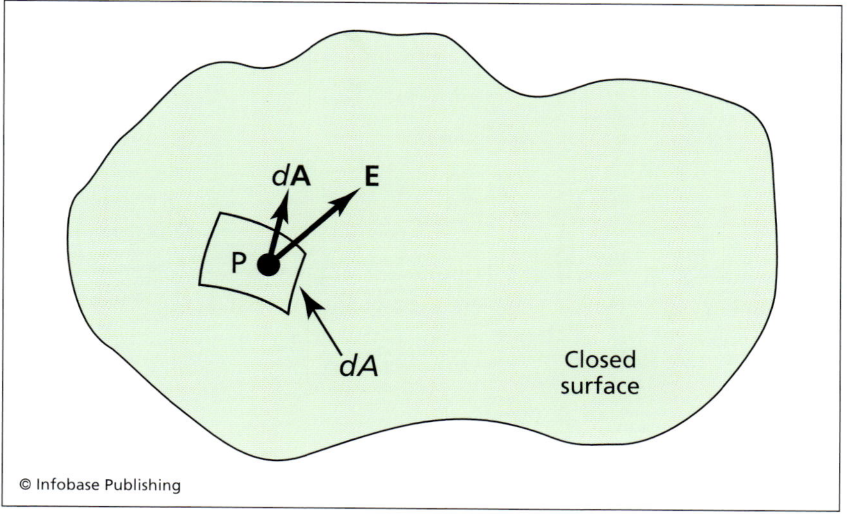

The figure shows a closed surface, called a Gaussian surface, to which Gauss's law is to be applied. An infinitesimal surface area element of magnitude dA is shown at an arbitrary point **P** on the surface. $d\mathbf{A}$ denotes the outward-pointing vector perpendicular to the surface at **P** with magnitude dA. **E** denotes the value of the electric field, a vector, at **P.** The outward electric flux through the surface is the integral of the scalar product $\mathbf{E}\cdot d\mathbf{A}$ over the whole surface. According to Gauss's law, this electric flux equals the algebraic sum of all the electric charges enclosed within the surface, divided by the permittivity of the medium.

the surface enter Gauss's law. The effects of all the charges outside the Gaussian surface cancel out.

Although Gauss's law holds for arbitrary surfaces, in many applications of the law, certain Gaussian surfaces are more useful than others. As an example, use Gauss's law to find the electric field caused by a single point charge. Consider the case of a single positive electric point charge q. What is the value (magnitude and direction) of the electric field **E** produced by the charge at some point? First, apply the following symmetry argument. The lone source charge, as the cause of the electric field, is symmetric under all rotations about all axes through itself. By the symmetry principle, the electric field, which is an effect of the charge, must possess the same symmetry. As a result, avoiding details, the magnitude of the electric field, E, can depend only on the distance of the field point from the charge, r, and the direction of the electric field is either everywhere radially away from the charge or everywhere radially toward it. In order to be able to solve for E, it must be possible to extract it from the integral in Gauss's law. This forces one to take for the Gaussian surface a spherical surface centered on the charge. Then at all points on the surface, E has the same value and **E** has the same direction with respect to $d\mathbf{A}$: either **E** points outward, $\theta = 0$, at all points, or **E** points inward, $\theta = 180°$.

Now assume that **E** points outward, $\theta = 0$, at all points on the spherical Gaussian surface. The left-hand side of Gauss's law then becomes as follows:

$$\begin{aligned}\Phi_e &= \oint_{\text{surface}} \mathbf{E}\cdot d\mathbf{A}\\ &= \oint_{\text{surface}} E\,dA\cos 0\\ &= E\oint_{\text{surface}} dA\end{aligned}$$

The integral in the last line is simply the total area of the spherical surface, $4\pi r^2$, where r is the radius of the sphere. So the left-hand side of Gauss's law in this case equals

$$\Phi_e = 4\pi r^2 E$$

The right-hand side for a single charge q becomes q/ε. Gauss's law then gives

$$4\pi r^2 E = \frac{q}{\varepsilon}$$

Solve for E to obtain

$$E = \frac{1}{4\pi\varepsilon}\frac{q}{r^2}$$

which is consistent with the relation given earlier, since both q and E are positive. Thus, for positive q, the electric field at any point is directed radially away from the source charge and its magnitude is given by the last equation, where r is the distance of the field point from the charge.

In the case of negative q, assuming that **E** points away from the source charge results in a sign inconsistency, which is resolved by assuming, instead, that **E** points toward the charge. The final result is that for a point charge q of any sign, the magnitude of the electric field produced by the charge at distance r from it is given by

$$E = \frac{1}{4\pi\varepsilon}\frac{|q|}{r^2}$$

where $|q|$ denotes the absolute value of q. The direction of the electric field is radially away from the source charge, if the charge is positive, and radially toward the charge, if it is negative. (Note that the magnitude and direction of the electric field can also be deduced from the definition of the electric field together with Coulomb's law.)

ENERGY AND POTENTIAL

The electric force is a conservative force (see FORCE), so a potential energy can be defined for it: the electric potential energy of a charge at any location is the work required to move the charge against the electric field from infinity (which is taken as the reference for zero electric potential energy) to that location. From this the electric potential, or simply the potential, at every location is defined as the electric potential energy of a positive unit test charge at that location. The electric potential is thus a scalar field. The potential difference, or voltage, between two locations is the algebraic difference of the values of the potential at the locations. The SI unit of potential and of potential difference is the volt (V), equivalent to joule per coulomb (J/C).

The potential energy E_p, in joules (J), of a test charge q at a location where the potential has the value V is

$$E_p = qV$$

The contribution V that a source charge q makes to the potential at any field point is

$$V = \frac{1}{4\pi\kappa\varepsilon_0}\frac{q}{r}$$

where the symbols have the same meanings as earlier. Note that the signed charge appears in this formula, so a positive source charge makes a positive contribution to the potential and a negative source charge makes a negative contribution. The total potential is the algebraic sum of the contributions from all source charges. As in the case of the electric field and its effect, here, too, a test charge cannot also serve as a source charge at the same time.

A useful device for describing the spatial configuration of the electric potential is the equipotential surface. This is an abstract surface in space such that at all points on it, the electric potential has the same value. The equipotential surfaces of a point source charge, as an example, are spherical shells centered on the charge. For any point in space, the electric field line passing through the point is perpendicular to the equipotential surface containing the point. A quantitative relation between the electric field and equipotential surfaces is that for any pair of close equipotential surfaces: (1) the direction of the electric field at any point between them is perpendicular to them and pointing from the higher-potential surface to the lower, and (2) the magnitude of the electric field E at such a point is given by the following:

$$E = \left|\frac{\Delta V}{\Delta s}\right|$$

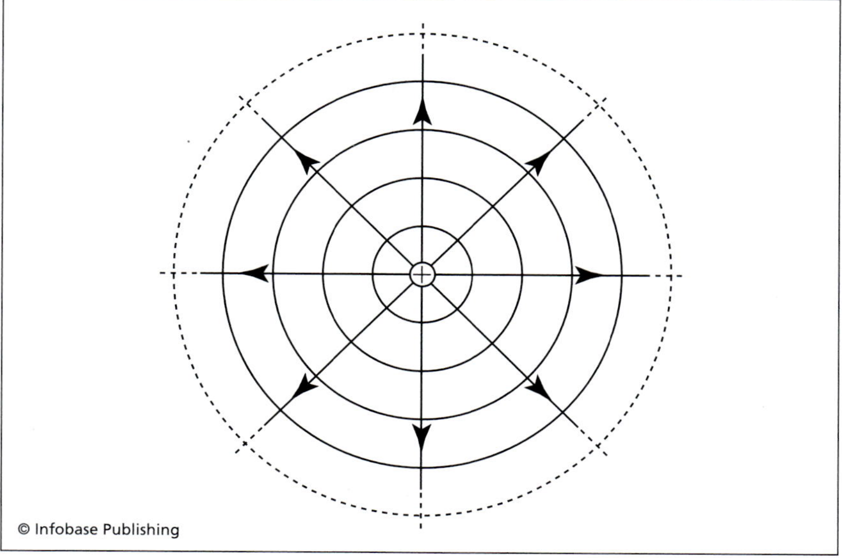

Electric field lines and equipotential surfaces for a positive point source charge, viewed in cross section. The outward-pointing radii indicate a number of field lines, and the concentric circles represent several equipotential surfaces, which are in reality concentric spherical surfaces. For a negative point source charge the picture is similar, but with the field lines pointing inward.

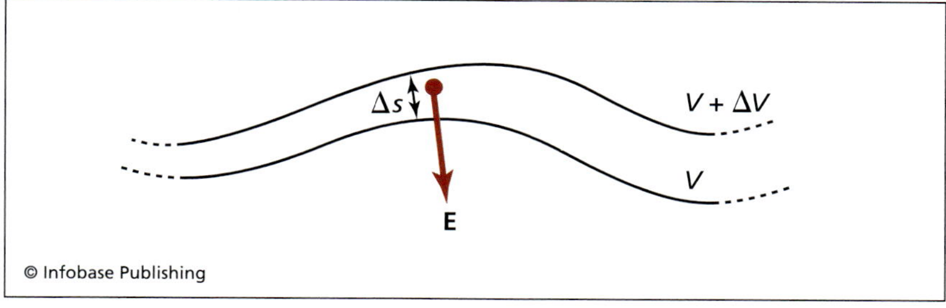

Parts of two close equipotential surfaces are shown in cross section. Their potentials are V (lower) and $V + \Delta V$ (higher). **E** is the electric field vector at a point between the surfaces. It is perpendicular to the surfaces and points from the higher-potential surface to the lower-potential one. Its magnitude, E, is given by $E = |\Delta V/\Delta s|$, where Δs is the (small) perpendicular distance between the surfaces.

where ΔV and Δs are, respectively, the potential difference between the equipotential surfaces and the (small) perpendicular distance between them in meters (m). Moving a test charge along an equipotential surface requires no work.

Energy is stored in the electric field itself. The energy density (i.e., the energy per unit volume), in joules per cubic meter (J/m^3), at a location where the magnitude of the electric field has the value E is given by

$$\text{Energy density of electric field} = \tfrac{1}{2}\varepsilon_0 E^2$$

VOLTAGE

A potential difference, also called a voltage, which is the difference between the potential at one location and that at another, relates to the work done on a charge—either by an external agent or by the electric field—when the charge moves from one location to the other. Denote the potentials at points A and B by V_A and V_B, respectively. Let a point charge q move from rest at point A to a state of rest at point B. By the work-energy theorem, the work required of an external agent to move the charge equals the change in the charge's potential energy (plus the change in

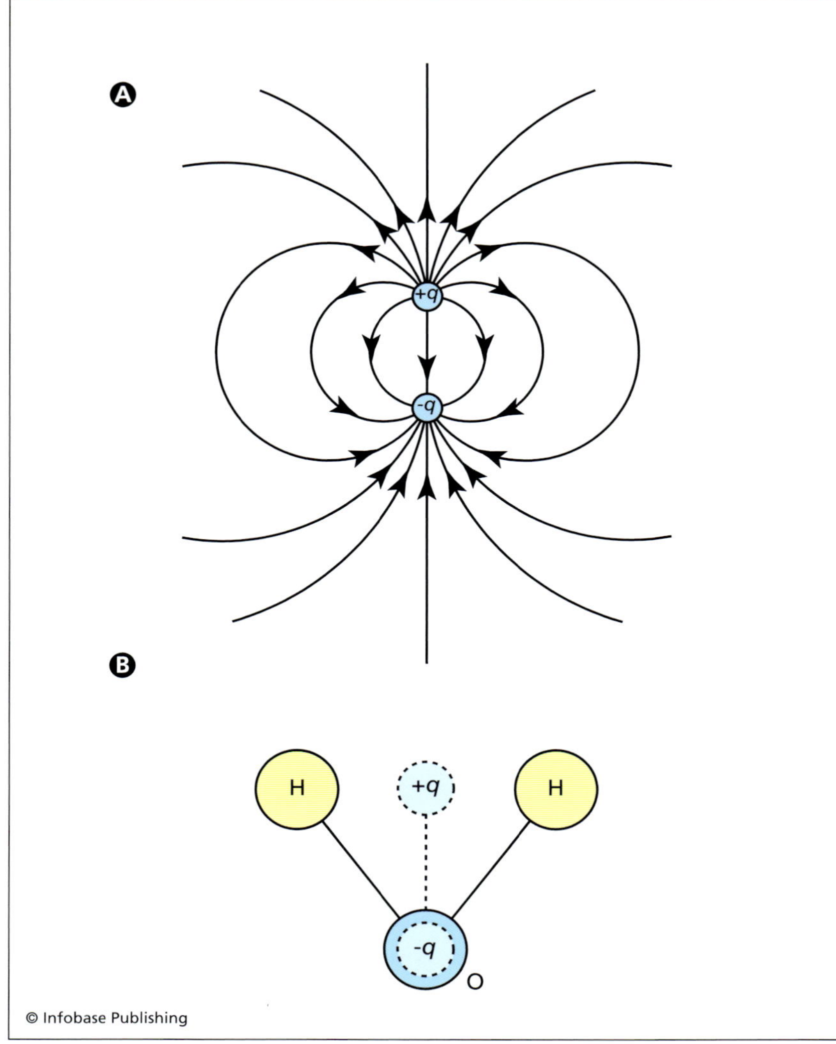

An electric dipole is a pair of separated equal-magnitude opposite electric charges or an equivalent situation. (a) An idealized electric dipole comprising charges +*q* and -*q*. A number of electric field lines are shown for it. (b) A water molecule, H_2O, which is a natural electric dipole. The oxygen atom forms a concentration of negative charge, while a center of positive charge is located between the hydrogen atoms. The equivalent electric dipole is shown in dashed lines.

its kinetic energy, but that is zero here). Denote the respective potential energies by E_{pA} and E_{pB}. Then

$$\text{Work done by external agent} = E_{pB} - E_{pA}$$
$$= qV_B - qV_A$$
$$= q(V_B - V_A)$$
$$= q\Delta V$$

where ΔV denotes the potential difference, or voltage, between points B and A. The external agent performs work against the force of the electric field. Thus, the work done by the electric field in this case is simply the negative of the work done by the external agent:

$$\text{Work done by electric field} = -q\Delta V$$

Devices such as batteries and electric generators produce voltages. In such cases, a voltage is often referred to as electromotive force, or *emf* for short. The term *electromotive force* for a quantity that is not a force is a relic of history. It remains from a time when scientists did not understand the nature of electricity as well as they do today and perhaps used the term *force* (and its translation in other languages) more loosely than they now do.

ELECTRIC DIPOLE

A pair of equal-magnitude and opposite electric charges at some distance from each other or a charge configuration that is equivalent to that is called an electric dipole. A water molecule is an electric dipole, since the oxygen atom's affinity for electrons brings about a concentration of negative charge in its vicinity, leaving a region of net positive charge where the hydrogen atoms are located. Because of the geometric structure of the water molecule (specifically, that the three atoms do not lie on a straight line), charge separation creates a natural electric dipole. In situations where an electric dipole does not normally exist, one might be created by an electric field, which tends to pull positive charges one way and negative charges the other way, resulting in charge separation. Since an electric dipole is electrically neutral, it suffers no net force in a uniform electric field. Rather, a uniform electric field exerts a torque on an electric dipole, whose magnitude depends on the dipole's orientation in the field. In a nonuniform field, however, a net force can act on a dipole, in addition to the torque.

An electric dipole is characterized by its electric dipole moment, which is a measure of the strength of an electric dipole. Electric dipole moment is a vector quantity, often denoted $\boldsymbol{\mu}_e$. For a pair of equal-magnitude, opposite electric charges, both of magnitude q, in coulombs (C), and separated by distance d, in meters (m), the magnitude of the electric dipole moment μ_e, in coulomb-meters (C·m), is given by

$$\mu_e = qd$$

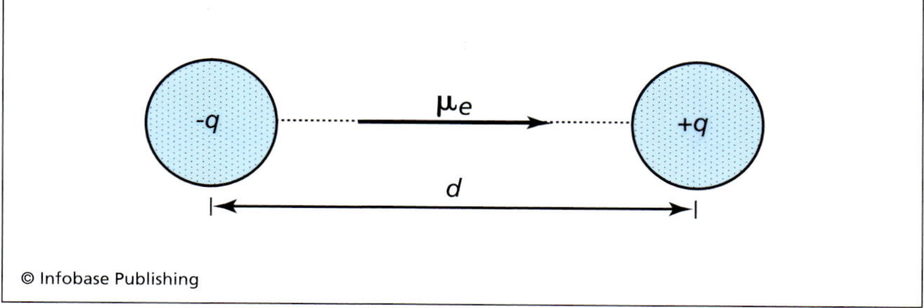

The electric dipole moment of an electric dipole, consisting of a pair of separated equal-magnitude opposite electric charges, is a vector $\boldsymbol{\mu}_e$ that points from the negative charge $-q$ to the positive one $+q$. Its magnitude μ_e is given by $\mu_e = qd$, where q is the magnitude of either charge and d the distance between the centers of the charges.

The direction of the electric dipole moment vector is along the line connecting the centers of the charges and pointing from the negative to the positive charge. Neutral charge configurations for which the center of positive charge does not coincide with the center of negative charge are similarly characterized by an electric dipole moment.

An electric field exerts a torque on an electric dipole, since it pulls the positive and negative charges in opposite directions, the positive charge in the direction of the field and the negative in the opposite direction. The torque tends to align the electric dipole moment with the field. The vector relation of torque to electric dipole moment and electric field is

$$\boldsymbol{\tau} = \boldsymbol{\mu}_e \times \mathbf{E}$$

where $\boldsymbol{\tau}$ denotes the torque vector, in newton-meters (N·m), and $\mathbf{E}$ the electric field vector at the location of the dipole. The magnitude of the torque τ is

$$\tau = \mu_e E \sin \phi$$

where E denotes the magnitude of the electric field and ϕ is the smaller angle (less than 180°) between $\boldsymbol{\mu}_e$ and $\mathbf{E}$. The direction of $\boldsymbol{\tau}$ is perpendicular to both $\boldsymbol{\mu}_e$ and $\mathbf{E}$ and points such that, if the right thumb is aimed in the same direction, the curved fingers of the right hand will indicate a rotation from the direction of $\boldsymbol{\mu}_e$ to that of $\mathbf{E}$ through the smaller angle between them. Although the sole effect of a *uniform* electric field on an electric dipole is a torque, as just described, in a *nonuniform* field, a net force acts on it as well, since then the forces on the positive and negative charges are not equal and opposite.

A potential energy is associated with a dipole in an electric field, since work is involved in rotating the dipole. The potential energy, in joules (J), is

$$\text{Potential energy} = -\boldsymbol{\mu}_e \cdot \mathbf{E} = -\mu_e E \cos \phi$$

It follows that the potential energy of the system is lowest when the electric dipole moment and the field are parallel ($\phi = 0$, $\cos \phi = 1$), which is thus the orientation of stable equilibrium, and highest when they are antiparallel ($\phi = 180°$, $\cos \phi = -1$). The potential-energy difference between the two states is $2\mu_e E$.

The importance of the study of dipoles and their behavior in electric fields lies in the fact that many molecules are electric dipoles and are affected by electric fields. Such situations occur in connection with capacitance.

CAPACITANCE

The ability of objects to store electric charge or to maintain a charge separation is referred to as capacitance. In the first case, when a body is charged with electric charge Q, it acquires electric potential V. It is found that V is proportional to Q, so their ratio

$$C = \frac{Q}{V}$$

is constant. This ratio, C, is the capacitance of the body. Note that a (relatively) high capacitance means that the body can take much charge with little increase of its electric potential, while a low-capacitance body needs only a small charge to cause it to reach a high potential.

In the case of charge separation, one is concerned with devices that possess two conducting parts, commonly called plates, which are electrically insulated from each other so no charge can flow between them. When charge Q is transferred from one plate to the other, a potential difference, or voltage, V, develops between the plates. In the present context, it is customary to denote a voltage by V rather than by ΔV, the notation introduced earlier. As before, V is proportional to Q, and their ratio C, defined as earlier and called the capacitance of the device, is constant. In this case, rather than becoming charged, the device as a whole remains electrically neutral, since charge is merely moved from one of its plates to the other. The device maintains a charge separation (rather than a charge), with one plate bearing a positive charge and the other carrying an equal-magnitude negative charge. Similarly to the first case, a (relatively) high capacitance means that the body can take a large charge separation with little increase of voltage, while a low-capacitance device needs only a small charge separation to cause it to develop a considerable voltage.

The SI unit of capacitance is the farad (F), equivalent to coulomb per volt (C/V). For many practical purposes, the farad is a very large unit, so the microfarad ($\mu F = 10^{-6}$ F) and picofarad ($pF = 10^{-12}$ F) are often used.

Any electric device that is constructed for the purpose of possessing a definite capacitance for charge separation is called a capacitor. An electric conductor allows the relatively easy movement of electric charge through itself. This is in opposition to an insulator, through which the movement of charge is extremely difficult. Every capacitor consists of two electric conductors (plates) separated by an insulator, called the dielectric. In use, electric charge is transferred, or separated, from one plate to the other, whereby the plates develop a voltage, or potential difference, between them, which is proportional to the transferred charge. The capacitance C of a capacitor is defined as the (constant) ratio of the transferred charge Q to the voltage V between the plates, as described earlier. When charged, a capacitor is actually electrically neutral, with one plate positively charged and the other carrying an equal-magnitude negative charge. The dielectric maintains a constant, uniform, controlled distance between the plates and, as an insulator, keeps the positive and negative charges separated. The dielectric contributes to the capacitor's capacitance also through its dielectric constant, as explained later.

The plates of a capacitor are almost always indeed plates in the usual sense: that is, they are thin compared to their other dimensions and are of uniform thinness. If they are parallel to each other and are large compared to the distance between them, as is very often the case, the capacitance of the capacitor is given by the following:

$$C = \frac{\kappa \varepsilon_0 A}{d}$$

where A is the area of each plate, in square meters (m^2); d the distance between the plates, in meters (m); κ is the dielectric constant of the dielectric; and ε_0 is the permittivity of the vacuum, whose value is $8.85418782 \times 10^{-12}$ $C^2/(N \cdot m^2)$, rounded to 8.85×10^{-12} $C^2/(N \cdot m^2)$. Thus the capacitance can be increased by increasing the plate area, decreasing the distance between the plates, and using a dielectric with higher dielectric constant.

Capacitors are normally specified by and labeled with two numbers: their capacitance and their maximal allowed voltage. The latter is the highest voltage at which the capacitor can be safely operated, without danger of its breaking down and burning out. It is derived from the dielectric strength of the dielectric, which is the greatest magnitude of electric field that the dielectric can tolerate without breaking down. Capacitors are essential for electronic circuits (see later discussion) and are ubiquitous in them.

As explained earlier, the ratio of the permittivity of a material to the permittivity of the vacuum is called the dielectric constant of the material. It is the factor by which the material reduces the magnitude of the electric force between charges immersed in it. Accordingly, it is the factor by which the material reduces the magnitude of an external electric field within itself. As a result, when a dielectric is used to fill the gap in a capacitor, the capacitance of the capacitor is increased by the factor of the dielectric constant from the value it would have without the dielectric, that is, in vacuum.

This effect of a dielectric is achieved by means of its electric polarization, whereby its atomic or molecular electric dipoles are aligned by an external electric field. This produces an internal electric field that is oppositely directed to the external field and partially cancels it inside the dielectric. A material that does not normally possess atomic or molecular dipoles might still be polarized by an external electric field. That happens through the field's creating dipoles by causing a separation of positive and negative electric charges in each of the material's atoms or molecules.

Air under ordinary conditions, for example, has a dielectric constant very close to 1, the dielectric constant of the vacuum. While materials such as rubber, paper, and glass possess dielectric constants in the 2–10 range, water has an especially high dielectric constant of around 80. See the following table.

ELECTRIC CURRENT

The subfield of electricity that deals with charges at rest is called electrostatics. Charges may be in motion, however, and moving charges form electric currents. The additional forces, over and above the electric forces, that affect moving charges, or currents, as a result of the charges' motion are the subject of magnetism.

For the flow of electric charge, the electric current is defined as the net amount of charge flowing past a point per unit time. If net charge Δq passes a

DIELECTRIC CONSTANTS OF VARIOUS MATERIALS

Material	Dielectric Constant κ
vacuum	1.00000
air	1.00054
paper	2–4
teflon	2.1
polyethylene	2.3
polystyrene	2.6
mica	3–6
mylar	3.1
Plexiglas	3.40
glass	5–10
germanium	16
ethanol	24.3
methanol	33.6
water	80.4

point during time interval Δt, in seconds (s), the average current during the time interval is as follows:

$$i = \frac{\Delta q}{\Delta t}$$

The SI unit of electric current is the ampere (A), equivalent to coulomb per second (C/s). In the limit of an infinitesimal time interval, the instantaneous current is given by the following:

$$i = \frac{dq}{dt}$$

Any net flow of charge forms an electric current. The most commonly known is the flow of electrons in a metal, such as a copper wire. Examples of other forms of electric current are the flow of ions in an electrolyte, an ion beam in a vacuum, the flow of electrons and ions in a corona discharge and in lightning, and the movement of electrons and holes in a semiconductor.

Electric devices, such as batteries, electric generators, resistors, inductors, capacitors, and transistors, connected by electric conductors, through which currents might flow, make up electric circuits. The analysis of electric circuits is performed with the help of tools such as Ohm's law and Kirchhoff's rules (see later discussion). Direct current (DC) is the situation when the currents flowing in a circuit maintain a constant direction, or, if currents are not flowing, the voltages (potential differences) involved maintain a constant sign. When the voltages and currents in a circuit undergo regular reversals, the situation is one of alternating current (AC). In more detail, alternating current involves a sinusoidal variation in time of the voltages and currents of a circuit at some frequency f, in hertz (Hz), whereby any voltage V or current i at time t is given by the following:

$$V = V_0 \sin(2\pi ft + \alpha)$$

$$i = i_0 \sin(2\pi ft + \beta)$$

Here V_0 and i_0 denote the maximal values of the time-varying quantities V and i, respectively, which thus vary from V_0 and i_0 to $-V_0$ and $-i_0$ and back again f times per second. The parameters a and β, in radians (rad), denote possible phase shifts, whereby the varying voltage or current reaches its maximum at a different time in the cycle, or at a different phase, than does some reference voltage or current in the circuit.

OHM'S LAW

Named for the German physicist Georg Simon Ohm, Ohm's law states that the electric potential difference, or voltage, V, across an electric circuit component and the current, i, through the component are proportional to each other:

$$V = iR$$

where R denotes the component's resistance. Here V is in volts (V), i in amperes (A), and R in ohms (Ω). This relation is valid for direct current as well as for alternating current. Ohm's law is not a law in the usual sense, but rather a description of the behavior of a class of materials for a limited range of voltages. A material that obeys Ohm's law for some range of voltages is termed an ohmic conductor. A nonohmic conductor conducts electricity but does not exhibit proportionality between the current and the voltage.

RESISTANCE

Electric resistance is the hindrance that matter sets to the passage of an electric current through it. That might be due to the unavailability of current carriers (which in solids are electrons and holes) or to impediments to the motion of current carriers (such as, again in solids, impurities and lattice defects and vibrations). As a physical quantity, the resistance of a sample of material, R, in ohms (Ω), is the ratio of the voltage across the sample, V, in volts (V), to the current through the sample, i, in amperes (A):

$$R = \frac{V}{i}$$

The resistance of a sample depends both on the nature of the material and on the sample's size and shape. In the case of a homogeneous material with a shape that has definite length and cross section, the resistance is given by the following:

$$R = \frac{\rho L}{A}$$

where L denotes the length, in meters (m); A the cross-sectional area, in square meters (m^2); and ρ the resistivity of the material from which the sample is made, in ohm-meters (Ω·m). Resistivity is a measure of the characteristic opposition of a material to the flow of electric current. It is defined as the electric resistance of a one-meter cube of the material when a voltage is maintained between two opposite faces of the cube. Its SI unit is the ohm-meter (Ω·m). So a resistivity of 1 Ω·m for a material means that when a voltage of 1 V is applied to a cube of the material as described, a current of 1 A flows between the opposite faces.

Resistivity is temperature dependent in general. For almost all solids, the resistivity increases with increase of temperature. For small temperature changes, the change in resistivity is approximately proportional to the change in temperature. That can be expressed as

$$\rho = \rho_0[1 + \alpha(T - T_0)]$$

where ρ_0 denotes the resistivity at temperature T_0 and ρ is the resistivity of the material at temperature T, with T close to T_0. The temperatures are either both in kelvins (K) or both in degrees Celsius (°C). The symbol α represents the temperature coefficient of resistivity, in units of inverse Celsius degree [$(°C)^{-1}$]. The following table presents a sampling of values of these quantities.

At sufficiently low temperatures, some materials enter a superconducting state, in which their resistivity is precisely zero.

When a current passes through matter, electric energy is converted to heat. The rate of heat generation, or power, P, in watts (W) is given by the following:

$$P = Vi$$

where i denotes the current through the sample and V is the voltage across the sample. This can be expressed equivalently in terms of the resistance of the sample as follows:

RESISTIVITIES AND TEMPERATURE COEFFICIENTS OF RESISTIVITY AT ROOM TEMPERATURE (20°C) FOR VARIOUS MATERIALS

Substance	Resistivity ρ ($\times 10^{-8}\ \Omega$·m)	Temperature Coefficient of Resistivity α [$(°C)^{-1}$]
Aluminum	2.75	0.0039
Carbon (graphite)	3.5×10^3	–0.0005
Constantan (Cu 60%, Ni 40%)	49	0.00001
Copper	1.72	0.00393
Iron	20	0.0050
Lead	22	0.0043
Manganin (Cu 84%, Mn 12%, Ni 4%)	44	0.00000
Mercury	95	0.00088
Nichrome (Ni 80%, Cr 20%)	100	0.0004
Silver	1.47	0.0038
Tungsten	5.25	0.0045

$$P = i^2R = \frac{V^2}{R}$$

This result is known as Joule's law, named for the British physicist James Prescott Joule, and the heat thus produced as Joule heat.

An electric device that is designed to possess a definite resistance and be able to function properly under a definite rate of heat production is called a resistor. A resistor is characterized by the value of its resistance in ohms (Ω) and by the value of its maximal allowed rate of heat production in watts (W). When resistors are connected in series (i.e., end to end in linear sequence, so that the same current flows through them all), the resistance of the combination—their equivalent resistance—equals the sum of their individual resistances. So if one denotes the resistances of the resistors by R_1, R_2, . . . and the equivalent resistance of the series combination by R, then

$$R = R_1 + R_2 + \ldots$$

In a parallel connection of resistors, one end of each is connected to one end of all the others, and the other end of each one is connected to the other end of all the others. In that way, the same voltage acts on all of them. The equivalent resistance R of such a combination is given by

$$\frac{1}{R} = \frac{1}{R_1} + \frac{1}{R_2} + \ldots$$

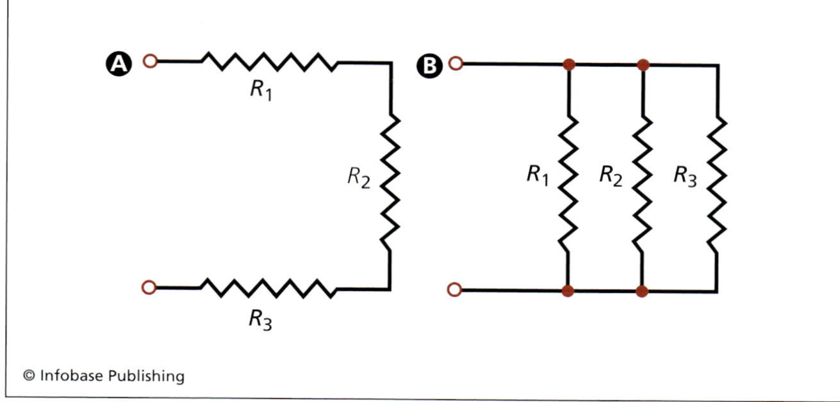

Resistors in combination. (a) Three resistors R_1, R_2, and R_3 connected in series, so that the same current flows through each. The equivalent resistance of the combination equals $R_1 + R_2 + R_3$. (b) Three resistors in a parallel connection, causing the resistors to have the same voltage across each of them. The equivalent resistance in this case equals $1/(1/R_1 + 1/R_2 + 1/R_3)$.

KIRCHHOFF'S RULES

Kirchhoff's two rules for the analysis of electric circuits, or networks, allow one to find the currents in all branches of any circuit and the voltages between all pairs of points of the circuit. Although in practice other methods of analysis might be used—and engineers do indeed use other methods—they are equivalent to and derived from Kirchhoff's rules. Introductory physics courses and textbooks present Kirchhoff's rules, rather than any of the other methods, since they are most directly based on fundamental physics principles, as explained later. The rules are named for the German physicist Gustav Robert Kirchhoff.

First, consider some definitions. A branch of a circuit is a single conducting path through which a definite current flows. Branches join at junctions, or nodes. Thus, a branch runs from one junction to another and has no other junctions in between. A loop in a circuit is a closed conducting path. We consider the application of Kirchhoff's rules to DC circuits, consisting of resistors (or other components possessing electric resistance) and sources of emf (voltage), such as batteries.

Next, prepare the given circuit diagram (with given emf sources and resistances) for the application of Kirchhoff's rules as follows. Identify the branches of the circuit, arbitrarily choose a direction of current in each branch, and assign a symbol to each current. (If one chooses the wrong direction, that is of no concern, since the algebra will straighten things out at the end.) Now apply the junction rule at each junction.

Kirchhoff's Junction Rule

The algebraic sum of all currents at a junction equals zero. Equivalently, the sum of currents entering a junction equals the sum of currents leaving it. The significance of this rule is basically conservation of electric charge, that charge is neither created nor destroyed at a junction: what goes in must come out.

The application of the junction rule at all junctions gives a number of equations for the unknown currents. Use the equations to eliminate algebraically as many unknown currents as possible. One is now left with some irreducible set of unknown currents. The equations obtained from the junction rule express all the other currents in terms of those in the irreducible set.

At this point, arbitrarily choose a number of independent loops in the circuit. (About "independent," see later discussion.) The number of loops should equal the number of unknown currents in the irreducible set. Apply the loop rule to every chosen loop.

Kirchhoff's Loop Rule

The algebraic sum of potential rises and drops encountered as the loop is completely traversed equals zero. Equivalently, the sum of potential rises around a loop equals the sum of potential drops. The loop rule is none other than conservation of energy together with the conservative nature of the electric force. It tells us that if a test charge is carried around a loop, it will end up with the same energy it had initially.

In detail, apply the loop rule in this manner. On each loop choose a point of departure. Traverse the loop in whichever direction you like. As you encounter circuit components, algebraically sum their potential changes. Crossing a battery from negative terminal to positive is a potential rise (positive change); from positive to negative, a drop (negative change). The emf of each battery is given, so no unknowns are involved in this. However, in crossing a resistor or the internal resistance of a battery, the magnitude of potential change equals the product of the unknown current and the known resistance, by Ohm's law. Crossing in the direction of the current gives a potential drop (negative change); against the current, a rise (positive change). (That is because a current flows from higher to lower potential.) When the traversal takes one back to the point of departure, equate the accumulated sum to zero.

If one chose independent loops, the equations one has are algebraically independent, and one has as many equations for the unknown currents in the irreducible set as there are currents in the set. Solve the equations for those currents. If one cannot solve for all of them, the equations are not independent, and one should modify the choice of loops and apply the loop rule again. When one has solved for the currents in the irreducible set, the junction rule equations serve to find all the other currents. A negative value for a current simply indicates that the guess of its direction was wrong; reverse the corresponding arrow in the circuit diagram and note a positive value instead. In this way one now knows the magnitudes and directions of the currents in all the branches of the circuit. Then, with the help of Ohm's law, one can find the voltage between any pair of points of the circuit.

Kirchhoff's rules are applicable to AC circuits as well.

SOURCES OF EMF

The following is a selection of sources of emf, means and devices for producing and maintaining voltages.

Battery

Any device for storing chemical energy and allowing its release in the form of an electric current is called a battery. The current is obtained from a pair of terminals on the battery, designated positive (+) and negative (–), where the positive terminal is at a higher electric potential than the negative. A battery may be composed of one or more individual units, called cells. Batteries and cells are characterized by their electromotive force (emf, commonly called voltage in this connection), designated in volts (V), and their charge-carrying capability, indicated in terms of current × time, such as ampere-hours (A·h) or milliampere-hours (mA·h). This indication of charge capability means that the battery is capable of supplying a steady current I for time interval t, where the product It equals the indicated rating. An automobile battery normally supplies an electric current at a voltage of 12 V and might be rated at 500 A·h. Such a battery could supply a current of 250 A, say, for two hours. Some batteries are intended for one-time use, while others are rechargeable. Attempts to recharge a nonrechargeable battery can be dangerous, possibly resulting in explosion.

Similar batteries or cells can be connected in series (the positive terminal of one to the negative terminal of another), giving a combined battery whose emf equals the sum of individual emf's (equivalently,

Batteries of various voltages, sizes, and types *(Andrew Lambert Photography/Photo Researchers, Inc.)*

the emf of one unit times the number of units). Four 1.5-volt batteries in series are equivalent to a six-volt battery. The charge-carrying capability of the combination equals that of an individual unit. Alternatively, they can be connected in parallel (all positive terminals connected and all negative terminals connected), whereby the combination has the emf of an individual unit and a charge capability that is the sum of those of all the units (or, that of one unit times the number of units). Four 1.5-volt batteries in parallel are equivalent to a 1.5-volt battery, but one that can supply a current that is four times the current that can be drawn from any one of the four. (Dissimilar batteries can be combined in series but should *never* be connected in parallel.)

A battery is equivalent to an ideal (zero-resistance) emf source in series with a resistor, whose resistance equals the internal resistance of the battery. If a battery possesses an emf of E and internal resistance r, the actual potential difference V, correctly called voltage, between its terminals is given by

$$V = \mathrm{E} - Ir$$

when the circuit is drawing current i from the battery.

Electric Generator

Any device for converting mechanical energy to electric energy is called an electric generator. In the operation of a generator, a coil, the rotor, rotates in a magnetic field, whereby a varying electromotive force (emf) of alternating polarity (i.e., an AC voltage) is induced in the coil. By suitable design, a generator can produce an emf of constant polarity (i.e., a DC voltage). Whenever an electric current is drawn from a generator, a torque is needed to maintain the coil's rotation. Thus work performed on the coil is converted into electric energy. The source of the magnetic field can be one or more permanent magnets or a fixed coil, the stator, in which the current that produces the field is taken from the rotor. In principle, a generator is an electric motor operated in reverse. Depending on the design and size of an electric generator, it might be referred to variously as an alternator, dynamo, or generator.

Piezoelectricity

The production of electricity from pressure, and vice versa, are termed piezoelectricity. The creation of an electric potential difference, or voltage or emf, across a crystal when a mechanical force is applied to a crystal is called the direct piezoelectric effect. The converse piezoelectric effect occurs when an applied voltage generates a mechanical distortion of a crystal. The effects are the result of the mutual dependence of electric dipole moment and mechanical strain that can occur for crystals of certain classes.

These are electric generators in the power station of the Hoover Dam at Lake Meade. They are powered by turbines, which are caused to rotate by water passing through the dam from the lake. *(Bryan Busovicki, 2008, used under license from Shutterstock, Inc.)*

Devices based on the direct piezoelectric effect include transducers that convert strain into an electric signal, such as microphones and pressure gauges, as well as certain ignition devices. The inverse piezoelectric effect underlies the operation of some acoustic sources and headphones. Both effects come into play when a vibrating crystal is used as a frequency reference.

Pyroelectricity

The production of an electric potential difference, or voltage or emf, across a crystal when the crystal is heated is termed pyroelectricity. The heat causes a rearrangement of the permanent electric dipoles in the crystal, which creates an electric dipole moment for the whole crystal. That brings about a positive electrically charged surface and an oppositely located negative surface, between which a voltage exists. Over time, ions from the air collect on the charged surfaces and cancel their surface charge. Then another applied temperature change can again rearrange the electric dipoles and produce fresh surface charges.

Thermocouple

A pair of wires of different metals that have both their ends, respectively, welded together serves as a thermocouple. When the two junctions are at different temperatures, an electromotive force (emf) is produced in the circuit and an electric current flows. The emf and current increase with an increase in temperature difference. This effect is named the Seebeck effect, for the German physicist Thomas Seebeck. By introducing a suitable electric or electronic device into the circuit, one can measure the current or emf. When such a device is calibrated, it can serve to measure temperature differences between the two junctions. By maintaining one of the junctions at a known temperature, the device then becomes a thermometer. Since the thermocouple wires can be extremely fine and the heat capacity of the probe junction consequently very small, a thermocouple can be suitable for measuring the temperature of very small objects.

An inverse effect is the Peltier effect, named for the French physicist Jean Charles Peltier. This is the creation of a temperature difference between the two junctions of a thermocouple as a result of an electric current flowing in the circuit.

See also ACOUSTICS; CONSERVATION LAWS; ELASTICITY; ELECTROMAGNETISM; ENERGY AND WORK; EQUILIBRIUM; FARADAY, MICHAEL; FORCE; HARMONIC MOTION; HEAT AND THERMODYNAMICS; MAGNETISM; MAXWELL, JAMES CLERK; PARTICLE PHYSICS; POWER; PRESSURE; ROTATIONAL MOTION; SUPERCONDUCTIVITY; SYMMETRY; VECTORS AND SCALARS.

FURTHER READING

Gibilisco, Stan. *Electricity Demystified: A Self-Teaching Guide*. New York: McGraw-Hill, 2005.

Parker, Steve, and Laura Buller. *Electricity*. New York: DK, 2005.

Young, Hugh D., and Roger A. Freedman. *University Physics*, 12th ed. San Francisco: Addison Wesley, 2007.

electrochemistry Electrochemistry is the branch of chemistry concerned with the movement of electrons and the relationship between electric energy and chemical reactions that can be utilized to create an electric current. Major topics in the study of electrochemistry include oxidation-reduction reactions, galvanic cells and batteries, and electrolytic cells. When spontaneous chemical reactions occur, they can be used to create an electric current, and that electric current can be used to drive chemical reactions. Electrochemical principles lead to the production of batteries and explain the chemical process of metal corrosion and rust.

OXIDATION-REDUCTION REACTIONS

The simplest definition of oxidation-reduction reactions is based on the gain and loss of electrons. An oxidation-reduction reaction occurs when one substance in the reaction gains one or more electrons (reduction) and one substance in the reaction loses one or more electrons (oxidation). When one substance is oxidized, another substance must be reduced. The reactions always occur in pairs—the lost electrons must go somewhere. The acronym *OIL RIG* (which stands for *o*xidation *i*s *l*oss, and *r*eduction *i*s *g*ain) is useful in remembering the distinction between oxidation and reduction. The substance that is responsible for oxidizing the other substance becomes reduced in the process and is known as the oxidizing agent. The substance that reduces the other substance becomes oxidized in the process and is known as the reducing agent.

In order to determine which substance is oxidized or reduced, it is necessary to keep track of the oxidation numbers of the elements involved. An oxidation number is the effective charge that an atom of an element would have if it were an ion, even if the compound is not actually ionic. The electron configuration of the element determines the oxidation number. The number of electrons necessary to complete the outer (valence) shell of the element determines the number of electrons that are gained or lost. When an electron is lost, the element has a positive oxidation number. When the element gains electrons, the element assumes a negative charge. If an element has three or fewer electrons, it will lose its valence electrons rather than gain five or more to fill the valence shell. This means that elements in Group IA have a +1 oxidation number, Group IIA has a +2 oxidation number, and Group IIIA has a +3 oxidation number. Elements with five, six, or seven electrons will gain electrons in order to fill their valence shell, meaning that elements in group VA will have an oxidation number of -3, elements in group VIA will have an oxidation number of -2, and elements in VIIA will have an oxidation number of -1. Elements in group VIIIA, the noble gases, already have eight valence electrons, so these elements have an oxidation number of 0 as they will neither gain nor lose electrons.

A more general definition of oxidation-reduction reactions involves the change in oxidation number of the elements of the participating compounds. A substance is oxidized when its oxidation number increases (becomes more positive as a result of the loss of electrons), and a substance is reduced when its oxidation number decreases (becomes more negative as a result of the gain of electrons).

In order to study oxidation-reduction reactions, it is helpful to separate the two parts of the reaction into an oxidation half-reaction and a reduction half-reaction. Once the two halves of the reaction are identified, the atom number and the charge need to be balanced.

ELECTROCHEMICAL CELLS

Electrochemical cells were originally developed by the Italian physicists Luigi Galvani and Alessandro Volta and are therefore also known as galvanic cells or voltaic cells. An electrochemical cell is an experimental container in which the two oxidation-reduction half-reactions are separated. The electrons that are formed in the oxidation side of the electrochemical cell travel to the reduction side of the electrochemical cell. This movement of electrons creates an electric current. The setup of electrochemical cells involves two electrodes, bars made of oxidizable and reducible elements. The electrodes are known as anodes and cathodes. Anodes are the electrodes at which oxidation occurs, and cathodes are the electrodes at which reduction occurs.

One type of electrochemical cell uses zinc as the oxidized component and copper as the reduced component. Therefore, the anode is the zinc bar and the cathode is the copper bar. The two electrodes are separated from one another, and the electrons released from the zinc travel through a wire that is connected to the copper cathode, leaving behind Zn^{2+}. As the electrons leave the zinc, they travel to the copper side and are used to reduce Cu^{2+} ions in solution to Cu, which becomes a part of the copper bar. The anode has a high concentration of electrons (due to the loss of electrons by zinc) and it is generally labeled with a negative sign. The cathode has a lower concentration of free electrons (due to the gain of electrons by copper) so it is labeled with a positive sign. These positive and negative designations are commonly observed on the ends of batteries. The flow of electrons is an electrical current that can be measured in volts using a voltmeter.

When oxidation-reduction occurs, there is an overall decrease in the amount of zinc present in the anode and an overall increase in the amount of copper in the cathode. The zinc bar decreases in size while the copper bar increases in size. The solution concentration of zinc increases, while the solution concentration of copper decreases. If this continued, the process would come to a halt. A porous barrier or a salt bridge is often used to correct the charge problem in the two compartments. A salt bridge is a U-shaped tube that is filled with an electrolyte solution that does not participate in the oxidation-reduction reaction. The electrolyte is often in a gel form to prevent it from leaving the tube. Two common salt bridge components are KCl and $NaNO_3$. The anions (Cl^- and NO_3^-) will leave the salt bridge and travel into the zinc side of the electrochemical cell. The cations (K^+ and Na^+) will leave the salt bridge and travel into the copper side of the cell. This movement of ions neutralizes the positive charge that builds up in the zinc compartment and the negative charge that builds up in the copper compartment. Without this neutralization, the oxidation-reduction reactions would come to a halt and the current would stop flowing.

STANDARD ELECTRODE POTENTIAL

Electrons move from the anode to the cathode in an electrochemical cell as a result of the potential difference between the electron concentrations at the anode versus the cathode. This difference is known as the cell potential. The electrons readily leave the negative anode to travel to the relatively positive

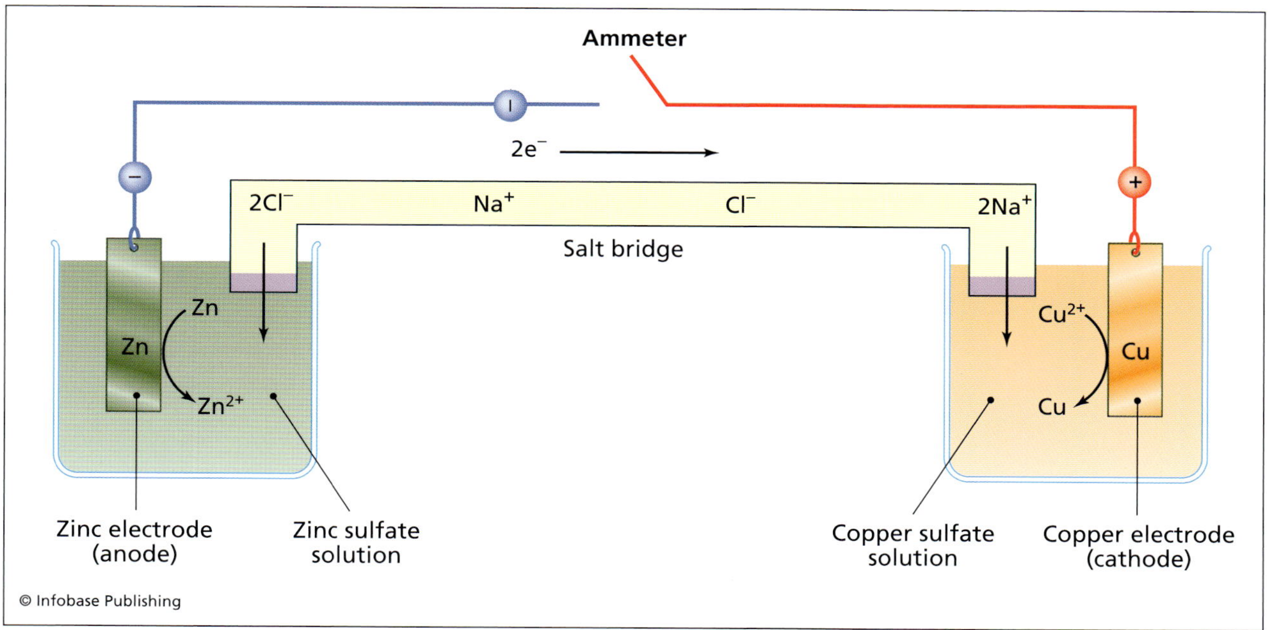

The zinc electrode (anode) is immersed in a zinc sulfate solution. The oxidation of zinc (Zn) leads forms positive zinc ions (Zn^{2+}) and electrons that can travel through the wire to the copper electrode (cathode), where copper (Cu^{2+}) is reduced to form copper (Cu). The salt bridge helps maintain the charge difference between the anode and the cathode to allow the electron current to continue to flow.

cathode. The potential difference creates a push for these electrons known as the electromotive force, or emf. The standard emf, also called the standard cell potential, is a measure of the voltage difference of a particular cell. Standard conditions are 1 M concentrations for reactants and products, a pressure of 1 atm for all gases, and a temperature of 77°F (25°C).

The cell potential will be positive for spontaneous oxidation-reduction reactions. The cell potential of a particular cell consists of the potential of the oxidation half-reaction as well as the reduction half-reaction. Rather than calculate the standard cell potential of every combination of half-reactions, the potential of the amount of reduction that occurs in each half-reaction is used and is conventionally known as the standard reduction potential. As reduction does not occur at the anode, the value for the reduction of that half-reaction is subtracted to give the total potential for the cell, as shown in the following formula:

$$E°_{cell} = E°_{red} \text{ at cathode} - E°_{red} \text{ at anode}$$

where $E°_{cell}$ stands for the total standard cell potential and $E°_{red}$ stands for the standard reduction potential at either the anode or the cathode under standard conditions.

CONCENTRATION EFFECT ON EMF

Not all reactions occur under standard conditions. It becomes necessary to be able to calculate potential differences under different temperatures and concentrations. To do this, one must employ the Nernst equation, named after the German chemist Walther Nernst. The formula that follows demonstrates how the values found on a table of standard reduction potentials can be utilized to convert to a potential difference under differing conditions.

$$E = E° - (RT/nF)\ln Q$$

where E is the electric potential at the experimental conditions, $E°$ is the standard potential obtained from a table, R is the ideal gas constant (equal to 8.31 J/K × mol), T is temperature in kelvins, n is the number of electrons being transferred, F is the Faraday constant (equal to 9.65×10^4 J/V × mol). Q is determined by the following equation:

$$Q = C^cD^d/A^aB^b$$

where the capital letters represent the actual concentration of each of the reaction components and the lowercase exponents are the coefficients of the balanced equation

$$aA + bB \rightarrow cC + dD$$

BATTERIES

A battery is an electrochemical cell or several connected electrochemical cells. The oxidation-reduction reactions that occur in these electrochemical cells can create an electric current that can power everything from telephones to computers, cameras, and cars. Different types of batteries have been developed for specific purposes.

Dry cell batteries, so called because they do not have a fluid portion in the battery, are an older version of alkaline batteries. The anode in this type of battery is often made of zinc. The cathode component often includes ammonium ions or manganese dioxide.

Alkaline batteries are an improvement on the dry cell battery in performance. They commonly use zinc and potassium hydroxide (KOH) at their anode and manganese dioxide (MnO_2) at the cathode. These types of batteries are used every day in devices such as flashlights, radios, games, and remote controls. They have a variety of sizes including AAA, AA, C, and D.

Twelve-volt lead-storage batteries are used in cars and contain six voltaic cells in a series. The cathode is made of lead dioxide, and the anode is made of lead. Sulfuric acid surrounds both the anode and the cathode, making the lead storage battery dangerous. The cathode and anode reactions that occur in such batteries are shown in the following.

$$PbO_2 + HSO_4^- + 3H^+ + 2e^- \rightarrow PbSO_4 + 2H_2O$$
(cathode)

$$Pb + HSO_4^- \rightarrow PbSO_4 + H^+ + 2e^- \text{ (anode)}$$

A major advantage of the lead storage battery is its ability to be recharged. While the engine of the car is running, a generator recharges the battery by reversing the oxidation-reduction process and turning the $PbSO_4$ back into Pb at one electrode and PbO_2 at the other.

CORROSION

The deterioration of metal in bridges, buildings, and cars is an undesirable effect of oxidation-reduction reactions. When metals react with components in their environment that can cause them to oxidize, they can turn into rust or other unwanted compounds. The oxidation-reduction of metals, especially iron, can be economically devastating. Protection of metal products from air can be accomplished by paints and finishes. The oxidation of iron by oxygen is summarized as follows:

$$O_2 + 4H^+ + 4e^- \rightarrow 2H_2O \qquad E°_{red} = 1.23 \text{ V}$$

$$Fe \rightarrow Fe^{2+} + 2e^- \qquad E°_{\_red} = -0.44 \text{ V}$$

As the iron half-reaction occurs at the anode, the reduction potential is much lower than that of oxygen at the cathode. Therefore, iron is reduced to Fe^{2+}, which can further be converted into Fe_2O_3 (rust). Coating the iron with something that is more easily oxidized causes the coating rather than the iron to oxidize, thereby protecting the metal product.

The corrosion of metal objects increases with higher moisture content, lower pH, and higher salt concentration. Leaving a bike outside in the Midwest does not have the same consequences as leaving a bike out on the beach. The salty air in coastal regions encourages corrosion.

ELECTROLYSIS

Electrolysis is an application of electrochemistry that involves using electricity to cause nonspontaneous oxidation-reduction reactions to occur. Electrolysis can be performed in an electrolytic cell and utilizes the same principles as an electrochemical cell.

The simplest case of electrolysis is running an electrical current through water. The products of this reaction are hydrogen gas and oxygen gas, as shown.

$$2H_2O \rightarrow 2H_2 + O_2$$

The change in free energy for this reaction is positive ($\Delta G = +237$ kilojoules), indicating that it is not a spontaneous reaction. To perform this electrolysis, one uses electrodes, usually made of platinum, to carry the electrical current. Since pure distilled water does not conduct an electric current well, a dilute sulfuric acid solution can be used. As the current begins to flow from the cathode to the anode, the water breaks down according to the following oxidation-reduction half-reactions:

At the anode, the hydrogen in the water molecule is oxidized.

$$\text{Anode: } 2H_2O \rightarrow O_2 + 4H^+ + 4e^-$$

At the cathode, the hydrogen ions are reduced to produce hydrogen gas.

$$\text{Cathode: } H^+ + e^- \rightarrow H_2$$

See also ELECTRICITY; MEMBRANE POTENTIAL; OXIDATION-REDUCTION REACTIONS.

FURTHER READING

Chang, Raymond. *Chemistry*, 9th ed. Boston: McGraw-Hill, 2007.

electromagnetic waves Electromagnetic radiation is transmitted through space in the form of electromagnetic waves, or disturbances in an electromagnetic field. Radios, television sets, microwave ovens, and X-ray machines all utilize electromagnetic radiation to function. An electric field is the mediator through which an electric charge exerts its force on another electric charge. Likewise, a magnetic field is the mediator through which a magnet exerts its force. The electromagnetic field consists of the electric and magnetic fields and serves as the mediator for all electromagnetic effects.

An oscillating electric charge, in particular the rapid movement of an electron back and forth, produces waves in the electromagnetic field. Such waves can travel through matter (such as water, air, and walls) and even in a vacuum. The frequency of the wave is the number of cycles of the wave that pass through a given point per unit time. Its unit is the hertz (Hz). Different frequencies of oscillation of the source result in different wavelengths of the emitted wave, where the wavelength is the length of one cycle of the wave as measured at any instant along the direction of propagation of the wave. For example, the distance from a crest to the adjacent crest of a wave is the wave's wavelength.

James Clerk Maxwell, a 19th-century Scottish mathematical physicist, discovered that electromagnetic waves traveled at the speed of light, 1.86×10^5 miles per second (3.00×10^8 m/s) in vacuum, demonstrating that light, therefore, must be a form of electromagnetic radiation. The relationship among the propagation speed of an electromagnetic wave, the wave's frequency, and its wavelength is described by the following equation:

$$c = \lambda f$$

where c is the speed of light (3.00×10^8 m/s in vacuum), λ is the wavelength in meters, and f is the frequency in hertz. Thus, lower frequencies produce longer wavelengths and higher frequencies, shorter wavelengths.

The range of wavelengths and corresponding frequencies of electromagnetic waves is huge and runs from about 10^{-13} meter to 10 meters, corresponding to approximately 10^{22} hertz to 10^8 hertz. This range is referred to as the electromagnetic spectrum. It is arbitrarily divided into subranges, or bands, in order of increasing wavelength: gamma rays, X-rays, ultraviolet light, visible light, infrared rays, microwaves, and radio waves. There is some overlap among bands.

Quantum physics offers an alternate description of an electromagnetic wave in terms of a flow of particles, called photons. The energy of each photon

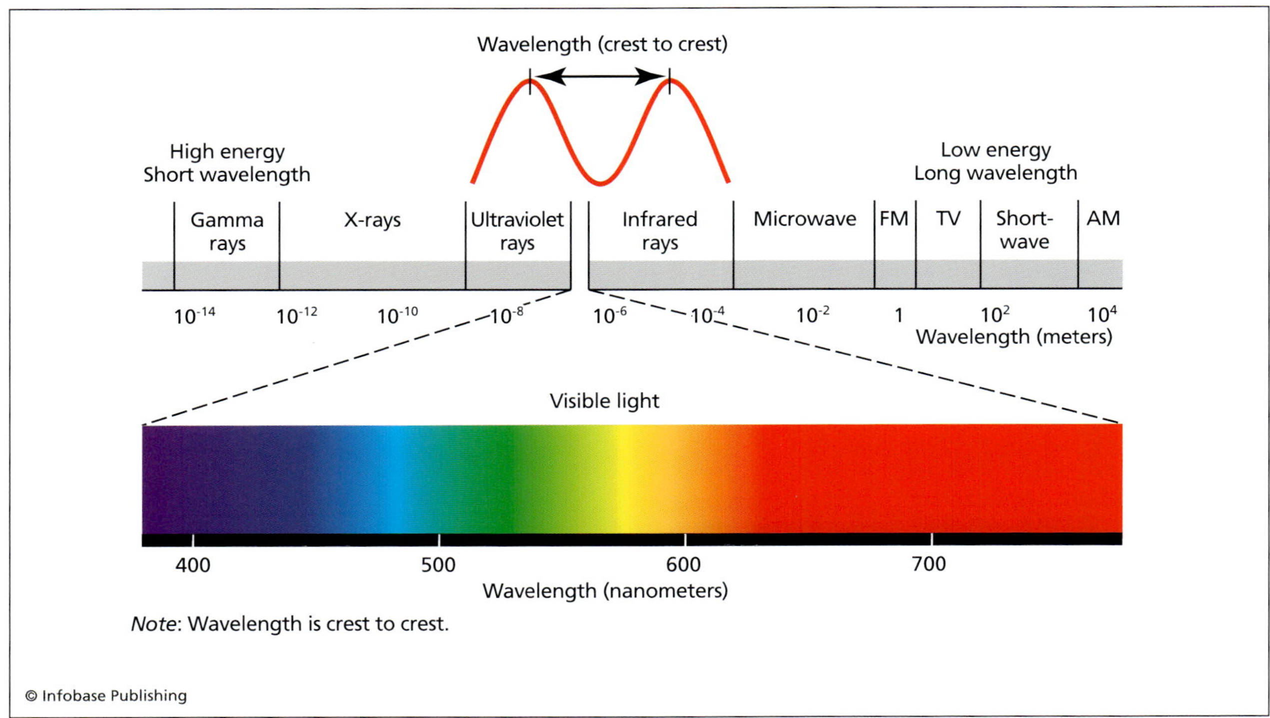

The electromagnetic spectrum encompasses the entire range of wavelengths or frequencies of electromagnetic radiation.

is proportional to the frequency of the wave and is given by

$$E = hf$$

where E is the photon's energy in joules (J), f is the frequency of the wave in hertz (Hz), and h stands for Planck's constant, whose value is 6.63×10^{-34} J·s.

Technology utilizes electromagnetic waves of different wavelengths for different functions. Of the forms of electromagnetic radiation, gamma rays have the shortest wavelength and their photons are the most energetic, having energies greater than several hundred thousand electron volts (eV). (The electron volt is a convenient unit of energy for certain purposes and is the amount of energy gained by one electron as it passes from a point of low potential to a point one volt higher in potential. One electron volt equals 1.60×10^{-19} joule.) Astronomical objects such as the Sun, galaxies, supernovae, and pulsars give off gamma radiation, and so also do certain elements, in a process called radioactive decay. Gamma radiation is a form of ionizing radiation, meaning that it induces the formation of ions as it passes through a substance. The energy carried by a photon bumps an electron out of its atom, thus converting a neutral atom into an ion, that is, into a charged particle. The photon is absorbed in the process. So as the radiation moves through an object and interacts with the electrons and with nuclei, it loses energy through depletion of its photons. The environment contains many naturally occurring sources of gamma radiation; thus the body is constantly bombarded by low doses. In higher doses the radiation can be directed to treat cancerous tumors.

X-rays consist of less-energetic photons than do gamma rays and are often used to diagnose medical conditions such as broken bones or cavities in teeth, because they can penetrate the flesh. Cancer treatments sometimes involve the use of X-rays to kill tumor cells. Because they can also harm healthy cells, health care professionals must monitor their use to prevent overexposure. Biophysicsts and materials science engineers utilize X-rays to probe the structure of substances at the atomic level on the basis of the way the substance diffracts the rays.

Ultraviolet rays, whose photons are less energetic than those of X-rays, effectively kill bacteria and viruses on surfaces of objects and are often used for this purpose in hospital or laboratory settings. The major natural source of ultraviolet radiation is the Sun, and overexposure can cause burns and lead to skin cancers. Ultraviolet rays are nonionizing, and plastic or glass sufficiently blocks most wavelengths. Electronics manufacturing also uses ultraviolet rays to produce integrated circuits.

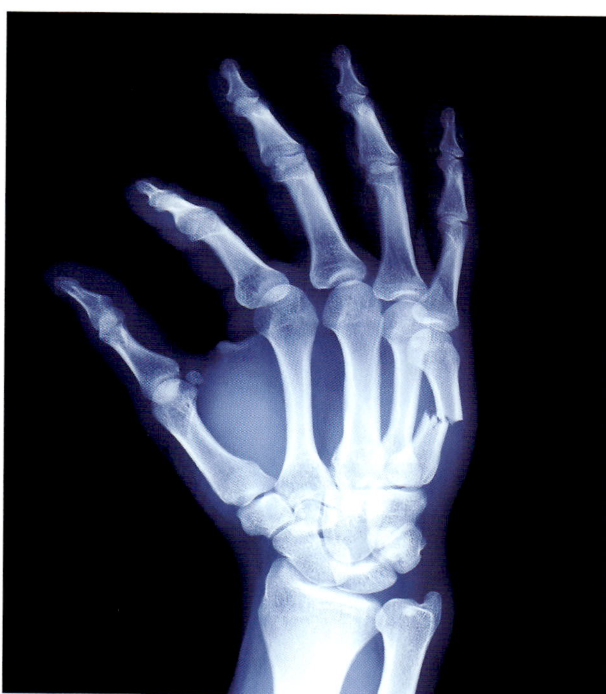

X-rays penetrate the flesh, allowing medical personnel to diagnose conditions by observing internal structures, such as this broken hand bone. *(Zephyr/ Photo Researchers, Inc.)*

A microwave oven heats and bakes food with electromagnetic waves that are confined inside the volume of the oven. Energy from the waves is absorbed by water molecules in the food, which heat up and share their heat with the rest of the food. *(Peter Widmann/Alamy)*

The visible range of electromagnetic radiation occurs at wavelengths from 390 nm to 750 nm. (Wavelengths of visible light are often specified in nanometers, where one nanometer [nm] equals 10^{-9} meter.) These are wavelengths that the human eye can detect and perceive as different colors. Red occurs at around 680 nm and violet at about 400 nm.

Infrared radiation, also called thermal radiation, is sensed by the body as heat. Infrared waves range from about 750 nm to one millimeter. Because warm-blooded animals, including humans, give off heat, police can use devices that detect infrared rays to locate people who are hiding at nighttime, and the military can use infrared detectors to see heat exhaust from vehicles or tanks traveling in the dark or in fog. Astronomers look for infrared radiation as evidence of newborn stars hidden behind gas or dust. More common devices that depend on the emission or detection of infrared radiation include electronic thermometers, toasters, and television remote controls.

Microwaves range between one and 300 millimeters in wavelength. The most familiar application of microwaves is cooking food, but because they can pass through rain and fog and the ionosphere (a region of the atmosphere that contains many charged particles), communications systems employ them for navigation, satellite transmissions, and space communications.

The antennas on the communications mast atop the Rockefeller Building in New York City send and receive electromagnetic waves for communication purposes, such as police and emergency radio networks and mobile phone use. *(Jeff Gynane, 2008, used under license from Shutterstock, Inc.)*

The type of electromagnetic waves with the longest wavelengths are radio waves, commonly used to communicate over long distances. Radio, television, and cell phones are examples of devices that utilize radio waves. In the broadcast process, sounds or other signals are converted into radio waves, a transmitter emits the electromagnetic waves through space, and a receiver detects the signals and converts them back into the original sounds or signals.

See also ELECTROMAGNETISM; FARADAY, MICHAEL; MAXWELL, JAMES CLERK; PHOTOELECTRIC EFFECT; WAVES.

FURTHER READING

Fritzsche, Hellmut. "Electromagnetic Radiation." In *Encyclopaedia Britannica*. 2006. From Encyclopedia Britannica Online School Edition. Available online. URL: http://school.eb.com/ed/article-59189. Accessed July 25, 2008.

National Aeronautics and Space Administration. "The Electromagnetic Spectrum." Available online. URL: http://science.hq.nasa.gov/kids/imagers/ems/index.html. Accessed July 27, 2008.

electromagnetism Electromagnetism is the natural phenomenon that encompasses electricity, magnetism, and their effects on each other. While many everyday manifestations of electricity and magnetism give no indication that the two are related, they can affect each other strongly. For example, an electric current flowing through a coiled wire becomes a magnet, called an electromagnet, which might unlock a car door or spin a computer's hard drive. On the other hand, in electric power plant generators and in automobile alternators, wire coils rotate near electromagnets and thereby produce electricity. These examples show that electricity and magnetism are not independent phenomena but are intimately linked as constituents of the broader phenomenon of electromagnetism.

Albert Einstein's special theory of relativity states that an observer will view a different mix of electric and magnetic effects, depending on her state of motion. If Amy is studying a purely electric effect, for example, then Bert, moving past Amy with sufficient speed, will see the same effect as a combination of electricity and magnetism. This, too, points to the deep connection between electricity and magnetism.

Electricity is described in terms of electric charges and the electric field. Magnetism involves no magnetic charges (at least as is presently known) and is described in terms of the magnetic field. Each field is a vector quantity that extends over all space and at every location possesses some value (possibly zero), which might change over time. The electric and magnetic fields together constitute the electromagnetic field and are components of it.

The electric field is produced by electric charges, independently of whether the charges are at rest or in motion. The electric field in turn causes forces to act on electric charges, again independently of whether the affected charges are at rest or in motion. The magnetic field is produced only by moving electric charges and causes forces only on electric charges in motion. So electric charges affect (i.e., cause forces on) each other through the mediation of the electromagnetic field.

The example of Amy and Bert can now be restated in terms of fields. Amy measures the forces on her test electric charge, both while keeping it at rest and while moving it in various ways. She finds a force acting on her test charge and discovers that it does not depend on the charge's motion. So she concludes that in her vicinity the electric field has some nonzero value, while the magnetic field is zero. Bert speeds by Amy and, as he passes her and while she is conducting her measurements, conducts similar experiments with his test charge. His finding is that in Amy's vicinity, *both* fields possess nonzero values. The values of the electric and magnetic components of the electromagnetic field at the same time and place can have different values for observers in different states of motion.

UNIFICATION AND WAVES

A full and unified understanding of classical electromagnetism (i.e., electromagnetism without quantum effects) was achieved in the 19th century by James Clerk Maxwell and Hendrik Antoon Lorentz. On the basis of discoveries of earlier physicists (such as André-Marie Ampère, Charles-Augustin de Coulomb, Michael Faraday, and Hans Christian Ørsted), Maxwell formulated a set of equations that show how the electric field is produced by electric charges and by the magnetic field's varying in time and how the magnetic field is produced by moving electric charges (electric currents) and by the electric field's changing over time. Lorentz's contribution was a formula giving the force on an electric charge in any state of motion due to the electromagnetic field (i.e., due to the simultaneous effects of the electric and magnetic fields). This is called the Lorentz force.

One of the most noteworthy features of Maxwell's equations is that they predict the existence and describe the properties of waves in the electromagnetic field, called electromagnetic waves or electromagnetic radiation. These waves span a very wide range, or spectrum, of radiations, which includes radio and TV waves, microwaves, infrared radiation, visible light, ultraviolet radiation, X-rays, and gamma rays. What differentiates the various radiations is their frequency, or, correspondingly, their wavelength.

Electromagnetic waves are transverse waves. As such, they can be polarized, that is, made to oscillate in a particular direction. In a vacuum, all electromagnetic waves propagate at the same speed, called the speed of light and conventionally denoted by *c*, with the value 2.99792458×10^8 meters per second (m/s). According to Maxwell's equations, the speed of light is related to the permittivity of the vacuum, ε_0, and the permeability of the vacuum, μ_0, by

$$c = \frac{1}{\sqrt{\varepsilon_0 \mu_0}}$$

That is indeed the case, with $E_0 = 8.85418781762 \times 10^{-12}\ C^2/(N \cdot m^2)$ and $\mu_0 = 4\pi \times 10^{-7}\ T \cdot m/A$. The speed of light is found to have the same value no matter what the state of motion of the observer. This seemingly paradoxical effect led Einstein to the special theory of relativity and serves as one of the theory's cornerstones.

Quantum electrodynamics (QED) is the name of the complete theory of electromagnetism, with the inclusion of quantum effects, which was developed in the 20th century. QED describes electromagnetic radiation in terms of flows of particles, called photons, and explains electric charges' affecting each other electromagnetically by exchanges of photons between them.

HISTORY

Studies on electromagnetism began in 1820 when the Danish physicist Hans Christian Ørsted discovered that electricity has a magnetic effect. He was setting up some materials for a lecture and noticed that a nearby compass needle deflected from the north position when he turned a battery on and off. A battery is an apparatus consisting of two connected cells through which an electrical current flows. As the current flowed, it produced a magnetic field that moved the compass needle, and the magnetic field emanated from the wire in all directions. The magnetic force acted in a circle rather than a straight line. After hearing of Ørsted's demonstration that electricity could produce magnetism, the French physicist André-Marie Ampère, who is best known for developing a way to measure the strength of an electric current, laid the mathematical groundwork for the development of what he called electrodynamics.

Stimulated by a colleague who was attempting to rotate a wire that carried a current by using a magnet, the English chemist and physicist Michael Faraday conceived of a device that could convert electrical energy into mechanical motion. He devised an apparatus consisting of two bowls of mercury, which is a good conductor of electricity, with metal rods inserted into each bowl and connected to a battery. One bowl had a movable magnet with a wire hanging into the mercury to complete the circuit, and the other had a fixed magnet and a movable hanging wire. When the battery was connected, a current passed through the wire, and the wire rotated about the fixed magnet in one bowl, and the movable magnet rotated in a circular motion around the stationary wire in the other bowl. Faraday achieved what he called electromagnetic rotation. The magnetic field made the current-carrying wire move in a circle around the magnet. After succeeding in building this first electric motor, he published in 1821 his findings that a field of force encircled a current-carrying wire. His findings were significant because until that time, scientists thought electricity was a fluid through which particles carrying electrical energy flowed. The notion of a force field eliminated the need for a medium.

In 1931 Faraday conducted investigations to determine whether magnetism could produce electricity. To attempt this, he wrapped one side of an iron ring with a copper wire attached to a battery and the other side of the iron ring with a wire attached to a galvanometer, a device for measuring electric current. When the circuit was completed, the needle jumped and then returned to its original position. When he disrupted the circuit to turn off the current, the needle jumped again, and then returned to its original position. Though he was unable to produce a steady current, changes (such as starting or stopping) in the magnetic field generated an electric current. A current in the wire attached to the battery generated a magnetic field that produced a second current in the wire attached to the galvanometer. Similarly, when he moved a bar magnet in and out of a coil formed from copper wire, the galvanometer detected a current but it did not detect any current when the apparatus was held stationary. He imagined lines of force surrounding the magnet, creating a magnetic field interacting with a force created by the traveling current. He next set up a copper disk between the north and south poles of a horseshoe-shaped magnet. A copper wire extended from the central axis around which the disk spun, through a galvanometer, and touched the edge of the disk. Mechanical rotation of the disk caused the galvanometer to register a current. Faraday had created the first dynamo, or electric generator, and given rise to the electromagnetic theory.

The generation of an electrical current by a changing magnetic field is called electromagnetic induction, and it is the foundation of the modern electrical industry. Electromagnetic induction is used to convert continuous motion into an electrical current. The American physicist Joseph Henry also independently discovered this phenomenon, but Faraday published his results first, and therefore he receives credit for this discovery. Electromagnetic induction

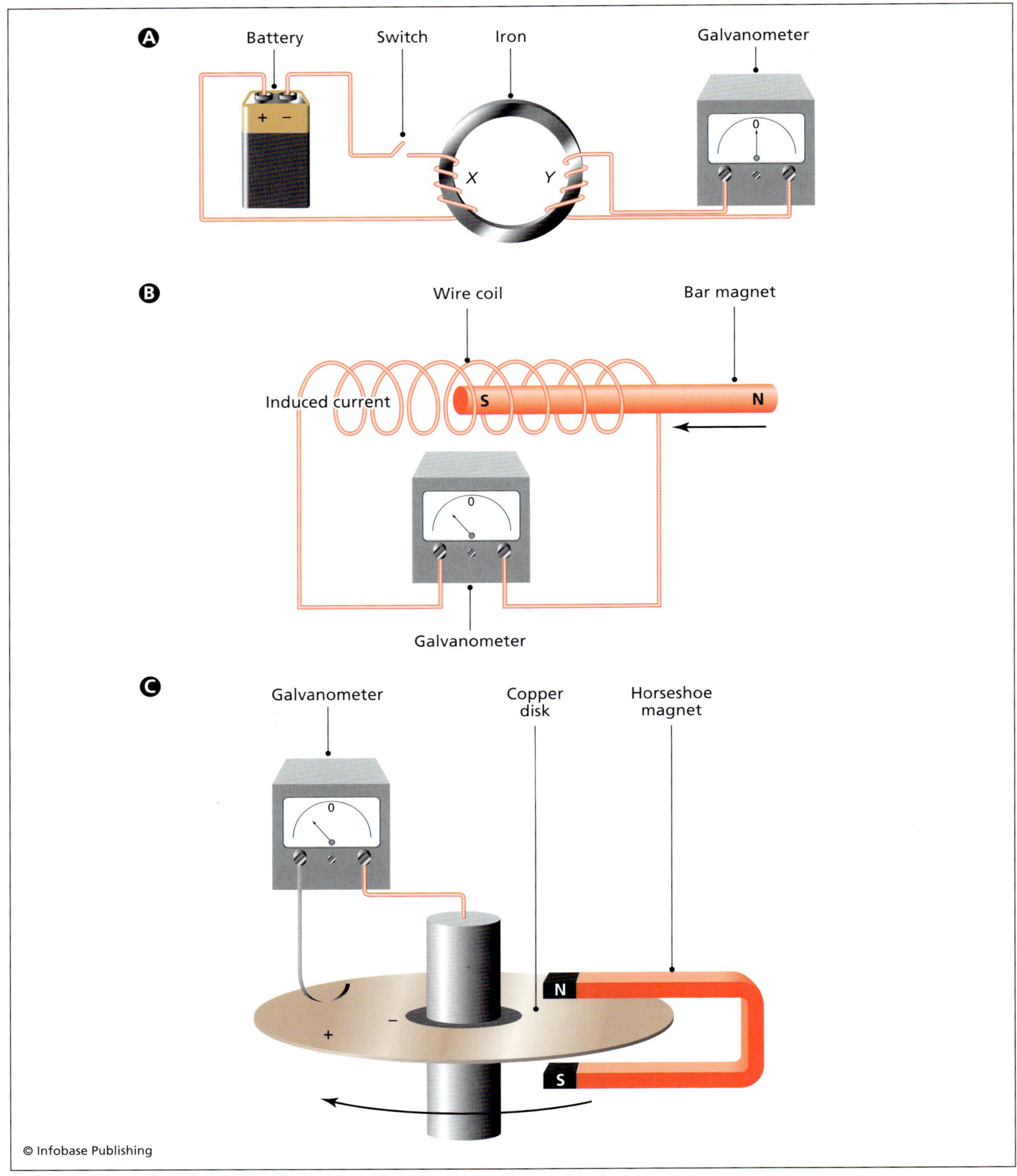

(a) A current produced by a battery magnetized an iron ring wrapped in a copper wire, which, in turn, induced an electric current in a second wire also wrapped around the iron ring. (b) Electromagnetic induction is the process of generating an electric current by changing a magnetic field. (c) In the first dynamo, the mechanical motion of rotating a copper disk produced an electrical current.

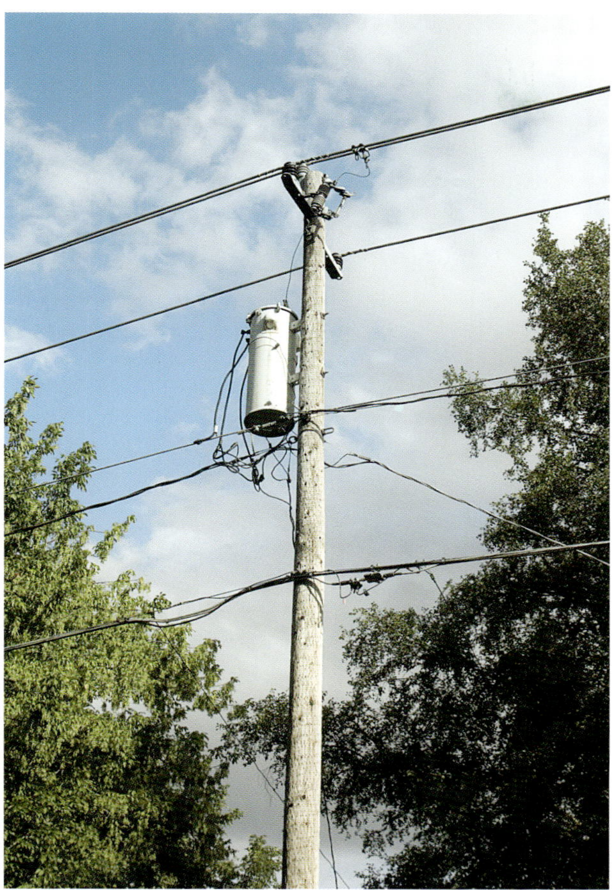

By means of electromagnetic induction this step-down transformer reduces the voltage of the electricity entering the neighborhood from several hundred volts, the voltage at which it is transmitted into the neighborhood, to 110 volts for home use. *(David Gaylor, 2008, used under license from Shutterstock, Inc.)*

is the basis for the function of generators, machines that convert mechanical energy into electrical energy that can be sent into the community. Transformers, devices that increase or decrease voltage, also operate by electromagnetic induction.

While lecturing at the Royal Institution in the 1840s, Faraday suggested that atoms, and therefore objects, be considered areas of concentrated forces that reached into space from charged, magnetic, or gravitating bodies. Electricity resulted from the interactions of the object's individual forces. The force field surrounding magnets can be visualized by sprinkling iron filings on a piece of paper placed over a bar magnet. Lines of filings will extend in distinct pathways to connect the two poles. The interactions between similar or opposite poles cause different patterns to form. This concept of fields—interactions that occur through space without a medium—is also used to explain the strong and weak interactions. Faraday also proposed that light was a vibration in a line of force, an idea that the renowned Scottish physicist James Clerk Maxwell would later demonstrate mathematically.

Maxwell built on and unified the work of other 19th-century scientists. While Faraday first conceived of an electromagnetic field, Maxwell quantified it, putting it into mathematical form. In 1865 he published his proposal of a theory that unified electricity and magnetism known as classical electromagnetism, a breakthrough in 19th-century science. The principles of electromagnetism dictate the manner in which charged particles interact. The forces are strong but weaken as distance increases. One prediction from Maxwell's theory was the existence of electromagnetic waves that could travel through space, a phenomenon that was later verified and has transformed the communications industry. Maxwell's theory of electromagnetism also helped scientists understand the nature of light as a disturbance in the electromagnetic field. These disturbances, called electromagnetic waves, are characterized by their frequencies, amplitudes, and orientation. Waves of different frequencies are used in different applications: some frequencies are used to transmit radio and television signals; others are used to take X-ray photographs. Maxwell summarized everything that was known about electricity and magnetism in four fundamental equations that came to bear his name—Maxwell's equations. In summary, his equations 1) related electric field to electric charge, a relationship called Gauss's law; 2) related magnetic field to its source, showing that magnetic field lines are continuous, whereas electric field lines begin and end at a charge; 3) showed that a changing magnetic field produces an electric field, called Faraday's law; and 4) demonstrated that an electric current or a changing electric field produces a magnetic field, a combination of contributions from Ørsted, Ampère, and Maxwell.

As beautiful as Maxwell's theory was, it contradicted classical Newtonian mechanics. The 20th-century physicist Albert Einstein took it upon himself to resolve the conflict between electromagnetism and Newton's laws of motion and, in doing so, developed his special theory of relativity in 1905. Einstein corrected Newton's laws as they applied to speeds approaching the speed of light so they could flawlessly merge with Maxwell's equations, which already held the speed of light in vacuum to be a universal constant. Relativity theory also helped scientists understand that magnetic fields and electric fields are really "two sides of the same coin." The quantum mechanical approach to the electromagnetic force had been invented by the 1940s and is called quantum electrodynamics (QED), often described as physics's most perfect theory.

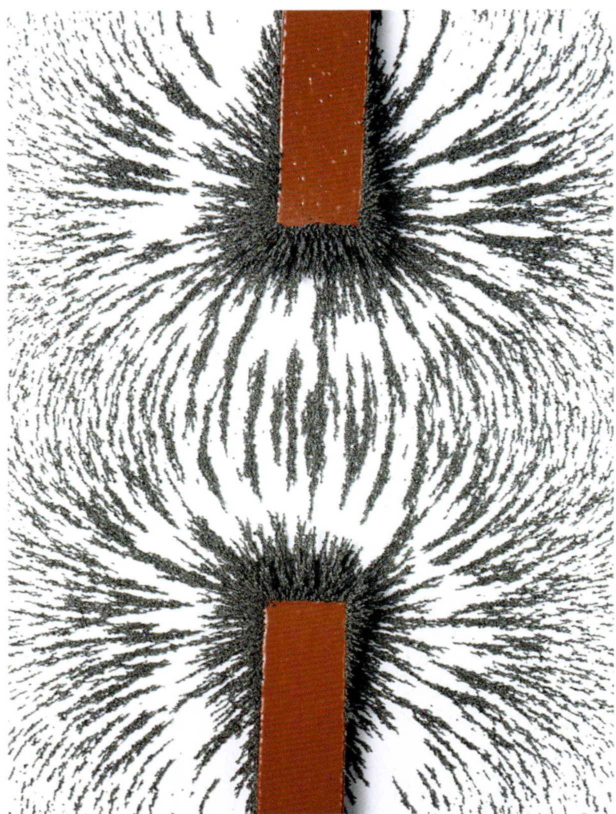

Michael Faraday conceived of the idea of a continuous field surrounding magnets. Sprinkling iron filings helps one visualize the attractive forces between these opposite magnetic poles, but defined lines of force do not really exist. *(Cordelia Molloy/ Photo Researchers, Inc.)*

Since the unification of electricity and magnetism, physicists have unified the weak nuclear force with the electromagnetic force, calling it the electroweak force. Many physicists have devoted their careers to developing a theory that also encompasses the strong nuclear force, but the proposed theories, such as superstring theory, have not been able to incorporate gravity successfully. Of the four forces, electromagnetism is the best understood.

Numerous inventions have electromagnets incorporated in their mechanisms. Ordinary magnets, such as people place on their refrigerator doors, have poles that attract opposite magnetic poles. Consisting of a coil of wire wrapped around a piece of magnetic metal, an electromagnet also has poles that attract one another if near an opposite pole and repel one another if alike, except that the electromagnet only operates when it is turned on, usually by an electric switch. Turning on an electromagnet means running an electric current through it, producing a circular magnetic field around the wire through which the current flows. Turning the switch to the off position stops the electron flow, and the magnetic field disappears. Numerous modern-day conveniences rely on electromagnets, including power locks on automobiles, cones inside stereo speakers, electric motors, read/write heads for hard disks that store digital information, and machinery used to lift tons of steel.

See also CLASSICAL PHYSICS; EINSTEIN, ALBERT; ELECTRICITY; ELECTROMAGNETIC WAVES; ENERGY AND WORK; FARADAY, MICHAEL; FORCE; MAGNETISM; MAXWELL, JAMES CLERK; PHOTOELECTRIC EFFECT; QUANTUM MECHANICS; SPECIAL RELATIVITY; VECTORS AND SCALARS.

FURTHER READING

Royal Institution of Great Britain. "Heritage: Faraday Page." Available online. URL: http://www.rigb.org/rimain/heritage/faradaypage.jsp. Accessed July 25, 2008.

electron configurations

electron configurations Electron configurations describe the probable locations of all the electrons in an atom, a particle composed of protons, neutrons, and electrons. The positively charged protons and the neutral neutrons are located in the nucleus and form the main contributors to the mass of the atom. The accurate description of locations of the negatively charged electrons has evolved from the plum pudding atomic model proposed by the English physicist Sir J. J. Thomson, in which the electrons were thought to be stuck in a positively charged mass of protons. In the early 1900s, the New Zealand–born British scientist Sir Ernest Rutherford performed experiments that led to the discovery of a positively charged nucleus that was densely packed in the center of the atom. These findings provided the Danish physicist Niels Bohr with information necessary for the development of his planetary model of the atom. Bohr's model suggested that the electrons in an atom orbited the nucleus much as planets orbit the Sun.

Contemporaries of Bohr's had observed while studying hydrogen that the amount and type of electromagnetic radiation given off by the element was always the same. The wavelengths of the visible light given off were 410 nm, 434 nm, 486 nm, and 656 nm. This set of spectral lines is known as the Balmer series. The portion of the line spectra that fell into the ultraviolet (UV) range is known as the Lyman series, and portions of the line spectra in the infrared range are known as the Paschen series. Scientists studied other elements and discovered that each gave off its own set of spectral lines, creating a fingerprint of the element. Since the elements always gave a characteristic pattern, the pattern could be used for identification.

The Bohr planetary model of the atom was unable to correlate the line spectra that were observed for each element with the location of the electrons in the orbits. Bohr studied the work of the German physicist Max Planck, who determined that energy is given off as electromagnetic radiation in the form of packets called quanta, or photons. These were always released according to the formula $E = hf$, where E is the photon energy in joules (J), h is a constant (Planck's constant) equal to 6.63×10^{-34} joule-second (J·s), and f is the frequency of the wave in hertz (Hz). Bohr used this knowledge to explain that if the element always releases the same amount of energy and absorbs the same amount of energy, the locations of the electrons must be fixed or limited. When the electron receives energy, it jumps up to a higher energy orbital, and, as it loses energy, it falls back down to the lower energy orbital and releases the energy. His model resembles a ladder—a person cannot stand between the rungs. The higher up one steps on the ladder, the higher the energy. Similarly, the electrons at higher levels have higher energy.

In 1926 the Austrian physicist Erwin Schrödinger developed an equation that describes the wavelike character of particles and opened the field of quantum mechanics, or quantum physics. In the quantum mechanical model of the atom, the electron locations are considered to be probable locations rather than exact locations. The German physicist Werner Heisenberg determined that one cannot know both the speed and the position of an electron at the same time. In determining the position of the electron, one cannot know how fast it is going, and while measuring its speed, one cannot know its exact position. The quantum mechanical model of the atom involved the use of probabilities, represented by electron clouds. Four quantum numbers describe an electron's position: principal quantum number (n), angular momentum quantum number (l), magnetic quantum number (m_l), and spin quantum number (m_s). Electrons with the same value for n are positioned in the same energy level, equal values for n and l indicate the same sublevel, and orbitals are the same if electrons share common values for n, l, and m_l. According to the Pauli exclusion principle developed by Wolfgang Pauli in 1925, no two electrons can have the same set of four quantum numbers. If two electrons have the same values for principal quantum number, momentum quantum number, and magnetic quantum number (n, l, and m_l), then they must have opposite spin (m_s).

ASSIGNING ELECTRON CONFIGURATIONS

Electrons that have the same value for n are in the same energy level. The higher the value of n, the farther from the nucleus the electron is, and the greater the energy of the electron. Electrons cannot exist between energy levels. The seven energy levels correspond to the seven periods of the periodic table. The values for n are thus 1, 2, 3, 4, 5, 6, and 7. The number of electrons that can be found in any energy level is given by $2n^2$, where n is the principal quantum number.

After determining the energy level of the electron, the next part of the electron's "address" is the sublevel (l), or area of specific energy located within the energy level. The number of sublevels is equal to the value of the principal quantum number of that energy level. Thus, the first energy level has one sublevel, the second energy level has two sublevels, and so on. Sublevels are named *s, p, d,* and *f* from lowest to highest energy. An *s* sublevel can hold two electrons, a *p* sublevel can hold six electrons, a *d* sublevel can hold 10 electrons, and an *f* sublevel can hold 14 electrons. Electrons fill the sublevels from lowest energy first with a few exceptions noted later.

Orbitals, or the actual locations of electrons, are found within sublevels. Two electrons fit in any orbital. The *s* sublevel has one orbital, the *p* sublevel has three orbitals, the *d* sublevel has five orbitals, and the *f* sublevel has seven orbitals. One must not confuse quantum mechanical orbitals with Bohr's orbits.

When assigning electrons to locations within the atom, one must first determine the total number of electrons in the atom, a quantity equal to the atomic number in a neutral atom. The Aufbau principle explains the distribution of electrons into the orbitals from the lowest to the highest energy. The energy of the orbitals is represented by the accompanying figure.

The octet rule, which states that every atom wants to have a filled outer shell and that the outer shell of any atom (except hydrogen and helium) can only hold up to eight electrons, determines the order of orbitals. An *s* sublevel holding two electrons combined with a *p* sublevel holding six electrons,

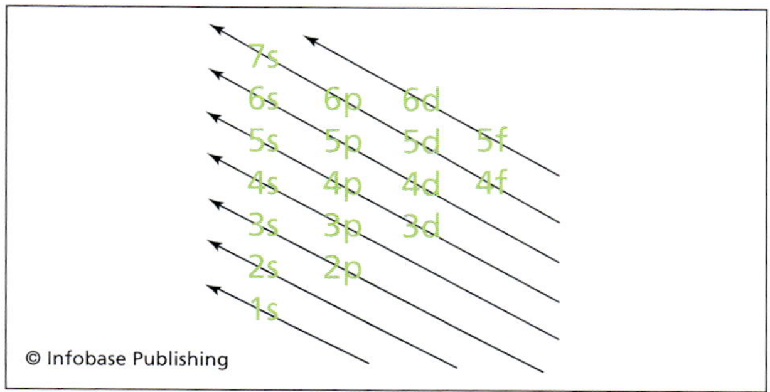

The order of increasing sublevel energy can be determined by following the arrow's head to tail. Electrons fill the sublevels from low energy to high energy.

constitute the eight electrons that satisfy the octet rule. Every element with 18 or fewer electrons, up to and including argon, will easily fit its electrons into the *s* and *p* sublevels of energy levels one, two, and three. After period three, each period has either 18 or 32 electrons to add that are accommodated in the *d* and *f* sublevels. To adhere to the octet rule, the atoms will not place electrons in any *d* or *f* sublevel until it has opened up the *s* sublevel of the next highest energy level. For example, prior to placing any electrons in the 3*d* sublevel, some must be placed in the 4*s* sublevel. This allows the fourth energy level to be the outer shell of the atom, and the octet rule can still be satisfied. Then the atom can have as many electrons in the 3*d* sublevel as it will hold.

Although every orbital can hold two electrons, some orbitals are degenerate, meaning they are of equivalent energy. The *p* sublevels have three degenerate orbitals oriented around the *x*-, *y*-, and *z*-axes. The *d* sublevels have five degenerate orbitals, and the *f* sublevels have seven degenerate orbitals. When filling degenerate orbitals, Hund's rule states that the electrons will fill each orbital singly before any one of a set of degenerate orbitals receives two. Once each of the degenerate orbitals has at least one electron, the orbitals will fill up doubly. An analogy to Hund's rule is a dormitory that has three equivalent rooms that hold two people each. According to Hund's rule, the rooms would each be filled with one student before pairing up any of the students with roommates. Once each equivalent room (degenerate orbital) houses one student (electron), then a second student (electron) would move into each room. Unpaired electrons cause the atom, known as paramagnetic, to be slightly attracted to a magnetic field. When all the electrons in an atom are paired (diamagnetic), a magnetic field will slightly repel the atom.

The shorthand method of writing electron configurations includes noting the energy level, sublevel letter, and number of electrons found in that sublevel as a superscript. For example, an atom that has five electrons would place two electrons in the 1*s* sublevel, two electrons in the 2*s* sublevel, and one electron in the 2*p* sublevel and would have an electron configuration of

$$1s^2 2s^2 2p^1$$

Because the number of electrons in the atoms increases as the atomic number increases, even this shorthand method becomes cumbersome. In order to assign the electron configuration of the larger elements, the symbol of the previous noble gas is written in brackets followed by the ending configuration of the electrons. For example,

$$[\mathrm{Ar}]\ 4s^2 3d^{10} 4p^3$$

is the electron configuration of arsenic (As) with an atomic number of 33. The [Ar] symbol represents the first 18 electrons, the number present in one atom of the noble gas argon, giving a total number of 33 electrons when combined with the remaining 15 electrons.

EXCEPTIONS TO AUFBAU PRINCIPLE

The Aufbau principle and orbital-filling rules work remarkably well for nearly all elements. Only two exceptions to this rule exist in the first 40 elements, chromium and copper. Chromium has an atomic number of 24 and would be expected to have the following electron configuration based on the energy of the orbitals:

$$[\mathrm{Ar}]\ 4s^2 3d^4$$

but its experimentally determined electron configuration is

$$[\mathrm{Ar}]\ 4s^1 3d^5$$

Copper has an atomic number of 29 and an expected electron configuration of

$$[\mathrm{Ar}]\ 4s^2 3d^9$$

but it has a real configuration of

$$[\mathrm{Ar}]\ 4s^1 3d^{10}$$

Even though the 3*d* is higher energy than the 4*s*, atoms are most stable when they have a full sublevel. If that is not possible, they prefer a half-filled sublevel to no arrangement. Therefore, in the predicted configuration, chromium started with one full sublevel ($4s^2$) and one with no arrangement ($3d^4$), while in the experimentally determined configuration of chromium, it had two half-filled sublevels ($4s^1$ and $3d^5$), a configuration that is energetically more stable. Copper has a similar situation, starting with one filled sublevel ($4s^2$) and one with no arrangement ($3d^9$) in the predicted configuration, while the experimentally determined configuration has one half-filled sublevel ($4s^1$) and one filled sublevel ($3d^{10}$) that is more stable.

ELECTRON CONFIGURATIONS AND THE PERIODIC TABLE OF THE ELEMENTS

The periodic table of the elements is a graphic representation of the electron configurations of the atoms. The periods (rows) of the periodic table correspond to the principal quantum numbers (*n*) of the electron configurations. Period one has all of its electrons in

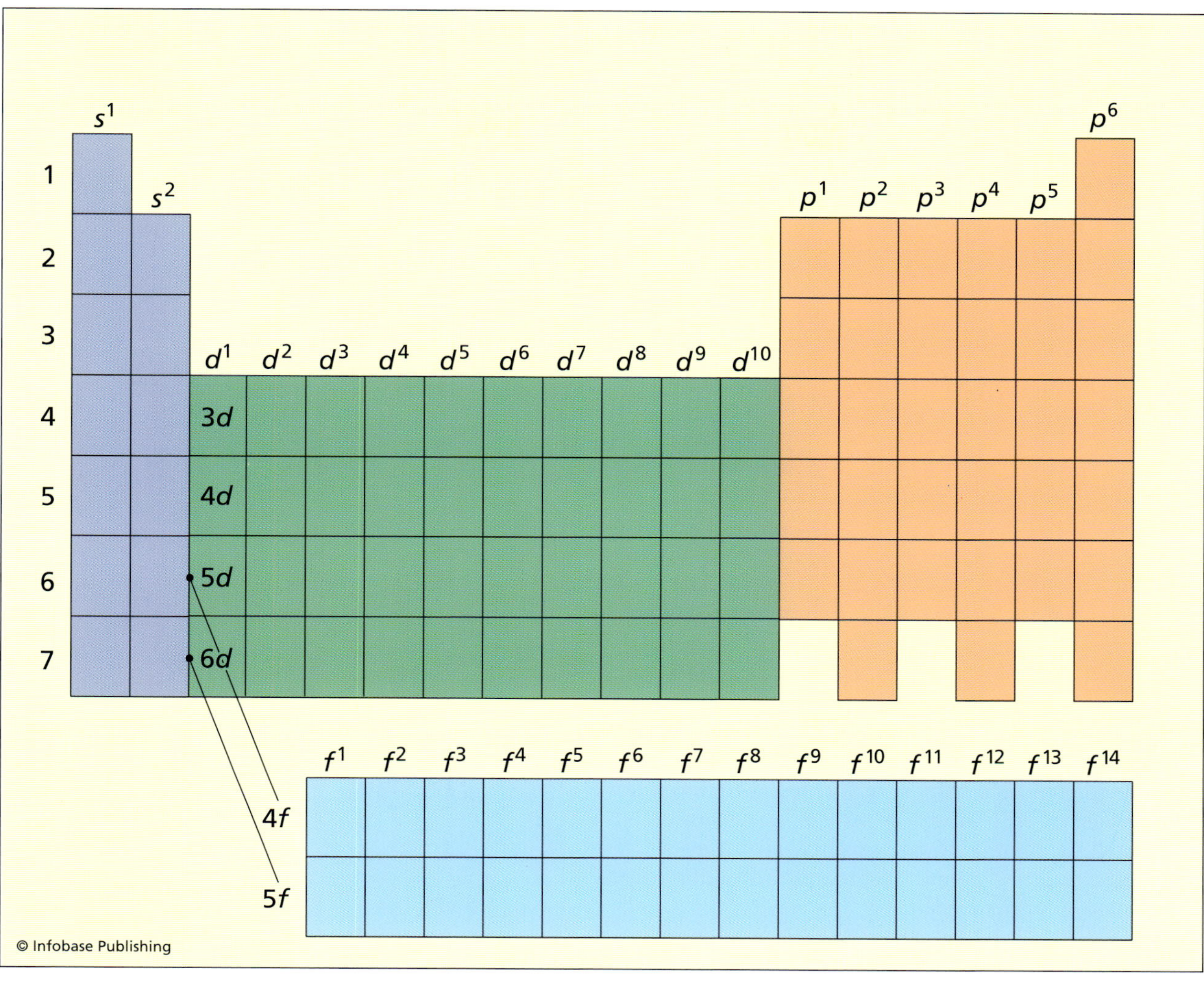

The ending electron configuration of an element can be determined by the location of the element on the periodic table.

the first energy level ($n = 1$); period two has all of its electrons in the second energy level ($n = 2$); the same is true for period three and energy level three. For periods four through seven, the period corresponds to the outer energy level with the exception of the transition metals and inner transition metals. The transition metals have their final electrons in the *d* sublevels of the period one energy level, meaning if the transition metal is in the fourth period on the periodic table, the highest-energy electrons would be added to the 3*d* sublevel. The inner transition metals have their final electrons added to the period two energy level. Therefore, if the inner transition metal were part of energy level seven, the highest energy electrons would be added to the 5*d* sublevel. This difference for transition metals and inner transition metals corresponds precisely to the ribbon diagram on page 207.

See also ATOMIC STRUCTURE; BOHR, NIELS; PERIODIC TABLE OF THE ELEMENTS; QUANTUM MECHANICS; THOMSON, SIR J. J.

FURTHER READING

Brown, Theodore, H. LeMay, and B. Bursten. *Chemistry: The Central Science,* 10th ed. Upper Saddle River, N.J.: Prentice Hall, 2006.

Chang, Raymond. *Chemistry,* 9th ed. New York: McGraw Hill, 2006.

electron emission This term refers to the emission of electrons from the surface of a conductor, especially of a metal. Three methods of achieving electron emission are the photoelectric effect, thermionic emission, and field emission. A separate entry (on page 527) discusses the photoelectric effect.

The present entry deals with thermionic emission and field emission.

THERMIONIC EMISSION

The spontaneous emission of electrons from the surface of a sufficiently hot metal is called thermionic emission. The emission occurs when a significant fraction of the free electrons in the metal possess sufficient thermal energy to overcome the forces they encounter at the metal's surface, which normally confine them to the volume of the metal. The thermal energy of those electrons must be greater than what is called the work function of the metal, which is the minimal amount of energy needed for an electron in the material to escape. As the temperature of the metal increases, both the rate of emission and the average energy of the emitted electrons increase as well.

Thermionic emission from a hot cathode serves as the source of electrons for the electron beam in television picture tubes, computer monitor tubes, and oscilloscope tubes, as examples. Such tubes are called cathode-ray tubes, or CRTs. Once emitted, the electrons are accelerated by an electric field, associated with an applied voltage, and collimated to form a beam. The beam is shaped, focused, and directed by magnetic fields. When the beam strikes the surface of the glass tube that serves as the screen, light is emitted at the spot of incidence through the process of fluorescence by a suitable material coating the inside surface of the screen.

The image tube for television sets and computer monitors is of the type called cathode-ray tube (CRT). The image is formed when a beam of electrons hits the inside of the screen surface in a raster pattern and excites phosphors coating the inside surface, causing them to emit light. The source of electrons in these tubes is thermionic emission from a heated cathode at the rear of the tube. (LCD and plasma screens, which operate by different principles from CRT tubes, are rapidly replacing CRT tubes in television sets and computer monitors.) *(Helene Rogers/Alamy)*

FIELD EMISSION

Field emission is the method of causing electrons to be emitted by forcing them out of matter by a strong electric field. The electric field in the vicinity of an electrically charged conductor is especially strong near regions of high curvature of the surface of the conductor. To maximize the effect, a conducting wire is sharpened to a very fine point. The conductor is placed in a vacuum chamber and given a negative potential (i.e., is charged negatively). For sufficient potential, the electric field at the tip of the point exerts forces on the electrons at the conductor's surface that overcome the forces confining them to the material and "tears" them out of the conductor. Quantum effects also play a role in this process, allowing the emission of electrons that otherwise would not be released by the electric field.

The field emission microscope is based on that process. The tip of the conductor's point is given a regular, round shape. Since electrons in the conductor are most likely to be in the neighborhood of atoms, the electrons pulled from the conductor by field emission emanate mostly from the atoms that are near the surface. These electrons leave the surface perpendicularly to the surface and travel in straight lines to the chamber's walls, which are designed to fluoresce when electrons strike them or otherwise record the electrons' arrival. Thus, the microscope creates an enlarged image of the atomic structure of the material from which the conductor is made.

See also ELECTRICITY; ENERGY AND WORK; HEAT AND THERMODYNAMICS; MAGNETISM; PHOTOELECTRIC EFFECT; QUANTUM MECHANICS.

FURTHER READING

Serway, Raymond A., Clement J. Moses, and Curt A. Moyer. *Modern Physics,* 3rd ed. Belmont, Calif.: Thomson Brooks/Cole, 2004.

Young, Hugh D., and Roger A. Freedman. *University Physics,* 12th ed. San Francisco: Addison Wesley, 2007.

electron transport system Embedded in the inner mitochondrial membrane of eukaryotic cells and in the plasma membrane of prokaryotic cells, the electron transport system consists of a chain of respiratory enzyme complexes that use energy obtained from a series of alternating oxidation-reduction reactions to create a gradient of hydrogen ions (H^+), or protons, across the membrane. Dissipation of the H^+ gradient powers the synthesis of adenosine triphosphate (ATP), the most abundant activated carrier molecule of energy in the cell, in a process called chemiosmosis.

Glycolysis is the catabolic pathway that degrades glucose into two molecules of pyruvate. The catabolic pathway generates small amounts of ATP by substrate-level phosphorylation, an energetically unfavorable reaction (requires energy) in which a phosphorylated metabolic intermediate transfers its phosphate group to an adenosine diphosphate (ADP), creating a molecule of ATP. The citric acid cycle, which further oxidizes the end products of glycolysis, also creates small amounts of ATP by substrate-level phosphorylation. In respiration, organisms produce ATP by oxidizing molecules and ultimately transferring the extracted electrons to an inorganic compound. In aerobic respiration, molecular oxygen (O_2) acts as the final electron acceptor, and in anaerobic respiration, oxidized inorganic compounds such as nitrate or sulfate act as the final electron acceptor. Electron transport systems produce the majority of ATP in a cell. Specialized molecules store the high-energy electrons released from the molecular energy source. Two oxidized molecules, nicotinamide adenine dinucleotide (NAD^+) and flavin adenine dinucleotide (FAD), carry the high-energy electrons extracted during the oxidation of glucose to the electron transport chain to convert the energy stored in the electrons to a high-energy bond between the second and third phosphate groups of ATP.

ELECTRON TRANSPORT

Many endergonic, or energy-requiring, biochemical reactions must couple with a complementary exergonic reaction to proceed. The free energy released from the exergonic reaction (ΔG) must exceed that required by the endergonic reaction. Oxidation-reduction, or redox, reactions involve the transfer of electrons from one molecule to another. Oxidation involves the loss of electrons, and reduction is the gain of electrons. Oxidizing a glucose molecule is an energetically favorable reaction (gives off energy), and by coupling it with a reduction reaction, some of the energy can be captured and stored in an activated carrier molecule. In this manner, redox reactions often supply the energy needed to drive otherwise energetically unfavorable reactions.

Hydrogen ions (H^+) are simply hydrogen atoms that have lost an electron (e^-). When one molecule donates electrons to reduce another molecule, the hydrogen ion often accompanies the electron in the move. This addition of a hydrogen atom during reduction is called a hydrogenation reaction, and the reverse is called a dehydrogenation reaction.

Oxidation-reduction or redox potential (E'_0) indicates a molecule's susceptibility to donate or accept electrons and is measured in voltage relative to the voltage required to remove or add an electron to H_2 (–0.42 V). For example, the redox potential for the pair $NAD^+/NADH + H^+$ is –0.32 V, and the redox potential for $FAD/FADH_2$ is –0.18 V. Compounds that have high redox potentials, meaning more negative voltages, are very reduced and generally have a lot of energy available for release by oxidation.

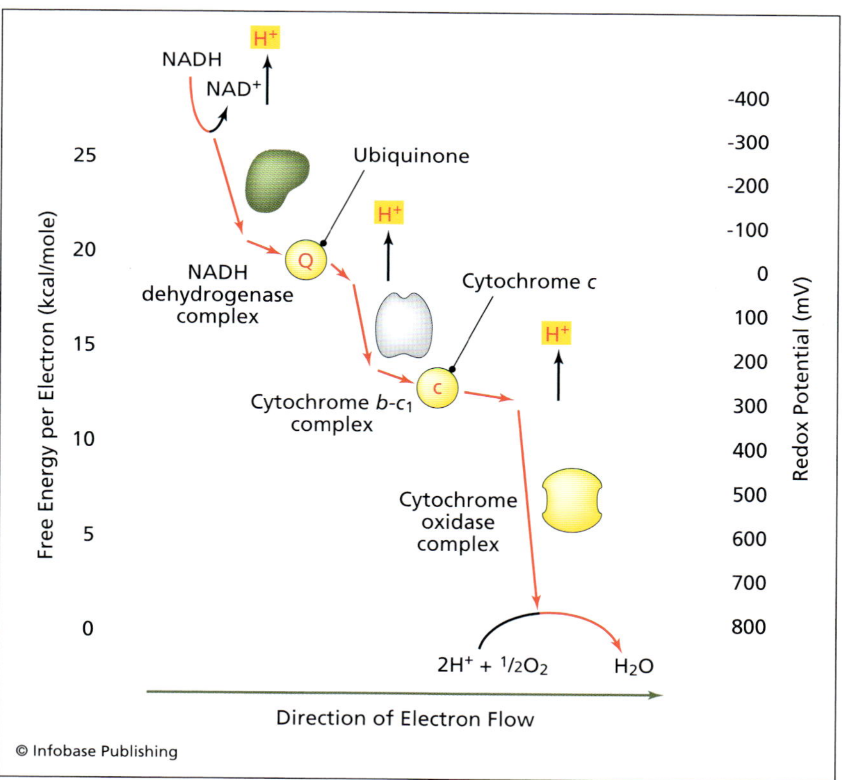

The electron transport chain passes high-energy electrons to carriers at lower energy levels and with successively higher reduction potentials.

The reduced carriers NADH + H^+ and $FADH_2$ carry the high-energy electrons to the electron transport chain. The specific molecules that serve as transporters in the chain differ among organisms but in mitochondria generally include several large protein complexes: the reduced nicotinamide adenine dinucleotide (NADH) dehydrogenase complex, ubiquinone, the cytochrome b-c_1 complex, and the cytochrome oxidase complex. Electron transport begins when NADH dehydrogenase removes a hydride ion (H^-) from NADH and subsequently splits it into a proton and two high-energy electrons.

$$H^- \rightarrow H^+ + 2e^-$$

This reaction occurs on the matrix side of the inner mitochondrial membrane. NAD^+ is in its oxidized form and can pick up more electrons released in another biochemical reaction. Inside the complex, the electrons pass from a flavin group to a set of iron-sulfur centers, and then to ubiquinone (Q), a lipid-soluble carrier that accepts the electrons and carries them in its reduced form, ubiquinol (QH_2), through the phospholipid bilayer to the next respiratory complex. At each step, the electrons move to a lower energy state on a carrier with an increased redox potential, or a higher affinity for the electrons, and a lower energy state. Ubiquinone donates the electrons to the next respiratory complex, cytochrome complex b-c_1, and the small protein cytochrome c carries them to the next complex, the cytochrome oxidase complex, until the electrons reach their final destination, molecular oxygen (O_2), which has the very high redox potential of +0.82 V. Cytochrome oxidase catalyzes the transfer of the electrons to the final electron acceptor of aerobic respiration—O_2—which has a very high affinity for electrons. Each O_2 molecule combines with four protons and four electrons from cytochrome c.

$$4e^- + 4H^+ + O_2 \rightarrow 2H_2O$$

ATP SYNTHESIS

Meanwhile the respiratory complexes have been using bits of energy along the way to pump protons from the mitochondrial matrix through the inner membrane into the intermembrane space, located between the inner and outer mitochondrial membranes. Moving the protons across the membrane creates an electrochemical gradient, also called a membrane potential. The electrical gradient results from the separation of charge, with the intermembrane space more positively charged than the matrix. The chemical contribution to the gradient is from the higher hydrogen ion concentration in the inner membrane space. The pH in the intermembrane space is approximately 7, compared to a pH of 8 in the matrix. Both gradients work together to create a steep electrochemical gradient.

The gradient serves as potential energy that the cell can utilize to perform useful work. The phospholipid composition of the membrane prevents the gradient of charged particles from dissipating in an uncontrolled manner. The gradient drives the H^+ back across the inner membrane into the matrix through the only channels that permit passage, ATP synthase. As H^+ ions flow through the transmembrane ATP synthase complex, the enzyme synthesizes ATP from ADP and an inorganic phosphate (P_i) in the matrix of the mitochondria. Because electrons derived from the oxidation of foodstuffs generated the gradient used to power the synthesis of ATP, the process is called oxidative phosphorylation.

The entire process of oxidizing biomolecules to generate ATP is quite efficient, almost 40 percent. The cell can only harvest energy in little packets; thus, proceeding to lower energy levels in a stepwise manner allows the activated carrier molecules to harness more of the available energy, rather than its being lost to heat.

See also CITRIC ACID CYCLE; GLYCOLYSIS; METABOLISM; OXIDATION-REDUCTION REACTIONS.

FURTHER READING

Alberts, Bruce, Dennis Bray, Karen Hopkin, Alexander Johnson, Julian Lewis, Martin Raff, Keith Roberts, and Peter Walter. *Essential Cell Biology,* 2nd ed. New York: Garland Science, 2004.

Rosen, Barry. "Chemiosmosis." In AccessScience@McGraw-Hill. Available online. URL: http://www.accessscience.com, DOI 10.1036/1097-8542.128350. Accessed July 25, 2008.

Roskoski, Robert Jr. "Biological oxidation." In AccessScience@McGraw-Hill. Available online. URL: http://www.accessscience.com, DOI 10.1036/1097-8542.082700. Accessed July 25, 2008.

energy and work

energy and work This entry discusses work, energy, and the relation between them, which includes the work-energy theorem and conservation of energy. The entry also presents various kinds and forms of energy, including kinetic, potential, internal, and external energies, and discusses energy quantization.

WORK

A scalar quantity (i.e., a quantity that is not characterized by direction), work is defined as a result of a force acting along a displacement in the direction of the force. Quantitatively, when a force **F** acts along an infinitesimal displacement ds, the infinitesimal amount of work performed dW is given by the scalar product

$$dW = \mathbf{F} \cdot d\mathbf{s}$$

This can also be expressed as by the following equation

$$dW = F\,ds \cos\theta$$

where F and ds are the magnitudes of $\mathbf{F}$ and $d\mathbf{s}$, respectively, and θ is the smaller angle (less than 180°) between the two vectors. Work is in joules (J), equivalent to newton-meters (N·m); force in newtons (N); and distance in meters (m).

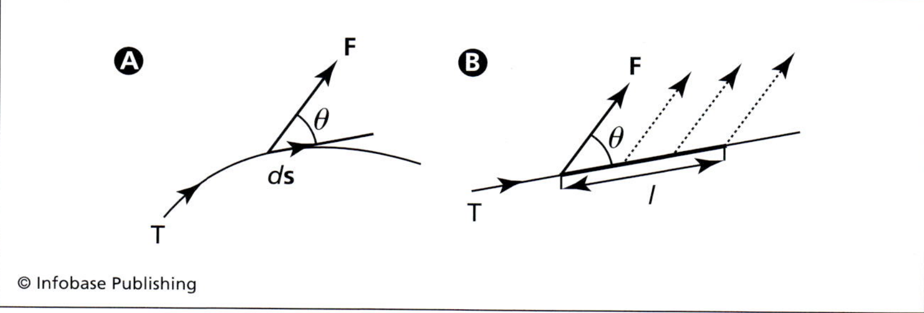

(a) The infinitesimal work done when a force **F** acts over an infinitesimal directed distance, or displacement, $d\mathbf{s}$, is given by $dW = \mathbf{F} \cdot d\mathbf{s}$. This can also be expressed as $dW = F\,ds \cos\theta$, where F and ds denote the magnitudes of the vectors **F** and $d\mathbf{s}$, respectively, and θ is the angle between the vectors. In the figure, the particle upon which the force acts is moving along a trajectory T. (b) In the special case when the force is constant and the trajectory straight, the work done by the force acting along path length l is $W = Fl \cos\theta$, where θ is the angle between the force **F** and the direction of motion.

When the force is in the direction of motion (i.e., when $\theta = 0$ and $\cos\theta = 1$), the work done equals simply $F\,ds$. When the force is opposite to the direction of motion (such as a braking force, with $\theta = 180°$ and $\cos\theta = -1$), the work is negative and equals $-F\,ds$. When the force is perpendicular to the direction of motion (such as in the case of a centripetal force or a magnetic force), then $\theta = 90°$ and $\cos\theta = 0$, and no work is performed.

The total work performed when a (possibly varying) force acts along a path is found by summing up (i.e., by integrating) dW along the path. A constant force of magnitude F, acting along a straight line segment of length l at constant angle θ from the direction of motion, performs work W such that

$$W = Fl \cos\theta$$

As an example, imagine someone pushing a stalled car to get it moving. If the pusher applies a force of 25 N, yet the car remains at rest, there is no displacement and no work is done. If an applied force of 100 N moves the car, the force is in the direction of motion ($\cos\theta = 1$), and the car rolls 10 m, then the work performed on the car by the pusher is given by the following calculation

$$(100 \text{ N}) \times (10 \text{ m}) \times 1 = 1{,}000 \text{ J}$$

Now, let the car be rolling and the pusher, obviously confused, pushes against the door of the car, perpendicularly to the direction of motion. Then $\cos\theta = 0$ and no work is done. Finally, let the car be rolling and the pusher trying to stop it by pushing on it backward ($\cos\theta = -1$) with a force of 90 N while the car rolls forward 8 m. Then the pusher performs

$$(90 \text{ N}) \times (8 \text{ m}) \times (-1) = -720 \text{ J}$$

of work on the car.

From the basic definition of work, presented above, it is possible to derive expressions for work performed in various additional specific cases. For instance, when a constant torque of magnitude τ acts through angle ϕ around an axis of rotation, the amount of work performed is given by the following equation

$$W = \tau\phi$$

ENERGY

The energy of a system is a measure of, and numerically equal to, the work that the system can perform. It, too, is a scalar quantity, and its SI unit is the joule (J), which is the same as that of work. Another unit of energy that is often used is the electron volt (eV). It is the amount of energy gained by an electron as it accelerates through a potential difference of one volt and equals 1.602176×10^{-19} J, rounded to 1.60×10^{-19} J. This unit and its multiples are useful and commonly used for the microscopic domain of atoms, molecules, nuclei, and elementary particles. An additional, common unit of energy is the kilowatt-hour (kW·h, often written kWh), used mostly in connection with electric energy. It equals 3.6×10^6 J.

Energy has two basic forms, kinetic energy and potential energy. For a system with structure (i.e., a system composed of microscopic components such as atoms, molecules, and ions) it is useful to distinguish between two manifestations of each form, internal energy and external energy. Energy is created as a result of work performed on a system. In addition, energy can be converted between its kinetic and potential forms (through the performance of work) as well as among the various forms of potential energy.

WORK-ENERGY THEOREM

Work is intimately related to energy through the work-energy theorem, which states that the work performed on an otherwise isolated system equals the increase of total energy content (in all forms) of the system. Conversely, when an otherwise isolated system performs work on its surroundings, the work equals the decrease of the system's total energy content. In this manner, work can be viewed as a "currency" through which energy of one system is converted to energy of another system. It then follows that when a system is completely isolated, so no work is done on it and it does no work on its surroundings, the system's total energy content remains constant. This is the law of conservation of energy. The energy of such a system can change form over time. For instance, the energy might convert from kinetic energy to potential energy and back again, but the total of all forms of energy of an isolated system remains constant.

As an example, consider the internal cup of a laboratory calorimeter and the contents of this cup as an approximately isolated system. Work can be done on the system through the stirring rod. The work causes an increase of the system's energy, in this case, its thermal energy. If the system is stirred sufficiently, the resulting increase in its temperature can be detected. On the other hand, if no work is performed, then the energy content of the calorimeter will remain (ideally) constant. Nevertheless, energies might change form within the calorimeter, as long as their total remains (ideally) constant. For instance, solids (such as ice) might melt, gases (such as water vapor) might condense, and solid bodies and liquids might warm or cool.

KINETIC ENERGY

Kinetic energy is energy due to motion. The kinetic energy of a point particle possessing mass m, in kilograms (kg), and speed v, in meters per second (m/s), is given by the following formula

$$\text{Kinetic energy} = \tfrac{1}{2}\, mv^2$$

Note that kinetic energy is relative, in that observers in different states of motion assign different values to the kinetic energy of the same particle, since they observe it as having different speeds.

Consider now a system with structure, such as a bowling ball. Its external kinetic energy is the sum of its translational kinetic energy and its rotational kinetic energy. The former is the system's kinetic energy due to the motion of its center of mass, whereby the system is equivalent to a point particle of the same mass located at its center of mass. So for a system with total mass m, whose center of mass is moving with speed v, the translational kinetic energy is given by the following equation:

$$\text{Translational kinetic energy} = \tfrac{1}{2}\, mv^2$$

If the system is rotating about its center of mass, it additionally possesses rotational kinetic energy, given by the following formula:

$$\text{Rotational kinetic energy} = \tfrac{1}{2}\, I\omega^2$$

where Ω denotes the magnitude of the system's angular velocity, its angular speed, in radians per second (rad/s), and I is the system's moment of inertia, in kilogram-meter^2 (kg·m^2), with respect to the axis of rotation.

A system with structure can possess also internal kinetic energy, which appears as thermal energy, or heat, of the system. Heat is the kinetic energy of the random motion of the system's microscopic constituents. Of all forms of energy, heat is an exception to the definition of energy as a measure of the work a system can perform, since, according to the second law of thermodynamics, heat is not fully convertible to work.

POTENTIAL ENERGY

Potential energy is energy of any form that is not kinetic. Equivalently, it is energy of a system that is due to any property of the system except motion. The term *potential* indicates that the energy possesses the potential to produce motion (through the system's performing work), thus converting to kinetic energy. A form of potential energy can be associated with any conservative force. There are many and diverse forms of potential energy. Here are some of them.

Elastic potential energy is the potential energy of a body due to its deformation. A stretched or compressed spring, for example, possesses this form of potential energy, whose value is given by the following equation

$$\text{Elastic potential energy} = \tfrac{1}{2}\, kx^2$$

where k is the spring constant, in newtons per meter (N/m), and x the amount of stretching or compression, in meters (m). Elastic potential energy is an internal energy.

Gravitational potential energy is the potential energy associated with the gravitational force. Two particles that are attracting each other gravitationally possess gravitational potential energy, given by the following equation

$$\text{Gravitational potential energy} = -\frac{Gm_1m_2}{r}$$

where G denotes the gravitational constant, whose value is 6.67259×10^{-11} N·m²/kg², rounded to 6.67×10^{-11} N·m²/kg²; m_1 and m_2 are the masses of the particles, in kilograms (kg); and r is the distance between them, in meters (m). (Here the reference separation, the separation for which the gravitational potential energy is defined as zero, is taken to be infinity. Hence the gravitational potential energy is negative for finite separations.) In the case of a body near the surface of Earth, the gravitational potential energy is given by the following equation

$$\text{Gravitational potential energy} = mgh$$

where m is the mass of the body; g the acceleration due to gravity, whose value is approximately 9.8 meters per second per second (m/s²); and h denotes the vertical position, in meters (m), of the body with respect to any reference level (at which the gravitational potential energy is taken to be zero).

For a system of bodies, the gravitational potential energy associated with their mutual attraction is an internal energy. However, the gravitational potential energy of a body in the gravitational field can be viewed as an external energy of the body.

Electric potential energy is the potential energy associated with the electric force. For two electrically charged particles, for instance, that are attracting or repelling each other electrically, the electric potential energy is given by the following formula:

$$\text{Electric potential energy} = \frac{1}{4\pi\kappa\varepsilon_0}\frac{q_1q_2}{r}$$

Here q_1 and q_2 are the (signed) charges of the particles, in coulombs (C); r is the distance between them, in meters (m); ε_0 is the permittivity of the vacuum, whose value is $8.85418782 \times 10^{-12}$ C²/(N·m²), rounded to 8.85×10^{-12} C²/(N·m²); and κ is the dielectric constant of the medium in which the particles are located ($\kappa = 1$ for the vacuum). (As for the gravitational potential energy of two particles, the

The hydroelectric power station in Hoover Dam (on the left) converts the gravitational potential energy of the water in Lake Meade behind the dam (not visible in this photograph) to electric energy. The water "falls" from lake level to output level (on the right) and in doing so rotates turbines, which in turn drive electric generators. The electric energy is carried to consumers over high-tension power lines (on the top of the ridge). *(Rafael Ramirez Lee, 2008, used under license from Shutterstock, Inc.)*

reference separation is here taken to be infinity.) The electric potential energy of a system of charged particles is an internal energy.

The electric potential energy of a charged particle in the electric field is given by the following relation

$$\text{Electric potential energy} = qV$$

where q denotes the particle's charge, in coulombs (C), and V is the electric potential, in volts (V), at the location of the particle. For a charged body in the electric field, this is an external energy.

Chemical potential energy is the energy that is stored in the chemical bonds of the materials that constitute a system and can be released and converted to other forms of energy through chemical reactions. The batteries in flashlights, electronic devices, and automobiles, as an example, contain chemical potential energy, which, as we use the batteries, is converted into the electric potential energy for which we buy the batteries. The recharging of rechargeable batteries is the reconversion of electric potential energy to chemical. We exploit the chemical potential energy of gasoline by converting it into heat by combustion (oxidation) in internal combustion engines. Similarly, food contains chemical potential energy, which is released for use by physiological oxidation in the body. Chemical potential energy is an internal energy.

Electromagnetic energy is the energy carried by electromagnetic waves. Earth receives energy from the Sun in this form, which is largely absorbed and converted to heat. Alternatively, solar energy might be converted to electric energy in solar cells. Electromagnetic energy is also the energy that is converted to heat in microwave ovens. Electromagnetic energy can be viewed as the energy of photons. This form of energy is not readily categorizable as internal or external.

OTHER FORMS OF ENERGY

Mass energy is the energy equivalent of mass, according to Albert Einstein's special theory of relativity. The value of the energy equivalent of mass m, in kilograms (kg), is given by the following formula

$$\text{Mass energy} = mc^2$$

where c denotes the speed of light in a vacuum, whose value is 2.99792458×10^8 meters per second (m/s), rounded to 3.00×10^8 m/s. One way of obtaining this energy is by means of nuclear reactions, when the sum of masses of the reaction products is less than the sum of masses of the initial reaction participants. The "disappearing" mass is converted to kinetic energy of the products according to Einstein's formula. Radioactivity demonstrates this phenomenon. Another manifestation of mass energy is through particle-antiparticle annihilation. When an electron collides with a positron, its antiparticle, they totally annihilate each other, and their mass is converted to electromagnetic energy in the form of a pair of photons. Mass energy is an internal energy.

The term *mechanical energy* is used to refer to the sum of all forms of external kinetic and potential energy. That excludes heat, chemical potential energy, electromagnetic energy, and mass energy.

As mentioned earlier, energy obeys a conservation law, whereby in any isolated system, the total amount of energy, of all forms, remains constant over time. In such a case, the system's energy can change form over time, for example, from kinetic to potential and back again, or from chemical to electric to heat, but the total amount does not change.

ENERGY QUANTIZATION

In certain situations in the quantum domain, energy is quantized: it cannot take any value but is constrained to a set of values. An example is a quantum harmonic oscillator, whose allowed energy values are given by the following formula:

$$E_n = (n + \tfrac{1}{2})hf \quad \text{with } n = 1, 2, \ldots$$

Here E_n denotes the value of the energy, in joules (J), of the harmonic oscillator's nth energy level; f is the classical frequency of the oscillator, in hertz (Hz); and h represents the Planck constant, whose value is $6.62606876 \times 10^{-34}$ joule-second (J·s), rounded to 6.63×10^{-34} J·s. Another example of a system whose energy is quantized is the hydrogen atom. Its energy levels, in electron volts, are given by the following equation:

$$E_n = -\frac{13.6}{n^2}\text{ eV} \quad \text{with } n = 1, 2, \ldots$$

A further example of energy quantization is the fact that electromagnetic radiation cannot exchange energy with matter in any arbitrary amount, but only in integer multiples of the quantity hf, where f denotes the frequency of the electromagnetic wave. That is related to the quantum picture of electromagnetic radiation as consisting of photons—each of which possesses energy hf—and interacting with matter through the emission and absorption of photons.

See also ACCELERATION; ATOMIC STRUCTURE; CONSERVATION LAWS; EINSTEIN, ALBERT; ELASTICITY; ELECTRICITY; ELECTROMAGNETIC WAVES; FORCE; GRAVITY; HARMONIC MOTION; HEAT AND THERMODYNAMICS; MASS; MATTER AND ANTIMATTER; MOTION; NUCLEAR PHYSICS; QUANTUM MECHANICS; RADIOAC-

TIVITY; ROTATIONAL MOTION; SPECIAL RELATIVITY; SPEED AND VELOCITY; VECTORS AND SCALARS.

FURTHER READING

Serway, Raymond A., Clement J. Moses, and Curt A. Moyer. *Modern Physics,* 3rd ed. Belmont, Calif.: Thomson Brooks/Cole, 2004.

Young, Hugh D., and Roger A. Freedman. *University Physics,* 12th ed. San Francisco: Addison Wesley, 2007.

engines An engine is any device that converts energy into mechanical force, torque, or motion, particularly when the energy source is a fuel. For example, a gasoline engine burns gasoline and uses the resulting heat to produce the torque that moves the car. Similarly for a diesel engine, where the fuel in this case is diesel oil. In a steam engine, it is coal, fuel oil, or gas that is burned to produce the heat that creates the steam that drives the engine. An engine is distinguished from a motor in ordinary speech, where a motor does not make use of primary fuel. An example of a motor is an electric motor or a hydraulic motor (which uses fluid pressure).

One way of categorizing engines is according to the location of fuel combustion: internal combustion or external combustion. In an internal-combustion engine, the fuel is burned inside the engine. Automotive engines are internal-combustion engines. Whether gasoline or diesel oil, the fuel burns in a precisely controlled manner inside cylinders that form part of the engine. The pressure from the hot combustion products drives pistons, which turn a crankshaft, which in turn applies torque to the transmission, from which torque is transmitted to the vehicle's wheels. Steam engines typify external-combustion engines. Their fuel is burned outside the engine proper. The resulting heat boils water and creates pressurized steam, which flows to the engine through pipes. There the steam might drive pistons inside cylinders, as described for automotive engines, or rotate turbines, in both cases creating torque.

Another way in which engines are classified is by the character of their mechanism: reciprocating, rotary, or reaction. Engines of the reciprocating variety operate with a back-and-forth motion of an essential part. All piston engines, such as automotive engines, are of this type. The back-and-forth motion of the pistons inside the engine's cylinders is converted to rotary motion of the crankshaft through the action of the piston rods on the crankshaft. Rotary engines solely use rotational motion for their entire operation. A common type of rotary engine is the turbine engine, such as the steam turbine. A stream of pressurized steam flows at high speed through a tube and exerts torque on fan wheels mounted on the engine's shaft. This causes the shaft to rotate. One finds such engines in electric generating plants, for example, where they drive the electric generators. Reaction engines operate by ejecting combustion products in the rearward direction, causing a forward-directed reaction force on the engine and a consequent forward acceleration of the engine and the vehicle attached to it, according to Newton's second and third laws. Jet and rocket engines are reaction engines. They are used to propel aircraft, spacecraft, and missiles.

Since automotive engines are so common in modern society, a brief description of the operation of the most common type of such an engine, the four-stroke gasoline engine, is relevant. This is an internal-combustion reciprocating engine. In each cylinder of such an engine, a piston moves back and forth inside a cylinder that is closed at one end by the cylinder head. The piston causes rotation of a crankshaft by means of a piston rod that links them. Valves in the cylinder head allow or prevent the passage of gases into or out of the cylinder. The number of cylinders in an engine varies, but the engines of many cars possess four to six cylinders. A full cycle of operation involves the following four motions, or strokes, of the piston:

1. Start with the piston moving from its position of smallest cylinder volume to that of largest volume. During this stroke, a mixture of gasoline and air is drawn or injected into the cylinder. This is the intake stroke.
2. The next stroke is the compression stroke. The piston moves from largest to smallest cylinder volume and compresses the fuel-air mixture.
3. When the mixture is compressed maximally, a spark plug produces a spark that ignites the mixture, which burns extremely rapidly. The very high pressure of the hot combustion products drives the piston from smallest to largest volume; that action applies torque to the crankshaft through the piston rod and causes rotation of the crankshaft. This stroke, called the power stroke, is the main purpose of the whole cycle of operation; during it, the chemical energy of the gasoline is converted to rotational motion of the rotating parts of the engine, then to the rotation of the wheels, and then to the linear motion of the vehicle, which is the intended function of the engine.
4. In the fourth stroke, the piston moves from largest to smallest volume and drives the combustion products from the cylinder. This is the exhaust stroke. It is followed by

the intake stroke, described earlier, and the cycle repeats.

One often characterizes such engines by their power rating, in units of horsepower (hp) or kilowatts (kW). This value gives the maximal power output—the greatest mechanical energy production per unit time—that the engine is capable of producing. Another specification of an engine is the torque it generates under various conditions, stated in units of foot-pound (ft·lb) or newton-meter (N·m). A further specification is an engine's displacement, given in units of cubic inch (in^3), liter (L), or cubic centimeter (cm^3). One derives this figure as follows: Consider the increase in volume of a single cylinder as its piston moves from smallest cylinder volume to largest. This is the displacement of one cylinder. Multiply it by the number of cylinders to obtain the engine's displacement. For example, if the single-cylinder displacement of a certain four-cylinder gasoline engine is 300 cm^3, then the engine's displacement is $4 \times (300\ cm^3) = 1{,}200\ cm^3$, or 1.2 L.

For a brief description of the operation of a very different kind of engine, consider a steam turbine engine, mentioned earlier, such as is commonly found in electric-generating plants. This is a rotary external-combustion engine. The purpose of combustion in this type of engine is to vaporize water and raise the pressure of the resulting vapor. Coal, fuel oil, or natural gas serves as fuel for this combustion. One might also include here nuclear electric-generating plants, where there is no chemical combustion, but heat is generated in a nuclear reactor from processes of nuclear fission—nuclear "burning," one might say—and this heat vaporizes the water. However it is produced, the high-pressure water vapor is led from the boiler via a pipe to the turbine engine proper. There the vapor flows at high speed along the engine's shaft. Fan wheels are mounted on the shaft, and the moving vapor applies forces to the vanes of these wheels, generating torque on the shaft. The shaft and its fan wheels are termed the engine's rotor. Stationary vanes positioned between the successive fan wheels redirect the motion of the moving vapor in order to increase its effectiveness in creating torque at the fan wheels. The engine's shaft connects directly to an electric generator. In this way, a steam turbine engine converts the chemical energy of the conventional fuel or the nuclear energy of the reactor fuel (uranium) to electric energy.

In addition to automobile engines and those in electric-generating plants, many other kinds of engines are used in vehicles, machines, and other devices. Small gasoline engines power lawn mowers, edgers, and leaf blowers. Larger gasoline engines serve as inboard and outboard engines for boats and high power motorcycles. Even larger gasoline and diesel engines move buses and trucks. Huge automotive engines power trains and ships. Reaction engines driving aircraft, spacecraft, rockets, and missiles were mentioned earlier.

See also ELECTRICITY; ENERGY AND WORK; FISSION; FORCE; HEAT AND THERMODYNAMICS; MACHINES; MOTION; NUCLEAR PHYSICS; POWER; ROTATIONAL MOTION.

FURTHER READING

Darlington, Roy, and Keith Strong. *Stirling and Hot Air Engines.* Ramsbury, England: Crowood Press, 2005.

Lumley, John L. *Engines: An Introduction.* Cambridge: Cambridge University Press, 1999.

Newton, Tom. *How Cars Work.* Vallejo, Calif.: Black Apple Press, 1999.

An automobile engine is an example of an engine that converts thermal energy—heat—to mechanical energy and work. In this case, the heat source is the combustion of fuel. *(Oleksiy Maksymenko/Alamy)*

enzymes Enzymes are biological molecules, usually proteins, that act as catalysts to accelerate chemical reactions in the cell. Chemical reactions, in and out of the cell, proceed only in the direction that results in a loss of free energy (energy that can be harnessed to work or drive a chemical reaction); in other words, the spontaneous direction for any reaction tends to proceed "downhill," toward the energetically favorable direction (gives off energy). Most cellular molecules exist in a stable state: that means that they cannot be changed to a state of lower energy without an accompanying input of energy. A molecule requires activation energy (a kick over its associated energy barrier) before it undergoes any chemical reaction, even if the reaction leaves the changed molecule in a lower, more stable state. As all catalysts do, enzymes provide chemical reactions an alternative path to a lower activation energy for the reactants; consequently, a dramatic increase in the rate of the reaction occurs when an enzyme is involved. On average, the rate of a cellular reaction in the presence of an enzyme specific to the reaction is millions of times faster than the same reaction without a catalyst. Each enzyme binds tightly to one or two molecules, called substrates or ligands, and holds the molecule(s) in a conformation that greatly reduces the activation energy required for the reaction to yield products. Enzymes, like other catalysts, are recyclable (meaning they are not changed over the course of the reactions they catalyze, so they can function over and over again), and they have no effect on the equilibrium of the reactants and products of a reaction. Enzymes are much more specific in nature than other catalysts—not only in the substrate that they bind, but in the specific reaction of the several possible reactions that their substrate might undergo that they catalyze. In this way, enzymes direct each of the many different molecules in a cell along specific reaction pathways. To date, biochemists have identified enzymes that affect the rates of more than 4,000 biochemical reactions, they have also discovered that not all biochemical catalysts are proteins—some ribonucleic acid (RNA) molecules, referred to as ribozymes, have catalytic properties. Enzyme activity is affected by other molecules and conditions in the cell. Inhibitors, one type of molecule, decrease enzyme activity, examples of inhibitor molecules include many drugs and poisons. Activators, a second type of molecule, increase enzyme activity. Temperature, pH, and concentration of substrate (three cellular conditions) affect enzyme activity as well.

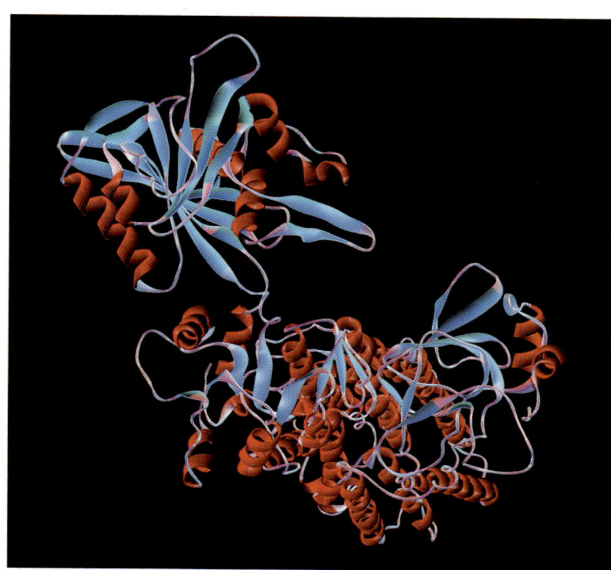

An enzyme is a catalyst that speeds up the rate of a reaction without being consumed itself. The adenosine triphosphatase (ATPase) shown in this ribbon diagram is an enzyme found in muscle that breaks down adenosine triphosphate (ATP), the primary energy source of the cell. The energy that is released is used to fuel the movement of calcium ions across membranes, a process important in the regulation of muscle contraction. *(Alfred Pasieka/ Photo Researchers, Inc.)*

HISTORY AND NOMENCLATURE

In the 1800s, Louis Pasteur studied fermentation, the conversion of sugar to alcohol. Pasteur observed a "vital force" in yeast cells that drove the process of fermentation, and he named this cell component *ferments*. Initially scientists believed that ferments must be within a living organism to be active. In 1878 the German physiologist Wilhelm Kühne renamed ferments *enzymes*, a name derived from Greek meaning "in leaven." Eduard Buchner, working at the University of Berlin, showed that an extract isolated from yeast was able to ferment sugar in the absence of living cells. Buchner called the enzyme *zymase*, and he received the 1907 Nobel Prize in chemistry for his biochemical research and the discovery of cell-free fermentation. As scientists isolated new enzymes, they named the molecules by following Buchner's example with zymase. One adds the suffix, *-ase*, to the name of the enzyme's substrate (see the accompanying table). For example, the lactase enzyme cleaves lactose. Enzymes are also given names that describe the type of reaction they catalyze. Scientists call the enzyme that synthesizes deoxyribonucleic acid (DNA) polymers *DNA polymerase*. After showing that enzymes could function outside living cells, scientists set out to determine the biochemical nature of these molecules. In 1926 James B. Sumner purified and crystallized the enzyme urease for structural analyses, suggesting that these catalytic molecules were proteins, and he later purified and crystallized a second enzyme, catalase (1937). After isolating and crystallizing three digestive enzymes (pepsin, trypsin, and chymotrypsin), John H. Northrop and Wendell M. Stanley proved conclusively that most enzymes were protein molecules with catalytic function, earning them and Sumner the 1946 Nobel Prize in chemistry.

STRUCTURES AND MECHANISMS

As proteins, enzymes consist of long, linear chains of amino acids that fold in unique three-dimensional shapes, and the three-dimensional structure of an enzyme determines its substrate (ligand) specificity. Enzymes are much larger than the substrate that they bind, with only a small portion (three or four amino acid residues) of the molecule interacting

CLASSES OF ENZYMES

Enzyme	Reaction Catalyzed
Hydrolases	catalyze a hydrolytic cleavage reaction
Nucleases	break down nucleic acids by hydrolyzing bonds between nucleotides
Proteases	break down proteins by hydrolyzing the peptide bonds between amino acids
Synthases	synthesize molecules by condensing two smaller molecules together
Isomerases	catalyze the rearrangement of bonds within the same molecule
Polymerases	catalyze reactions in which polymers are synthesized
Kinases	catalyze the addition of phosphate groups to molecules
Phosphatases	catalyze the hydrolytic removal of a phosphate group from a molecule
Oxidoreductases	catalyze reactions in which one molecule is oxidized while a second molecule is reduced. Enzymes of this type are often called oxidases, reductases, or dehydrogenases
ATPases	hydrolyze adenosine triphosphate (ATP)

with the substrate and participating in the catalytic events. The active site (containing the catalytic residues) usually consists of a cavity in the protein surface formed by a particular arrangement of amino acids. These amino acids often belong to widely separated regions of the amino acid chain that are placed together when the protein folds. The ability of an enzyme to bind selectively and with high affinity to the substrate is due to the formation of weak, noncovalent bonds—hydrogen bonds, ionic bonds, van der Waals interactions, and favorable hydrophobic interactions—between the two molecules. Each individual bond is weak, but the cumulative effect of many weak bonds gives the enzyme its high affinity for a specific substrate. The class of polymerases, enzymes involved in copying and expression of the genome, and being endowed with proofreading mechanisms, has the highest specificity and accuracy of all enzymes. Some enzymes may even have binding sites for small molecules that are by-products of the reaction being catalyzed; these small molecules may increase or decrease the enzyme's activity upon binding to form a feedback loop.

In the late 19th century and the first half of the 20th century, scientists tried to elucidate a general mechanism by which enzymes worked. Emil Fischer proposed the lock-and-key model of enzyme-substrate interactions in 1894. In the lock-and-key model, the enzyme and the substrate possess specific complementary geometric shapes that fit exactly into one another as a key fits into a lock. The lock-and-key model gave a feasible explanation for enzyme specificity, but it failed to address the means by which enzymes stabilized a substrate's transition state, permitting a lower activation energy for the reaction to proceed. In 1958 Daniel Koshland proposed the induced fit model of enzyme-substrate interactions, a modified lock-and-key model. Since enzymes are flexible structures, Koshland proposed that the active site changes conformation upon interaction with the substrate. Amino acid side chains take on precise positions within the active site, permitting the enzyme to perform its catalytic function. In some instances, as in glycosidases, the substrate changes shape, too.

The most recent insights into the mechanism for specificity involve the connection between the internal dynamics of the enzyme and the process of catalysis. The internal dynamics of an enzyme involve movement of the polypeptide parts (single amino acids, groups of amino acids, a loop region in the protein, shifts of secondary structures such as alpha helices, and even whole domains). The amount of time required for these movements to take place and affect a catalytic event may be on the order of femtoseconds (one-quadrillionth of a second, or 10^{-15} second) to seconds. Regardless, networks of residues within the structure of the enzyme contribute to the catalytic reactions. Allosteric effects, in which the binding of a molecule induces a conformation change of a second molecule, are an example of the consequences of the internal dynamics of an enzyme.

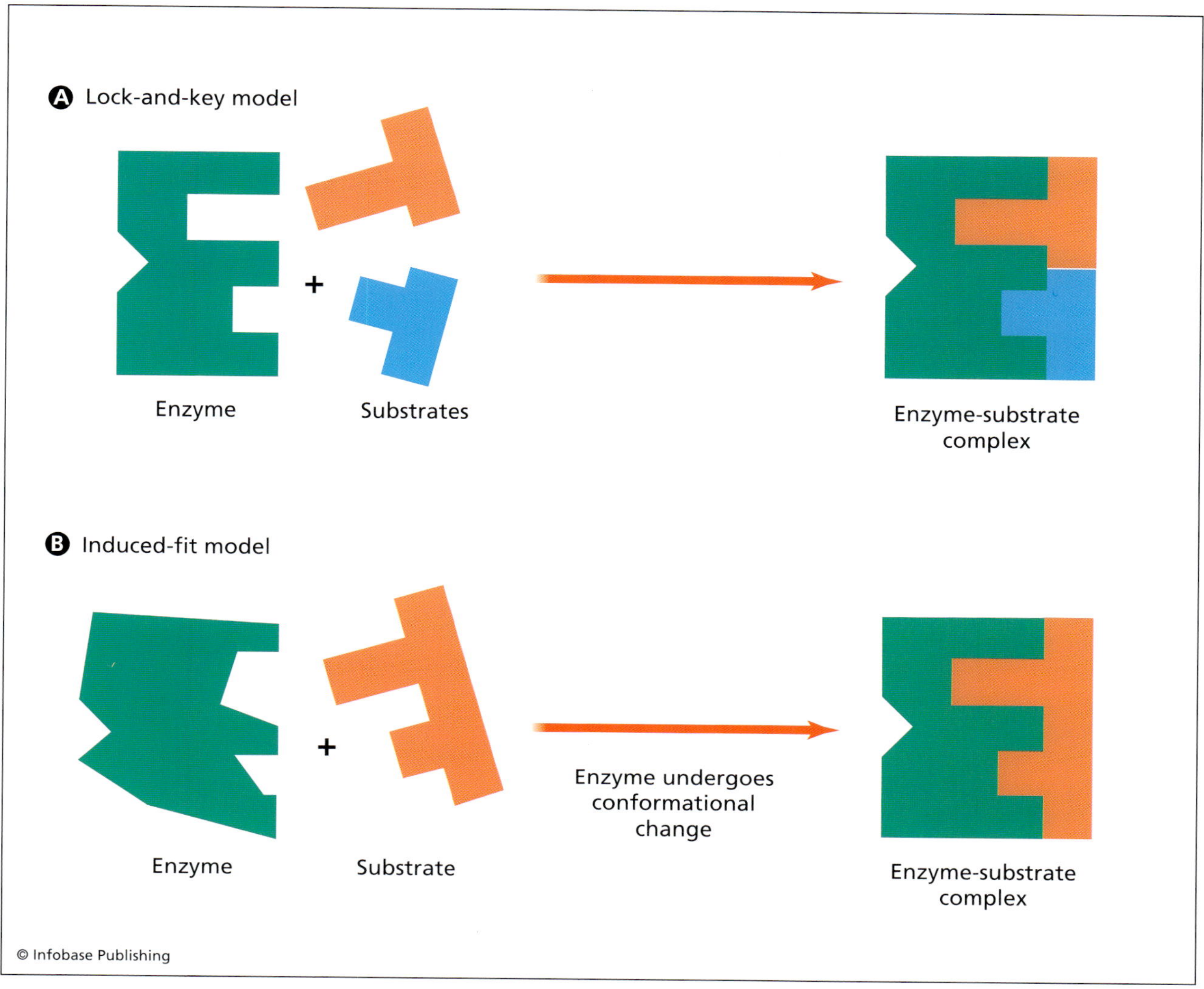

When an enzyme binds to a substrate, they together form an enzyme-substrate complex. (a) In the lock-and-key model the substrate fits into a complementary geometric shape on the surface of the enzyme. (b) In the induced fit model the binding of the substrate induces a conformational change in the enzyme that increases binding affinity.

COFACTORS AND COENZYMES

Cofactors associate with certain enzymes for maximal catalytic activity. Both inorganic cofactors and organic cofactors exist; inorganic cofactors include metal ions and iron-sulfur clusters, and organic cofactors include flavin and heme, two non-protein groups that are tightly bound to the enzymes they assist. These organic compounds are distinguished from coenzymes such as NADH, a reduced form of nicotinamide adenine dinucleotide, because NADH is not released from the active site during the reaction. Enzymes that require a factor but lack one are referred to as apoenzymes. When an apoenzyme binds to its cofactor via noncovalent interactions, the complex is called a holoenzyme (the enzyme's active form).

The energy released by metabolic processes in the cell can be stored temporarily in coenzymes before it is used to drive other reactions, and ultimately events. In most cases, coenzymes, or these small carrier molecules, store the energy in the form of one or more energy-rich covalent bonds. These molecules diffuse rapidly through the cell, carrying the stored energy from one enzyme to another, where it is released to drive the second catalytic event. Activated coenzymes carry the stored energy in an easily transferable form, either as a functional group (acetyl, formyl, and methyl groups are common) or as high-energy electrons (hydride ion with H^+ and two electrons), and they serve a dual role as a source of energy and a functional group donor for cellular reactions. Coenzymes undergo a chemical change after interactions with enzymes; many different enzymes require coenzymes for activity, which are sometimes referred to as second substrates. The most important of the activated coenzymes are ATP (adenosine 5′-triphosphate) and two molecules that are closely related to one another, reduced nicotinamide adenine dinucleotide (NADH) and NADPH (the reduced form of nicotinamide adenine dinucleotide phosphate). Coenzymes are usually regenerated by taking them through a metabolic pathway (NADPH is regenerated by the pentose phosphate pathway); thus, the cellular concentration of these molecules remains constant.

THERMODYNAMICS AND KINETICS

Like other catalysts, reactions catalyzed by enzymes must be spontaneous, meaning that the reaction proceeds toward a net negative Gibbs free energy. The reactions catalyzed by enzymes run in the same direction as uncatalyzed reactions, but enzymes increase the rate of the reaction. Enzymes eliminate many spontaneous reactions that occur between the reactants if permitted to interact over a longer period, preventing the formation of undesirable products. Depending on the concentration of reactants and products, enzymes catalyze reactions forward and backward. Enzymes also do not change the equilibrium of a reaction, simply its rate of completion. By holding the bound substrate in a specific conformation, called the transition state, the amount of energy required (activation energy) to generate a product is lowered. Given enough time (albeit millions of years in some cases), the substrate would assume this transition state eventually of its own accord; enzymes merely speed up the conformation change needed for the reaction to proceed.

When a fuel molecule, such as glucose, is oxidized in the cell, enzyme-catalyzed reactions ensure that a large part of the free energy released by the oxidation is captured in a chemically useful form, rather than lost as heat. Enzymes achieve this task through coupled reactions, in which thermodynamically favorable reactions drive thermodynamically unfavorable ones through carrier molecules, or coenzymes. Driving a variety of chemical reactions, the most important and versatile of the activated carriers in cells is ATP. An energetically unfavorable phosphorylation reaction synthesizes ATP by adding a phosphate group to ADP (adenosine 5′-diphosphate). When needed, ATP releases an inorganic phosphate, and the energy stored in that covalent bond fuels energetically unfavorable reactions. Many of the coupled reactions involve the transfer of the terminal phosphate in ATP to another molecule. Specific reactions coupled to hydrolysis of ATP include pumps that transport substances into and out of the cell, molecular motors that enable muscles to contract, and transport of materials from one end of a nerve cell to the other.

To catalyze a reaction, an enzyme first binds its substrate. The substrate then undergoes a reaction to form product molecules that initially remain bound to the enzyme. Finally, the product diffuses away, leaving the enzyme free to bind another substrate and to catalyze another reaction. The rates of the three steps vary greatly from one enzyme to another; enzyme kinetics studies the differences in each of the steps by performing enzyme assays. (Assays involve mixing purified enzymes and substrates together under carefully defined conditions.)

In 1913 Leonor Michaelis and Maud Menten developed a quantitative theory of enzyme kinetics, referred to as Michaelis-Menten kinetics. G. E. Briggs and J. B. S. Haldane expanded on this theory, deriving equations that are still widely used today. Michaelis-Menten kinetics divides enzyme reactions into two stages. In the first stage, the substrate binds reversibly to the enzyme, forming an enzyme-substrate complex (called a Michaelis-Menten complex); in the second stage, the enzyme catalyzes the chemical reaction and releases the product. The two stages can be represented by the following equation:

$$E + S \rightleftharpoons ES \rightarrow E + P$$

where E stands for enzyme, S stands for substrate, and P stands for product. An enzyme's effect on reaction rate is clearly shown in the reaction catalyzed by orotidine-5′-phosphate decarboxylase. If no enzyme is present, the reaction uses half of its substrate after 78 million years. When the enzyme is present, the reaction uses half of its substrate in a little more than 25 milliseconds (one millisecond equals one-thousandth of a second, or 10^{-3} second).

If the concentration of the substrate gradually increases from a low value, the concentration of the enzyme-substrate complex and the rate at which product forms initially increase in a linear fashion in direct proportion to substrate concentration. As more and more enzyme molecules bind substrate, the rate of product formation tapers off, until it reaches a maximal value, V_{max}, where the concentration of substrate is very high. At this point, substrate occupies the active sites of all available enzyme molecules (E = ES), and the rate of product formation depends on how quickly the enzyme catalyzes the reaction. For many enzymes, the turnover number is about 1,000 substrate molecules per second, although turnover numbers may vary from one to 10,000 substrate molecules per second.

The concentration of substrate needed to make an enzyme work efficiently is measured by the Michaelis-

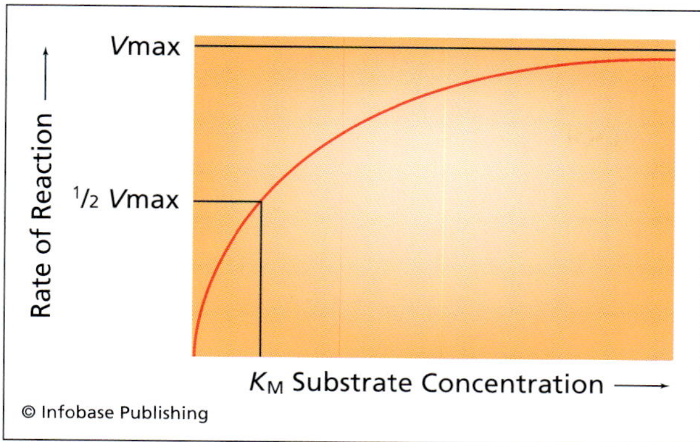

The rate of an enzyme reaction (V) increases as substrate concentration increases until a maximum value (V_{max}) is reached. At this point, substrate occupies all active sites, and the enzyme's catalytic activity determines the rate of reaction. For most enzymes the K_m, which measures the concentration of substrate needed for the enzyme to work most efficiently, directly correlates with the tightness with which the substrate binds. A high K_m corresponds to lower affinity for the enzyme.

Menten constant, K_m, the concentration of substrate at which the enzyme works at half its maximal speed (0.5 V_{max}). Each enzyme has a characteristic K_m, indicating how tightly the E and S bind. In general, a low K_m suggests that a substrate binds very tightly to an enzyme, and a high K_m suggests that a substrate binds very weakly to an enzyme. Another useful constant in enzyme kinetics is K_{cat}, which indicates the number of substrate molecules handled by one active site per second. The efficiency of an enzyme is often expressed in terms of K_{cat}/K_m, or the specificity constant. The specificity constant incorporates the rate constants for all the steps in an enzymatic reaction, and it reflects both the affinity of an enzyme for a particular substrate as well as its catalytic ability. Specificity constants are especially useful when one wants to compare different enzymes or compare the same enzyme with different substrates.

BIOLOGICAL FUNCTIONS

Enzymes carry out a wide variety of functions in the cell. Kinases and phosphatases regulate many aspects of signal transduction. Enzymes generate movement, with myosin hydrolyzing adenosine triphosphate (ATP) to induce muscle contraction. ATPases (enzymes that break down ATP) in the cell membrane act as ion pumps involved in active transport. Enzymes also carry out specialized functions, such as lucerifase, which is responsible for the light produced by fireflies.

Enzymes play a key role in the digestive systems of animals. Enzymes, such as amylases and proteases, break down large molecules of starch and proteins, respectively, so that the smaller pieces might be absorbed more easily by the intestines. Different enzymes digest different foods. In ruminants (e.g., cows), which have herbivorous diets, microorganisms in the gut synthesize the enzyme cellulose, an enzyme able to break down the cellulose cell walls of plants.

Often many enzymes work in concert, creating metabolic pathways. In a metabolic pathway, one enzyme takes the product of another and uses it as its substrate. After the catalytic reaction, another enzyme uses the product as its substrate. Enzymes determine which steps occur in these pathways. Without enzymes, metabolism would neither progress along the same steps each time, nor be fast enough to serve the needs of the cell. The intricate pathway of glycolysis could not exist without enzymes. Glucose, for example, can react directly with ATP, leading to phosphorylation of any one of its carbons. If hexokinase is present, glucose 6-phosphate is the only product. Thus, the type of metabolic pathway within each cell depends on the repertoire of enzymes that are present.

REGULATION OF ENZYMES

The enzyme that catalyzes the first step in a metabolic pathway is often inhibited by the ultimate product in the pathway. The biosynthesis of isoleucine in bacteria is an example of this type of regulation, which is called feedback inhibition. For example, threonine deaminase is the first enzyme used in a five-step pathway converting threonine to isoleucine. When the concentration of isoleucine reaches a sufficiently high level, the amino acid binds to the enzyme at a regulatory site distinct from the active site. This inhibition is sometimes referred to as noncompetitive inhibition. The binding of isoleucine produces an allosteric effect, reversibly changing the shape of the enzyme and consequently inactivating it. When the isoleucine concentration decreases, the amino acid no longer binds to the enzyme, and threonine deaminase becomes active again, allowing more isoleucine to be synthesized. Another form of inhibition, competitive inhibition, involves an inhibitor binding to the active site of the enzyme. In competitive inhibition, the inhibitor usually strongly resembles the substrate. Methotrexate looks very similar to folic acid, and it is a competitive inhibitor of the enzyme dihydrofolate reductase that catalyzes the reduction of dihydrofolate to tetrahydrofolate.

Enzymes are also controlled by regulatory proteins that can either stimulate or inhibit. The activities of many enzymes are regulated by calmodulin, a protein that acts as calcium sensor in many eukaryotic cells. The binding of calcium ions (Ca^{2+}) to four sites in calmodulin induces the formation of an alpha helix, converting the protein from an inactive to an active form. Activated calmodulin then binds to many enzymes, as well as other cellular proteins, and modifies their activities.

Covalent modification is a third mechanism of enzyme regulation. For example, the attachment of a phosphoryl group to a specific serine residue regulates the activities of the enzymes that synthesize and degrade glycogen. The phosphorylation of threonine and tyrosine controls other enzymes. Hydrolysis of the phosphate ester linkage reverses these modifications. Other enzymes catalyze the attachment and removal of phosphoryl and other modifying groups.

Some enzymes are synthesized in an inactive precursor form that is activated at the physiologically appropriate time and place. The digestive enzymes exemplify this kind of control, called proteolytic activation. For example, pancreatic cells synthesize trypsinogen, a precursor to the active form trypsin that a peptide-bond cleavage in the small intestine releases. Cells use similar means of control with enzymes associated with blood clotting. The enzymatically inactive precursors of proteolytic enzymes are called zymogens or proenzymes.

RIBOZYMES (CATALYTIC RNA)

Ribozymes (ribonucleic acid enzymes) are ribonucleic acid (RNA) molecules that catalyze chemical reactions. Many ribozymes catalyze their own cleavage (breaking) or cleavage of other RNA molecules; ribozymes also catalyze the aminotransferase activity of ribosomes, the molecular machine that synthesizes proteins. Before the discovery of ribozymes, proteins were the only known biological catalysts. In the late 1960s Carl Woese, Francis Crick, and Leslie Orgel suggested that RNA could act as a catalyst because of its ability to form complex secondary structures. In the 1980s Thomas Cech discovered the first ribozyme while studying RNA splicing in the ciliated protozoon *Tetrahymena thermophila,* and Sidney Altman discovered other ribozymes while working on bacterial ribonuclease (RNase) P complex. The ribozyme found by Altman existed in an intron (area in a messenger RNA [mRNA] that does not code for a protein). This ribozyme removed itself from the rest of the mRNA. Altman also discovered ribozymes associated with the maturation of pretransfer RNA molecules. In 1989 Cech and Altman shared the Nobel Prize in chemistry for their discovery of the catalytic properties of RNA.

Although rare, ribozymes play an essential role in cells. For example, the functional part of ribosomes is a ribozyme. Ribozymes have the potential to behave as a hereditary molecule, a fact that encouraged Walter Gilbert to propose the "RNA world hypothesis" of the origin of life, arguing that the cell may have used RNA as the genetic material as well as the structural and catalytic model, rather than dividing these functions between DNA and protein as they are today. Some known ribozymes include RNase P, Group I and Group II introns, hairpin ribozyme, hammerhead ribozyme, and tetrahymena ribozyme.

See also BIOCHEMISTRY; METABOLISM; PROTEINS; RATE LAWS/REACTION RATES; RIBOZYMES (CATALYTIC RNA).

FURTHER READING

Copeland, Robert A. *Enzymes: A Practical Introduction to Structure, Mechanisms, and Data Analysis.* New York: John Wiley & Sons, 2000.

EPR

EPR The abbreviation *EPR* stands for *Einstein-Podolsky-Rosen,* the 20th-century physicists: the German-Swiss-American Albert Einstein, the American Boris Podolsky, and the American-Israeli Nathan Rosen. In 1935 Einstein, Podolsky, and Rosen published a paper entitled "Can Quantum-Mechanical Description of Physical Reality Be Considered Complete?" which became the most cited paper in physics for its implications, described later. In the paper the authors proposed an experiment to test what seemed to be paradoxical aspects of quantum theory. This paper is referred to as EPR, and *EPR* often appears in expressions, such as *EPR experiment* and *EPR paradox.*

The "paradox" comprises two components. First, according to quantum theory, a system does not possess a measurable physical characteristic until it is actually measured. If position, for instance, is measured, then a result is obtained, and the system can be said to have such and such a position at the time of measurement. But prior to the measurement, the state of the system might have been such that it could not be thought of as even possessing a position. A measurement, according to quantum physics, does not generally merely reveal the existing, but unknown, value of a physical quantity characterizing the system, as is the case in classical physics. Rather, a measurement actually endows the system with the property being measured by putting it in a state characterized by the property. This complicated idea was revolutionary at the time, and many physicists are still uncomfortable with it. The second component of the EPR paradox is that as long as a system is isolated, all parts of it participate in its quantum state, no matter how spread out the system becomes.

EPR proposed to have two particles created by one source, so that they are part of the same state, and fly apart at high speed. Then, when the particles are far apart, perform a measurement on one and immediately afterward perform a measurement on the other. According to quantum theory, the first measurement should endow the whole system, including the other particle, with the property being measured. The second measurement, if designed properly, should detect whether or not the system is indeed in the state brought about by the first measurement. (Actually, the whole procedure must be repeated many times, and the statistical correlation between the first and second measurement results must be examined.) The paradox is this: For the second particle measured to "know" it is in a state that it was not in a very short time earlier, it would seem that some "influence" must have reached it from the first measurement, an influence that must have propagated from the location of the first measurement to the location of the second. However, if one performs the measurements sufficiently closely in time, that "influence" must travel faster than the speed of light in a vacuum, and according to Einstein's special theory of relativity, no signal or information can do that.

The British physicist John Bell proposed in 1964 a statistical criterion for analyzing EPR-type experiments in the form of what is known as Bell's theorem, or Bell's inequalities. In 1982 the French physicist Alain Aspect and collaborators reported that they performed a version of the EPR experiment for the first time, and since then, the experiment has been performed very often. The result is clear: according to Bell's theorem, the predictions of quantum theory are confirmed. The two particles, produced by the same source and part of the same quantum state, are forever linked in what is called quantum entanglement no matter how far apart they are. The change of state brought about by the first measurement does indeed immediately affect the result of the second measurement through quantum entanglement. Contradiction with the special theory of relativity is avoided, however, because quantum entanglement does not involve the faster-than-light transmission of a signal or of information. Quantum entanglement underlies, for example, quantum encryption (the secure transmission of secrets) and quantum cloning (reproducing the quantum state of one system in another system).

See also CLASSICAL PHYSICS; EINSTEIN, ALBERT; QUANTUM MECHANICS; SPECIAL RELATIVITY.

FURTHER READING

Aspect, Alain, Jean Dalibard, and Gérard Roger. "Experimental Test of Bell's Inequalities Using Time-Varying Analyzers." *Physical Review Letters* 49 (1982): 1804–1807.

Bell, John S. "On the Einstein Podolsky Rosen Paradox." *Physics* 1 (1964): 195–200.

Einstein, Albert, Boris Podolsky, and Nathan Rosen. "Can Quantum-Mechanical Description of Physical Reality Be Considered Complete?" *Physical Review* 47 (1935): 777–780.

Ferris, Timothy. *The Whole Shebang: A State-of-the-Universe(s) Report.* New York: Simon & Schuster, 1997.

Smolin, Lee. *The Life of the Cosmos.* New York: Oxford University Press, 1997.

equilibrium

A system is said to be in equilibrium when it does not spontaneously evolve but rather maintains itself as it is. A pendulum at rest in its lowest position is in equilibrium, for example. If the pendulum is displaced from its lowest position by, say, an angle of 5°, it is not in a state of equilibrium, since, if not prevented from doing so, it will immediately and spontaneously start moving (toward its lowest position).

One usefully distinguishes among three kinds of equilibrium: stable, labile, and unstable. The distinction is based on the behavior of the system when it is displaced slightly from its equilibrium state. If the system tends to return to its equilibrium state, the latter is a state of stable equilibrium. The lowest position of a pendulum is an obvious example of stable equilibrium. If, on the other hand, the system tends to evolve away from its equilibrium state when it is displaced slightly from it, the state is said to be one of unstable equilibrium. With a rigid suspension rod, a pendulum in its highest position, when the bob is situated directly over the pivot, is in such a state. In everyday language, the pendulum exists in a state of precarious equilibrium. The bob can balance there, at least in principle. But the slightest nudge, a tiny breeze, or even a molecular motion can cause the bob spontaneously to move away from that position. Labile equilibrium is intermediate between stable and unstable equilibrium. When a system is displaced slightly from a state of labile equilibrium, it is still in equilibrium and does not tend to change at all with respect to the new state. A ball resting on a level floor is an example of labile equilibrium. Displace it slightly to the side, and it is just as happy to remain in its new position as it was in its previous one. Labile equilibrium can be viewed as a continuous set of equilibrium states.

MECHANICAL EQUILIBRIUM

For mechanical systems, equilibrium is the situation in which the net external force and the net external torque acting on the system are both zero. Then, according to Newton's second law of motion, the system will remain at rest, if it is at rest, or will move with constant velocity and rotate about its center of mass with constant angular velocity. In symbols, the conditions for mechanical equilibrium are summarized by the vector equations

$$\Sigma \mathbf{F} = 0, \quad \Sigma \boldsymbol{\tau} = 0$$

In textbook problems, in order to emphasize the principles and preclude excessive complications, the systems under consideration are mostly constrained to move in a plane, say, the *xy* plane, with all forces parallel to that plane. Then the conditions for equilibrium reduce to the vanishing of the *x*- and *y*-components of the net external force on the system and the vanishing of the net external torque on the system with respect to any single pivot point O in the *xy* plane,

$$\Sigma F_x = 0, \quad \Sigma F_y = 0, \quad \Sigma \tau_0 = 0$$

The treatment of such problems is best done with the help of a free-body diagram.

Mechanical equilibrium is not necessarily a static situation, one in which the system is at rest with regard to both translation and rotation. Indeed, engineers study statics as a distinct subject. Although their buildings and bridges should maintain equilibrium under a range of conditions, they should also remain at rest (and not be moving, even at constant velocity).

ELECTROMAGNETIC EQUILIBRIUM

An example of stable equilibrium in electromagnetic systems is the following. Lenz's law states that the emf (voltage) induced in an electric circuit by a change of magnetic flux through the circuit is such that, if the circuit is closed, the magnetic field produced by the resulting current tends to oppose the change of flux. Thus, electromagnetic systems "attempt" to maintain the status quo of currents and magnetic fluxes.

This stability law applies even at the level of magnetic forces. In a common physics demonstration, a small nonmagnetic cylinder is dropped down a vertical metal tube. The time it takes to reappear at the bottom end of the tube appears quite in line with normal free-fall times. Then a similar-appearing cylinder, this time a magnet, is dropped down the tube. Its fall time is so much longer than the pervious one that the observers begin to wonder what is happening to it inside the tube. What is happening is this: The circumference of the tube at any location along the tube can be considered a closed electric circuit. As the magnet falls inside the tube, the flux of its magnetic field through the part of the tube beneath it increases, while the flux in the part of the tube above it decreases. The currents induced in the circumferences of the tube at all locations along it produce magnetic fields that tend to oppose these changes. The magnetic fields thus produced are equivalent to the fields that would be produced by two magnets facing the falling magnetic cylinder, one beneath the cylinder and one above it. To make the explanation simpler, assume the magnet falls with its north pole first. Then the equivalent magnet that the falling magnet "sees" beneath it has its north pole on top, facing the falling magnet's north pole, while the equivalent magnet that is above the falling

At the instant this photograph was taken, the rope was in equilibrium, since the two teams were pulling it with equal-magnitude and opposite forces, resulting in zero net force. A moment later, however, the team on the right succeeded in increasing its force, causing the rope to accelerate to the right and that team to win the contest. *(Boris Horvat/AFP/Getty Images)*

magnet has its north pole on its bottom, facing the falling magnet's south pole. Since like poles repel and opposite poles attract, the falling magnet is repelled by the equivalent magnet beneath it and attracted to the one above it. Both those forces oppose the force of gravity, with the result that the acceleration of the falling magnet is noticeably less than the acceleration due to gravity. The stability aspect of the situation is that the system is "attempting" to maintain the original state of an empty tube with the magnetic cylinder outside it and no magnetic flux inside it.

CHEMICAL EQUILIBRIUM

In chemistry, equilibrium for a chemical reaction means that the concentrations of the reactants and the products of the reaction remain constant in time. This also means that the reaction rates for the forward and inverse reactions must be equal. Take, as an example, magnesium sulfate ($MgSO_4$) in aqueous solution. On the one hand, the $MgSO_4$ molecules dissociate into ions,

$$MgSO_4 \rightarrow Mg^{2+} + SO_4{}^{2-}$$

at a rate that is proportional to the concentration of $MgSO_4$. On the other hand, the ions Mg^{2+} and $SO_4{}^{2-}$ recombine to $MgSO_4$,

$$Mg^{2+} + SO_4{}^{2-} \rightarrow MgSO_4$$

at a rate that is proportional to the product of the ions' concentrations. For the total reaction

$$MgSO_4 \leftrightharpoons Mg^{2+} + SO_4{}^{2-}$$

equilibrium will be achieved, and the concentrations of $MgSO_4$, Mg^{2+}, and $SO_4{}^{2-}$ will remain constant, when the concentrations are such that the rates equalize. Then as many magnesium sulfate molecules dissociate as are created per unit time.

Chemical equilibrium is stable equilibrium, as expressed by Le Chatelier's principle: if a chemical system is in equilibrium and the conditions (such as one or more concentrations) change so that it is no longer in equilibrium, the system will react to the change by reaching a new equilibrium in a way that tends to counteract the change. In the previous example, for instance, let magnesium hydroxide [$Mg(OH)_2$] be added to the mix. This substance will dissociate, and the concentration of Mg^{2+} ions in the solution will thus increase. That will cause the $MgSO_4$ equilibrium to shift to the left: the concentration of $MgSO_4$ will increase, while the concentrations of Mg^{2+} and $SO_4{}^{2-}$ will decrease. $MgSO_4$ will "devour," so to speak, some of the excess Mg^{2+} ions that appear in the solution as a result of introducing $Mg(OH)_2$.

THERMODYNAMIC EQUILIBRIUM

For isolated systems to which thermodynamic considerations apply, states of stable equilibrium are states of maximal entropy, and vice versa. That is due to the second law of thermodynamics, according to which isolated systems tend to increase their entropy. The uniform distribution of an isolated gas in its container is an example of a state of stable equilibrium and of maximal entropy. Any nonuniform state, such as a higher density on one side of the container than on the other, possesses less entropy than the uniform state and spontaneously evolves to it in a very short time.

Such systems might also be in unstable equilibrium. For instance, the condition of a superheated liquid is a state of unstable equilibrium. That is when the liquid's temperature and pressure are such that it normally boils, but the onset of boiling is prevented by the absence of nucleating sites, which are irregularities (such as surface roughness or particles in the liquid) at which vapor bubbles can form. The liquid is indeed in equilibrium, since, with no interference, it remains in its superheated state and does not tend to change. But a small disturbance, such as jarring the vessel or introducing a rough grain of sand, can immediately bring about sudden, sometimes violent boiling.

The significance of equilibrium for such systems is that their macroscopic state does not change. Nevertheless, changes are taking place at the microscopic level. As an example, consider a mixture of liquid water and ice at the freezing point in a closed system. The mixture is in equilibrium, as neither the liquid nor the solid phase increases at the expense of the other. Yet the individual water molecules are hardly in equilibrium. Molecules are continuously leaving the solid for the liquid, and vice versa. Such equilibrium is called dynamic equilibrium, or a steady state.

See also CHEMICAL REACTIONS; ELECTROMAGNETISM; FLUX; FORCE; HEAT AND THERMODYNAMICS; MAGNETISM; MOTION; NEWTON, SIR ISAAC; ROTATIONAL MOTION; VECTORS AND SCALARS.

FURTHER READING

Serway, Raymond A., and John W. Jewett. *Physics for Scientists and Engineers,* 7th ed. Belmont, Calif.: Thomson Brooks/Cole, 2008.

Young, Hugh D., and Roger A. Freedman. *University Physics,* 12th ed. San Francisco: Addison Wesley, 2007.

Ertl, Gerhard (1936–) German *Physical Chemist* Gerhard Ertl is a German-born chemist best known for his pioneering work in the field of surface chemistry—the study of the atomic and molecular interactions at the interface between two

Gerhard Ertl is a German chemist who is best known for discovering the mechanism of the Haber-Bosch process to produce ammonia. He was awarded the 2007 Nobel Prize in chemistry for his work. Ertl is shown here in his lab at the Fritz Haber Institute of the Max Planck Society in Berlin. *(Johannes Eisele/dpa/Landov)*

surfaces. In the 1960s Ertl demonstrated that reaction mechanisms can be determined using meticulous experimental techniques at the gas-solid interface that allow for the visualization of the precise atomic interactions. Ertl worked specifically on elucidating the mechanism of the well-known and much used Haber-Bosch process of ammonia production. His work also elucidated the mechanism of the conversion of carbon monoxide to carbon dioxide at the platinum catalyst of the catalytic converter in automobiles. Ertl's work on hydrogen catalysis at the interface between gaseous hydrogen and a solid catalyst is helping to pave the way for the development of hydrogen fuel cells. For his work in the establishment and development of the field of surface chemistry, Ertl received the Nobel Prize in chemistry in 2007.

EARLY YEARS AND EDUCATION

Gerhard Ertl was born on October 10, 1936, in Stuttgart, Germany. From a young age, he had a strong inclination toward science, and chemistry in particular. As a teen, he developed an interest in physics as well. Ertl attended the Technical University in Stuttgart from 1955 until 1957, when he entered the University of Paris. In 1958–59 he attended Ludwig-Maximilian University in Munich. As an undergraduate, Ertl studied the branch of physical chemistry known as electrochemistry. He worked on the interaction of solids and liquids and the transfer of electrons between them. Upon completing his degree and beginning work for his doctorate at the Technical University in Munich, Ertl began studying the solid-gas interface rather than the solid-liquid interface familiar to electrochemistry. At the time, little was known about the interactions that occurred on the surface of a solid catalyst. Ertl received a Ph.D. from the Technical University in Munich in 1965.

After the pioneering work of Ertl and his colleagues led to the development of a new field called surface chemistry, Ertl stayed on as a professor at the Technical University from 1965 to 1968. From 1968 to 1973 Ertl served as a chemistry professor as well as the director of the Institute for Physical Chemistry at the Technical University in Hanover. From 1973 to 1986 Ertl was a professor of chemistry and the director of the Institute for Physical Chemistry at Ludwig-Maximilian University in Munich. During this time, Ertl held several visiting professor positions in the United States including at the California Institute of Technology, University of Wisconsin in Milwaukee, and University of California at Berkeley.

SURFACE CHEMISTRY

The field of surface chemistry allows scientists to get a closer look at reaction mechanisms, or the step-by-step processes by which a reaction proceeds. Traditional obstacles to studying reaction mechanisms occur because the reacting species typically interact with other substances in a mixture. Studying chemical reaction mechanisms is a complicated process but is simplified by considering first the reacting species in the gas form. If both reactants are in the gas phase, the gas molecules only interact with the individual molecules that are part of the reaction. When the reaction takes place in solution (usually water), the movement of the water and the other particles complicates the analysis of the mechanism of the reaction. Isolation of the reaction components is required in order to understand the mechanism of the reaction, and this is where surface chemistry plays a role.

Ertl's contributions to surface chemistry were both theoretical and technical. Progress in the study of atomic interactions at solid surfaces had lagged since the 1930s. The difficulties in studying atomic interactions on a solid catalyst were twofold: no mechanism existed to maintain a clean surface on the solid catalyst, and the surface of the solid was not smooth and even. Ertl demonstrated that if a reaction took place between a gas and a solid under a vacuum, then atmospheric gases would not interfere with the reaction. This process enabled the surface of the solid to remain "clean." Prevention of contamination within the reaction allowed Ertl to study the reacting species unhindered. To solve the smooth surface problem, Ertl used a single crystal surface to enable the reaction to proceed.

HYDROGEN CATALYSIS

One of the first areas of research for Ertl in the newly emerging field of surface chemistry was hydrogen catalysis. Ertl first demonstrated the arrangement of hydrogen atoms on the surface of such metals as platinum and lead. Using an experimental technique known as LEED, low-energy electron diffraction, Ertl demonstrated that the individual hydrogen atoms were singly arranged in a monolayer on the surface of the metal. This description of the arrangement of the hydrogen atoms enabled scientists to explain the proper mechanism for hydrogenation reactions. The single layer of hydrogen atoms explained how the hydrogen atoms could be added singly rather than reacting with a hydrogen molecule, which consists of two covalently bonded hydrogen atoms.

HABER-BOSCH SYNTHESIS

At the time, the most economically significant discovery by Ertl was the precise reaction method of the ammonia-producing reaction developed by the German chemist Fritz Haber that had been in use for a long time. Chemists had employed this process since the early part of the 1900s, yet Ertl was the first to explain how the reaction occurred. While people understood that the metal sped up the process of turning molecular nitrogen from the atmosphere into ammonia that could be useful in fertilizers, no evidence had been produced that explained the mechanism. Several different reaction mechanisms had been proposed, but no accurate experimental data supported them. Ertl utilized his surface chemistry techniques to determine the mechanism.

The reaction occurs in a series of steps in which the nitrogen and hydrogen gases are adsorbed onto the metal surface. Many doubted that nitrogen would dissociate into atomic nitrogen, which would be necessary for the reaction to occur because of the strong triple bond found between the nitrogen atoms in molecular nitrogen. Ertl's research brought to light several factors of the reaction that confirmed the mechanism of the reaction. First he used auger electron spectroscopy (AES) to demonstrate that under high pressure, atomic nitrogen was evident at the surface of the metal. This led the way for the determination of the following reaction mechanism, where *adsorption* means the accumulation of a substance on the surface of a solid or liquid:

1. Adsorption of molecular hydrogen to the surface of the metal

$$H_2 \rightarrow 2H_{adsorbed}$$

2. Adsorption of molecular nitrogen to the surface of the metal

$$N_2 \rightarrow 2N_{adsorbed}$$

3. Reaction of the atomic nitrogen and atomic hydrogen

$$N_{adsorbed} + H_{adsorbed} \rightarrow NH_{adsorbed}$$

4. Addition of hydrogen to produce NH_2

$$NH_{adsorbed} + H_{adsorbed} \rightarrow NH_{2adsorbed}$$

5. Addition of one more hydrogen atom

$$NH_{2adsorbed} + H_{adsorbed} \rightarrow NH_3$$

6. Removal of ammonia from the metal

$$NH_{3adsorbed} \rightarrow NH_3$$

Determination of the mechanism for ammonia production was a financially important achievement. Fertilizers are a lucrative business, and by understanding the mechanism of the reaction for the production of one of its main components, one can optimize the process. Surface chemistry studies allow one to examine the reaction mechanism so scientists can determine how chemical reactions are taking place. Ertl's demonstration of the rate-limiting step (the dissociation of molecular nitrogen) allowed industries to develop appropriate reaction conditions and kinetics to increase their yield of ammonia.

THE CATALYTIC CONVERSION OF CARBON MONOXIDE TO CARBON DIOXIDE

Gerhard Ertl also applied his understanding of catalytic mechanisms on solid surfaces to investigate the function of automobile catalytic converters. Carbon monoxide, a by-product of the incomplete combustion of fossil fuels, exists as an odorless, colorless, and tasteless gas that is toxic to humans and other animals. The production of carbon monoxide in automobiles is polluting to the environment as well as potentially toxic to those in the vehicle. In a catalytic converter, installed in the exhaust system of automobiles, platinum metal catalyzes the conversion of carbon monoxide to carbon dioxide. Ertl studied this catalysis process and elucidated the mechanism for the conversion on the platinum catalyst noted as follows:

1. The catalytic converter contains transition metals oxides (usually platinum) adsorbed to a matrix.
2. Exhaust gases containing carbon monoxide (CO) pass through the catalytic converter.
3. The carbon monoxide reacts with the adsorbed platinum oxide and becomes oxidized.
4. The presence of the platinum oxide catalyst lowers the activation energy of the oxidation reaction of toxic carbon monoxide to carbon dioxide.

Understanding the reaction mechanism once again gave the scientific community the opportunity to make informed decisions when developing catalytic converters.

PRESENT RESEARCH INTERSETS

A unique trait of Gerhard Ertl is his willingness to revisit topics he has previously studied as new information and technology arises. This holds true for his three most important research areas: hydrogen catalysis, ammonia production, and the catalytic conversion of carbon monoxide to carbon dioxide. Although he studied the catalysis of hydrogen in his early years of surface chemistry research, he revisited the topic near the end of his career as a method of reducing the environmental impact of the carbon dioxide produced from the catalytic conversion of carbon monoxide to carbon dioxide.

Gerhard Ertl changed chemistry forever by developing the experimental techniques that enabled chemists to understand the reaction mechanisms of many well-studied reactions. Ertl is best known for his study of hydrogen catalysis, the optimization of ammonia production for use in fertilizers, and his study of the reactions involved in catalytic conversion in automobiles. The development of the field of surface chemistry in the 1960s occurred in large part as a result of the work of Gerhard Ertl. After serving as the director of the Department of Physical Chemistry at the Fritz-Haber-Institute of the Max-Planck Society from 1986 to 2004, he retired, becoming a professor emeritus in 2004.

See also ANALYTICAL CHEMISTRY; CHEMICAL REACTIONS; PHYSICAL CHEMISTRY; RATE LAWS/REACTION RATES; SURFACE CHEMISTRY.

FURTHER READING

Ertl, Gerhard. *Environmental Catalysis.* New York: Wiley-VCH, 1999.

Faraday, Michael (1791–1867) English *Physicist* Michael Faraday is well known for inventing the electric motor, the generator, and the transformer, three devices upon which the electrical industry is built, but he also made important contributions to the fields of chemistry and physics. He was a simple man born to an unknown family, but his creative intelligence and desire to learn led him out of the slums of London into the most respected intellectual societies in Europe. His research led the industrial revolution out of the factories of Great Britain and into the households of common people.

CHILDHOOD

Michael Faraday was born on September 22, 1791, to James and Margaret Faraday in London. James Faraday was a skilled blacksmith but suffered from rheumatism and was often too ill to work, and their family lived in poverty. One of Faraday's jobs as a child was delivering papers from one customer to another, as many families were too poor to buy their own copy.

At age 14, Faraday worked as an apprentice at a local bookbindery owned by George Riebau. While learning his trade, Michael devoured every book that he bound. He took notes from many of them and bound them in a special notebook, and he asked lots of questions. Hearing the answers was not enough for young Michael. He needed to test them and prove for himself that the explanation worked.

Michael was particularly curious about electricity. He had read a treatise on electricity in the *Encyclopaedia Britannica* that had passed through Riebau's shop. He was also engrossed by Jane Marcet's *Conversations on Chemistry* text. With permission from his employer, he set up a little lab in his bedroom to duplicate experiments he read about. To quench his thirst for scientific knowledge, in 1810 he started attending meetings of the City Philosophical Society. This organization was a sort of intellectual club, where ordinary men could assemble to discuss scientific matters. Sometimes they sponsored lectures by famous scientists. Riebau's customers knew Michael. One was so impressed by Michael's desire to learn that, in 1812, he offered him tickets to the final four lectures in a series given by a famous English chemist, Humphry Davy.

ASSISTANT TO DAVY

Davy was a professor of chemistry at the Royal Institution (RI) of Great Britain, an organization that supports scientific research and teaching. He was well known for discovering that electricity could be used to liberate metals from compounds and for discovering the gas nitrous oxide. He was a popular, entertaining lecturer. Faraday was engrossed by his discourses and recopied all of his lecture notes neatly onto fresh paper, illustrated them, and bound them. The end of his apprenticeship was drawing near, and he worried about his future. He had a job lined up with a bookbinder in town, but by now Faraday knew he wanted to be a scientist. Faraday believed that scientists were pure, that they were somehow morally superior to other people, especially businessmen, and he wanted to join their faction. By a strange twist of fate, Davy was temporarily blinded in a lab explosion in 1812, and by recommendation, Faraday served as his secretary for a few days. Shortly thereafter, Faraday sent Davy the nearly-400-page bound volume of his lecture notes with a request for a position. Davy was impressed but disappointed Faraday as no position was available.

Michael Faraday's electrical inventions—including the electric motor, the generator, and the transformer—led the way for the industrial revolution. *(National Library of Medicine)*

Several months later, a laboratory assistant at the RI was fired for fighting. Davy recommended Faraday for the spot, and in March 1813, at the age of 21 years, Michael Faraday started working for the RI as a chemical assistant. His duties included assisting Davy in his laboratory research, maintaining the equipment, and helping professors in their lecture preparation. The pay was less than he made as a bookbinder, but Faraday felt he was in heaven. He remained associated with the RI until his retirement.

After only six months, Faraday left London to tour Europe with Davy and his wife. During the extended trip, Davy discovered the new element iodine. Faraday witnessed Davy's genius firsthand and had the opportunity to meet with many other scientists. Upon his return to London in 1815, he worked very long hours. He certainly did not make time to socialize. In 1816 he started lecturing at the City Philosophical Society.

One of Faraday's first true scientific explorations was to help Davy study the dangerous problem of improving coal miner safety from gas explosions. The problem stemmed from the workers' illuminating the mines by using naked flames. They found that by enclosing the flame in wire gauze, the *heat* from the flame would be absorbed by the wire and not ignite the methane. Thus the Davy lamp was invented. Shortly thereafter, in 1816, Faraday published his first of many scientific papers, entitled "Analysis of Native Caustic Lime of Tuscany," in the *Quarterly Journal of Science*. In 1821 Faraday married the sister of a friend, Sarah Barnard. The couple moved into the RI, where they lived for 46 years. Michael and Sarah Faraday did not have any children of their own, but they did raise his niece from age 10.

His chemical researches early on involved producing and analyzing new chemical compounds, such as those consisting of carbon and chlorine, and studying metallurgy. He worked to make higher-grade steel alloys. This task was laborious, but Faraday kept current with the scientific literature. Two decades prior, the Italian physicist Alessandro Volta had invented the first electric battery, called a voltaic pile, by stacking disks of copper, zinc, and cardboard soaked in salt water, then attaching a wire from the top to bottom. This produced a steady flow of electric current. One important more recent discovery by the Danish physicist Hans Christian Ørsted was that electricity had a magnetic field. This was exceptional because the needle of a magnetic compass was deflected at right angles to the wire carrying an electric current, suggesting that the magnetic force produced by an electric current acted in a circle rather than a straight line.

THE ELECTRIC MOTOR

In 1821 another English scientist, William Wollaston, visited Davy's laboratory. He attempted to rotate a wire that was carrying an electric current on its axis by using a magnet. This inspired Faraday, who devised an apparatus that would convert electric energy into mechanical motion. The device consisted of two bowls of mercury, which is an electric conductor. A battery was attached below the bowls to metal rods inserted into the bottom of each bowl. A metal bar dipped into the tops of both bowls to complete the circuit. One bowl had a fixed upright magnet and a movable wire hanging into the mercury, and the other bowl had a fixed upright wire and a movable magnet hanging into the bowl. When the current was turned on, the movable wire circled the stationary magnet in the first bowl, and the movable magnet rotated about the fixed wire in the second bowl. Faraday had transformed an electric current into a continuous mechanical motion, creating the first electric motor. His publication, "On Some New Electro-Magnetical Motions, and on the Theory of Magnetism," in the *Quarterly Journal of Science* (1821) reported these investigations and contained the first mention of a line of force that encircled the current-carrying wire. This viewpoint was different from that of most scientists at the time, who thought of electricity as a fluid through which particles carrying electrical energy flowed.

Unfortunately, this amazing accomplishment was darkened by a mistake that Faraday later regretted. When he reported his results, he failed to give any credit to Davy or Wollaston. Faraday defended himself by saying that Wollaston had been trying to make a current-carrying wire rotate on its own axis when held near a magnet, and, besides, Wollaston had failed. Faraday was successful. Wollaston was more forgiving than Davy, but Faraday was always extremely careful from then on to give a credit where it was rightfully due.

BECOMES RESPECTED SCIENTIST

In 1823 Faraday was working with chlorine hydrate, and Davy suggested he heat the crystals in a closed tube. When he did so, the crystals melted, and as the pressure built up inside the enclosed tube, an oily liquid collected in the opposite arm. He had produced liquid chlorine. He attempted this procedure on a variety of other gases such as carbon dioxide with similar results. While doing this, he became the first scientist to create temperatures below 0°F (-17.8°C) in the laboratory. Incidentally, he injured his eye in an explosion during these experiments. Faraday reported his results in two papers submitted to the *Philosophical Transactions of the Royal Society*. Before publishing them, Davy added a statement crediting himself with the suggestion. These results were extraordinary because scientists were not sure some gases could exist in more than one state. Even though Davy was Faraday's mentor, he was jealous of all the attention that his protégé was receiving. The next year, when Faraday was elected as a fellow of the Royal Society, Davy (then the president of the Royal Society) opposed his nomination.

Faraday often did private consulting work and donated his earnings to the RI's research funds. He was often solicited for advice on chemical matters by the Royal Engineers, the Royal Artillery, and the Admiralty. He also taught chemistry at the Royal Military Academy from 1829 to 1853. During the decades he worked for the RI, he served as a court expert witness, helped lighthouses burn fuel more efficiently, and performed chemical analyses for gas companies. In 1825 he was sent a sample for chemical analysis. After skillfully investigating the substance, he concluded it was a compound made of carbon and hydrogen, which he called bicarburet of hydrogen. This was Faraday's greatest contribution to the field of chemistry. Today we call this compound benzene, and its utility in organic chemistry is immeasurable. It is utilized in the production of nylon, polystyrene, and rubber.

Also in 1825, the Royal Society assigned Faraday to a project aimed at improving the quality of optical glass, such as glass used to make telescope lenses. He worked on this for about five years, during which he made modest advances and obtained an assistant named Sergeant Charles Anderson, who remained Faraday's only assistant for 40 years.

Davy's health deteriorated, and he died in 1827. Despite their differences, Faraday had great respect for Davy and always spoke of him with reverence and fondness. He was deeply saddened by his passing. He felt a great loyalty to Davy and to the institution. When he was offered a professorship in chemistry at the new London University, he declined.

RESEARCH ON ELECTROMAGNETISM

By 1831 Faraday felt confident enough that after completing his report on optical glass, he could work on subjects of his own choosing. He was eager to experiment further with electromagnetism. Physicists knew that electricity could produce magnetism, but he wondered whether the reverse were also true—could magnetism be used to create an electric current? After much thought on how to test this, he wrapped one side of an iron ring with an insulated copper wire, attached to a voltaic battery. On the other side, he wound another wire, attached to a galvanometer to detect an electrical current. When the final connection was made, the needle jumped and then rested again. Starting a current in one coil magnetized the iron ring, which then induced an electrical current in the other wire, as detected by the galvanometer. When the current was turned off, the needle momentarily jumped again, then rested. To make sure what he saw was real, he repeated his experiment using a more sensitive measuring device, ensuring there were not leaks of current or metallic connections between the two wires. It appeared that only changes in the magnetic field generated an electric current. Also, the voltage in the second coil depended not only on the power from the battery, but also on the number of coils in the wire. One could alter the voltage by changing the number of turns in the coil.

A few months later, Faraday performed a similar experiment using a magnetic rod, which he moved in and out of a solenoid formed from a copper wire. The ends of the wire were connected to a galvanometer. When the magnet or the solenoid was moved, a current was detected. When the relationship between the two objects was stationary, no current was detected. Though he was successful in inducing a current, he expected it to be continuous. Faraday's thoughts returned to lines of force that he imagined surrounded the magnet. These lines are invisible but can be visualized by sprinkling iron shavings onto a piece of paper held over a magnet. The shavings line up in definite patterns surrounding the magnet. Only when the lines of force moved and crossed the wire did they create an electric current in it.

He next set up a copper disk so that its edge passed between the poles of a horseshoe-shaped magnet. Wires and sliders connected the axle at the center of the disk and the edge of the disk to a galvanometer, to detect the presence of a current. When the disk was spun, a continuous current was produced between the center of the disk and its edge. Motion had been converted into electricity.

Faraday presented these results to the Royal Society in November 1831. These experiments allowed Faraday to conclude that a changing magnetic field can produce an electric current, a process called electromagnetic induction (or electrical induction). Around the same time, the American physicist Joseph Henry also discovered electrical induction, but Faraday published his results first, so he is given credit. (Faraday's paper appeared in *Philosophical Transactions* in 1832.) One of the most common uses of electromagnetic induction today is in generators (also called dynamos). A magnetic field inside the generator creates an electric current in a rapidly spinning wire coil; the current is then sent to homes or other buildings that use electricity. Thus generators convert mechanical energy (motion) into electric energy. Transformers also operate on the principle of induction. Transformers are devices used to increase or decrease voltage. For example, electric power plants create extremely high-voltage currents, but only low-voltage currents are needed to run ordinary household appliances such as lamps or toasters.

Faraday wondered whether the sort of electricity that he created through the use of magnets was the same as static electricity or electricity produced from a voltaic pile. He searched the literature and performed a series of tests that showed the same effects resulted no matter what the electrical source: thus electricity was electricity. While the intensity might vary, the nature remained the same.

ELECTROCHEMISTRY AND LIGHT

Over the next few years, Faraday merged his knowledge of chemistry with the developing field of electricity. Years before, Davy had figured out how to separate metals by passing an electric current through them. Faraday called this process electrolysis, the chemical decomposition of a compound by sending an electric current through it. In 1833 he established two basic laws of electrolysis. The first law stated that the quantity of metal liberated was proportional to the quantity of electricity used. The second law stated that the electricity required to liberate the unit equivalent mass of any element was precisely the same for all elements. In order to describe his research, he had to make up several new words, including *anode, cathode, electrode, electrolyte, anion,* and *cation.* These are now an essential part of scientific vocabulary.

In 1833 Michael Faraday was named the first Fullerian Professor of Chemistry at the RI for his advances in electrochemistry. He began to collect many other honors and medals at this time, including two Copley Medals from the Royal Society (1832, 1838), an honorary doctorate from Oxford, and membership in the Senate of the University of London. In 1836 he was named scientific adviser for Trinity House, which was in charge of maintaining safe waterways in England and Wales and overseeing the lighthouses.

By 1835 his research in electrochemistry led him to consider electrostatics. He had found that electrochemical forces were intermolecular, resulting from strains passed along a series of molecular partners. Could electrostatic discharges be explained similarly? From experiments designed to answer this, he formulated his theory of electricity. He described static discharges as being the result of a release of a strain caused by an electrical force. Currents were made of strains on particles of matter and were passed without any of the matter itself being transferred.

As he developed his theory of electricity, his mind began to fail him. He suffered memory losses and acted giddy at inappropriate times. Though his wife tried to protect him from embarrassment, by 1840 he could no longer work and did not fully resume his activities until 1844. It is possible that he was suffering from mild chemical poisoning.

After he resumed his research, he became interested in the relationship among light, electricity, and magnetism. Were there a connection among them? If so, what was the nature of the relationship? He tried many different experiments to determine whether there was such a relationship, and his persistence paid off when he placed heavy glass (which, incidentally, he had produced when working on optical glass years before) across the poles of two powerful electromagnets placed side by side. Then he shined polarized light (light whose waves are all oscillating in the same direction) through the glass at one end. When the magnet was turned on, the direction of the light's polarization was rotated. This phenomenon came to be called the Faraday effect. Furthermore, the glass itself was affected. To test this more directly, he hung a bar of heavy glass between the poles of a strong horseshoe-shaped electromagnet, and the glass aligned itself perpendicularly across the lines of force of the magnet. That is, instead of the ends' of the bar each reaching toward a magnetic pole, the bar aligned itself transversely between the poles. After further experimentation, he found that other substances acted similarly. He called this phenomenon "diamagnetism" and concluded that all substances were either magnetic or diamagnetic. These findings earned him two medals from the Royal Society in 1846, the Rumford and Royal Medals.

Though again suffering temporary confusion and memory troubles, in 1846 Faraday published "Thoughts on Ray Vibrations," a short paper that, as the title suggests, was a series of musings on the subject of radiation. In it, he proposed but could not define a relationship between light and magnetism. Throughout the late 1840s and the 1850s, Faraday's skill as a lecturer was perfected. He worked hard to improve his delivery and was very popular among a variety of always crowded audiences. He delighted the attendees with carefully planned experiments and explained the concepts so they could be easily understood. In 1848 he delivered the famous lecture "The Chemical History of the Candle," and he gave the Christmas Lecture every year between 1851 and 1860.

He also continued his research during this time. He was prolific, but his experiments were not as awe-inspiring as his earlier discoveries. He examined the effect of magnetism on gases and found oxygen to be magnetic. He believed in the unity of all natural forces (gravity, light, heat, magnetism, and electricity) and wanted to find a relationship between electricity and gravity but was unsuccessful.

PLAIN MICHAEL FARADAY

In 1857 Faraday was offered the presidency of the Royal Society but declined it, reportedly because he feared the effect on his mind of assuming any new responsibilities. He also turned down an offer of knighthood from Queen Victoria, desiring instead to remain plain Michael Faraday. He did accept the offer of a house and garden in Hampton Court by the queen in 1858. Though his laboratory research had suffered since his breakdown in the early 1840s, he did not resign as a lecturer at the RI until 1861. He was offered the presidency of the RI, but he refused that as well. He did continue advising the staff of Trinity House until 1865, particularly in the matter of electric lighting for the lighthouses. By 1864 he had resigned from all his duties, and he and Sarah moved to their new house permanently in 1865. By the time he died on August 25, 1867, he could barely move or speak. He was buried in the Sandemanian plot in the Highgate Cemetery under a gravestone that, at his request, simply read, *Michael Faraday*.

His name lives on. In his honor, a unit of capacitance was named a *farad*. The Christmas Lectures that Faraday initiated are now a flagship of the Royal Institution and are broadcast internationally. The Institution of Electrical Engineers awards a Faraday Medal for notable achievement in the field of electrical engineering and holds annual Faraday Lectures. He was a pure scientist, interested in gaining a better understanding of natural forces but always leaving the applications and widespread fame to others. He was a pioneer who made important foundational discoveries upon which others built new sciences. Years later the work of James Clerk Maxwell helped support the work of Faraday. His achievements were appreciated in his own time; after all, he started so young and accomplished so much. Faraday was honored by more than 50 academic societies during his lifetime. Even so, much of the importance of what he discovered was not fully appreciated until the impact of the applications resulting from his research became apparent years later. Faraday never had any formal training. He was particularly deficient in mathematics and relied on others to translate his ideas mathematically into concrete theories, but he had an uncanny sense of the concepts and theories he researched. The application of the phenomena he discovered developed into the modern electrical industry.

See also ELECTRICITY; ELECTROCHEMISTRY; ELECTROMAGNETISM, MAXWELL, JAMES CLERK.

FURTHER READING

Hirshfield, Alan. *The Electric Life of Michael Faraday*. New York: Walker, 2006.

The Royal Institution of Great Britain. "Heritage Faraday Page." Available online. URL: http://www.rigb.org/rimain/heritage/faradaypage.jsp. Accessed January 6, 2008.

fatty acid metabolism Lipids are one of the four main classes of biomolecules. Fatty acids, a major component of lipids, are long carbon chains, typically having between 14 and 24 carbons, with a terminal carboxyl group (-COOH). The length can vary, as can the number and position of double or triple bonds. Three fatty acids covalently linked to one molecule of glycerol constitutes a triacylglycerol (also called a triglyceride) or, simply, a fat, an energy-rich molecule that cells use for long-term energy stores. Phospholipids and glycolipids, the structural building blocks for biological membranes, both contain fatty acids. A typical phospholipid consists of one glycerol molecule covalently linked to two fatty acids and one phosphate group esterified to a hydroxyl group of an alcohol. Glycolipids are derived from sphingosine, an amino alcohol that contains a long fatty acid unit and, as the name implies, also contains sugar moieties. Some hormones are also derived from fatty acids.

Fatty acid nomenclature depends mainly on the parent hydrocarbon chain, with a few modifications

A Fatty Acid

$$\overset{3}{H_3C\text{-}(CH_2)_n\text{-}}\overset{2}{CH_2}\text{-}\overset{1}{CH_2}\text{-}C{\overset{\displaystyle O}{\underset{\displaystyle OH}{\lessgtr}}}$$

(ω) (β) (α)

R

B Triacylglycerol

$$\begin{array}{l} \qquad\qquad\quad CH_2\text{-}O\text{-}\overset{\overset{O}{\|}}{C}\text{-}R_1 \\ \qquad\qquad\quad | \\ R_2\text{-}\overset{\overset{O}{\|}}{C}\text{-}O\text{-}CH \\ \qquad\qquad\quad | \\ \qquad\qquad\quad CH_2\text{-}O\text{-}\overset{\overset{O}{\|}}{C}\text{-}R_3 \end{array}$$

(a) Fatty acids consist of a long hydrocarbon chain and a terminal carboxyl group. (b) Triacylglycerols (also called triglycerides) store fatty acids as uncharged esters of glycerol, with one glycerol bonded to three separate fatty acids.

based on whether or not the molecule contains any double or triple bonds. Fatty acids that contain the maximal possible number of hydrogens and whose carbons are all connected to one another by single bonds are said to be saturated. If one or more double or triple bonds exist, the fatty acid is unsaturated. The number of double bonds is variable, but at least one methylene group separates them. Having more double bonds and being shorter enhance the fluidity of fatty acids. Numbering of the carbon atoms begins at the carboxy terminus and proceeds in order until reaching the carbon at the chain's end, which is called the ω carbon. Carbons numbered 2 and 3 are referred to as the α and β carbons, respectively.

Because fatty acids are highly reduced, they release a lot of energy when oxidized. Fatty acids have the potential to yield about nine kilocalories per gram (9 kcal/g) whereas carbohydrates and proteins yield about 4 kcal/g. Thus fatty acids can store more than twice the amount of energy in the same quantity of mass, making them much more efficient molecules for storing energy than carbohydrates and proteins. Also, because fatty acids are nonpolar, they contain little water and can store more energy. In mammals, triacylglycerol accumulates in the cytoplasm of adipose cells, specialized fat-storage cells. The fat droplets merge to form large lipid globules inside the cell and make up most of the cell's volume.

FATTY ACID DEGRADATION

Adipose cells are specialized to mobilize the fat stores when the body needs energy. Hormones, including epinephrine, norepinephrine, glucagon, and adrenocorticotropic hormone, secreted by the adrenals, the pancreas, and the anterior pituitary stimulate this process. Adipose cells have receptors on the surface on their cell membranes that specifically recognize and bind these hormones. The hormone binding sites of the receptor proteins face the cell's exterior; the proteins also have nonpolar regions that span the phospholipid membrane and regions that contact the cytoplasmic side of the membrane. Hormone binding stimulates changes inside the cell, such as activation of the enzyme adenylate cyclase, which produces cyclic adenosine monophosphate (cAMP), a second messenger molecule that mediates numerous cellular activities. An increase in cAMP levels activates the enzyme protein kinase A, which in turn activates the enzyme lipase. Lipase activation leads to lipolysis, or the hydrolysis of lipids, a process that converts triacylglycerols into fuel molecules that circulation transports to the body's tissues. The cells in the body's tissues then uptake the fuel molecules so they can synthesize adenosine triphosphate (ATP), a molecule containing high-energy bonds that, when broken, release energy for use in many other biochemical reactions.

One of the responses of adipose cells to the hormones mentioned is the activation of the enzyme lipase, which catalyzes the hydrolysis of ester bonds such as the bonds that link fatty acid chains to glycerol in triacylglycerols, releasing the individual fatty acids. After being released into the blood, the free fatty acids are bound by serum albumin and transported to the body tissues. The liver absorbs the freed glycerol and converts it into glyceraldehyde 3-phosphate, an inter-

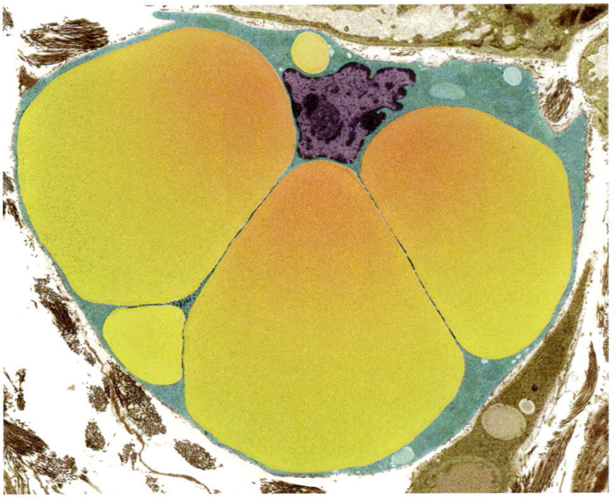

Adipose cells are specialized cells that store fat. The lipid droplets (yellow), apparent in this colored transmission electron micrograph (TEM), of a developing fat cell can take up practically the entire volume of the cell. The cell's nucleus is shown in purple. *(Steve Gschmeissner/Photo Researchers, Inc.)*

mediate of the biochemical pathways glycolysis and gluconeogenesis. Liver and muscle cells can metabolize the fatty acids as a source of energy.

Fatty acid oxidation occurs in the mitochondria, but before entering the mitochondrial matrix, fatty acids must be activated by the enzyme acyl CoA (coenzyme A) synthetase, also called fatty acid thiokinase. Located on the outer mitochondrial membrane, this enzyme catalyzes the formation of acyl CoA using the energy released by successively breaking both of the high-energy bonds of an ATP molecule. Because two high-energy bonds are broken, and only one is formed (the one attaching the acyl moiety to coenzyme A), this reaction is irreversible. In order to move into the mitochondrial matrix for oxidation, long-chain acyl CoAs must drop the CoA and temporarily bind to a compound called carnitine that carries it through the membrane transport protein translocase. After reaching the mitochondrial matrix, another CoA exchanges with the carnitine, and the carnitine moves back into the cytoplasm to pick up another fatty acid for transport. The acyl CoA is now ready to proceed with oxidation inside the mitochondrial matrix.

Fatty acid metabolism occurs by a process called beta-oxidation, in which the carbon chains are degraded two carbons at a time. The name *beta-oxidation* is due to the characteristic that the beta (β) carbon is oxidized as the carboxyl and the α carbons are removed. Oxidation of saturated fatty acids occurs in four main steps: oxidation, hydration, oxidation again, and thiolysis. During the first step, the enzyme acyl CoA dehydrogenase oxidizes acyl CoA, reducing flavin adenine dinucleotide (FAD) to $FADH_2$ in the process and forming a *trans*-double bond between carbons number two and three. The second step is the hydration of the newly formed *trans*-double bond, a reaction catalyzed by enoyl CoA hydratase. This allows for the subsequent oxidation of the third carbon (the β carbon) by the enzyme L-3-hydroxyacyl CoA dehydrogenase, generating NADH + H^+. Finally, with the assistance of CoA, β-ketothiolase breaks the linkage between carbons number two and three, releasing an acetyl CoA from the shortened acyl CoA chain. In summary, the completion of these four steps results in the release of one molecule of acetyl CoA from an acyl CoA and the reduction of one FAD to $FADH_2$ and one NAD^+ to NADH + H^+. The acetyl CoA can enter the citric acid cycle to release more energy, and the reduced forms of NADH and $FADH_2$ can carry electrons to the electron transport chain for ATP synthesis. Oxidation of the acyl CoA continues two carbons at a time.

If the fatty acid is not saturated, degradation proceeds in the same manner as in the case of saturated fatty acids until reaching a double bond. The presence of a *cis*-double bond between carbons three and four will prevent the formation of a double bond between carbons number two and

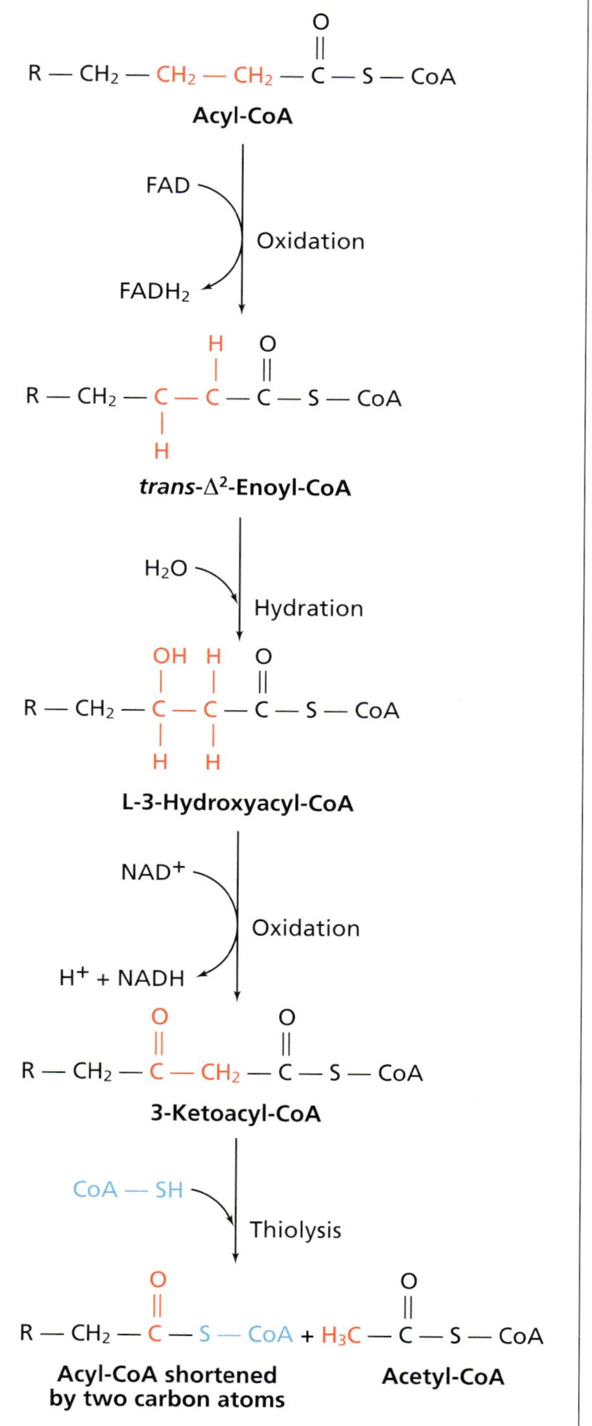

Degradation of fatty acids occurs by a process called β-oxidation, in which a series of four steps removes two carbons at a time from the acyl group of a fatty acid.

three during the first oxidation step of degradation. To overcome this, an isomerase converts the *cis*-double bond between carbons three and four to a *trans*-double bond between carbon number two and three, preparing it for the subsequent hydration step. If the problem is the presence of a *cis*-double bond at an even numbered carbon, after the first oxidation step, an intermediate that contains two double bonds forms, one between carbon two and carbon three and one between carbon four and carbon five. The enzyme 2,4-dienoyl CoA reductase remedies this by using a reduced nicotinamide adenine dinucleotide phosphate (NADPH) to reduce the intermediate to form another intermediate that has one *cis*-double bond between carbon number two and three. Isomerase converts the *cis*-double bond to a *trans*-double bond, and degradation can continue at the hydration step.

When high concentrations of acetyl CoA are produced in the liver, such as when fat breakdown predominates, not all of it enters the citric acid cycle. Under fasting conditions, or in the case of diabetes, ketogenesis, the formation of ketone bodies, occurs. The joining of two acetyl CoA molecules forms acetoacetic acid, which is converted to β-hydroxybutyric acid and acetone. These three molecules, called ketone bodies, travel through blood circulation to other tissues, such as muscle, whose cells convert the ketone bodies back into acetyl CoA, which feeds into the citric acid cycle and leads to the synthesis of more ATP. The presence of low levels of ketone bodies in the blood is not damaging; in fact, heart muscle and the renal cortex prefer to burn acetoacetate rather than glucose. High levels of ketone bodies in the blood can overwhelm the buffering capacity of the blood, causing it to become acidic.

FATTY ACID SYNTHESIS

The biochemical pathways that synthesize fatty acids are distinct from the catabolic pathway. One difference is that synthesis occurs in the cytoplasm of cells, whereas degradation occurs in the mitochondrial matrix. As does fatty acid degradation, synthesis occurs by a four-step process. Intermediates are carried by an acyl carrier protein (ACP) rather than CoA, and chains elongate two carbons at a time. The four main steps are condensation, reduction, dehydration, and another reduction.

A single, large, multifunctional polypeptide chain that forms a complex called fatty acid synthase catalyzes the reactions of fatty acid synthesis. The precursors for fatty acid synthesis are acetyl CoA (acetyl groups contain two carbons), which is made in the mitochondria and transported to the cytoplasm in the form of citrate, and malonyl CoA, a three-carbon derivative formed by the carboxylation of acetyl CoA. Acetyl transferase and malonyl transferase attach these molecules to ACP to form acetyl-ACP and malonyl-ACP. The acyl-malonyl-

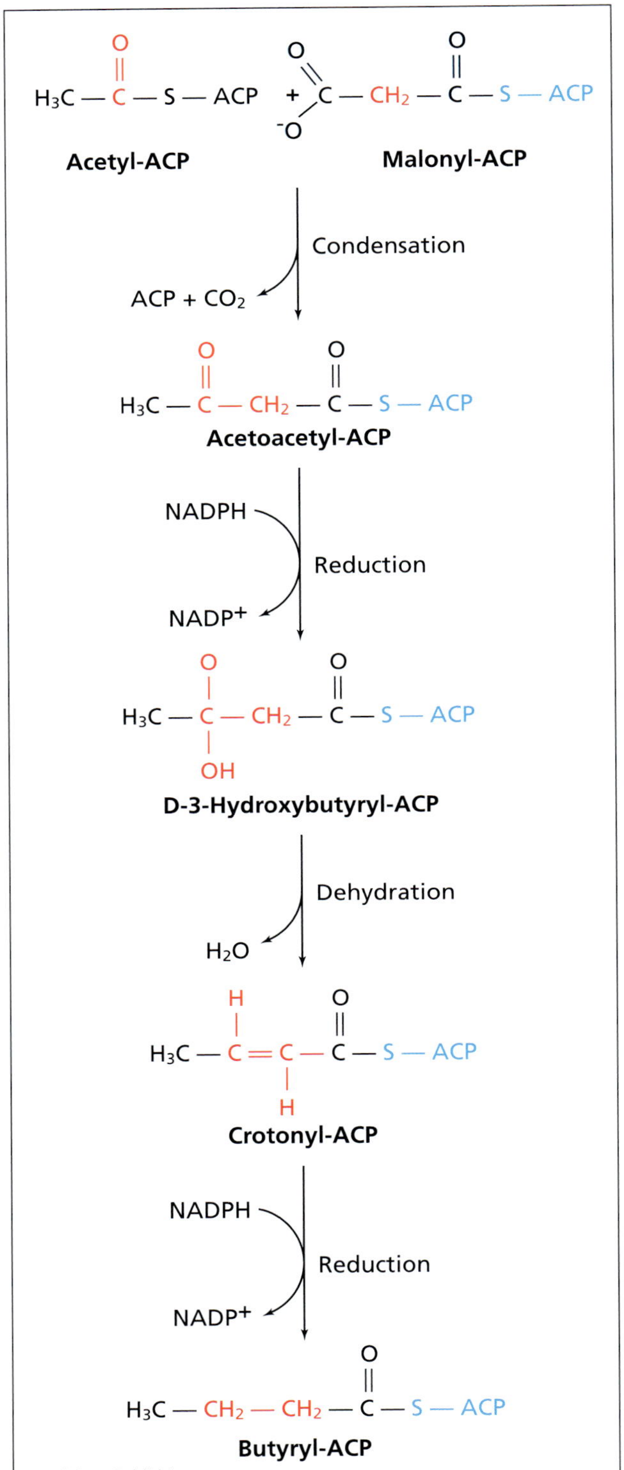

Fatty acids elongate two carbons at a time through a series of four-step sequences.

ACP condensing enzyme combines acetyl-ACP with malonyl-ACP to form the four-carbon acetoacetyl-ACP, releasing ACP and carbon dioxide (CO_2) in the process. From this point on, the number of carbons will be even. (The synthesis of fatty acids with odd numbers begins with the three-carbon precursor, priopionyl-ACP.) During the second step, β-ketoacyl-ACP reductase reduces the keto group using NADPH, forming D-3-hydroxybutyryl-ACP. Next, dehydration of D-3-hydroxybutyryl-ACP by 3-dehydroxyacyl-ACP dehydrase results in crotonyl-ACP, which contains a *trans*-double bond. In the fourth step, enoyl-ACP reductase reduces crotonyl-ACP into butyryl-ACP using another NADPH, thus completing the first round of elongation.

To continue with a second round of elongation, malonyl-ACP combines with butyryl-ACP in a condensation reaction, followed by reduction, dehydration, and another reduction step. This occurs as many times as necessary until the 16-carbon fatty acid palmitate results. The net reaction for the synthesis of palmitate is as follows:

$$8\text{acetyl CoA} + 7\text{ATP} + 14\text{NADPH} + 6\text{H}^+ \rightarrow \text{palmitate} + 14\text{NADP}^+ + 8\text{CoA} + 6\text{H}_2\text{O} + 7\text{ADP} + 7\text{P}_i$$

After the synthesis of palmitate, other enzymes may add more two-carbon units to elongate the hydrocarbon chain further, and still others may introduce double bonds into long-chain acyl CoAs.

Genetic mutations in the genes for enzymes involved in fatty acid metabolism or insufficient levels of these enzymes can cause disorders called lipid storage disorders. Without the ability to oxidize fatty acids, over time, the storage of excess fats can cause harm to body tissues, particularly the brain and peripheral nervous system, since myelin, the substance that coats and protects some nerve cells, is composed of lipids. The liver, spleen, and bone marrow may also be damaged. In addition, defects in the transport protein that translocates acyl CoA bound to carnitine into the mitochondrial matrix can cause muscular cramps as a result of fasting, exercise, or consumption of a high-fat meal because of the body's inability to degrade the fatty acids into acetyl CoA molecules. Symptoms of lipid storage disorders include enlarged spleen and liver, liver malfunctions, neurological symptoms, developmental delays, and impaired kidney function. No cures or even treatments exist for many of these disorders, and some lead to death during childhood. Enzyme replacement therapy effectively treats a few types.

See also CARBOHYDRATES; CITRIC ACID CYCLE; ELECTRON TRANSPORT SYSTEM; ENZYMES; GLYCOLYSIS; LIPIDS; METABOLISM.

FURTHER READING

Berg, Jeremy M., John L. Tymoczko, and Lubert Stryer. *Biochemistry,* 6th ed. New York: W. H. Freeman, 2006.

Fermi, Enrico (1901–1954) Italian-American *Physicist* Enrico Fermi was an Italian-born physicist who is best known for work on development of the first nuclear chain reaction. Fermi developed the principle of slow neutrons and is also known for his contributions to quantum theory with Fermi-Dirac statistics and fermions relating the properties of particles obeying the Pauli exclusion principle. Enrico Fermi received the Nobel Prize in physics in 1938 for his development of slow neutrons for nuclear chain reactions.

Enrico Fermi was born in Rome on September 29, 1901, to Alberto Fermi and Ida de Gattis. He attended school in 1918 at the Scuola Normale Superiore in Pisa. He earned a Ph.D. in physics in 1922 from the University of Pisa. After completion of his degree, Fermi worked with the German physicist Max Born at Göttingen, Germany, and then became a professor of theoretical physics at the University of Rome in 1927.

HISTORY OF RADIOACTIVITY BEFORE FERMI

Radioactivity was first discovered by the French physicist Henri Becquerel. Since his discovery, numerous scientists have worked to understand the process and characteristics of the radioactive elements.

Lord Ernest Rutherford laid the groundwork for Fermi's work on neutron bombardment in 1907 with his gold foil experiment. He directed alpha particles, which had been identified as helium atoms lacking their two electrons, at targets of gold foil. If an atom consisted of a cloud of positive charge dotted with negatively charged electrons, Rutherford predicted that most of the alpha particles in his studies would pass through the gold atoms in the foil. The majority of them did, but a small fraction of the alpha particles did not pass through the gold foil. This fraction of particles deflected at large angles away from the foil and, in some cases, deflected straight back toward the source of the particles. Rutherford proposed that an atom has a small dense core, or nucleus, composed of positively charged particles that makes up most of an atom's mass and that negatively charged electrons float around in the remaining space. When alpha particles hit this dense core, they bounced back. These experimental results led to his development of an atomic model characterized by a nucleus containing protons with the rest of the atom made of mostly empty space filled with fast-moving electrons. Rutherford also

Enrico Fermi was an Italian physicist responsible for the successful completion of the first nuclear chain reaction. This work was known as the Chicago Pile and was completed at an underground squash court at the University of Chicago on December 2, 1942. Fermi is shown here at the University of Chicago, November 25, 1946. *(AP Images)*

determined that some of the alpha particles were trapped by the nucleus, and a new particle (in most cases, a proton) was given off. This was one of the first examples of an atom of one element being turned into another element. Many cases showed the new product formed to be one of the already established elements, while several examples of unstable nuclei that had not previously been discovered were the first examples of artificial radioactivity as established by Irène Joliot-Curie and Frédéric Joliot-Curie in 1933. This had been the goal of alchemists for centuries, and via chemical means it could not be done. Fermi utilized this concept of nuclear decay to explain atomic structure further.

Creating radioactivity by bombardment of nuclei with hydrogen and helium nuclei (a type of radiation) only works for elements that have an atomic number less than 20. Protons have an inherent positive charge that causes them to be repelled when they are in close proximity with the nucleus, causing them to lose speed. The discovery of neutrons in the early 1930s led the way for Fermi and others to study radioactivity using neutrons to bombard. A neutron does not have a repulsion problem because it is an uncharged particle. Direct collisions with nuclei are the only way to slow traveling neutrons. As the size of an atom is very large relative to the size of a nucleus, these collisions are relatively rare.

Fermi used beryllium (atomic number 4) and radon (atomic number 86) as sources of neutrons to begin the bombardment process. These radioactive substances spontaneously emit neutrons, which can then bombard other substances. He experimentally determined the radioactivity given off as well as chemical properties of all of the elements of the periodic table in response to this neutron bombardment. During his career, Fermi was responsible for the identification of several previously unknown elements, with atomic numbers greater than that of uranium.

SLOW NEUTRONS

As Fermi was carrying out his systematic evaluation of the potential to produce radioactivity from different elements, he varied the materials between the neutron source and the element being tested. He found that placing paraffin between the beryllium tube and the sample to be tested dramatically increased the amount of radioactivity emitted. Paraffin is a hydrocarbon that is solid at room temperature. Fermi discovered that when the neutrons were passed through the hydrogen nuclei in the paraffin, the kinetic energy of the original neutron was split between the neutron and the proton in hydrogen. Therefore, the resulting neutron was termed a *slow* neutron. The slow neutron has the potential for more collisions with nuclei and therefore could can more radioactive emissions from the samples. The radioactive emissions of elements of higher mass number increased considerably with slow neutrons rather than fast-moving neutrons. In 1938 Fermi received the Nobel Prize in physics for his work on artificial radioactivity produced by slow neutrons. After receiving the Nobel Prize, Fermi moved to the United States, where he became a professor of physics at Columbia University.

After Fermi's work, a team of German chemists, Otto Hahn and Fritz Strassman, demonstrated that when uranium was bombarded with neutrons from a radon-beryllium source, they found residual barium after the reaction. Barium was a much less massive element than uranium. They believed the atom split (fissioned) when it was bombarded with the neutrons. They shared their work with the German physicist Lise Meitner, who was working with Niels Bohr in Copenhagen in 1938. As Meitner further studied the reaction, she found that the amount of mass recovered from the splitting of uranium was not completely recovered in the products. Using the equation proposed by Albert Einstein,

$$E = mc^2$$

which showed matter (mass, denoted m) could be converted into energy (E), with c the speed of light, Meitner proposed that the missing matter had been converted into energy. Meitner's prediction quickly spread throughout the scientific community, and verification of her prediction by experimental results soon followed. The discovery that nuclear fission could produce a large amount of energy had been made. Otto Hahn received the Nobel Prize in chemistry in 1944 for his work on nuclear fission of heavy nuclei. Many believed Meitner deserved to share this award.

After the discovery of nuclear fission, Fermi saw its potential to produce more neutrons in order to create a chain reaction of nuclear decay. He immediately got to work to determine whether a self-sustaining nuclear fission reaction was possible.

THE CHICAGO PILE

After the discovery of nuclear fission and the possibility of a chain reaction, a number of countries began the race to produce an atomic bomb. Albert Einstein sent a letter to President Franklin D. Roosevelt describing the importance of the research into nuclear fission by American scientists. Rumors surfaced that the Germans had already begun work on an atomic bomb program. So in 1939 Roosevelt set up the Uranium Committee, which included Fermi.

In order to explore the possibility of a chain reaction, one must have a good source of neutrons, housing for the reaction, and a way to control the reaction. The amount of uranium required to produce a self-sustaining nuclear reaction is known as the critical mass. Fermi developed the science and procedures responsible for the first controlled nuclear fission reaction. This work took place on a squash court beneath Stagg Field at the University of Chicago. The components of the pile were graphite, uranium, and cadmium control rods. The pile itself was approximately 26 feet (7.9 m) in diameter and approximately 20 feet (6.1 m) high. The completed pile contained approximately 771,000 pounds (349,720 kg) of graphite, 81,000 pounds (36,741 kg) of uranium oxide, and 12,400 pounds (5,624 kg) of uranium metal. Graphite bricks that had uranium pieces built into them served as the neutron source. Fermi and his team then put in neutron-absorbing cadmium bars (control rods) that could stop the reaction if need be. When the control rods were pulled out, the neutron source could begin bombarding the uranium and create a controlled reaction. At 3:25 P.M. on December 2, 1942, they succeeded in creating the first self-sustaining nuclear reaction.

NUCLEAR FISSION

Nuclear energy is the energy found within the attraction between subatomic particles in an atomic nucleus known as protons and neutrons. When the energy of these interactions is released by splitting the nucleus of the atom such that smaller nuclei are produced, the process is known as nuclear fission. In order for fission to occur, the nucleus of the atom being broken apart (the fissionable material) must be made unstable: the fissionable material is bombarded with particles in order to break it apart. On average, two to three neutrons are released from a fission reaction cycle. The fission reaction is known as a chain reaction, meaning that once started, the reaction continues on its own until all of the uranium 235 atoms are used up. Because two neutrons are produced from the reaction of one neutron, each reaction that happens is able to fuel subsequent reactions.

It did not escape attention that this nuclear fission reaction could be used as both a source of energy as well as weaponry. Fermi's work was incorporated in the Manhattan Project, which was responsible for the development of the atomic bomb. After his work at the University of Chicago, in 1942, Fermi became director of what became known as Argonne National Laboratory back when it was the University of Chicago's "Argonne Lab." In 1944 he moved to Los Alamos, New Mexico, having become a naturalized citizen of the United States, and headed the Manhattan Project's "F" division, which solved special problems as they arose. After the war, Fermi returned to the University of Chicago and became the Charles H. Swift Distinguished Service Professor of Physics. He received many additional medals and awards in recognition of his contributions to physics and the Manhattan Project.

Fermi was married in 1928 to Laura Capon, and they had one son and one daughter. He died of cancer on November 28, 1954, at the age of 53.

Enrico Fermi was an unparalleled scientist whose discovery of slow neutrons advanced both theoretical and experimental physics. As one of the foremost experts on neutrons, in the late 1930s, Fermi did work that was crucial to the achievement of the first controlled nuclear fission reaction in 1942. The creation of the atomic bomb changed the course of history. Without the work of Fermi, this could not have been accomplished.

See also Bohr, Niels; Fermi-Dirac statistics; fission; fusion; Meitner, Lise; nuclear physics; Oppenheimer, J. Robert; radioactivity.

FURTHER READING

Bernardini, Carlo, and Luisa Bonolis. *Enrico Fermi: His Work and Legacy*. Bologna: SIF, 2001.

The Nobel Foundation. "The Nobel Prize in Physics 1938." Available online. URL: http://nobelprize.org/nobel_prizes/physics/laureates/1938/index.html. Accessed February 1, 2008.

Segrè, Emilio. *Enrico Fermi, Physicist*. Chicago: University of Chicago Press, 1970.

Stux, Erica. *Enrico Fermi: Trailblazer in Nuclear Physics*. Berkeley Heights, N.J.: Enslow, 2004.

Fermi-Dirac statistics The statistical rules governing any collection of identical fermions are called Fermi-Dirac statistics, named for the 20th-century Italian-American physicist Enrico Fermi and the English physicist Paul Dirac. A fermion is any particle—whether an elementary particle or a composite particle, such as an atom or a nucleus—whose spin has a half-integer value (i.e., 1/2, 3/2, 5/2, . . .). All the elementary matter particles are fermions. In particular, they comprise the quarks and the leptons (the electron, muon, tau, and their corresponding neutrinos). These are all spin-1/2 fermions. At the level of atomic nuclei, the nucleons (neutrons and protons) constituting nuclei are also spin-1/2 fermions.

Fermi-Dirac statistics is based on (1) the absolute indistinguishability of the identical particles and (2) no more than a single such particle being allowed to exist simultaneously in the same quantum state. The latter is known as the Pauli exclusion principle.

One result of Fermi-Dirac statistics is that in a system of identical fermions in thermal equilibrium at absolute temperature T, the probability for a particle to possess energy in the small range from E to $E + dE$ is proportional to $f(E)\,dE$, where $f(E)$, the probability distribution function, is

$$f(E) = \frac{1}{e^{\frac{E-E_F}{kT}} + 1}$$

Here E and E_F are in joules (J), T is in kelvins (K), f is dimensionless (i.e., is a pure number and has no unit), and k denotes the Boltzmann constant, whose value is $1.3806503 \times 10^{-23}$ joule per kelvin (J/K), rounded to 1.38×10^{-23} J/K. The quantity E_F is the

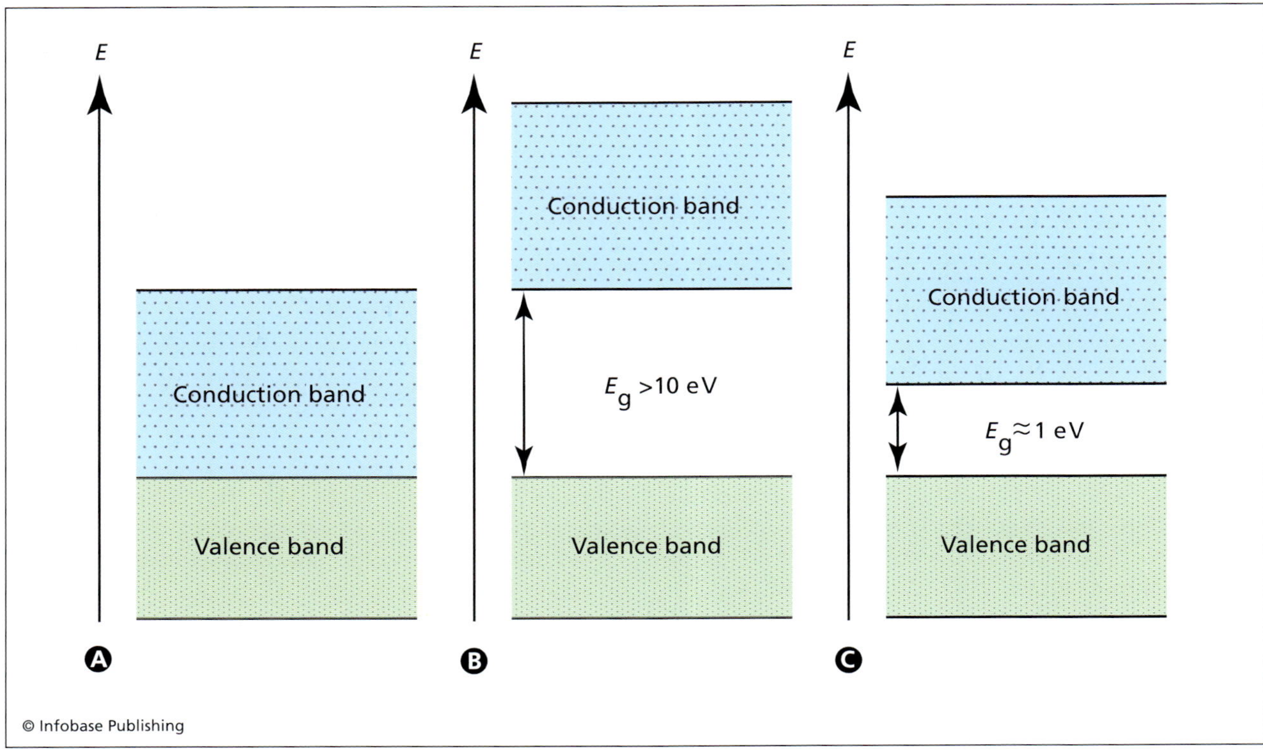

The conduction band of a solid represents the range of energies at which electrons can move about freely in the volume of the solid. The valence band is the range of energies of electrons in the ground state. The energy axis is labeled *E*. (a) When the bands overlap or almost overlap, the material is a conductor. (b) When the energy gap E_g between the bands is large, greater than about 10 eV, the material is an insulator. (c) Semiconductors have intermediate energy gaps, on the order of 1 eV.

Fermi energy, the energy for which the probability distribution function takes the value ½:

$$f(E_F) = ½$$

Note that this distribution function has its maximal value (less than 1) at zero energy and decreases to zero as the energy increases. For temperatures close to absolute zero, the function is quite flat with the value 1 for energies below the Fermi energy, drops abruptly to zero at the Fermi energy, and remains at this value for higher energies. That reflects the uniform filling of energy states from zero energy to the Fermi energy, while the higher-energy states are unoccupied. The states are filled uniformly from zero energy up, because only a single particle is allowed in any state. The Fermi energy can then be understood as the highest particle energy in the system, when the system is close to absolute zero temperature.

Fermi-Dirac statistics is crucial for the understanding of, for instance, solid conductors, semiconductors, and insulators. Solids possess a very large number of energy levels available to the electrons (which are fermions) of the material's atoms. The energy levels possess what is known as a band structure, whereby, depending on the material, they group into ranges, or "bands," with gaps of forbidden energy between the bands. Electrons in sufficiently high-energy states, states of the conduction band, can move freely throughout the solid and conduct electricity, while electrons in the lowest-energy band, called the valence band, are immobile. Without the constraint of Fermi-Dirac statistics, the material's electrons would tend to populate the lowest-energy state available to them, and much energy would be required to take electrons from the valence band into the conduction band. All materials then would be insulators. As it is, however, the electrons fill the available states one electron per state (where each of the two spin directions corresponds to a different state), allowing the highest-energy electrons to be in the conduction band (in conductors) or in states from which relatively small amounts of energy, around one electron volt, can raise them to the conduction band (in semiconductors). (One electron volt is the amount of kinetic energy an electron gains when it spontaneously moves to a location where the electric potential is greater by one volt than at its original location. One electron volt [eV] equals 1.60×10^{-19} joule [J].) In insulators, the gap between the conduction band and the valence band is wide, greater than about 10 electron volts, and the highest-energy electrons are below the gap. Thus, in insulators a large amount of energy is needed to raise electrons to the conduction band.

See also ATOMIC STRUCTURE; BOSE-EINSTEIN STATISTICS; ELECTRICITY; ELECTRON CONFIGURATIONS; ENERGY AND WORK; FERMI, ENRICO; HEAT AND THERMODYNAMICS; MAXWELL-BOLTZMANN STATISTICS; PARTICLE PHYSICS; QUANTUM MECHANICS; STATISTICAL MECHANICS.

FURTHER READING

Serway, Raymond A., Clement J. Moses, and Curt A. Moyer. *Modern Physics*, 3rd ed. Belmont, Calif.: Thomson Brooks/Cole, 2004.

Feynman, Richard (1918–1988) American *Theoretical Physicist* Richard Feynman epitomized a new type of modern theoretical physicist to whom the common man could relate. A maverick with a charismatic personality and a love for adventure, Feynman began his professional career devising a mathematical formula that predicted the energy yield of an atomic explosion for the Manhattan Project. He went on to explain the concept of superfluidity, a critical point where gas and liquid behave the same way, as it pertained to helium. His seminal work occurred during his tenure at the California Institute of Technology (Cal Tech) in the early 1950s with his novel approach to quantum electrodynamics (QED), an area of physics that combines theories of quantum mechanics and electromagnetic radiation in a single theory of matter and light. After years of inadequate means of visualizing the events occurring at the subatomic level, Feynman developed diagrams, now called Feynman diagrams, that provide a visual way to follow interactions among various subatomic particles in both space and time. During this time, other physicists, namely, Julian Schwinger and Sin-Itiro Tomonaga, used traditional mathematical approaches based on previous work to develop a valid theory of QED, but Feynman chose to reinvent quantum mechanics from the ground up and ultimately arrived at the same conclusion. In 1965 he shared the Nobel Prize in physics with Schwinger and Tomonaga for their collective contributions to QED theory.

Born May 11, 1918, to Melville and Lucille Feynman, Richard Feynman grew up in Far Rockaway, New York, a seaside suburb of New York City. Melville Feynman always encouraged his son to identify not what he knew but what he did not know. His father also believed that the best way to seek knowledge was through a well-asked question. Along with his mother's sense of humor, Richard took his father's sage pieces of advice to heart, and the combination served him well throughout his life. By the time he was 10 years old, Richard busied himself assembling and disassembling crystal radio sets, and he often used his younger sister, Joan, as the hapless laboratory assistant in his hair-raising bedroom laboratory

The inventor of the Feynman diagram, Richard Feynman was a colorful character who is famous for the development of quantum electrodynamics and for his work in elementary-particle physics. Feynman shared the 1965 Nobel Prize in physics for his work. *(Photograph by Robert Paz, California Institute of Technology, courtesy AIP Emilio Segrè Visual Archives)*

experiments. With a natural propensity toward anything to do with mathematics or science, Feynman taught himself differential and integral calculus by the time he was 15. Before entering college, he frequently experimented with esoteric mathematical concepts like half-integrals. Feynman enrolled at the Massachusetts Institute of Technology, where he received a bachelor's degree in physics in 1939. Feynman chose Princeton University to seek a doctoral degree in physics under John Wheeler. His doctoral thesis was theoretical in nature—dealing with a part of the theory of electromagnetic radiation concerning waves that travel backward in time. Feynman's approach to this problem in quantum mechanics laid the groundwork for his quantum electrodynamic theory and the Feynman diagrams. Icons in the field, Albert Einstein, Wolfgang Pauli, and John von Neumann, thought enough of Feynman's work that they attended the seminar when he defended it to obtain his doctorate in 1942.

After Feynman's graduation from Princeton, another mentor, Robert Wilson, encouraged him to participate in the Manhattan Project, which took nuclear physics from the laboratory into the battlefield with the goal to design and detonate an entirely new weapon, the atomic bomb. Initially Feynman hesitated to take part in the endeavor, not only for its ethical ramifications but also because his first wife, Arline Greenbaum, whom he had married in 1942, had contracted tuberculosis. After moving his wife to a sanitorium close to Los Alamos, Feynman joined the project, serving as group leader in the theoretical division. At Los Alamos, Feynman developed formulae to measure how much radioactivity could be stored safely in one place and to predict the amount of energy released by a nuclear weapon. Feynman witnessed the detonation of the world's first atomic bomb at the Trinity Bomb Test in July 1945. After the death of his wife that same month, Feynman immersed himself in his work until Hans Bethe transferred him to the Oak Ridge, Tennessee, facilities to help engineers with safety issues. Before Feynman left Los Alamos, he perpetually joked about military security, and he pointed out its weaknesses by learning to crack safes containing top-secret information throughout the facility.

After World War II, Feynman accepted a position at Cornell University as professor of theoretical physics. Feynman doubted his abilities as a physicist during this period of his professional career, yet countless universities continued to offer him opportunities at their institutions. After a brief sojourn to Brazil in 1951, Feynman joined the faculty of Cal Tech, where over the course of his tenure he addressed problems in superfluidity, models of weak nuclear decay, and QED. One of his first projects at Cal Tech examined the physics of the superfluidity of supercooled liquid helium. At the point where helium gas and helium liquid exist in one homogeneous fluid state (called superfluidity), helium experiences no frictional resistance as it flows. Feynman successfully applied Schrödinger's equation describing quantum mechanical behavior to this phenomenon occurring at a macroscopic level. Then he and his colleague Murray Gell-Mann worked collaboratively to develop a description of the weak interactions that result in the decay of a free neutron into an electron, a proton, and an antineutrino.

Feynman next turned his energies toward the development of a quantum mechanical theory for electromagnetism, called quantum electrodynamics (QED). Electrons interact with one another, with positively charged nuclei, and with electromagnetic radiation. When two charged particles, such as two electrons or an electron and a proton, are near each other, the particles exchange a photon, or quantum of light, with one another. As two negatively charged electrons approach one another, one electron donates and the other electron accepts the light energy, yet both particles are deflected onto new paths after the photon exchange because like charges repel. Feynman showed this phenomenon mathematically as an inverse-square law that fit nicely into known aspects of the emerging QED theory. Feynman also described strong and weak nuclear forces in terms of similar exchanges of force particles. To help visualize these events, Feynman created simple diagrams—graphic bookkeeping devices that permitted one to visualize and subsequently calculate interactions of particles in space and time. These diagrams, called Feynman diagrams, have a horizontal axis representing space and a vertical axis representing time. Feynman used lines to depict all particles, whether matter or antimatter, with straight lines symbolizing matter particles such as electrons and wavy lines symbolizing force particles such as photons. He showed real particles (detectable entities such as electrons, protons, and neutrons) as entering and leaving the diagrams with these particles strictly adhering to all known laws of quantum mechanics. Feynman represented virtual particles (particlelike abstractions used in quantum mechanics to visualize interactions among real particles) as intermediate lines that stayed inside the diagram.

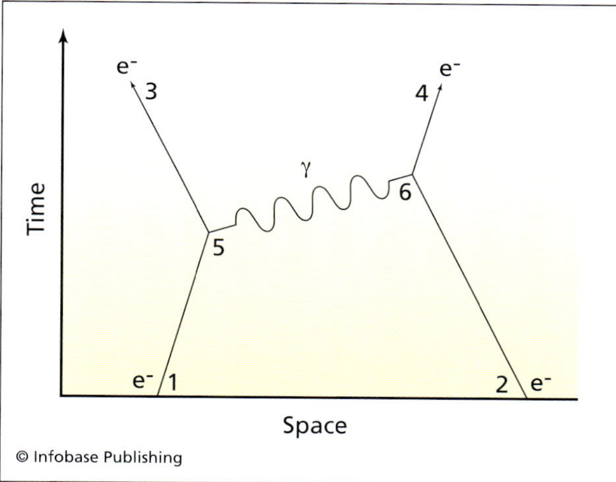

This Feynman diagram represents the simplest interaction between two electrons. The horizontal axis represents space and the vertical axis represents time. The first electron starts at 1 and, as time progresses, moves through space to point 5, where it emits a photon (symbolized by the wiggly line) and alters its path. The electron then travels to point 3 and off the graph. A second electron starts at point 2 and, as time advances, travels through space to point 6, where it absorbs the photon emitted by the first electron, which causes its path to change. The second electron moves along its new path to point 4 and off the graph.

Before Feynman's effort to understand QED, an infinite series of correction terms in the mathematical equations made it difficult to predict the state of a particle accurately. The goal of the calculations was to predict the probability of transition by a particle from one state to some other subsequent state. In theory, every possible path from one state to another is equally likely; therefore, the final path by which the transition occurs is the sum of all possible paths. In his typical cavalier way, Feynman reinvented quantum mechanics from scratch. During these prolific Cal Tech years, he developed rules that all quantum field theories obey, renormalized the theory of QED, and invented a means to represent quantum interactions schematically. In the end, theoretical values calculated using Feynman's formulas differed from experimental values by a minute ±0.00000001 percent. Feynman compared the accuracy of the theoretical calculations using QED theory to measuring the distance from New York City to Los Angeles within the thickness of a single human hair. Peers recognized the significance of his work in the field, and he shared the Nobel Prize in physics in 1965 with Julian Schwinger and Sin-Itiro Tomonaga for their cumulative work on QED theory.

Outside the laboratory, Richard Feynman expressed a deep concern about the education of future scientists and nonscientists alike. He served for two years on a California council that evaluated mathematics and physics textbooks for primary and secondary schools. Approached by several members of the physics faculty at Cal Tech, Feynman agreed to teach a series of undergraduate physics courses that enlivened the institution's physics program. Two faculty members, Robert Leighton and Matthew Sands, diligently recorded and transcribed the lecture series, and the three-volume work *The Feynman Lectures on Physics* (1963–65) remains popular today among students and teachers for its unique approach to explain many difficult concepts. Later Feynman made physics accessible to the general public when he wrote *QED—The Strange Theory of Light and Matter* (1986).

In 1960 Richard Feynman married Gweneth Howarth, a British woman, with whom he had two children, a son named Carl Richard and a daughter named Michelle Catherine. When he was not playing the bongos or playing a practical joke on someone, Feynman enjoyed weaving stories about his many personal escapades. His humorous autobiography, *Surely You're Joking, Mr. Feynman!* (1985), became a best seller as people seemed enamored with this wisecracking physicist from Queens. Near the end of his illustrious career, Feynman served on the committee that investigated the space shuttle *Challenger* explosion, which led to the death of the entire seven-person crew. In typical showman style, Feynman demonstrated that O-rings on the shuttle failed because of the cold temperatures on the day of the launch by dunking an O-ring into ice water and compromising its integrity during the televised committee meeting. Considered by many to be the greatest and most popular physicist of modern times, Richard Feynman lost a five-year battle with a rare abdominal cancer at the age of 69 on February 15, 1988, refusing any additional treatments so that he might die with dignity and on his own terms.

See also Einstein, Albert; electromagnetic waves; electromagnetism; fluid mechanics; Gell-Mann, Murray; general relativity; matter and antimatter; Oppenheimer, J. Robert; quantum mechanics; theory of everything.

FURTHER READING

Feynman, Richard P., Paul Davies, Robert B. Leighton, and Matthew Sands. *Six Easy Pieces: Essentials of Physics Explained by Its Most Brilliant Teacher*. Reading, Mass.: Addison-Wesley, 1995.

Feynman, Richard P., and Ralph Leighton. *Surely You're Joking, Mr. Feynman! (Adventures of a Curious Character)*. New York: W. W. Norton, 1984.

Gleick, James. *Genius: The Life and Science of Richard Feynman*. New York: Pantheon Books, 1992.

The Nobel Foundation. "The Nobel Prize in Physics 1965." Available online. URL: http://nobelprize.org/nobel_prizes/physics/laureates/1965/index.html. Accessed July 25, 2008.

fission Nuclear energy is the energy found within the attraction between subatomic particles in an atomic nucleus, the protons and neutrons. When the energy of these interactions is released by splitting the nucleus of the atom such that smaller atoms are produced, the process is known as nuclear fission. In order for fission to occur, the nucleus of the atom being broken apart (the fissionable material) must be unstable. The fissionable material is bombarded with particles in order to break it apart. The process can be visualized by thinking of billiard balls on a pool table. When the balls are racked, they are packed closely together. When they are struck by the cue ball, the collisions among them cause them to bounce into each other and be broken off from the group. When nuclear fission occurs in a controlled environment, the energy from the collisions can be harnessed and used to create electricity, propel submarines and ships, and be employed in research purposes. An atomic bomb is the result of uncontrolled nuclear fission.

The particle that bombards the fissionable material and causes it to break apart is a neutron. Since it is a neutral particle, repulsive forces that a positively charged particle would have with the nucleus do not present a problem. The most common fissionable materials for fission reactions are uranium 235 and plutonium 239. Uranium 235 reacts with an incoming neutron in several ways, one of which is demonstrated by the following process:

$$\text{uranium 235} + \text{1 neutron} \rightarrow$$
$$\text{2 neutrons} + \text{krypton 92} + \text{barium 142} + \text{energy}$$

The number of neutrons produced in this reaction is significant because there are more neutrons produced

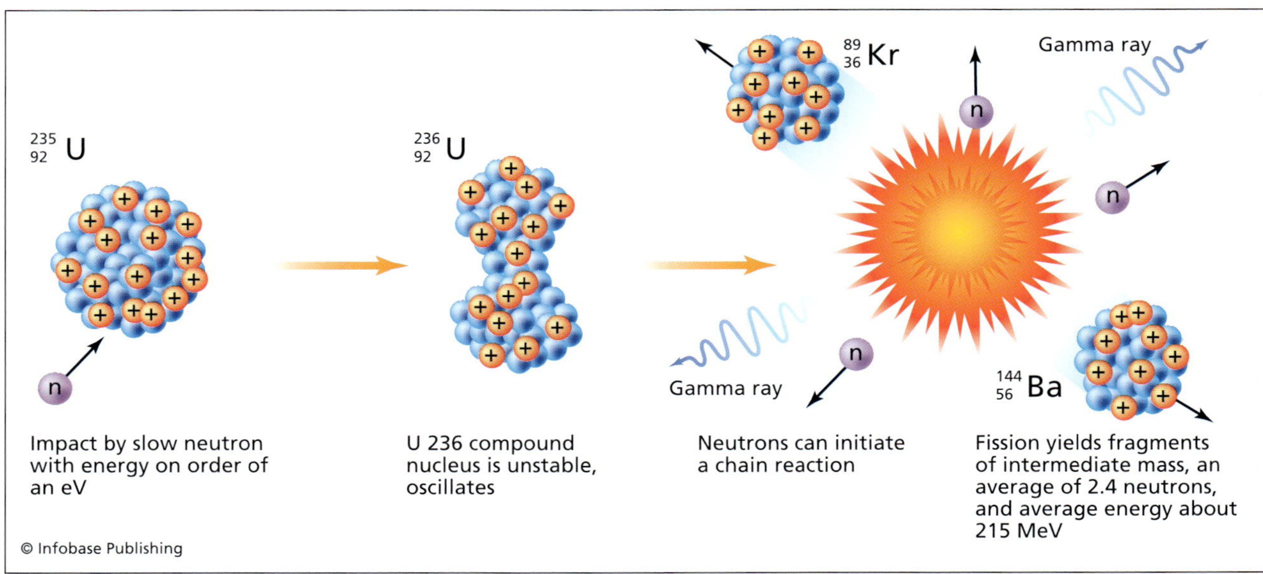

When uranium 235 is bombarded with a neutron, energy is released along with two to three neutrons that can induce additional reactions.

A test of a nuclear fission bomb, popularly known as an atomic bomb, took place in Bikini lagoon July 1, 1946, creating a characteristic mushroom cloud. The bomb was dropped from Super Fortress "Dave's Dream" and a remote controlled camera on Bikini Island took this photograph. *(AP Images)*

than used up in the reaction. On average two to three neutrons are released from a fission reaction cycle. The fission reaction is known as a chain reaction, meaning that once started, the reaction will continue on its own until all of the uranium 235 atoms are used up. Each newly released neutron can fuel subsequent reactions. Because the fission reaction released two neutrons, though only one bombarded the nucleus originally, the total number of neutrons released during each cycle, or generation, of reactions will increase exponentially.

Chain reaction:

1st generation: 1 neutron →
  2 neutrons (can fuel 2 new reactions)
2nd generation: 2 neutrons →
  4 neutrons (can fuel 4 new reactions)
3rd generation: 4 neutrons →
  8 neutrons (can fuel 8 new reactions)

The theoretical number of neutrons created can be approximated by raising 2 to the number of reaction cycles. A fission reaction occurs very rapidly: one reaction cycle can take as little as one millisecond. If 10 reaction cycles occur, the number of neutrons produced could reach 2^{10}, or 1,024, neutrons to power 1,024 new reactions. If 100 reaction cycles occur, 1.25×10^{30} neutrons could be produced with each of the exothermic reactions in every cycle, contributing to the overall energy being produced. This demonstrates the power of the exponential chain reaction. Uncontrolled, this reaction powers atomic bombs like those dropped at Hiroshima and Nagasaki during World War II.

The development of the nuclear reactor allowed for controlled and contained nuclear reactions. The nuclear reactor was patented in 1955 by the Italian American physicist Enrico Fermi and American physicist Leo Szilard for work they did in 1942. They were able to create a controlled nuclear chain reaction in the Chicago Pile reactor. Their work later became part of the Manhattan Project, a coordinated project with the goal of developing nuclear weapons during World War II. Nuclear reactors can produce plutonium for nuclear weapons, energy to run ships and submarines, as well as electrical power. Prior to 1946 and the Atomic Energy Act, nuclear energy was used primarily for weapons. This act allowed for the development of nuclear technology for peacetime purposes.

Nuclear power plants harness the energy from the exothermic fission reaction of uranium 235 and use it to heat water that turns a turbine in a generator that creates electricity. The Experimental Breeder Reactor (EBR-1) was the first reactor to produce electrical power, in Idaho, in 1951. The first true nuclear power plant was built in Obininsk, USSR, in 1954. The first commercial grade nuclear power plant, the Shippingport Reactor, was opened in 1957 in Shippingport, Pennsylvania. The Nuclear Regulatory Commission (NRC) regulates and inspects commercial nuclear reactors. The core of a nuclear power plant contains the fissionable material (uranium 235 in most reactors). Uranium is formed into pellets, and its arrangement is based on the following diagram:

Uranium pellets
↓
many pellets make fuel tubes
↓
many fuel tubes make fuel assemblies
↓
many assemblies make up the reactor core

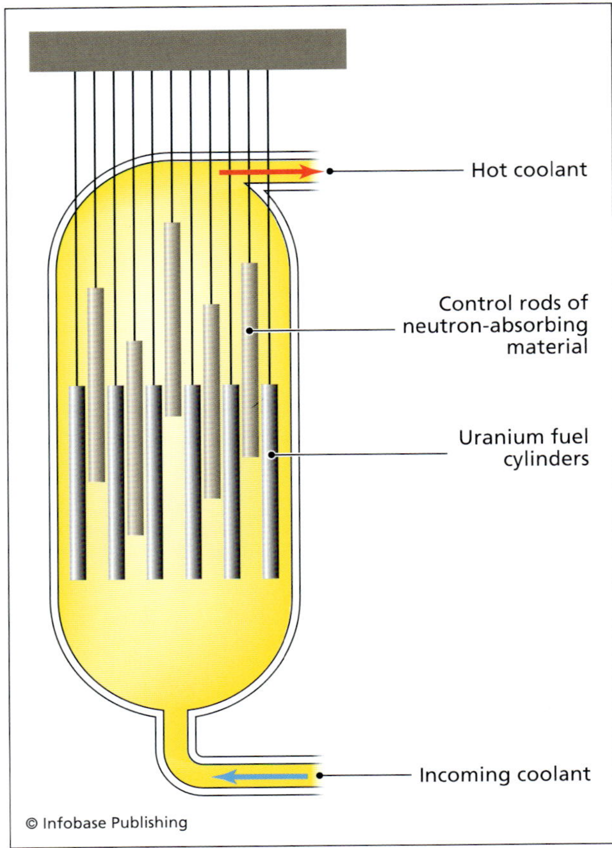

The uranium fuel tubes in the core of the reactor are separated from each other by neutron-absorbing control rods.

In order to control the rate of the reaction, rods made out of a substance that absorbs neutrons are placed between the pellets of uranium 235. Many neutrons catalyze many fission reactions, with the result that much heat is produced. The speed of the incoming neutron is another important factor in controlling the fission reaction. If the neutron is moving too quickly, it will not cause fission of uranium 235. The neutron must be slowed (moderated), and most reactors use water for this purpose. These types of reactors are known as light water reactors. The water serves two purposes: it slows the neutrons so that the number of fission reactions increases, and it serves as a cooling mechanism for the reactor itself. A containment building surrounds the entire reactor core to minimize the risk of radioactivity's escaping the reactor.

There are two types of nuclear power plants in the United States: pressurized water reactors (PWRs) and boiling water reactors (BWRs). In the United States, 104 operating nuclear power plants presently produce approximately 20 percent of the nation's total electric energy supply. Of these plants, 69 are PWRs and 35 are BWRs. In PWRs the power plant is made up of two loops. The primary loop is where water under high pressure is heated by the fission reactions at the core of the plant. The heated water carries its energy to the steam generator, which then heats the water in a secondary loop that turns a turbine to create electricity. The steam in the secondary loop is then condensed back into water to be reused. The water in the primary loop of the pressurized water reactor never mixes with the water from the secondary loop; therefore the possibility of radioactivity's being released is reduced. In a boiling water reactor, the same water that is heated by the reactor core is used directly to turn the turbine that creates electricity in a generator. This means that the water that was in contact with the core of the reactor does leave the containment building, increasing the possibility of radiation's escaping.

The major advantage of nuclear power plants is that they can create a large amount of energy with a relatively small amount of input reactants. The chain reaction nature of the process allows for much energy to be produced. With the future of fossil fuel supplies uncertain, it is important to maximize the energy production of alternative energy sources, including nuclear fission reactors. Nuclear power plants possess many disadvantages as well. The potential damage created by nuclear power plants' melting down could include large-scale contamination of the surrounding environment. In 1979 an accident at a nuclear power plant at the Three Mile Island reactor near Middletown, Pennsylvania, created much public scrutiny of nuclear power plants. A malfunctioning pressure release valve allowed cooling water to be released into the environment. The cooling water was not contaminated since this was a pressurized

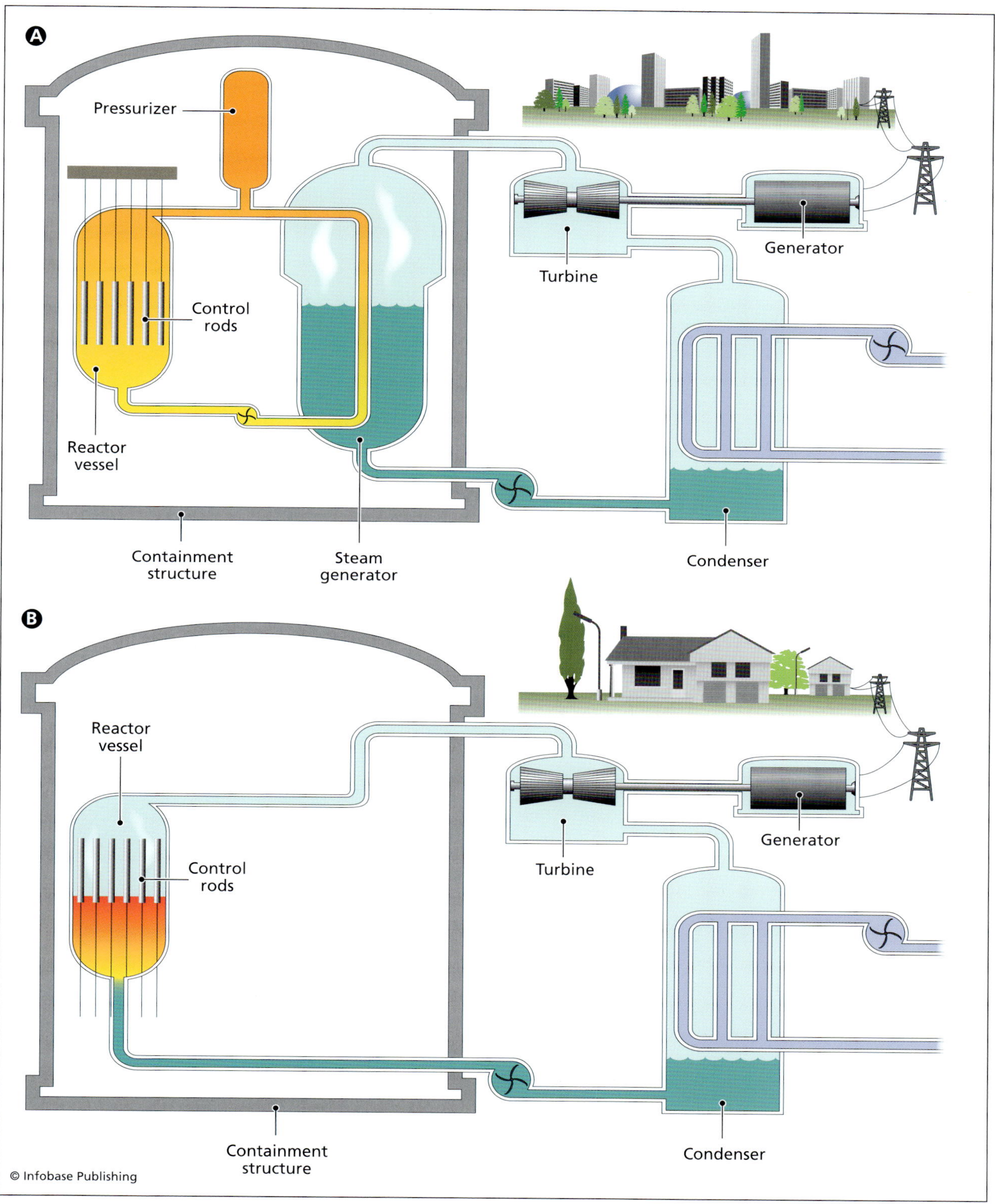

The two basic designs of nuclear reactors—(a) pressurized water reactor and (b) boiling water reactor—use fission reactions to heat water to steam that turns a turbine to generate electricity.

water reactor with a two-loop system. The loss of the cooling water was not immediately apparent to the workers of the plant, so the plant continued operating without proper coolant. This caused a meltdown of nearly half of the reactor core, leading to some environmental leakage of radiation. Assessments conducted after the fact by the NRC and the Environmental Protection Agency (EPA) determined that the amount of radiation exposure to individuals outside the plant was very low, less than that of a chest X-ray. Despite these findings, the meltdown of the reactor caused much public concern over the safety of the reactors in general and led the NRC to modify many safety procedures to prevent these types of accidents in the future.

The situation was very different with the Chernobyl nuclear power plant disaster near Kiev, USSR, in 1986. The design of the reactor (RBFK) type was flawed (including the absence of a containment building) and had been shown to be flawed prior to this accident. The safety procedures that should have been used were disregarded. A reactor test was performed with well under the minimal number of control rods to prevent the reaction from occurring too quickly, and several safety features of the reactor were intentionally disabled prior to the disaster. On the morning of April 26, 1986, the entire reactor core melted down and created an explosion that tore the roof of the reactor. Environmental consequences of the Chernobyl disaster are still not fully understood. Areas within a six-mile (10-km) radius were destroyed, and the radiation that was released could have reached the entire Northern Hemisphere. This accident caused 30 immediate deaths and innumerable radiation-related deaths, illnesses, and diseases. Unlike Three Mile Island, where much was learned and positive measures that improved safety and operating procedures were taken after the accident, Chernobyl did not offer much more information than what was already known at the time. The safety rules were known, but they were just not implemented at this plant.

Nuclear waste, including the spent fuel rods from nuclear reactor cores, pose a very serious and as of yet unanswered environmental question. Immediately after use, the spent fuel rods can be stored in a cooling tank of water that allows them to continue cooling and to decay radioactively. The waste then needs to be stored until it decays to the point that it is safe. Some believe aboveground storage is sufficient. Protection for the surrounding community would need to be carefully monitored so that no one would be exposed to the damaging effects of the radiation. Some proposals exist for burying the nuclear waste under the surface of the earth. The plans for nuclear waste storage at Yucca Mountain, in Nevada, have come under much criticism for the unknown effects on the environment. One point is certain: future generations will still be dealing with the nuclear waste produced in nuclear reactors 50 years ago.

See also ALTERNATIVE ENERGY SOURCES; ATOMIC STRUCTURE; ENERGY AND WORK; FERMI, ENRICO; FUSION; RADIOACTIVITY.

FURTHER READING

Dahl, Per F., ed. *From Nuclear Transmutation to Nuclear Fission.* Philadelphia: Institute of Physics, 2002.

fluid mechanics A fluid is matter in any state that is not solid (i.e., either liquid or gas). In other words, a fluid is matter that can flow. What differentiates a fluid from a solid is that whereas in the latter the constituent particles (atoms, molecules, or ions) are constrained to fixed positions relative to each other, in a fluid the particles are sufficiently free that flow can occur. In a liquid the particles move rather freely about, maintaining only a constant average distance from each other. A sample of liquid maintains a fixed volume, while flowing to take the shape of its container. The particles of a gas are even less constrained, and a gas flows so as to fill the volume of its container completely. The mechanics of fluids is separated into hydrostatics, which is the study of fluids at rest, and hydrodynamics, dealing with fluids in motion.

HYDROSTATICS

Hydrostatics is the field of physics that studies the behavior of fluids at rest. Among the main principles of hydrostatics are direction of force, hydrostatic pressure, transmission of pressure, and buoyancy.

Direction of Force

The force exerted upon a solid surface by a fluid at rest is perpendicular to the surface. For example, the water in an aquarium presses outward on the aquarium's walls, exactly perpendicularly to the glass at every point.

Hydrostatic Pressure

The pressure p at depth h beneath the surface of a liquid at rest is given by

$$p = p_0 + \rho g h$$

where p_0 is the pressure on the surface of the liquid (often atmospheric pressure), ρ is the density of the liquid, and g is the acceleration due to gravity. Here the pressures are in pascals (Pa), equivalent to newtons per square meter (N/m^2); density is in kilograms per cubic meter (kg/m^3); g is in meters per second per second (m/s^2) with nominal value of 9.8 m/s^2; and h is in meters (m). This relation is felt very noticeably by any diver into deep water.

Transmission of Pressure

Pascal's principle, named for the 17th-century French physicist and mathematician Blaise Pascal, states that an external pressure applied to a fluid that is confined in a closed container is transmitted at the same value throughout the entire volume of the fluid. This principle underlies the operation of hydraulic car lifts, for example, in which a pump raises the pressure in the operating fluid (a liquid, in this case), which is transmitted to the piston, which pushes the car upward.

Buoyancy

Named for the ancient Greek scientist, Archimedes' principle states that a body immersed in a fluid, whether wholly or partially immersed, is buoyed up by a force whose magnitude equals that of the weight of the fluid that the body displaces. This force is called the buoyant force. If volume V of the body (which might be the whole volume of the body or only part of it) is submerged, then V is also the volume of displaced fluid. The mass, in kilograms (kg), of that volume of displaced fluid is $V\rho$, where ρ denotes the density of the fluid in kilograms per cubic meter (kg/m^3) and V is in cubic meters (m^3). The magnitude of the weight of the displaced fluid, in newtons (N), and thus the magnitude of the buoyant force, is $V\rho g$.

If the magnitude of the buoyant force acting on a totally submerged object is less than the magnitude of the object's weight, the object will sink. This occurs when the density of the object is greater than that of the fluid. A stone dropped into a lake behaves in this way. The density of the stone is greater than the density of the lake water. When the object's density is less than the fluid's, the object will rise if totally submerged. If the fluid is a liquid, the object will then float partially submerged at the liquid's surface, in such a manner that the submerged part of the object displaces a quantity of liquid whose weight equals the body's weight. This is the behavior of a piece of wood immersed in a lake. When released from under water, the wood will rise and eventually float with only part of its volume submerged. If the density of the object exactly equals the density of the fluid, the object will hover at any height in the fluid, neither sinking nor rising.

HYDRODYNAMICS

Hydrodynamics is the field of physics that studies the flow of fluids. Fluid flow is generally either laminar or turbulent. Laminar flow, also called streamline flow, or steady flow, is a smooth flow, in which the path of any particle of the fluid is a smooth line, a streamline. Turbulent flow, on the other hand, is chaotic and rapidly changing, with shifting eddies and tightly curled lines of flow. Observe the smoke rising from a smoking ember, for example. Close to the ember it ascends in smooth, laminar flow; as it rises higher, it breaks up into eddies and continues ascending in turbulent flow.

The flow of fluids is, in general and even in laminar flow, complex. Viscosity plays its role, acting in the manner of friction by removing kinetic energy from the system and converting it to heat.

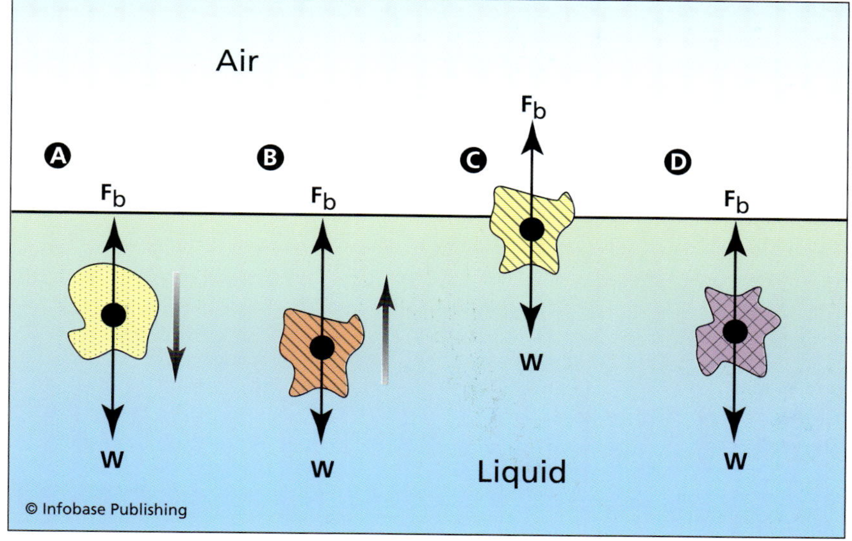

Buoyancy is the effect of a lifting force—the buoyant force—acting on an object that is wholly or partially submerged in a fluid. The magnitude of the buoyant force equals that of the weight of the fluid that the object displaces, by Archimedes' principle. The figure shows an object in a liquid. **W** denotes the object's weight, and $\mathbf{F_b}$ represents the buoyant force acting on the object. (a) The density of the object is greater than the density of the liquid, so the magnitude of the object's weight W is greater than the magnitude of the buoyant force F_b, $W > F_b$, and the object sinks. (b) The density of the object is less than the density of the liquid. Then when the object is fully submerged, $F_b > W$ and the object rises. (c) It reaches equilibrium when it is floating on the surface and partially submerged to the extent that $F_b = W$. (d) If the object's density equals that of the liquid, the object will be in equilibrium when wholly submerged, since then $F_b = W$.

The density of the fluid can change along the flow. Moreover, the particles of the fluid might attain rotational kinetic energy about their centers of mass. And the flow, even if laminar and steady, might vary in time as the causes of the flow vary. The following discussion makes a number of simplifying assumptions, which nevertheless allow the treatment and understanding of various useful phenomena. First, assume one is dealing with flow that is laminar and stationary: the flow is smooth and steady and, moreover, at each point in the fluid, the velocity of the flow does not change in time. Next, assume that the fluid, is incompressible, that its density is the same at all points and at all times. Further, assume that the effect of viscosity is negligible, that the flow is nonviscous (i.e., no internal friction). Finally, assume irrotational flow, that the fluid particles do not rotate around their centers of mass (although the fluid as a whole might flow in a circular path).

With those assumptions, consider any flow tube, an imaginary tubelike surface whose walls are parallel to streamlines. The streamlines that are inside a flow tube stay inside it and those that are outside it do not penetrate and enter it. It is as if the fluid is flowing through a tube, except there is no actual tube. Let A denote the cross-sectional area, in square meters (m^2), at any position along a flow tube and ν the flow speed there, in meters per second (m/s). The product $A\nu$ is the volume flux, or volume flow rate: the volume of fluid flowing through the cross section per unit time. Its SI unit is cubic meters per second (m^3/s). Under the preceding assumptions, the volume flow rate is constant along a flow tube: the volume of fluid flowing past any location along a flow tube during some time interval is the same as that passing any other location along the same flow tube during the same time interval. This is expressed by the continuity equation,

$$A_1\nu_1 = A_2\nu_2$$

where the subscripts indicate any two locations along the flow tube. The continuity equation tells one, for instance, that where the flow narrows (and the streamlines bunch together), the flow speed increases, and where the flow broadens, it slows. One use of this effect is to increase the range of the water stream from a garden hose by narrowing the end of the hose, either with the thumb or by means of a nozzle.

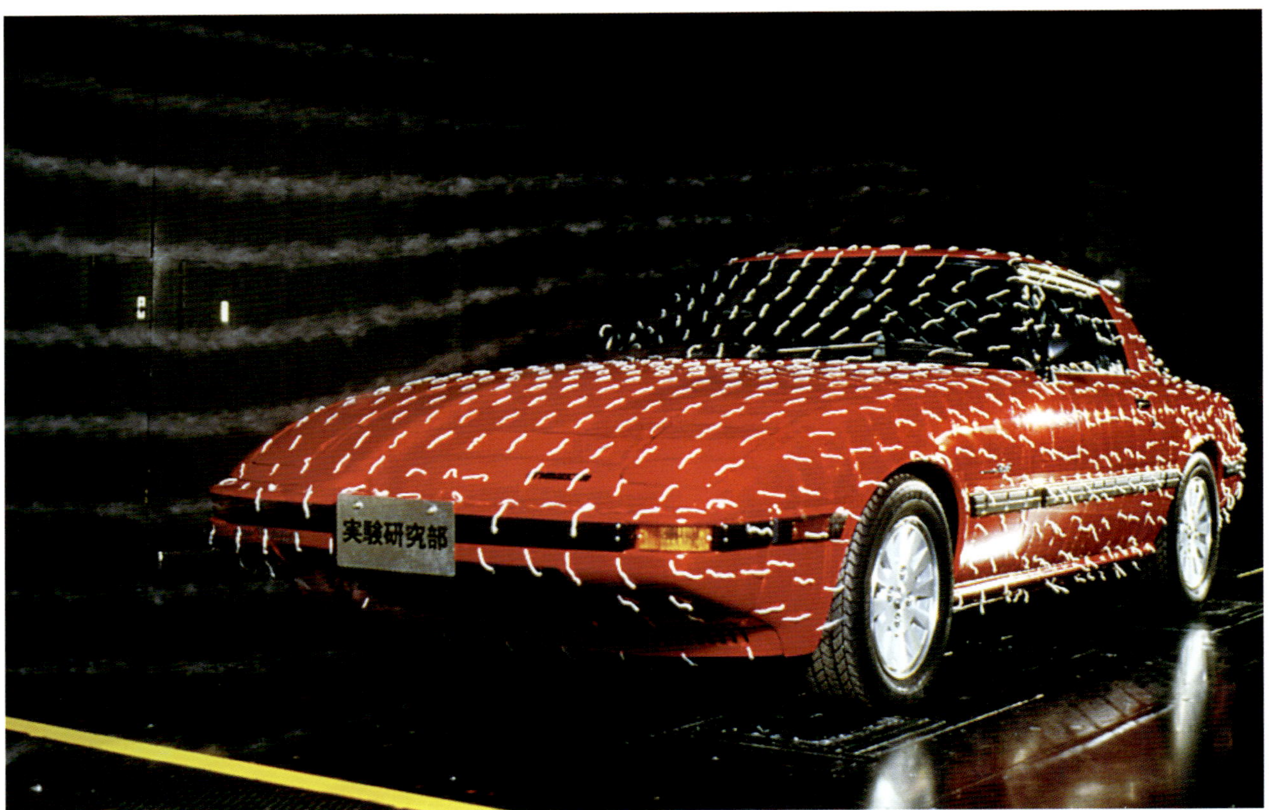

Engineers study the aerodynamic properties of this car with a wind tunnel, in which the flow of air around the car is visualized by means of smoke streams. *(Takeshi Takahara/Photo Researchers, Inc.)*

Consider a small volume of fluid that flows along a streamline within a flow tube, changing in shape and in speed as it flows, in accord with the continuity equation. The work-energy theorem states that the work done on such a volume by the pressure of the fluid, as the volume flows from one location along the flow tube to another, equals the sum of its gains in kinetic energy and gravitational potential energy. That results in Bernoulli's equation, named for the 18th-century Swiss mathematician and physicist Daniel Bernoulli. One form of this equation is that as the small volume of fluid flows along, the following quantity remains constant:

$$p + ½ \rho v^2 + \rho g h$$

In this expression p denotes the pressure of the fluid, in pascals (Pa), equivalent to newtons per square meter (N/m²); ρ is the density of the fluid, in kilograms per cubic meter (kg/m³); g is the acceleration due to gravity, whose nominal value is 9.8 meters per second per second (m/s²); and h denotes the height of the fluid volume, in meters (m), from some reference level. This is expressed as

$$p + ½ \rho v^2 + \rho g h = \text{constant}$$

One can equivalently cast Bernoulli's equation in the following form:

$$p_1 + ½ \rho v^2{}_1 + \rho g h_1 = p_2 + ½ \rho v^2{}_2 + \rho g h_2$$

where the subscripts 1 and 2 refer to the values of the quantities at any two locations along the flow.

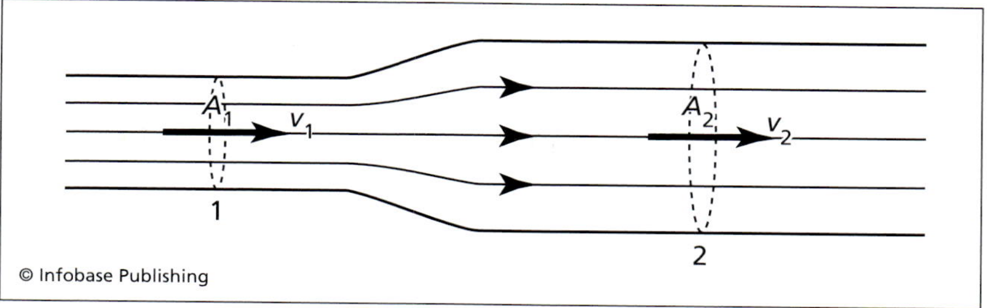

The continuity equation for an incompressible fluid is an expression of the fact that in a confined flow, or in an equivalent situation, the volume of fluid entering the flow during any time equals the volume exiting it during that time. The equation is $A_1v_1 = A_2v_2$, where A_1 and A_2 denote the cross-sectional areas of the flow at positions 1 and 2, respectively, while v_1 and v_2, represent the respective flow speeds. It follows that the fluid flows faster in narrower regions than in wider ones.

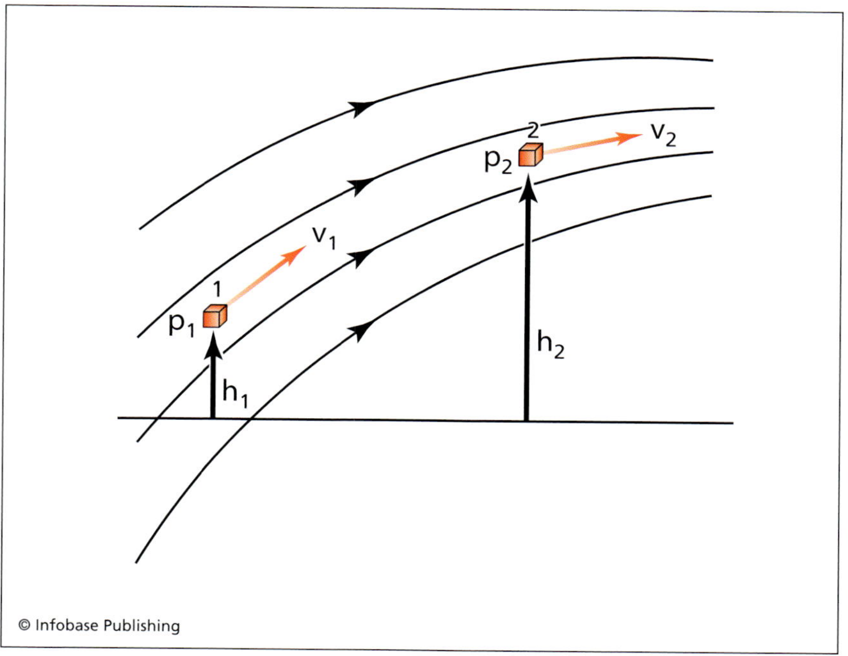

Bernoulli's equation: for a small volume of an incompressible, nonviscous fluid in smooth, steady, irrotational flow, the quantity $p + {}^1/_2\rho v^2 + \rho gh$ remains constant along the flow. The variable p denotes the pressure of the fluid, ρ the fluid density, v the flow speed, g the acceleration due to gravity, and h the height from some reference level. The figure shows stream lines describing the flow of a fluid and a small volume of fluid that flows from location 1 to location 2. For this volume, $p_1 + {}^1/_2\rho v_1{}^2 + \rho gh_1 = p_2 + {}^1/_2\rho v_2{}^2 + \rho gh_2$.

An example of application of Bernoulli's equation is this. Let a fluid flow through a horizontal pipe that is narrower in some region than in the rest of the pipe. The earlier example of use of the continuity equation shows that the fluid flows faster in the narrower region than in the wider sections. Putting $h_1 = h_2$ and $v_1 > v_2$ in Bernoulli's equation, one obtains $p_1 < p_2$. In other words, the pressure of the fluid in the narrower region, where it is flowing faster, is less than in the wider region, where its speed is lower. This is referred to as the Venturi effect, named for the 18–19th-century Italian physicist Giovanni Battista Venturi.

LIFT, DRAG, AND TURBULENCE

As an additional example, the air flow around a horizontal airplane wing can be treated with the continuity equation and Bernoulli's equation similarly to the derivation of the Venturi effect. The result is that the air pressure is greater beneath the wing than above it. This is one source of the upward force, called lift, that keeps airplanes in the air.

In general, lift, also called aerodynamic lift or hydrodynamic lift, forms a component of the force acting on a body due to the flow of a fluid past the body. It is the component in the direction perpendicular to the direction of flow if the body were absent. It is just such lift due to moving air that keeps kites aloft, for instance. It is the relative motion that matters; the body might in fact be moving through a stationary fluid, as in the case of an airplane. One source of lift is a pressure difference above and below the body resulting from faster flow above than below, according to Bernoulli's equation, as mentioned earlier. Another source is fluid deflection by the body, when the body causes the flow to change direction. In such a situation, the force on the body is the reaction force, according to Newton's second law of motion, to the force the body exerts on the fluid to change its direction of flow. The analysis of lift forces is complicated, especially because of turbulence and viscosity.

Lift forces caused by moving air keep aloft heavier-than-air aircraft, as well as birds and flying insects, for example. Submarines are designed to exploit a combination of lift and buoyancy forces to control their depth in water.

The retarding effect of forces of turbulence and viscosity acting on a body moving through a fluid is called drag. Such forces are called drag forces. Those forces are dissipative, performing negative work on the body, and thus reducing its mechanical energy (the sum of its kinetic and potential energies) by converting it irreversibly to heat.

As an example, for a blunt body moving sufficiently rapidly through air, the magnitude of the drag force is given by

$$\text{Drag force} = \tfrac{1}{2} D\rho A\nu^2$$

where the drag force is in newtons (N); ρ denotes the density of the air, in kilograms per cubic meter (kg/m^3); A is the effective cross-sectional area of the body, in square meters (m^2); and ν is the body's speed, in meters per second (m/s). The symbol D denotes the drag coefficient, which is dimensionless and so has no unit.

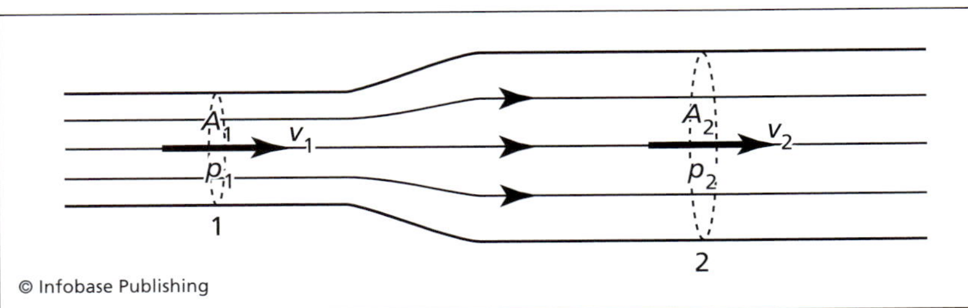

The Venturi effect in fluid flow, which follows from Bernoulli's equation, is the reduction of pressure in regions of higher flow speed. These are regions of smaller cross-sectional area, according to the continuity equation. A_1, v_1, and p_1 denote the cross-sectional area, flow speed, and pressure, respectively, at position 1 along the flow, while A_2, v_2, and p_2 denote the corresponding quantities at position 2. Since in this example $A_1 < A_2$, then $v_1 > v_2$, by the continuity equation, and $p_1 < p_2$, by Bernoulli's equation.

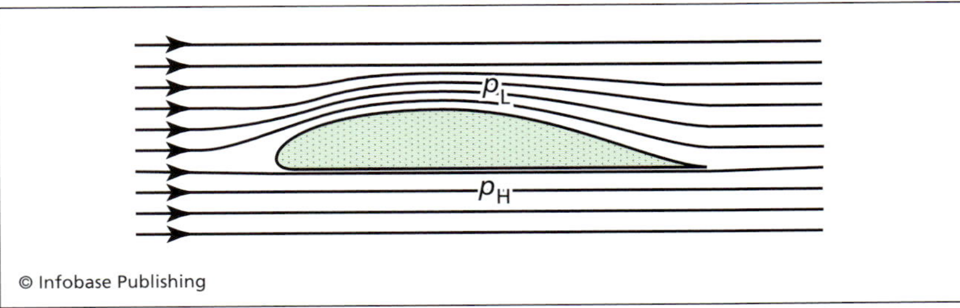

When air flows over and under a horizontal aircraft wing (or when the wing moves through air), the continuity equation and Bernoulli's equation show that the pressure beneath the wing is greater than the pressure above it. That results in a net upward force on the wing, which is a form of hydrodynamic lift. The symbols p_H and p_L denote the higher and lower pressures, respectively.

Named for its discoverer, the 19–20th-century Irish-British physicist and mathematician George Stokes, Stokes's law gives the magnitude of the viscous force that acts on a solid sphere moving at sufficiently low speed through a fluid. If we denote the magnitude of the force by F_v, Stokes's law gives

$$F_v = 6\pi r\eta\nu$$

where F_v is in newtons (N); r denotes the radius of the sphere, in meters (m); η is the viscosity of the fluid, in newton-second per square meter (N·s/m^2); and ν represents the speed of the sphere in the fluid (or the speed of the fluid flowing past the sphere), in meters per second (m/s).

If a sphere whose density is greater than that of the fluid is released from rest in a fluid, it falls and accelerates until it reaches its terminal speed, the speed at which the viscous force balances the net downward force. The sphere then continues to fall at the terminal speed. The magnitude of the net downward force acting on the sphere equals the magnitude of the force of gravity (the weight) less the magnitude of the buoyant force. For a homogeneous sphere, this quantity is

$$F_{\mathrm{d}} = \frac{4\pi r^3}{3}(\rho - \rho_{\mathrm{f}})g$$

where $4\pi r^3/3$ is the volume of the sphere, and ρ and ρ_{f} denote the densities of the sphere and the fluid, respectively, in kilograms per cubic meter (kg/m^3). The net downward force is independent of the sphere's speed. At terminal speed, $F_{\mathrm{v}} = F_{\mathrm{d}}$. Equating the two expressions, one obtains the terminal speed, v_{t}:

$$v_{\mathrm{t}} = \frac{2r^2(\rho - \rho_{\mathrm{f}})g}{9\eta}$$

Alternatively, one can find the fluid's viscosity η by measuring the terminal speed:

$$\eta = \frac{2r^2(\rho - \rho_{\mathrm{f}})g}{9v_{\mathrm{t}}}$$

Another law that is relevant to this discussion is Poiseuille's law, named for the 19th-century French physician Jean Poiseuille. This law deals with the steady flow of a viscous fluid through a cylindrical pipe. Denote the inside radius of the pipe by r and the pipe's length by L. Both are in meters (m). Let the difference in fluid pressures at the ends of pipe be Δp, in pascals (Pa), and denote the viscosity of the fluid, in newton-second per square meter (N·s/m^2), by η. Then Poiseuille's law states that the volume flow rate, or volume flux, of the fluid through the pipe (i.e., the volume of fluid passing through the pipe per unit time), in cubic meters per second (m^3/s), is given by

$$\text{Volume flow rate} = \frac{\pi r^4}{8\eta}\frac{\Delta p}{L}$$

Note that a greater pressure difference or a greater tube radius brings about a greater volume flow rate, while increasing the tube's length or the viscosity of the fluid causes a decrease in volume flow rate.

The situation in fluid flow where the flow velocity and fluid pressure at every point vary wildly in time is called turbulence. Compare this with laminar flow, which is a steady flow in which the velocity and pressure remain constant in time, or vary only slowly, throughout the fluid. The Reynolds number, named for the 19–20th-century British engineer Osborne Reynolds, is a parameter that characterizes the flow of incompressible, viscous fluids. It is a dimensionless quantity defined as

$$\text{Reynolds number} = \frac{\rho v l}{\eta}$$

where ρ denotes the fluid's density, in kilograms per cubic meter (kg/m^3); v is the flow speed, in meters per second (m/s); l is a length, in meters (m), that is characteristic of the geometry of the system; and η represents the fluid's viscosity, in newton-second per square meter (N·s/m^2). When the value of the Reynolds number is less than about 2,000, the flow is smooth. For values of Reynolds number greater than

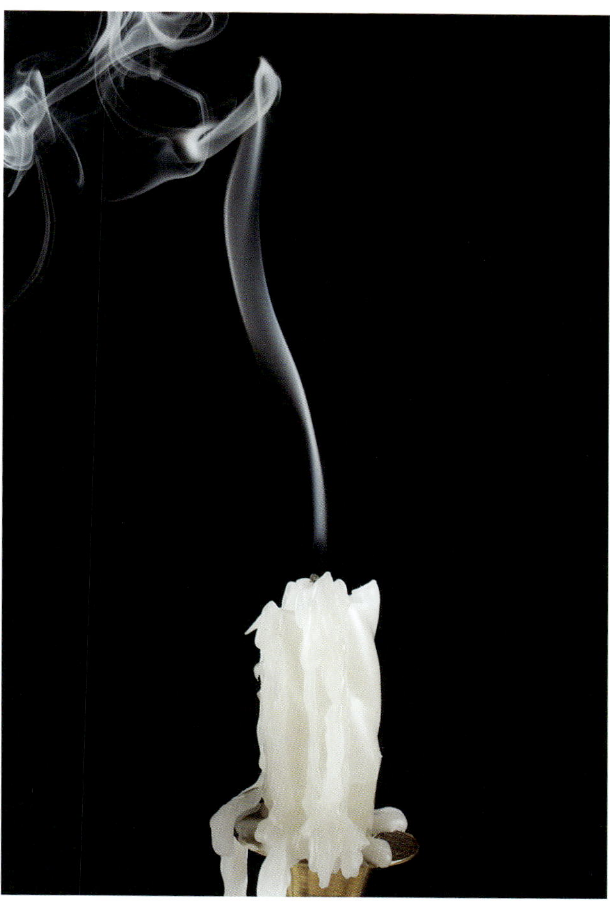

Turbulence in air flow. Smoke carried by the upward flow of warm air from the wick of a freshly extinguished candle shows initial laminar flow (center) becoming turbulent (upper left). *(Adam Gryko, 2008, used under license from Shutterstock, Inc.)*

around 3,000, the flow is turbulent, with velocity and pressure varying wildly in time.

Note that for a given incompressible fluid, the density ρ and viscosity η are fixed. The geometry of the situation determines l. Then the character of flow is controlled solely by the flow speed, v. Accordingly, for sufficiently low flow speeds, the flow will be laminar, while turbulent flow develops at high speeds.

SURFACE TENSION

One needs to perform positive work in order to increase the area of the free surface of a liquid (i.e., the surface separating the liquid from vacuum or from gas). This property of liquids is known as surface tension. Accordingly, a liquid minimizes its potential energy by minimizing its free surface, all other effects being equal. Surface tension comes about in this way. Every molecule of a liquid is attracted by its nearby neighbors, an effect called cohesion. For a molecule inside the volume of the liquid, the attractive forces cancel out. A molecule at a free surface suffers a net force pulling it into the liquid. Increasing the area of a free surface draws more molecules to the surface and requires overcoming the force that pulls those molecules back into the liquid. In this way one performs work when increasing the surface area of a liquid. The surface of a liquid acts as a membrane under tension. But the force required to stretch the "membrane" is independent of the "membrane's" extension, whereas for a real elastic membrane, the force increases with extension. The surface tension of a liquid is quantified and defined as the magnitude of the force required to extend the surface, per unit length—perpendicular to the force—along which the force acts. It is determined by stretching the surface and measuring the required force and the length of the edge of the stretching device, then dividing the former by the latter. Its SI unit is newton per meter (N/m). Surface tension is temperature dependent, tending to decrease with rising temperature.

A water strider is able to walk on water because of water's surface tension. Note the depressions in the water's surface under the insect's six feet. *(© Dennis Drenner/Visuals Unlimited)*

One effect of surface tension is that objects that are not very heavy, but have a density greater than that of water, can be made to float on the surface of water, whereas ordinarily they sink in water. One might try that with small pieces of aluminum foil. While floating, the object slightly depresses and stretches the water's surface at its points of contact. Some insects exploit this phenomenon to walk on the surface of water. Another effect is the tendency of drops of a liquid to take a spherical shape. For a given volume, the shape of a sphere gives it the smallest surface area.

CAPILLARY FLOW

Capillary flow is the spontaneous flow of liquids in very narrow tubes, called capillaries. It is caused by an imbalance between the attraction of the liquid's molecules at the liquid-tube-air boundary to the tube (adhesion) and their attraction to the liquid (surface tension, or cohesion). If the former dominates the latter, the liquid will tend to be drawn into the tube, such as water in a glass capillary. If the latter is greater, however, the liquid will tend to be ejected from the tube. This is exemplified by liquid mercury in a glass capillary.

In the situation of a vertical capillary whose end is dipped into a liquid, capillary flow can cause the level of the liquid in the tube to be different from its level outside, either higher or lower, according to whether the adhesion is greater or lesser than the surface tension, respectively, as described earlier. The surface of the liquid in the tube, called the meniscus, will not be flat in those cases; it will be concave or convex, respectively, as viewed from its top. For a round capillary tube of radius r, in meters (m), the height h, in meters (m), that a liquid of density ρ, in kilograms per cubic meter (kg/m^3), and surface tension γ, in newtons per meter (N/m), rises is given by

$$h = \frac{2\gamma \cos \theta}{\rho g r}$$

where θ is the angle between the surface of the meniscus at its edge and the surface of the capillary. This angle is measured below the meniscus. For a concave meniscus (such as forms on water in glass),

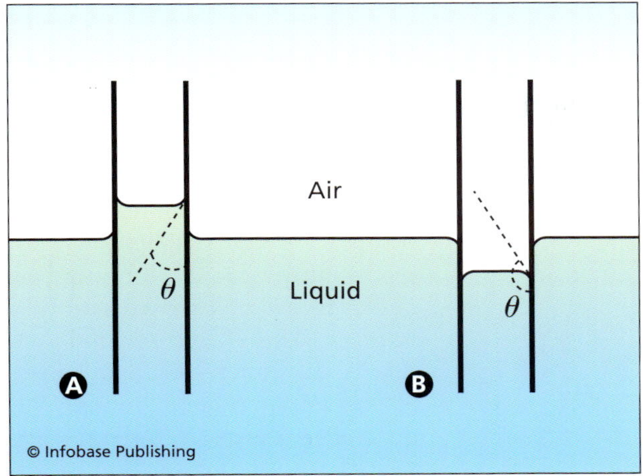

Two cases of capillary flow in a vertical tube as viewed in vertical cross section. (a) The contact angle θ between the edge of the meniscus and the capillary wall is less than 90°, the meniscus is concave as viewed from above, and the liquid rises in the capillary. Water and glass behave in this way. (b) When the contact angle is greater than 90°, the meniscus is convex, and the liquid in the capillary descends. An example of this is mercury and glass. Capillary tubes are very narrow, and the scale of the figure is exaggerated for the sake of clarity.

$0 \le \theta < 90°$, giving a positive cosine and a positive h, indicating a rise of the liquid in the capillary. For a convex meniscus (in the example of mercury in glass), $90° < \theta \le 180°$, so the cosine and h are both negative, with the liquid lower in the tube than outside. In the intermediate case of a flat meniscus, when $\theta = 90°$ and $h = 0$, adhesion and surface tension balance and there is no capillary flow.

SUPERFLUIDITY

Superfluidity is a low-temperature state of liquid matter that flows with no viscosity. Superfluidity has been discovered in the isotope helium 4 below about 2.2 K, where it is a form of Bose-Einstein condensate. This state has also been detected in helium 3 at temperatures of thousandths of a kelvin. A superfluid can flow through tiny pores with no friction and is extremely mobile over surfaces with which it is in contact. The superfluid state possesses unusual thermal and acoustical properties as well. It flows under the effect of a temperature difference between different locations and transmits temperature waves, as examples.

See also ACCELERATION; BOSE-EINSTEIN STATISTICS; CENTER OF MASS; ELASTICITY; ENERGY AND WORK; FLUX; FORCE; GRAVITY; HEAT AND THERMODYNAMICS; MOTION; PRESSURE; ROTATIONAL MOTION; SPEED AND VELOCITY; STATES OF MATTER.

FURTHER READING

Young, Hugh D., and Roger A. Freedman. *University Physics,* 12th ed. San Francisco: Addison Wesley, 2007.

flux This is a physical quantity that is related to vector fields, such as the magnetic and electric fields or the velocity field of a flowing fluid. The flux of a vector field through a surface is the integral over the surface of the component of the field perpendicular to the surface. In detail: Consider a vector field **F** and a surface. (The surface might be bounded by a curve, in which case it possesses an edge. Alternatively, the surface might be closed, such as the surface of a sphere, and have no edge. Then it will possess an inside and an outside.) Define a direction for the surface by thinking of one side of the surface as "from" and the other side as "to" and, at each point of the surface, imagining an arrow pointing from the "from" side to the "to" side. If the surface is closed, it is common to have these arrows pointing outward. At each point of the surface, consider an infinitesimal element of surface area of magnitude dA and define a corresponding vector $d\mathbf{A}$ as the vector whose magnitude equals dA and that is perpendicular to the surface element in the direction of the "from-to" arrow. At each point of the surface form the scalar product of the field vector **F** and the surface element vector $d\mathbf{A}$, $\mathbf{F} \cdot d\mathbf{A}$ (which equals $F\,dA\cos\theta$, where F is the magnitude of **F** and θ is the smaller angle between **F** and $d\mathbf{A}$). Now, integrate this expression over the whole surface to obtain the flux of the field through the surface in the defined direction, Φ:

$$\Phi = \int_{\text{surface}} \mathbf{F} \cdot d\mathbf{A}$$

To help obtain a sense of what flux is about, imagine that the vector field **F** describes some flow (which is the origin of the term *flux*) by giving the velocity of a flowing liquid at every point. (Of course, in general, fields do not represent flows, but this model might be helpful.) To make matters simple, let the flow be uniform and the surface bounded and flat. In other words, let **F** have the same value at every point, so that everywhere in the liquid its magnitude (the flow speed) F is the same and the direction of flow forms the same angle θ with the "from-to" arrow (perpendicular to the surface). Denote the area of the surface by A. Then the flux becomes simply

$$\Phi = FA\cos\theta$$

This is the volume of liquid that flows through the surface per unit time. Note that the flux is directly proportional to both F and A: double the amount of liquid flowing through an opening by

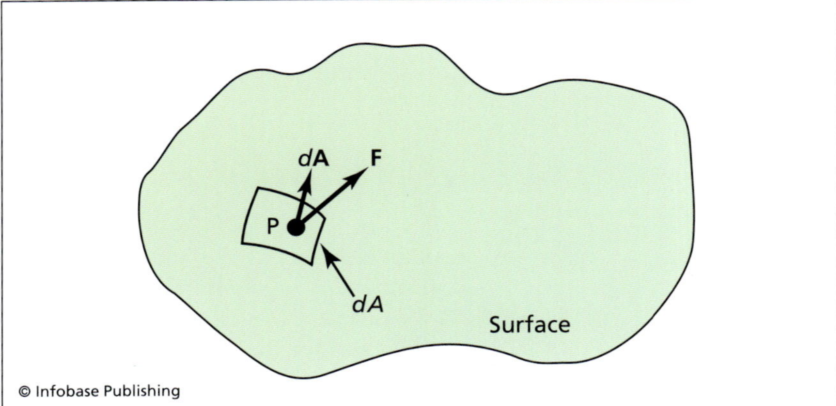

The figure shows a surface through which the flux of a vector field is to be calculated. An infinitesimal surface area element of magnitude dA is shown at an arbitrary point P on the surface. $d\mathbf{A}$ denotes the vector perpendicular to the surface at point **P** with magnitude dA. Its chosen sense, from one side of the surface to the other, is the same for all points on the surface. **F** represents the value of the vector field at P. The flux of the vector field through the surface is the integral of the scalar product $\mathbf{F}\cdot d\mathbf{A}$ over the whole surface.

doubling the flow speed or doubling the area of the opening. Also note the effect of direction: for given flow speed and area, the amount of liquid passing through the opening is maximal for flow that is perpendicular to the surface ($\theta = 0$, $\cos\theta = 1$), decreases as the flow is slanted away from the perpendicular, and vanishes for flow parallel to the surface ($\theta = 90°$, $\cos\theta = 0$). For angles greater than 90° and up to 180°, the flux is increasingly negative, indicating flow in the negative sense, against the "from-to" arrow. For = 180° ($\cos\theta = -1$), the flux is maximally negative.

Flux is an important concept in electricity and magnetism. In particular, Gauss's law in electricity involves the electric flux, the flux of the electric field **E** through a closed surface. This flux is

$$\Phi_e = \int_{\text{surface}} \mathbf{E} \cdot d\mathbf{A}$$

The SI unit of electric flux is newton-meter² per coulomb ($\text{N}\cdot\text{m}^2/\text{C}$), where **E** is in newtons per coulomb (N/C) and $d\mathbf{A}$ is in square meters (m^2).

In magnetism, Faraday's law is related to the rate of change of the magnetic flux, the flux of the magnetic field **B** through a bounded surface. Magnetic flux is

$$\Phi_m = \int_{\text{surface}} \mathbf{B} \cdot d\mathbf{A}$$

The SI unit of magnetic flux is the weber (Wb), equivalent to tesla-meter² ($\text{T}\cdot\text{m}^2$), and the magnetic field is in teslas (T).

Other fluxes are used in physics. One is radiant flux, which is the power transported by light or other electromagnetic radiation. Its SI unit is that of power, the watt (W). Another is luminous flux, the power transported through a surface by light, as corrected and evaluated for the light's ability to cause visual sensation in humans. The SI unit of luminous flux is the lumen (lm), equivalent to candela-steradian (cd·sr). The difference between radiant flux and luminous flux is that while the former involves all radiation energy, the latter weighs the energy according to the strength of its visual effect. For instance, humans are visually insensitive to infrared and ultraviolet radiation. For visible light, people are most sensitive to a wavelength of about 560 nm, which produces the sensation of yellow-green. (Note that the radiant intensity of the Sun is maximal for this wavelength.)

Each of those fluxes has an associated surface density. The radiant flux density, also called irradiance, or radiant power density, on a surface is the radiant flux through or into the surface per unit area of surface. Its SI unit is watt per square meter (W/m^2). Similarly, luminous flux density, also illuminance or illumination, is the luminous flux through or into the surface per unit area of surface. Its SI unit is lumen per square meter (lm/m^2).

See also ELECTRICITY; ELECTROMAGNETIC WAVES; FLUID MECHANICS; MAGNETISM; POWER; SPEED AND VELOCITY; VECTORS AND SCALARS.

FURTHER READING

Young, Hugh D., and Roger A. Freedman. *University Physics*, 12th ed. San Francisco: Addison Wesley, 2007.

force Force is often defined as that which causes a body's acceleration or causes change of a body's momentum. This is based on Newton's second law of motion, which is expressed by

$$\mathbf{F} = m\mathbf{a}$$

or

$$\mathbf{F} = \frac{d\mathbf{p}}{dt}$$

Here **F** denotes force, which is a vector quantity whose SI unit is the newton (N); *m* is the mass, in kilograms (kg), of the body on which the force acts; **a** the acceleration of the body, in meters per second per second (m/s^2); **p** the momentum of the body, in kilogram-meters per second (kg·m/s); and *t* the time, in seconds (s). (Both the acceleration **a** and the momentum **p** are vector quantities.) The equations express the effect of a force acting on a body. The first equation, in the usual formulation of Newton's second law, states that a body's acceleration produced by a force acting on the body is in the direction of the force, and its magnitude is proportional to the magnitude of the force and inversely proportional to the body's mass. The second equation is more general, while the first is valid only for speeds that are sufficiently small compared to the speed of light and thus forms a sufficiently good approximation for everyday situations.

As an example of Newton's second law in its usual form, consider forces of varying magnitude acting on the same body and the magnitude of the resulting accelerations. Say one gives a soccer ball, initially at rest on the pavement, kicks of increasing force, all kicks being parallel to the pavement. Newton's law states that the magnitude of the acceleration is proportional to the magnitude of the force. So it is expected that as the force on the soccer ball increases, so does the ball's acceleration, as estimated by the speed that results from the kick. Indeed, a gentle tap with the toe barely gets the ball rolling. A light kick results in the ball's sliding a short distance, then rolling faster than from the tap. A medium kick makes the ball slide for a longer distance and then roll at an even higher speed than from the light kick. And a full-strength smash sends the ball zipping past the goalie and into the goal.

For another example, consider the effect of forces of equal magnitude acting on objects of different mass. Newton's second law states that the magnitudes of the resulting accelerations will be inversely proportional to the masses. In other words, forces of equal magnitude give smaller masses higher accelerations and larger masses lower accelerations. We can assure equality of force magnitudes by using Newton's third law of motion, the law of action and reaction: when a body exerts a force on another body, the second body exerts on the first a force of equal magnitude and of opposite direction. So consider the unfortunate case of a motorcycle colliding head-on with a semitrailer. The result is the smashed motorcycle stuck to the smashed grille of the semi. During the collision, the magnitude of the motorcycle's force on the truck is the same as the magnitude of the truck's force on the motorcycle. But the massive semitrailer continues on its way as if nothing much happens to it. It suffers almost no change of velocity, so the collision causes it only a very small acceleration (actually, a deceleration, since the direction of the force and of the acceleration—backward—is opposite the direction of the truck's motion). On the other hand, the motorcycle, whose mass is much less than the semi's, reverses the direction of its velocity and adopts that of the truck. This large change of velocity indicates a large acceleration in the collision. Newton's second law is demonstrated again.

CONCEPTUAL CONSIDERATIONS

It would appear, then, that a force acting on a particle is detected and identified by the particle's acceleration, or change of momentum. No acceleration, no force; the momentum changed, a force acted. However, Newton's second law of motion is valid only in inertial reference frames, which are those reference frames in which Newton's first law of motion is valid. Newton's first law states that in the absence of forces acting on it, a body will remain at rest or in motion at constant velocity (i.e., at constant speed in a straight line). Inertial reference frames are then those reference frames in which a particle that has no net force acting on it moves at constant speed in a straight line (or remains at rest, as the special case of zero speed), and vice versa. So the validity of the common definition of force, based on Newton's second law, hinges on inertial reference frames, whose definition involves force, according to Newton's first law. As a result, the common definition of force is flawed in that it refers to what it is defining.

As an example of that, consider an accelerating jet airplane building up speed for takeoff. The plane serves as a convenient reference frame for the passengers. In this reference frame, during the acceleration process, the passengers appear to be accelerated backward and would crash into the lavatories at the rear of the plane if they were not restrained by the backs of their seats. Yet there is no physical cause for this acceleration in the form of a backward force acting on the passengers. Nothing is pushing them and nothing is pulling them. No ropes, springs, or giant pushing hands cause this apparent acceleration. That is a violation of Newton's first law of motion, according to which any change of state of motion (i.e., acceleration) must be caused by a force. Or conversely, in the absence of force, as is the case here, a passenger must remain at rest or move at constant speed in a straight line. The reference frame of the airplane is not an inertial reference frame, and in this reference frame Newton's second law does not hold. For instance, according to the second law a 10-N force should cause a body of mass 5 kg to accelerate at 2 m/s^2 ($a = F/m$, where a and F denote the magnitudes of the acceleration and the force, respectively). In the airplane reference frame, that simply does not hold. Indeed, even in the absence of force, there is still a backward acceleration.

Newton's first law seems to require a primitive, underived understanding of a force as an obvious, readily identifiable physical effect. It appears that Newton's thinking was that if a body is not moving at constant speed in a straight line or is spontaneously abandoning its state of rest, yet our examination of it reveals no physical influence affecting it, then we are not observing the body in an inertial reference frame. That is exactly the case of the preceding example of an accelerating airplane. Unrestrained passengers appear to be accelerating toward the rear of the plane, but no physical influence can be found to cause this. In fact, an accelerating airplane is not an inertial reference frame. Once inertial reference frames have thus been identified, only then can the second law be applied. But there is no point in then defining force by the second law, since the concept has already been used in identifying inertial reference frames.

The definition of a force is somewhat arbitrary. That should not be of concern in elementary applications but needs to be taken into account in advanced physics. Albert Einstein's general theory of relativity, for instance, abolishes the gravitational force. Instead, Newton's first law is modified to the effect that in the absence of a net nongravitational force, a body moves along a geodesic, which is a curve in space-time that is the closest to a straight line. For situations in which Newton's law of gravitation is valid, the resulting motion of a body according to Einstein is that which would have been found nonrelativistically by using Newton's second law to calculate the effect of the gravitational force on the body's motion. What is a dynamic effect (involving force) in nonrelativistic physics is an inertial effect (force-free motion according to the geometry of space-time) in the general theory of relativity.

As an example, consider the motion of Earth around the Sun. In the hypothetical absence of the gravitational force of the Sun (and ignoring all other solar-system objects), Earth would move inertially (i.e., in accordance with Newton's first law) in a straight line at constant speed. The gravitational force of the Sun on Earth makes the situation dynamic, however, and the second law applies. The resulting dynamic motion turns out to be motion in an elliptical orbit around the Sun. The general theory of relativity describes matters differently. According to it, the Sun causes a deformation of space-time. No forces act on Earth, which therefore moves inertially. Inertial motion is geodesic motion in space-time, which is the space-time analog of straight-line motion at constant speed in ordinary space. When Earth moves inertially along a geodesic in space-time, it *appears* to be moving dynamically along an ellipse in ordinary space.

REACTION FORCE

A reaction force is a force that comes into being as a result of an action force, according to Newton's third law of motion, also called the law of action and reaction. The law states that when a body exerts a force on another body, the second body exerts on the first a force of equal magnitude and of opposite direction. Forces occur in pairs, where one is the action force and the other the reaction force. Note that the forces of an action-reaction pair *act on different bodies*. Either of such a pair can be considered the action force, although one usually thinks of the force one has more control over as the action, with the other serving as its reaction. If someone pushes on a wall, for instance, the wall pushes back on the person with equal magnitude and in opposite direction. One usually thinks of the person's push as the action and the wall's push as its reaction. But the reaction force is no less real than the action force. If one is standing on a skateboard and pushes on a wall, he will be set into motion away from the wall by the wall's force on him, according to the Newton's second law of motion.

FRICTION

The term *friction* is a broad one and covers mechanisms and forces that resist and retard existing motion and that oppose imminent motion. They include forces of adhesion and cohesion. The former is the attraction between different materials, such as between water and glass, causing the capillary flow of water in a narrow glass tube. Cohesive forces are those that hold matter together, such as the forces that constrain the molecules of a liquid to remain within the volume of the liquid or maintain the shape of a solid. In the capillary example, adhesive forces hold the water to the glass, while cohesive forces act among the water molecules, allowing the higher molecules to pull up the lower ones. Both kinds of force are manifestations of the mutual electromagnetic attraction of the atoms or molecules constituting the materials. Friction also includes the effects of turbulence and viscosity in fluids (i.e., liquids and gases). Turbulence is an extremely irregular fluid flow, while viscosity is the resistance of fluids to flow. When resisting actual motion, friction forces are dissipative, in that they convert nonthermal forms of energy into heat. In spite of the great practical importance of friction, a full understanding of it and its reduction has not yet been achieved.

Two commonly considered types of friction are solid-solid friction and solid-fluid friction.

SOLID-SOLID FRICTION

Solid-solid friction is friction between two bodies in contact. Its origin seems to be adhesion and impeding surface irregularities. Imagine two bodies, such as a book and a table top, in contact along flat faces, and imagine an applied force, parallel to the contact surface and tending to cause one body to move relative to the other. For a sufficiently weak applied force, there is no motion: the friction force matches the applied force in magnitude and opposes it in direction. This is static friction. But static friction has its limit. For some magnitude of the applied force, motion commences. The maximal value of the magnitude of the force of static friction is found, to a good approximation, to be proportional to the magnitude of the normal force, the force that is perpendicular to the contact surface and pushes the bodies together. For the book-on-table example, the normal force is the weight of the book. This maximal magnitude is also found to be practically independent of the area of contact. The proportionality constant is the coefficient of static friction. Thus

$$F_f \leq F_{f\,max} = \mu_s F_N$$

where F_f denotes the magnitude of the force of static friction, $F_{f\,max}$ the maximal magnitude of the static friction force, F_N the magnitude of the normal force, and μ_s the dimensionless coefficient of static friction for the two materials involved. This coefficient also depends on the condition of the contact surfaces, such as whether they are smooth, rough, dry, wet, and so on.

Once motion is achieved, a friction force then acts to retard the motion. This friction is called kinetic friction or sliding friction. The direction of the force is parallel to the contact surface and opposite to the direction of the velocity. Its magnitude is found, to a good approximation, to be independent of the speed and contact area and to be proportional to the normal force. Thus

$$F_f = \mu_k F_N$$

where F_f is the magnitude of the force of kinetic friction and μ_k is the dimensionless coefficient of kinetic friction for the two materials in contact and depends on the condition of the surfaces. Normally, for the same pair of surfaces in contact, $\mu_k < \mu_s$.

For the example of a book resting on a table, let the weight of the book be 30 N. On a horizontal surface, as in this example, that is also the magnitude of the normal force, $F_N = 30$ N. Assume that the coefficients of static and kinetic friction between the book and the table top have the values be $\mu_s = 0.5$ and $\mu_k = 0.3$, respectively. Then the maximal magnitude of static friction force is

$$\begin{aligned} F_{f\,max} &= 0.5 \times (30 \text{ N}) \\ &= 15 \text{ N} \end{aligned}$$

Now, imagine pushing the book horizontally, starting with a very small force and gradually increasing it. As long as the pushing force is less than $F_{f\,max} = 15$ N, the force of static friction balances the pushing force, the net horizontal force on the book is zero, and the book remains at rest. When the pushing force reaches 15 N, the opposing static friction force reaches its maximum. The book does not yet move. When the pushing force exceeds 15 N, even slightly, the static friction force can no longer balance it, and the book begins to move. Then the friction between the book and the table top is no longer static friction, but kinetic. Then the friction force that opposes the motion is that of kinetic friction and its magnitude is

$$\begin{aligned} F_f &= 0.3 \times (30 \text{ N}) \\ &= 9 \text{ N} \end{aligned}$$

less than the maximal force of static friction. So once the book is set in motion, the pusher will notice a sudden reduction in resistance to motion: instead of a 15-N resisting force, when the force of static friction just reaches its maximum and the book is not yet in motion, the resisting force suddenly jumps down to 9 N.

In this connection, the advantage of an antilock braking system in a car becomes clear. When a car is braking, as long as the wheels are turning, the force of *static* friction between the tires and the road decelerates the car. The maximal magnitude of this force is proportional to the coefficient of *static* friction. If the wheels lock, the tires slide on the pavement and the braking force is then the force of *kinetic* friction, whose magnitude is proportional to the coefficient of *kinetic* friction. The kinetic friction force is less than the maximal static friction force, since the coefficient of kinetic friction—about 0.7 for tires on dry concrete—is less than the coefficient of static friction—around 1.0 in the same case. Thus, the braking force when the car is skidding is less than the maximal braking force when the wheels are rolling, and the breaking distance is concomitantly greater in skidding than in rolling. In an emergency, a skilled driver controls the brakes in such a way that the car is on the verge of skidding but is still rolling. In this way, she achieves the greatest possible braking force and shortest braking distance. Ordinary drivers in an emergency tend to push the brake pedal so hard that the wheels lock and the car skids. The antilock braking system then intervenes by alternately releasing and applying the brakes at a rather high rate, allowing the locked wheels to unlock part of the time and the braking force to be, at least partly, the greater force of static friction. While the performance of an antilock braking system does not match that of a

APPROXIMATE COEFFICIENTS OF FRICTION		
Materials	**Coefficient of Static Friction, μ_s**	**Coefficient of Kinetic Friction, μ_k**
Steel on ice	0.1	0.05
Steel on steel—dry	0.74	0.57
Steel on steel—greased	0.1	0.05
Aluminum on steel	0.61	0.47
Copper on steel	0.53	0.36
Brass on steel	0.51	0.44
Shoes on ice	0.1	0.05
Tires on dry concrete	1.0	0.7
Tires on wet concrete	0.7	0.5
Tires on icy concrete	0.3	0.02
Rubber on asphalt	0.6	0.4
Teflon on Teflon	0.04	0.04
Teflon on steel	0.04	0.04
Glass on glass	0.9	0.4
Wood on wood	0.5	0.3

skilled driver, it is noticeably better than the performance of an ordinary driver.

When a round body rolls on a surface, there is rolling friction. The origin of rolling friction is this. The weight of the rolling body causes distortions in the shape of the surface and the body by compression at the area of contact. When the body rolls and the area of contact advances on the surface and around the body, the distortions are continuously both created anew ahead and relieved behind. In the absence of any internal friction within the materials of the body and the surface, the decompression would aid the motion to the same extent that the compression hinders it. But the presence of internal friction, which cannot be avoided, makes the retarding force greater than the assisting one, and the net result is what is called rolling friction. As is the case for kinetic friction, the magnitude of the rolling friction force is proportional to the magnitude of the normal force:

$$F_f = \mu_r F_N$$

where μ_r denotes the dimensionless coefficient of rolling friction for the two materials involved, which is normally much smaller than the other two coefficients of friction.

For hard surfaces, which undergo little deformation, the coefficient of rolling friction is small. For instance, for steel wheels rolling on steel tracks, its value is around 0.001. For more flexible materials, the coefficient is larger. For example, for rubber tires on concrete, the coefficient might be in the range 0.01 to 0.02.

Rolling friction needs to be overcome, at the expense of engine power, in order to keep a vehicle moving. To increase fuel efficiency for motor vehicles, tires might be made less flexible, giving a lower coefficient of rolling friction and accordingly a smaller force of rolling friction. The downside of that is a harder and noisier ride, however.

SOLID-FLUID FRICTION

Solid-fluid friction must be overcome, for example, in order to force blood through an artery or an aircraft through the air. Phenomena such as adhesion, viscosity, and turbulence cause this type of friction. Poiseuille's law describes the effect in the case of a liquid flowing through a uniform circular pipe, for instance. This law relates the volume of liquid flowing through the pipe per unit time, called the volume flux, or flow rate, to the dimensions of the pipe, the viscosity of the liquid, and the difference of pressures

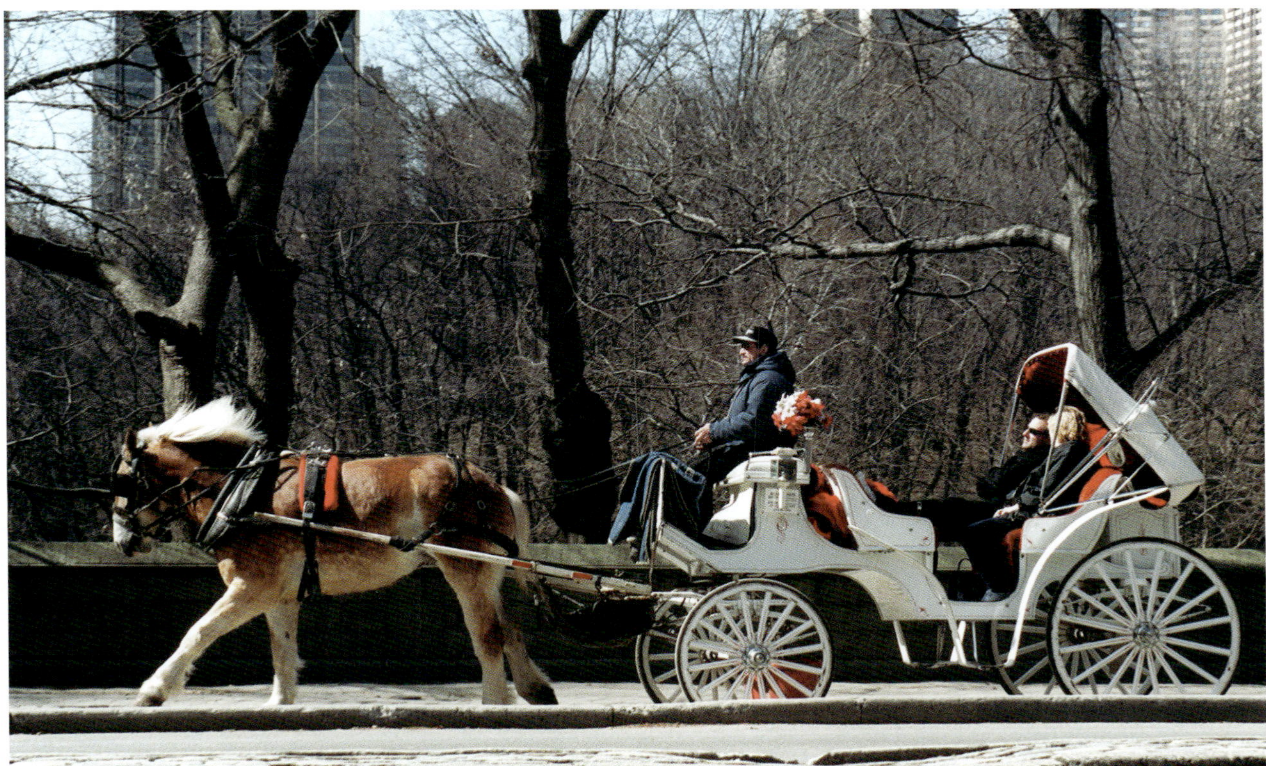

The horse is exerting a forward force on the carriage, keeping it in motion through New York City's Central Park. This force must overcome the backward-directed forces acting on the carriage: air resistance (which is not very great at walking speeds), rolling friction between the wheels and the road, and kinetic friction between the wheels and their axles. *(AP Images)*

at the ends of the pipe. The relation is approximately valid in many useful situations:

$$\text{Volume flux} = \frac{\pi}{8}\frac{R^4}{L}\frac{\Delta p}{\eta}$$

In this relation, the volume flux is in cubic meters per second (m^3/s); R and L denote the pipe's radius and length, respectively, in meters (m); Δp is the difference of the pressures at the ends of the pipe, in pascals (Pa), equivalent to newtons per square meter (N/m^2); and η represents the liquid's viscosity, in newton-second per square meter ($N\cdot s/m^2$). The relation demonstrates that the fluid in the pipe does not simply accelerate under the net force of the pressure difference, according to Newton's second law of motion. Rather, an opposing friction force balances the applied force and causes the fluid to flow at constant speed.

The formula shows that the flow rate increases with greater diameter and with higher pressure difference and decreases for longer pipes. And all other factors being equal, a more viscous—"thicker"—liquid will flow at a lower rate. As an example, the viscosity of whole blood is about twice that of water at body temperature. The same pressure difference at the ends of the same pipe, perhaps a blood vessel, will push twice the volume of water as of blood through the pipe during the same time interval.

Another manifestation of solid-fluid friction is the drag force that acts on a solid body moving through a fluid, such as a dolphin through water or an aircraft through air. For a blunt body moving sufficiently rapidly through air, as an example, the magnitude of the drag force is given by

$$\text{Drag force} = \tfrac{1}{2} D\rho A v^2$$

where the drag force is in newtons (N); ρ denotes the density of the air, in kilograms per cubic meter (kg/m^3); A is the effective cross-sectional area of the body, in square meters (m^2); and v is the body's speed, in meters per second (m/s). The symbol D denotes the drag coefficient, which is dimensionless (does not have a unit) and whose value depends on the shape of the body.

According to this relation, for example, at the same speed an aircraft encounters less air resistance at higher altitudes, where the air density is lower—

VISCOSITIES OF SOME LIQUIDS

Liquid	Temperature (°F/°C)	Viscosity, η ($N \cdot s/m^2$)
Water	212/100	0.000283
Gasoline	68/20	0.0006
Water	68/20	0.00100
Ethanol	77/25	0.000110
Blood plasma	98.6/37	0.0013
Whole blood	98.6/37	0.00208
Light oil	68/20	0.11
Heavy oil	68/20	0.66
Glycerine	68/20	1.49

the air is "thinner"—than at lower altitudes. Clearly, for the same aircraft at the same altitude, the air resistance increases with greater speed. And, for the same shape, a larger aircraft presents a greater cross section and accordingly suffers a greater drag force than does a smaller aircraft at the same altitude and speed.

FREE-BODY DIAGRAM

Drawing what is called a free-body diagram is useful in applying Newton's first and second laws to bodies. In this diagram the only body that appears is the body under consideration, upon which the effect of Newton's laws needs to be found. Then labeled arrows are added to the diagram to indicate all the forces acting on this body, and only on this body, and their directions. If rotations are being considered, then the arrows should be drawn so as to show also the points of application of the forces, to allow their torques to be found. Otherwise, indicating only the direction of the forces is sufficient.

In order to take account of all the forces acting on the body, it is useful to consider forces as being of two types: contact forces and field forces. Contact forces are forces that result from direct contact with other bodies, such as a hand pushing, a rope pulling, or a rough surface producing friction. Field forces are forces acting over a distance and are either gravitation forces, electric forces, or magnetic forces. Actually, viewed microscopically, even contact forces are field forces, since they are really electromagnetic forces. But for dealing with macroscopic bodies, the contact-field distinction is useful. So in setting up a free-body diagram, every object that is in contact with the body under consideration is a potential source of a (contact) force acting on the body. Each such force should be indicated in the diagram. Then distant bodies, such as Earth or a magnet, should be considered as potential sources of field forces, and those forces should be indicated as well. The most common field force that is encountered is the weight of the body, in other words, the force by which the body is attracted gravitationally to Earth. This force acts at the body's center of gravity.

CONSERVATIVE AND NONCONSERVATIVE FORCES

A force that performs zero total work on a body moving in a closed path is termed a conservative force. An equivalent formulation is that the work a conservative force does on a body that moves between two points is independent of the path the body follows between those points. As a result, a potential energy can be associated with a conservative force. (A third definition of a conservative force is a force with which a potential energy can be associated.) What that means is this. Imagine, for example, that a force does some positive work on a particle, thus increasing the particle's kinetic energy. By the law of conservation of energy, the particle's gain in kinetic energy is at the expense of an energy decrease of some source. For a conservative force, and only for such a force, the source's energy can be regained in full by having the particle return to its initial state. Then the force will do negative work on the particle of the same magnitude as before; the particle will lose the same amount of kinetic energy it previously gained, thus replenishing the source. In the potential-energy picture, the potential energy serves as the source of the particle's kinetic energy. The process can then be viewed as an initial conversion of potential energy to kinetic, followed by a total reconversion of kinetic energy back to potential.

Two examples of conservative forces are the gravitational force and the electric force. For the former, there is gravitational potential energy, and for the latter electric potential energy. From these one can derive the gravitational and electric potentials. As a demonstration of the gravitational case, consider an ideal pendulum displaced from equilibrium and released from rest. As the pendulum descends, the gravitational force between it and Earth performs positive work on it, whereby it gains kinetic energy and achieves maximal kinetic energy as it passes through its equilibrium point. When the pendulum is ascending, however, the gravitational force does negative work, causing the pendulum to lose kinetic energy until it comes to momentary rest at its initial height. In terms of gravitational potential energy, the pendulum starts with zero kinetic energy and some value of potential energy. As it descends, its potential energy decreases and its kinetic energy increases such that their sum remains constant (conservation of energy). When the pendulum passes through its equilibrium point, its potential energy is lowest and its kinetic energy highest. As the pendulum ascends, its kinetic energy converts back into potential energy, and at the end of its swing it again has zero kinetic energy and the same amount of potential energy with which it started.

Any force that is not conservative is called a nonconservative force. Some such forces only perform negative work and convert other forms of energy into heat. Common examples of those are friction and viscosity forces. In the preceding pendulum example, if the pendulum is realistic rather than ideal, so that friction in the suspension bearing and air resistance are in play, the pendulum will eventually come to rest at its equilibrium position. The energy given it by the work that was done in initially displacing it from its equilibrium position eventually dissipates as heat.

See also ACCELERATION; CENTER OF MASS; CONSERVATION LAWS; EINSTEIN, ALBERT; ELECTRICITY; ELECTROMAGNETISM; ENERGY AND WORK; EQUILIBRIUM; FLUX; GENERAL RELATIVITY; GRAVITY; HEAT AND THERMODYNAMICS; MAGNETISM; MOMENTUM AND COLLISIONS; MOTION; NEWTON, SIR ISAAC; PRESSURE; ROTATIONAL MOTION; SPEED AND VELOCITY; VECTORS AND SCALARS.

FURTHER READING

Jammer, Max. *Concepts of Force.* Mineola, N.Y.: Dover, 1998.

Young, Hugh D., and Roger A. Freedman. *University Physics,* 12th ed. San Francisco: Addison Wesley, 2007.

fossil fuels Fossil fuels are energy sources formed from the decomposition of organic compounds over a long period. Examples of fossil fuels include energy sources such as oil, natural gas, and coal. According to the U.S. Department of Energy, presently 85 percent of the U.S. energy supply is from the combustion of fossil fuels. Fossil fuels have been used since ancient times to generate light and for heat. Burning candles and waxes to produce light and burning peat for heat are some of the first examples of fossil fuel use by humans. Today people depend on fossil fuels to heat homes and businesses, produce electricity, and power transportation, including automobiles and airplanes. Imagining a world without fossil fuels is difficult, but alternatives need to be found as fossil fuels are nonrenewable resources that are rapidly becoming depleted. Estimates consider the production of petroleum and natural gas supplies from decomposing organisms that lived millions of years ago. Prior to the beginning of fossil fuel extraction, the estimated amount of fossil fuels present in the earth was 2.2 trillion barrels, with one barrel equivalent to 42 U.S. gallons. The amount of fossil fuels left is presently estimated at less than 1.1 trillion barrels.

HYDROCARBONS

Fossil fuels are all hydrocarbons, the simplest form of organic compounds that are created by the covalent bonding of carbon and hydrogen atoms. Hydrocarbons are divided into two categories: aliphatic hydrocarbons and aromatic hydrocarbons. Aliphatic hydrocarbons take their name from the Greek for "fat," and they are straight-chained or branched compounds. Aromatic hydrocarbons, named for their original isolation from aromatic plant mixtures, contain ring structures made up of benzene moieties. Aliphatic hydrocarbons are distinguished by the length of the carbon chain, the branching of the chain, and the type of bonds formed between the carbon atoms. Hydrocarbons that contain all single bonds are called alkanes. When the hydrocarbon has at least one double bond, it is known as an alkene, and hydrocarbon compounds containing triple bonds are known as alkynes.

The name of a hydrocarbon is based on the length of the carbon chain and the type of bonds found in the compound. The length of the longest carbon chain in the compound determines the prefix given to the compound.

HYDROCARBON PREFIXES

Number of Carbons	Prefix	Number of Carbons	Prefix
1	*meth-*	2	*eth-*
3	*prop-*	4	*but-*
5	*pent-*	6	*hex-*
7	*hept-*	8	*oct-*
9	*non-*	10	*deca-*

The ending of the hydrocarbon name is related to the type of bonds, with alkanes ending in *-ane,* alkenes ending in *-ene,* and alkynes ending in *-yne.* For example, the molecular formulas for propane, 1-propene, and 1-propyne would be

$CH_3CH_2CH_3$ propane
CH_2CHCH_3 propene (double bond after carbon 1)
$CHCCH_3$ propyne (triple bond after carbon 1)

The chain length (and therefore molecular weight) of hydrocarbons determines many physical characteristics, such as boiling point, physical state, and solubility. The shorter the carbon chain, the lower the boiling point of the compound. The smallest hydrocarbons are gases, the longer hydrocarbons (more than five carbons) are generally liquids, and even longer hydrocarbons are waxy. Hydrocarbons are generally insoluble in water; however, lower-molecular-weight molecules are able to dissolve in water to some extent.

The properties of hydrocarbons are altered by the addition of functional groups, atoms or groups of atoms that contribute to the chemical and physical properties of a compound, to the carbon chain. Functional groups added to hydrocarbons include alcohols, alkyl halides, ether, aldehyde, ketones, carboxylic acid, esters, and amide groups. Each of these alters the reactivity of the compound to which it is attached. When naming hydrocarbons that have functional groups attached, one must first number the longest carbon chain from the end that

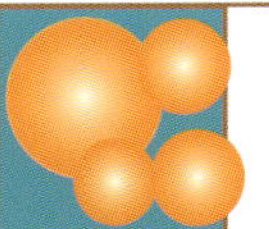

COMPACT FLUORESCENT LIGHTBULBS

by Lisa Quinn Gothard

In 2008 customers had a selection of several different types of lightbulbs: the incandescent lightbulb that screws into a normal light socket, the compact fluorescent lightbulb that also uses a normal light socket, and the traditional fluorescent (tubular) bulb that requires special receptacles. Another type of lightbulb that is emerging as new technology is the LED (light emitting diode) bulb. As of yet, the cost of LED bulbs cannot compete with that of either incandescent or compact fluorescents. The light that these bulbs emit is also substandard compared to that of the first two types. Thus, consumers truly have two choices when it comes to lighting in their home that does not require special fixtures—the incandescent bulb and the compact fluorescent bulb.

The first incandescent lightbulb was perfected by the American inventor Thomas Edison in 1879 and reigned in the lighting world for nearly 130 years. The pleasant glow given off by this type of lightbulb is due to the fact that it produces all wavelengths of visible light. The filament of incandescent lightbulbs is made of tungsten, and it burns at a temperature of 3,992°F (2,200°C). The time of the incandescent lightbulb is coming to an end, however. In a world that is becoming more and more concerned with energy consumption, this type of bulb is very inefficient. Incandescent bulbs only utilize approximately 10 percent of the energy produced by the lightbulb to create light. The rest, approximately 90 percent, is lost as heat. This inefficiency is becoming noticeably less acceptable in a world that is attempting to become more focused on energy conservation. In December 2007 the U.S. Congress passed new energy efficiency standards for lighting, requiring that all 100-W. bulbs be 30 percent more energy efficient than today's 100-W. bulbs, by the year 2012. This will be virtually impossible for makers of incandescent bulbs, leading the way for the emergence of compact fluorescent bulbs to take control of the market.

Ed Hammer, an American engineer who worked for General Electric, developed the first compact fluorescent bulb (CFL) in the 1970s. These types of bulbs already meet the 2012 efficiency standards, giving them a competitive advantage compared to incandescent bulbs. Compact fluorescent bulbs have ballasts that send the electrical current through a tungsten electrode. This creates an arc that stimulates mercury vapor in the lightbulb to create ultraviolet (UV) light. The glass of the lightbulb has a phosphor coating that when struck by the UV light begins to fluoresce, or give off visible light. Many consider the light from these bulbs too harsh since CFLs do not give off all wavelengths. Scientists are working to correct this problem.

Consumers usually consider two parameters when selecting which type of bulb to purchase: energy efficiency and cost. The quantity of light intensity given off by a bulb is measured in lumens. The wattage of the bulb is a measure of the amount of energy the bulb consumes when it is turned on. The spiral design of the compact fluorescent lightbulb has practically become synonymous with the green movement of energy conservation. Compact fluorescent bulbs create the same amount of light while utilizing nearly four times less energy. That means that the lighting power achieved from a 60-W. incandescent lightbulb can be achieved by a 15-W. CFL. The average American household has more than 30 lightbulbs. If each of these bulbs were 60 W. the total wattage to power these lightbulbs would be 1,800 W. If the same household contained only CFLs contributing the same amount of light, the total energy usage would be 450 W. Simply changing the type of lightbulbs can lead to significant energy savings.

The cost to run the bulb depends on the wattage. Running a 60-W. bulb for one hour consumes 0.06 kilowatt of electricity. Electric companies charge by the kilowatt/hour or kWh. The cost of running a 15-W. CFL for that same hour would only be 0.015 kWh. According to the U.S. Energy Information Administration, the average cost of residential energy in the United States in December 2007 was 10.81

is closest to the functional group. The location of the functional group is written in numbers and separated from the name of the compound by a dash. The addition of the functional group -OH changes a hydrocarbon to an alcohol, and the name of the hydrocarbon changes to reflect the change in reactivity.

$CH_3CH_2CH_3$ propane

CH_3CH_2CHOH 1-propanol

COAL

Coal is formed from the decomposition of organic materials. All coal contains carbon, hydrogen, oxygen, and varying degrees of sulfur. The original plant and animal matter is formed into a soft, spongy material known as peat. This was used as a fuel source for many generations. Burning of peat provided little heat and the fires were relatively smoky, but peat was readily available and less expensive than other heating sources. When subjected to increased pressure and temperature within the earth's crust, the peat

cents/kilowatt hour. That means that the incandescent bulb would cost (0.060 kWh × 10.81 cents) = $0.65 to run for one hour. The CFL would utilize (0.015 kWh × 10.81 cents) = $0.16 to run for one hour.

The cost and efficiency of two lightbulbs available for purchase are compared in the table below with regard of their cost and efficiency.

The cost difference between the incandescent lightbulb and the compact fluorescent bulb is significant. One must consider the lifetime of the bulb when considering price. The incandescent bulb, in the example above, costs $0.00037 per hour of life. The compact fluorescent bulb lasts 10,000 hours and costs $0.000799 per hour of life.

Combining the two examples, one can determine the total price of purchasing and running incandescent bulbs versus compact fluorescent bulbs.

Incandescent 60-W. bulb cost per hour =
price per hour + energy cost per hour
$0.00037 + $0.65 = $0.65037/hour

Compact fluorescent 15-W. bulb cost
per hour = price per hour + energy cost
per hour
$0.000799 + $0.16 = $0.160799/hour

The total energy savings is clearly in favor of CFLs, as is the financial saving over the long term.

One question frequently asked by consumers of CFLs relates to their most efficient method of use. Should they be left on, or should they be turned off every time someone leaves the room? The question arises because the initial energy surge to start up this type of lightbulb is large relative to the amount of energy required to run the bulb. Also that initial surge decreases the life span of the bulb. If one leaves the lightbulb on continuously, the bulb does not experience that energy surge. The recommendation is that if one is leaving a room for more than five minutes, it is still beneficial to turn off the lightbulbs. The overall life span of the bulb is not decreased relative to the reduction in the number of hours that the bulb is used. This means that the overall savings on the life of the bulb is greater if you leave it off when not in use.

As the number of compact fluorescent bulbs increases, one problem emerges, and that is of disposal. The CFLs contain mercury vapor; thus they do present an environmental disposal problem. The accidental breakage of lightbulbs, an event that occurs often in a household, could pose a health hazard, especially for infants and children. Exposure to mercury vapors can cause serious respiratory and neurological damage. The disposal of mercury-containing bulbs is also difficult, as landfills generally do not accept them. They should be collected and recycled at designated hazardous waste recycling centers. The total number of these bulbs being recycled is estimated to be as low as 2 percent. Improvements in recycling efforts for these types of bulbs will become necessary as the number of the bulbs used continues to increase.

Efforts are under way to incorporate components other than mercury in the compact fluorescents in order to reduce the risk. Unfortunately, no compound or element tried to date works as well as mercury. Scientists are currently working on developing new technology to reduce the hazardous risk of mercury disposal.

Bulb	Cost (Per Bulb)	Life (Hours)	Lumens	Watts
Incandescent Soft White	37 cents	1,000 hours	840	60
Compact Fluorescent	7.99	10,000 hours	840	15

FURTHER READING

Energy Star (A joint project of the U.S. Environmental Protection Agency and the U.S. Department of Energy). "Compact Fluorescent Lightbulbs." Available online. URL: http://www.energystar.gov/index.cfm?c=cfls.pr_cfls. Accessed April 20, 2008.

GE Lighting. "Compact Fluorescent Lightbulb FAQs." Available online. URL: http://www.gelighting.com/na/home_lighting/ask_us/faq_compact.htm. Accessed April 20, 2008.

solidified more, creating coal of various consistencies. The lowest grade of coal is lignite, which has the lowest carbon content and the highest sulfur content. Subbituminous coal has a higher carbon content than lignite, but less than bituminous. The highest-quality coal is anthracite coal, which is also the cleanest-burning coal.

Coal can be removed from the earth in two ways, underground mining and surface mining (strip mining). In underground mining, shafts can be bored into the earth and workers are transported down into the shafts and caves via elevators and tracks to work in the removal of coal from the mine. Coal mining is a dangerous occupation. Inhalation of coal dust over a long period can lead to pneumoconiosis (CWP), also known as black lung for the color that the coal deposits turn the workers' lungs. In 1969 the federal government passed the Federal Coal Mine Safety and Protection Act, which regulated the amount of coal dust exposure allowable in coal mines. The legislation has led to a decrease in the incidence of black lung, although it is still a concern. Fatalities in the coal mine are also possible. The U.S. Department of Labor Mine Safety and Health Administration is responsible for regulating safety in coal mines. They reported that in the 10-year span from 1992 to 2002, there were more than 400 fatalities at coal mines with most of these fatalities of miners who had been on the job for less than one year.

The second type of mining to remove coal from the earth is known as surface mining or strip mining. In this type of mining the surface layers of the earth are stripped away to reveal the coal below. The process of strip mining is devastating to the environment around the mines. The land that is left behind is destroyed and is not able to support plant or animal life.

The first coal reserves and coal mines were found in China. Present-day production of coal in China is still at record levels. In the United States, the top five states with the highest percentage of coal production are Wyoming, West Virginia, Pennsylvania, Kentucky, and Texas. Coal is a major environmental pollutant. The sulfur content within the coal is damaging, as the sulfur dioxides released from coal plants contribute to acid rain. When sulfur dioxides react with water, the product formed is sulfuric acid, H_2SO_4. The Clean Coal Power Initiative was introduced in 2002 by President Bush to promote the development of new technologies that will reduce the polluting factors of coal as a fuel source.

NATURAL GAS

Not to be confused with gasoline used in cars, natural gas is a hydrocarbon that is primarily made of methane, CH_4. Methane, as are all fossil fuels, is formed by organic deposits that were trapped underneath the surface of the earth long ago. Methane is also produced from animal digestion. Natural gas is generally an odorless gas, and it is used in household heating and in running of household appliances. The natural gas pipelines transfer the gas from the source to individual homes or businesses. Combustion of methane is relatively clean compared to that of other fossil fuels. Natural gas is often found in areas near oil reserves. Natural gas prices in 2008 were at record levels, with an April 2008 price of $14.30 per thousand cubic feet for residential gas. The majority of natural gas reserves in the United States are located in the Gulf of Mexico and Texas.

PETROLEUM PRODUCTS

The production of oil or petroleum is divided into four areas: location of oil reserves, extraction of the crude oil, refining of crude oil into the different components, and distribution of the final products. The cost for crude oil in July 2008 was $147.27 per barrel. This was more than triple the cost of a barrel of crude oil in 2003. The cost of crude oil in December 2008 fell 77 percent to $44.80 per barrel from the record high of July 2008.

The countries with the largest production of oil include Saudi Arabia, Iran, Russia, United States, Mexico, China, and Europe's North Sea, with Saudi Arabia producing double the amount of the next-closest country, Iran. In the United States, the largest oil reserves are in California, Alaska, the Gulf of Mexico, and West Texas.

Crude oil that is removed from the earth contains many different types of hydrocarbons that must be further purified before use. The first step in oil refining is fractional distillation, which separates the hydrocarbons of different carbon chain lengths on the basis of their differing boiling points. When the boiling point of a particular compound is reached, it will evaporate and be collected. Boiling points of nearly of 1,112°F (600°C) remove the hydrocarbons with the highest boiling point, those that contain more than 80 carbons in the chain. At approximately 572°F (300°C), the heavy gas oil of 44 carbons is isolated. Boiling temperatures of 392°F (200°C) isolate lubricating oil with approximately 36 carbons. As the boiling point decreases, gas oil or diesel (which has approximately 16 carbons) is removed. Twelve-carbon compounds, including kerosene, boil at approximately 158°F (70°C). Gasoline (automobile fuel), made of eight carbons, boils and evaporates next. Naphtha is the mixture of the remaining hydrocarbons with varying numbers of carbon atoms, generally five to nine carbons. This mixture is a by-product of the distillation process and moves to the reformer, the next stage of the purification process, leading to the lowest-boiling-point compounds from oil refining, gases with fewer than four carbons. The series of separation techniques are able to take crude oil and purify it into useful components such as oil, gasoline, and natural gas.

The smallest carbon compounds collectively make up petroleum gas, which is used for heating, cooking, and the synthesis of some types of plastics. These types of compounds are often liquefied with pressure to create liquefied petroleum (LP) gas used in home heating and cooking.

Kerosene is a precursor for many other petroleum products. It is a liquid that ranges from 10 to 18 carbon atoms. Kerosene is sometimes used for heating in kerosene space heaters and is also used as jet fuel.

Gasoline is used as automobile fuel, known commonly as gas, and is rated according to its octane level. Naphtha, the mixture of refined hydrocarbons with five to nine carbons that have a boiling point lower than 212°F (100°C), generally consists of unbranched alkanes and cycloalkanes. This is what is processed into gasoline. Gasoline, which contains hydrocarbons with five to 12 carbons, is used to power automobiles and boils at temperatures below 392°F (200°C). The ignition properties of these hydrocarbons lead to undesirable engine knocking. The standard of octane is the measure of the amount of knocking that occurs in a sample of gasoline relative to the same standard sample of heptane, which has an octane value of 0, and 2,2,4-trimethylpentane, which has an octane value of 100. The ratio of 2,2,4-trimethylpentane to heptane gives the octane value and the same amount of knocking relative to that sample of gas. Highly branched forms of alkanes give less knocking. Therefore, naphtha is further processed after distillation in order to produce a higher octane gasoline. This process is known as reforming, in which unbranched alkanes and cycloalkanes in naphtha are converted to highly branched alkanes, or aromatic hydrocarbons, thus decreasing the knocking produced from that sample of gasoline.

Cracking is the process of cleaving the carbon-carbon bonds in high-molecular-weight hydrocarbons found in petroleum in order to produce more of the low-carbon-number molecules that are used as heating and fuel sources. Cracking can be divided into two types: thermal cracking and catalytic cracking. Thermal cracking involves high temperature and pressure to break the high-molecular-weight compounds into smaller compounds. Catalytic compounds achieve this cleavage by using a chemical catalyst.

PREDICTED AMOUNT OF FOSSIL FUEL RESERVES

Estimates in 2005 placed the amount of fossil fuels consumed to that point at approximately 1.1 trillion barrels of oil, meaning that nearly half of the total reserves had been used up. At a rate of 83 million barrels per year, this predicts that the remaining fuel will last until approximately 2036. Given that fossil fuel consumption has been increasing steadily over the last 50 years, these predictions could be overestimates. For example, in June 2008, the U.S. Energy Information Administration projected energy consumption to increase by 57 percent from 2004 to 2030.

The most credible predictions of fossil fuel expenditures and lifetime were made by the Shell Oil geophysicist Marion King Hubbert in 1957. The curve he developed, called Hubbert's curve, correctly predicted in 1956 the peak of oil production from 1965 to 1970. The true peak occurred in 1970. This added credibility to Hubbert's peak and to the prediction that there is less than 50 years of oil left at present consumption rates.

No matter what level of fossil fuel consumption exists, the fact of the matter is that fossil fuel reserves will eventually be depleted and cannot be replaced. It required nearly 300 million years to produce the reserves that humans are using today, and they are not being replaced. The dependence on fossil fuels is a political, economic, and environmental concern. Viable alternatives to fossil fuel consumption need to be developed in order to reduce society's dependence on coal, oil, and natural gas. Present alternatives include conservation, alternative energy sources, and more improved drilling and isolation techniques.

See also ACID RAIN; ALTERNATIVE ENERGY SOURCES; ORGANIC CHEMISTRY; SEPARATING MIXTURES.

FURTHER READING

Goodell, Jeff. *Big Coal: The Dirty Secret behind America's Energy Future.* New York: Mariner Books, 2007.

U.S. Department of Energy. "Fossil Energy." Available online. URL: http://fossil.energy.gov/education/. Accessed July 25, 2008.

Franklin, Rosalind (1920–1958) British *Physical Chemist* Rosalind Franklin is one of the most controversial figures in modern biochemistry because her studies played a major role in the elucidation of the structure of deoxyribonucleic acid (DNA), yet she received nominal, at best, recognition during her lifetime. Trained as a physical chemist with expertise in X-ray crystallography, Franklin made important contributions to the understanding of the physical structures of DNA, viruses, and coal. Although her father intended Franklin to enter social work, an acceptable occupation for women of her social standing, the strong-willed Franklin aspired to be a scientist and set out to achieve her dream with or without her family's blessing. Peers described Franklin as tenacious, logical, and meticulous—qualities essential to the tedious field of X-ray crystallography that she entered. During her graduate studies, she studied the physical properties of coal and discovered important properties that helped optimize its use during World War II. She further honed her skills in X-ray crystallography in France before accepting a research associate position in the laboratory of Sir John Randall at King's College in London, Great Britain. From 1951 to 1953, she identified the physiological form of DNA (B-form) and gathered physical data, including a specific X-ray diffraction photograph identified as photograph #51, that her peers James Watson, Francis Crick, and Maurice Wilkins incorporated with other experimental results to elucidate a working model of DNA. Watson and Crick proposed that DNA is a double helix with two sugar-phosphate backbones oriented in opposite directions (antiparallel), linked through nitrogenous base pairs: adenine-thymine and guanine-cytosine. In the famous April 1953 publication of *Nature* containing the theoretical model of Crick and Watson, Franklin coauthored another paper outlining the X-ray diffraction results from her studies of B-form DNA, yet her role in the discovery of the DNA structure appeared minor to those outside the field despite the significance of her data and measurements. She continued to play a major role in the field of crystallography, determining the structure of tobacco mosaic virus before she succumbed to ovarian cancer at the age of 37. Because scientists cannot be posthumously nominated, only Crick, Watson, and Wilkins received the prestigious 1962 Nobel Prize in physiology or medicine for the determination of DNA structure and its properties. The missing acknowledgment of Rosalind Franklin's key role in this pivotal discovery is a source of ongoing controversy among many biochemists in the field.

Rosalind Franklin was an English physical chemist best known for her X-ray crystallography patterns of DNA. *(Science Source/Photo Researchers, Inc.)*

A member of an affluent Jewish family living in London, Rosalind Elsie Franklin was born on July 20, 1920, to Muriel Waley Franklin and Ellis Franklin. A merchant banker by profession, Rosalind's father expected her to focus her education, talents, and skills on political, educational, and charitable forms of community service. Franklin attended St. Paul's Girl School, where she excelled in chemistry, Latin, physics, and sports. Franklin decided to be a scientist at the age of 15, but her father refused to pay for continuing university studies until an aunt offered to support Rosalind's studies. Grudgingly, Franklin's father relented, permitting his daughter to enter Newnham College, a college of Cambridge University, in 1938. Franklin passed her final exams in physical chemistry in 1941, but she never received a bachelor's degree because women were not eligible for Cambridge degrees at the time. For the next four years, Franklin worked at the British Coal Utilization Research Association studying the physical properties of coal, specifically examining the idea of high-strength carbon fibers, which later became the basis for her dissertation research. Franklin earned her Ph.D. from Cambridge in 1945, and then she joined the laboratory of Jacque Mering at the Laboratoire Central des Services Chimiques de l'État in Paris, France.

During her time in Paris, she developed the skills necessary to perform X-ray crystallography studies on imperfect crystals, such as coal and DNA. X-ray crystallography studies yield results in the form of X-ray diffraction data, a means of visualizing a molecule too small to photograph by regular photography. X-rays create pictures of tiny structures because these wavelengths are so short that they bounce off atoms themselves. When X-rays bombard a molecule, the X-rays ricochet off atoms in their path, like balls in pinball machines, leaving a pattern that hints at the shape of the molecule by the way that the rays have been scattered, or diffracted. Photographic film captures these patterns for crystallographers to study. Using X-ray diffraction methodologies, Franklin acquired enough results from her studies of graphitizing and nongraphitizing carbons that she authored five coal-related papers within a three-year period. Her work was sufficiently significant that researchers in the field continue to cite it today. Franklin immensely enjoyed her time in France, flourishing in its people, language, cooking, and intellectual liberalism. In 1950 she decided that the time had come to return to Great Britain, and she accepted a position in the Medical Research Council Biophysics Unit at King's College in London.

Rosalind Franklin joined the laboratory of Sir John Randall in 1951. Randall hired Franklin as a research associate, whose project would be the study of protein structures using X-ray crystallography. When she arrived, Randall reassigned Franklin to another project, the ongoing structural analyses of another important macromolecule, DNA. When Franklin and the graduate student Raymond Gosling began their studies of DNA, biochemists knew that each nucleotide (the building block of DNA) consisted of a deoxyribose sugar, a phosphate group, and one of four nitrogenous bases: the purines adenine (A) and guanine (G), which are double-ringed carbon and nitrogen molecules, or the pyrimdines thymine (T) and cytosine (C), which are single-ringed carbon and nitrogen molecules. Another research associate by the name of Maurice Wilkins had spent the previous year performing X-ray crystallographic studies on A-form DNA. More ordered and crystalline in nature, A-form DNA appears under conditions of low humidity and water. Franklin quickly identified an alternative configuration of DNA, B-form DNA. With less ordered crystalline structures, the B-form DNA existed at high humidity and water content similar to conditions found in vivo. Extra hydration makes it easier for the B-form DNA to assume a lowest-energy helical configuration. The increased water content also keeps the two helical backbone chains farther apart than in A-form, allowing the bases to orient themselves perpendicularly to the backbone.

Using a nondenatured DNA sample from Rudolph Signer of the University of Bern in Switzerland, Franklin extracted thin single fibers of DNA. She controlled the humidity of the fibers by flooding the specimen chamber with a mixture of hydrogen gas and salt solutions. In order to obtain the best X-ray pictures possible, she designed a tilting multifocus camera to achieve better images of diffraction patterns, and she developed a technique to improve the orientation of the DNA fibers for X-ray exposure. Franklin slowly and precisely transformed the crystalline A-form to the hydrated B-form and back again, and she obtained X-ray photographs of such clarity with her labors that she easily determined both forms' density, unit cell size, and water content. From her knowledge of chemistry, Franklin suggested that the hydrophilic sugar and phosphate groups were located on the external part of the DNA molecules and the hydrophobic nitrogenous bases were located on the internal part of the molecules. Franklin corroborated the placement of the sugar and phosphate groups on the external part of both forms through mathematical calculations derived from the X-ray diffraction data in a progress report she submitted to the Medical Research Council (MRC).

In May 1952, Rosalind Franklin developed a photograph, referred to as photograph #51, from X-ray diffraction studies of B-form DNA. Laws of diffraction hold that X-rays moving through a helix diffract at angles perpendicular to the object. If two strands twist around each other in an arrangement like a spiral staircase, an X shape appears in the diffraction results. The high resolution of photograph #51 clearly showed an X-shaped pattern, suggesting that the structure of DNA was a double helix. The four white diamond shapes surrounding the X in the photograph indicated that the helix continues below and above this cross section of the DNA. At horizontal points (layer lines) along the X, the absence of black makes it appear that atoms are not present at these points, but the atoms of one strand simply cancel out the diffraction pattern of the atoms in the second strand as they wrap around one another. These layer lines also suggest that B-form DNA contains two types of grooves along the outer edge of the double helix: deeper and wider major grooves and shallower and thinner minor grooves. Using trigonometry, Franklin calculated the dimensions of B-form DNA to be 1.0 nanometer (nm, one-billionth of a meter) across (the radius). She also confirmed earlier measurements made in the 1930s by William Astbury and Florence Bell that one full twist of DNA is 3.4 nm and a distance of 0.34 nm exists between each base pair, meaning that each twist of DNA contains 10 nucleotides, or base pairs. Franklin approached the Oxford crystallographer Dorothy Hodgkin with photograph #51, and Hodgkin helped Franklin to determine the correct crystallographic space group of B-form DNA. At this date, Franklin was only unclear about the way the nitrogen bases form pairs with each other and the orientation of the two strands with respect to each other (parallel or antiparallel). With this information, she could have built a theoretical model of DNA's structure, as Linus Pauling had with the alpha helix of proteins, but Franklin played it safe, deciding that experimental structural determination should precede model building.

Until Franklin joined Randall's laboratory, Maurice Wilkins headed up the DNA structural studies. While Wilkins vacationed, Franklin started in the laboratory. Because Randall never defined the roles of the two peers in the project, conflict between them immediately arose upon Wilkins's return. Wilkins mistakenly

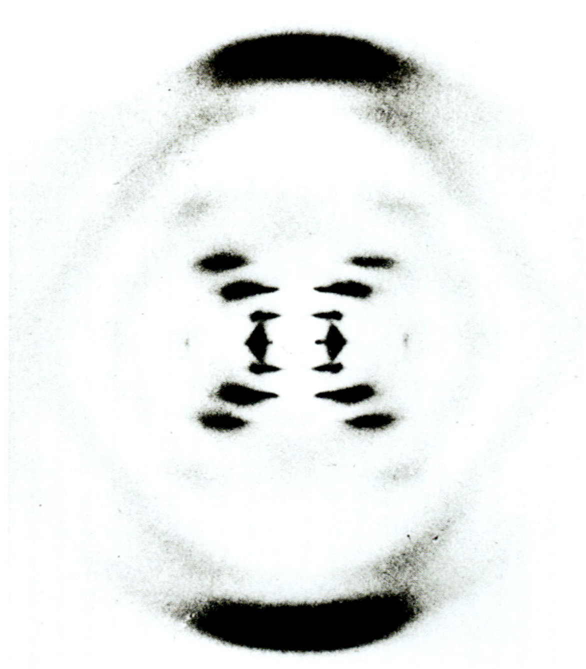

Shown here is one of Franklin's famous X-ray diffraction photos—the 1953 X-ray diffraction photo of the B-form of DNA utilized by James Watson and Francis Crick in their determination of the structure of DNA. Although Franklin was not given due credit for her contributions at the time, her structural determinations were vital to solving the structure of DNA. *(Omikron/Photo Researchers, Inc.)*

believed that Franklin was a technical assistant, not a peer, and treated her as such. The misunderstanding came to light, but the incident permanently damaged their professional relationship. Personality conflicts between the mild-mannered, reserved Wilkins and the strong-willed, assertive Franklin only aggravated their strained relationship. Franklin decided to transfer her fellowship to Birbeck, another institution in Great Britain, in 1953. Sir John Randall made Franklin promise not to perform additional studies on DNA and to leave her laboratory notebooks, which contained her measurements and calculations, as well as all diffraction photographs, including photograph #51.

During this period, the director of the Cavendish Laboratory and a member of the MRC, Max Perutz, permitted two research associates in his laboratory, Francis Crick and James Watson, to read Franklin's MRC progress report detailing the placement of the sugar and phosphate groups externally along the DNA molecule. A disgruntled Wilkins also showed Watson, whom he met through his friend Crick, the infamous photograph #51. Watson, a zoologist by training, only recognized that an X indicated a helix, but Wilkins helped Watson interpret the data. On March 18, 1953, Wilkins wrote a letter acknowledging receipt of a Watson and Crick manuscript that described the structure of DNA. One day earlier, on March 17, 1953, Franklin finished a written draft that concluded that B-form DNA exists as a double helix. If this manuscript had not been simultaneously published with the Crick and Watson paper in the April issue of *Nature,* they would have had to cite Franklin's work as a seminal piece of the puzzle.

After leaving King's College, Rosalind Franklin changed her focus to ribonucleic acid (RNA) viruses and virus particles. During this time, she began a long and fruitful collaboration with Aaron Klug. While Klug studied spherical viruses, Franklin focused on rodlike viruses, such as tobacco mosaic virus (TMV). In 1955, against the ideas of the preeminent virologist of the day, Norman Pirie, Franklin published results in *Nature* correctly proposing that the building blocks (virus particles) of TMV were the same length. Working with the graduate student Kenneth Holmes, Franklin elucidated the structure of TMV. Later, she and Don Casper separately demonstrated that the RNA in TMV wound along the inner surface of the hollow virus.

Rosalind Franklin was an outstanding and accomplished scientist—a fascinating person with a strong personality that left a lasting impression on everyone she met. Outside the laboratory, friends found her bright, witty, and fun. But in the laboratory, she displayed no lightheartedness as she focused exclusively on her work, often bypassing the morning coffeepot and afternoon tea, social heresy in Great Britain, where she worked. In summer 1956, on a work-related trip to the United States, Franklin fell ill. An operation performed in September 1956 found two tumors in her abdomen. She convalesced at Francis Crick's home, The Golden Helix, until she returned to work. In that year, she published seven papers, and she continued working in 1957, preparing six more publications. A promotion to research associate in biophysics occurred at Birbeck in January 1958. She fell ill again at the end of March and subsequently died of ovarian cancer on April 16, 1958. Long exposure to X-ray radiation may have contributed to her early death, although she behaved no more recklessly than other X-ray crystallographers of her time.

See also BIOCHEMISTRY; CRICK, FRANCIS; DNA REPLICATION AND REPAIR; NUCLEIC ACIDS; PAULING, LINUS; WATSON, JAMES; WILKINS, MAURICE; X-RAY CRYSTALLOGRAPHY.

FURTHER READING

Maddox, Brenda. *Rosalind Franklin: The Dark Lady of DNA.* New York: HarperCollins, 2002.

Sayre, Anne. *Rosalind Franklin and DNA,* 2nd ed. New York: W. W. Norton, 2000.

fullerenes In 1985 scientists knew that the element carbon existed in two pure forms, the three-dimensional crystalline array of carbons assumed by diamonds and the two-dimensional stacked sheets of hexagonal arrays of carbon atoms assumed by graphite. Trying to prove that long-chain hydrocarbons existed in interstellar dust, Harold Kroto, Robert Curl, and Richard Smalley set up a collaboration; results from their studies yielded discrete peaks, corresponding to two molecules with the exact masses of 60 or 70 carbon atoms, respectively, in data obtained by mass spectrometry (a technique using the properties of light to determine components of a substance on the basis of their mass). Kroto and the others initially believed that these molecules were the long-chained molecules for which they were searching, but further analysis revealed that they had discovered a new type of molecule that took the form of a hollow, closed cage. They named the carbon molecule with 60 carbons *buckminsterfullerene,* after the architect and philosopher Richard Buckminster Fuller, who built geodesic domes (seen at places like Epcot at Walt Disney World) that had the same shape as the newly identified molecule. Scientists working on this third form of pure carbon molecules shortened the name to *fullerene,* and the general public nicknamed these molecules "buckyballs," one of the most significant discoveries in the physical sciences since superconductors. The renowned journal *Science* went so far as to name the buckyball the molecule of the year in 1991, saying that its discovery would shape the course of scientific research for years. The C60 fullerenes, a specific type of buckyballs, consist of a mixture of pentagon and hexagon rings bound together into round, hollow molecular cages resembling the classic black-and-white-patterned soccer ball. The C70 fullerenes consist of additional hexagon rings, resulting in a contortion of the sphere into a less stable, elliptical shape that resembles a rugby ball. By the early 1990s, scientists had discovered hundreds of combinations of these interlocking pentagon/hexagon molecules such as buckybabies (spheroid molecules with fewer than 60 carbon atoms), fuzzyballs (buckyballs with 60 hydrogen atoms attached), giant fullerenes (fullerenes with hundreds of carbon atoms), nanotubes (tubes of ring-shaped building blocks), and dopeyballs (buckyballs that have metals inserted into their hollow core). All of these fullerenes are relatively stable: that is, they retain their structural integrity despite their interaction with other atoms or molecules and their exposure to high temperatures and pressures. Because they are versatile, modifiable, and stable, fullerenes have enormous practical applications in electronics, medicine, metallurgy, optics, and nanotechnology. Considered a virtual playground of possibilities for chemists, physicists, and material scientists, the discovery of fullerenes led to Curl, Kroto, and Smalley's 1996 Nobel Prize in chemistry.

BUCKYBALLS

Fullerenes are a class of discrete molecules that are similar in structure to graphite (another pure carbon molecule), not only composed of sheets of hexagonal rings (as graphite) but also containing pentagonal rings (unique to fullerenes) that cause the sheet to bend in toward itself to form a spherelike structure. The symmetry of the soccer ball–shaped C60 buckyballs contributes to the fascination with the shape of the molecule, which artists and mathematicians have exhibited over the centuries. As early as 250 B.C.E., the shape assumed by the C60 molecule was appreciated—the great scientist and mathematician Archimedes deemed it important enough to be one of the "Archimedean solids." The German astronomer Johannes Kepler coined the name *truncated icosahedron* for the C60 buckyball. Since finding that these molecules naturally exist in and out of the laboratory, physical scientists have let their imaginations run wild about the possible uses of buckyballs. (Candle soot contains measurable quantities of fullerenes, and helium-associated fullerenes have been found at geologic sites for meteroids.) These molecules might be used in anything from controlled drug release of anticancer agents to creation of nanostructures with tailored electronic properties for large-scale industrial applications. So stable is this entire family of naturally hollow molecules that they will certainly stand up to the rigors of the manufacturing process, making them attractive subjects for research and development in multiple industries. With the ability in hand to synthesize these molecules in substantial quantities, a new branch of chemistry, fullerene chemistry, has come into being, continues to identify new members of the fullerene family, and studies the structural, thermodynamic, mechanical, and electrical properties of fullerenes.

One C60 buckyball measures approximately one-billionth of a meter in diameter. To put this measurement in perspective, if the C60 buckyball were the size of an actual soccer ball, a soccer ball would be the size of the Earth. The C60 buckyball is the roundest molecule known. As do all geodesic structures, this buckyball consists of 12 pentagon rings that transform the network of atoms into a spheroid, as first predicted by the 18th-century Swiss mathematician Leonhard Euler. The remaining carbons in the C60 buckyball form 20 hexagon rings that fill in the gaps to make a continuous hollow sphere resembling a soccer ball. In both the pentagon rings and the hexagon rings, carbon atoms exist at each of the vertices, and bonds between carbon atoms create the edges of the rings. Double bonds link two carbon atoms within a single hexagon, but single bonds link a carbon atom from a hexagon ring to a carbon atom from a pentagon ring. The C60 buckyball is also the most symmetric molecule known. Within this fullerene, 120 identical points of symmetry exist within the molecule. The symmetry and roundness of the C60 molecules confer a great deal of stability to the molecular structure. With more or fewer than 20 hexagonal rings, fullerenes lose their roundness. The C70 buckyball, with 25 hexagons, looks more like an elliptically-shaped rugby ball. Giant fullerenes, like the C960 buckyball with hundreds of hexagons, take on a pentagon-shaped structure. Smaller fullerenes, with fewer than 20 hexagons, look like asteroids. With the roundness lost, these other fullerenes lose their intrinsic stability as well.

To measure the physical properties of this family of molecules, the German scientist Wolfgang Krätschmer at the Max Planck University developed a method for synthesizing large amounts of fullerenes. To synthesize fullerenes, researchers generate a plasma arc between two graphite rods in the presence of helium gas. The heat from the arc vaporizes the rods into individual carbon atoms that coalesce to form sheets. Scientists believe that the inert helium gas holds the carbon atoms close enough to the arc and to each other that they close in on themselves, forming millions and millions of predominantly C60 and C70 fullerene spheres. To obtain pure crystals of these fullerenes, one uses chromatography techniques to separate the two prevalent fullerenes, C60 and C70, from each other, and condenses the purified fullerenes into powder form. Chemists then vaporize the powder of buckyballs at high temperature inside a quartz tube. With the right conditions of temperature, concentration, and gas inside the quartz tube, fullerene crystals grow on the walls of the tubes. To synthesize fullerenes to their powdered form, the process takes 10 minutes; to make pure fullerene crystals, the process takes 10 days or more. Once a pure crystal is in hand, fullerenes can be doped, meaning that single atoms such as potassium or rubidium may be inserted into the hollow core of individual bucky balls.

Fullerenes are stable molecules yet possess a certain degree of reactivity. Fullerenes tend to act as electrophiles, accepting hydrogen atoms at the unsaturated double bonds between adjacent hexagonal carbons. Before saturation, electrons of the unsaturated carbon atoms are in a sp^2-hybridized orbital with a bond angle of 120 degrees. When the double bond between the hexagonal carbons becomes saturated, the electrons move to a sp^3-hybridized orbital with a bond angle of 109.5 degrees. The decrease in bond angles in this saturated state leads to a subsequent decrease in angle strain when closing the sphere, which leads to a more stable molecule. The hexagonal carbons exhibit no other aromatic properties: that is, the electrons in these orbitals never delocalize over the rest of the molecule but stay within the confines of the hexagonal arrays.

Fullerenes possess other intrinsic properties that chemists have elucidated since their discovery. C60 and C70 fullerenes will dissolve in 1,2,4-trichlorobenzene, carbon disulfide, toluene, and benzene, even at room temperature. Solutions of C60 fullerenes take on a purple color, and solutions of C70 fullerenes are reddish brown. Other fullerenes, both smaller fullerenes and most larger fullerenes, are insoluble, with the exception of C72 fullerene when a metal is associated with it.

When chemists add other atoms and molecules into the hollow core of fullerenes, they create endohedral fullerenes, a new class of fullerenes that has additional intrinsic properties. When an alkali metal (such as potassium or rubidium) becomes internalized into a fullerene core, the fullerene conducts electricity. When chemists insert an organic reducing agent, the endohedral fullerene created exhibits magnetic properties despite the absence of a metal within the molecular structure, an unprecedented phenomenon.

When fullerenes form crystal structures such as diamond and graphite, they can serve not only as insulators but also as conductors, semiconductors, or even superconductors. The superconductivity of endohedral fullerenes that contain alkali metals and alkaline earth metals particularly interests physical scientists because they have a high transition temperature (T_c) at 40 K (-387°F or -233°C). Below this critical temperature, these molecules conduct electrical current without any resistance. Vibrations within the crystal lattice and the resulting photons (quanta or particles associated with these lattice vibrations) attract electrons within individual fullerene molecules, and the attraction causes the electrons to pair up. With pairs of electrons in the crystal lattice, the motion of one electron is precisely countered by its partner, so no overall resistance to the flow of an electric current through the crystal exists. The C60 fullerene is the best superconductor because its spherical shape increases the opportunity for interactions of electrons and phonons to occur as they bounce off the curved surfaces of the molecule.

NANOTUBES

Another type of fullerene exists, fullerene tubes or nanotubes, one-dimensional cylinders with a wider range of potential applications than the buckyballs.

First discovered in Japan, these cylinders are composed of hexagonal and pentagonal arrays of carbon atoms that form tubes with rounded ends. These ends are highly reactive with other atoms because of their curvature. These nanotubes are typically a few nanometers wide in diameter, and they can be a micrometer to a millimeter in length. As buckyballs do, these molecules behave as insulators and also as conductors or semiconductors, depending on how the tube is closed. To increase the versatility of these tubes further, one could place an atom into the core of the tube, in a manner similar to inserting atoms into other fullerenes. To achieve this end, scientists continue their efforts to grow a carbon cage around an atom with the intention of forcibly injecting the caged atom into the nanotube using a scanning tunneling microscope to breach the carbon walls. The unique cylindrical structure of the molecule imparts several significant features to the nanotube, including a high tensile strength, high electrical conductivity, and high resistance to heat. Possibly the world's strongest wire, a one-meter nanotube has 50 to 100 times the strength of a similar length of steel, at one-sixth of the weight. In the future, nanotubes may replace traditional steel as the structural support of buildings and the skeletons of airports and cars. The electrical conductivity of nanotubes is four orders higher than the current-carrying capacity of copper, a traditional conductor. Commercial use of nanotubes as conductors could revolutionize the electronics industry, specifically computer electronics and molecular electronics. The limiting factor to nanotubes' usefulness to date surrounds the accurate synthesis of the molecule desired. When the synthesis protocols become accurate, the practical applications of both nanotubes and buckyballs will be realized.

See also BONDING THEORIES; ELECTROCHEMISTRY; ELECTRON CONFIGURATIONS; MAGNETISM; MATERIALS SCIENCE; NANOTECHNOLOGY; ORGANIC CHEMISTRY.

FURTHER READING

Aldersey-Williams, Hugh. *The Most Beautiful Molecule: The Discovery of the Buckyball.* New York: John Wiley & Sons, 1997.

Hirsch, Andreas, Michael Brettreich, and Fred Wudl. *Fullerenes: Chemistry and Reactions.* New York: John Wiley & Sons, 2005.

Koshland, Daniel E., and Elizabeth Culotta. "Buckyballs: Wide Open Playing Field for Chemists (Molecule of the Year; Includes Information on Other Significant Stories of 1991)." *Science* 254 (1991): 1706–1710.

fusion The energy associated with the attraction between subatomic particles is known as nuclear energy. The energy produced by nuclear reactions is nearly a million times greater than the amount of energy stored in an atomic bond. If harnessed, that energy could be used to produce electricity, heat, and fuel, replacing the need for fossil fuels in the production of energy for the world. Models predict that fossil fuel supplies (coal, natural gas, and oil) will only last approximately 50 more years. Fossil fuels are presently the main source of energy used for heating, transportation, and production of electricity. Two types of nuclear reactions release the energy found between subatomic particles: nuclear fission and nuclear fusion. Fission results in the splitting of atomic nuclei, and fusion results in the joining of two atomic nuclei. Fusion, although not yet attainable, is being heavily researched as a safe, alternative energy source.

Nuclear fusion is a method of producing energy by combining two smaller atomic nuclei into one larger nucleus. The fusion of any nuclei with a lower mass than iron (with an atomic mass of 26) will lead to the release of energy, as long as the mass of the resulting nucleus is less than the sum of the two individual nuclei prior to fusion. The extra mass is converted to energy, a phenomenon explained by Albert Einstein's famous equation

$$E = mc^2$$

where E represents the energy equivalent in joules (J) of mass m in kilograms (kg), and c is the speed of light in vacuum, whose value is 3.00×10^8 meters per second (m/s).

Fusion requires extremely high temperatures to overcome the natural repulsion between the positively charged nuclei, temperatures that naturally occur on the surface of the Sun, catalyzing the fusion of hydrogen and helium nuclei there with the concomitant production of solar energy. These high temperatures are necessary in order to increase the kinetic energy of the nuclei in order to overcome the natural repulsive forces between the protons known as Coulomb forces. Hydrogen has the smallest possible charge of all nuclei (+1), thus requires less-energetic collisions to cause fusion than other nuclei, but it still requires temperatures approaching millions of degrees Celsius, conditions unattainable on the surface of the Earth except in specialized reactors. Even if temperatures that high are attainable, the new problem of what material could withstand and contain the matter undergoing fusion arises. The only state of matter that exists at temperatures this high is plasma, which possesses such high kinetic energy that the matter is stripped of all of its electrons. Since the atoms of matter in the

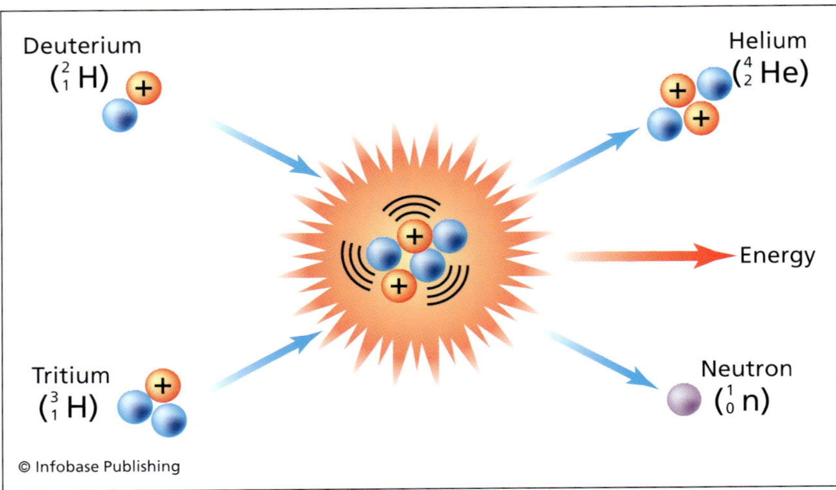

In deuterium-tritium fusion a helium atom, a neutron, and energy are produced.

plasma state are stripped of their electrons, they become charged and are referred to as ions in this state.

An isotope is a form of an atom that contains the same number of protons but a different number of neutrons and therefore has a different atomic mass. Several isotopes of hydrogen exist: the most common form of hydrogen has an atomic mass of 1 and contains one proton and one electron; deuterium has a mass of 2 and contains one proton, one neutron, and one electron; tritium, an atomic mass of 3 and one proton, two neutrons, and one electron. All of the hydrogen isotopes contain one proton, or they would not be hydrogen, because the number of protons defines the element. The fusion of the hydrogen isotopes deuterium and tritium is the most promising type of reaction for successfully achieving fusion. The combined total of particles would be two protons and three neutrons. The nuclear fusion results in the production of one new helium nucleus that contains two protons and two neutrons, and the release of one free neutron and a large amount of energy (see the following equations). The atomic number of the isotope is written as a subscript preceding the symbol, and the mass number of the element is written as a superscript preceding the symbol. The energy released in each process is given in units of a million electron volts (MeV), where 1 million electron volts equals 1.60×10^{-13} joule (J).

Deuterium-deuterium fusion occurs by the following reactions:

$$^2_1\text{H} + ^2_1\text{H} \rightarrow ^3_2\text{He} + ^1_0\text{n} + 3.27 \text{ MeV}$$

$$^2_1\text{H} + ^2_1\text{H} \rightarrow ^3_1\text{H} + ^1_1\text{H} + 4.03 \text{ MeV}$$

Deuterium-tritium fusion occurs by the following reaction:

$$^2_1\text{H} + ^3_1\text{H} \rightarrow ^4_2\text{He} + ^1_0\text{n} + 17.59 \text{ MeV}$$

The symbol $^2_1$H represents deuterium; $^3_1$H represents tritium; $^1_1$H represents hydrogen; $^4_2$H represents helium; and $^1_0$n represents a neutron.

Using fusion as an energy source has several advantages. The amounts of air pollution, thermal pollution, and nuclear waste produced are low, and the reactants are readily available. The main end product of the reaction, helium, is an inert gas that does not damage the environment or harm humans. Seawater naturally contains enough deuterium to provide an essentially limitless supply, since most of the Earth is covered with water. Deuterium can be isolated from ocean water by electrolysis, but the tritium required for the most productive deuterium-tritium fusion reaction must be produced from lithium in a process known as breeding, shown in the following reaction.

$$^6_3\text{Li} + ^1_0\text{n} \rightarrow ^4_2\text{He} + ^3_1\text{H} + 4.8 \text{ MeV}$$

where lithium ($^6_3$Li) reacts with a free neutron ($^1_0$n) to produce helium ($^4_2$He) and tritium ($^3_1$H) that can then be used in the deuterium-tritium fusion reaction. Breeder reactors specifically synthesize the tritium necessary to run fusion experiments. Lithium is abundant in Earth's surface, and the deuterium supply is plentiful, so tritium production is the limiting factor that determines the amount of fusion product formed.

Another advantage of generating energy by fusion is that the fusion process does not function by a chain reaction; therefore, turning off a reactor stops the reaction, eliminating the potential for a reaction's getting out of control and leaking radioactivity into the surrounding community. The production of radioactive tritium by the breeder reaction poses a problem with containment and disposal, but the half-life of tritium is only about 10 years, so long-term storage is not a problem. Because of these advantages, deuterium-tritium fusion is the most promising option for designing functional fusion reactors.

The major problem with fusion reactors is that no country has been successful in creating one that works. A functional fusion reactor must surpass the breakeven point, the point where the amount of energy produced exceeds the amount of energy necessary to run the reactor.

The procedure for fusion requires a thermonuclear reaction at extremely high temperatures and pressures. Three factors that need to be controlled in a fusion reaction are temperature, time, and ion concentration. The deuterium-tritium fusion reaction produces 17.6 MeV but requires temperatures in the millions of degrees Celsius to be performed. The temperature of the reaction must reach a critical ignition temperature, and it must be held there for the critical length of time in order for the plasma to be able to react. If the concentration of the components is not sufficiently high to induce a significant number of collisions, then there is really no point producing the high temperatures to create the plasma: no appreciable amount of energy will be produced. In 1957 J. D. Lawson determined the minimal criterion predicted for the success of a fusion reaction. Used in evaluating reactors, Lawson's criterion includes the length of time the critical ignition temperature must be maintained (the confinement time) and the ion density required in order to achieve fusion.

Part of the energy released during a fusion reaction must heat other nuclei to provide energy needed for fusion to continue, to be self-sustained. Fusion reactors generally fall into one of two categories based on the method of confinement, the means of maintaining the conditions necessary for the plasma to undergo fusion: magnetic confinement reactors and inertial confinement reactors. The magnetic confinement reactors prevent the plasma from ever touching the containment material by surrounding it with a circular magnetic field in two directions, causing the plasma to move in a circular path without touching the surface of the container. This allows the temperature to be maintained high enough (critical temperature) for a long enough period (confinement

Fusion reactions must take place within a contained environment. A tokamak maintains the plasma separate from the surface by utilizing magnetic fields. The Tokamak Fusion Test Reactor (TFTR) shown here was in operation from 1982 to 1997 at the Princeton Plasma Physics Laboratory. *(Princeton Plasma Physics Laboratory)*

This mushroom cloud is the result of the explosion of a thermonuclear bomb, commonly called a hydrogen bomb, in the Mike hydrogen bomb test in 1952 on the island of Elugelap in the Marshall Islands. Such an explosion is an uncontrolled nuclear fusion reaction, triggered by the explosion of a fission (atomic) bomb. *(Courtesy of the Los Alamos National Laboratory Archives)*

time) without destroying the confinement materials. The tokamak is one such system, which supplies a magnetic field in two directions, allowing for the confinement and heating of the plasma. The Tokamak Fusion Test Reactor (TFTR) that operated at the Princeton Plasma Physics Laboratory from 1982 to 1997 was a very successful magnetic fusion device. TFTR set records for producing 10.7 million watts of power using a fuel mix of half deuterium and half tritium in 1994, and for reaching a temperature of 510 million degrees Celsius in 1995. While the tokamak has reached the critical ignition temperature and almost fulfilled Lawson's criterion, scientists have not been able to accomplish both feats simultaneously.

Presently the largest tokamak in the world is JET (Joint European Torus facility) in Abingdon, in the United Kingdom, prepared by the European Fusion Development Agreement (EFDA), a collaboration among European fusion research institutes. Information obtained using JET will facilitate development of the newest collaborative effort, known as ITER (International Thermonuclear Experimental Reactor). Seven partners have joined ITER: the European Union, United States of America, Russia, China, South Korea, India, and Japan. The goal for ITER, which will be much larger than JET but share the same magnetic construction, is to achieve sustained burn, a point at which the reactor becomes self-heating and productive. ITER will be built at Cadarache, France, and should be finished by 2015.

The second strategy for confinement is called inertial confinement. In inertial confinement reactors, the nuclei need to be in close enough contact to undergo fusion. Though there is no physical means of confinement in this method, the entire process happens very quickly, so the nuclei cannot escape their own inertia, and they undergo fusion before the energy escapes. Rather than maintaining the temperature for an extended period, these reactors raise the temperature on the nuclei incredibly quickly so they are forced to react with one another and fuse. The best known of the inertial confinement reactors are found at Lawrence Livermore National Laboratories (LLNL) and include the Shiva and the more powerful Nova. LLNL has been run by the University of California since its inception in 1952, for the U.S. Department of Energy.

An uncontrolled fusion reaction drives the hydrogen bomb, also known as a thermonuclear bomb. The bomb is designed with an atomic bomb (fission bomb) at the center surrounded by lithium and deuterium that will participate in the events leading up to fusion. Detonation of the atomic bomb creates the temperature and pressure required for fusion as well as providing the neutrons required by lithium to create the tritium necessary for fusion. The fusion reaction releases an enormous quantity of heat and pressure, causing waves of damage. The first hydrogen bomb was created by the United States in 1953. Today many countries maintain similar nuclear weapons capabilities.

With the difficulty and expense of using traditional fusion production, one hopes for the development of an easier, less costly, and more usable type of fusion experiment. This process was thought to have occurred in March 23, 1989, when two scientists, Stanley Pons and Martin Fleischmann, claimed to have successfully completed a "cold fusion" experiment. The reaction did not really occur in the cold—it was actually a room temperature experiment—but relative to the temperatures in fusion reactors, it was cold. Much controversy surrounds these results. While many scientists claim that the experiments were falsified, many others claim that they were successful. Much more research needs to be done to determine whether cold fusion is a real possibility.

Because fusion power does not generate high levels of radioactive waste and because it is renewable (sustainable), fusion is an attractive alternative energy source. With current technological limitations, fusion power is not yet feasible or practical, but physicists and engineers are actively working to remedy this.

See also ALTERNATIVE ENERGY SOURCES; ENERGY AND WORK; FISSION; RADIOACTIVITY.

FURTHER READING

Hsu, Tom. *Foundations of Physical Science with Earth and Space Science.* Peabody, Mass.: Cambridge Physics Outlet, 2003.

Galilei, Galileo (1564–1642) Italian *Physicist, Mathematician, and Astronomer* Sir Isaac Newton is famously quoted as stating, "If I have seen farther it is by standing on the shoulders of Giants." The giants to whom he was referring included Galileo Galilei and Johannes Kepler, especially Galileo. Today's scientists recognize Newton and Galileo together as the fathers of modern science. It was Galileo who introduced unwavering, consistent dependence on observation and measurement in the evaluation of scientific theories. Before him, there was no serious observation or measurement at all, only metaphysics and dogma. The ancient Greek thinker Aristotle had made some claims about the natural world. The Catholic Church, a powerful force in Europe at that time, supported and defended those claims that were consistent with its then-literal interpretation of the Bible. The most important of those claims was the cosmological scheme called geocentrism. In this scheme Earth holds the position as the center of the universe, while all the heavenly bodies, consisting of the Sun, the Moon, the planets, and the stars, revolve around Earth on "spheres." Earth and its inhabitants were deemed imperfect, while the heavenly spheres and their heavenly bodies were supposed to be perfect. Woe to anybody who questioned that accepted picture of the world. Such a person was in danger of being ostracized from the community of natural philosophers and, moreover, exposed himself to the wrath of the Inquisition, that powerful body of the church that "defended" the faith from heresy. Galileo not only questioned the prevailing dogmas about the natural world, but proved some of them wrong.

GALILEO'S LIFE

Galileo was born on February 15, 1564, in Pisa in what is now Italy, and was the first of the six children of his father, Vincenzio Galilei, and mother, Giulia Ammannati. Galileo's father was a musician and music theorist, who studied such matters as the relation of the pitch of a plucked string to the force stretching the string, the string's tension. This interest exposed Galileo at an early age to the investigation of phenomena through measurement. As a young man, Galileo considered becoming a priest, but at his father's urging, instead started studying medicine at the University of Pisa. In the middle of his course of study, he switched direction and studied mathematics instead, obtaining his degree in that field.

In 1589, at the age of 25, Galileo was hired to teach mathematics at the University of Pisa, and three years later he took a position at the University of Padua, where he taught astronomy and mechanics, in addition to mathematics, for 18 years, until 1610. Although never married, Galileo had three children with Marina Gamba, who were born during 1600–06. The two oldest were girls, Virginia and Livia, born in 1600 and 1601, respectively. They were both sent to a convent, where they took the names *Maria Celeste* and *Suor Arcangela* and remained for their whole lives. The son, Vincenzio, was born in 1606. While at the University of Padua, Galileo made important discoveries in pure and applied science, including the study of motion, astronomy, and telescope construction.

From 1610 on, Galileo was in continuous interaction and controversy with the Catholic Church and the Inquisition concerning his discoveries and ideas and about their publication. Matters came to a head in 1633, when Galileo stood trial before the Inquisition for heresy. He was sentenced to recant his heliocentric (Sun-centered) view of the solar system; to go to prison, later eased to house arrest, for life; and it was ordered that a recently published offending book of his, *Dialog Concerning the Two Chief World Systems,* be banned and that none of his past or future works be published. But Galileo continued his investigations and writing and in 1637 completed another work, his last, *Discourses on Two New Sciences,* which was smuggled out of Italy and published in Holland the following year.

By 1638 Galileo had lost his vision. He died on January 8, 1642.

DISCOVERIES IN PHYSICS AND ASTRONOMY

In science Galileo was an innovator in his insistence on performing measurements and describing natural laws mathematically. It was reported that he dropped balls from the Leaning Tower of Pisa, demonstrating that, neglecting air resistance, all bodies fall the same distance during the same time interval. That clearly refuted Aristotle's claim, generally accepted at that time as dogma, that heavier objects fall faster than lighter ones. Although it is in doubt as to whether he actually performed the experiment, there is no doubt that Galileo performed rigorous experiments involving balls rolling down inclined planes. He was the first to demonstrate that the acceleration of similar rolling bodies is independent of their mass. Since rolling is falling in slow motion, Galileo's result refuted Aristotle's claim. From his investigations, Galileo determined the correct relation between the distance covered by a body falling from rest and the elapsed time: that the distance is proportional to the square of the time. In addition, he discovered his principle of inertia: a body moving on a level surface will continue in the same direction at constant speed unless disturbed. This principle later developed into Newton's first law of motion.

Another discovery that Galileo made was that the period—the time of one complete swing—of a pendulum is independent of the amplitude of the pendulum's swing. Thus, for instance, as a freely swinging pendulum loses energy (a concept yet unknown to Galileo) as a result of air resistance and friction in the bearing, and its amplitude concomitantly decreases, its swing should describe equal time intervals throughout. Legend has it that Galileo made his discovery while watching the swinging chandelier in Pisa's cathedral, using his pulse as a timer. On the basis of this result, Galileo seems to have been the first to suggest using a pendulum to control a clock. After Galileo's time, mathematicians

Galileo was a 16th–17th-century scientist whose ideas and experimental results laid a foundation for later understanding of mechanics and gravitation. He was infamously persecuted by the religious authorities for publicly claiming (correctly) that Earth and the planets revolve around the Sun (rather than that the planets and the Sun revolve around Earth). *(Scala/Art Resource, NY)*

proved that Galileo's conclusion is strictly valid only for small amplitudes of swing.

In 1608 a report reached Galileo that a telescope had been invented in the Netherlands. Although the report was sketchy on details, Galileo constructed a working telescope that same year and then improved it, attaining a magnification of 32-fold. His telescopes performed well for use on land and also for astronomical observations. Using his telescope, in 1610 Galileo discovered four of Jupiter's moons. They are Io, Europa, Callisto, and Ganymede and are now named the *Galilean satellites* in honor of Galileo. The concept of a planet with bodies orbiting it was at odds with the geocentric model of the universe, which had everything revolving around Earth.

Later Galileo observed that the planet Venus went through a complete cycle of phases, just as the Moon does. This observation refuted the geocentric idea that Venus and the Sun revolve around Earth with Venus always between the Sun and Earth and showed that Venus orbits the Sun—another nail in the coffin of geocentrism.

In addition to his other astronomical observations, Galileo observed sunspots on the Sun and mountains and craters on the Moon. Both phenomena clashed with the dogma of perfection of the heavenly bodies, which were held to be unblemished.

In 1638 Galileo proposed a method for measuring the speed of light, which involved two observers with lanterns. When the experiment was carried out, though, the results were inconclusive. Galileo achieved an understanding of the frequency of sound and proposed a theory for the tides, which was wrong.

Galileo also made contributions to technology. He devised a compass for artillery and surveying and made a thermometer that was based on the expansion and contraction of air. He also constructed a compound microscope, devised a mechanism for a pendulum clock, and designed various devices, including an automatic tomato picker.

CONFLICT WITH THE CATHOLIC CHURCH

As Galileo's discoveries discredited the geocentric model of the universe more and more, he became convinced of the correctness of the heliocentric model, which Nicolaus Copernicus had proposed in 1543. In this scheme, the Sun and stars were fixed; Earth revolved around the Sun, as did all the other planets; and the Moon revolved around Earth. The theological problem with the heliocentric model, as perceived by some in the Catholic Church, was that while humanity was believed to hold a central position in creation, heliocentrism did not respect this tenet geometrically. The geocentric model indeed realized this belief in its geometry, with Earth and humanity situated at the center of things. The heliocentric picture, on the other hand, had the Sun, rather than Earth, at the geometric center of the solar system (if not of the universe), with Earth forming just another planet and Earth's inhabitants not centrally located. Further, the idea that Earth moved and the Sun was stationary seemed to contradict the literal reading of certain biblical passages.

The evidence that Galileo had amassed put the geocentric model in doubt and raised the credibility of the heliocentric model. Galileo did not hesitate to express his opinions in the matter. Some in the Catholic Church were angered by this and pushed to have Galileo reined in. In 1616 the Inquisition decreed that Galileo was no longer to hold or defend the idea that the Sun stands still while Earth moves. Galileo did not consider this decree as forbidding him to investigate and discuss heliocentrism as a hypothetical alternative to geocentrism. He refrained from controversy for a while. When a pope who was favorable to Galileo was elected in 1623, Galileo continued working on his book *Dialog Concerning the Two Chief World Systems*. This book was supposed to confine itself to presenting arguments for and against heliocentrism, but not to advocate that model. It was granted permission to be published by the pope and the Inquisition.

But when the book was published in 1632, the pope and other powerful figures in the church considered it biased, and the Inquisition ordered Galileo to stand trial for heresy in 1633. The Inquisition found Galileo guilty and imposed the following sentence:

- He must recant the heliocentric view, which the Inquisition declared heretical.
- He was sentenced to life imprisonment, which was later commuted to house arrest.
- The Inquisition banned his offending book and forbade the publishing of his past and future works.

Popular legend has it that as Galileo was leaving the courtroom after publicly declaring before the Inquisition that Earth stood still, as he was required to do, he muttered under his breath, "Eppur si muove" (and yet it moves). A nice tale, but apparently unfounded.

From that time, the Catholic Church was viewed askance by the community of scientists and objective thinkers for its treatment of Galileo, for its ban of heliocentrism, and for its inclusion of Galileo's and Copernicus's books in its index of forbidden works. From 1718 on, however, the church gradually eased its restrictions and had completely abandoned its antiheliocentric position by 1835, when Galileo's and Copernicus's books were finally dropped from the index. In 1992 the pope officially expressed regret for the Catholic Church's treatment of Galileo.

GALILEO'S LEGACY

Galileo gave subsequent generations the telescope, the pendulum clock, and other devices. He made astronomical discoveries and advanced the heliocentric model. He laid a firm foundation for the study of motion (kinematics), upon which Newton constructed his magnificent theoretical edifice. He fought blind adherence to dogma in understanding the natural world and got into deep trouble for doing so.

But most important of all, from the viewpoint of today's science, was Galileo's separation of physics from metaphysics. Before Galileo, the people attempting to understand the natural world were philosophers, who mainly applied metaphysical considerations and very little observation. For example, objects fall as a result of their natural tendency to be at the center of Earth and of the universe. More massive objects must fall faster, since they have a greater such tendency. Since the heavenly spheres are perfect, heavenly bodies must possess perfectly round shapes and be unblemished. That was not science. Galileo gave birth to true science by being the first to insist that the only valid criterion for the truth of any theory about the natural world was its consistency with observation and measurement—not its consistency with the Bible or with theological considerations or with Aristotle's ideas, but *solely* its consistency with measurement and observational results. Facts must take precedence over ideology, if any meaningful and useful understanding is to be achieved.

See also ACCELERATION; ENERGY AND WORK; FORCE; HARMONIC MOTION; KEPLER, JOHANNES; MASS; MOTION; NEWTON, SIR ISAAC; OPTICS; TELESCOPES.

FURTHER READING

Brecht, Bertolt. *Galileo.* New York: Grove, 1994.

Cropper, William H. *Great Physicists: The Life and Times of Leading Physicists from Galileo to Hawking.* New York: Oxford University Press, 2001.

Galilei, Galileo. *Dialog Concerning the Two Chief World Systems.* New York: Modern Library, 2001.

Shea, William R., and Mariano Artigas. *Galileo in Rome: The Rise and Fall of a Troublesome Genius.* New York: Oxford University Press, 2003.

Sobel, Dava. *Galileo's Daughter: A Historical Memoir of Science, Faith, and Love.* New York: Penguin, 2000.

gas laws

A gas is a state of matter that has no definite shape or definite volume. The molecules of a gas move rapidly, and the kinetic molecular theory determines the behavior of a gas. The first principle of the kinetic molecular theory of gases states that all matter is made up of atoms and molecules that have a very small volume relative to the entire volume occupied by the gas and that have negligible attractive or repulsive forces for each other. The second point of the kinetic molecular theory of gases is that the particles of a gas are in constant random motion and the collisions are perfectly elastic, meaning that no kinetic energy is lost when the collisions occur. Four factors that affect a gas are pressure (P), volume (V), temperature (T), and number of moles of a gas (n). The laws governing the behavior of gases include the relationship between volume and pressure, described by Boyle's law; the relationship between temperature and volume, known as Charles's law; the relationship between the number of moles and the volume of a gas, referred to as Avogadro's law; and the relationship between the pressure and temperature of a gas, called Gay-Lussac's law.

Pressure is the force per unit area. The pressure of a gas can be measured by the force exerted on the walls of a container by the gas molecules as they collide with the walls of the container. The barometer is the tool used to measure the pressure of a gas. The SI unit of pressure is the pascal (Pa), equal to one newton per square meter (N/m^2); other commonly used units are atmosphere (atm), kilopascal (kPa), and millimeter of mercury (mm Hg). The relationships among these units of pressure are given by the following equation:

$$\begin{aligned} 1 \text{ atm} &= 760 \text{ mm Hg} = 101.3 \text{ kPa} \\ &= 1.013 \times 10^5 \text{ Pa} \end{aligned}$$

Temperature is a measure of the amount of random kinetic energy that the particles of matter possess. The higher the random kinetic energy is, the higher the temperature of the matter. The Fahrenheit temperature scale was named after the German physicist Daniel Gabriel Fahrenheit. Degrees Fahrenheit are related to degrees on the Celsius temperature scale, named for the Swedish astronomer Anders Celsius, by the following formula:

$$°\text{C} = \frac{5}{9}(°\text{F} - 32)$$

The boiling point of water on the Celsius temperature scale is 100°C and is 212°F on the Fahrenheit scale. The temperature scale commonly used when measuring the characteristics of a gas is the Kelvin scale, named after the Scottish physicist Lord William Thomson Kelvin. The Kelvin scale is based on absolute zero, the theoretical temperature at which all motion of particles is predicted to cease. Since the temperature of the substance relates to the amount of random kinetic energy in a linear relationship, the linear relationship can be extrapolated in order to predict this temperature. The American physicists Carl Weiman and Eric Cornell and the German physicist Wolfgang Ketterle won the Nobel Prize in physics in 2001 for measuring a temperature within a billionth of a degree of absolute zero. The Kelvin temperature scale uses absolute zero as the zero point and uses the same degree increments as the Celsius scale. This causes the Celsius and Kelvin temperature scales to be offset by 273° and related by the following formula

$$\text{K} = 273 + °\text{C}$$

Robert Boyle, an Irish-born scientist, examined the relationship between volume and pressure of a gas in 1662. Boyle's law demonstrates the relationship between pressure and volume when the amount of the gas and the temperature remain constant. The

Hot air balloons are made to fly by increasing the temperature of the gases within the balloon. As the air becomes warmer, it takes up more space and becomes less dense, causing the balloon to inflate and to rise as a result of buoyant force. *(Rod Beverley, 2008, used under license from Shutterstock, Inc.)*

relationship between volume and pressure is given by

$$PV = \text{constant}$$

where the constant is a proportionality constant. By this formula, the relationship between volume and pressure is an inverse relationship. When the volume of a gas decreases, the pressure increases by the same factor, and vice versa. According to kinetic molecular theory, as the volume that the gas occupies decreases, the molecules are forced closer together and will have more collisions. The more collisions that gas particles have with the sides of the container, the higher the pressure of a gas. When the volume occupied by a quantity of a gas is increased, the pressure of the gas decreases. The larger volume allows the particles to spread out, and, as a result, they will have fewer collisions with each other and with the sides of the container, thus reducing the pressure of the gas.

Boyle's law, described by the following formula, allows for the calculation of the pressure or volume change in a gas when the temperature and number of moles of the gas remain constant:

$$V_1P_1 = V_2P_2$$

V_1 is the initial volume, P_1 is the initial pressure, V_2 is the final volume, and P_2 is the final volume. Any volume unit can serve here, as long as both the initial and final volumes are expressed in the same unit. Similarly for pressure. To illustrate, consider the following problem. If the initial pressure of a gas is 100.36 kPa and it fills a volume of 2.5 liters (L), what is the new pressure of the gas if the volume is expanded to 5.0 L?

$$V_1 = 2.5 \text{ L}$$

$$P_1 = 100.36 \text{ kPa}$$

$$V_2 = 5.0 \text{ L}$$

$$P_2 = ?$$

$$(2.5\text{ L})(100.36\text{ kPa}) = (5.0\text{ L})(P_2)$$

$$P_2 = 50.18\text{ kPa}$$

When the volume doubled, the pressure decreased by a factor of 2, demonstrating the inverse relationship between pressure and volume in Boyle's law.

Charles's law shows the relationship between volume and temperature of a gas as

$$V = \text{constant} \times T$$

where the constant is a proportionality constant. Charles's law was developed in the 1800s by the French scientist Jacques Charles. When the pressure and number of moles of a gas are held constant, the relationship between the volume and temperature is directly proportional. The linear relationship between volume and temperature shows that when the temperature rises, the volume of a gas increases at the same rate. The relationship shown between volume and temperature is utilized to control the altitude of hot air balloons. When the air inside the balloon is heated, the volume of the air expands and inflates the balloon. When the balloon needs to rise higher, the air is heated to expand the volume. Since the mass of air remains constant, increasing the volume lowers the density (mass per volume) compared to the surrounding air, and the balloon rises.

The following formula allows one to use the Charles's law relationship to calculate the change in volume that results from heating a gas, assuming the pressure and the number of moles are kept constant:

$$\frac{V_1}{T_1} = \frac{V_2}{T_2}$$

where V_1 is the initial volume, T_1 is the initial temperature in kelvins (K), V_2 is the final volume, and T_2 is the final temperature in kelvins. Both volumes must be expressed in the same unit, but any volume unit can serve. Here is an example. If the temperature of 2.5 L of a gas under constant pressure and moles is changed from 37°C to 55°C, what is the new volume of the gas? Since the temperature of the gas must be used in kelvins, that conversion must be done first.

$$T_1 = 37°\text{C} + 273 = 310\text{ K}$$

$$T_2 = 55°\text{C} + 273 = 328\text{ K}$$

$$V_1 = 2.5\text{ L}$$

$$V_2 = ?$$

$$(2.5\text{ L})/(310\text{ K}) = V_2/(328\text{ K})$$

$$V_2 = 2.7\text{ L}$$

As stated in Charles's law, as the temperature increased, the volume also increased.

Joseph-Louis Gay-Lussac, a 19th-century French scientist, established the relationship between temperature and pressure of a gas when the volume of the system and the number of moles of the gas samples are constant.

$$P = \text{constant} \times T$$

where the constant is a proportionality constant. The relationship described by Gay-Lussac's law is the reason why aerosol cans display warnings to avoid extreme temperatures. The gas added to the can as a propellant would tend to expand when the temperature was increased; however, the volume of the container is fixed and cannot be changed. Therefore, the relationship becomes one between temperature and pressure. As the temperature increases, the pressure exerted by the expanding gas on the can will also increase. If the pressure increases beyond the limit that the container can withstand, it will explode.

When calculating the relationship between the pressure and temperature of a gas sample using Gay-Lussac's law the following formula is used:

$$\frac{P_1}{T_1} = \frac{P_2}{T_2}$$

where P_1 is the initial pressure, T_1 is the initial temperature in kelvins, P_2 is the final pressure, and T_2 is the final temperature in kelvins. Any pressure unit can be used, as long as the same unit is used for both pressures. As an example, if an aerosol can at room temperature of 72°F and pressure of 101.3 kPa is caught in a fire and the temperature reaches 250°F, what is the new pressure of the gas on the inside of the can? Since the temperature needs to be in kelvins before it can be used in these formulas, the Fahrenheit temperatures need to be converted to degrees Celsius and then to kelvins.

$$T_1 = 72°\text{F}$$

$$T_1(°\text{C}) = \frac{5}{9}\,[T_1(°\text{F}) - 32] = \frac{5}{9}\,(72 - 32) = 22.2°\text{C}$$

$$T_1(\text{K}) = T_1(°\text{C}) + 273 = 22.2 + 273 = 295.2\text{ K}$$

Using the same set of calculations to convert T_2 into kelvins,

$$T_2 = 121.1°\text{C} = 394.1\text{ K}$$

Then,

$$(101.3 \text{ kPa})/(295.2 \text{ K}) = P_2/(394.1 \text{ K})$$

$$P_2 = 135.2 \text{ kPa}$$

As described by Gay-Lussac's law, when the temperature inside the can increased, the pressure inside the can also increased because of the linear relationship of these two factors. Since the volume of the container could not expand, the can could potentially explode.

Avogadro's law, developed by Amedeo Avogadro in the early 1800s, demonstrates the relationship between the number of particles of a gas and the volume of the gas. Avogadro stated that the volume taken up by the same number of moles of a gas will always be the same at a constant temperature and pressure. The relationship between volume and number of moles is

$$V = \text{constant} \times n$$

where the constant is a proportionality constant and n is the number of moles. The volume filled by one mole of a gas at the temperature of 0°C and pressure of 1 atm, called standard temperature and pressure (STP), is 22.4 liters. Because the size difference of different types of gas molecules is small relative to the large amount of space between the gas molecules, the amount of space that the gas samples take up is virtually independent of the type of gas that it is.

The volume occupied by the number of moles of the gas at constant temperature and pressure is given by the following relationship:

$$\frac{V_1}{n_1} = \frac{V_2}{n_2}$$

where V_1 is the volume occupied by n_1 moles of a gas, and V_2 is the volume occupied by n_2 moles of the gas. The same unit of volume must be used for both volumes, but any unit is allowed. Consider this example. What volume will 3.6 moles of a gas occupy at STP?

$$V_1 = 22.4 \text{ L}$$

$$n_1 = 1 \text{ mol}$$

$$V_2 = ?$$

$$n_2 = 3.6 \text{ mol}$$

$$(22.4 \text{ L})/(1 \text{ mol}) = V_2/(3.6 \text{ mol})$$

$$V_2 = 80.6 \text{ L}$$

As described by the direct proportionality between volume and number of moles, when the number of moles was nearly quadrupled, the volume of the gas increased by the same factor.

Gases often exist in situations where two of the factors cannot be kept constant, and therefore Boyle's law, Charles's law, Gay-Lussac's law, and Avogadro's law are not very useful per se. The combined gas law shows the relationship among the temperature, volume, and pressure of a gas when all three of these factors can vary. The combined gas law incorporates Boyle's law, Charles's law, and Gay-Lussac's law all in the following formula:

$$\frac{P_1 V_1}{T_1} = \frac{P_2 V_2}{T_2}$$

where P_1 and P_2 represent the initial and final pressures, V_1 and V_2 represent the initial and final volumes, and T_1 and T_2 represent the initial and final temperatures in kelvins. The unit for volume is arbitrary but must be the same for both volumes. Similarly for the pressure. For example, when a gas that has a temperature of 35°C, pressure of 102.3 kPa, and volume of 2.3 L changes to 45°C and 101.3 kPa, what will the new volume of the gas sample be?

$$P_1 = 102.3 \text{ kPa}$$

$$V_1 = 2.3 \text{ L}$$

$$T_1 = 35°\text{C} + 273 = 308 \text{ K}$$

$$P_2 = 101.3 \text{ kPa}$$

$$T_2 = 45°\text{C} + 273 = 318 \text{ K}$$

$$V_2 = ?$$

$$(102.3 \text{ kPa})(2.3 \text{ L})/(308 \text{ K}) = (101.3 \text{ kPa})V_2/(318 \text{ K})$$

$$V_2 = 2.40 \text{ L}$$

The relationship that allows calculation for any of the four factors affecting a gas while permitting all of the factors to change is the ideal gas law. An ideal gas is one that behaves as predicted by the ideal gas formula. Ideal gases are thought of as consisting of molecules that are point masses and do not have any interactions with each other. When a gas deviates from the predicted ideal behavior by having intermolecular forces attracting or repelling the gas molecules from each other and having nonzero volume for its molecules, it is called a real gas. The ideal gas law is given by the formula

$$PV = nRT$$

where P is the pressure of the gas, V is its volume, n is the number of moles, T is the temperature in kelvins, and R is the ideal gas constant. The value of R depends on the units being used. If liter (L) and kilopascal (kPa) are used for volume and pressure, respectively, then

$$R = 8.31 \frac{\text{L} \cdot \text{kpa}}{\text{mol} \cdot \text{K}}$$

The same value is obtained for the SI units of cubic meter (m^3) and pascal (Pa)

$$R = 8.31 \frac{\text{m}^3 \cdot \text{Pa}}{\text{mol} \cdot \text{K}}$$

Chemists often use the units of liter (L) and atmosphere (atm) for volume and pressure, in which case the value of R is

$$R = 0.08206 \frac{\text{L} \cdot \text{atm}}{\text{mol} \cdot \text{K}}$$

The ideal gas law incorporates each of the four previous gas law relationships.

When more than one gas is present in a mixture, the pressures of each of the gases can be used to determine the overall pressure of the mixture of gases. The relationship is given by Dalton's law of partial pressure of gases, established by the English scientist John Dalton in 1801. Dalton's law of partial pressures states that the pressure of a mixture of gases is equal to the sum of the individual pressures of each of the component gases, an important consideration when studying or using air that is made up of a mixture of many gases. Knowing the partial pressure of each component gas is important in scuba diving for regulating the pressure of a diver's tank in order to receive the proper amount of oxygen and prevent decompression sickness, also called the bends. Dalton's law of partial pressures is given by the formula

$$P_{\text{total}} = P_1 + P_2 + P_3 \ldots$$

where P_{total} represents the combined pressure of all the gases, and P_1, P_2, P_3, and so on, represent the partial pressures of the component gases.

See also Boyle, Robert; Dalton, John; pressure; states of matter.

FURTHER READING

Silberberg, Martin. *Chemistry: The Molecular Nature of Matter and Change,* 5th ed. New York: McGraw Hill, 2008.

Gell-Mann, Murray (1929–) American *Physicist* Until the late 19th century, scientists thought atoms were the smallest units of matter; the root of the word *atom* even means "indivisible." Then in 1897, Sir J. J. Thomson discovered the first subatomic particle, the negatively charged electron. In 1911 the British physicist Sir Ernest Rutherford described the structure of the atom as a dense nucleus surrounded by orbiting electrons, and three years later he announced the existence of protons, positively charged nuclear particles. When another British physicist, Sir James Chadwick, discovered the neutron in 1932, physicists confidently believed they had identified the fundamental particles of matter, the building blocks that composed everything in the universe. The electron, proton, and neutron were actually just the leaders of a parade of hundreds of subatomic particles that were discovered by the middle of the 20th century. Perplexed scientists called the procession the "particle zoo" and searched frantically for a means to organize the hodgepodge of minute morsels. The American physicist Murray Gell-Mann found the connection in an abstract mathematical

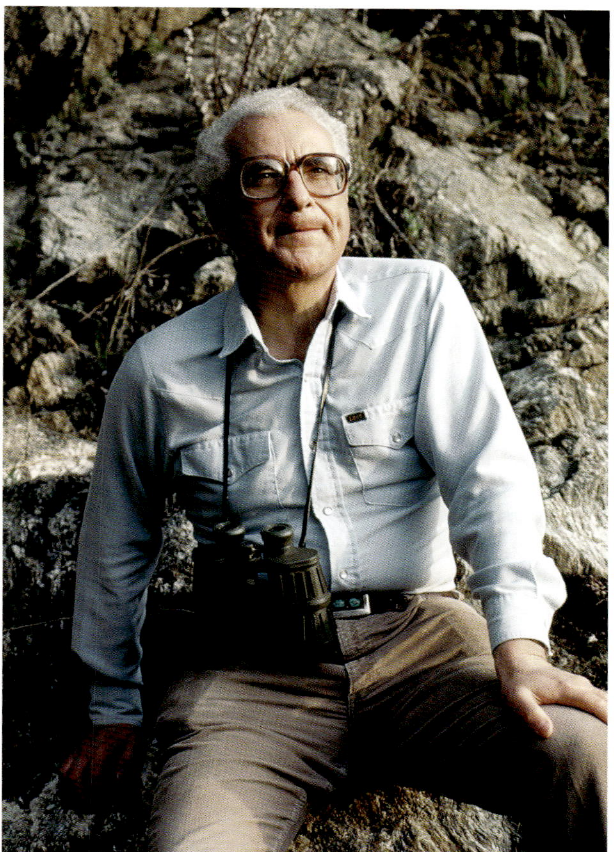

Murray Gell-Mann has worked mainly in theoretical elementary-particle physics and is known especially for his particle classification scheme, called the eightfold way, for which he was awarded the 1961 Nobel Prize in physics. *(© Kevin Fleming/CORBIS)*

model and classified elementary particles on the basis of their symmetry properties. Gell-Mann's scheme, the eightfold way, led to his proposal of quarks as the ultimate building blocks of matter and to the development of quantum chromodynamics as the field theory explaining the force responsible for holding atomic nuclei together.

CHILD GENIUS

Murray Gell-Mann was born on September 15, 1929, to Arthur Isidore and Pauline Reichstein Gell-Mann in New York City. Murray's father, an immigrant from Austria-Hungary, ran a language school that closed down during the Great Depression, forcing him to obtain a position as a bank custodian. The couple already had a nine-year-old son, Ben, who took Murray under his wing and introduced him to science through trips to the local museums and bike rides in the nearby park, where they identified birds, trees, and insects. By age three, Murray could read and carry out multiplication in his head, and by age seven he beat 12-year-old competitors in a spelling bee. Murray enrolled in a private school on a full scholarship and quickly skipped several grades. His older peers called the tiny genius the school's "pet genius" and sometimes teased him during gym class. Graduated by the time he was 14, Murray received a full scholarship to Yale University, where youthful awkwardness gradually disappeared, and he flourished academically and socially.

Murray wanted to study archaeology or linguistics, but his father encouraged him to major in a field that he believed was more financially dependable, such as engineering. They agreed on physics, though Murray hated the subject in high school. At Yale, he became entranced by theoretical physics and sometimes annoyed the other students with his ability to find the answers to problems without even working through the steps. After obtaining a bachelor's degree in physics in 1948, Gell-Mann proceeded to the Massachusetts Institute of Technology (MIT) and earned a doctoral degree in physics in only three years. Victor Weisskopf, a nuclear physicist who had served as a leader of the Manhattan Project during World War II, advised Gell-Mann on his dissertation topic, events that follow neutron bombardment of an atomic nucleus. Learning physics and mathematics came easily to him, but writing did not. For his entire life, Gell-Mann procrastinated at summarizing his ideas, and a fear of being wrong in intractable print prevented him from submitting his papers for publication in a timely manner.

J. Robert Oppenheimer, who headed the construction of the atomic bomb during World War II and was the director of the Institute for Advanced Study at Princeton, gave Gell-Mann a temporary appointment. Gell-Mann collaborated with his office mate, Francis Low, on his first scientific paper, "Bound States in Quantum Field Theory," published in *Physical Review* in 1951. The University of Chicago hired him as an instructor at the Institute for Nuclear Studies in 1952, and his growing reputation as a particle physicist earned him a promotion to assistant professor in 1953. After he held a visiting professorship at Columbia University for one year, the California Institute of Technology lured him away in 1955 with an impressive salary offer and an associate professorship with full tenure. The following year, Gell-Mann became the youngest person to be named a full professor at the institution.

Before moving to Pasadena that summer, he married a British archaeology student, J. Margaret Dow. They had a daughter, Lisa, born in 1956 and a son, Nicholas, born in 1963.

CLASSIFICATION OF ELEMENTARY PARTICLES

Until scientists discovered unstable particles in cosmic rays in the mid-20th century, they believed that the electron, proton, and neutron were the most fundamental particles of matter. Electrons were very tiny negatively charged particles that orbited around the atomic nucleus, which contained the protons and the neutrons. About the time that Gell-Mann entered academia, physicists were rapidly discovering many new subatomic particles—positrons in 1932, followed shortly thereafter by muons and pions. Positrons were the antimatter equivalent of electrons; muons were similar to electrons but much more massive. The use of particle accelerators beginning in the 1930s eliminated the need to rely on cosmic rays to collect particles for analysis. The ability to produce these particles on demand led to the rapid discovery of more than 100 kinds of subatomic particles, and every particle had its own antimatter particle with an identical mass but an opposite electrical charge.

By the early 1950s, scientists faced a chaotic conglomeration that Oppenheimer termed the subatomic "particle zoo." Scientists looked for patterns to organize the particles into groups, originally using mass as the primary determinant. Leptons were the lightest group and contained electrons, positrons, and neutrinos. The middleweight mesons included pions and kaons. Baryons were the heaviest and included protons and neutrons. There was an abundance of schemes for sorting the particles; Gell-Mann thought it was best to use the interactions in which the particles participated.

The different types of forces that act among particles included gravitational, electromagnetic, strong, and weak forces. The concept of a quantum field described interactions of elementary particles as exchanges of other elementary particles. Gravitons are the hypothetical carriers of gravitation, the attractive interaction between all particles; however, this interaction is so weak that elementary-particle physicists do not give it much consideration. Electrically charged particles, such as electrons, emit and absorb particles called photons that carry the electromagnetic force, which the theory of quantum electrodynamics (QED) elegantly explained. The nature of the nuclear interactions, the strong and weak forces, was still a mystery.

Gell-Mann grouped together elementary particles that participated in the strong interactions, forming a family called the hadrons that was further divided according to spin, a quantum number that represents the intrinsic angular momentum of an elementary particle. The baryons had half-integer spins, and mesons had integer units of spin. Leptons do not participate in strong interactions, but leptons and hadrons both participate in weak interactions.

Among the newest species to join the particle zoo were some unstable particles that only lived for 10^{-23} second (1 over a denominator of 1 with 23 zeros following it, or one-hundred-thousandth of a billionth of a billionth) and others that remained for 10^{-10} second, longer than existing strong-force models predicted. To explain the resistance of some particles to decay, Gell-Mann proposed the existence of a new quantum number that he called strangeness and concluded that, as must charge, it must be conserved during a strong or electromagnetic interaction. That is, if a particle with a strangeness value of +1 decayed, the total sum of the strangeness values for the decay products must also equal 1. He assigned ordinary particles such as protons and neutrons a strangeness of 0 and the so-called strange particles, including the kaons and sigmas, either +1 or −1 (others discovered later had other whole-number values). When nucleons and pions collided, strange particles formed. Since the strangeness of the ordinary parent particles had to be 0 (by definition of ordinary), then the creation of a particle with a strangeness of +1 necessitated that another strange particle of strangeness −1 also form to conserve the total strangeness. (This production in pairs was called associated production.) If either of the newly created strange particles decayed back into an ordinary particle by the strong force, then the strangeness would not be conserved. Decay via the strong force occurred rapidly, while weak decays did not conserve strangeness and occurred more slowly; thus, the particles with lifetimes longer than expected decayed by the weak force. This system, proposed around 1953 by Gell-Mann and independently by Kazuhiko Nishijima, predicted that some associated pairs should be produced more often than others, a prediction confirmed by Brookhaven National Laboratory accelerator experiments.

Though the strangeness theory explained prolonged decay and associated production, others initially had difficulty accepting some aspects of Gell-Mann's proposal. Within a few years, however, the strangeness quantum number had become deeply rooted. Researchers continued to find new particles that the theory predicted.

COLLABORATION WITH FEYNMAN

In Pasadena, Gell-Mann flourished with the brilliant American physicist Richard Feynman as his stimulating yet exasperating theory-sparring partner and with his new wife at home. He researched a problem related to the long-lived particles from which he developed the concept of strangeness and the conservation law of parity, related to a certain symmetry. Symmetry exists when some change can take place while some aspect of the situation remains unchanged. To illustrate symmetry, Feynman explained that an alarm clock acts the same way no matter where or when it is; its behavior is said to be symmetric in space and time. If taken to a different time zone, the hands move around the face of the clock at the same speed, and it will work the same way next week or next year as it works today. The symmetry Gell-Mann studied was called parity symmetry, where parity was a quantum number related to reflection, as an object appears in a mirror image. Even in a mirror image, a clock reads the same, the arrows still point to the same numbers, and the hands still pass through number 1 on the way to number 2; only the right and left are exchanged. Physicists long assumed that nature did not recognize right- versus left-handedness—in other words, parity was conserved—but the behavior of strange particles suggested that parity was not conserved. Gell-Mann solved this puzzle but was hesitant to publish his solution in case his idea was wrong. In early 1956, he was horrified to see a preprint (a prepublication version) of an article by T. D. Lee and Frank Yang proposing the same idea. The idea eventually proved false, but Gell-Mann resolved never to be scooped again.

That spring Gell-Mann attended a conference where physicists discussed the violation of parity symmetry and suggested that the weak force violated this symmetry. The Italian-born American physicist Enrico Fermi had introduced the weak force as a mediator of beta decay in 1933. In this process, a neutron transforms into a proton and emits an electron and an antineutrino. The creation of the electron conserved charge during the reaction, and the antineutrino conserved energy and momentum. More than 20 years later, the nature of the weak force was still a mystery. Lee and Yang suggested experiments to determine whether the weak force conserved parity. In January 1957, researchers found convincing evidence that parity was broken during beta decay—electrons were emitted preferentially in one direction.

Five possible forms of particle interaction could potentially explain the weak force in quantum field theory. Gell-Mann wanted to determine which two of the five explained all weak interactions, including beta decay. Recent experiments suggested two forms for one interaction and two different forms for a second interaction, and physicists wondered whether no universal weak interaction existed. After discussing the problem with E. C. George Sudarshan and Robert Marshak, who were visiting California from the University of Rochester, Gell-Mann recognized the pattern followed by the weak force and planned to include it in another paper he was going to write, but then he found out Feynman had formulated the same conclusions. Gell-Mann was annoyed at Feynman's arrogance and did not want him to get credit for the discovery. Their department head suggested that joint authorship on a paper was in the best interest of the school. "Theory of the Fermi Interaction," published in *Physical Review* in 1958, generated praise for both Feynman and Gell-Mann, and though they acknowledged Sudarshan and Marshak, history chiefly credits the Caltech professors.

Quantum field theory did not seem to explain the strong and weak nuclear forces as elegantly as it described electromagnetism. Infinities plagued equations for nuclear interactions, even after Gell-Mann and Feynman developed their theory of the weak force. Physicists believed pions carried the strong force and had no idea what carried the weak force.

In a weak interaction, two particles entered with one set of quantum numbers and two emerged with another set. Gell-Mann suggested "uxyls," also referred to as X particles, as the intermediaries of the weak force. At least two charged intermediaries were needed to carry away the charge difference when a weak interaction changed the charge of a particle (as when a neutron decayed into a proton in beta decay), and Gell-Mann suspected two additional neutral particles existed. He thought unification of the weak force with electromagnetism might help solve the problem, but was quantum field theory the right framework? And why was strangeness sometimes conserved and sometimes not?

THE EIGHTFOLD WAY

Frustrated, Gell-Mann switched his focus to the strong force, and he started looking for patterns in hadrons, the group of particles that were affected by the strong force. A Caltech mathematician working on group theory reminded Gell-Mann about a grouping system based on symmetry. Realizing this may be the key to uniting the hadrons, he started arranging the particles by strangeness and electric charge into higher orders. When everything finally came together, he found a pattern complete with eight baryons and a similar pattern with seven mesons. In 1961 Gell-Mann described an abstract mathematical method, which he named the eightfold way, for arranging hadrons into families based on properties of symmetry. The Israeli physicist Yuval Ne'eman independently proposed the same method. One could view what appeared to be several separate particles as different states, with different configurations of quantum numbers, of one particle. Baryons with the same spin and parity composed one group, and mesons with identical spin and parity belonged to another. The scheme suggested that heavier baryons should form a group of 10 members. Nine of these 10 particles were identified already.

When Gell-Mann devised this classification scheme, he predicted the existence and properties of several undiscovered particles, including the baryon called omega-minus to complete the group of 10 heavy hadrons. A newly discovered eighth meson had properties that fit it into a pattern similar to the eightfold baryon pattern. Experiments conducted in 1963 at Brookhaven National Laboratory produced an omega-minus particle, confirming the eightfold way.

QUARKS

The symmetry patterns of the eightfold way suggested that hadrons were not elementary, but were composed of other, more fundamental particles. In 1964 Gell-Mann published a short paper in *Physics Letters,* "A Schematic Model of Baryons and Mesons," proposing that subatomic particles he named quarks were the stable, smallest building blocks of all hadrons, tremendously simplifying the field of elementary particle physics. At the same time, a former graduate student of Gell-Mann and Feynman's, George Zweig from CERN (the European Organization for Nuclear Research), arrived at the same hypothesis. Quarks were unusual because their assigned electric charges were only a fraction of the electron charge. He initially described three "flavors" (a fanciful term for "kinds") of quarks: up (u, with a fractional electric charge of +2/3), down (d, with a charge of -1/3), and strange (s, with a charge of -1/3). Gell-Mann was able to describe the composition of many hadrons from these three building blocks. Experiments performed by others substantiated Gell-Mann's quark theory. When experiments suggested the three-quark model was no longer sufficient to explain everything, physicists proposed a fourth flavor, charm (c, with a charge of +2/3). Later, when additional, more massive hadrons were discovered, top (t, charge of +2/3) and bottom (b, charge of -1/3) flavors were also predicted and eventually discovered, making the total number of quarks six. It turned out that the family of leptons comprised the same number of members, six: electron, electron neutrino, muon, muon neutrino, tau, and tau neutrino. Particle physicists generally assume a deep relationship between the two families, as different as they are, which is still far from being understood.

A proton consisted of two up quarks and one down quark, with total electric charge of

$$(+2/3) + (+2/3) + (-1/3) = +1$$

One up and two down quarks made up an electrically neutral neutron, with total charge

$$(+2/3) + (-1/3) + (-1/3) = 0$$

When particles decayed, the weak force changed the flavor of quarks; for instance, when a neutron (ddu) decayed into a proton (duu), one down quark changed into an up quark.

The quark theory seemed to work, but were the quarks real or contrived mathematical entities? No other particle was known to possess fractional charges, and experimenters were unsuccessful in tracking down quarks in cosmic rays, accelerators, or anywhere else. In 1968 physicists began to accumulate evidence of electrons' hitting and bouncing off something inside a proton, but quarks cannot normally be separated from each other and therefore have not yet been directly observed.

Gell-Mann received the 1969 Nobel Prize in physics for his contributions and discoveries concerning the classification of elementary particles and their forces. Though he shared credit with his codiscoverers of strangeness, the eightfold way, and quarks (Nishijima, Ne'eman, and Zweig, respectively), this award belonged to Gell-Mann alone. After impressing the Swedes by speaking in their native language for part of his Nobel acceptance speech, he procrastinated in submitting his lecture to the celebratory publication *Le Prix Nobel,* and despite numerous requests, he never sent one.

QUANTUM CHROMODYNAMICS

Others accepted Gell-Mann's field theory of the strong force involving quarks and bosons long before he was convinced it was more than a mathematical contrivance. By 1995 particle physicists had obtained experimental evidence for all six quarks, which, as it turned out, not only had six flavors: each flavor had three possible states, whimsically called "colors" and labeled *red, blue,* and *green.* The addition of this quantum number was necessary to respect Pauli's exclusion principle. Colored particles (quarks) attracted one another by exchanging gluons, carriers of the strong force (also called the color force), just as electrons interacted by exchanging photons, carriers of the electromagnetic force. This exchange of gluons between quarks from different protons held the protons together tightly even though their electrical charges simultaneously pushed them apart electrostatically. Whereas individual quarks carried the color charge, hadrons did not. Baryons contained three quarks of the three different colors: red, blue, and green. Combined, these colors cancelled each other out, just as these primary colors of visible light combined to form white light. Mesons were composed of one quark and one antiquark, so if the quark was red, the antired antiquark eliminated the color charge in the meson.

One aspect of the strong force theory that continued to puzzle Gell-Mann and others was that individual quarks were never observed. In 1973 the presentation of a concept called asymptotic freedom explained this puzzle, stating that the force holding the quarks together intensified at greater distances, just the opposite of what occurred in electromagnetism or gravity. In 2004 David J. Gross, H. David Politzer, and Frank Wilczek shared the Nobel Prize in physics for the discovery of asymptotic freedom in the theory of the strong interaction.

The property of asymptotic freedom completed the theory of the strong force called quantum chromodynamics (QCD). Hadrons were composed of quarks, or a quark and an antiquark, which were held together by the strong force, whose intermediaries were a set of eight massless bosons called "gluons" that differed only in their color properties. The color gluons affected only color-carrying particles, that is, the quarks. QCD also explained how the strong force held nuclear particles together so tightly, and the quark scheme explained all the possible interactions among nuclear particles and the symmetries within groups. Many brilliant minds contributed to the development of all the elements of QCD, and Gell-Mann's fingerprints covered many of the pieces.

RETIREMENT YEARS

After his Nobel Prize, many distractions kept Gell-Mann on the fringes of physics; in particular, his wife, Margaret, died of cancer in 1981, and in 1992 he married a poet and English professor, Marcia Southwick. He served as a regent of the Smithsonian Institution (1974–88) and as director of the MacArthur Foundation (1979–2002). Gell-Mann has appreciated nature since his days of bird watching in the park with his older brother, and in 1993 he received the Lindbergh Award for his efforts in promoting balance between advancing technology and conservation. In 1984 Gell-Mann helped found the Santa Fe Institute, a research and education center where theorists collaborate on topics ranging from quantum mechanics to economics in order to understand better the simple mechanisms underlying complex phenomena. Having retired from Caltech in 1993, he currently serves on the board of trustees of the Santa Fe Institute and has served as chairman of the board (1984–85) and cochair of the science board (1985–2000). He has written books about the topic of complexity for a popular audience, such as *The Quark and the Jaguar* (1994). Living in Santa Fe, he still enjoys linguistics, ornithology, and archaeology. He teaches part time at the University of New Mexico and continues to feed his hunger for patterns by researching the basis of complex phenomena.

Gell-Mann belongs to the American Physical Society, which awarded him the Dannie Heineman Prize for mathematical physics in 1959, and the National Academy of Sciences, which awarded him the John J. Carty Medal in 1968. He also belongs to the Royal Society of London as a foreign member and the American Academy of Arts and Sciences, and he has received awards from the Franklin Institute and the Atomic Energy Commission. More than 10 academic institutions have awarded Gell-Mann honorary doctorates.

Gell-Mann's proposal of the eightfold way provided theoretical physicists a framework for classifying the plethora of subatomic particles discovered during the 20th century. His ability to recognize relationships among fundamental particles and the forces that affect them led to a better understanding of the structure of matter. Strangeness, the eightfold way, and quarks, three of Gell-Mann's most influential ideas, all contributed to the development of the standard model that combines the electromagnetic, weak, and strong forces to describe particles and their interactions. This model may lead to final resolution in the form of an all-encompassing grand unified theory during the 21st century.

See also ATOMIC STRUCTURE; ELECTRICITY; ELECTROMAGNETISM; FERMI, ENRICO; FEYNMAN, RICHARD; GRAVITY; INVARIANCE PRINCIPLES; OPPENHEIMER, J. ROBERT; PARTICLE PHYSICS; RUTHERFORD, SIR ERNEST; SYMMETRY; THOMSON, SIR J. J.

FURTHER READING

Gell-Mann, Murray. *The Quark and the Jaguar: Adventures in the Simple and the Complex*. New York: W. H. Freeman, 1994.

Johnson, George. *Strange Beauty: Murray Gell-Mann and the Revolution in Twentieth-Century Physics*. New York: Alfred A. Knopf, 1999.

Murray Gell-Mann home page. Santa Fe Institute. Available online. URL: http://www.santafe.edu/~mgm/. Accessed January 13, 2008.

The Nobel Foundation. "The Nobel Prize in Physics 1969." Available online. URL: http://nobelprize.org/physics/laureates/1969/. Accessed July 25, 2008.

general relativity Albert Einstein's general theory of relativity, known in short as "general relativity," is the theory of gravitation that Einstein proposed in 1915. A generalization of the special theory of relativity ("special relativity"), this theory, too, is formulated in terms of four-dimensional space-time. In contrast to the way it is in the special theory, space-time is not flat in the general theory: rather, it curves as an effect of and in the vicinity of masses and energies. According to the theory, gravitation is not a force, like, say, the magnetic force, but rather is an effect of the curved geometry of space-time. That comes about in this way. In the theory, force-free bodies obey a generalization of Newton's first law of motion. Their free motion, or inertial motion, however, is not in a straight line, but follows a path that is the closest to a straight line in curved space-time, called a geodesic. This is referred to as geodesic motion, motion that is the analog in curved space-time of straight-line motion at constant speed in ordinary space. The result is motion that appears in ordinary space as neither in a straight line nor at constant speed, in general, and manifests the effect of gravitation on the bodies.

As an example, consider the motion of Earth around the Sun. Viewed nonrelativistically, in the hypothetical absence of the gravitational force of the Sun (and ignoring all other solar-system objects), Earth would move inertially (i.e., in accordance with Newton's first law) in a straight line at constant speed. The gravitational force of the Sun on Earth makes the situation dynamic, however, and Newton's second law applies. The resulting dynamic motion turns out to be motion in an elliptical orbit around the Sun. The general theory of relativity describes conditions differently. According to the general relativistic view, the Sun causes a deformation of space-time. No forces act on Earth, which therefore moves inertially. Inertial motion is geodesic motion in space-time, as described earlier. When Earth moves inertially (freely) along a geodesic in space-time, it appears to be moving dynamically (under the effect of a force) along an ellipse in ordinary space.

Here is a description of an often-presented model, to give some idea of what is going on. Set up

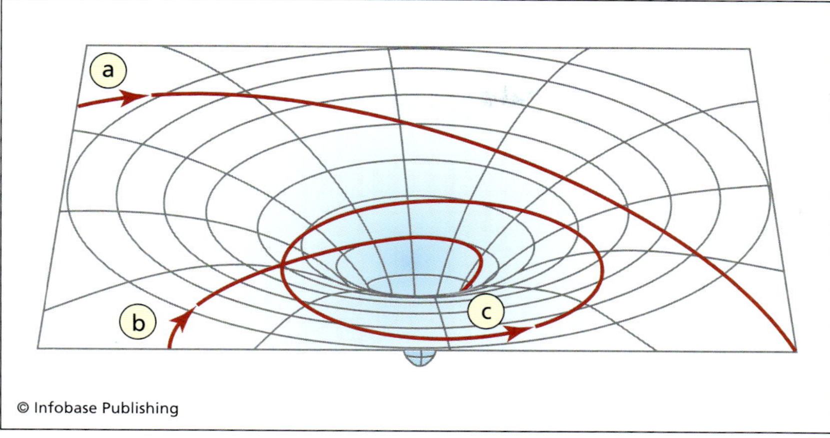

The figure shows a rubber sheet model of gravitation according to the general theory of relativity. A heavy object, such as a marble, placed on a stretched horizontal rubber sheet deforms it as in the figure. The motion of freely rolling ball bearings, or other light spherical objects, across the sheet is no longer the straight line it would have been in the absence of the distortion. Three sample trajectories are shown: (a) the rolling object approaches the marble and leaves it, (b) the rolling object falls onto the marble, and (c) the rolling object orbits the marble.

a horizontal stretched sheet of rubber to represent flat space-time. Place a heavy marble on the sheet. The marble represents a massive body. Because of the marble's weight, the rubber sheet is stretched downward at the location of the marble and assumes a somewhat conical shape with the marble at the apex. This shape represents curved space-time in the presence of and in the vicinity of mass. Now, toss light marbles or ball bearings onto the rubber sheet so they roll at various speeds and in various directions. Some might orbit the heavy marble. Low-speed ones will spiral into the marble. High-speed ones will escape capture but still have their trajectory affected. This behavior represents the effect of space-time curvature on bodies as an explanation of gravitation.

EQUIVALENCE PRINCIPLE

The equivalence principle was important for Einstein's development of the general theory of relativity and is central to the theory. This principle states that mass as it affects inertia—such as in Newton's second law of motion—is equal to mass as it is affected by gravitation—following Newton's law of gravitation, for instance. Experiments by the Hungarian physicist Lóránt Eötvös and later by others for the purpose of detecting a violation of this principle have produced negative results.

Another formulation of the equivalence principle is this. One cannot determine, by means of experiments carried out solely within a laboratory (i.e., no peeking outside the lab, for example), whether the room is at rest on the surface of a planet, say (so the masses in the lab are being affected by gravitation), or is undergoing constant acceleration (whereby the masses are exhibiting their inertial character) in a region where there is no net gravitational force. In both cases, if a body is attached to a spring anchored to the wall, the spring is stretched. An additional formulation is that it is similarly impossible to distinguish between the lab's being in a state of free fall under the influence of gravitation and the lab's being at rest in a region where there is no net gravitational force acting. In both cases, a body floats freely around the laboratory. The occupants of a space station in orbit around the Earth experience a free-fall situation.

MACH'S PRINCIPLE

In addition, Einstein was strongly influenced by Mach's principle. This principle, named for the Austrian physicist and philosopher Ernst Mach, deals with the origin of mechanical inertia. Mechanical inertia is expressed by Newton's first and second laws of motion. According to the first law, in the absence of forces acting on a body (or if the forces on it cancel out), the body remains at rest or moves with constant velocity (i.e., at constant speed in a straight line). The second law states how a body resists changes in its force-free motion, or inertial motion. Such a change is an acceleration, and according to the second law, the acceleration resulting from a force is inversely proportional to a body's mass: the greater the mass, the less change a force can effect. Thus mass serves as a measure of a body's inertia.

The question immediately arises, In what way do inertial motion and accelerated motion differ? Since in an appropriately moving reference frame, any motion can appear to be inertial (in particular, the frame only needs to track the moving object, which then appears at rest), the question can be restated as: What distinguishes inertial reference frames—those in which Newton's laws of motion are valid—from noninertial ones?

A few words at this point about inertial and noninertial reference frames: Consider an accelerating jet airplane building up speed for takeoff. The plane serves as a convenient reference frame for the passengers. In this reference frame, during the acceleration process, the passengers appear to be accelerated backward and would crash into the lavatories at the rear of the plane if they were not restrained by the backs of their seats. Yet there is no physical cause for this acceleration in the form of a rearward force acting on the passengers. There is nothing pushing them and nothing pulling them—no ropes, springs, or giant pushing hands causing this apparent acceleration. That is a violation of Newton's first law of motion, according to which any change of state of motion (i.e., acceleration) must be caused by a force. Conversely, in the absence of force, as is the case here, a passenger must remain at rest or move at constant speed in a straight line. The reference frame of the airplane is not an inertial reference frame, and in this reference frame, Newton's second law does not hold.

According to Newton, the difference between inertial and noninertial reference frames lies in the state of motion relative to absolute space. Newton assumed the existence of an absolute space (as well as an absolute time) that is unobservable except for its inertial effects. Inertial and accelerated motions occur with respect to absolute space. Accordingly, inertial reference frames are reference frames moving at constant velocity (or at rest) with respect to absolute space.

Mach's philosophical approach to science did not allow entities that are inherently unobservable, including absolute space. But in the absence of absolute space, how are inertial motion or inertial reference frames different from all other motions or reference frames? Considerations such as of the fundamental meaning of motion for a single body, two bodies, and three bodies in an otherwise empty universe led Mach to his principle: The origin of inertia lies in all the matter of the universe. Since the matter of the universe consists overwhelmingly of distant stars, galaxies, and so forth, one can interpret Mach's principle to mean that the origin of inertia lies in the distant stars.

So, when the motions of the distant stars are averaged out, they define a reference frame for rest. Stated in more detail, this reference frame—which might be called the universal reference frame—is that in which the average linear momentum of all the matter in the universe is zero. Then, according to Mach, inertial reference frames are those, and only those, that are at rest or moving at constant velocity relative to the just-defined reference frame. All other reference frames, which are in relative acceleration or rotation, are noninertial. In terms of motions of bodies, the rest, velocity, and acceleration with which Newton's first two laws of motion are concerned are relative to the universal rest frame defined by the matter in the universe.

Although Einstein, in his development of the general theory of relativity, was greatly influenced by Mach's thinking and principle, the theory does not succeed in fully implementing the principle, in showing just *how* distant matter engenders inertial effects. Neither has Mach's principle yet been successfully implemented by other theories in a way that is generally accepted. So at present Mach's principle can best be viewed as a guiding principle.

THE GENERAL THEORY

One of the basic postulates of the general theory, which is a straightforward generalization of a basic postulate of the special theory, is that the laws of physics are the same in all reference frames, *no matter what their relative motion.* (In the special theory, the reference frames are restricted to relative motion at constant velocity, i.e., at constant speed in a straight line.) Einstein designed the general theory to produce Newton's law of gravitation, under appropriate conditions.

The general theory of relativity has passed all the tests it has been given. It explains the deviation of the orbit of the planet Mercury from the prediction of Newtonian mechanics, and it correctly gives the amount of deflection from a straight-line trajectory of starlight as it passes near the Sun. Its prediction of the gravitational redshift—the change in frequency of light or of any electromagnetic wave as it passes through the gravitational field—has been confirmed. The theory also predicts the existence of gravitational waves. There is indirect evidence for such waves, although physicists have not yet detected them with the special detectors designed for that purpose. The general theory of relativity predicts black holes. Astronomical observations indicate their existence as a common phenomenon. The theory predicts gravitational lensing of light, and the effect is observed. This is the bending of light rays from a distant astronomical source by an intervening massive astronomical object, such as a star, galaxy, or galaxy cluster, on their way to an observer on Earth, causing the source to appear distorted—as several sources or as a (possibly incomplete) ring. The general theory of relativity is incorporated in the standard theoretical framework for the study of cosmology, called relativistic cosmology. As for practical applications, the theory is routinely invoked in connection with the navigation of spacecraft and in the design and operation of the Global Positioning System (GPS).

One of the early confirmations of the general theory of relativity, the observation of the predicted deflection of starlight by the Sun, involved adventure and travel to distant places. In order to observe this deflection, one must measure the apparent position of a star when it appears close to the edge of the Sun in the sky and compare it with the star's usual observed position. The Sun's deflection of the light

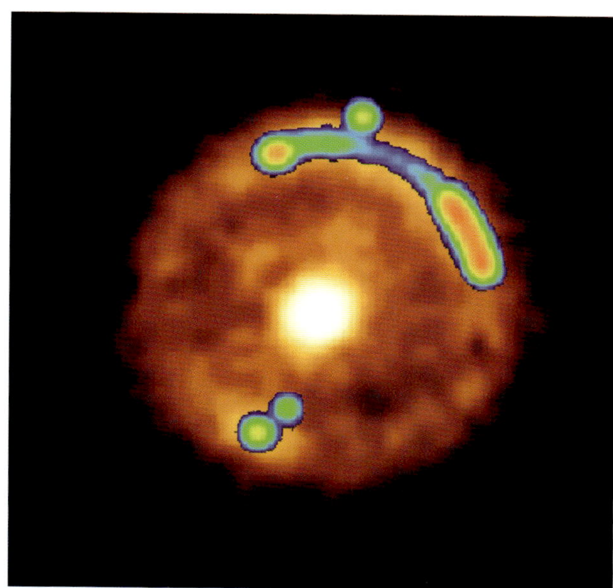

When one galaxy is situated in front of another, as viewed from Earth, the closer galaxy can bend the light from the farther one that is passing close to it, according to the general theory of relativity. Thus, the farther galaxy, whose light would otherwise be blocked by the closer galaxy from reaching Earth, might appear to astronomers as a full or partial ring surrounding the closer galaxy, called an Einstein ring. The photograph, a combination image in false color, shows a huge galaxy (white, at center) positioned between Earth and a distant galaxy. Infrared (orange) and radio frequency (blue, green) radiation from the distant galaxy are seen to form both a full Einstein ring in infrared and a partial ring in radio (consisting of an arc and three blobs). *(Space Telescope Science Institute/NASA/Science Source/Photo Researchers, Inc.)*

on its way from the star to Earth causes a difference between the two measurements, and general relativity predicts the magnitude of difference. Astronomers routinely measure a star's usual position at night; however, measurement of the apparent position of a star, when it is near the edge of the Sun, requires observation during the daytime, when the stars cannot be seen. That problem is solved by waiting for a total solar eclipse, when the Moon completely covers the Sun. Then, for a few minutes the sky becomes dark and one can observe the stars. Arthur Eddington, an English astronomer and physicist, proposed exploiting for this purpose a solar eclipse that was predicted for May 29, 1919. Unfortunately, the eclipse would not be total in England, where excellent telescopes and equipment were available. The best locations for observing the total eclipse and making the necessary measurements were two small islands, Sobral off the coast of Brazil and Principe in the Gulf of Guinea along the West African coast, where there were no telescopes or other observational equipment. Eddington participated in organizing and leading expeditions to those islands. The weather cooperated during the eclipse, although just barely for the Principe group under Eddington, and both groups photographed the star field near the Sun. Upon return to England, Eddington compared the stars' apparent positions during the eclipse with their usual positions. The result confirmed the prediction of the general theory of relativity. Since then, additional such observations have strengthened confirmation of the theory.

See also ACCELERATION; BLACK HOLE; EINSTEIN, ALBERT; ELECTROMAGNETIC WAVES; ENERGY AND WORK; FORCE; GRAVITY; MACH'S PRINCIPLE; MASS; MOMENTUM AND COLLISIONS; MOTION; NEWTON, SIR ISAAC; SPECIAL RELATIVITY; SPEED AND VELOCITY; TIME.

FURTHER READING

Geroch, Robert. *General Relativity from A to B.* Chicago: University of Chicago Press, 1981.

Schutz, Bernard. *Gravity from the Ground Up: An Introductory Guide to Gravity and General Relativity.* Cambridge: Cambridge University Press, 2003.

glycogen metabolism Carbohydrates serve as the body's major source of energy. After ingestion, food is chemically broken down into simpler components that the body can oxidize to extract energy for adenosine triphosphate (ATP) synthesis or use to build macromolecules needed for cell maintenance or reproduction. Carbohydrates, organic molecules composed of carbon, hydrogen, and oxygen, are ultimately digested into monosaccharides, individual subunits used to build larger, more complex carbohydrates. Glucose ($C_6H_{12}O_6$), a widely occurring six-carbon monosaccharide, circulates throughout the body in the blood, and because of its important role in carrying energy to all the body's tissues, its levels must be maintained within a certain range.

Animal cells store glucose as glycogen, a highly branched polymer of glucose subunits that can easily be removed when the body needs energy. Plants typically store glucose as starch, another complex carbohydrate. Though most cells in the body can store glucose as glycogen, liver and muscle cells play an especially important role in glycogen metabolism, the synthesis and breakdown of glycogen. Numerous enzymes play a role in glycogen synthesis and degradation and, when deficient, can cause a variety of dangerous metabolic disorders.

Glycogen is composed of numerous glucose subunits linked by α-1,4-glycosidic bonds. A molecule of glycogen resembles a tree with extensive branching. In

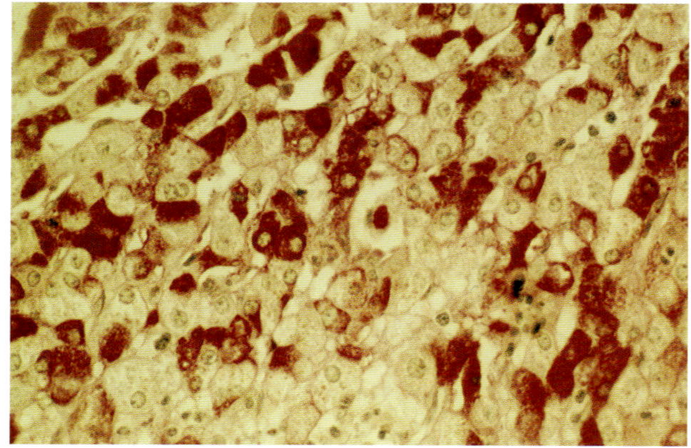

Glycogen is the storage form of glucose in animals. The liver and muscle cells are primarily responsible for the synthesis and degradation of glycogen in the body. Shown here is a light micrograph of a section through liver cells showing stored glycogen, which appears as bright pink. *(SPL/Photo Researchers, Inc.)*

the interior portion of the glycogen molecule, branches linked by α-1,6-glycosidic bonds occur every four or so residues, compared to every eight or 10 residues for the exterior branches. The number combinations 1,4 and 1,6 indicate the specific carbon atoms that participate in the glycosidic linkage, the name for the type of covalent linkage that forms between monosaccharides. The branched structure allows the cell to release glucose subunits from the glycogen stores rapidly and to synthesize glycogen rapidly, as several enzymes can act at multiple chain ends simultaneously. Branching also increases the solubility of the large molecule. The ends of the branches are called nonreducing ends, and these are the site of synthesis and degradation reactions. One glucose subunit of a glycogen molecule is called the reducing end and is attached to a protein called glycogenin.

GLYCOGEN SYNTHESIS AND DEGRADATION

Glycogen synthesis and degradation occur by different biochemical pathways, a strategy that confers more flexibility with respect to control and energetics. One pathway can be activated at the same time as the other is inhibited. In most cells, the enzyme hexokinase converts glucose to glucose-6-phosphate immediately after it enters the cell. In the liver, glucokinase performs this function. The presence of excess glucose favors the synthesis of glucose-1-phosphate from glucose-6-phosphate by phosphoglucomutase. In order to synthesize glycogen, the enzyme uridine disphosphate (UDP)-glucose phosphorylase must first activate glucose molecules by forming a high-energy bond between glucose and UDP to form UDP-glucose. Most glycogen synthesis involves the addition of new residues to the nonreducing ends of a preexisting glycogen primer. The enzyme glycogen synthase transfers the glucosyl groups from UDP-glucose to the nonreducing ends, forming α-1,4 linkages. After the chain grows to 11 residues in length, a separate branching enzyme removes a fragment of about six to eight residues, then reattaches these residues by an α-1,6 linkage to a different residue, forming a new branch.

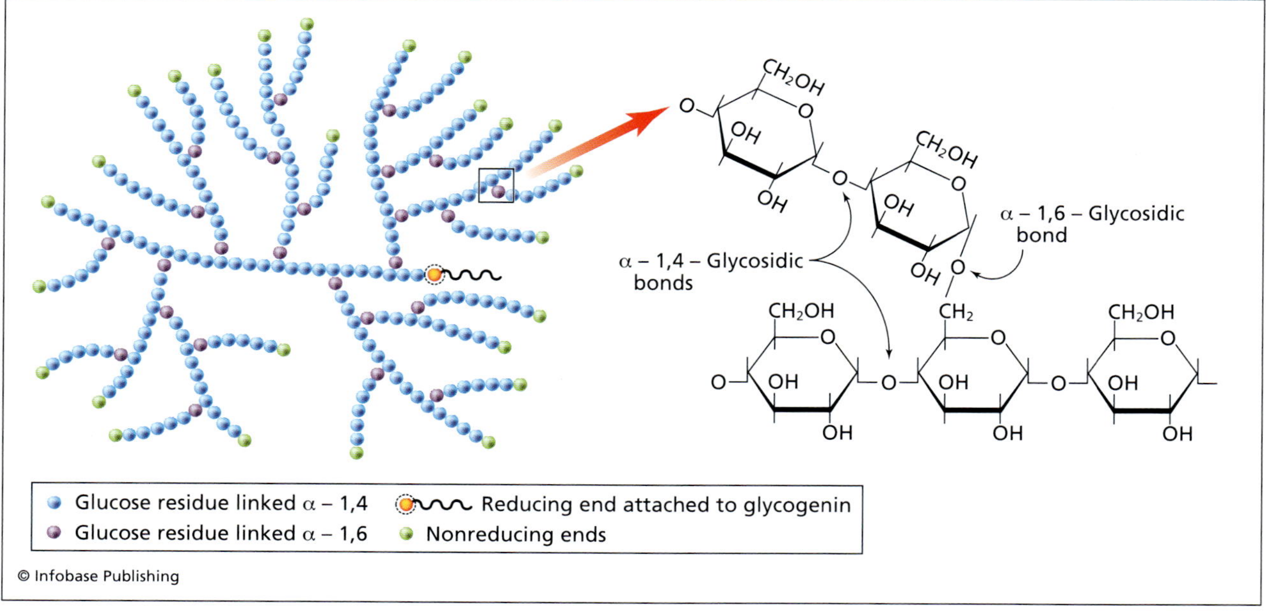

Glycogen is a branched polysaccharide of glucose containing α-1,4-glycosidic bonds between each subunit with branches attached by α-1,6-glycosidic bonds every eight to 10 residues.

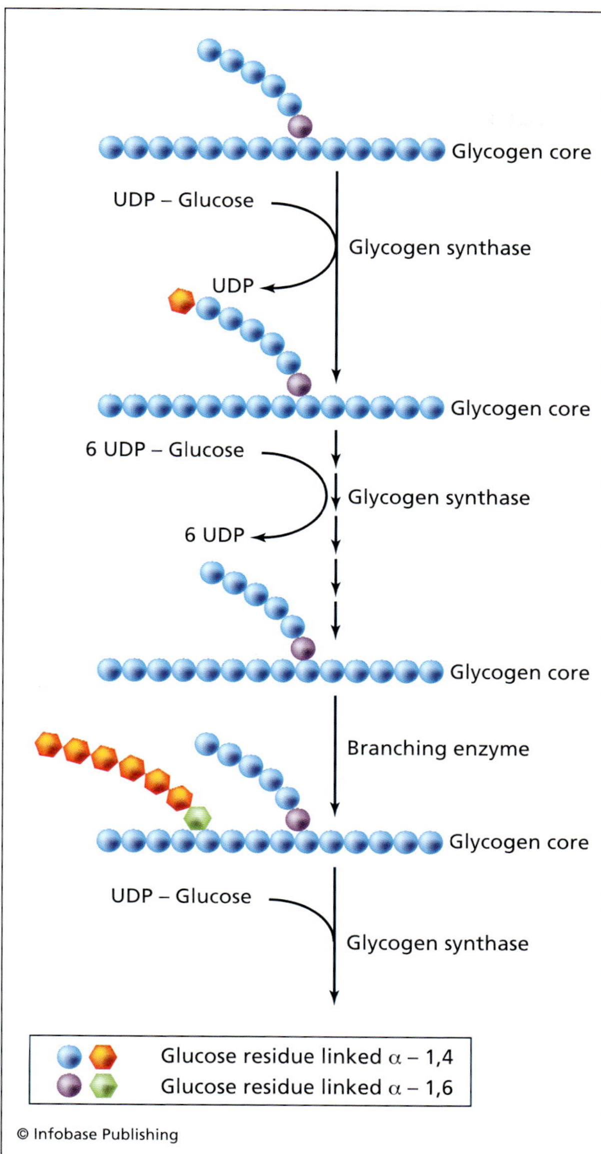

When the energy charge of the cell is high, glucose residues are stored in the form of highly branched glycogen molecules.

Glycogen primers form when the protein glycogenin attaches a glucosyl residue onto itself, then adds more glucosyl residues until the chain is long enough to serve as a glycogen primer for glycogen synthase.

The breakdown of glycogen requires two different enzymes, a debrancher enzyme to cleave the α-1,6 bonds and glycogen phosphorylase to break the α-1,4 bonds. Glycogen phosphorylase cleaves individual residues from the end of the chain one at a time, adding a phosphate in its place. The released molecules of glucose 1-phosphate can be converted to glucose 6-phosphate and feed into glycolysis when energy is needed. Glycogen phosphorylase cannot fit close enough to cleave subunits closer than four away from the branch points, and a second enzyme is necessary. The debrancher enzyme performs two jobs: it transfers three of the remaining residues to the end of a longer chain by forming an α-1,4 linkage, and it cleaves the α-1,6 bond that still holds a single residue at the branch point. This process yields approximately eight free glucose 1-phosphate residues from every branch point broken down.

REGULATION OF GLYCOGEN METABOLISM

The liver is responsible for maintaining blood glucose levels within a certain range. The antagonistic hormones insulin and glucagon regulate the activity

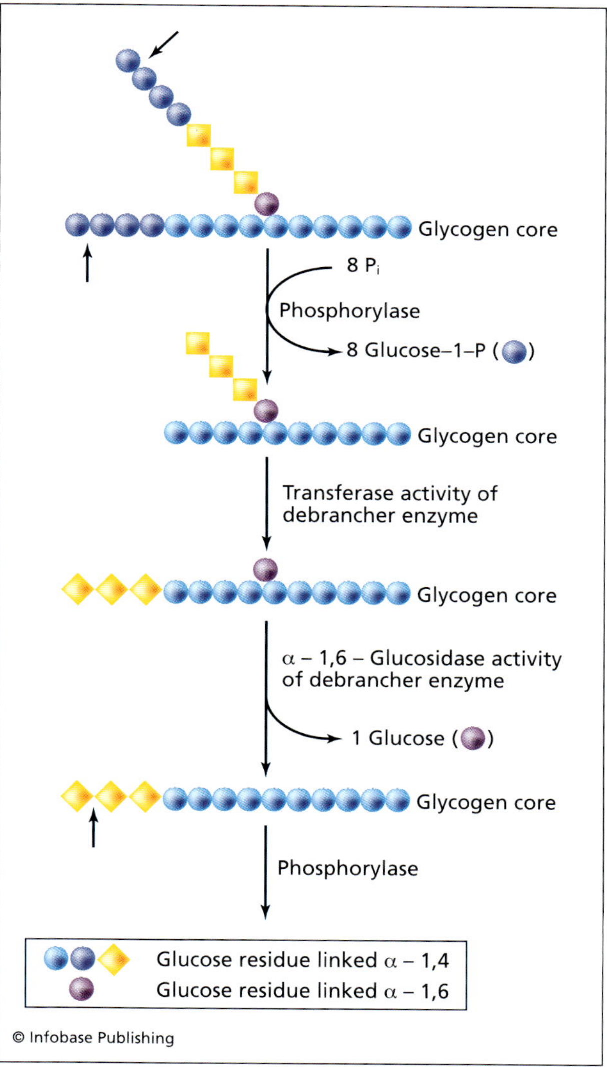

The breakdown of glycogen into individual glucose residues occurs by a stepwise process involving several enzymes.

of liver cells in response to changing levels of blood glucose due to exercise or fasting, and the liver monitors the levels of blood glucose and imports or exports glucose accordingly. These hormones act by inducing alterations in the phosphorylation state of the enzymes glycogen phosphorylase and glycogen synthase.

When the stomach is empty, glucagon levels rise and insulin levels drop. This results in the phosphorylation of glycogen phosphorylase, an event that activates the enzyme and initiates glycogen breakdown. The glucose 1-phoshate residues released from the liver glycogen stores are converted to glucose 6-phosphate, from which glucose 6-phosphatase (an enzyme that only liver cells possess) removes the phosphate. This generates glucose molecules that are transported out of the liver cells and into blood circulation. By breaking down liver glycogen stores to release glucose for export, relatively constant blood glucose levels are maintained even when a long period of time has passed since someone has eaten.

Muscle cells express two interconvertible forms (*a* and *b*) of the enzyme glycogen phosphorylase because they need to break down glycogen to release glucose for their own use, not to restore circulating blood levels to supply the rest of the body cells. The addition of a covalently linked phosphate group converts the *b* form into the *a* form, and both forms have active and inactive conformations. Phosphorylase *a* most often is present in its active conformation, whereas the inactive conformation of phosphorylase *b* dominates unless the energy charge of the muscle cell is low and the levels of adenosine monophosphate (AMP) are high. These conditions occur when the muscles are metabolically active such as during exercise. In resting muscle cells, most of the enzyme exists in the inactive conformation of the *b* form. Hormonal action, such as a burst in epinephrine (a hormone that the body secretes during stressful situations), causes most of the *b* form to convert to the *a* form, which does not depend on cellular levels of ATP, AMP, or glucose in order to be activated. With most of the muscle glycogen phosphorylase present in the active *a* form, the cells rapidly break down glycogen to provide the cells with an immediate source of fuel to burn.

The reverse situation occurs when someone ingests more food fuel than the body can immediately utilize. After ingesting a meal high in carbohydrates, the digestive system will break down the carbohydrates into simple sugars. These glucose molecules make their way to the circulatory system, which carries them to tissues throughout the body, which responds by coordinating an inhibition of glucagon secretion and an increase in insulin production. The action of insulin will eventually assist the body cells in glucose uptake, but the rise in glucose levels also has a more immediate effect. Glucose stimulates the dephosphorylation of the liver enzyme glycogen phosphorylase, inhibiting it so glycogen degradation will halt. Being able to stop the breakdown of glycogen at the same time as stimulating its synthesis is necessary to avoid wasting energy. Continuing to break down glycogen into glucose residues at the same time as the cell is trying to synthesize glycogen to store excess glucose would result in no net change in glucose levels. Phosphorylation controls the activity of glycogen synthase just as it does the activity of glycogen phosphorylase. Phosphorylation inactivates glycogen synthase, in contrast to activating glycogen phosphorylase.

GLYCOGEN STORAGE DISORDERS

The glycogen storage diseases result from genetic mutations leading to deficiencies in specific enzymes involved in glycogen metabolism. Most are inherited as autosomal recessive conditions, meaning both parents pass on a defective copy of the gene for the responsible enzyme. Von Gierke's disease, named after the physician who first identified this condition in 1929, and characterized by the married biochemists Gerty and Carl Cori in 1952, results when the enzyme glucose-6-phosphatase is defective. These individuals cannot convert glucose-6-phosphate into glucose for transport from the liver to blood circulation; thus they are severely hypoglycemic (having very low blood sugar levels) with an increased amount of glycogen in enlarged livers. This disease can also result from a nonfunctional glucose-6-phosphate transport protein, resulting in the inability of cells to move this carbohydrate into the endoplasmic reticulum, the location of the enzyme that converts glucose-6-phosphate into glucose. Cori's disease causes milder similar symptoms but results from a defective debranching enzyme. Examination of the glycogen molecules from individuals who have Cori's disease reveals short outer branches, and this abnormal glycogen does not serve as an efficient storage form of glucose, since only the outermost branches can be utilized. Individuals who have McArdle's disease have faulty muscle phosphorylase activity. Without being able to efficiently break down glycogen during exercise, the muscle cells do not have an adequate fuel supply, and severe cramps result, limiting the person's ability to perform strenuous exercise. Other glycogen storage diseases resulting from defects in other enzymatic activities include Pompe's disease, Andersen's disease, and Hers disease.

See also BIOCHEMISTRY; CARBOHYDRATES; ENERGY AND WORK; ENZYMES; GLYCOLYSIS; METABOLISM.

FURTHER READING

Berg, Jeremy M., John L. Tymoczko, and Lubert Stryer. *Biochemistry,* 6th ed. New York: W. H. Freeman, 2006.

Champe, Pamela C., Richard A. Harvey, and Denise R. Ferrier. *Lippincott's Illustrated Reviews: Biochemistry,* 3rd ed. Philadelphia: Lippincott, Williams, & Wilkins, 2004.

glycolysis Glycolysis is a central pathway in metabolism that breaks down a molecule of glucose into two molecules of pyruvate, releasing a small amount of adenosine triphosphate (ATP) in the process. The net overall reaction, which occurs in the cytosol, is as follows:

$$\text{glucose} + 2P_i + 2\text{ADP} + 2\text{NAD}^+ \rightarrow$$
$$2\text{ pyruvate} + 2\text{ATP} + 2\text{NADH} + 2H^+ + 2H_2O$$

where P_i represents inorganic phosphate, ADP represents adenosine diphosphate, NAD^+ represents the oxidized form of nicotinamide adenine dinucleotide, NADH + H^+ represents the reduced form of nicotinamide adenine dinucleotide, and H_2O represents water. In aerobic respiration, glycolysis precedes the citric acid cycle and electron transport system. When oxygen levels are low or oxygen is absent, fermentation follows glycolysis. During fermentation, the cell reduces pyruvate into lactate, ethanol, or another organic acid or alcohol, depending on the species, in order to regenerate the oxidized form of NAD^+ so glycolysis can continue.

In 1860 Louis Pasteur, a French chemist, discovered that the process of fermentation, as used in wine making, was carried out by living microorganisms. Glycolysis, the first step in fermentation, was discovered by accident in 1897, when Hans Buchner and Eduard Buchner were searching for a safe substance to add to yeast extracts to preserve them for potential therapeutic use. Many foods, such as jams and jellies, contain high concentrations of sucrose as a preservative, so they tried adding sucrose to the yeast extracts, but the extracts rapidly fermented the sucrose into alcohol. These findings meant that fermentation was a chemical process; it did not need to occur inside a living cell, a revelation that opened the door to the modern field of biochemistry. Many other pioneering biochemists contributed to the complete elucidation of the glycolytic pathway: Arthur Harden, William Young, Gustav Embden, Otto Meyerhof, Carl Neuberg, Jacob Parnas, Otto Warburg, Gerty Cori, and Carl Cori.

REACTIONS OF GLYCOLYSIS

Glycolysis converts a six-carbon molecule of glucose into two three-carbon molecules of pyruvate. The intermediates are either six-carbon or three-carbon derivatives, all of which are phosphorylated, meaning phosphoryl groups ($-PO_3^{2-}$) are attached as esters or anhydrides.

In order to extract energy from the glucose molecule, the cell must first trap it inside the cell and convert it into a form that can be cleaved into two three-carbon units. Because cells transport glucose through the cell membrane via specific carrier proteins, if the glucose molecule is modified, the glucose-specific carrier protein no longer recognizes the molecule, and it remains inside the cell. The enzyme hexokinase phosphorylates glucose in the following reaction

$$\text{glucose} + \text{ATP} \rightarrow \text{glucose-6-phosphate} + \text{ADP} + H^+$$

The divalent metal ion Mg^{2+} acts as a cofactor for hexokinase, which replaces a hydroxyl group from the sixth carbon of glucose with a phosphoryl group from ATP.

The second step of glycolysis is the isomerization of the glucose-6-phosphate into fructose-6-phosphate. Isomerization is the change in the structural arrangement of a molecule without a change in the number of atoms of the same elements. Phosphoglucose isomerase performs this reversible reaction by rearranging the glucose from an aldose form to a ketose form.

$$\text{Glucose-6-phosphate} \rightarrow \text{fructose-6-phosphate}$$

The enzyme phosphofructokinase adds another phosphoryl from a molecule of ATP to the sugar, creating fructose-1,6-bisphosphate. This step is one of the main regulatory steps of glycolysis.

$$\text{Fructose-6-phosphate} + \text{ATP} \rightarrow$$
$$\text{fructose-1,6-bisphosphate} + \text{ADP} + H^+$$

Thus far into glycolysis, two molecules of ATP have been utilized, but none has been produced. The next group of reactions generates ATP, a main goal of glycolysis. The enzyme aldolase splits fructose 1,6-bisphosphate into two three-carbon molecules: dihydroxyacetone phosphate and glyceraldehyde 3-phosphate. The same enzyme catalyzes the reverse reaction, an aldol condensation, in which two carbonyl compounds join to form a compound called an aldol.

$$\text{fructose-1,6-bisphosphate} \rightarrow \text{dihydroxyacetone}$$
$$\text{phosphate} + \text{glyceraldehyde-3-phosphate}$$

Glyceraldehyde-3-phosphate continues through the glycolytic pathway, but dihydroxyacetone phosphate does not. Triose phosphate isomerase readily

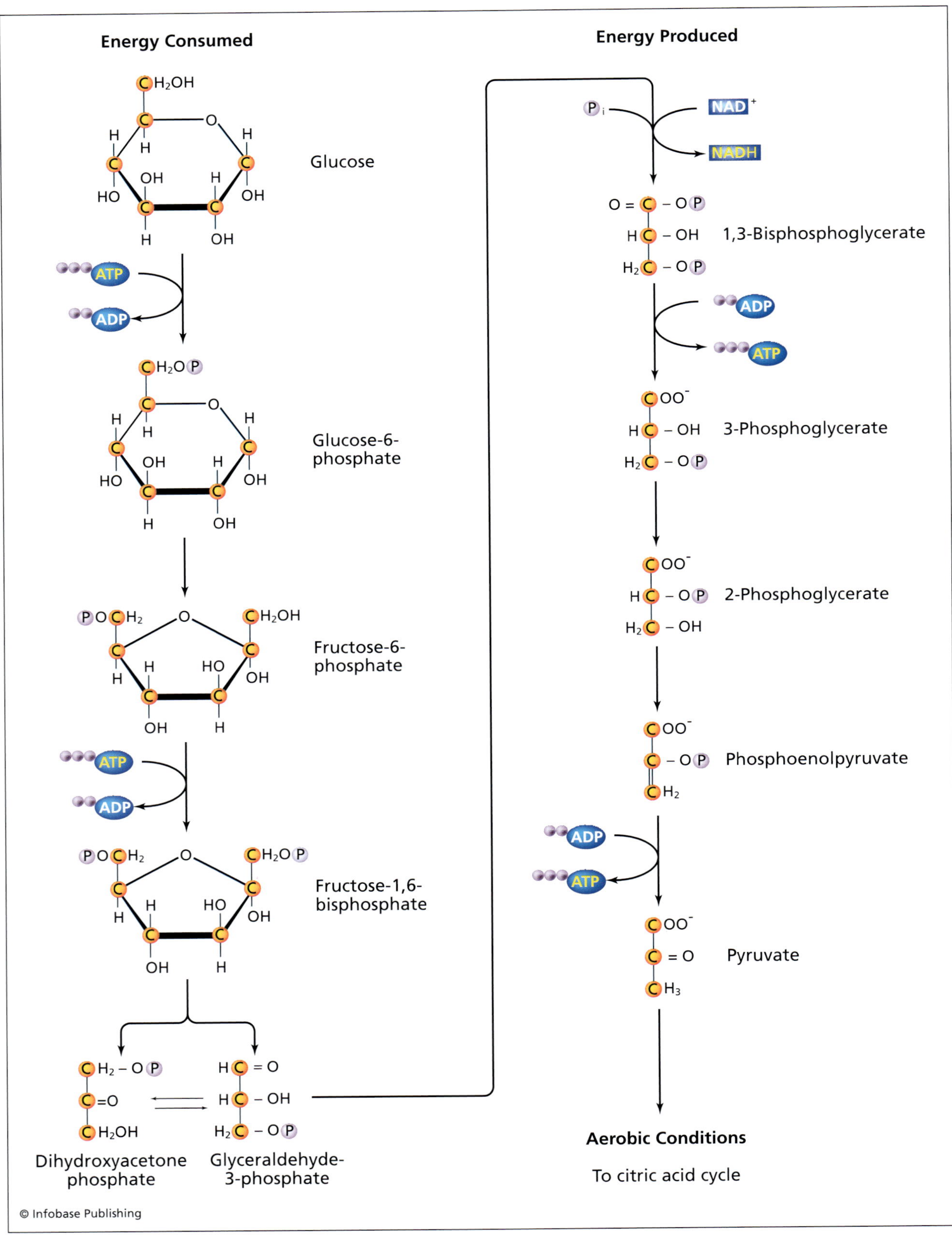

During glycolysis, a six-carbon molecule of glucose is broken down into two three-carbon molecules of pyruvate, generating a net of two molecules of adenosine triphosphate (ATP).

converts the ketose dihydroxyacetone phosphate into the aldose glyceraldehyde-3-phosphate; thus, when performing stoichiometric calculations, from this point on, all the reactions occur in duplicate for a single molecule entering the glycolytic pathway.

dihydroxyacetone phosphate →
glyceraldehyde-3-phosphate

The next step is the addition of a phosphoryl group, resulting in the reversible conversion of glyceraldehyde 3-phosphate into 1,3-bisphosphoglycerate. Glyceraldehyde 3-phosphate dehydrogenase catalyzes the conversion of the aldehyde group on the first carbon into what is called an acyl phosphate group, which has a high transfer potential. The oxidation of the aldehyde group provides the energy necessary to drive the formation of the new bond. NAD^+ picks up the electrons, generating NADH + H^+.

glyceraldehyde-3-phosphate + NAD^+ + P_i →
1,3-bisphosphoglycerate + NADH + H^+

Phosphoglycerate kinase transfers the phosphoryl group from 1,3-bisphosphoglycerate to an ADP, generating the first ATP of glycolysis and releasing 3-phoshoglycerate.

1,3-bisphosphoglycerate + ADP →
3-phosphoglycerate + ATP

The last few reactions of glycolysis convert the three-carbon intermediate into pyruvate and form additional ATP. First phosphoglycerate mutase rearranges 3-phosphoglycerate into 2-phosphoglycerate by moving the phosphoryl group from carbon 3 to carbon 2.

3-phosphoglycerate → 2-phosphoglycerate

An enolase catalyzes the subsequent dehydration reaction, forming phosphoenolpyruvate and releasing a molecule of water. The purpose of this step is to create a phosphoryl with a high transfer potential in preparation for the synthesis of another ATP.

2-phosphoglycerate →
phosphoenolpyruvate + H_2O

In the final step, pyruvate kinase transfers the phosphoryl group to ADP to generate another ATP.

phosphoenolpyruvate + ADP + H^+ →
pyruvate + ATP

The cell uses five different types of reactions in the controlled, stepwise degradation of glucose into pyruvate: phosphoryl transfer, phosphoryl shift, isomerization, dehydration, and aldol cleavage. For each molecule of glucose entering the pathway, glycolysis consumes two molecules of ATP, generates four molecules of ATP, and forms two reduced molecules of NADH + H^+, which either carry the high-energy electrons to the electron transport system for ATP synthesis or transfer them back to an organic intermediate if fermentation is to follow rather than the citric acid cycle.

Other carbohydrates besides glucose can feed into the glycolytic pathway. The three most abundant monosaccharides are glucose, fructose, and galactose. Sucrose, common table sugar, consists of a molecule of glucose linked to a molecule of fructose. The enzyme hexokinase can phosphorylate fructose to create fructose-6-phosphate, but the affinity of hexokinase for fructose is much weaker than for glucose. Instead, the enzyme phosphofructokinase converts fructose into fructose-1-phosphate, which fructose-1-phosphate aldolase converts into glyceraldehyde and dihydroxyacetone phosphate. A triose kinase phosphorylates the glyceraldehyde using an ATP to form glyceraldehyde-3-phosphate, an intermediate of glycolysis, and triose phosphate isomerase converts dihydroxyacetone phosphate into glyceraldehyde-3-phosphate, so both three-carbon molecules can enter the glycolytic pathway. Lactose, the disaccharide found in milk, consists of one molecule of glucose and one molecule of galactose. Galactose can also feed into glycolysis through a series of four reactions that convert it into glucose-6-phosphate. Galactokinase phosphorylates galactose, using an ATP to make galactose-1-phosphate. Galactose-1-phosphate uridyl transferse uses UDP-glucose to make UDP-galactose and glucose-1-phosphate. Then UDP-galactose-4-epimerase converts the UDP-galactose into glucose-1-phosphate. Phosphoglucomutase isomerizes glucose-1-phosphate into glucose-6-phosphate.

REGULATION OF GLYCOLYSIS

Several key steps participate in the regulation of glycolysis. The most important is the step catalyzed by phosphofructokinase, the enzyme that transfers a phosphoryl group from ATP to fructose-6-phosphate to form fructose-1,6-bisphosphate. High levels of ATP inhibit this enzyme by binding to a region of the enzyme other than its active site and changing its shape in a manner that decreases the enzyme's affinity for the substrate. This form of regulation is called allosteric control. When plenty of ATP is present, the cell does not need to undergo glycolysis to produce more. A slightly lowered pH also inhibits phosphofructokinase, which decreases the production of lactic acid by fermentation. When the energy charge of the cell is low and ATP levels drop, the inhibition lifts, and glycolysis resumes.

The synthesis of ATP is one goal of glycolysis, but glycolysis also provides metabolic intermediates that feed into other biochemical pathways. If the cell has plenty of these other intermediates to use as building blocks to synthesize other biomolecules, then glycolysis is not necessary. A sufficient concentration of citrate, one of the intermediates of the citric acid cycle that follows glycolysis in aerobic respiration, enhances the inhibitory effect of ATP on phosphofructokinase. Another allosteric regulator of phosphofructokinase is fructose-2,6-bisphosphate, a molecule that activates the enzyme.

Two other regulatory steps in glycolysis are the first and the last. Glucose-6-phosphate, the end product of the reaction catalyzed by hexokinase, inhibits the enzyme that creates it, in a process called feedback inhibition. If phosphofructokinase activity is inhibited, then the levels of fructose-6-phosphate build up because the next reaction does not proceed. Since fructose-6-phosphate and glucose-6-phosphate are in equilibrium with one another, the levels of glucose-6-phosphate will also rise and inhibit hexokinase. (The liver, which is responsible for regulating the levels of glucose in the blood, has a different enzyme, glucokinase, which glucose-6-phosphate does not inhibit, so the liver can continue to generate glucose-6-phosphate, a precursor to glycogen, a stored form of glucose.)

Pyruvate kinase, the third enzyme in glycolysis that catalyzes an irreversible reaction, actually exists in several different forms called isozymes that perform similar functions but are regulated by different methods. Fructose-1,6-bisphosphate, an earlier intermediate of glycolysis and the product of the reaction catalyzed by phosphofructokinase, activates pyruvate kinase. When the energy charge in the cell is high, ATP allosterically inhibits pyruvate kinase. The amino acid alanine is made from pyruvate, and it also allosterically inhibits pyruvate kinase. In the liver, phosphorylation controlled by circulating hormone levels acts as another mechanism to regulate this enzyme's activity.

See also BIOCHEMISTRY; CARBOHYDRATES; CITRIC ACID CYCLE; ELECTRON TRANSPORT SYSTEM; ENERGY AND WORK; ENZYMES; GLYCOGEN METABOLISM; METABOLISM; OXIDATION-REDUCTION REACTIONS.

FURTHER READING

Berg, Jeremy M., John L. Tymoczko, and Lubert Stryer. *Biochemistry*, 6th ed. New York: W. H. Freeman, 2006.

Boiteux, A., and B. Hess. "Design of Glycolysis." *Philosophical Transactions of the Royal Society of London B* 293 (1981): 5–22.

Fruton, Joseph S. *Molecules and Life: Historical Essays on the Interplay of Chemistry and Biology*. New York: Wiley-Interscience, 1972.

King, Michael. From the Medical Biochemistry Page. Available online. URL: http://themedicalbiochemistrypage.org/. Last modified July 11, 2008.

graphing calculators Mechanical calculating methods originated as early as the 13th century. The abacus, the slide rule, and the mechanical calculating machine were all sequential improvements made in an effort to simplify the calculating process. These original devices began to be replaced by electronic calculators in the 1960s. Calculating devices have become so commonplace that most students do not remember a time without them. Gone are the days of the simple four-function calculator. Today's calculators have computing and graphing ability beyond what people imagined even for computers 50 years ago. The evolution of the electronic graphing calculator is a prime example of scientific and technological advances building on previous designs.

ABACUS AND SLIDE RULE

The earliest known formalized calculating device is the abacus. Chinese and Japanese versions of the abacus had slight variations, but all types of the abacus have a series of strings or wires and beads. The abacus simply helps the user keep track of the number of items that have been counted in order to allow totaling. A simple abacus has two different sets of strings. One set of wires has five beads and allows the user to count from 1 to 5. The other set of wires holds two beads representing items in groups of five or 10. When one counts an item, a bead from the set of five is moved to the other side. Once 5 is reached, the user can move the corresponding bead in the set of two to represent a total of 5. This allows a cumulative total to be reached.

The invention of logarithms by the Scottish mathematician John Napier in 1617 simplified calculations by transforming the process of multiplication into one of addition. Scientific notation requires a coefficient greater than or equal to 1 and less than 10. The second part of scientific notation is 10 raised to a power. For example,

$$1 \times 10^3 = 1{,}000$$

The exponent in this measurement describes the location of the decimal point in order to return to the

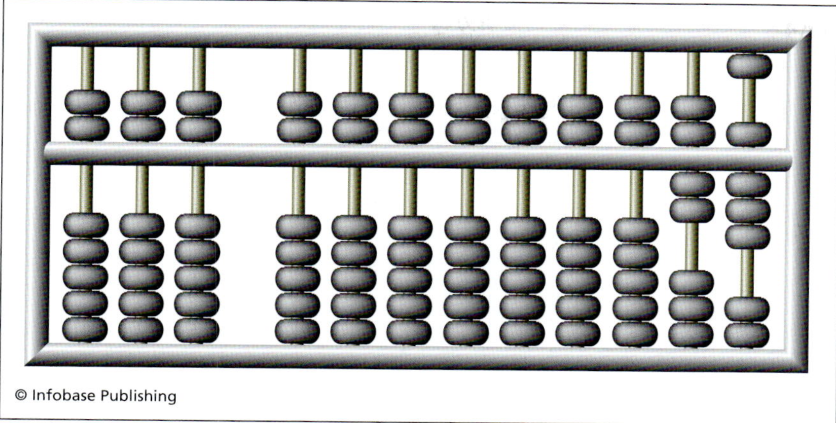

An abacus is the earliest known calculating device. The abacus has beads on strings in sets of five and two. Five beads on the bottom of the abacus shown are used to count numbers from 1 to 5. Once 5 is reached, a bead on the top is moved and the lower beads are counted again.

original number. Base 10 logarithms are simply the exponent to which 10 is raised. The logarithm of 100 equals 2 because 10^2 is equal to 100. The general formulas for logarithms are as follows:

$$\log ab = \log a + \log b$$

$$\log a/b = \log a - \log b$$

$$\log a^b = b \log a$$

When performing calculations using logarithms, the log of any two multiplied numbers is equal to the sum of each individual log. The log of any two divided numbers is equal to the difference between their logs.

Utilizing logarithms was still tedious because the user had to determine the logarithm of both of the numbers and then add them together. To simplify this process, the English mathematician William Oughtred developed the slide rule in 1633. A slide rule is a device with two parallel rulers that are able to slide past one another. Each of the bars has a number line based on the logarithms of the numbers. Sliding the bars past one another allows for the distance to be measured for each of the values, revealing the solution. The slide rule continued to be a significant calculating tool well into the 1900s.

EARLY MECHANICAL CALCULATING MACHINES

The first documented drawings for a mechanical calculating device were found in the notes of the Italian scientist and mathematician Leonardo da Vinci from the early 1500s. The end of the 1600s saw the development of the first mechanical calculating machine. Although neither of the devices was particularly usable for the general public, the devices designed by the German mathematician Wilhelm Schickard and the French mathematician Blaise Pascal are noteworthy. Both of these developments influenced the design of calculating machines well after their production.

Schickard developed a device known as the calculating clock in 1623. His machine involved base 10 wheels that completed a full turn every 10 numbers added. This adding machine facilitated the addition of large numbers. However, no actual version of the calculating clock exists.

A child prodigy, Blaise Pascal, at the young age of 21 developed and produced a calculating device known as the Pascaline in 1643. The wheels in this device were only able to add numbers easily; subtraction was a different story. Although unwieldy and difficult to use, the Pascaline helped Pascal's father, a tax collector, complete his arithmetic calculations more quickly.

In 1672 a German mathematician named Gottfried Leibniz created the first mechanical calculating device that could successfully carry out addition, subtraction, multiplication, and division. His apparatus, called the stepped drum, contained nine notches of increasing length around the drum. Only a few versions of this calculating machine were made, but subsequent computing machines utilized the stepped drum technology.

The 1700s produced minimal advances in computational technology. None of these devices was developed commercially. This changed in 1820, when the French mathematician Thomas de Colmar utilized Leibniz's stepped drum mechanism to create the first mass-produced calculating machine, the arithmometer. This machine was produced at greater quantities than any previous model. By the 1930s, more than 1,500 arithmometers were produced. As the 1800s proceeded, the quest for new and improved calculators led to increasingly improved designs as well as the development of keys rather than levers for entering numbers. The innovation of the printout from the calculator was developed in 1888, when William Seward Burroughs received a patent for the first calculating machine that had a keyboard and a printout. In 1886 Burroughs founded the American Arithmometer Company, which originated in St. Louis and later moved to Detroit. By 1905 the company was renamed the Burroughs Adding Machine

Company, and by 1925 the company had sold more than 1 million devices.

ELECTRONIC CALCULATORS

By the 1920s the mechanical calculating machine was on the way out, and the development of the electronic calculator had begun. Jerry Merryman, James Van Tassel, and Jack St. Clair Kilby at Texas Instruments invented the first electronic handheld pocket calculator in 1966. What followed over the next 30 years was a rapid series of improvements that transformed the original simple four-function calculator into an electronic programmable calculator.

Companies such as Hewlett-Packard and Texas Instruments led the developments. Hewlett-Packard started as a partnership in 1930 between Bill Hewlett and Dave Packard, and the company's HP-35 was one of the world's first handheld calculators. The Hewlett-Packard 9100 series that followed were quite expensive because of the newness of the technology—some cost $1,000 to $2,000. In the early 1970s companies began using microchips in calculators. This led to the introduction of the following:

- desktop versions of calculators
- calculators with scientific functions such as logarithms
- programmable calculators
- electronic calculating toys including such consumer products as the Texas Instruments' Little Professor in 1976
- solar-powered calculators
- graphing calculators

Graphing calculators such as the Texas Instruments TI-84 Plus Silver Edition shown here are common tools in classrooms today. The technology required to produce these calculators has developed rapidly. *(Texas Instruments images used with permission)*

The innovative graphing calculator was introduced in the 1990s. This calculator not only performed all of the arithmetic and scientific functions, but could also graph data points and perform curve-fitting functions. The field of education recognized the implications of these types of calculators. Only scientists and researchers had previously had access to computers that were sophisticated enough to perform these complicated tasks. As a result of financial constraints, schools and students did not have this same access. The production of the first handheld graphing calculator, the TI81, by Texas Instruments in 1990, gave educators the opportunity to implement higher-level mathematics skills in everyday classrooms as students now had the calculating capabilities to perform them. Today fitting a curve to a plot of data points using such calculators is standard practice in high school math classes. Graphing calculators became increasingly sophisticated, and in the early 1990s even more advanced graphing calculators, the TI83 and TI85, were introduced. With each upgrade in technology, the calculators gained more functions, more memory, and more capabilities.

The graphing calculator became even more useful with the advent of data collecting devices that were capable of taking readings and transferring the data to the calculator for graphing. In 1994 the CBL (calculator-based laboratory) was created, followed by the CBR (calculator-based ranger) in 1996. These devices allowed students to gather data such as motion and speed of an object and transfer these data directly to the calculator, where information about the velocity and acceleration of the moving object could be calculated and graphed. Improvements of the graphing calculators continued through the late 1990s, including upgradable functions and new programs that could be added to the calculator, resulting in the development of calculators such as the TI-89, the TI-83 Silver Edition, the HP48gll, and the HP39gs. By the year 2003, 25 million graphing calculators had been shipped by Texas Instruments. Hewlett-Packard offers their latest line of graphing calculators including the HP50g. The technology continues to improve.

From the abacus as a counting device, the slide rule, numerous variations of mechanical calculating devices, to the electronic calculator, and finally to the graphing calculator, the evolution of computing devices is a shining example of the development of science and technology in action. Each scientist tried to develop devices building on the successes of the previous device to create a more useful product. The story demonstrates how the success of one innovation can stimulate rapid succession in the development of many similar and improved devices. Examples of this are seen in the change from the first rough Pascaline to the multiple improvements that followed it in the 1600s. The same is seen from the 1960s, when the first electronic calculator was developed, to the rapid achievements seen today.

See also SCIENTIFIC METHOD.

FURTHER READING

Hewlett-Packard. "History." Available online. URL: http://www.hp.com/hpinfo/abouthp/histnfacts/. Accessed February 20, 2008.

gravity Gravity is the force of attraction between objects due to their mass. Gravity is also referred to as gravitation, and some physicists choose to restrict the term *gravity* to mean the gravitational effect of an astronomical body, such as Earth. Gravity is most familiar in its effect on objects and people on the surface of Earth: they tend to stay on the surface or fall to it. But gravity is a mutual effect. When Earth is pulling an apple toward it, the apple is pulling Earth toward *it* with the same magnitude of force. As a result of the tremendous difference in mass, these forces affect the apple much more than they affect Earth. Beyond preventing houses, horses, and harmonicas from flying off into space, gravity appears to be the dominant force in cosmological and astronomical phenomena, holding stars' planets in their orbits and planets' moons in theirs. The solar system and the Earth-Moon system serve as examples. Gravity binds stars into galaxies, galaxies into clusters, and the latter into superclusters. And gravity appears to operate at the scale of the whole universe, where it is thought to govern the rate of expansion of the universe and even to cause what seems to be an accelerating rate of expansion. Under certain conditions, gravity can cause repulsion, rather than the much more common attraction.

For a large range of conditions, Sir Isaac Newton's law of gravitation adequately describes gravity. The law states that two point particles of masses m_1 and m_2 attract each other with forces of magnitude

$$F = G\frac{m_1 m_2}{r^2}$$

where F is in newtons (N); m_1 and m_2 in kilograms (kg); r denotes the distance between the particles, in meters (m); and G is the gravitational constant,

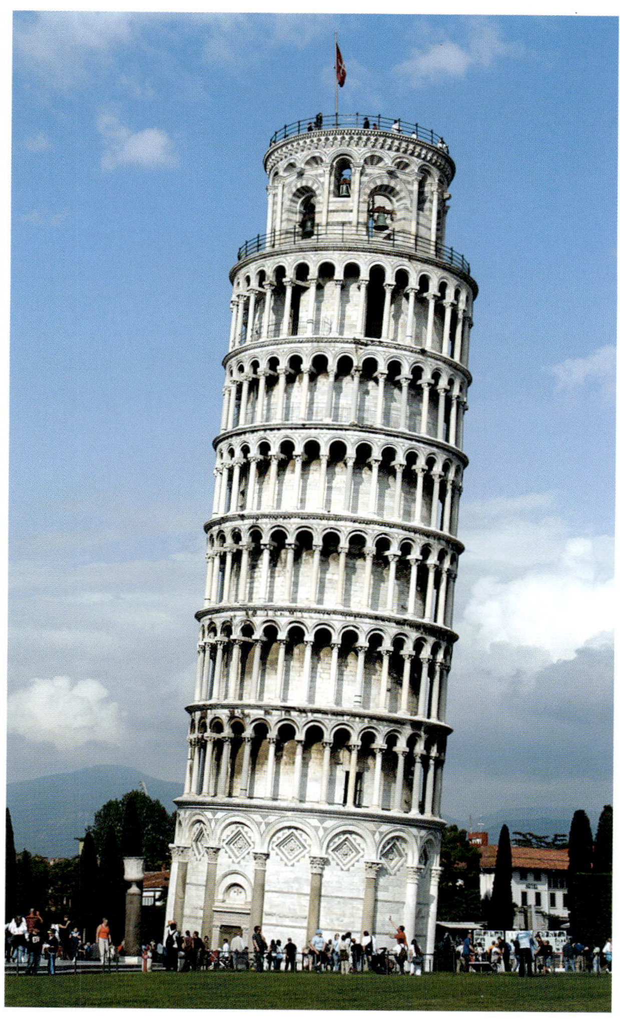

Galileo was among the first to investigate gravity experimentally. He is reported to have dropped cannonballs from the Leaning Tower of Pisa, shown here, and found that their time of fall was independent of their mass. *(Helen Shorey, 2008, used under license from Shutterstock, Inc.)*

whose value is 6.67259×10^{-11} N·m²/kg², rounded to 6.67×10^{-11} N·m²/kg². The gravitational forces of attraction are proportional to each of the masses of the mutually attracting bodies and are inversely proportional to the square of the bodies' separation. Double the separation, and the bodies are attracted to each other with only one-fourth of the original forces. One can determine the gravitational constant experimentally by measuring the forces between two bodies. This measurement is called the Cavendish experiment, after the 18th-century English chemist Henry Cavendish, who first performed it, and can be carried out in student laboratories.

Gravity is a conservative force (*see* FORCE) and thus possesses a defined potential energy. The gravitational potential energy of two point particles as earlier is given, in joules (J), by

$$\text{Gravitational potential energy} = -G\frac{m_1 m_2}{r}$$

Here the reference level of zero potential energy is defined as that of infinite separation. For finite separations, the potential energy is negative and becomes more negative as the gravitating bodies are closer. When two gravitating bodies are pulled apart so that their separation increases, the positive work that is done appears as increased (i.e., less negative) gravitational potential energy of the bodies (according to the work-energy theorem). For very small changes in the separation distance, when a small body moves small vertical distances near the surface of a planet, such as Earth, the gravitational potential energy can be conveniently approximated by

$$\text{Gravitational potential energy} = mgh$$

where m denotes the mass of the body, in kilograms (kg); h is the body's altitude, in meters (m), above an arbitrary reference height near the planet's surface (h is positive or negative when the body is above or below the reference level, respectively); and g denotes the acceleration due to gravity, the free-fall acceleration, of objects at the surface of the planet, in meters per second per second (m/s²). At the surface of Earth, g has the approximate value of 9.8 m/s².

As an example of application of this formula, consider this problem. Let a body fall from rest near the surface of Earth. When the body has fallen 5.0 m, what is its speed? (Ignore air resistance.) For this problem to serve as an example of gravitational potential energy, its solution should involve the use of an energy method (rather than a kinematic approach). Let the original altitude of the body serve as the reference level. So in the body's initial state its gravitational potential energy is zero, and, since it is at rest, so is its kinetic energy, giving zero total mechanical energy. In the body's final state, its altitude is h = -5.0 m with respect to the reference level, and it has unknown speed v. No external work is being done on the body-Earth system, so mechanical energy is conserved:

$$\tfrac{1}{2} mv^2 + mgh = 0$$

where m denotes the mass of the body. The first term on the left-hand side is the body's final kinetic energy, the second term denotes its final gravitational potential energy, and their sum is the body's total mechanical energy in its final state. This sum equals zero, since mechanical energy is conserved and the body's mechanical energy in its initial state is zero. Now the mass m cancels out of the equation—the result will be independent of the body's mass—and the solution for v is

$$\begin{aligned} v &= \sqrt{-2gh} \\ &= \sqrt{-2 \times (9.8\ \text{m/s}^2) \times (-5.0\ \text{m})} \\ &= 9.9\ \text{m/s} \end{aligned}$$

When the body has fallen 5.0 m from rest, its speed is 9.9 m/s.

As is the case for electricity and magnetism, so too can gravity be described in terms of a field, the gravitational field. The sources of this field are all masses, while it affects all masses by causing forces to act on them.

Newton's law of gravitation is not adequate for all situations. The orbit of the planet Mercury, for example, is not consistent with Newton's law. Neither is the amount of deviation of a light ray as it passes near a massive object, such as the Sun, correctly predicted by Newton's law. A more widely applicable theory of gravitation is Albert Einstein's general theory of relativity. This theory has so far successfully passed all experimental tests.

Gravity is one of the four fundamental forces among matter, together with electromagnetism, the weak force, and the strong force. The force of gravity is the most universal among them in the sense that it affects all the fundamental constituents of matter. Electromagnetism affects only electrically charged elementary particles, and the weak and strong forces each affects only a certain class of elementary particles. Gravity alone affects *all* the elementary matter particles, since they all have mass, although at the atomic and subatomic scales, gravity is the weakest of all four forces. While electromagnetism and the weak force have been unified as the electroweak force, which, in turn, is expected to be unified with the strong force in a grand unified theory (GUT), the gravitational force is problematic in that regard. The theories of the other three forces can easily be quantized (i.e., made consistent with the quantum character of nature). Gravity is not so amenable. This seems to be intimately related to the fact that gravity, as described by the general theory of relativity, involves space-time in an essential way, while the other three forces "merely" appear to play their roles on the space-time stage, so to speak, without affecting or being affected by space-time. So a quantization of gravitation is tantamount to a quantization of space-time itself. At present it is not clear how that might be accomplished.

According to the quantum picture of nature, the gravitational force is assumed to be mediated by the exchange of force particles called gravitons. Present understanding holds that the graviton should be a massless, spin-2 particle, but it has not yet been discovered experimentally.

GRAVITATIONAL LENS

The effect whereby light rays from a distant astronomical source are bent by an intervening massive astronomical object, such as a galaxy or a galaxy cluster, on their way to an observer on Earth, causing the source to appear distorted—as several sources or as a possibly incomplete ring around the intervening object, called an Einstein ring—is gravitational lensing, with the intervening object serving as a gravitational lens. Einstein's general theory of relativity requires the bending of a light ray passing near a massive object. Since a gravitational lens tends to bring to an observer light rays from the source that would otherwise not reach him, it tends to brighten the source's image and can even reveal sources that would otherwise be hidden by the intervening object. On the other hand, gravitational lensing, called microlensing in this case, can detect the presence of dark massive astronomical objects by the temporary apparent brightening of a distant light source as a dark object passes through the line of sight to the source. Gravitational lensing is a small effect and requires powerful telescopes for its observation.

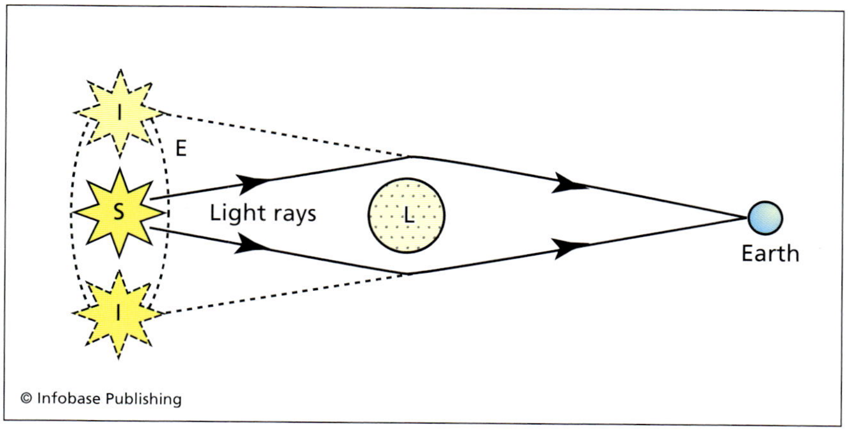

GRAVITATIONAL WAVES

A wave in the gravitational field is termed a gravitational wave. The general theory of relativity shows that time-varying mass configurations can produce waves in the gravitational field. Gravitational waves travel at the speed of light in vacuum (3.00×10^8 m/s) and have other similarities to electromagnetic waves. They should be detectable through the time-varying forces they cause on bodies. Physicists are currently carrying out an ongoing effort to construct gravitational-wave detectors of sufficient sensitivity to detect such waves. One such effort is the Laser Interferometer Gravitational-Wave Observatory (LIGO), being constructed in the United States and consisting of two widely separated installations—one in Hanford, Washington, and the other in Livingston, Louisiana—operated in unison as a single observatory. In addition, a space-based gravitational-wave detector, the Laser Interferometer Space Antenna (LISA), is in the planning stage. The most intense gravitational waves are candidates for earliest detection, when detectors become sensitive enough. Large masses undergoing high accelerations, such as supernovas, massive stars falling into black holes, and even the big bang, are expected to generate them.

See also ACCELERATION; BIG BANG THEORY; BLACK HOLE; CONSERVATION LAWS; EINSTEIN, ALBERT; ELECTRICITY; ELECTROMAGNETIC WAVES; ELECTROMAGNETISM; ENERGY AND WORK; FORCE; GENERAL RELATIVITY; MAGNETISM; MASS; MOTION; NEWTON, SIR ISAAC; QUANTUM MECHANICS; WAVES.

A gravitational lens is a massive astronomical object, such as a galaxy, star, or dark object, that is located between a distant luminous object, such as a galaxy, and Earth and bends light rays emanating from the distant object on their way to Earth. In this manner the light source, as viewed on Earth, might appear distorted in various ways. The figure (not to scale) represents two light rays in an ideally symmetric lensing situation, in which the source S would appear to an observer as a ring E, called an Einstein ring, with the lensing (bending) object L at the ring's center. To an observer on Earth, the two light rays appear as two images I of the source, which form part of the Einstein ring.

FURTHER READING

Serway, Raymond A., Clement J. Moses, and Curt A. Moyer. *Modern Physics,* 3rd ed. Belmont, Calif.: Thomson Brooks/Cole, 2004.

Young, Hugh D., and Roger A. Freedman. *University Physics,* 12th ed. San Francisco: Addison Wesley, 2007.

green chemistry Green chemistry is a field of chemistry that is a direct outgrowth of the Pollution Prevention Act passed by Congress in 1990, which requires the Environmental Protection Agency to promote the prevention or reduction of pollution from industry whenever possible. Historically manufacturers did not worry about the consequences of the production of medicines, plastics, fertilizers, and synthetic fabrics on the environment. This disregard led to the release of an increasing number of toxins into the dirt, air, and water over the years, affecting both the immediate environment and the people living in the vicinity. The only solution was to treat or to clean up the problems after they were present. Green chemistry prevents waste at the outset by reducing or eliminating the use or creation of toxic substances in the design and manufacture of chemical products. This growing branch of chemistry promotes designing products that are nontoxic, using reactants that are not hazardous, minimizing waste and unwanted by-products, and educating both chemists and the general public about alternatives. Green chemistry encompasses many subdisciplines of chemistry including analytical chemistry, biochemistry, inorganic, organic, and physical chemistry. A collaborative partnership among scientists, scientific societies, academia, industry, government, and private agencies has led to several early achievements in this growing field. Accomplishments to date include the introduction of nontoxic supercritical fluids (substances that behave as a gas and a liquid at certain temperatures and pressures) as alternative solvents, the use of metathesis chemistry to synthesize drugs and plastics with little waste, and the creation of synthetic molecules called tetra-amido macrocyclic ligand activators (TAMLs) that act in concert with hydrogen peroxide to break down toxins in water.

TWELVE PRINCIPLES OF GREEN CHEMISTRY

When trying to address a problem with green chemistry, individuals and groups rely on the 12 principles of green chemistry, a concept that Paul Anastas and John Warner introduced in their book, *Green Chemistry: Theory and Practice,* when the field was in its infancy. Issues addressed in these guidelines can be grouped into four categories: using safer starting materials, using renewable resources whenever possible in manufacturing, producing materials that are nontoxic and degrade rapidly, and designing chemical syntheses that require the least amount of reactants, generate the minimal amount of waste by-products, and manufacture products in the most energy efficient manner. Together these principles serve as an instruction manual for designing synthetic materials that are "benign by design."

Safety to people and the environment is an important aspect of green chemistry. In the past, the starting materials for many chemical processes were hazardous because they were poisonous to humans, they could lead to fires and explosions, or they released fumes into the environment. Green chemistry addresses these issues by designing and implementing chemical processes that use nontoxic starting materials that are safer to handle. For example, the goal of a common laboratory exercise performed in undergraduate organic chemistry courses is to create adipic acid, a chemical used to make nylon, using nitric acid as an oxidant in the reaction. A by-product of this reaction is nitrous oxide, a chemical that depletes the ozone layer. At the University of Oregon, a green organic chemistry laboratory now uses hydrogen peroxide, an innocuous oxidizing agent, to yield the same product. Green chemists have also identified safer solvents to dissolve reactants in the industrial sector. The dry cleaning industry has traditionally used perchloroethylene to dissolve dirt and grease from clothes; an increasing portion of the dry cleaning industry now uses a nontoxic alternative cleaning solvent called supercritical carbon dioxide.

Green chemistry also encourages the design of chemical syntheses that use renewable resources, resources replaced by natural ecological cycles or natural chemical or physical processes. The plastics industry traditionally has used fossil fuels, such as petroleum, as a starting material, resulting in the depletion of natural resources. Green chemists design processes that use renewable resources from plants, such as corncobs and fallen leaves, to manufacture products. Natureworks LLC recently built a plant in Nebraska that uses plant sugars to manufacture polylactic acid (PLA) pellets that are used to make water bottles and packing material.

A third principle that green chemistry follows is designing chemical syntheses that use the least amount of starting material, produce the least waste, and synthesize materials in the most energy-efficient manner. One way to accomplish this goal is to minimize the number of steps in the synthetic process. In the 1960s the pharmaceutical industry synthesized ibuprofen, an over-the-counter pain-relief medication, using a six-step process that produced a larger percentage of waste products than desired end product. In 1991 chemists determined that the number of steps in ibuprofen production could be reduced to three, with less than one-fourth of the products generated considered waste. The new synthesis protocol also significantly diminished the amount of required reactants and produced millions fewer pounds of waste. To decrease the necessary energy input for a synthetic reaction, green chemists often use catalysts to drive the desired reactions. Using catalysts reduces the number and quantity of reactants needed to drive the various steps in the syntheses. Performing syntheses at ambient temperature and pressure also reduces the amount of energy required for materials to be synthesized.

The last element of the design process to consider is the product that is manufactured using these principles. According to the 12 principles of green chemistry, materials produced must not be toxic to people, and they must be able to degrade rapidly in order to prevent a buildup of synthetic material in the environment. Manufacturers formerly made packing material using petroleum-derived polystyrene pellets that fill up landfills because they do not rapidly degrade; in contrast, green chemists synthesize starch packing pellets, nontoxic products that dissolve readily in water.

ADVANCEMENTS IN GREEN CHEMISTRY

As more groups in the scientific community and the industrial sector embrace the 12 principles of green chemistry, manufacturers are forsaking long-established processes that generated toxic by-products for environmentally friendly alternatives. In some cases, the newer green technology employs products, such as supercritical fluids, that were identified over 100 years ago, but whose use as alternative solvents has just recently been investigated. Many new synthesis methods that were previously unavailable have been implemented, and these processes yield more product and less waste than older methods of synthesis. Metathesis is an example of these new types of syntheses. Green chemists have also studied natural processes to design synthetic molecules, like TAMLs, that will degrade pollutants into harmless molecules.

Supercritical fluids are substances that behave as both a gas and a liquid, meaning these fluids can diffuse through solids as a gas does and dissolve solute material as a liquid does, after they reach a critical temperature or pressure. At this critical point, the density of this molecule in its liquid form equals the density of its gas form. The two most common supercritical fluids used in widespread applications are supercritical water and supercritical carbon dioxide.

Supercritical carbon dioxide possesses many favorable characteristics. Like carbon dioxide, supercritical carbon dioxide is nonflammable, is nontoxic, and does not release any emissions that contribute to global warming. From a manufacturer's perspective, supercritical carbon dioxide is a feasible alternative to traditional materials because it is highly abundant and consequently inexpensive to use. Commercial uses of supercritical carbon dioxide include decaffeinating coffee, extracting oils from plant materials for the perfume industry, and serving as an alternative solvent for metal parts and garment cleaning.

The dry cleaning industry traditionally has used a chlorinated solvent called perchloroethylene, or tetrachloroethylene. This particular solvent is toxic and depletes the ozone layer. The conventional process using this solvent requires water and results in a waste stream of solvent, sludge, and debris in the wastewater. Supercritical carbon dioxide is an environmentally friendly solvent, but it does not have the same cleaning ability as perchlorothylene; therefore, some dirt and stains are not removed using the supercritical fluid alone. A breakthrough in green chemistry resulted when Dr. Joseph DeSimone at the University of North Carolina successfully designed and synthesized a group of surfactants, substances that act as detergents and readily dissolve in supercritical carbon dioxide. These surfactants and the supercritical carbon dioxide combine to form spherical structures called micelles, with the surfactants forming an external shell that encloses the supercritical fluid in the center. These micelles have soaplike properties because they trap dirt inside and carry it away from the garment being cleaned. When dry cleaning is complete, the supercritical carbon dioxide reverts to the gas state and the surfactants can be collected and reused. The entire process is performed in the absence of water, so natural resources are neither depleted nor contaminated. The dry cleaning industry also saves money and energy as garments do not have to be dried after this alternative process.

In 2005 Yves Chavin, Robert H. Grubbs, and Richard R. Schrock won the Nobel Prize in chemistry for the development of the metathesis method in organic syntheses. *Metathesis* means "to change places." In metathesis syntheses, two reactants join in the presence of a catalyst, bonds within individual reactants break, parts of one molecule swap with parts from the second molecule, and then bonds reform, resulting in products that are a mix of the original reactants. Chemists sometimes refer to this type of reaction as a double-replacement reaction. These reactions require catalysts such as the metals nickel, tungsten, ruthenium, and molybdenum. Currently different sectors of industry use metathesis syntheses to produce a variety of products including medicines, polymers, and enhanced fuels. Metathesis syntheses are more efficient than other syntheses because they require fewer steps, consume fewer resources, and produce less waste. Since metathesis syntheses can be performed at ambient temperature and pressure, and because the products are stable in air, these syntheses are simple to use. Green chemists also favor metathesis syntheses because they use nontoxic solvents and create little, if any, hazardous waste. Research and industry have both embraced this type of synthesis as a means to introduce green chemistry on a large, commercial scale.

Every year billions of gallons of waste from industry end up in streams and rivers. The water becomes contaminated with dyes, plastics, pesticides, and other toxic substances. Pollution is not limited to rivers and streams; traces of pesticides, cosmetics, and even birth control hormones also contaminate drinking water. Not only is the environmental impact apparent as fish and other animals that have contact with these pollutants decline in number, but scientists also suspect that human exposure to these contaminants affects development, immunity, and reproduction.

To address the problem of water pollution, a group of investigators at Carnegie Mellon University's Institute for Green Oxidation Chemistry have developed a synthetic group of catalyst molecules called TAMLs that bind to specific pollutants and, in concert with oxygen or hydrogen peroxide, break down these pollutants. The scientists used naturally occurring enzymes, such as peroxidases and cytochrome p450s, as inspiration for the design of these TAMLs. In nature, fungi use peroxidases and hydrogen peroxide to degrade lignin, a molecule that helps wood stick together, so that the wood breaks down into smaller components that the fungi can use as a nutritional source. Rather than imitating the entire protein structure of these large enzymes, scientists focused on recreating a functional catalytic site in the laboratory. These synthetic molecules needed to be strong enough to withstand a violent reaction between hydrogen peroxide and the targeted pollutant. Using nature as a guide, the catalytic portions of these TAMLs were designed with an iron atom at their center, surrounded by four nitrogen atoms that were bound to the iron through covalent bonds, forming a square. Placement of organic groups around the center formed a large outer ring called a macrocycle, designed to attract and bind specific pollutants. To ensure that these TAMLs did not ultimately become a pollutant, the scientists devised TAMLs that would decompose in a few minutes to a couple of hours.

Initial laboratory studies and field tests suggest that the TAMLs effectively eliminate a wide range of pollutants. Evidence shows that TAMLs can bind and destroy anthraxlike bacteria in a water supply. In field trials, TAMLs and peroxide added to wastewater from the paper and wood pulp industries resulted in a significant decrease in the brownness of the treated water, suggesting that lignin was being degraded by these catalysts. A reduction in toxic by-products of paper manufacturing, such as organochlorines, was also observed in field trials using different TAMLs specific to these pollutants. From a corporate perspective, TAMLs are appealing because they are relatively inexpensive to make and can be injected directly into wastewater, eliminating the need to purchase additional equipment.

See also ACID RAIN; ENZYMES; GREENHOUSE EFFECT.

FURTHER READING

Collins, Terrence J., and Chip Walter. "Little Green Molecules." *Scientific American,* March 2006, 82–90.

United States Environmental Protection Agency. Green Chemistry homepage. Available online. URL: http://www.epa.gov/greenchemistry. Last updated July 25, 2008.

greenhouse effect The greenhouse effect is a natural phenomenon that traps heat from the Sun in the atmosphere and creates a comfortable life on Earth. Many natural gases in the atmosphere absorb energy (in the form of infrared radiation) that has been reradiated from the surface of the Earth into the atmosphere. When the energy is absorbed, it is not released back into space, thus raising the global temperature. Without this insulating blanket of gases in the atmosphere the temperature on the Earth could be as low as 30°F (16.7°C) colder than it is today. The greenhouse gases include carbon dioxide (CO_2), nitrous oxide (N_2O), methane (CH_4), perfluorocarbons (PFCs), hydrofluorocarbons (HFCs), and sulfur hexafluoride (SF_6).

The largest gaseous components of the Earth's atmosphere, oxygen and nitrogen, are not involved in atmospheric temperature regulation. In order to be able to absorb the infrared radiation, the gas molecules must contain more than two atoms and have a relatively flexible bond structure. Molecules that can absorb infrared radiation (IR) are known as IR-active; those molecules that cannot absorb infrared radiation are known as IR-inactive. The bonds of IR-active molecules have a dipole moment (an asymmetric charge distribution), and their bonds are able to absorb the energy of the infrared radiation. These molecules then release the energy in the form of infrared radiation that can increase the temperature of the planet or possibly excite another molecule of a greenhouse gas. Ultimately these molecules prevent the heat from leaving the atmosphere and moving back into space.

Heat transfer can occur by three mechanisms: radiation, convection, and conduction. Radiation is the transmission of energy in the form of waves through space. Conduction is the transfer of energy in the form of heat from one object to another through direct contact. Convection is the transfer of energy by moving air. The energy from the Sun is transferred to the Earth through space in the form of electromagnetic radiation with wavelengths between 100 nm and 5,000 nm. The wavelength of a light wave is a measure of the distance from the crest of one wave to the crest of the next wave or from the trough of one wave to the trough of the next wave. The greater the distance between the crests or troughs, the longer the wavelength of the wave. Electromagnetic radiation with long wavelengths has low energy. Conversely, waves with short wavelengths have higher energy. The visible region of the electromagnetic spectrum includes radiation with wavelengths ranging from the most energetic form of visible light (violet) at 400 nm to the lowest energy form of visible light (red) around 700 nm. Visible light makes up approximately half of the electromagnetic radiation from the Sun that reaches the Earth. Infrared specifically refers to radiation having wavelengths longer than visible light up to about 1 millimeter.

When electromagnetic radiation hits the surface of the Earth, much of it is reflected back by gases in the atmosphere, some is reflected back into space by the clouds, and some is absorbed by the clouds. The rest is absorbed by the Earth, and then a portion of that is reflected back in the form of infrared radiation. The percentage of the Sun's energy that is reflected back from the surface is known as albedo. The surface that the sunlight hits determines whether the energy is absorbed or returned into the atmosphere. The bright white surfaces of glaciers or ice reflect most of the energy back into the atmosphere. Brown or dark surfaces like the desert absorb much energy and radiate very little back.

The greenhouse effect is found on any planet that has an atmosphere, including Mars and Venus. The greenhouse gases of the atmosphere on Mars are not present in large enough quantities to sustain a livable temperature on the surface of Mars. The

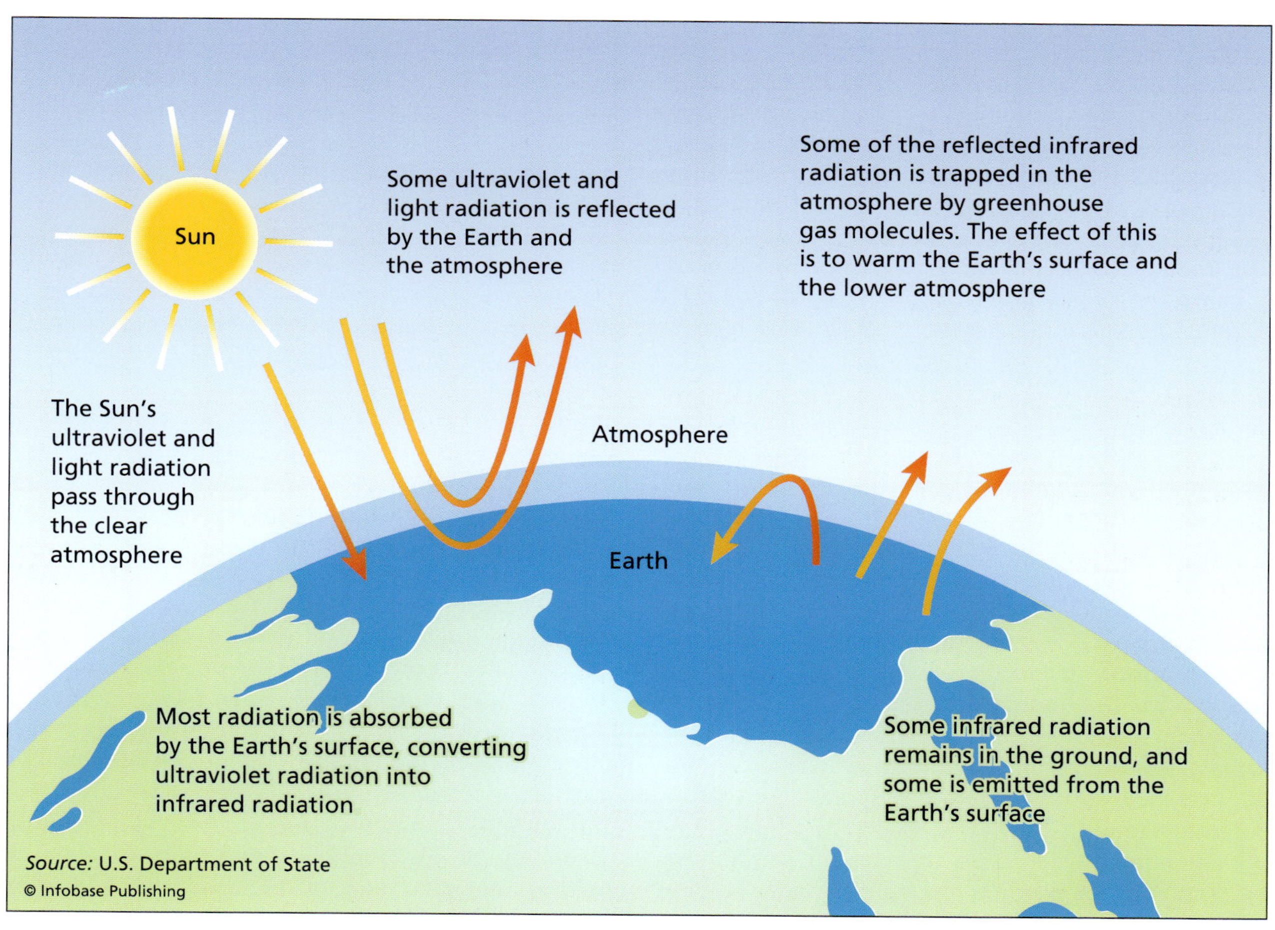
Sun
Some ultraviolet and light radiation is reflected by the Earth and the atmosphere
Some of the reflected infrared radiation is trapped in the atmosphere by greenhouse gas molecules. The effect of this is to warm the Earth's surface and the lower atmosphere
The Sun's ultraviolet and light radiation pass through the clear atmosphere
Atmosphere
Earth
Most radiation is absorbed by the Earth's surface, converting ultraviolet radiation into infrared radiation
Some infrared radiation remains in the ground, and some is emitted from the Earth's surface
Source: U.S. Department of State
© Infobase Publishing

Greenhouse gases trap solar radiation and prevent it from escaping the Earth's atmosphere.

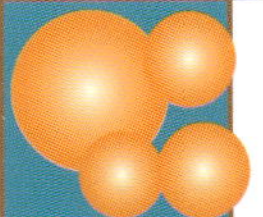

USES OF CARBON DIOXIDE: REMOVAL OF HARMFUL GREENHOUSE GASES

by Lisa Quinn Gothard

Carbon dioxide is a simple molecule that has received a lot of attention due to the dramatic environmental impact it has as a greenhouse gas. The greenhouse effect is a natural phenomenon caused by the presence of gases in Earth's atmosphere that prevent reradiated solar energy from escaping into outer space. This allows the temperature of the planet to remain stable and warm enough to maintain life on Earth.

The natural recycling of carbon-containing compounds occurs via the carbon cycle. Plants and other photosynthetic organisms take in atmospheric carbon dioxide and create oxygen, while the release of carbon dioxide occurs during the processes of cellular respiration (carried out by many life-forms) and the combustion of carbon-containing compounds.

In the past-century, human activities have disturbed the carbon cycle. The dramatic increase of the amount of greenhouse gases, including carbon dioxide, being produced in the world today is creating an environmental condition known as global warming. The actual impact of global warming is not fully understood, but scientists agree that the dramatic increase of carbon dioxide and other greenhouse gases will have negative consequences on the planet.

The burning of fossil fuels including natural gas, gasoline, fuel oil, and coal, in addition to burning of wood, produces a large quantity of carbon dioxide. The net amount of atmospheric carbon dioxide has also increased as a result of the widespread removal of plants and trees that are capable of removing carbon dioxide from the atmosphere. The U.S. carbon dioxide production in 2003 was 25 billion metric tons. More than one-third of that carbon dioxide was from coal.

At present, the focus for reducing carbon dioxide levels has been on the prevention of carbon dioxide emissions. Alternative energy sources such as geothermal, solar, wind, and nuclear energy are being researched in order to find a nonpolluting replacement for fossil fuels. The market demand for hybrid vehicles that are more energy efficient and burn cleaner is on the rise. One new approach to help reduce emissions is the utilization of plants instead of fossil fuels in the creation of plastics and as an energy source. The American chemist Conrad Zhang has shown that in the presence of chromium (II) chloride catalysts, glucose and fructose can be converted into plastics and fuels. Prevention of carbon dioxide emissions by limiting the amount of fossil fuels consumed must continue if atmospheric carbon dioxide levels are to be returned to acceptable levels.

New research has begun looking at the potential uses and removal methods of carbon dioxide that is already present in the atmosphere. Until recently, the process of photosynthesis has maintained the natural balance of carbon dioxide, keeping the atmospheric levels sufficiently low. Unfortunately the dramatic increase in emissions of carbon dioxide has disrupted the balance of the carbon cycle. Scientists have begun working on methods to put the excess carbon dioxide in the atmosphere to good use by turning carbon dioxide into plastics and fuels in addition to developing a process known as carbon capture and storage.

Carbon dioxide captured from smokestacks or other sources has been successfully used to manufacture polycarbonate plastics by applying large amounts of pressure and heat in the presence of a catalyst. The process allows for the formation of long-chain polymers. The German chemist Thomas Müller and the Japanese chemist Toshiyasu Sakakura

surface of Venus has a large amount of greenhouse gases in its atmosphere that insulate the planet and do not let much of the Sun's energy escape from the atmosphere back into space. This makes the surface of Venus incredibly hot. Greenhouse gases surrounding Earth have been present throughout the history of the atmosphere and act to insulate the planet from severe cold. The increased production and release of greenhouse gases, especially carbon dioxide, into the atmosphere by human activities have created concern about the effect of increased greenhouse gases in the environment.

The human activities that create greenhouse gases include the production of electricity for everyday use (estimated to be the highest producer of greenhouse gases), burning of fossil fuels in cars and trucks (the next highest contributor), production from factories and industries, followed by energy consumption for home heating. Deforestation and the burning of trees also contribute to greenhouse gases because the burning of trees not only causes an increase in the production of carbon dioxide released into the environment but also removes a primary user of carbon dioxide from the Earth. During the process of photosynthesis, trees and other green plants take in and metabolize carbon dioxide to produce sugar. When trees are cut down and burned, greenhouse gases increase as a result of carbon dioxide production from burning in addition to the lowered utilization of atmospheric carbon dioxide by reducing the number of photosynthetic organisms.

have researched this topic. This research is still in its developmental stages, as the amount of energy presently required in order for the carbon dioxide to be turned into plastic would require more coal to be burned, releasing more carbon dioxide than it actually captured. Scientists are looking for methods to fuel this process using nonpolluting sources such as geothermal power and wind or solar treatments. If the production of plastics could remove carbon dioxide in an energy-efficient manner, this could become a viable method of carbon dioxide removal from the atmosphere. In November 2007 the New York–based company Novomer received $6.6 billion in funding to begin production of a biodegradable plastic formed from carbon dioxide.

Another alternative to help remove the carbon dioxide from the atmosphere is the conversion of carbon dioxide into fuels. The Italian chemist Gabriele Centi has shown that it is possible to ionize liquid CO_2 that when mixed with water produces hydrocarbons that include such gases as methane (natural gas). Centi has also worked out a method to use gaseous carbon dioxide in the presence of catalysts that more reliably produce hydrocarbons. This new research might enable the production of fuels when traditional fuels are not available. Chemical engineers are presently performing research to optimize this technique, which could lead to a solution to two problems—the limited supply of fossil fuels and the increased levels of carbon dioxide in the atmosphere.

Removal of carbon dioxide at the source is known as carbon capture. The principle behind carbon capture includes capturing the carbon dioxide from coal-fired plants and compressing the carbon dioxide back into a liquid. This liquid carbon dioxide can then be sent into the earth. The carbon capture technique has been utilized in the North Sea at a location known as the Sleipner gas field. This carbon capture site has been used to sequester up to 1 million tons of carbon dioxide per year. The true environmental impact of such a technique is not fully understood. Measures must be taken in order to prevent the carbon dioxide from seeping back out, and monitoring is difficult.

This method also has several other problems. Scientists have discovered that the carbon dioxide can mix with water and form carbonic acid, which leads to destruction of the surrounding rock. The geological implications of this phenomenon are yet unknown. Also, many types of rocks allow the carbon dioxide to escape, making the process inefficient. Many believe the carbon capture technique is just burying the problems of today and leaving them for future generations to solve. Proponents of increased burning of coal as an alternative to oil and natural gas strongly support this method as an important part of their carbon emissions control. Much more research needs to be done on this technique to determine whether it could become a valid method of carbon dioxide removal.

Carbon dioxide levels have reached an all-time high and are continuing to increase. Scientific and technological advances in reducing carbon dioxide emissions as well as controlling the emissions that are present are needed. Scientists continue to perfect these and other techniques to help control the problem of greenhouse gases.

FURTHER READING

Marshall, Jessica. "Carbon Dioxide: Good for Something." *Discovery News,* April 10, 2008. Available online. URL: http://dsc.discovery.com/news/2008/04/10/carbon-dioxide-plastic.html. Accessed April 20, 2008.

Patel-Predd, Prachi. "Carbon Dioxide Plastic Gets Funding: A Startup Is Moving Ahead with an Efficient Method to Make Biodegradable Plastic." *Technology Review,* November 14, 2007. Available online. URL: http://www.technologyreview.com/Biztech/19697/. Accessed April 20, 2008.

U.S. Department of Energy. "Carbon Capture Research." Available online. URL: http://fossil.energy.gov/sequestration/capture/index.html. Accessed April 20, 2008.

Prior to the industrial revolution, the concentrations of carbon dioxide did not vary dramatically over time though periodic fluctuations did occur. The atmospheric levels of carbon dioxide have been officially measured annually since 1958. Since the 1950s, the atmospheric levels of carbon dioxide have increased approximately 10 percent. Scientists use ice cores from deep inside glaciers to measure the levels of atmospheric carbon dioxide throughout Earth's history.

According to these measurements, the atmospheric levels of carbon dioxide have increased 30 percent since the beginning of the industrial revolution. The 2006 levels of carbon dioxide stood at 380 parts per million (ppm). Levels of carbon dioxide prior to the industrial revolution were approximately 280 ppm. Present models predict, on the basis of the level of energy use today, that the atmospheric carbon dioxide levels will double in the next 100 years.

The ramifications of these estimations are not completely clear. Greenhouse gases trap the infrared radiation, and this affects the temperature of the planet; therefore, it is completely reasonable to predict that the climate and temperature will change on the basis of the amount of carbon dioxide present in the atmosphere. The logical consequence of this increased carbon dioxide concentration would be a dramatic rise in the temperature on Earth. The Earth's temperature has not differed appreciably over thousands of years, but the last 100 years demonstrate an increase of 1°F (0.6°C). The planet's average temperature has changed more in the last 100

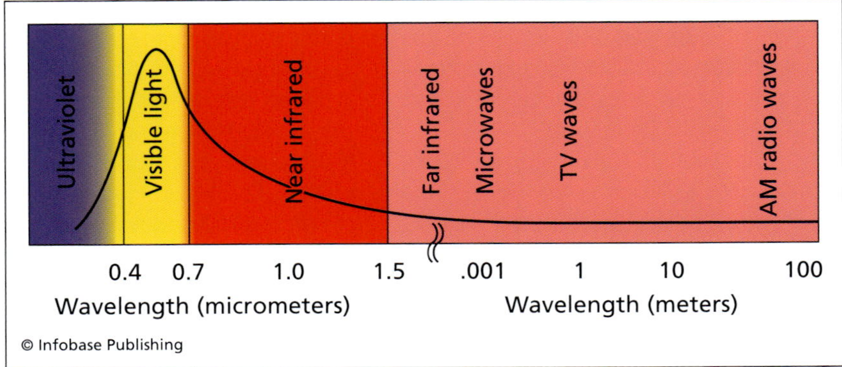

Incoming solar radiation has wavelengths across the electromagnetic spectrum.

years than it has in the 1,000 years prior. Models showing the doubling of carbon dioxide levels predict that this will lead to a change in the temperature of Earth of 1.8–5.4°F (1–3°C) in the next 100 years if there is not a change in the amount of carbon dioxide produced by human activities. Humans could be impacting the entire climate of the planet Earth by the production of greenhouse gases.

What is the effect of increased overall average temperatures? One major consequence is the melting of glaciers (large blocks of ice on the surface of the Earth that move very slowly). Glaciers in the United States and all over the world are decreasing. As they melt, the water level of the oceans increases. While this does not sound like a serious consequence, areas that were beachfront could become underwater, and areas that were once dry and farmable could be submerged or at least very wet and unfarmable. In the same respect, on the basis of the increased temperatures, areas that were once tillable farmland could become dry desolate wastelands, and the people would be unable to feed themselves. The hardest hit countries in the world would be the nations that are most dependent on farming, including the poorest countries, which are least capable of absorbing the economic impact of a changing climate. The economic instability that could be created by a small change in global temperature cannot be predicted.

What needs to be done to correct this climate problem has been a matter of discussion among scientists and politicians for some time now. Many believe that since there has not yet been a serious detrimental effect of global warming, the planet will not experience a major overall effect. Others warn that Earth may reach a point of no return, where the levels of greenhouse gases climb so high that the problem cannot be corrected. The easiest way to reduce the emission of greenhouse gases and reduce the amount of sunlight energy that the atmosphere retains involves a decreased dependence on fossil fuels for our energy sources. Alternative energy sources such as nuclear energy, solar energy, and hybrid and fuel cell vehicles require more research and implementation to reduce the levels of greenhouse gases significantly. An immediate end to deforestation programs and an increase in reforestation programs would also decrease greenhouse gases. Methane production could be curbed by capping landfills to prevent the release of methane into the atmosphere. The promotion of recycling, composting, and reducing consumption will also help prevent landfills from filling so quickly. Industrial plants that convert waste into energy by burning the trash and harvesting the energy released can also contribute to the solution of landfill overcrowding and methane overproduction. As of January 2000 the Environmental Protection Agency stated that the production of CFCs is illegal. CFCs have already proven to be an environmental hazard by their destruction of the ozone layer of the Earth's atmosphere. CFCs have been taken off the market, and the release of CFCs from older appliances into the atmosphere is unlawful.

In 1992 the United Nations Framework Convention on Climate Change (UNFCCC) put forth an international treaty to reduce the production of greenhouse gases. The treaty was ratified by 198 countries and had as its mission the gathering and sharing of information on greenhouse gas emissions and the study of how to reduce these emissions and thus reduce the impact on the environment. In 1997 38 countries agreed to the Kyoto Protocol, an addition to the UNFCCC treaty that included a legally binding aspect requiring that the emission of the most abundant greenhouse gases (CO_2, N_2O, CH_4, and CFCs) be reduced by at least 5 percent relative to 1990 levels in the years 2008–12. The levels that the United States must reach represent a 7 percent reduction relative to 1990 levels. The United States is the largest contributor to the greenhouse gas problem. According to the UNFCCC reports, the U.S. levels of carbon dioxide are more than double those of the next closest country. In 1998 President William Clinton, on behalf of the United States, signed the Kyoto Protocol, agreeing to meet the 7 percent reduction expected for the United States. In 2001 President George W. Bush withdrew the U.S. agreement to the Kyoto Protocol. Because the Kyoto Protocol requires ratification by enough countries to constitute 55 percent of the total greenhouse gas emissions, the loss of the United States, the largest contributor of greenhouse gases, creates a problem for the treaty.

See also ATMOSPHERIC AND ENVIRONMENTAL CHEMISTRY; CRUTZEN, PAUL; ELECTROMAGNETIC WAVES; MOLINA, MARIO; PHOTOSYNTHESIS.

FURTHER READING

Gore, Al. *An Inconvenient Truth.* New York: Rodale, 2006.

United Nations Framework Convention on Climate Change home page. Available online. URL: http://unfccc.int/2860.php. Accessed July 25, 2008.

harmonic motion Commonly called simple harmonic motion, harmonic motion is oscillatory (i.e., repetitive) motion in which the displacement from the center of oscillation is described by a sinusoidal function of time. What is undergoing this motion might be anything. For instance, a weight hanging from a vertical spring will perform harmonic motion when pulled down and released. A swinging pendulum undergoes approximate harmonic motion, as might the atoms or molecules forming a crystal. Harmonic "motion" might be performed figuratively by physical quantities that do not actually move. Electric voltage, for example, might and often does have a sinusoidal time dependence. Such a situation is designated "alternating current," or AC for short. The electricity used in households is of this kind.

If one denotes the displacement by x, then the most general harmonic motion is represented by this function

$$x(t) = A \cos(2\pi ft + \alpha)$$

where A denotes the (positive) amplitude of the oscillation, f its (positive) frequency (number of cycles per unit time), t the time, and α is a constant, the phase constant, or phase shift. Note that *phase,* in this context, refers to the stage in the cycle. The motion takes place between the extremes of $x = A$ and $x = -A$. As for units, x and A are in meters (m) most commonly, although any length unit can serve (as long as both are in the same unit), or for figurative harmonic "motion" might be in units such as volt (V) for voltage or ampere (A) for electric current. Frequency f is in hertz (Hz), t in seconds (s), and α in radians (rad). Note that the period of oscillation, which is the time of a single cycle, denoted by T and specified in seconds (s), relates to the frequency by

$$T = \frac{1}{f}$$

Successive derivatives of $x(t)$ give the velocity, $v(t)$, in meters per second (m/s) and acceleration, $a(t)$, in meters per second per second (m/s^2) of the motion as follows:

$$v(t) = \frac{dx(t)}{dt} = -2\pi fA \sin(2\pi ft + \alpha)$$

$$a(t) = \frac{dv(t)}{dt} = \frac{d^2x(t)}{dt^2} = -(2\pi f)^2 A \cos(2\pi ft + \alpha)$$

Note the proportionality, with negative coefficient, between the displacement and the acceleration that is characteristic of simple harmonic motion,

$$a = -(2\pi f)^2 x$$

The preceding relationships show that at the points of maximal displacement ($x = \pm A$) the speed is zero, and the acceleration is maximal, equal to $(2\pi f)^2 A$ in magnitude, and directed toward the center. At the center ($x = 0$) the speed is maximal and equals $2\pi fA$, while the acceleration is zero.

In the particular case of motion starting at the center, $x = 0$, at time $t = 0$ and initially heading for increasing values of x, the expressions for displacement, velocity, and acceleration take the form

$$x = A \sin 2\pi ft$$

$$v = 2\pi fA \cos 2\pi ft$$

$$a = -(2\pi f)^2 A \sin 2\pi ft$$

For motion starting at maximal positive displacement, $x = A$, at time $t = 0$, the corresponding expressions are

$$x = A \cos 2\pi ft$$

$$v = -2\pi fA \sin 2\pi ft$$

$$a = -(2\pi f)^2 A \cos 2\pi ft$$

Harmonic motion is seen, for example, when the position of a point in circular motion at constant angular speed around the origin is projected onto the x- or y-axis.

In physical systems, simple harmonic motion results when a system in stable equilibrium and obeying Hooke's law (that the restoring force is proportional to the displacement) is displaced from stable equilibrium and allowed to evolve from that state. Let x denote the displacement from equilibrium for such a system and F the restoring force in newtons (N). Hooke's law is then

$$F = -kx$$

where k is the force constant (called the spring constant in the case of a spring), in newtons per meter (N/m), and the minus sign indicates that the restoring force is in the opposite direction from that of the displacement. Newton's second law of motion states that

$$F = ma$$

where m denotes the mass of the moving part of the system, in kilograms (kg), and a its acceleration. From the last two equations we obtain that the acceleration is proportional, with negative coefficient, to the displacement,

$$a = -\frac{k}{m}x$$

As shown, this relation is characteristic of simple harmonic motion, with

$$(2\pi f)^2 = \frac{k}{m}$$

From this one sees that such a system can undergo simple harmonic motion at the frequency

$$f = \frac{1}{2\pi}\sqrt{\frac{k}{m}}$$

The mechanical energy of such a system remains constant (in the ideal, frictionless case) and is continually undergoing conversion from elastic potential energy, $\frac{1}{2}kx^2$, to kinetic energy, $\frac{1}{2}mv^2$, and back, so that

$$\text{Mechanical energy} = \tfrac{1}{2}kx^2 + \tfrac{1}{2}mv^2 = \tfrac{1}{2}kA^2$$

where energies are in joules (J).

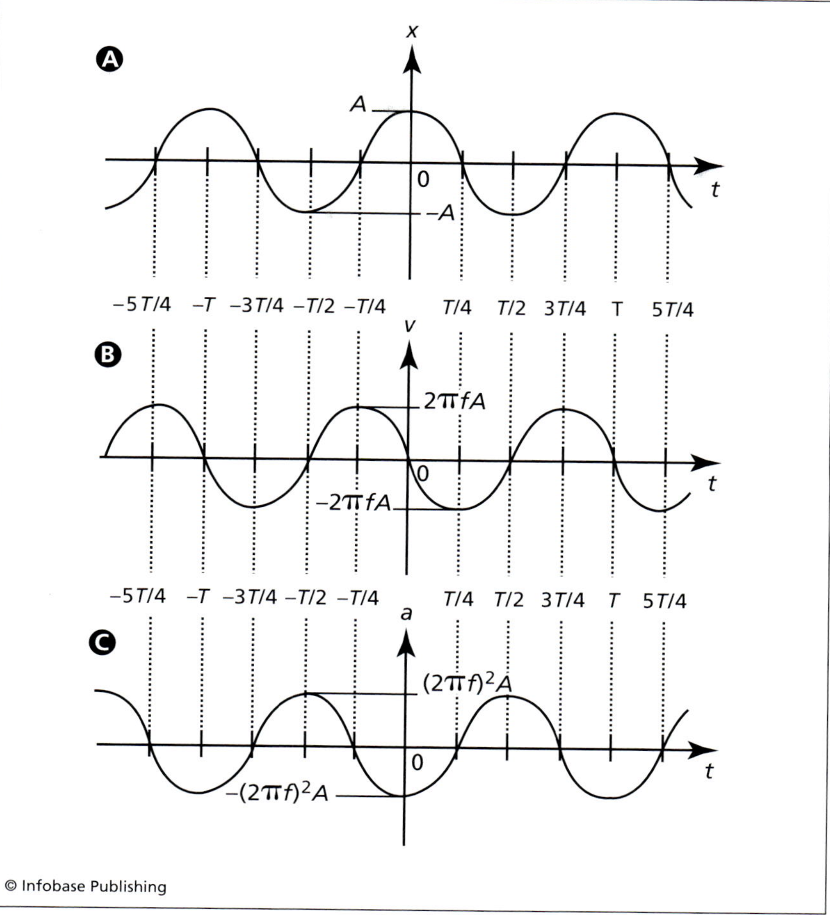

Harmonic motion is described by a sinusoidal function of time. (a) The figure depicts $x = A \cos 2\pi ft$, where x denotes the displacement from equilibrium, A the amplitude, f the frequency, and t the time. The period is $T = 1/f$. (b) The velocity of the motion as a function of time, $v = -2\pi fA \sin 2\pi ft$. (c) The acceleration as a function of time, $a = -(2\pi f)^2 A \cos 2\pi ft$.

In general, almost any system in stable equilibrium, when displaced from equilibrium, responds with a restoring force that, for sufficiently small displacements, is approximately proportional to the displacement. The proportionality coefficient might differ for different ways of displacing the system from equilibrium, such as in different directions. So systems in stable equilibrium can, in general, oscillate harmonically at a number of frequencies.

The preceding discussion also applies to nonmechanical systems, in which motion, displacement, velocity, acceleration, and so forth, refer to change in the value of some not necessarily spatial physical quantity, and force is generalized appropriately. Simple harmonic variation can be undergone, for instance, by voltages and currents in electric circuits, by reagent concentrations in chemical reactions, and by air pressure in musical instruments. In AC (alternating-current) circuits, as mentioned earlier, the voltages, as well as the currents, have harmonic time dependence. That occurs in all household appliances that plug into wall sockets. The concentrations of certain reagents in a prepared chemical solution can oscillate harmonically, resulting in impressive color oscillations of the solution. Musical instruments produce sounds that consist of harmonic oscillations, as described later in this article. When a listener hears sound, she is actually sensing air pressure oscillations acting on her eardrums, as a result of sound waves impinging on and entering her ears. The sound source might be a wind instrument, such as a tuba, saxophone, or clarinet. In such instruments the air pressure oscillates harmonically, and that generates the sound waves emanating from them.

When struck by a rubber mallet or by the hand, the tines of a tuning fork oscillate in harmonic motion and produce a pure tone. This tuning fork oscillates at the frequency of 512 hertz. The symbol C indicates that its musical pitch is C, in this case one octave above middle C. It is tuned low, however, since the standard frequency of C above middle C is 523 Hz. *(Jason Nemeth, 2008, used under license from Shutterstock, Inc.)*

HARMONIC OSCILLATOR

Any device that is designed to oscillate in harmonic motion, either literally or figuratively, is called a harmonic oscillator. As an example, a weight suspended from a spring is a harmonic oscillator: when displaced from equilibrium, it oscillates in literal harmonic motion, with its vertical displacement varying as a sinusoidal function of time. For a figurative example, an electric circuit can be designed so that it produces a voltage that varies sinusoidally in time. Such a circuit is thus a harmonic oscillator. The essential physical property of a harmonic oscillator is that it possess a state of stable equilibrium, such that when it is displaced from that state, a restoring force (or suitable generalization of force) whose magnitude is proportional to the magnitude of the displacement (or of its appropriate generalization) from equilibrium arises.

Whereas according to classical physics a harmonic oscillator may possess energy in a continuous range of values, the energy of a quantum harmonic oscillator is constrained to a sequence of discrete values:

$$E_n = (n + {}^1/_2)\, hf \qquad \text{for } n = 1, 2, \ldots$$

Here E_n denotes the value of the energy, in joules (J), of the harmonic oscillator's nth energy level; f is the classical frequency of the oscillator, in hertz (Hz); and h represents the Planck constant, whose value is $6.62606876 \times 10^{-34}$ joule-second (J·s), rounded to 6.63×10^{-34} J·s. Note that the quantum harmonic oscillator cannot possess zero energy. That fact is related to the Heisenberg uncertainty principle. A diatomic molecule, such as molecular oxygen O_2 or carbon monoxide CO, serves as an example of a quantum harmonic oscillator. The oscillations are in the separation distance of the atoms. The molecule can be modeled by a pair of masses connected by a compressible spring. The oscillations involve alternate stretching and compression of the spring.

HARMONICS

The term *harmonics* refers to an infinite sequence of frequencies that are all integer multiples of the lowest among them:

$$f_n = nf_1 \qquad \text{for } n = 1, 2, \ldots$$

where f_n denotes the nth frequency of the sequence, called the nth harmonic. Such a sequence is called a harmonic sequence, or harmonic series. The lowest frequency of the series, f_1, the first harmonic, is called the fundamental frequency, or simply the fundamental. The higher frequencies of the series can be referred to also as the harmonics of the fundamental.

The relevance of harmonics to harmonic motion is that certain systems can oscillate at any or some of a harmonic sequence of frequencies and, moreover, can oscillate at any number of such frequencies simultaneously. This is especially important for pitched musical instruments. The name *harmonic* originated from the fact that the frequencies of musical tones that are generally perceived to sound pleasant, or harmonious, when heard together are members of such a sequence. Because of this, musical instruments that produce definite pitches were developed so that their natural frequencies are members of a harmonic sequence. The musical pitch of such an instrument is determined by the fundamental of the sequence. For example, the natural frequencies of a stretched string are (for small displacements from equilibrium)

$$f_n = \frac{n}{2L}\sqrt{\frac{F}{\mu}} \quad \text{for } n = 1, 2, \ldots$$

which is clearly a harmonic sequence, with

$$f_1 = \frac{1}{2L}\sqrt{\frac{F}{\mu}}$$

Here F denotes the tension of the string, in newtons (N); L is the length of the string, in meters (m); and μ is the string's linear density (mass per unit length), in kilograms per meter (kg/m). The fundamental frequency, which determines the pitch, is f_1. This is the acoustic foundation of all stringed instruments, such as the violin, guitar, and piano. Such a sequence of harmonics might be this one, as an example: Let the fundamental, or first harmonic, be the tone C two octaves below middle C. The second harmonic is then an octave above the fundamental (doubling the frequency always raises a tone by an octave), which in this example is the C one octave below middle C. The third harmonic is G below middle C. The rest of the sequence consists of, in ascending order, middle C, E above it, G above that, a tone between the next A and B-flat that does not exist on the piano, the C above middle C, D above it, E above that, a tone between the next F and F-sharp, G above that, and a theoretically infinite sequence of increasingly higher tones that become closer and closer together in pitch.

When a pitched musical instrument is played at a designated pitch corresponding to frequency f_1, all or some of the harmonics of f_1 sound simultaneously. In the parlance of musical acoustics, the higher members of the harmonic sequence of f_1 are called overtones. The first overtone is the second harmonic, the second overtone the third harmonic, and so on. The relative intensities of the simultaneously sounding harmonics determine the tonal character, or timbre, of the instrument. In double-reed instruments, such as the oboe and the bassoon, for example, the first overtone (the second harmonic) contains more acoustic energy than does the fundamental. The clarinet, on the other hand, lacks the odd overtones (even harmonics) altogether and possesses a rather strong fundamental. The flute is very weak in all its overtones compared to its fundamental and thus has a very strong fundamental. A tone that consists of the fundamental alone and that completely lacks overtones is referred to as a pure tone. On the other hand, a psychoacoustic effect exists that allows a listener to *perceive* the pitch of the fundamental when the fundamental is lacking from a tone when higher harmonics are present.

See also ACCELERATION; ACOUSTICS; CLASSICAL PHYSICS; ELASTICITY; ELECTRICITY; ENERGY AND WORK; EQUILIBRIUM; FORCE; MASS; MOTION; NEWTON, SIR ISAAC; PRESSURE; QUANTUM MECHANICS; ROTATIONAL MOTION; SPEED AND VELOCITY.

FURTHER READING

Hall, Donald E. *Musical Acoustics,* 3rd ed. Pacific Grove, Calif.: Brooks/Cole, 2002.

Young, Hugh D., and Roger A. Freedman. *University Physics,* 12th ed. San Francisco: Addison Wesley, 2007.

Hawking, Stephen (1942–) British *Physicist*

At the age of 21 Stephen Hawking learned he would die of amyotrophic lateral sclerosis (known in the United States as Lou Gehrig's disease) within a few years. His engagement and marriage to his girlfriend helped pull him out of depression, he did not die as predicted, and he proceeded to become one of the most famous scientists of the late 20th and early 21st centuries, specializing in theoretical physics and particularly in black holes, cosmology, and quantum gravity. Hawking's image, slumped in his wheelchair, is familiar to many, and he has become an icon for successfully overcoming severe adversity. His first book written for a general readership, *A Brief History of Time,* became a runaway best seller. Hawking's appearance on many television shows has added to his celebrity.

Although he is now almost completely paralyzed and able to speak only by means of a computer-operated voice synthesizer, Hawking's wry sense of humor endears him to all. He placed a bet with a physicist colleague against the success of one of his own ideas and happily lost the bet. And he was the first and only paralyzed person to float weightless.

Against all odds, Stephen Hawking, who proved that a black hole is not really black, continues his research into the most fundamental aspects of nature.

HAWKING'S LIFE

Stephen William Hawking was born on January 8, 1942, in Oxford, England. (Hawking likes to point out that his birth date followed the death of Galileo by 300 years to the day.) His parents, Isobel and Frank, were living in London. World War II was in full fury, and the city was a dangerous place as a result of heavy bombing by the German air force. The parents-to-be chose to move temporarily to safer Oxford for the birth of their first child. Over time, Stephen gained two sisters, Philippa and Mary, and a brother, Edward. In 1950, the Hawkings moved to St. Albans, where Stephen attended elementary and high schools. An exceptional mathematics teacher inspired him to specialize in mathematics in high school, but Stephen's father insisted that chemistry was a better specialization. Even as a world-renowned physicist, Stephen Hawking maintains contact with his high school.

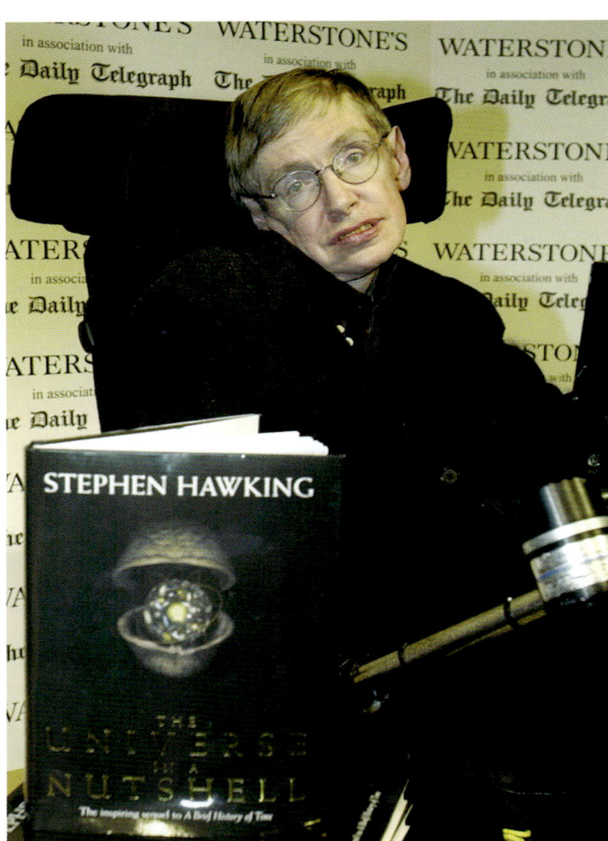

Stephen Hawking is a cosmologist who is investigating the implications of the unification of the general theory of relativity and quantum physics. Here he is launching his popular book *The Universe in a Nutshell.* *(AP Images)*

Upon graduation from high school in 1959 at the age of 17, Hawking enrolled at University College of Oxford University with a scholarship, intending to study science. This was the college his father had attended. Although Hawking was especially interested in mathematics, there was no mathematics specialist at the college. He instead chose to major in physics, in which he proved exceedingly talented. Hawking's teachers recognized his exceptional mind. He did not work very hard at his studies, however, and the result of his final exam put him on the borderline between graduating with a first-class or a second-class degree. The atmosphere at the college was one of an antiwork attitude. The prevailing student opinion was that one got the degree one deserved with no effort, according to one's natural brilliance or lack thereof. It was simply not done to work to improve one's academic standing through effort. For Hawking an oral exam determined first-class honors and he graduated in 1962, at the age of 20.

Hawking immediately went to Cambridge University for graduate studies in general relativity and cosmology. His research supervisor there was Dennis Sciama, a noted astrophysicist and cosmologist. Around that time, Hawking developed noticeable clumsiness. Two weeks of tests in a hospital revealed that he was suffering from amyotrophic lateral sclerosis (ALS, also known as Lou Gehrig's disease), which would eventually deprive him of almost all muscular control. Hawking's physicians did not expect him to live to finish his Ph.D. This news was a tremendous shock, and Hawking initially found himself unmotivated to continue his studies, since he expected to die soon. But the advance of his disease slowed and, motivated especially by his engagement and marriage, in 1965, to Jane Wilde, a student of languages at Cambridge, Hawking buckled down, worked hard, and completed his Ph.D. in 1966, at age 24. Jane and Stephen had three children, Robert, Lucy, and Tim, born in 1967, 1970, and 1979, respectively.

Hawking's first positions after earning his Ph.D. were at Gonville and Caius College, in Cambridge, as a research fellow and then a professorial fellow. In 1973, he joined the department of applied mathematics and theoretical physics of Cambridge University and was appointed professor of gravitational physics in 1977. In 1979, Cambridge University appointed Hawking Lucasian Professor of Mathematics, the position he has occupied since. He is the latest in a line of 17 distinguished mathematicians and scientists who have held the Lucasian Chair of Mathematics at Cambridge University, which includes the theoretical physicist Paul Dirac, the physicists/mathematicians George Stokes and Joseph Larmor, and, perhaps most significantly, the physicist/mathematician/astronomer Sir Isaac Newton, who was the second to hold the position.

In 1975, Hawking was working on the theory of black holes. At that time, astronomers had not yet established the existence of such objects. (They are now known to be very common.) In order to gain some consolation in the event black holes are found not to exist, Hawking bet against their existence with a colleague, the astrophysicist Kip Thorne at the California Institute of Technology in Pasadena. Upon proof of the existence of a black hole, Hawking would give Thorne a one-year subscription to the magazine *Penthouse.* If black holes were found not to exist, Thorne would give Hawking four years of *Private Eye.* The four-to-one ratio derived from the estimate at that time of an 80 percent chance (four-to-one odds) that the star Cygnus X-1 would turn out to be a black hole. Eventually, the existence of black holes became incontrovertible, and Hawking was glad to concede the bet.

After a respiratory crisis in 1985, Hawking underwent a tracheotomy in order to ensure his breathing. That put an end to his speech, which had in any case become very slurred and difficult to understand. Since then, Hawking has been using a computer-operated speech synthesizer for communication. He has become very proficient in its use and jokingly complains that the only problem with the synthesizer is its American accent.

In 1988, six years after he started work on the project, Hawking's book *A Brief History of Time,* with the subtitle *From the Big Bang to Black Holes,* appeared in print. The book was intended for the general public, his first of that type. The book was an amazing success, easily breaking sales records. It gently introduced the reader to extremely esoteric aspects of modern physics and the universe and became something of a fad, although rumors have it that apparently very few of its readers actually managed to finish it. *A Brief History of Time* called Hawking to the public's attention, which he has not left since. Hawking has subsequently written other books for the general public, including *A Briefer History of Time* with Leonard Mlodinow, published in 2005. Additional books of general interest that he wrote are *Black Holes and Baby Universes and Other Essays* (1993), *The Universe in a Nutshell* (2001), and *On the Shoulders of Giants: The Great Works of Physics and Astronomy* (2002).

Hawking celebrated his 65th birthday in 2007 by participating in a zero-gravity flight and became the first quadriplegic to float weightless. He enjoyed weightlessness eight times during the flight. (The pilot accomplishes that by guiding the aircraft in a carefully designed, alternately ascending and descending flight path, following a parabolic trajectory for some of each segment of the path. Prospective National Aeronautic and Space Administration [NASA] astronauts routinely train in this way to prepare for the weightless state in space.) One of the reasons Hawking did this was to encourage public interest in space, since he believes that the future of humankind is in space, given the increasing risk of its being destroyed on Earth.

In 1991 Stephen and Jane separated. Hawking married Elaine Mason, his nurse, in 1995 and in 2006 filed for divorce. He reconciled with Jane and their children in 2007.

Hawking has been the recipient of many prestigious awards and honors. Perhaps the most prestigious among them was the 1988 Wolf Prize in Physics, which he shared with Roger Penrose at Oxford University "for their brilliant development of the theory of general relativity, in which they have shown the necessity for cosmological singularities and have elucidated the physics of black holes. In this work they have greatly enlarged our understanding of the origin and possible fate of the universe." Other awards include the Albert Einstein Medal (1979), the Order of the British Empire (1982), and the Copley Medal of the Royal Society (2006). In 1974, when he was 32, the Royal Society elected Hawking as one of its youngest fellows.

HAWKING'S WORK

A theoretical physicist, Hawking has as his main areas of research the general theory of relativity, cosmology, and quantum gravity.

Much of Hawking's earliest work was done in collaboration with Roger Penrose, then at Birkbeck College in London. He investigated general relativity with regard to the occurrence of singularities. A singularity is a situation in which the mathematics used to describe the physics gives an infinite value for one or more physical quantities. Physicists take this as a "red flag," indicating that the theory behind the mathematics has exceeded its limits of validity, or, stated more vividly, the theory breaks down. Hawking investigated the conditions under which, according to general relativity, singularities must develop in the universe. One result was that the origin of the universe must be singular. Physicists generally understand this as indicating that classical physics, at least in the form of the general theory of relativity, cannot deal with the origin of the universe and that quantum physics must enter the picture. Since general relativity is a theory of gravitation, the merger of quantum physics with general relativity is called quantum gravity.

Hawking then focused his attention on black holes. General relativity was known to allow the existence of such objects, which are so dense and compact that nothing, not even light, can escape from their vicinity because of their immense gravitational attraction. Since these objects absorb all radiation, as well as everything else, that impinges on them, they were given the name *black holes*. The general relativistic description of a black hole involves singularities. Hawking applied quantum concepts and thermodynamics to black holes and surprised the science community with a number of results. One can assign a temperature to a black hole, a temperature that is inversely proportional to its mass. A black hole is in no way black, as it radiates similarly to a blackbody at the same temperature. This radiation is called Hawking radiation. For both a blackbody and a black hole, the rate of radiation, their emitted power, is proportional to the fourth power of their temperature. As a black hole radiates away its energy, its mass decreases, according to Albert Einstein's mass-energy equivalence relation, and its temperature increases, according to Hawking. Its rate of mass loss by radiation then increases greatly (through the fourth power of the temperature). This is a runaway process, leading to a black hole's ending its life in an explosion. Thus, the black holes that are the end product of the evolution of sufficiently massive stars and the gigantic black holes at the centers of most galaxies possess low temperatures, radiate slowly, and live long and prosper. They postpone their demise by swallowing matter and increasing their mass, thus lowering their temperature and reducing their rate of radiation. On the other hand, low-mass black holes, called mini black holes, such as some researchers, including Hawking himself, suggest might have been produced after the big bang, should all have evaporated by now.

Hawking's fruitful investigations range over general relativity, classical cosmology, quantum cosmology, structure of the universe, quantum entanglement, the nature of space and time, gravitational radiation, wormholes, and much, much more. He continues making progress in and important contributions to the understanding of the universe in which humankind finds itself. Many view Hawking as Einstein's successor, continuing along the path that Einstein blazed. And both share the distinction of serving as popular cultural icons as well as being the two scientists who are best known to the general public.

See also BIG BANG THEORY; BLACKBODY; BLACK HOLE; CLASSICAL PHYSICS; EINSTEIN, ALBERT; GALILEI, GALILEO; GENERAL RELATIVITY; GRAVITY; HEAT AND THERMODYNAMICS; NEWTON, SIR ISAAC; QUANTUM MECHANICS; SPECIAL RELATIVITY.

FURTHER READING

Cropper, William H. *Great Physicists: The Life and Times of Leading Physicists from Galileo to Hawking*. New York: Oxford University Press, 2001.

Hawking, Stephen. *Black Holes and Baby Universes and Other Essays*. New York: Bantam, 1994.

———. *A Brief History of Time: The Updated and Expanded Tenth Anniversary Edition*. New York: Bantam, 1998.

———. *On the Shoulders of Giants: The Great Works of Physics and Astronomy*. Philadelphia: Running Press, 2002.

———. *The Universe in a Nutshell*. New York: Bantam, 2001.

Hawking, Stephen, and Leonard Mlodinow. *A Briefer History of Time*. New York: Bantam, 2005.

Stephen Hawking home page. Available online. URL: http://www.hawking.org.uk/. Accessed February 23, 2008.

White, Michael, and John Gribbin: *Stephen Hawking: A Life in Science*. Washington, D.C.: John Henry, 2002.

"The Wolf Foundation's Prizes in Physics." Available online. URL: http://www.wolffund.org.il/cat.asp?id=25&cat_title=PHYSICS. Accessed February 23, 2008.

heat and thermodynamics

Also called thermal energy, heat is the internal energy of matter, the energy of the random motion of matter's microscopic constituents (its atoms, molecules, and ions). In a gas, the molecules or atoms are confined only by the walls of the container. The form of heat for a gas is then the random, relatively free flying about of its molecules or atoms, occasionally bouncing off the walls and off each other. The microscopic constituents of a liquid are confined to the volume of the liquid. So for a liquid, heat is the random dance of its constituents as they collide often with each other and occasionally with the liquid surface and with the container walls. In a solid, the constituents are confined to more or less fixed positions. Thus, heat in solids is the random oscillation of its constituents about their positions.

Of all forms of energy, heat is unique in its random nature and in its not being fully convertible to work and, through work, to other forms of energy. Thermodynamics is the field of physics that deals with heat and its transformation. It is the second law of thermodynamics that expresses the impossibility of total conversion of heat to work. Thermodynamics is treated later in this entry.

TEMPERATURE

Temperature is a measure of the average random energy of a particle of matter. One of the most important properties of temperature is that if two systems possess the same temperature, they will be in thermal equilibrium when they are in contact with each other. In other words, when two systems are brought into contact so that heat might flow from one to the other (but with no transfer of matter or work), their properties do not change if they are at the same temperature. Another of temperature's most important properties is that heat spontaneously flows from a region or body of higher temperature to one of lower temperature; never the opposite. The second law of thermodynamics deals with this.

Absolute temperature, or thermodynamic temperature, is defined by the behavior of an ideal gas, or equivalently, by thermodynamic considerations. For the Kelvin absolute scale, a property of water is utilized as well. The lowest conceivable temperature, called absolute zero, is assigned the value 0 kelvin (K) in SI units. Another temperature that is assigned to define this temperature scale is that of the triple point of water. The triple point of any substance is the unique set of conditions under which the substance's solid, liquid, and gas phases are in equilibrium and can coexist with no tendency for any phase to transform to another. The triple point of water is at the pressure of 0.619×10^3 pascals (Pa), and its temperature is given the value 273.16 K. (Note that the degree sign ° is not used with kelvins.) The freezing point of water, the temperature at which ice and liquid water are in equilibrium at a pressure of one atmosphere (1 atm = 1.013×10^5 Pa), is one-hundredth of a kelvin lower than the triple point, and is thus 273.15 K. In the commonly used Celsius temperature scale, the freezing point of water is taken as 0°C. In this scale the boiling point of water, the temperature at which liquid water and water vapor are in equilibrium at a pressure of 1 atm, is 100°C and absolute zero is -273.15°C. The magnitude of a kelvin is the same as one Celsius degree, which is one-hundredth of the temperature difference between water's freezing and boiling points. The subdivision of this temperature difference is based on the behavior of an ideal gas, or equivalently, on thermodynamic considerations.

The conversion between the Kelvin and Celsius scales is given by

$$T = T_C + 273.15$$

where T and T_C denote the same temperature in the Kelvin and Celsius scales, respectively. In the Fahrenheit scale, still used in the United States and only in the United States, the freezing point of water is set at 32 degrees Fahrenheit (°F), and a Fahrenheit degree is defined as 5/9 of a Celsius degree. As a result, the conversion between the Fahrenheit and Celsius scales is given by

$$T_F = \frac{9}{5} T_C + 32°$$

$$T_C = \frac{5}{9}(T_F - 32°)$$

where T_F and T_C denote the same temperature in the Fahrenheit and Celsius scales, respectively.

Absolute zero temperature is not attainable, although it can be approached as closely as desired. As it is approached, the random energy of the particles of matter approaches a minimum. The minimum is not zero, because of a quantum effect involving the Heisenberg uncertainty principle, whereby some residual energy, called zero-point energy, remains. For general temperatures, the average random energy of a matter particle has the order of magnitude of the value of kT, where T is the absolute temperature, in kelvins (K), and k denotes the Boltzmann constant, whose value is $1.3806503 \times 10^{-23}$ joule per kelvin (J/K), rounded to 1.38×10^{-23} J/K.

THERMAL EXPANSION

Thermal expansion is the expansion that most materials undergo as their temperature rises. Consider an unconfined piece of solid material. When its temperature changes by a small amount from T to $T + \Delta T$ and any of its linear dimensions increases concomitantly from L to $L + \Delta L$, the relative change in length is approximately proportional to the temperature change:

$$\frac{\Delta L}{L} = \alpha \, \Delta T$$

Here L and ΔL are both in the same unit of length, ΔT is either in kelvins or in Celsius degrees (which are the same for a temperature *difference*), and α, in inverse kelvins (K^{-1}), denotes the coefficient of linear expansion of the material. This coefficient is generally temperature dependent. The absolute change of length ΔT is then

$$\Delta L = \alpha L \, \Delta T$$

As the solid expands, any area on its surface expands also, from A to $A + \Delta A$ (both A and ΔA in the same unit of area). Then, to a good approximation, the relative change in area is

$$\frac{\Delta A}{A} = 2\alpha \, \Delta T$$

and the absolute area change

$$\Delta A = 2\alpha A \, \Delta T$$

The coefficient 2α is the coefficient of area expansion. Similarly, any volume of the material undergoes expansion as well, from V to $V + \Delta V$ (both V and ΔV

in the same unit of volume), where the relative volume change is

$$\frac{\Delta V}{V} = 3\alpha\,\Delta T$$

and the absolute change

$$\Delta V = 3\alpha V\,\Delta T$$

So the coefficient 3α is the coefficient of volumetric expansion.

For liquids and gases only the volume expansion has significance, and the coefficient of volumetric expansion is denoted by β in the expressions

$$\frac{\Delta V}{V} = \beta\,\Delta T$$

and

$$\Delta V = \beta V\,\Delta T$$

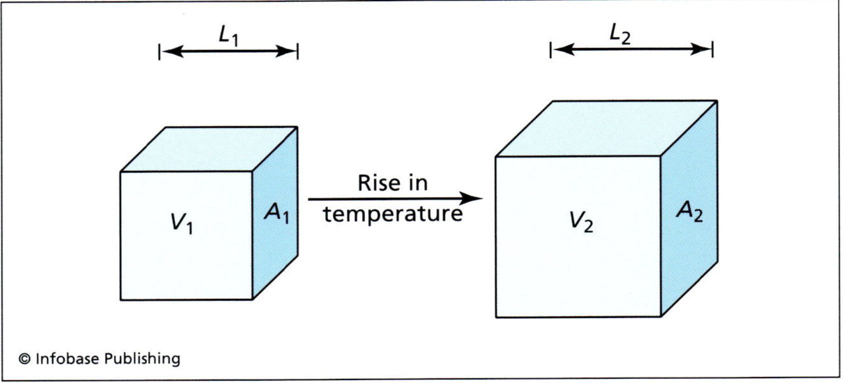

Most materials expand when their temperature rises. For a solid body, every length, area, and volume increases. The figure shows this, with the length of one edge of the body expanding from L_1 to L_2, the area of one of the body's surfaces increasing from A_1 to A_2, and the body's volume expanding from V_1 to V_2. In every case, the relative change is proportional to the change in temperature (for small temperature changes).

A gas at densities that are not too high and temperatures that are not too low can be approximated by an ideal gas. For temperature change of a sample of ideal gas at fixed pressure, this relation holds

$$\frac{\Delta V}{V} = \frac{\Delta T}{T}$$

with temperature in kelvins, giving for the coefficient of volumetric expansion

$$\beta = \frac{1}{T}$$

In all cases where an increase of temperature (positive ΔT) causes expansion, a decrease of temperature (negative ΔT) brings about contraction, termed thermal contraction.

As a result of the particular importance of water for life on Earth, the abnormal behavior of liquid water should be pointed out: in the range 0–4°C water possesses a *negative* coefficient of volumetric expansion. So, as the water temperature rises within that range, the liquid *contracts,* with concomitant *increase* of density. When liquid water freezes, it expands, an unusual reaction among materials. Thus the density of ice is less than that of liquid water. As a result, ice floats on liquid water. As bodies of water freeze in cold weather, the ice forms and remains at the top, allowing fish to survive the winter in the liquid water under the ice. And since still liquids tend to stratify by density, with the denser ones lying beneath the less dense ones, in freezing weather a lake might have this state: a layer of ice floats on top, with the coldest water, at 0°C, immediately beneath it. Deeper water is increasingly warmer up to 4°C. The special behavior of water relates to the structure of the water molecule and of the ice crystal.

Here is an example of a thermal-expansion calculation. Consider a 10.00-m aluminum rod at initial temperature 15°C. To what temperature does the rod need to be heated in order for it to fit snugly into a gap of 10.01 m, that is, for it to expand by 1 cm. The coefficient of linear expansion for aluminum is 2.4×10^{-5} K^{-1}. The data for this problem are $\Delta L = 0.01$ m and $\alpha = 2.4 \times 10^{-5}$ K^{-1}. In order to obtain the answer, which is $(15 + \Delta T)$°C, it is necessary to solve for ΔT. Using the preceding formulas for linear expansion, one finds

$$\begin{aligned}\Delta T &= \frac{1}{\alpha}\frac{\Delta L}{L}\\ &= \frac{1}{2.4\times 10^{-5}\ \mathrm{K}^{-1}}\frac{0.01\ \mathrm{m}}{10.00\ \mathrm{m}}\\ &= 42 \text{ celsius or kelvin degrees}\end{aligned}$$

As a result, the rod needs to be heated to a temperature of (15 + 42)°C = 67°C.

The following is a table of coefficients of linear and volumetric expansion of a number of materials.

QUANTITY OF HEAT

Since heat is a form of energy, it is measured in the SI energy unit of joule (J). A commonly used non-SI unit for heat is the calorie (cal), defined as the quantity of thermal energy required to raise the tem-

COEFFICIENTS OF LINEAR AND VOLUMETRIC EXPANSION*

Material	Coefficient of Linear Expansion, α (× 10^{-5} K^{-1})	Coefficient of Volumetric Expansion, 3α or β (× 10^{-5} K^{-1})
Solids		
Aluminum	2.4	7.2
Brass	2.0	6.0
Copper	1.7	5.1
Diamond	0.1	0.3
Glass (ordinary)	0.9–1.2	2.7–3.6
Glass (Pyrex)	0.5	1.5
Gold	1.3	3.9
Granite	0.8	2.4
Invar (iron-nickel alloy)	0.09	0.27
Quartz (fused)	0.04	0.12
Steel (structural)	1.2	3.6
Steel (stainless)	1.7	5.1
Liquids		
Acetone		149
Ethanol		75
Gasoline		95
Glycerin		49
Mercury		18
Water		21

*Approximate values at room temperatures (around 20°C).

perature of one gram of water by one Celsius degree from 14.5°C to 15.5°C. The kilocalorie (kcal, 1,000 calories) is also used. In fact, what is called a *Calorie* (written with a capital C) in the business of food and diet is really a kilocalorie. An additional unit, used only in the United States, is the British thermal unit (Btu), defined as the amount of heat needed to raise the temperature of one pound of water by one Fahrenheit degree from 63°F to 64°F. These units are related as follows:

$$1 \text{ cal} = 4.186 \text{ J}$$

$$1 \text{ kcal} = 1 \text{ dietary Calorie} = 1{,}000 \text{ cal} = 4{,}186 \text{ J}$$

$$1 \text{ Btu} = 252 \text{ cal} = 1{,}055 \text{ J}$$

Heat capacity is the amount of heat required to raise the temperature of a body by a single unit of temperature. Its SI unit is joule per kelvin (J/K). The heat capacity of a body depends both on the type of material from which the body is made and the amount of material in the body. Specific heat capacity and molar heat capacity are related to heat capacity. They both are properties of the materials themselves and are independent of the amount of material.

Specific heat capacity, also called specific heat, is the amount of heat required to raise the temperature of one unit of mass by one unit of temperature. Its SI unit is joule per kelvin per kilogram [J/(kg·K), also written J/kg·K]. Specific heat is commonly denoted by c. So for a sample consisting of m kilograms of a substance whose specific heat is c in J/(kg·K), the amount of heat Q, in joules, required to raise its temperature by ΔT Kelvin or Celsius degrees is

$$Q = mc\,\Delta T$$

Alternatively, if Q joules of heat are required to raise the temperature of m kilograms of a material by ΔT Kelvin or Celsius degrees, then the material's specific heat capacity in J/(kg·K) is given by

$$c = \frac{1}{m}\frac{Q}{\Delta T}$$

The heat capacity of the sample is

$$\text{Heat capacity} = \frac{Q}{\Delta T} = mc$$

Molar heat capacity, commonly denoted by C, is the amount of heat required to raise the temperature of one mole of a substance by one unit of temperature. Its SI unit is joule per kelvin per mole [J/(mol·K), also J/mol·K]. The molar mass of a substance is the mass of a single mole of that substance. Usually denoted by M, its SI unit is kilogram per mole (kg/mol). So the mass of a sample of material of molar mass M that contains n moles of the material is given by

$$m = nM$$

The amount of heat needed to raise the temperature of this sample is then

$$\begin{aligned} Q &= mc\,\Delta T \\ &= nMc\,\Delta T \end{aligned}$$

On the other hand, by the definition of molar heat capacity C,

$$Q = nC\,\Delta T$$

giving

$$c = \frac{1}{n}\frac{Q}{\Delta T}$$

and

$$C = Mc$$

So the molar heat capacity of a body equals the product of the molar mass of the substance from which the body is made and the specific heat capacity of the same substance. In molar terms, the heat capacity of an object containing n moles of a substance whose molar heat capacity is C is

$$\text{Heat capacity} = nC$$

Here is a table listing the specific heats, molar masses, and molar heat capacities for a number of materials. Note the exceptionally large value for the specific heat of liquid water, another unusual and unique characteristic of water. As a result of this property, which allows water to absorb and then give off relatively large quantities of heat, bodies of water of considerable size, such as oceans, seas, and large lakes, have a moderating influence on climate: at seaside or lakeside, the winters are warmer and the summers cooler than they are inland. In addition, water's high specific heat makes it useful as a coolant, such as in a nuclear power generating plant, where it absorbs heat from a high-temperature source and discharges it to a lower-temperature heat absorber.

As an example, the specific heat of solid iron is 4.70×10^2 J/(kg·K), while its molar mass is 0.0559 kg/mol. Accordingly, the molar heat capacity of iron is

$$\begin{aligned} C &= Mc \\ &= (0.0559 \text{ kg/mol}) \times [4.70 \times 10^2 \text{ J/(kg·mol)}] \\ &= 26.3 \text{ J/(mol·K)} \end{aligned}$$

The number of moles of iron in a five-kilogram iron dumbbell is

$$\begin{aligned} n &= \frac{m}{M} \\ &= \frac{5.00 \text{ kg}}{0.0559 \text{ kg/mol}} \\ &= 89.4 \text{ mol} \end{aligned}$$

The heat capacity of this object can then be calculated by

$$\begin{aligned} \text{Heat capacity} &= mc \\ &= (5.00 \text{ kg}) \times [4.70 \times 10^2 \text{ J/(kg·K)}] \\ &= 2.35 \times 10^3 \text{ J/K} \end{aligned}$$

or equivalently by

$$\begin{aligned} \text{Heat capacity} &= nC \\ &= (89.4 \text{ mol}) \times [26.3 \text{ J/(mol·K)}] \\ &= 2.35 \times 10^3 \text{ J/K} \end{aligned}$$

The amount of thermal energy required to raise the temperature of the dumbbell by 17.2 Kelvin or Celsius degrees, should anybody desire to do so, is calculated by

SPECIFIC HEATS, MOLAR MASSES, AND MOLAR HEAT CAPACITIES*			
Substance	**c [J/(kg·K)]**	**M (kg/mol)**	**C [J/(mol·K)]**
Aluminum	910	0.0270	24.6
Beryllium	1,970	0.00901	17.7
Copper	390	0.0635	24.8
Ethanol	2,428	0.0461	111.9
Ice (near 0°C)	2,100	0.0180	37.8
Iron	470	0.0559	26.3
Lead	130	0.207	26.9
Marble ($CaCO_3$)	879	0.100	87.9
Mercury	138	0.201	27.7
Salt (NaCl)	879	0.0585	51.4
Silver	234	0.108	25.3
Water (liquid)	4,190	0.0180	75.4

*Approximate specific and molar heat capacities, at constant pressure and at ordinary temperatures.

$$Q = (\text{Heat capacity})\,\Delta T = (2.35 \times 10^3 \text{ J/K}) \times (17.2 \text{ K}) = 40.4 \text{ J}$$

HEAT OF TRANSFORMATION

The quantity of heat per unit mass that is absorbed or released when a substance undergoes a phase transition is known as the material's heat of transformation for that transition. Its SI unit is joule per kilogram (J/kg). The most familiar phase transitions are melting and its inverse, freezing, and vaporization and its inverse, condensation. For melting, the heat of transformation is called heat of fusion and is the amount of heat per unit mass required to melt the material in its solid state without changing its temperature (which is then at the material's melting point). The same quantity of heat per unit mass needs to be removed from the material in the liquid state in order to freeze it at constant temperature (which is again at the material's melting point). The amount of heat Q in joules (J) needed to melt, by adding it, or freeze, by removing it, mass m, in kilograms (kg), of a material whose heat of fusion is L_f is

$$Q = mL_f$$

The heat of vaporization of a substance is the amount of heat per unit mass needed to convert it at its boiling point from its liquid state to a gas. In the inverse process of condensation, the same amount of heat per unit mass must be removed from the gas to liquefy it at the boiling point. So the amount of heat Q needed to be added or removed from mass m of a substance with heat of vaporization L_v in order to vaporize or condense it, respectively, is given by

$$Q = mL_v$$

The following table lists these values for a number of substances.

In addition to the phase transitions just mentioned, there exist also solid-solid phase transitions, such as from one crystalline structure to another, and they too involve heats of transformation.

In order to illustrate the ideas of this section and the previous one, let us consider the process of continually supplying heat to a sample of a substance originally in its solid state until the substance reaches its gaseous state, and let us consider all the heat exchanges along the way. Let the sample have mass m, initial temperature T_{init} (below the freezing point), and final temperature T_{fin} (above the boiling point), and let the material be characterized by freezing point T_f, boiling point T_b, heat of fusion L_f, and heat of vaporization L_v. Further, let the specific heats of the solid, liquid, and gaseous states of the substance

MELTING AND BOILING POINTS AND HEATS OF FUSION AND VAPORIZATION*

Material	Melting Point (°C)	Heat of Fusion L_f (× 10^3 J/kg)	Boiling Point (°C)	Heat of Vaporization L_v (× 10^3 J/kg)
Antimony	630.5	165	1,380	561
Ethanol	-114	104	78	854
Copper	1,083	205	2,336	5,069
Gold	1,063	66.6	2,600	1,578
Helium	-269.65	5.23	-268.93	21
Hydrogen	-259.31	58.6	-252.89	452
Lead	327.4	22.9	1,620	871
Mercury	-38.87	11.8	356.58	296
Nitrogen	-209.86	25.5	-195.81	199
Oxygen	-218.4	13.8	-182.86	213
Silver	960.8	109	1,950	2,336
Water	0.0	333.7	100.0	2,259

*The heats of fusion and vaporization are approximate.

be c_s, c_l, and c_g, respectively, which for the purpose of this example are assumed to be constant throughout their relevant temperature range. In the first step of the process the sample warms from its initial temperature to its freezing point. The amount of heat Q_1 needed for this is

$$Q_1 = mc_s(T_f - T_{init})$$

In step two the sample is melted at its freezing point by the addition of amount of heat Q_2, where

$$Q_2 = mL_f$$

Step three consists of heating the liquid from the freezing point to the boiling point, which requires the quantity of heat Q_3,

$$Q_3 = mc_l(T_b - T_f)$$

Then, in step four the liquid is vaporized at the boiling point. That needs the amount of heat Q_4, where

$$Q_4 = mL_v$$

Step five, the final step, involves heating the gas from the boiling point to the final temperature, which requires the quantity of heat Q_5,

$$Q_5 = mc_g(T_{fin} - T_b)$$

The total amount of heat used in the process, Q, is the sum of the quantities of heat supplied in all five steps:

$$\begin{aligned} Q &= Q_1 + Q_2 + Q_3 + Q_4 + Q_5 \\ &= m[c_s(T_f - T_{init}) + L_f + c_l(T_b - T_f) + \\ &\quad L_v + c_g(T_{fin} - T_b)] \end{aligned}$$

For the inverse process, in which the sample is initially in its gaseous state at a high temperature and heat is continually removed from it until it reaches its solid state, all the quantities of heat found earlier remain the same, except that they are removed from the sample rather than supplied to it.

HEAT TRANSFER

The transfer of heat occurs in one or more of three ways: conduction, convection, and radiation. In the former, heat flows through matter without concomitant transfer of the matter. In convection the motion of the material itself accomplishes the transfer, such

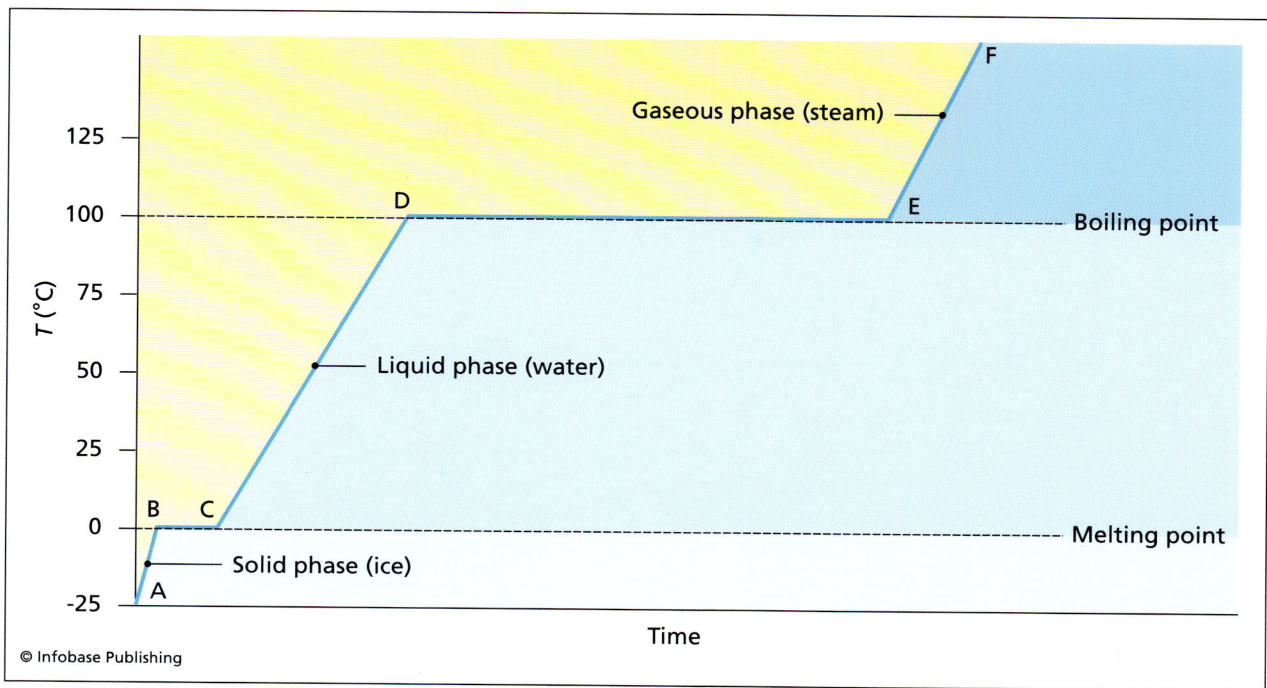

A graph of temperature versus time for a sample of water initially in its solid phase (ice) with heat being added to it at a constant rate. A to B: ice warms from its initial temperature to its melting point (0°C). B to C: the temperature remains constant (at 0°C) while all the ice melts. C to D: liquid water warms from the melting point (0°C) to its boiling point (100°C). D to E: the temperature remains constant (at 100°C) as all the water vaporizes. E to F: steam warms from the boiling point to its final temperature.

as when heated air from a furnace blows through ducts into the rooms of a house. Radiation is the transfer of heat by electromagnetic radiation, which is the way the Sun heats Earth.

The mechanism of heat conduction depends on the nature of the heat energy. In fluids (i.e., in gases and liquids) the constituent molecules or atoms are free to move about within the confines of the material. Heat is the random motion of the molecules or atoms, whereby they move more or less freely between collisions with the container walls and with each other. Heat conduction takes place through the diffusion of higher-speed molecules or atoms away from regions where they are concentrated (to be replaced by lower-speed ones, with no net transfer of matter). This type of conduction is relatively efficient compared to that typical of solids.

In solids the constituent molecules or atoms are to a large extent confined to fixed positions, and heat is the random oscillation of the constituents about their positions. Conduction of heat takes place by the more energetically oscillating molecules or atoms nudging their neighbors into greater activity. In some solids, typically in metals, there exist what are known as free electrons, which are electrons that are not bound to any particular atom and behave as a gas confined to the volume of the material. They contribute strongly to heat conduction in the manner described for a fluid and make metals typically relativity good conductors of heat.

The quantitative relation that is found to govern thermal conduction is the following. For a small, thin slab of material with area A, thickness Δx, and temperature difference ΔT between its two faces, the heat current H, which is the thermal energy passing through the slab's thickness per unit time, is given by

$$H = -kA\frac{\Delta T}{\Delta x}$$

Here H is in joules per second (J/s) or watts (W), A in square meters (m^2), ΔT in Kelvin or Celsius degrees, and Δx in meters (m). The symbol k denotes the material's thermal conductivity, which is a characteristic of every material. Its SI unit is joule per second per meter per kelvin [J/(s·m·K), also J/s·m·K] or watt per meter per kelvin [W/(m·K), or W/m·K]. A table of thermal conductivities of a number of materials is shown. The expression $\Delta T/\Delta x$ is called the temperature gradient and is positive when the temperature increases with increasing x, negative

THERMAL CONDUCTIVITIES	
Material	**k [W/(m·K)]**
Metals	
Aluminum	205.0
Brass	109.0
Copper	385.0
Lead	34.7
Mercury	8.3
Silver	406.0
Steel	50.2
Nominal values for certain solids	
Brick, insulating	0.15
Brick, construction	0.6
Concrete	0.8
Fiberglass	0.04
Glass	0.8
Ice	1.6
Styrofoam	0.01
Wood	0.12–0.04
Gases	
Air	0.024
Argon	0.016
Helium	0.14
Hydrogen	0.14
Oxygen	0.023

otherwise. The minus sign in the expression for H indicates that the direction of heat flow is opposite the temperature gradient: heat flows from higher to lower temperatures.

Good heat conductors are used, of course, where it is desirable to allow rapid heat flow from high-temperature regions or bodies to lower-temperature ones. For example, cooking pots and pans are almost always made of metal. On the other hand, poor heat conductors, or heat insulators, are used where heat flow needs to be minimized. For instance, oven pads and mitts are made of insulating materials, such as cotton batting, and if the handles of cooking pots and pans are to be held without the use of hand protectors, they are made of wood or plastic, not of metal. Homes might be thermally insulated by means of brick construction and sheets of fiberglass batting.

Heat energy transfer by convection, when the transfer is accomplished through the motion of matter, is a very complicated process. In general, though, for a body losing or gaining heat by means of convection of a fluid, such as a human body exposed to air, the heat current is proportional to the area exposed to the fluid. (In winter one tries to minimize one's skin area exposed to the cold air in order to keep warm.) The heat current increases with the speed of the fluid. (One gets cold faster in a cold wind than in still air at the same temperature.) And further, the heat current is approximately proportional to the 5/4 power of the temperature difference between the body and the volume of the fluid. (One chills faster in cold air than in cool air.)

Radiational heat transfer involves electromagnetic radiation. Every body, whatever its temperature, emits electromagnetic radiation to the environment and absorbs some of the electromagnetic radiation that is incident upon it from the environment. The heat current H, in joules per second (J/s), of the thermal energy lost by a body through radiation is proportional to the surface area of the body A, in square meters (m^2), and to the fourth power of the body's absolute temperature T, in kelvins (K). The expression for the heat current is

$$H = Ae\sigma T^4$$

where σ denotes the Stefan-Boltzmann constant, whose value is 5.670400×10^{-8} W/(m^4·K^4), rounded to 5.67×10^{-8} W/(m^4·K^4), and e is a number between 0 and 1, called the emissivity of the surface. This number has no unit and depends on the nature and treatment of the surface. Surfaces that absorb electromagnetic radiation well are also efficient radiators and have higher values of e. An ideal absorber, one that absorbs all radiation impinging upon it, is called a blackbody (also spelled *black body*) and is the most efficient radiator, with $e = 1$. The fourth-power dependence of heat current on absolute temperature is a very strong dependence, whereby small temperature differences result in large differences in radiational heat transfer.

As an example of radiational heat transfer, consider the effect that frost might form on the roof of a car outdoors overnight under a clear sky, while it does not form under similar conditions under a cloudy sky. The car roof both radiates heat to its surroundings and absorbs heat that is radiated by its surroundings. The sky forms a major component of the roof's surroundings. For radiation to which air is mostly transparent, the effective temperature of a cloudy sky is the temperature of the clouds, while the effective temperature of a clear sky is the temperature of outer space, which is around 3 K and is much lower than that of clouds, which is some 260–280 or so degrees higher. As a result of the fourth-power temperature dependence, the warming heat transfer from the clouds is very, very much greater than that from outer space. The result could possibly be that the car roof is warmed by the radiation absorbed from the clouds to above 0°C, so frost, which is ice, cannot form on it, while the radiation absorbed from outer space on a clear night might leave the car roof colder than 0°C, allowing frost to form.

KINETIC THEORY

The explanation of properties of bulk matter in terms of the motion of its constituent particles (atoms, ions, molecules) is called kinetic theory. In particular, the kinetic theory of gases derives the properties of pressure, temperature, and quantity from the particle nature of gases. Pressure is understood as the average perpendicular force on a unit area of container wall due to collisions of the gas particles with the wall. Quantity of gas is basically the number of particles in the container. And, of relevance to the present entry, temperature is related to the average kinetic energy of a particle of gas.

For an ideal gas consisting of a single species of particle, the kinetic theory straightforwardly gives some simple results. Note that the equation of state of an ideal gas is

$$pV = nRT$$

where p denotes the pressure, in pascals (Pa), equivalent to newtons per square meter (N/m^2); V is the volume of the gas, in cubic meters (m^3); T the absolute temperature, in kelvins (K), n the amount of gas, in moles (mol); and R the gas constant, whose value is 8.314472 joules per mole per kelvin [J/(mol·K), also J/mol·K], rounded to 8.31 J/(mol·K). (Other sets of units can be used as well.) This equation can be cast in the "microscopic" form

$$pV = NkT$$

where N denotes the number of particles of the gas in volume V and k is the Boltzmann constant, whose value is $1.3806503 \times 10^{-23}$ joule per kelvin (J/K), rounded to 1.38×10^{-23} J/K. The number of moles n and the number of particles N are related by

$$N = N_A n$$

where N_A denotes Avogadro's number, which is the number of particles per mole, and has the value $6.02214199 \times 10^{23}$ per mole (mol^{-1}), rounded to 6.02×10^{23} mol^{-1}. In words, the number of particles equals the number of moles multiplied by the number of particles per mole. Note that the gas constant R, the Boltzmann constant k, and Avogadro's number N_A are related by

$$R = N_A k$$

Kinetic theory shows that the pressure can be expressed as

$$p = \frac{2N(E_{kin})_{av}}{3V}$$

where $(E_{kin})_{av}$ represents the average kinetic energy of a gas particle, in joules (J). So, as an example, for a fixed volume of gas, say, the air in the tank of a scuba diver, the pressure increases when more air is introduced into the tank—N increases and more molecules are hitting the wall of the tank—or when the average kinetic energy of an air molecule is increased—so each collision exerts a greater force on the wall. This relation can be put in the form

$$pV = \tfrac{2}{3}N(E_{kin})_{av}$$

Now, equating the right-hand side of this equation with the right-hand side of the microscopic form of the ideal-gas equation shown earlier gives the relation between the absolute temperature T and the average kinetic energy of a gas particle $(E_{kin})_{av}$,

$$(E_{kin})_{av} = \tfrac{2}{3}kT$$

or

$$T = \frac{2}{3k}\,(E_{kin})_{av}$$

This relation reveals the microscopic significance of the macroscopic physical quantity temperature: in the case of an ideal gas, absolute temperature is proportional to the average random kinetic energy of a gas particle. From this relation one obtains an expression for v_{rms}, the root-mean-square speed (the square root of the average squared speed) of a gas particle, in terms of the temperature and either the mass of a gas particle m or the molar mass M, in kg/mol, of the gas. Using the well-known expression for kinetic energy, the average kinetic energy of a gas particle can be expressed as

$$(E_{kin})_{av} = \tfrac{1}{2}\,m(v^2)_{av}$$

where m is the mass of a single gas particle, in kilograms (kg), and $(v^2)_{av}$ denotes the average squared speed of a gas particle, with speed in meters per second (m/s). Then it follows that

$$v_{rms} = \sqrt{(v^2)_{av}} = \sqrt{\frac{3kT}{m}} = \sqrt{\frac{3RT}{M}}$$

For example, the molar mass of molecular hydrogen H_2 is 2.00×10^{-3} kg/mol. So the root-mean-square speed of a molecule in a sample of hydrogen gas at a temperature of 300 K (27°C) is

$$v_{rms} = \sqrt{\frac{3 \times 8.31 \times 300}{(2.00 \times 10^{-3})}} = 1.93 \times 10^3 \text{ m/s}$$

or almost 1.2 miles per second (2 km/s).

THERMODYNAMICS

Thermodynamics is the field of physics that deals with the conversion of energy from one form to another while taking temperature into account and derives relationships among macroscopic properties of matter. Thermodynamics is founded on four laws, which are based on and form generalizations of our experience with nature.

The Zeroth Law of Thermodynamics

This law states that physical systems may be endowed with the property of temperature, which is described by a numerical quantity. When two systems possess the same temperature, meaning that they have the same value for this quantity, they will then be in thermal equilibrium while in contact with each other. In other words, when two systems are put in contact so that heat might flow from one to the other (but with no transfer of matter or work), their properties do not change if they are at the same temperature. In thermodynamics, temperature is usually denoted T and is expressed on an absolute scale in the SI unit of the kelvin (K). This is a very technical and formal law. The need for this law was recognized only after the other laws of thermodynamics—the first through third—were known and numbered, in order to give thermodynamics a firm theoretical foundation: hence its strange designation. The zeroth law is not commonly included in introductory presentations of thermodynamics.

The notion of temperature is derived from the fact that if systems A and B are in thermal equilibrium when in contact and so are systems B and C, then invariably systems A and C are also found to be in thermal equilibrium when in contact. That suggests the existence of a property that is a physical quantity and has the same value for systems that are in thermal equilibrium, hence temperature.

The First Law of Thermodynamics

This law is equivalent to the work-energy theorem, or the law of conservation of energy, while recognizing that systems possess internal energy. The law states that the increase in the internal energy of a system equals the sum of work performed on the system and heat flowing into the system. In symbols:

$$\Delta U = W + Q$$

where ΔU denotes the increase of the system's internal energy, W is the work done on the system, and Q the heat flowing into the system. All quantities are in the same unit of energy, which in the SI is the joule (J).

Note that while internal energy is a property of a system, neither work nor heat flow is a property. The latter two are modes of energy transfer. Work is the transfer of orderly energy. An external force compressing a system is performing work on it. Heat flow is the transfer of disorderly energy, the random motion of the particle constituents of matter. Heat flows as a result of a temperature difference between a system and its surroundings. Heat flows from a hot flame to the cooler water in the pot on top of the flame. It follows, then, that for an isolated system, a system that exchanges neither heat nor work with its environment, the internal energy must remain constant. That is a constraint on processes occurring in isolated systems.

As an example, let a sample of gas confined to a cylinder with a movable piston expand, thereby pushing the piston outward and doing 162 J of work on the surroundings. Then, with no additional work done, heat in the amount of 98 J is made to flow into the gas by heating the cylinder wall with a propane torch. Since heat flows *into* the gas, Q in the preceding formula is positive and $Q = 98$ J. But as the gas is performing work on the environment, W in the formula is negative, $W = -162$ J. The first law of thermodynamics states that as a result of this two-step process the change of internal energy of the gas is

$$\begin{aligned}\Delta U &= W + Q \\ &= (-162 \text{ J}) + 98 \text{ J} \\ &= -64 \text{ J}\end{aligned}$$

So the gas's internal energy decreases by 64 J. In other words, the gas does more work on the surroundings (162 J) than can be "paid for" by the heat flowing into it (98 J), so some of the work (64 J) is at the expense of the gas's internal energy, which correspondingly decreases (by 64 J).

An analogy useful for demonstrating this concept likens a system's internal energy to a bank account. Deposits into and withdrawals from the account may be made in either of two forms, say, by cash or by check (representing heat flow and work). If you pay \$162 by check from the account but only deposit \$98 in cash, the account balance drops by \$64.

Nature does not allow any process that violates the first law. For instance, the example just presented describes an allowed process. For the same system, however, a hypothetical process in which the gas performs 200 J of work on its surroundings (W = –200 J), absorbs 98 J of heat (Q = 98 J), and still loses only 64 J of internal energy (ΔU = –64 J) is not allowed and will never take place. For this process

$$\Delta U \neq W + Q$$

since

$$-64 \text{ J} \neq (-200 \text{ J}) + 98 \text{ J}$$
$$= -102 \text{ J}$$

The Second Law of Thermodynamics

This law goes beyond the first law and adds a further constraint. Not all processes that are allowed by the first law will indeed occur. They may occur if they are consistent with the second law and will never occur if they violate it. One formulation of the second law is that heat does not flow spontaneously from a cooler body to a warmer one (i.e., from lower to higher temperature).

Another, equivalent formulation is in terms of a heat engine, which is any device that converts thermal energy, heat, into mechanical energy, from which other forms of energy might be derived, such as electric energy. The heat engine accomplishes this by taking in heat from an external source and performing work on its surroundings. In these terms, the second law can be stated thus: a cyclic heat engine must discharge heat at a lower temperature than the temperature at which it takes in heat. In other words, not all the heat input is convertible to work; some must be "wasted" by being discharged at a lower temperature. Since after a complete cycle a heat engine returns to its original state, its internal

A steam locomotive is a heat engine that converts some of the thermal energy of steam to mechanical energy. The source of the steam's thermal energy is the chemical energy of the coal that is burned to create the steam by boiling water. According to the second law of thermodynamics, not all of the steam's heat can be so converted, and some of it is discharged to the atmosphere. *(Louie Schoeman, 2008, used under license from Shutterstock, Inc.)*

energy at the end of a cycle has the same value as at the start of the cycle. Then, according to the first law, any work done by the engine on the surroundings during the cycle must equal the *net* heat flow into the engine—heat input less discharged heat—during the cycle. The second law states that some of the heat flowing in is discharged, so that not all the heat input is converted to work. In other words, the heat input is greater than the work performed. Since the thermal efficiency of a heat engine is the ratio of the work performed to the heat input, the second law determines that the thermal efficiency of a cyclic heat engine is less than 100 percent.

In order to quantify these notions, a physical quantity, called entropy and commonly denoted S, is introduced as a property of physical systems. Entropy is a measure of a system's disorder, of its randomness. As an example, a state of a gas in which all the molecules are confined to one half of the container is more orderly than the state in which the molecules are distributed throughout the container. Accordingly, the system has a lower value of entropy in the former state than in the latter one. One way that the entropy of a nonisolated system can be increased or decreased is by the flow of heat into the system or from it, respectively. The increase in entropy ΔS, in joules per kelvin (J/K), due to the amount of heat ΔQ, in joules (J), flowing reversibly into the system at absolute temperature T, in kelvins (K), is given by

$$\Delta S = \frac{\Delta Q}{T}$$

Note that a reversible process is one that can be run in reverse; thus, a reversible heat flow must take place at a single temperature, denoted T in the formula. That is because heat flow through a temperature difference, from higher temperature to lower, cannot be reversed, as heat cannot flow from lower to higher temperature. Since heat also does not flow between equal temperatures, there is no reversible heat flow in practice; the notion is a theoretical idealization. Reversible heat flow can be approximated by allowing heat to flow very slowly through a very small temperature difference. Heat flowing into a system brings about an increase in the system's entropy, while the entropy decreases when heat flows out. For example, in a process during which 3,000 J of heat leaves a system ($\Delta Q = -3{,}000$ J) very slowly while the system is at a temperature of 300 K ($T = 300$ K), the system's entropy changes by the amount

$$\Delta S = \frac{-3{,}000}{300} = -10 \text{ J/K}$$

In other words, the entropy of the system decreases by 10 J/K.

Internal processes can also effect change on the entropy of a system. That can happen for isolated as well as nonisolated systems.

The second law of thermodynamics is formulated in terms of entropy: the entropy of an isolated system cannot decrease over time. Spontaneous, irreversible evolution of an isolated system is always accompanied by an increase of the system's total entropy. That might involve a decrease of entropy in part of the system, but it will be more than compensated for by an increase in the rest of the system. On the other hand, for an isolated system that is undergoing a reversible process or is not evolving at all, the entropy remains constant, but in no case does the entropy of an isolated system decrease.

As an example, consider a sample of gas in a container of fixed volume. If the gas is evenly distributed in the container and its temperature is uniform throughout its volume, the gas remains in that state and does not evolve; it is in equilibrium, and its entropy remains constant. If the gas is initially distributed unevenly, say, all the gas is on one side of the container and none on the other, it very rapidly evolves to the uniform state just described and in doing so increases its entropy. Since systems evolve from nonequilibrium states, like the uneven distribution just described, to equilibrium states and then remain in those states, and since a system's entropy increases during such evolution but remains constant once equilibrium is attained, it follows that equilibrium states are states of maximal entropy.

In connection with the second law, the Carnot cycle, named for the 19th-century French engineer Nicolas Léonard Sadi Carnot, describes the operation of an ideal, reversible heat engine that takes in heat at higher temperature T_H and discharges heat at lower temperature T_L. In a single cycle of operation, heat Q_H is absorbed at the higher temperature and heat Q_L discharged at the lower. For absolute temperatures, the relation

$$\frac{T_L}{T_H} = \frac{Q_L}{Q_H}$$

holds for a Carnot engine. Carnot cycles are not realizable in practice, since they involve reversible processes, but can only be approximated. Their importance is mainly theoretical.

The thermal efficiency of any engine, not necessarily a Carnot engine, that is taking in and discharging heats Q_H and Q_L per cycle at temperatures T_H and T_L, respectively, is the ratio of work performed per cycle to the heat taken in per cycle, as mentioned earlier. Since at the end of a cycle the internal energy of the engine returns to its value at the start of the cycle, the first law of thermodynamics (equivalent to the work-energy theorem) gives that the work performed by the engine per cycle, $-W$ (in the earlier formulation of the first law, W is the work done *on* the engine), equals the difference between the heat taken in and the heat discharged per cycle:

$$-W = Q_H - Q_L$$

So the efficiency e is

$$\begin{aligned} e &= \frac{-W}{Q_H} \\ &= \frac{Q_H - Q_L}{Q_H} \\ &= 1 - \frac{Q_L}{Q_H} \end{aligned}$$

From the second law, it follows that the thermal efficiency of a Carnot engine e_C sets a theoretical upper limit for the efficiency of any heat engine operating between the same two temperatures. From the preceding temperature-heat relation for the Carnot engine, the efficiency of such an engine is given by

$$e_C = 1 - \frac{T_L}{T_H}$$

This shows that the efficiency of a Carnot engine, and thus the theoretical maximal efficiency of any engine operating between two temperatures, can be increased by lowering the discharge temperature or raising the intake temperature. In other words, engine efficiency is improved by having an engine operate over a greater temperature difference. This formula serves engineers in designing and estimating the efficiencies of heat engines.

The Third Law of Thermodynamics

This law is related to absolute zero temperature. One statement of it is that it is impossible to reach absolute zero temperature through any process in a finite number of steps. Equivalently, a Carnot engine cannot have perfect efficiency. Since $T_H = \infty$ is not possible, the only other way of achieving $e_C = 1$ is by taking $T_L = 0$. But the third law does not allow that. Thus, $e_C < 1$ and the efficiency is less than 100 percent.

Another formulation of the third law is that at absolute zero temperature all substances in a pure, perfect crystalline state possess zero entropy. This law sets the zero point for entropy—note that the second law deals only with changes of entropy. With a zero level now defined by the third law, the entropy of any substance can be measured and calculated with the help of the previous formula for entropy change due to reversible heat flow.

PERPETUAL MOTION

Various and diverse devices are occasionally proposed by inventors as sources of free or almost-free energy. Such devices seem too good to be true, and indeed, they *are* too good to be true. Some devices are claimed to produce more energy than needs to be invested or that produce energy with *no* need for investment. Their operation is termed perpetual motion of the first kind, which means they violate the first law of thermodynamics (i.e., the law of conservation of energy). Since, at least as far as energy is concerned, you cannot get something for nothing, such devices cannot work as a matter of principle, and there is no need to analyze the details of their operation further.

An example of putative perpetual motion of the first kind might be a wheel with bar magnets arrayed on it and stationary bar magnets distributed around a frame surrounding the wheel. When given an initial spin, such a wheel might be claimed to rotate forever. Even if no useful work is obtained from the device, it still violates the first law in that it is performing work on its surroundings through the friction forces at the bearings and the force of air resistance. That work is not being "paid for" by heat entering the system or by depletion of the system's internal energy. With no need to analyze the magnetic forces that are operating, it is clear from the first law of thermodynamics that this device cannot work as claimed.

Then there are those devices that obey the first law but violate the second law of thermodynamics. Their operation is called perpetual motion of the second kind. They do not create energy—such as mechanical or electric—or do work from nothing. But they are claimed to be able to convert thermal energy, heat, into such orderly forms of energy at 100-percent efficiency, or at least at efficiencies that exceed those that are allowable according to the second law. As an example, a tremendous amount of internal energy exists in the oceans. Some of it can be converted to orderly energy, as long as heat is discharged at a lower temperature. But that is neither a simple nor an inexpensive procedure, since warm and cold ocean water do not coexist sufficiently near each other and at a sufficient temperature difference to make the operation possible without considerable investment. A perpetual motion device of the second kind might be purported to dispense altogether with the discharge and simply take thermal energy from the ocean and convert all of it to electricity. Such claims violate the second law of thermodynamics and are thus invalid in principle, so again, it is not necessary to look further into the way the devices operate.

STATISTICAL MECHANICS

The properties of bulk matter that thermodynamics deals with—temperature, pressure, internal energy, entropy, and so on—are macroscopic manifestations of the situation that exists at the level of constituent particles, at the microscopic level. So thermodynamics should be, and is, translatable into microscopic terms. Kinetic theory, discussed earlier, is an example of that. The field of physics that fulfills this function in general is statistical mechanics, which studies the statistical properties of physical systems containing large numbers of constituents and relates them to the systems' macroscopic properties.

See also CALORIMETRY; CONSERVATION LAWS; ELECTROMAGNETIC WAVES; ENERGY AND WORK; EQUILIBRIUM; FORCE; GAS LAWS; HEISENBERG, WERNER; PRESSURE; QUANTUM MECHANICS; STATISTICAL MECHANICS.

FURTHER READING

Maxwell, James Clerk. *Theory of Heat.* Mineola, N.Y.: Dover, 2001.

Ord-Hume, Arthur W. J. G. *Perpetual Motion: The History of an Obsession.* Kempton, Ill.: Adventures Unlimited, 2006.

Serway, Raymond A., and John W. Jewett. *Physics for Scientists and Engineers,* 7th ed. Belmont, Calif.: Thomson Brooks/Cole, 2008.

Van Ness, H. C. *Understanding Thermodynamics.* Mineola, N.Y.: Dover, 1983.

Young, Hugh D., and Roger A. Freedman. *University Physics,* 12th ed. San Francisco: Addison Wesley, 2007.

Werner Heisenberg was a 20th-century German physicist. He received the Nobel Prize in physics in 1932 for the development of the field of quantum mechanics. The Heisenberg uncertainty principle contributed greatly to the development of the atomic model. *(AP Images)*

Heisenberg, Werner (1901–1976) German *Physicist*

Werner Heisenberg was a German-born physicist who is best known for his development of the Heisenberg uncertainty principle and his contributions to the modern model of the atom by the development of the quantum mechanical model. Many consider Heisenberg to be one of the most influential physicists of the 20th century. He played a large role in the development of Germany's nuclear weapons program and received the Nobel Prize in physics in 1932 for his development of quantum mechanics.

TRAINING AND CAREER

Heisenberg was born on December 5, 1901, in Würzburg, Germany, in Bavaria, to Dr. August Heisenberg and Annie Heisenberg. His father was a professor of Greek languages. Heisenberg was the younger of two sons, and his older brother was also interested in science and became a chemist.

Heisenberg's education and training led to a prolific career in academia—teaching and researching physics. He was interested in science and math at an early age and that interest influenced his studies. Heisenberg's family moved to Munich in 1911, and he attended school in Munich, at the Maximillian School. While in high school, Heisenberg was strongly affected by the involvement of Germany in World War I and joined the youth militia groups in support of the war effort. This early experience with the political process influenced Heisenberg and may have played a role in his later involvement with the German nuclear weapons program during World War II.

In 1920, Heisenberg entered the University of Munich, where he originally planned to major in mathematics but eventually chose theoretical physics. At the age of 21 in 1923, Heisenberg received a Ph.D. in physics working under the physicist Arnold Sommerfeld. Heisenberg was privileged to work with some of the greatest physicists at the time—in addition to Sommerfeld, he collaborated with Max Born, Niels Bohr, and his contemporary Wolfgang Pauli, who became a close friend and collaborator on the quantum theory work. Heisenberg then became an assistant professor at the University of Göttingen, Germany, with Max Born. While at Göttingen, Heisenberg began his work on the quantum mechanical model of the atom. In 1924–25, Heisenberg worked at the University of Copenhagen, where he formed a lasting collaboration with Bohr while they put the final touches on the quantum mechanical model.

In 1927, Heisenberg became professor of theoretical physics at the University of Leipzig, making him the youngest full professor in Germany at the time. In this same year, he developed what became known as the Heisenberg uncertainty principle (described later).

Heisenberg's career in theoretical physics continued at Leipzig. With the start of World War II, physicists all around the world were working on what was thought to be an impossible task, nuclear fission reactions that would be theoretically feasible for nuclear weapon production. In 1941, Heisenberg became a professor of physics at the University of Berlin and the director of the Kaiser Wilhelm Institute for Physics, where he worked on the development of the German nuclear weapons program. Heisenberg's involvement in the unsuccessful attempts of the German military to develop a nuclear weapon has been the cause for much speculation. Many believe that he intentionally slowed his work to prevent them from creating a nuclear bomb because of moral convictions. Others believe that he just could not make a nuclear bomb work. Correspondence and quotes from Heisenberg suggest he may have misled the military when he informed them that it was not feasible, causing them to pursue nuclear energy techniques rather than weapons.

After the war, German scientists, including Heisenberg, were rounded up and held in detention in England. Heisenberg returned to Germany in 1946 to the University of Göttingen, which became known as the Max Planck Institute for Physics in 1948. In 1937, Heisenberg married Elisabeth Schumacher, and they lived in Munich with their seven children.

QUANTUM MECHANICS

When Heisenberg was in his early 20s, the atomic theory of the day was put forward by Niels Bohr, with whom Heisenberg worked in Copenhagen. The Bohr model of the atom explained the locations of the electrons orbiting the positively charged nucleus in discrete orbits. Bohr's model is sometimes called the "planetary model" of the atom. There were several weaknesses of Bohr's atomic theory including its inability to explain spectroscopic results. Bohr's model did not fully explain the spectral lines found in atoms that were more massive than hydrogen. There were even lines in hydrogen's own spectrum that could not be accounted for by Bohr's model. His model also did not fully explain in which form the light traveled—waves or particles. As Heisenberg entered the field, Bohr's model was facing more and more criticism for its inability to explain these problems.

Heisenberg's work in Göttingen and Copenhagen resulted in his matrix mechanics model. The calculations in his model involved matrices using the tools of a different branch of math, matrix algebra, with which many were not familiar. Matrix algebra is a method to manipulate multiple equations with multiple variables that if understood leads to straightforward calculations. Most physicists at the time were experimental physicists and worked primarily in laboratories. Heisenberg was an originator of the field of theoretical physics, which many did not understand. The new field and the abstract nature of his matrix theory caused some to question its validity.

Around the same time, Erwin Schrödinger developed his wave mechanics model based on more familiar calculations and more familiar wave motions of light. The arguments between Heisenberg and his colleagues versus Schrödinger and his colleagues continued for several years. In the late 1920s, Schrödinger demonstrated that the matrix model put forth by Heisenberg and the wave mechanics model were equivalent mathematically. But Schrödinger and Albert Einstein still adamantly argued that the wave mechanics model was significantly more suitable than matrix mechanics.

An acceptable compromise to most, except Schrödinger and Einstein, was drafted by Pasqual Jordan and Paul Dirac, who had worked with Heisenberg at Göttingen and Copenhagen, respectively. Jordan-Dirac theory was based on the premise that large objects follow normal Newtonian mechanics while smaller objects such as subatomic particles are affected by the wavelength of the light. They generated a transformation theory that attempted to combine both versions, and it became known as quantum mechanics. In 1933, both Erwin Schrödinger and Paul Dirac received the Nobel Prize in physics for their contributions to atomic theory.

HEISENBERG UNCERTAINTY PRINCIPLE

Heisenberg's uncertainty principle helped to describe calculations of the position and motion of subatomic particles in the Bohr model of the atom. The summary statement by Heisenberg from his 1927 paper succinctly explains the uncertainty principle:

> The more precisely the position is known, the less precisely the momentum is known in this instant and vice versa.

In order to determine the location of any object, one must visualize the object to determine where it is. The resolution of an image is only as good as the wavelength of light that is used to observe it. Imagine that while camping out one wanted to determine where a tent was; it would be necessary to shine a flashlight on it. Because of the large size of the tent, the wavelength of the light from the flashlight would have little or no effect on the determination of the position of the tent. In contrast, when a smaller particle such as an electron is visualized, the wavelength of the light being used to determine its position can have an effect on the determination of the position of the electron. In order to determine the location of the electron accurately, one needs to use light with a short enough wavelength. The shorter the wavelength, the higher the energy of each photon in the light wave, and this higher energy has enough momentum to cause the small mass of the electron to increase its momentum. When light has a sufficiently long wavelength, such that the momentum of the light's photons is not high enough to affect the momentum of the electron appreciably, it will not be a high enough wavelength to allow significant resolution of the electron's position. When the location measurement of an electron becomes more and more precise, the momentum must be less precise. When the momentum measurement increases in precision, the position of the electron must become less precise.

If the position is labeled x, and the momentum is p, the mathematical relationship of the Heisenberg uncertainty principle can be summarized by the following equation:

$$\Delta x \, \Delta p \geq h/4\pi$$

where Δx is the uncertainty in the position of the particle, Δp is the uncertainty in the momentum of the particle, and h is Planck's constant, equal to 6.6×10^{-24} J·s. The product of the uncertainty of the position and the uncertainty of the momentum can be equal to or greater than but can never be less than the value of $h/4\pi$.

This principle, now known as the Heisenberg uncertainty principle, developed by Werner Heisenberg in 1927, paved the way for the quantum mechanical model of the atom based on the location of electrons. In 1932, Heisenberg received the Nobel Prize in physics for the Heisenberg uncertainty principle, the development of quantum mechanics, and the discovery of allotropic versions of hydrogen.

See also Bohr, Niels; Broglie, Louis de; Einstein, Albert; nuclear physics; Pauli, Wolfgang; quantum mechanics; Schrödinger, Erwin.

FURTHER READING

Cassidy, David C. *Uncertainty: The Life and Science of Werner Heisenberg.* New York: W. H. Freeman, 1992.

Hodgkin, Dorothy Crowfoot (1910–1994)

British *Physical Chemist* Dorothy Crowfoot Hodgkin was a pioneer in the field of X-ray crystallography, using her knowledge of chemistry, physics, and mathematics to perform structural analyses of biomolecules such as sterols, antibiotics, and vitamins. Over the course of her 70-year career, she solved increasingly complex structures that culminated in the elucidation of the structure of insulin in 1969. Prior to the determination of insulin's structure, Hodgkin received the Nobel Prize in chemistry in 1964 for her use of X-ray crystallography to determine the molecular structures of other molecules, including cholesterol, penicillin, and vitamin B_{12}. At the time of her death, the Nobel laureate and her fellow crystallographer Max Perutz remarked that Hodgkin would be remembered not only as a great chemist but also as a gentle, tolerant person and as a devoted proponent of world peace.

Born on May 12, 1910, in Cairo, Egypt, Dorothy Crowfoot was the daughter of John Winter Crowfoot and Grace Mary Hood Crowfoot. Her father served as an administrator in the Egyptian Education Service, and her mother worked at home with Dorothy and her sisters. Until the outbreak of World War I, Dorothy and her family traveled extensively in the Middle East. In 1916, her father accepted a position as the assistant director of education and of antiquities in Sudan, and Dorothy's parents sent her and two sisters to Worthing, England, to live with her aunt. She was reunited with her family in England at the age of eight. At 10, Dorothy was exposed to chemistry while taking a class at the National Educational Union, and she created her own makeshift laboratory in her parents' attic, where she grew copper sulfate and alum crystals. While attending the Sir John Leman School in Beccles, she took part in a chemistry class normally exclusive to boys at the school. When Dorothy was 15, her mother gave her a book by the Nobel laureate William H. Bragg, *Concerning the Nature of Things* (1925), in which he set forth the potential uses of X-ray crystallography to "see" atoms and molecules of interest. Because atoms are smaller than the wavelengths emitted by visible light, one cannot use traditional microscopy to visualize molecular structures. X-rays emit wavelengths smaller than atoms. When researchers bombard a molecule with X-rays, the atoms in the molecule diffract (bend) those X-rays, and a picture of the molecule of interest emerges from mathematical analyses of the pattern generated by the diffracted rays onto a piece of photographic paper. After Hodgkin enrolled (1928–32) in Somerville College (the women's college of Oxford University), where she majored in physics and chemistry, she took a special course in X-ray crystallography to become more familiar with the techniques involved. For her fourth-year project, she used these developing skills to perform an X-ray diffraction study of thallium dialkyl halide crystals with Dr. H. M. Powell as her adviser. A chance meeting on a train that involved her old mentor, Dr. A. F. Joseph, secured Hodgkin a position in the Cambridge University laboratory of John Desmond Bernal, a pioneer in the study of metals using X-ray crystallography, for her graduate studies. Upon learning that Bernal intended to shift his focus from metals to sterols, Hodgkin eagerly joined Bernal's laboratory. For her Ph.D. thesis in crystallography, Hodgkin performed crystallographic studies of steroid crystals, yet between 1933 and 1936, she also examined minerals, metals, inorganic molecules, and organic molecules such as proteins and viruses. During her short time in Bernal's laboratory, she coauthored with Bernal 12 papers including a paper describing the crystallographic analysis of pepsin, a digestive enzyme that hydrolyzes peptide bonds between amino acids of proteins in the stomach, the first reported analysis of that kind. Ironically the same afternoon in 1934 that Bernal captured the first good X-ray diffraction pattern of pepsin on film, a physician diagnosed the tenderness and inflammation in the joints of Hodgkin's hands as rheumatoid arthritis, a deforming and crippling autoimmune disease that ultimately left Hodgkin wheelchair bound.

As financial assistance from her aunt, Dorothy Hood, ran short, Hodgkin accepted a research fellowship in 1933 that required her to leave Cambridge after one year to complete her doctoral studies at Somerville College at Oxford University. She left Cambridge and returned to Oxford per the conditions of her fellowship, but she continued her work on sterols until receiving her Ph.D. in 1937. At Oxford, Hodgkin studied more than 100 different types of sterols, a group of organic molecules consisting primarily of unsaturated alcohols, including cholesterol, which is a precursor to many steroid hormones and is found in the cell membrane of animals. When Hodgkin began her structural studies of cholesterol, the chemistry of the molecule had been determined (i.e., chemists had already deciphered the basic chemical formula with the number and type of atoms making up the molecule), but no one knew the atomic arrangement of the molecule's carbon, hydrogen, and oxygen atoms. Using her growing expertise in X-ray crystallography, Hodgkin elucidated the molecular structure of cholesteryl iodide, the most complex biological crystal structure solved at that time, in 1945.

When World War II broke out in Europe and many scientists turned their attention to helping with the war effort, Bernal gave Hodgkin his crystallography equipment, making it possible for her to study more complex molecules. With the increase in the complexity of the diffraction patterns generated by these molecules and of the mathematical computations necessary for the analysis of the patterns, Hodgkin also fortuitously gained access to a Hollerith punch-card machine, an early type of computer that made the numerous calculations for the X-ray diffraction studies manageable. Trying to help the war effort, Hodgkin focused her research efforts on determining the structure of penicillin, an effective antibiotic made from the mold *Penicillium notatum.* With large-scale synthesis of penicillin in mind, Hodgkin believed that the process would be greatly expedited if the molecular structure of the antibiotic were in hand. Several different forms of the molecule existed, and the heterogeneous mixture of crystals made X-ray diffraction analyses difficult. Hodgkin

Dorothy Crowfoot Hodgkin was a 20th-century British chemist. She received the Nobel Prize in chemistry in 1964 for the determination of the crystal structures of penicillin and vitamin B_{12}. *(Harold Clements/Daily Express/Hulton Archive/Getty Images)*

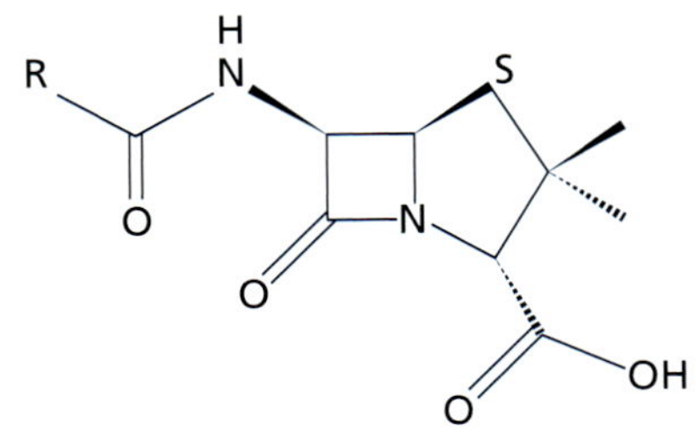

The structure of penicillin was discovered by Dorothy Crowfoot Hodgkin in the 1940s. The formula of penicillin is R–$C_9H_{11}N_2O_4S$. There are several side chains (R) in different forms of penicillin.

progressed slowly until she received a three-milligram sample of an American strain of the penicillin that crystallized in a more homogeneous manner. In 1945, using diffraction data, she elucidated that penicillin consisted of an unusual atomic arrangement central to its structure called a beta-lactam ring. The structure contains a square with two methyl groups, a carbonyl, and a nitrogen atom linked together, joined to a five-sided thiazolidine ring made up of a sulfur atom, carbon atoms, and a nitrogen atom shared with the beta-lactam ring.

In 1926, other researchers discovered the vitamin B_{12}, an essential nutrient necessary for red blood cell production, and the purified form of this vitamin became available 18 years later. Pharmaceutical companies found the molecular structure of vitamin B_{12}, whose absence results in pernicious anemia and subsequent death, too complex for traditional chemical analyses. One pharmaceutical company, Glaxo, approached Hodgkin to help solve the molecular structure of this molecule because they wanted to manufacture a synthetic version of the vitamin. For six years, Hodgkin plugged away at the structural analysis of vitamin B_{12}. Initial studies indicated that a ringed structure similar to porphyrin, a flat-ringed structure found in hemoglobin consisting of four smaller rings called pyrroles, created the core of the vitamin's molecular structure. Hodgkin made more than 2,500 X-ray photographs, and she sent the data via telegram and airmail to the University of California at Los Angeles to the American crystallographer Kenneth Trueblood, who used the National Bureau of Standards Western Automatic Computer, the most sophisticated computer of the time, to perform the necessary crystallographic calculations 100 times faster than standard methods using a program that he had designed. In 1955, Hodgkin described the structure of vitamin B_{12} in the journal *Nature*, including a new substructure called a corrin ring that is similar but not the same as a porphyrin ring. In 1960, the Royal Society of London recognized her accomplishments in the field of chemistry and awarded Hodgkin the first Wolfson Research Professorship, a position she held for 16 years. Because of the significant biochemical structures she had determined using X-ray crystallography, Hodgkin received the 1964 Nobel Prize in chemistry, becoming the first female recipient from Great Britain.

During her graduate studies, the Nobel laureate Sir Robert Robinson gave Hodgkin a crystal sample of insulin, a hormone that regulates blood sugar levels. More than 30 years later, Hodgkin resumed her attempts to determine the molecular structure of the most complex biomolecule she had studied to date with 777 atoms to arrange and rearrange in the analyses. In the late 1930s, she grew and regrew insulin crystals, gathering data from wet and dry samples; three decades later, she prepared heavy atom derivatives of insulin crystals to collect additional crystallographic data. The Hodgkin laboratory analyzed more than 70,000 X-ray diffraction spots and the corresponding unwieldy mathematical calculations. In 1969, she gave one of her postdoctoral students, Thomas Blundell, the honor of announcing the results of their efforts, the low-resolution determination of the structure of insulin, at an International Union of Crystallography (IUCr) (an organization that Hodgkin helped form in 1947 to encourage and facilitate the exchange of scientific information by crystallographers around the globe) meeting. Two years later, Hodgkin's group refined their model of insulin to a resolution of 1.9 angstroms (Å), but Hodgkin believed that this higher-resolution model could be refined further. In 1988, Hodgkin authored her last scientific paper, "The Structure of 2Zn Pig Insulin Crystals at 1.5 Å Resolution," in the *Philosophical Transactions of the Royal Society of London*. Hodgkin's diligence and perseverance not only led to a greater understanding of insulin's behavior in solution, its chemical properties, and its folding patterns, but also demonstrated the usefulness of X-ray crystallography to a new generation of organic chemists trying to understand chemical molecules.

Dorothy Crowfoot Hodgkin entered the field of X-ray crystallography during its infancy. Her keen intuition and dogged determination resulted in the successful elucidation of countless biomolecules as she tested the limits of the technique. A quiet, modest woman by nature yet an extremely powerful influence on those around her, Hodgkin paved the way for many crystallographers who followed by promoting open exchange of information among peers and mutual helpfulness and support. In her private life, Hodgkin married Thomas Hodgkin, a political historian who studied the politics of the African and

Arab worlds, in December 1937. Between 1938 and 1946, the couple had three children—two sons and a daughter; as in everything she did, Hodgkin juggled career and family with great success, being remembered by those around her as a devoted, loving, and patient mother and wife. In addition to the Nobel Prize, Hodgkin received countless other awards for her achievements including the Royal Society of London's Royal Medal for her work on vitamin B_{12} (1957), Great Britain's Order of Merit (1965), and the Copley Medal (1976). From 1976 to 1988, Hodgkin served as president of the Pugwash Conferences on Science and World Affairs, a set of meetings that invites scholars to find ways to reduce the danger of armed conflict. During the last decade of her life, the rheumatoid arthritis that plagued Hodgkin during her adult life confined her to a wheelchair, yet she continued to travel to scientific meetings and peace conferences until her death at home on July 29, 1994, at the age of 84.

See also BIOCHEMISTRY; ORGANIC CHEMISTRY; PERUTZ, MAX; X-RAY CRYSTALLOGRAPHY.

FURTHER READING

Ferry, Georgina. *Dorothy Hodgkin: A Life.* London: Granta Books, 1998.

holography In the most general sense of the term, *holography* is the recording and storage of three-dimensional information in a two-dimensional medium and its reconstruction from the medium. A

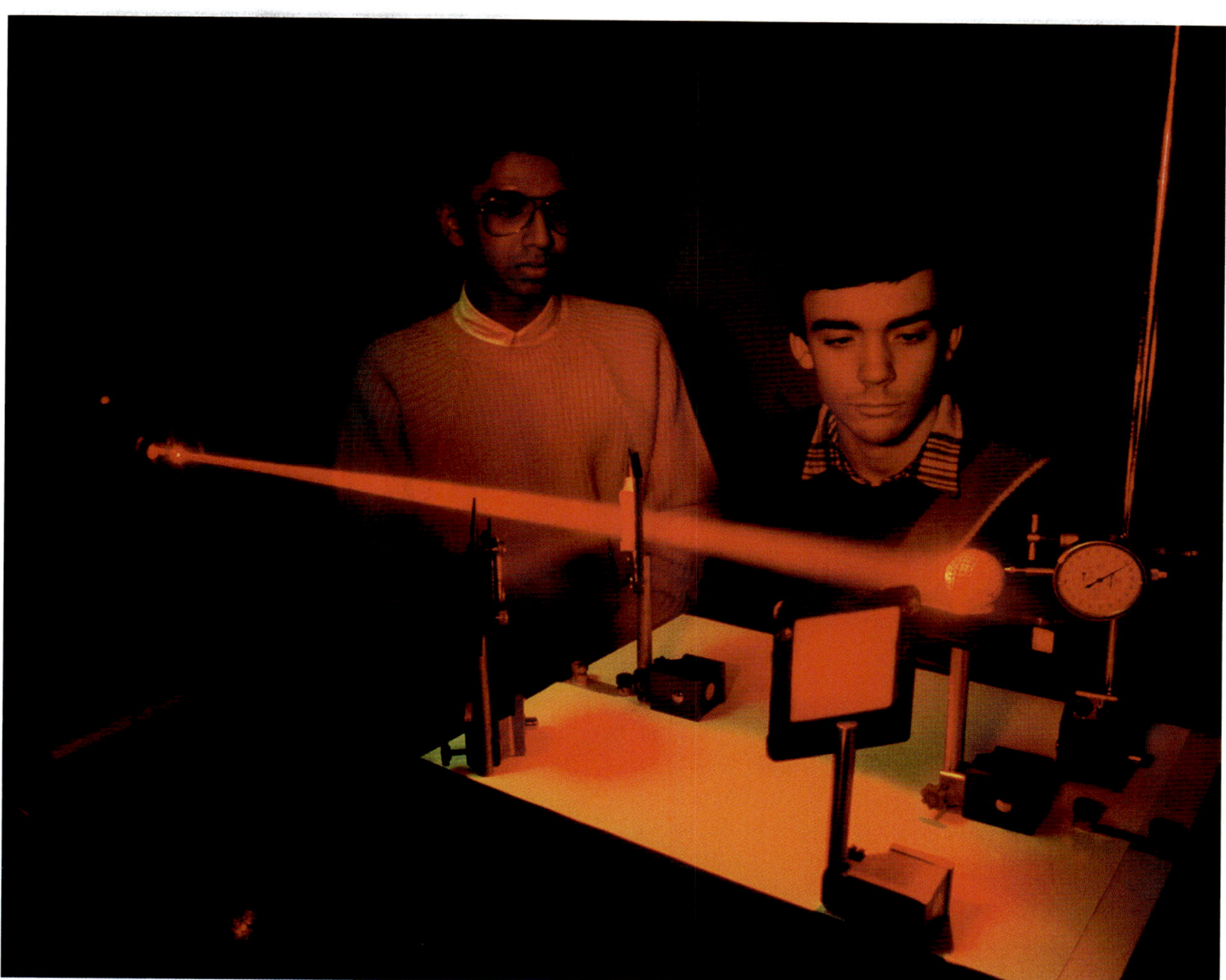

These undergraduate students are experimenting with the production of a holographic image of a golf ball from a hologram, in which the three-dimensional image of the golf ball is recorded. The light source, a laser, is on the left; the hologram is at the center of the photograph; and the image of the golf ball is at the right, just to the left of the dial. *(Department of Physics, Imperial College London/Photo Researchers, Inc.)*

two-dimensional record of three-dimensional information is called a hologram, or a holographic image. In particular, *holography* most commonly refers to the recording of three-dimensional scenes on photographic film in such a manner that the scenes can be viewed in their full three-dimensionality from the image on the film. The procedure for producing a hologram is to allow a beam of coherent light—light whose waves maintain a constant phase relation with each other—both to illuminate a scene and to fall directly on the film. The light reflected to the film from the scene and the light directly reaching the film interfere, and the resulting interference pattern is recorded on the film, thus creating a hologram. The holographic image contains information about both the amplitude, related to the intensity, and the phase of the light from the scene, thereby allowing a three-dimensional reconstruction of the scene. (In ordinary photography, on the other hand, it is only the intensity that is recorded in the image.) Reconstruction is achieved by shining light through or reflecting light from the hologram. The image thus generated can be viewed as if it were the original three-dimensional scene. As an observer moves about, different sides of the scene come into and disappear from view.

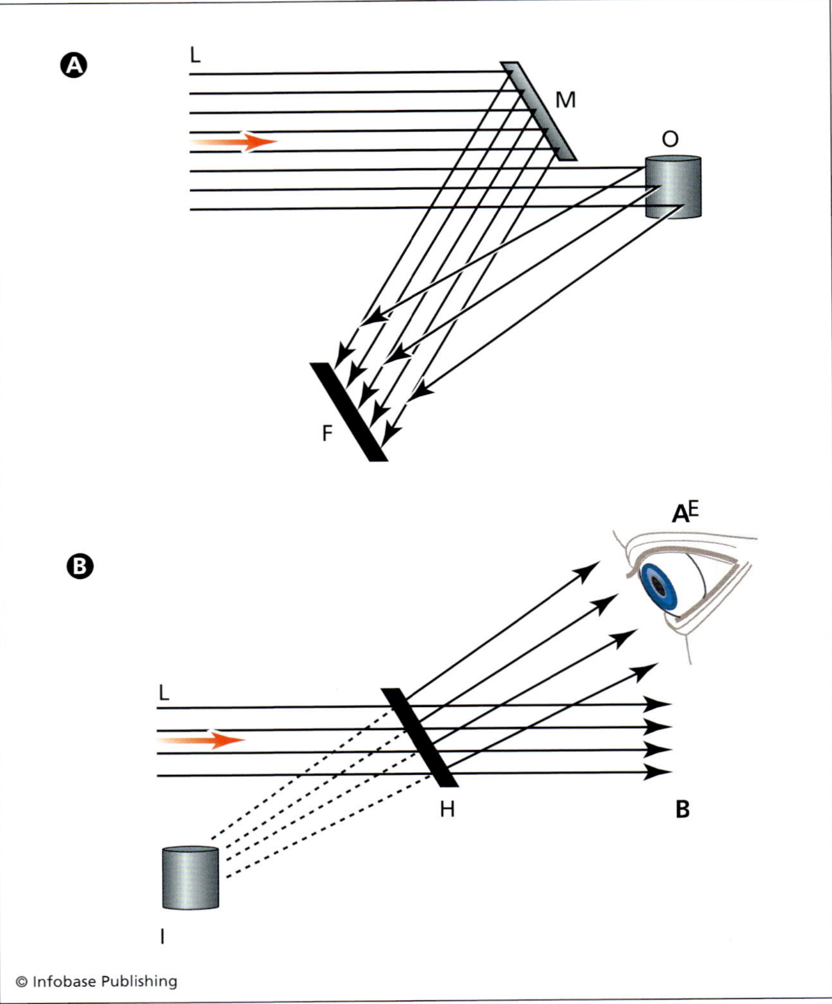

(a) A hologram is produced by having a coherent beam of light L reach photographic film F, both directly (with the help of a mirror M in this setup) and by reflection from an object O. (b) A holographic virtual image I of the object is viewed by shining light L through the hologram H and looking back through it (represented by an eye). As the observer moves around, different sides of the image disappear from and come into view.

Holography was invented by the 20th-century Hungarian-British physicist and electrical engineer Dennis Gabor. Gabor was awarded the 1971 Nobel Prize in physics "for his invention and development of the holographic method." Current uses of holography include art, security (such as on identity cards), and commercial purposes. Common examples of the latter are the holographic logos found on many credit cards and the holograms that Microsoft Corporation affixes to its products to ensure authenticity.

See also OPTICS; WAVES.

FURTHER READING

Biedermann, Klaus. *Lippmann's and Gabor's Revolutionary Approach to Imaging*. The Nobel Foundation. Available online. URL: http://nobelprize.org/nobel_prizes/physics/articles/biedermann/index.html. Accessed February 27, 2008.

The Nobel Foundation. "The Nobel Prize in Physics 1971." Available online. URL: http://nobelprize.org/physics/laureates/1971/. Accessed February 27, 2008.

Young, Hugh D., and Roger A. Freedman. *University Physics*, 12th ed. San Francisco: Addison Wesley, 2007.

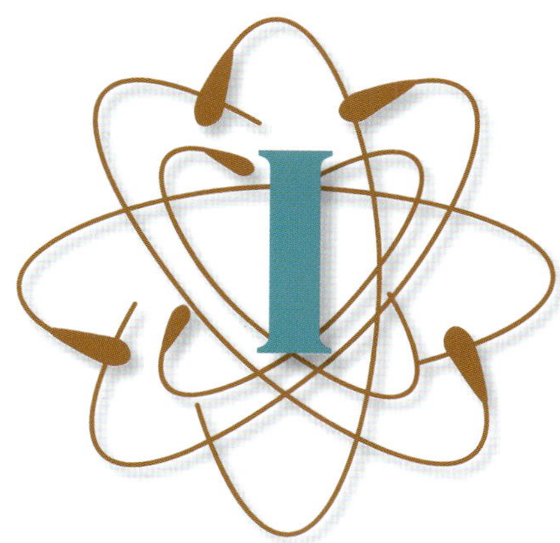

industrial chemistry Industrial chemistry is the application of chemical processes and understanding to all aspects of industrial processes. Industry is the labor or business that leads to the manufacturing of a product. All of the branches of chemistry, including organic, inorganic, physical, analytical, and biochemistry, can be applied to the field of industrial chemistry, which contributes to the manufacture of goods for the food, clothing, automobile, pharmaceutical, plastics, and environmental industries.

LIMESTONE-BASED PRODUCTS

As well as manufacturing final products, industrial chemists make many commercially important chemicals that are used by other industries in the creation of products. Several commercially important chemicals include those derived from limestone (calcium carbonate, $CaCO_3$). Limestone itself can be used for many industrial products including glass and concrete, as filler for roads, and in buildings. Pure calcium carbonate is white. Impurities add color and design options when limestone is used for building. Calcium carbonate is used as chalk and is also the primary component in egg shells.

Calcium carbonate can be chemically modified to produce other important compounds including calcium chloride and sodium carbonate. One of the first industrial chemical processes was the development of a method for producing sodium carbonate from calcium carbonate known as the Solvay process, named after two Belgian brothers, Ernest and Alfred Solvay. Their first factory was located in New York in 1861. Sodium carbonate (also called soda ash) is used for the production of glass, soaps, and paper. In the early 2000s, roughly 11 million tons of soda ash were produced per year in North America. The overall reaction (although the process requires many steps) is shown in the following:

$$CaCO_3 + 2NaCl \rightarrow Na_2CO_3 + CaCl_2$$

Calcium carbonate is readily available from limestone, and sodium chloride is available from brine solutions (seawater). This process for the production of sodium carbonate was the primary method used after its development in most parts of the world, including the United States. The discovery of natural sodium carbonate deposits in the Unites States has led to the closing of Solvay plants in North America.

As seen in the preceding reaction, the production of sodium carbonate generates a by-product called calcium chloride. One industrial use for calcium chloride is as the primary form of road deicer that reduces slippery surfaces on roadways in the winter. Calcium chloride can also speed the drying time of cements as it is hygroscopic (readily absorbs water) and is often used as a commercial drying agent.

SODIUM CHLORIDE–BASED PRODUCTS

Sodium chloride (NaCl), also known as table salt, is a water-soluble ionic compound that is readily abundant in halite (rock salt) deposits and seawater. Ocean water generally contains 2 to 3 percent sodium chloride. The isolation of sodium chloride is relatively straightforward. Mining can yield sodium chloride directly, or it can be removed from brine by evaporation. Uses of sodium chloride include as food flavoring, in the generation of sodium and chlorine for other industrial uses, and as a deicer (although it is not as effective a deicer as calcium chloride).

Sodium chloride–derived compounds are also commercially important. Such products as sodium hydroxide, chlorine, and hydrochloric acid are valuable chemicals that are used in many industries. Sodium hydroxide, an ionic compound with the formula NaOH, is the most widely used base in the world. Many cleaning products such as soap and drain cleaners contain this strong base, which is also used in paper synthesis. Sodium hydroxide can be produced from the electrolysis of sodium chloride according to the following steps:

1. NaCl in aqueous solution dissociates into Na^+ and Cl^-.
2. The Na^+ at the cathode leads to the reduction of water.
3. Hydrogen gas (H_2) and hydroxide ions are produced from the reduced water.
4. The hydroxide ions (OH^-) react with sodium ions (Na^+) to make sodium hydroxide (NaOH).

Chlorine is a by-product of the electrolysis of brine during sodium hydroxide production. One of the best known uses of chlorine is as a disinfectant. Added to water supplies and used in household cleaners such as bleach, chlorine disinfects surfaces and helps prevents waterborne illnesses. The creation of polymers such as polyvinyl chloride (PVC) and vinyl also depends on chlorine. PVC is a polymer that forms from addition reactions of vinyl monomers. PVC is one of the most commonly used plastics in the world, although there are concerns about the environmental and health risks of use of PVC. Rigid PVC, also known as vinyl, is used in siding for houses and household plumbing supplies. Plastic products also depend on chlorine production.

FERTILIZERS

Another example of chemical synthesis that leads to commercially valuable chemicals is the synthesis of ammonia from molecular nitrogen. Production of ammonia can occur in the presence of a metal catalyst by a process known as the Haber process. Here atmospheric nitrogen and hydrogen gas are adsorbed to the metal (i.e., accumulate on its surface) and dissociate into their atomic forms. These atoms then react to produce ammonia. This process is economically important because plants and animals require nitrogen to synthesize proteins and nucleic acids. Although the atmosphere is composed predominantly of nitrogen gas, plants cannot use this form of nitrogen. The fixation of nitrogen by bacteria in the soil results in the synthesis of nitrogen compounds that plants can use, but because the amount of available usable nitrogen present is often the limiting factor in plant growth, fertilizers containing synthetic nitrogen compounds are commonly used to replace nitrogen that is missing from the soil.

PLASTICS

Industrial chemistry does not just produce the chemicals themselves; it is also responsible for commercial products such as plastics and textiles. Examples of commercial polymers include such common substances as polyethylene, polyethylene terephthalate, polypropylene, polystyrene, polyvinyl chloride, polyurethane, and nylon. Each of these polymers has been developed by organic chemists and polymer scientists to fill the need of a commercial product.

Polyethylene formation results from a free radical reaction on the double bond of the ethylene molecule, causing the previously bonded electrons to form new bonds with new ethylene molecules. Billions of pounds of polyethylene are produced every year. Different types of polyethylene can be used in packaging and for plastic soda, water, and juice bottles. Polyethylene terephthalate (PET or PETE) is a polymer formed by condensation reactions. PET is utilized in the manufacture of polyester clothing. The most common and easily identifiable use of PET is in plastic soda and water bottles. Polypropylene, utilized in kitchenware, appliances, food packaging, and other containers, is formed by addition reactions and is made up of propylene units. Polyurethane is a polymer formed by condensation reactions and can be utilized in such applications as furniture upholstery foam, Spandex materials, and water-protective coatings for wood.

TEXTILES

The chemical industry contributes to the clothes people wear and the upholstery of the furniture people use. The creation of new and versatile fibers to supplement or replace natural components such as wool, silk, or cotton is the focus of the textile industry. Synthetic fibers such as rayon, polyester, acrylic, and acetates demonstrate another application of the chemical industry to everyday life. One of the fastest growing areas of textiles today is the production of intelligent textiles. These are synthetic textiles that are designed and engineered to fit a certain need or solve a particular problem. Some estimates put the intelligent textiles at as high as 38 percent of the textile market. Current research also aims to continue increasing the efficiency and environmental friendliness of the dyeing process as well as the chemical design of fabric-specific dyes that allow for long-lasting and penetrating color for all types of fabrics.

PHARMACEUTICALS

Another industrial application of chemistry is pharmaceutical drug development, the process by which new pharmaceutical medicines are designed or discovered. Chemists in the pharmaceutical industry perform experiments to determine the tolerability, safety, and efficacy of new drugs for a specified patient population. The chemical production of safe, pure, and cost-effective medications is critical to the health and longevity of the population.

FOSSIL FUELS

Fossil fuels are energy sources formed from the decomposition of organic compounds over a long period. Fossil fuels include such energy sources such as oil, natural gas, and coal. According to the U.S. Department of Energy, 85 percent of the U.S. energy supply presently results from the combustion of fossil fuels that people burn to heat homes and businesses, produce electricity, and power transportation, including automobiles and airplanes. Imagining a world without fossil fuels is difficult, although finding alternatives is necessary as fossil fuels are nonrenewable resources that are rapidly becoming depleted. Chemists are busily working on optimizing the extraction procedures of petroleum products, developing new fuel sources such as biodiesel that can begin to replace fossil fuels, and reducing societal dependence on the highly polluting fossil fuels.

ENVIRONMENTAL APPLICATIONS

Industries of every type face a new obstacle in addition to the one of simply manufacturing their product. That obstacle is the method of making their product in a manner that does not negatively impact the environment so that the product itself does not negatively impact the environment. The merging of economic and industrial interests with the protection of the environment is a daunting task. Most industrial processes create waste that must be handled, and many products have a negative impact on the environment. Two primary environmental impacts are acid rain and the greenhouse effect (which leads to global warming). Acid rain occurs when precipitation falls in a form such as rain, snow, or sleet that is more acidic than normal rainfall. Natural rainwater registers a pH in the acidic range. Carbon dioxide (CO_2) naturally released into the atmosphere reacts with water to create carbonic acid, leading to the lower than neutral pH of normal rainfall at approximately 5.0–5.6. Sulfur dioxide (SO_2) and nitrogen oxides (NO_x) contribute to the characteristically low pH of acid rain. Sulfur and oxygen react according to the following equation:

$$S + O_2 \rightarrow SO_2$$

Several nitrogen oxides, including nitrogen monoxide (NO) and nitrogen dioxide (NO_2), form according to the following equation:

$$N_2 + O_2 \rightarrow NO_x$$

When SO_2 and NO_2 react with the water present in the atmosphere, they ultimately form sulfuric acid (H_2SO_4) or nitric acid (HNO_3). These acids fall to earth in the form of acid precipitation that can drastically affect plants, animals, and buildings.

Many natural gases in the atmosphere absorb energy (in the form of infrared radiation) that has been reradiated from the surface of the Earth into the atmosphere. This process is known as the greenhouse effect. Many industrial processes today produce gases that prevent the energy from escaping the atmosphere. When the energy is absorbed, it is not released back into space and the average global temperature increases. The greenhouse gases include carbon dioxide (CO_2), nitrous oxide (N_2O), methane (CH_4), perfluorocarbons (PFCs), hydrofluorocarbons (HFCs), and sulfur hexafluoride (SF_6).

The focus of industrial chemistry in the 21st century is shifting from simply making a useful product to manufacturing a useful product without damaging the environment in the process.

See also ACID RAIN; ALTERNATIVE ENERGY SOURCES; FOSSIL FUELS; GREENHOUSE EFFECT; INORGANIC CHEMISTRY; IONIC COMPOUNDS; ORGANIC CHEMISTRY; PHARMACEUTICAL DRUG DEVELOPMENT; POLYMERS; TEXTILE CHEMISTRY.

FURTHER READING

Heaton, Alan. *An Introduction to Industrial Chemistry.* London: Chapman and Hall, 1996.

inorganic chemistry Inorganic chemistry is the branch of chemistry involving compounds that do not contain carbon. This type of chemistry is the general chemistry that is taught in high schools and colleges as it lays the foundation for all other types of chemistry. Inorganic chemistry affects every aspect of daily life. Medications, metallurgy, ceramics, cosmetics, geology, semiconductors, computers, environmental pollutants, agricultural products, and cleaning chemicals all involve inorganic chemistry. Biological applications of inorganic chemistry include processes such as photosynthesis, respiration, and oxygen transport by hemoglobin. Inorganic chemists study many different subjects, such as atomic structure, periodic properties, reaction types, bonding patterns, crystal structure, solution chemistry, acid-base chemistry, and reaction rates.

ATOMIC STRUCTURE

The basis of inorganic chemistry is atomic structure. The modern model of the atom was developed as a result of chemical discoveries that spanned several centuries. The number of protons within a nucleus determines the identity of the atom. The nucleus is composed of protons and neutrons that together contribute to the mass of an atom. In a neutral atom, the number of positively charged protons equals the number of negatively charged electrons. An electron is approximately 1/1,840 the size of a proton, so the mass of an electron is not considered when calculating the mass of an atom.

The location of the electrons within an atom determines the chemical and physical properties of that element. Electrons are localized within energy levels, sublevels, and orbitals that describe their distance from the nucleus, their likely position, and the associated energy of that position. Ground state electrons are in the lowest possible energy state, closest to the nucleus, and the most stable. When electrons are energized, they can be excited into higher energy levels. As these electrons fall back down to their ground state, they release energy in the form of electromagnetic radiation of a certain wavelength. Because specific wavelengths of visible light are associated with certain colors, inorganic chemists can identify elements on the basis of the characteristic color of light emitted during heating. Spectral analysis of atomic emissions is an important tool for inorganic chemists.

Several rules govern the distribution of electrons within an atom; these include the Pauli exclusion principle, the Aufbau principle, Hund's rule, and the octet rule. The Pauli exclusion principle, named after the Austrian physicist Wolfgang Pauli, states that no two electrons within an atom can exist at the exact same location. If two electrons are in the same energy level, the same sublevel, and the same orbital, they must be spinning in opposite directions. Consider a mailing address; no two houses can have exactly the same mailing address: If two residences are located in the same building, they must have different numbers

SILICONE: A VERSATILE POLYMER

by Lisa Quinn Gothard

Silicon is the second most abundant element on Earth; it has an atomic number of 14 and an atomic mass of 28. The elemental form of silicon is a brittle, shiny, dark gray solid. This group IV element makes up more than one-quarter of the Earth's crust and has even been found on the Moon. Silicon is a versatile element and generally exists in compounds, although the elemental form can be found in computer chips everywhere. The electron configuration of silicon has four valence electrons; thus it is similar to carbon in that respect. Some of the most common compounds that involve silicon are those formed with oxygen to create silicon oxide compounds. Silicon dioxide, for example, is the main component of sand and quartz.

A useful substance formed from silicon, known as silicone, forms when silicon bonds with two oxygen atoms and two methyl groups to form a polymer. Called silicone, this durable polymer resists damage due to high temperature, sunlight, moisture, and age. Silicone has applications in many fields, exists in various structures, and exhibits many functions based on the groups attached to the silicone backbone as well as the interactions between the backbones of other silicone polymers. The most common form of silicone is polydimethylsiloxane (PDMS). This type of silicone forms the basis of many commercially available products including elastomers, resins, and emulsions. Functions of silicones include the following:

- as an adhesive
- to provide stability
- as insulation
- as a lubricant
- to provide surface protection
- in the manufacture of cosmetics

Silicone adhesives are extremely durable and resistant to water, pressure, acids, and bases. They are able to form tight bonds between compounds such as glass and metal. Silicone adhesives are used for bonding glass windshields in vehicles, airplanes, buildings, and other structures that must resist a large amount of force from the wind. The silicone-based adhesive used to hold this glass in place creates a stronger bond than has been found with any other product. Silicone caulks and adhesives have changed the way buildings are designed as a result of their superior strength and durability under stressful conditions. The silicone-based release factors found on the back side of tape, the paper holding stickers, and other peel-off products are responsible for the easy removal of the paper while maintaining the sticky side of the product.

Thermal stability is a quality of silicones that allows them to withstand temperature fluctuations. This property makes silicones a good base for many fluids that must maintain their integrity under high temperatures.

Silicone-based products are also utilized for their insulative properties, as

or a letter following the address to indicate the specific residence. On the basis of this rule, each electron resides in its own space.

One task of inorganic chemistry is to determine the positions of these electrons. The Aufbau principle, named for the German term meaning "building up," states that electron orbitals are filled from the lowest to the highest energy states. Those orbitals that are closest to the nucleus have the lowest potential energy and are the most stable. Electrons fill these orbitals first. Orbitals of higher energy are filled only after those with lower energy contain electrons. To illustrate this principle, consider the following similar situation. Fans at a free concert fill the first rows of the concert venue first before the later rows are filled. One would not pass up a front-row seat for a fourth-row seat if it were available. The same principle applies to electrons filling energy levels.

Electrons being added to equivalent energy orbitals (for example, the three equivalent *p* orbitals) will add singly, one each into each orbital, before doubling up. This phenomenon is known as Hund's rule. When two equivalent bus seats are open, unrelated people fill the seats individually before they pair up.

The final rule, the octet rule, determines the reactivity and many of the chemical properties of atoms. The octet rule states no more than eight electrons (valence electrons) can be in the outer energy level of an atom. This means that atoms react the way they do in order to obtain a complete outer energy level with a total of eight electrons. These rules—Pauli exclusion principle, Aufbau, Hund's, and octet—describe why the electrons fill the sublevels the way they do.

PERIODIC TABLE OF THE ELEMENTS AND PERIODIC TRENDS

Organization of the elements into the periodic table is based on increasing atomic number, and the electron configuration of an element can be determined from

they are nonpolar molecules. The inability of silicone to conduct an electric current makes it a perfect candidate for the insulation around electrical wires, transformers, and generators.

Silicone also has a big impact on fields that require a successful lubricant that will hold up under high temperatures. For example, moving parts in engines require the grease of silicone-based products.

Silicones even offer protection against water damage. The water-repelling, nonpolar nature of silicone compounds in a highly cross-linked polymer create fantastic treatments for protection against exposure to water. Materials that depend on this include shingles, concrete, and ceramic tiles.

Along with the construction industry, one of the most productive applications involving silicones is in the cosmetic industry. Silicone-containing substances are found in lotions, creams, conditioners, and all types of makeup. The silicone component helps the product maintain its creaminess and keeps the skin soft and supple by acting as a waterproofing agent to prevent the skin from drying out.

Silicone-based products are incredibly useful and have thousands of applications to common substances and everyday life. However, the impact of some silicon products on the environment and on human health is not completely clear. One example involves the use of silicone breast implants. These types of implants are used for breast augmentation or reconstruction. They are made of a silicone elastomer shell and filled with silicone. When the implants fail, generally after approximately 10 years, or after a sharp blow, they can leak silicone into the surrounding scar tissue. This can create additional scar tissue that may cause pain, discomfort, or disfigurement. The health risks of ruptured silicone breast implants are still being debated. In 1992 class action lawsuits were filed on behalf of thousands of women who claimed that ruptured silicone implants had seriously affected their health. The U.S. Food and Drug Administration (FDA) banned the use of silicone-based implants after that. Years later, it still has not been definitively shown that silicone leakage from a ruptured breast implant causes any long-term consequences, but the research continues. Care must be taken when silicone-based products are used for applications involving long-term exposure to human tissues. Many breast implants now utilize saline solution rather than silicone to prevent any potential damage to the patient in the case of a rupture, but in 2006 the FDA again allowed the production and use of silicone-based implants.

Silicone is an incredibly versatile polymer that has improved the building and technological industries; it even makes our bandages stay on longer. The silicone industry is continuing to develop and find new and better ways to impact human life.

FURTHER READING

Center for Devices and Radiological Health, U.S. Food and Drug Administration. Breast Implants home page. Available online. URL: http://www.fda.gov/cdrh/breastimplants/. Last updated November 17, 2006.

Dow Corning. *The Basics of Silicone Chemistry.* Available online. URL: http://www.dowcorning.com/content/sitech/sitech-basics/silicones.asp. Accessed May 1, 2008.

the location of the element on the table. Inorganic chemists study the physical properties of the elements, which directly result from their atomic structure. Understanding physical and chemical properties that are related to the atomic structure and periodic table location of an element reveals much information about the element. Such periodic trend information includes shielding, electronegativity, electron affinity, ionization energy, atomic radius, and ionic radius.

Shielding occurs as more energy levels are added in larger atoms. The attraction of the nucleus for the electrons in the atomic orbitals decreases as the distance between the electrons and the nucleus increases. The electrons in the lower energy levels are said to shield the outer level electrons from the nucleus. This shielding causes less of an attraction for outer level electrons. Thus, as one travels down a group of the periodic table, the property of electron affinity, defined as the attraction for electrons, decreases. The related property of electronegativity, the desire of an atom to have more electrons, also decreases as one moves down a group.

The atomic radius of an element is measured as half the distance between the nuclei of two covalently bonded atoms of that element or two atoms in contact. The ionic radius is half the distance across an ion of that element. These distances increase as one travels down a group of the periodic table, as having more energy levels increases the size of the element. Surprisingly, the atomic radius and ionic radius decrease across a period of the periodic table. Elements in the same period have the same total number of energy levels, which are located at the same distance from the nucleus, but as the atomic number increases, the number of protons (in the nucleus) that attract the electrons is higher, creating a stronger attraction of the nucleus for the electrons. This explains the strong attraction of elements such as fluorine for their own electrons, a characteristic known as high ionization energy. The attraction for additional electrons is known as electronegativity, and fluorine is the most electronegative atom, a property that explains why it is extremely chemically reactive.

INORGANIC REACTION TYPES

There are four general types of reactions in inorganic chemistry: synthesis, decomposition, single-replacement, and double-replacement. Synthesis reactions, also called combination reactions, combine two or more separate reactants to form one product. Decomposition reactions involve the breakage of a single reactant into two or more products, one of which is usually a gas, and generally require the addition of heat. Products of decomposition reactions differ depending on the type of reactant involved. Single-replacement reactions involve the exchange of one positive ion, or cation, for another in an ionic compound. Double-replacement reactions occur when two ionic compounds switch their cations or anions.

BOND TYPES AND BONDING THEORIES

Two main types of chemical bonds exist in inorganic chemicals—ionic and covalent. Ionic bonds involve the removal of an electron from a less electronegative atom by a more electronegative atom. Once this electron is removed, the atom that lost the electron has a positive charge and is called a cation, and the atom that stole the electron has a negative charge and is called an anion. The electrostatic attraction between the cation and anion holds them together in an ionic bond.

A covalent bond forms when electrons are shared between two atoms that have similar electronegativities. Neither atom of the bond has complete possession of the electron, and no electrostatic interaction occurs. Covalent bonds that occur between atoms of slightly different electronegativities do not have equal sharing of bonded electrons, creating a charge separation and a polar covalent bond.

Several theories exist to describe the bonding in inorganic chemistry, including valence bond theory, valence shell electron repulsion theory, and molecular orbital theory. Valence bond theory is based on the overlap of electron orbitals, the allowed energy states of electrons. Chemists developed this bonding model to explain the differences in bond length and bond energy. The overlap of orbitals permits the sharing of electrons between the two atoms in the region of space that is shared between the two orbitals. Valence bond theory explains how bond length and bond energy can be different for similar molecules. The valence shell electron pair repulsion theory (VSEPR) predicts bonding in atoms on the basis of repulsion. One can accurately predict the shape of a molecule by applying this theory. The molecular orbital theory is a more sophisticated model of bonding that takes the three-dimensional shape of a molecule into account. The main principle of this theory involves the formation of molecular orbitals from atomic orbitals, similar to the valence bond theory overlap of atomic orbitals.

CRYSTAL STRUCTURE

Inorganic chemists often study the three-dimensional shape of compounds. Techniques such as nuclear magnetic resonance (NMR) and X-ray crystallography help chemists determine the shape of the compound. Molecular modeling programs can be utilized to determine the arrangements of the atoms in the molecule.

Inorganic chemists study the three-dimensional crystal shape of solids. Chemists classify crystals by their chemical and physical properties as well as their crystal structure. Seven categories of crystal classification are based on structure: cubic, hexagonal, orthorhombic, trigonal, monoclinic, triclinic, and tetragonal. Crystal structure influences the physical and chemical properties of a compound.

Different crystal forms of a compound can possess a property known as optical activity. White light travels in all directions, but plane-polarized light has passed through a polarizing filter and only the light that vibrates in the same direction passes through the polarizing filter. The filter blocks the passage of light traveling in other directions. Different crystal forms of a compound demonstrate optical activity by rotating plane-polarized light differently. Distinguishing among the different forms and the chemical reactivities of each form of a compound is the job of an inorganic chemist.

SOLUTION CHEMISTRY

Inorganic chemists often study reactions and compounds that are in a solution, a homogeneous mixture between two or more types of substances. The solvent is the substance that does the dissolving, and the solute is the substance that is dissolved. If the solvent used is water, the solution is referred to as an aqueous solution, and these types of solutions play an important role in biological and chemical systems. Chemical reactions in solution are dictated by the properties of the reactants as well as the properties of the solvent. When the solvent is a polar molecule, it dissolves polar solutes, such as water and salt. Nonpolar solvents dissolve nonpolar solutes, such as gasoline and oil.

ACID-BASE CHEMISTRY

The simplest definition of an acid is a substance that donates a proton (H^+) and of a base is a substance that accepts an H^+. One can calculate the concentration of protons in a solution using the pH formula:

$$pH = -\log[H^+]$$

The pH of the solution in which a chemical reaction occurs dramatically affects the outcome. Measurement of the pH is carried out by organic chemists using pH paper, chemical indicators, and pH probes that electronically calculate the pH of a solution. Acidic solutions have a pH below 7, neutral solutions have a pH of 7, and basic solutions have a pH above 7. Bases are found in many cleaning supplies, soaps, and compounds such as bleach and ammonia. Many types of food, such as lemons, oranges, vinegar, coffee, and soft drinks, contain acids.

REACTION RATES

The speed at which a chemical reaction takes place is known as the reaction rate. Many chemical reactions take place on the order of pico- (10^{-12}) or femto- (10^{-15}) seconds while others take place on the order of thousands of years. Inorganic chemists who specialize in the field called kinetics study the speed at which chemical reactions occur. Measuring the rate of disappearance of reactants or the rate of appearance of products allows the determination of the speed of a reaction. Chemical catalysts are often used to increase the rate of a chemical reaction by decreasing the amount of energy required to start the reaction, called the activation energy.

EDUCATION AND TRAINING

Education and training in the field of inorganic chemistry require comprehensive study of each of the topics already mentioned: atomic structure, periodic properties, reaction types, bonding patterns, crystal structure, solution chemistry, acid-base chemistry, and reaction rates. Laboratory training is critical in the field of inorganic chemistry, and inorganic chemistry education involves extensive study and training in laboratory techniques. Degrees in inorganic chemistry range from four-year bachelor of science degrees to research-based master of science degrees and doctoral degrees. In general, laboratory technician positions require a bachelor's or master's degree. Supervisory roles require either master's degrees or doctoral (Ph.D.) degrees. Full-time academic positions generally require a Ph.D., and many require additional postdoctoral work in the field.

Careers in chemistry are diverse and involve every aspect of life. Chemists work in research and development of numerous types of production processes. The production of goods requires chemical research to create the product.

See also ACIDS AND BASES; ATOMIC STRUCTURE; BONDING THEORIES; CHEMICAL REACTIONS; COVALENT COMPOUNDS; ELECTRON CONFIGURATIONS; IONIC COMPOUNDS; ORGANIC CHEMISTRY; PAULI, WOLFGANG; PERIODIC TABLE OF THE ELEMENTS; PERIODIC TRENDS; PH/POH.

FURTHER READING

Zumdahl, Stephen S., and Donald J. DeCoste. *Basic Chemistry,* 6th ed. Geneva, Ill.: Houghton Mifflin, 2008.

inorganic nomenclature Chemical nomenclature is the systematic naming of chemical compounds. The goals of the chemical naming system include making the name recognizable to people of all languages and levels of expertise and giving information about the composition and ratio of atoms in the compound. More recent developments in nomenclature include three-dimensional information about the compound. Alchemy-based systems historically named the compounds according to their physical properties or place of origin. This system led to several different names for a single compound, depending on the situation and location, and caused much confusion. The French chemist Guyton de Morveau first articulated the need for a systematic procedure for naming chemical compounds in 1782. Several scientists expanded on his system at the end of the 1700s and in the early 1800s including the French chemists Antoine-Laurent Lavoisier and Claude Louis Berthollet. In 1813, Jons Berzelius, a Swedish chemist, standardized the writing of formulas by using one- and two-letter symbols to represent the element's name, generally derived from the Latin form. In 1892, a meeting of chemical societies in Geneva developed a system for naming organic compounds. The Council of the International Association of Chemical Societies formed in 1913 to work on the systematic procedure for naming organic and inorganic compounds but was interrupted by World War I. Their work continued in 1921 after the establishment of the International Union of Pure and Applied Chemistry (IUPAC) in 1919. IUPAC continued the commissions on inorganic chemistry, organic chemistry, and biochemistry, publishing the first naming system in 1940. The current IUPAC naming systems for organic and inorganic compounds can be found in a series of IUPAC books commonly known as the "Red Book," which covers inorganic nomenclature, and the "Blue Book," which describes organic nomenclature.

NAMING BINARY IONIC COMPOUNDS

Ionic compounds that contain only two elements (binary compounds) are the simplest compounds to name. Binary ionic compounds form by the electrostatic interaction between a metal and a nonmetal. The nonmetal steals an electron from the less electronegative metal, resulting in a positively charged metal and a negatively charged nonmetal. The value of the charge is equivalent to the number of electrons lost or gained. The positively charged metal is known as a cation, and the negatively charged nonmetal is known as an anion. The name of the metal precedes the name of the nonmetal ending in *-ide.* Consider the formation of a binary ionic compound containing magnesium (a metal) and fluorine (a nonmetal), during which two fluorine atoms steal electrons from magnesium. The magnesium takes on a +2 charge, and each fluorine atom assumes a -1 charge. The interaction of the negative and positive charges causes the formation of the ionic compound.

$$Mg^{2+} + 2F^- \rightarrow MgF_2$$

The name of this compound is magnesium fluoride. If the metal were lithium and the nonmetal chlorine, the name of the compound would be lithium chloride.

$$LI^+ + Cl^- \rightarrow LiCl$$

If the metal involved has more than one potential oxidation number, then it is necessary to indicate which oxidation state of the metal is reacting to form the compound. For instance, iron has two oxidation states, +2 and +3. If iron were to react with chlorine to form an ionic compound known as iron chloride, it would not be clear whether it was iron with an oxidation state of +2 or iron with an oxidation state of +3.

$$Fe^{2+} + 2Cl^- \rightarrow FeCl_2$$

$$Fe^{3+} + 3Cl^- \rightarrow FeCl_3$$

The use of roman numerals solves this problem. When a metal has more than one possible oxidation state, a roman numeral set in parentheses after the name of the metal indicates its oxidation state in the compound. The two iron compounds formed in the preceding examples are known as iron (II) chloride and iron (III) chloride, respectively. These two compounds have different chemical formulas and different chemical reactivity. Most of the transition metals, tin, and lead have more than one oxidation state and require the use of the roman numeral system.

NAMING TERNARY IONIC COMPOUNDS

Naming ternary compounds, ionic compounds that contain more than two elements, is similar to naming binary ionic compounds. The name of the cation is first, followed by the name of the anion. In the case of ternary compounds, the anion is not a single element but a covalently bound group of atoms that make up a polyatomic ion. The formulas for many common polyatomic ions are shown in the following table.

When writing the names of ternary ionic compounds, the name of the metal appears unchanged and precedes the name of the polyatomic ion. The combination of potassium and carbonate yields the compound potassium carbonate.

$$2K^+ + CO_3^{2-} \rightarrow K_2CO_3$$

The use of roman numerals differentiates between metals that have more than one oxidation state, just as in binary ionic compounds. The reaction between

POLYATOMIC IONS

Name	Formula	Name	Formula
Ammonium	NH_4^+	Hydroxide	OH^-
Acetate	$C_2H_3O_2^-$	Peroxide	O_2^{2-}
Carbonate	CO_3^{2-}	Silicate	SiO_3^{2-}
Chlorate	ClO_3^-	Permanganate	MnO_4^-
Chlorite	ClO_2^-	Chromate	CrO_4^{2-}
Perchlorate	ClO_4^-	Dichromate	$Cr_2O_7^{2-}$
Hypochlorite	ClO^-	Tartrate	$C_4H_4O_6^{2-}$
Nitrate	NO_3^-	Oxalate	$C_2O_4^{2-}$
Nitrite	NO_2^-	Tetraborate	$B_4O_7^{2-}$
Sulfate	SO_4^{2-}	Iodate	IO_3^-
Sulfite	SO_3^{2-}	Cyanide	CN^-
Phosphate	PO_4^{3-}	Thiosulfate	$S_2O_3^{2-}$

tin (Sn^{2+}) and phosphate yields the compound tin (II) phosphate.

$$Sn^{2+} + PO_4^{3-} \rightarrow Sn_3(PO_4)_2$$

WRITING FORMULAS FOR IONIC COMPOUNDS

In order to use chemical nomenclature, one must also be able to write the formula when given the name of a compound. In the case of ionic compounds, the order of the anion and cation helps with identifying the elements involved in the compound. If the ionic compound contains only two element names and the second name ends in *-ide,* then the compound is a binary ionic compound. The symbols for the metal and nonmetal are written just as they appear on the periodic table of the elements. The next question is how many of each element react to form the compound. The relative number of each element depends on the oxidation number of each of the elements involved. The final overall charge of the compound must equal zero because the ionic compounds themselves are neutral. This means that the amount of positive charge contributed by the metals must balance the amount of negative charge contributed by the nonmetals. Consider the following reaction of potassium (K) with oxygen (O) to form potassium oxide:

$$K^+ + O^{2-} \rightarrow K_2O$$

Since potassium has an oxidation number of +1 and oxygen has an oxidation number of -2, the final compound must contain two potassium atoms in order to balance the charge of the oxygen ion and create a neutral compound.

The crisscross method is often useful when writing the formula of ionic compounds. The value (without the charge) of the oxidation number of the cation becomes the subscript of the anion, and vice versa. Thus, knowing the oxidation states allows for quick formula writing. One can determine the oxidation state for most elements by their position on the periodic table of the elements. All group I elements have an oxidation number of +1, group IIA elements have an oxidation number of +2, Group IIIA is +3, Group VA is -3, Group VIA is -2, Group VIIA is -1, and the halogens have an oxidation number of 0 because they do not form ions.

The formulas for ternary ionic compounds can be written as binary ionic compounds are except the atoms of the polyatomic ions are treated as an individual unit and their formulas must be memorized or looked up. The oxidation state of the anion is included in the formula of the polyatomic ion, and the crisscross, or balancing, charges method can be used to write the formula of the compound. When writing the formula for calcium sulfate, the +2 of calcium (because it is in group IIA) balances out the -2 of sulfate.

$$Ca^{2+} + SO_4^{2-} \rightarrow CaSO_4$$

If more than one polyatomic ion is necessary to balance the charges, then parentheses must enclose the formula for the polyatomic ion before adding the subscript. The parentheses indicate that the subscript applies to all of the atoms within the polyatomic ion. The formula for calcium phosphate requires two phosphate ions to balance out the three calcium ions, so the formula requires parentheses around the phosphate group to represent the number of atoms in the compound correctly.

$$Ca^{2+} + PO_4^{3-} \rightarrow Ca_3(PO_4)_2$$

NAMING MOLECULAR COMPOUNDS

Molecular compounds (covalent compounds) do not involve ions in their formation. The two nonmetals of a molecular compound are covalently bonded by sharing electrons. The easiest way to determine whether a compound is ionic or molecular when given a name or a formula is to look for a metal. If a metal and a nonmetal are present, the compound is generally ionic, and if a metal is not present, the compound is molecular. There are no charges to balance when writing molecular formulas and no roman numerals when writing their names, since nonmetals do not have multiple oxidation states.

In names of molecular compounds, prefixes represent the quantity of each atom type present in that compound. The following table lists some common prefixes.

The names of molecular compounds consist of two words with the second element ending in *-ide*. The formula for the molecular compound carbon dioxide is CO_2. The prefix *di-* indicates two oxygen atoms are present. The prefix *mono-* is not used when there is only one atom of the first element. For instance, CO_2 is not monocarbon dioxide. The prefix *mono-* is used when there is only one atom of the second element in the compound, as in carbon monoxide, CO.

MOLECULAR PREFIXES

Prefix	Number of Atoms	Prefix	Number of Atoms
mono-	1	hexa-	6
di-	2	hepta-	7
tri-	3	octa-	8
tetra-	4	nona-	9
penta-	5	deca-	10

NAMING ACIDS

Simply defined, an acid is an ionic compound that has a hydrogen ion as its cation or an ionic compound that donates a proton. Inorganic acids are not named as other ionic compounds are. Binary acids are named by adding the prefix *hydro-* to the name of the anion and ending the anion with *-ic acid*. The name *HF* represents hydrofluoric acid, and the name *HCl* represents hydrochloric acid.

If the acid contains a polyatomic ion, then the name of the acid depends on the ending of the polyatomic ion. For example, an acid containing a polyatomic ion that ends with *-ite* ends in *-ous acid*. An acid containing a polyatomic ion that ends with *-ate* will end in *-ic acid*. For example, when nitrite is the anion, as in HNO_2, the name of the acid is *nitrous acid*, and when nitrate is the anion, as in HNO_3, the acid is *nitric acid*. The following table summarizes the rules for naming inorganic acids.

See also ACIDS AND BASES; COVALENT COMPOUNDS; INORGANIC CHEMISTRY; IONIC COMPOUNDS; LAVOISIER, ANTOINE-LAURENT; ORGANIC CHEMISTRY; ORGANIC NOMENCLATURE.

FURTHER READING

Connelly, N. G., R. M. Hartshorn, T. Damhus, and A. T. Hutton, eds. *Nomenclature of Inorganic Chemistry:*

NAMING INORGANIC ACIDS

Ending of Anion	Name of Acid	Examples
-ide (e.g., chloride, fluoride)	hydro_____ic acid	HCl—hydrochloric acid, HF—hydrofluoric acid
-ite (e.g., nitrite, sulfite)	_____ous acid	HNO_2—nitrous acid H_2SO_3—sulfurous acid
-ate (e.g., nitrate, sulfate)	_____ic acid	HNO_3—nitric acid H_2SO_4—sulfuric acid

Recommendations 2005. London: Royal Society of Chemistry, 2005.
Koppenol, W. H. "Naming of New Elements, IUPAC Recommendations 2002." *Pure and Applied Chemistry* 74 (2002): 787–791.

invariance principles A statement that the laws of physics, or laws of nature, are invariant under a certain change is an invariance principle. In other words, if a certain change were imposed, the laws of physics would nevertheless remain the same. An invariance principle is a symmetry of nature and might be associated with a conservation law. Some of the best-known invariance principles are as follows:

- *Spatial-displacement invariance*. The laws of physics are the same at all locations. The same experiment performed here or there gives the same result.
- *Temporal-displacement invariance*. The laws of physics are the same at all times. The same experiment performed now or then gives the same result.
- *Rotation invariance*. The laws of physics are the same in all directions. The same experiment aimed one way or another gives the same result.
- *Boost invariance*. A boost is a change of velocity. The laws of physics are the same at all velocities. The same experiment performed at this or that velocity gives the same result.

Those invariance principles all deal with changes that involve space and time. They can also be expressed in the following way. No experiment carried out in an isolated laboratory can detect the lab's location or orientation in space, the reading of any clock outside the lab, the speed of the lab's straight-line motion, or the direction of that motion.

Two more invariance principles are obeyed by nature as long as the weak force is not involved, since it violates them. They are as follows:

- *Reflection invariance*. The laws of physics, except for the weak force, do not change under mirror reflection. Every experiment that does not involve the weak force gives the same result as its mirror-image experiment, if that result is reflected as well. The operation of reflection (or more precisely, the reversal of all three spatial directions) is conventionally denoted P (for parity).
- *Particle-antiparticle conjugation invariance. This is also called charge conjugation invariance*. The laws of physics, except for the weak force, are the same for elementary particles as for their respective antiparticles. Every experiment that does not involve the weak force gives the same result when all particles are replaced by their respective antiparticles, if in that result all particles are replaced by their respective antiparticles. The change involved in this invariance principle is denoted C (for charge).

The weak interaction *is* invariant under the combined operation of reflection and particle-antiparticle conjugation, denoted CP, even if not under each one separately. This invariance principle is nevertheless violated by a little-understood effect involving the elementary particle of type neutral kaon. An additional invariance principle, this one related to reversal of temporal ordering, is similarly valid for all of physics except the neutral kaon. The two invariance principles are as follows:

- *CP invariance*. The laws of physics are the same for elementary particles, except neutral kaons, as for their respective reflected antiparticles. Every experiment that does not involve neutral kaons gives the same result as its mirror-image experiment with all particles replaced by their respective antiparticles, if the result is also reflected and all particles in it are replaced by their respective antiparticles.
- *Time reversal invariance*. The name of this invariance principle is somewhat misleading, since time cannot be reversed. What is reversed is the temporal order of events making up a process. One compares any process with the process running in reverse order, from the final state to the initial state. Such pairs of processes are the time reversal images of each other. The invariance principle is that, except for processes involving neutral kaons, the time reversal image of any process that is allowed by nature is also a process allowed by nature. Except for experiments involving neutral kaons, if the result of any experiment is turned around and used as the input to an experiment, then the result of the latter experiment, when turned around, will be the same as the input to the original experiment. Time reversal is denoted by T.

In spite of the fact that P, C, CP, and T invariances are all violated in one way or another, there are compelling theoretical reasons to believe that the triple combined operation of CPT—particle-antiparticle conjugation, reflection, and time reversal, together—is one that all laws of physics are invariant under. No experimental evidence has as yet contradicted this invariance, which can be formulated as follows:

- *CPT invariance*. The laws of physics are the same for all elementary particle processes as for their respective particle-antiparticle conjugated, reflected, and time reversed image processes. For every experiment, if the particles in the result are replaced by their respective antiparticles, are reflected, and are turned around and used as input to an experiment, then the result of the latter experiment, when turned around, reflected, and with all its particles replaced by their respective antiparticles, will be the same as the input to the original experiment.

Additional invariance principles, called gauge symmetries, are found to hold but are beyond the scope of this encyclopedia.

The importance of invariance principles is twofold. First, they offer insight into the workings of nature at the most fundamental level. For instance, it is not obviously guaranteed that the same laws of physics that apply here and now on Earth were valid also for a galaxy millions of light-years away and millions of years ago. And yet they appear to be. A second reason for the importance of invariance principles is that they offer clues and hints for the development of theories. As an example, the knowledge that the strong force acts in the same way for particles and for antiparticles helped physicists find a successful theory of the strong force, called quantum chromodynamics.

See also CONSERVATION LAWS; MATTER AND ANTIMATTER; PARTICLE PHYSICS; SPEED AND VELOCITY; SYMMETRY.

FURTHER READING

Mittelstaedt, Peter, and Paul A. Weingartner. *Laws of Nature*. Berlin: Springer, 2005.

ionic compounds An ionic compound forms by an electrostatic interaction between positively charged and negatively charged ions. The electron configuration of atoms determines the types of chemical bonds an atom will form. The only electrons that are involved in bonding are those in the outer energy level, the valence electrons. The number of valence electrons determines whether an atom will gain or lose electrons to create ions. A neutral atom contains the same number of protons as electrons, balancing the electrical charge. When an atom has three or fewer valence electrons, it is classified as a metal and will lose its valence electrons when forming an ion. Cations (positively charged ions) result when metals lose electrons, leaving behind more protons than electrons and a positive charge. Anions are formed by nonmetals that take electrons from other atoms, causing them to have one additional electron relative to the number of protons in the nucleus.

The likelihood of a nonmetal's stealing an electron is determined by the electronegativity of the atom. Electronegativity is the attraction of an atom for additional electrons. In 1932, Linus Pauling developed a scale to compare the electronegativities of elements. This scale, known as the Pauling scale, gives the most electronegative element, fluorine, a value of 4. If two atoms have an electronegativity difference greater than or equal to 1.7, the nonmetal will "steal" the electron from the metal and the bond between them will be an ionic bond rather than a covalent bond.

The interaction between these two charges is based on the Coloumb interaction of the positive and negative ions. The association of the cation and anion with one another to form a neutral compound is more stable (i.e., has lower energy) than existence of the ions as separate entities. The comparison of the energy of the free ions to the stable ionic compound is shown in the following:

$$E = 2.31 \times 10^{-19}(\text{J})(\text{nm})\ (Q_1Q_2)/r$$

where E is energy in joules, r is the distance between ions, and Q_1 and Q_2 are the charges of the ions. As an example, if the compound being studied is sodium chloride, the chlorine will steal the electron from sodium as chlorine has a higher electronegativity. The positively charged potassium ion will have a +1 charge and the negatively charged fluorine ion will have a -1 charge. The distance between the ions in this compound is 0.276 nanometers.

$$E = 2.31 \times 10^{-19}(\text{J})(\text{nm})\ (+1) \times (-1)/(0.276\ \text{nm})$$

$$E = -8.37 \times 10^{-19}\ \text{J}$$

The negative value for E means that energy is released during the formation of this ionic bond. Thus, the ionically bonded compound has lower energy than the individual ions alone. This energetic attraction is the driving force behind the formation of an ionic bond.

PHYSICAL CHARACTERISTICS OF IONIC COMPOUNDS

A formula unit of a compound is the smallest representative unit of a substance. In an ionic compound, the formula unit indicates the lowest possible ratio of cations to anions. In covalent compounds, the term *molecule* is used to represent the smallest bonded unit. The physical characteristics of covalently bonded molecules and ionic compounds differ distinctly with respect to melting points, boiling points, hardness, solubility, and electrical conductivity.

Formula units of ionic compounds strongly attract one another, creating strong compounds. The boiling points and melting points of ionic compounds are, therefore, much higher than boiling points and melting points of covalent compounds. The increased heat necessary to reach these phase transitions is due to the increase in required kinetic energy to separate formula units of the ionic compound from one another. Ionic compounds are nearly all solids at room temperature, whereas covalent compounds can be solids, liquids, or gases. Most ionic compounds form crystalline solids that are very hard and brittle.

The solubility of ionic compounds is significantly higher than that of covalent compounds. When ionic compounds are placed in water, the ions dissociate into the original ions and become surrounded by polar water molecules. As an ionic compound dissolves, the cation associates with the partial negative charge of the oxygen end of the water molecules, and the anion associates with the partial positive charges of the hydrogen end of the water molecules. This creates a separation between the cations and the anions that is maintained until the water evaporates and the ions are able to reassociate.

Electrolytes are compounds that are capable of conducting an electric current when dissolved in aqueous solution. As most ionic compounds are soluble in water, ionic compounds are generally electrolytes. Covalent compounds, on the other hand, are poor electrolytes or nonelectrolytes.

NAMING IONIC COMPOUNDS

Ionic compounds are named differently on the basis of the number of atoms involved in the compound. The overall naming process includes first identifying the metal or the cation, then naming the anion.

Binary ionic compounds consist of only two atoms. The name of the metal is followed by the name of the nonmetal with the ending changed to *-ide*. For example, the ionic compound formed between the metal sodium ion (Na^+) and the nonmetal chlorine ion (Cl^-) is called sodium chloride ($NaCl$).

Ternary ionic compounds involve more than two atoms. The cation is generally a single metal atom (with a few exceptions noted later), and the anion is a polyatomic ion, a charged particle consisting of two or more covalently bonded atoms. When naming ternary ionic compounds, the name of the metal precedes the given name of the polyatomic ion. For example, the ternary ionic compound formed between the metal sodium ion (Na^+) and the polyatomic ion hydroxide (OH^-) is known as sodium hydroxide ($NaOH$). Several polyatomic cations also commonly participate in the formation of ionic compounds, such as ammonium (NH_4^+), mercury (I) (Hg_2^{2+}), and hydronium (H_3O^+). When naming compounds that contain these ions, the name of the polyatomic ion is stated first, followed by the name of the anion. For example, the combination of ammonium (NH_4^+) and a chlorine ion (Cl^-) creates the compound ammonium chloride (NH_4Cl), and the combination of ammonium (NH_4^+) with hydroxide (OH^-) creates the compound ammonium hydroxide (NH_4OH).

When ionic compounds are formed between metals that have more than one possible charge, the charge of the ion must be written as a roman numeral in parentheses after the name of the cation. For example, ions of iron can possess a charge of either +2 or +3. When writing the name of a compound containing iron, one must include the charge of iron in the compound. $FeCl_2$ is known as iron (II) chloride since the charge of each chloride ion is -1, and balancing the positive charge of the iron requires two chloride ions. If iron had a +3 charge in this compound, the formula would have to be written $FeCl_3$ and the name would be written iron (III) chloride.

WRITING FORMULAS OF IONIC COMPOUNDS

When the name of an ionic compound is known, the formula of the compound is based on the information in the formula. Covalent compounds have names that contain prefixes indicating the amount of each type of atom present in the compound. Names of ionic compounds do not contain such information. Rather, ionic compounds require that the charges of each of the ions involved in the compound be known. Once the charges of the ions are known, the positive and negative charges must be balanced in the final formula. For example, the compound with the name *sodium sulfate* would be written using the following symbols:

sodium ion = Na^+

sulfate ion = SO_4^{2-}

sodium sulfate = Na_2SO_4

As the negative charge on the sulfate ion is twice that of the positive sodium ion, the compound requires two sodium ions in the formula.

A simple method of determining the correct formula of ionic compounds is known as the crisscross method. When the ions and their respective charges are known, the numerical value of the charges for the cation become the subscripts representing the number of atoms for the anion, and the value of the charge for the anion becomes the subscript of the cation. For example, consider the formula of calcium phosphate. The calcium cation has a charge of +2, and the phosphate anion has a charge of -3. When writing the formula using the crisscross method phosphate takes 2 as a subscript and calcium takes 3 to give the formula $Ca_3(PO_4)_2$.

See also BONDING THEORIES; ELECTRON CONFIGURATIONS; PERIODIC TABLE OF THE ELEMENTS; PERIODIC TRENDS.

FURTHER READING

Hein, Morris, and Susan Arena. *Foundations of College Chemistry,* 11th ed. Hoboken, N.J.: John Wiley & Sons, 2004.

isomers Isomers are different geometric or spatial arrangements of compounds made of the same elements in the same ratio or the same molecular formula. Isomers can be divided into either constitutional (structural) isomers or stereoisomers. Constitutional isomers differ in the arrangement of the atoms in the molecules. The order in which the elements are bonded together differ. Stereoisomers have all of the same elements bonded together in the same order, but they differ in the spatial arrangements of the elements. Stereochemistry is the study of the three-dimensional arrangement of chemical compounds. Jacobus van't Hoff, a Dutch chemist, laid the foundation for the field of stereochemistry by his studies of the tetrahedral structure of carbon.

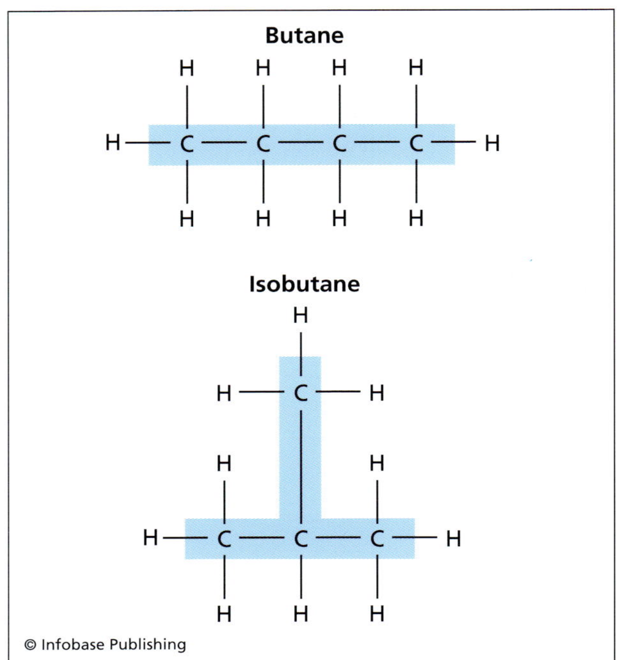

In constitutional isomers (such as butane and isobutene, shown here), the atoms are bonded to different partners, although the isomers have the same molecular formula.

Constitutional isomers have identical molecular formulas, but the bonding order of the elements differs. The position of double or triple bonds can differ as well as the location of any substituent present in a compound. As long as the two compounds contain the same number and type of elements, then the two compounds are considered constitutional isomers. Constitutional isomers can be divided into several categories including chain isomers, positional isomers, and functional group isomers. Chain isomers differ from each other by the way that their carbon atoms are connected. The carbon atoms can be put together in a straight chain, a branched chain, or a chain with multiple branches. Positional isomers differ in the location of functional groups added onto identical carbon chains. Functional group isomers, by definition of isomers, have the same chemical formula, but they contain different functional groups, chemically bonded groups that give a compound its reactivity. In order to have identical chemical formulas yet not contain the same functional group, some type of modification of the functional group must distribute the elements of the functional group throughout the molecule. Examples of functional groups that can be altered to give functional group isomers include aldehydes and ketones. Both of these functional groups contain a carbonyl group, a carbon atom double-bonded to an oxygen atom. In an aldehyde, the carbonyl group is located at the end of the carbon chain so the carbonyl carbon is bonded to one other carbon atom and a hydrogen atom on the other side. In a ketone, the carbonyl group is in the interior of the carbon chain, and the carbonyl carbon is bonded to two other carbon atoms.

Stereoisomers, unlike constitutional isomers, have the same elements bonded together and are divided into two categories known as enantiomers and diastereomers. Enantiomers (also known as optical isomers) are nonsuperimposable mirror images, meaning that they resemble one another as do a right and a left hand. Although a person's right and left hands contain the same number of fingers and thumbs and are virtually identical, one cannot put a right hand into a left-handed glove. The orientation of the atoms on the molecule prevents the superimposition of one onto the other without breaking bonds. The molecules in a pair of enantiomers have the same physical properties as each other, including melting point, boiling point, viscosity, colors, odors, and reactivity, but they can have different chemical properties. For example, many enzymes only recognize one isomer of a compound and will not be able to bind the other isomer as a substrate.

Plane-polarized light results from white light that has been sent through a polarizing filter and goes out in only one plane. A polarimeter is a device that contains a polarizing filter and a tube filled with the compounds of interest. The light passes through the filter, where it is polarized and then enters into the collecting tube and interacts with the compounds being studied. The amount of rotation of the polarized light depends on several variables including the wavelength of the light, the length of the collecting tube, the amount of the compound that is dissolved in the solution (its concentration), and the compound's ability to rotate light, known as its specific rotation. In 1827, Friedrich Woehler, a German chemist, discovered that some compounds had the same composition yet had different properties when he was studying uric acid. In 1848, Louis Pasteur manually separated two different versions of crystals of sodium ammonium tartrate. He then demonstrated that the two different versions of the compound rotated plane-polarized light in opposite directions when dissolved in aqueous solution.

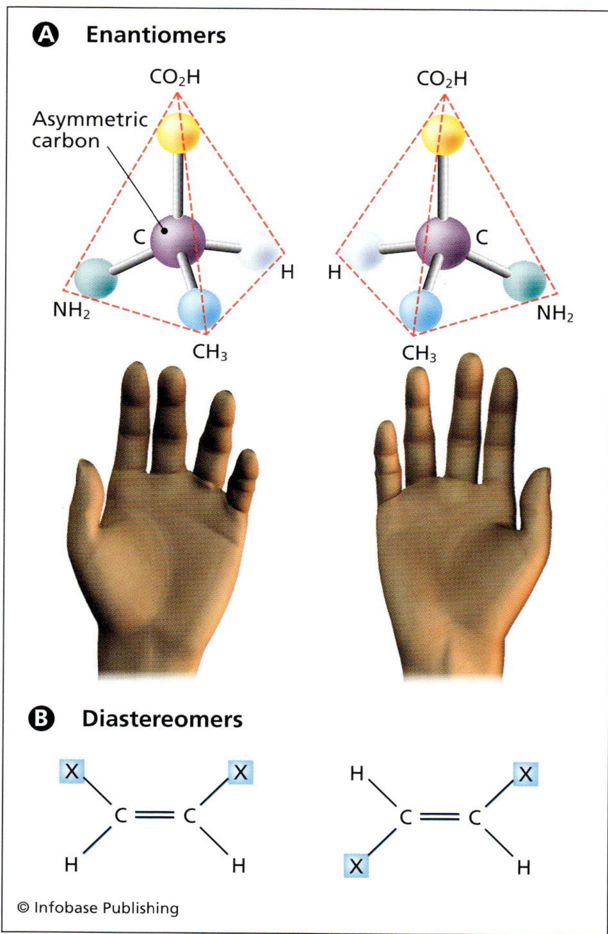

(a) Enantiomers are nonsuperimposable mirror images of each other. (b) Diastereomers are isomers that differ in their three-dimensional arrangement around a double bond.

Enantiomers are chiral molecules, meaning that they can rotate plane-polarized light and are therefore known to have optical activity. If chiral molecules are present in the tube, they are able to rotate the plane of polarized light. The presence of enantiomers was responsible for the rotation of polarized light seen by early scientists. When the direction of rotation of the light is to the left, the compound is designated (l), (–), or levorotatory. The enantiomer partner is able to rotate the plane-polarized light in the opposite direction, to the right, and is designated (d), (+), or dextrorotatory. Two enantiomers always have the same magnitude of rotation, just in opposite directions. When both the d and l versions of a compound are present in equal amounts, the mixture is known as racemic and the compounds are racemates. The right-handed rotation balances out the left-handed rotation and the mixture does not have optical activity. Chiral molecules contain stereogenic centers, which are atoms with four different substituents.

Diastereomers (also known as geometric isomers) are not mirror images but have a different geometric arrangement between the elements that are bonded in the same order. Generally geometric isomers contain a double bond or a ring structure in the molecule. Unlike single bonds, around which there is free rotation, atoms involved in a double bond are not able to rotate. This means that if two groups are attached to the carbon atoms involved in a double bond, the position of the groups relative to the double bond is significant. In order to demonstrate the location of these groups relative to the double bond, the prefixes *cis* and *trans* are used. If the substituents are on the same side of the double bond, then they are called *cis,* and if they are on opposite sides of the double bond, they are known as *trans*. In general, the *cis* and *trans* forms of a compound have significantly different physical and chemical properties such as boiling points, melting points, colors, odors, and reactivity. Diastereomers contain achiral centers that can be recognized by their plane of symmetry and the fact that they are superimposable mirror images.

The absolute configuration of a compound is dependent on the locations of the atoms of each element in space. The relative configuration depends upon a compound's relationship to another compound. The absolute configuration of organic compounds is indicated using the Cahn-Ingold-Prelog

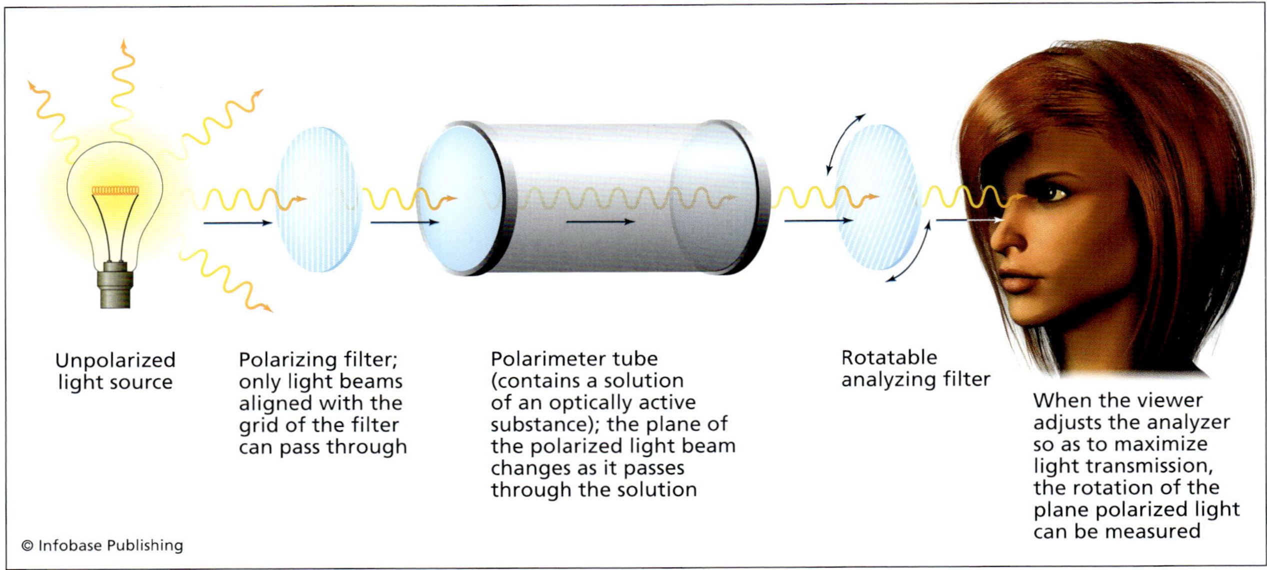

A polarimeter is used to measure the rotation of plane polarized light.

notational system, R and S, developed in 1956 and named after Robert Sidney Cahn, a British chemist; Christopher Ingold, a British chemist; and the Croatian chemist Vladimir Prelog. This system works by determining which direction, clockwise or counterclockwise, the substituents located on the stereogenic center will go when placed in rank order, from highest to lowest, based on the atomic masses of the element. They are ranked from the largest to the smallest atomic mass. If the first element attached to the stereogenic carbon is the same for two of the groups, then one must continue down the chain to the first point of difference and make an atomic mass determination. Once all of the substituents are ranked, point the lowest-ranking substituent away from the viewer. The three highest are drawn as they look when the lowest is pointed away from the viewer. Then the substituents are read from highest rank to lowest rank in order. If the direction that these substituents are read is clockwise, then the compound is given the designation R. If the direction of the three highest-ranking substituents is counterclockwise, the compound is designated S.

R and S designations are based on the structure of a single compound. The d or l, (+) or (–), designations are a characteristic of a group of compounds in solution. Both of these descriptions can apply to the same mixtures. The designations of R and S as well as (+) and (–) are written in front of the name of the compound in parentheses and are separated from the name of the compound by dashes.

(S)–(–)–2–bromobutane

Structures of isomers can be represented in a variety of ways. Ball and stick models and wedge and dash structures are common methods of representing the molecules. Another common method of drawing structural formulas of compounds is the Fischer projection drawing system, or Fischer projections. Horizontal lines represent elements that are directed out of the page, vertical lines represent elements that are directed into the page, and the central carbon is not shown. If the compound contains a chain of carbon atoms bonded together, then the chain is usually written vertically.

Often the same molecule can exist in different conformations and these compounds are called conformers. Conformers are represented by Newman projections, which were developed by the American chemist Melvin S. Newman of Ohio State University. This method uses a point to represent the front carbon, and the carbon that is behind is written as a circle. Two common conformations are the staggered and eclipsed conformations. In the staggered conformation, the carbon-hydrogen bonds on the front carbon are rotated relative to the carbon-hydrogen bonds on the back carbon. The eclipsed conformation is one in which the carbon-hydrogen bonds on the front carbon are lined up with the carbon-hydrogen bonds in the back carbon.

Most medications are chiral molecules. Many naturally occurring molecules are isolated as individual isomers rather than mixtures. This allows the particular isomer of interest to be used in the medication. Most organically synthesized medications are made as a racemic mixture of both isomers. The problem with this is that both of the isomers are not functional as a medication. Despite this fact, many medications are distributed as racemic mixtures. This presents the possibility of three different outcomes: the opposite isomer's being nonfunctional, the opposite isomer's being converted into the functional isomer, or the opposite isomer's having a deleterious effect. If the other enantiomer is nonfunctional, then the user will only obtain half of the medication he or she would with a completely pure medication consisting of one enantiomer. Asthma medication known as albuterol contains a racemic mixture with equal amounts of left and right isomers. The right-handed isomer produces more of the positive effects of the medication. The left-handed isomer contributes to many of the side effects of albuterol. Levalbuterol is a version of albuterol that contains only the right-handed isomer, thus maximizing the positive effects of the medication while removing some of the side effects. Ibuprofen is an example of a medication that is sold as a racemic mixture. Only the S-enantiomer is active, and it makes up 50 percent of the mixture. The R-enantiomer is able to be used because the body converts it to the S-enantiomer. Another medication, thalidomide, had the worst possible combination of isomers. Thalidomide was used to counteract morning sickness in pregnant women in the 1950s. Researchers later found out that the drug had teratogenic properties, meaning it caused birth defects. Many studies suggest that the S isomer was effective as a treatment for morning sickness and the R isomer caused the birth defects, possibly through binding to the deoxyribonucleic acid (DNA) of the developing fetus. Thalidomide was therefore pulled off the market. In May 2006, the U.S. Food and Drug Administration (FDA) approved the use of thalidomide in cases of multiple myeloma. As a result of the devastating teratogenic effects of thalidomide, the prescription and use of this drug require a registration process that regulates to whom and when the drug can be administered.

See also ORGANIC CHEMISTRY; PASTEUR, LOUIS.

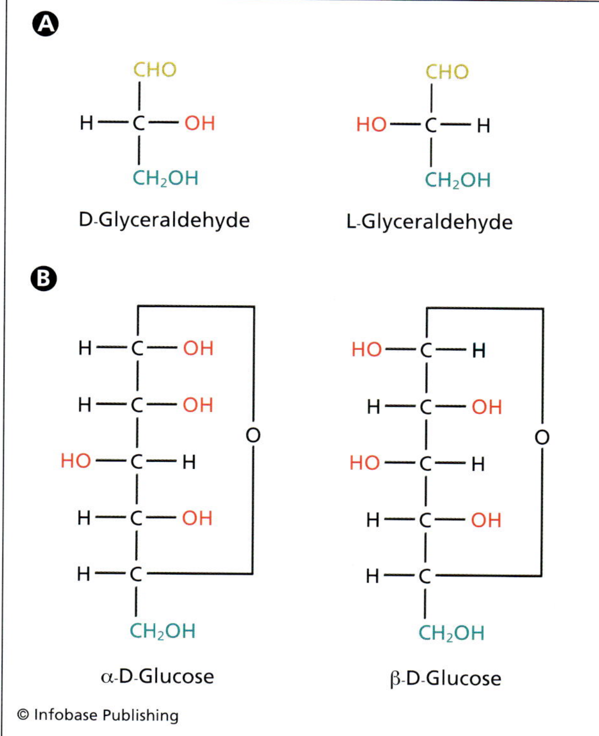

In a Fischer projection, the carbon chain is written vertically with the chiral carbon in the center. (a) Fischer projection formulas representing two isomers of glyceraldehyde. (b) Fischer projections representing two isomers of glucose.

FURTHER READING

Wade, Leroy G. *Organic Chemistry*, 6th ed. New York: Prentice Hall, 2006.

isotopes While a chemical element is characterized by its atomic number, an isotope is a version of a chemical element that has both a definite atomic number and a definite atomic mass number. All nuclei of an isotope possess the same number of protons and the same number of neutrons. Different isotopes of the same element possess the same electron structure and the same chemical properties. However, the reaction rates of heavier isotopes, those containing more neutrons in their nuclei, are somewhat slower than those of lighter ones of the same element. This is called the kinetic isotope effect. The physical properties of isotopes of the same element may differ as a result of the different masses of their atoms. For instance, the diffusion rate of an isotope and the speed of sound in it are different for different isotopes of the same element, as is the infrared absorption spectrum of molecules that include different isotopes of the same element but are otherwise identical. Most naturally occurring elements are mixtures of isotopes. Natural carbon, for instance, with atomic number 6, contains a mixture of three isotopes, having atomic mass numbers 12, 13, and 14 and containing in each of their nuclei, in addition to six protons, six, seven, and eight neutrons, respectively. The first, carbon 12, is by far the most abundant isotope. Both carbon 12 and carbon 13 are stable isotopes, while carbon 14 is radioactive: that is, its nucleus is unstable and eventually decays (to a nitrogen 14 nucleus, an electron, and an antineutrino of electron type). Isotopes that do not occur naturally might be produced in nuclear reactors or in particle accelerators. As an example, the radioactive isotope carbon 11, which does not occur naturally, can be produced by bombarding boron 11, the most abundant naturally occurring isotope of boron, with protons in an accelerator.

The actual mass of an atom of an isotope in atomic mass units, which is the isotope's atomic mass, generally differs from the isotope's atomic mass number because of the nuclear binding energy. In SI the atomic mass of the isotope carbon 12 is defined as 12 atomic mass units (amu) exactly, and the atomic masses in amu of all the other isotopes are scaled accordingly.

The number that is given as the atomic weight (also called atomic mass, depending on the book or table) of an element is a weighted average of the atomic masses of the naturally occurring isotopes of that element, where the weighting is according to the abundances of isotope atoms in the element. In the case of carbon, for instance, atoms of carbon 12 make up 98.89 percent of the atoms in the natural element, while carbon 13 atoms form 1.11 percent. The abundance of carbon 14 is negligible at this level of precision and in any case varies by source. The atomic mass of carbon 12 is defined as 12 amu exactly, as just mentioned, while that of carbon 13 is 13.003 amu. The calculation of carbon's atomic weight is

$$\begin{aligned}&\text{Atomic weight of C} =\\&(\text{atomic mass of C 12}) \times (\text{abundance of C 12})\\&+ (\text{atomic mass of C 13}) \times (\text{abundance of C 13})\\&= (12\text{ amu}) \times 0.9889 + (13.003\text{ amu}) \times 0.0111\\&= 12.011\text{ amu}\end{aligned}$$

This is the number (to five significant figures) that one finds in the literature and in periodic tables of the elements.

APPLICATIONS OF ISOTOPES

Medical applications of isotopes include imaging and therapy. One example of the use of isotopes for medical imaging is positron emission tomography (PET). This imaging procedure uses a radioactive isotope that is a positron (the antiparticle of the electron) emitter, that is, an isotope that decays radioactively via inverse beta decay, also called e^+ decay or β^+ decay. Carbon 11, mentioned earlier, is such an isotope, as are nitrogen 13, oxygen 15, and fluorine 18. The isotopes are incorporated into compounds that are absorbed by the body, such as glucose, water, or ammonia, and are injected into the body.

An example of use of isotopes for therapy is what is called brachytherapy, which is the placement of a radioactive source beside or inside a cancer tumor. Prostate cancer, for instance, can be treated this way. Tiny rods, called "seeds," of iodine 125 are inserted in the prostate for the purpose of destroying the cancerous tissue.

One of the most common applications is isotopic labeling, which is the use of radioactive isotopes as tracers in chemical reactions. For instance, if one needs to know where the oxygen, say, in the reactants is going, one can replace some of the oxygen atoms with oxygen 15 and then detect the presence of oxygen among the reaction products by the isotope's radioactivity.

See also ACCELERATORS; ATOMIC STRUCTURE; MATTER AND ANTIMATTER; NUCLEAR PHYSICS; RADIOACTIVITY.

FURTHER READING

Faure, Gunter, and Teresa M. Mensing. *Isotopes: Principles and Applications,* 3rd ed. Hoboken, N.J.: Wiley, 2004.

Firestone, Richard B., Coral M. Baglin, and S. Y. Frank Chu. *Table of Isotopes,* 8th ed. Hoboken, N.J.: Wiley, 1999.

Kepler, Johannes (1571–1630) *German Astronomer and Mathematician* Sir Isaac Newton is famously quoted as stating, "If I have seen farther it is by standing on the shoulders of Giants." The giants to whom he was referring included Galileo Galilei and Johannes Kepler. Kepler was a skilled mathematician. Using his remarkable abilities, he analyzed the precise (for that time) astronomical data of his employer, the astronomer Tycho Brahe, which he "liberated" from Brahe's home upon the latter's death before the heirs could lay their hands on the material. Kepler's arduous analysis led to his formulation of what are known as Kepler's three laws of planetary motion. These laws, together with Galileo's discoveries, constituted the foundation upon which Newton constructed his laws of motion, which form the fundamental laws of mechanics, and his law of gravitation, which describes the dominant force governing the motion of astronomical bodies.

KEPLER'S LIFE

Johannes Kepler was born on December 27, 1571, in Weil, in what is now Germany, to Heinrich Kepler and Katherine Guldenmann. By his own description, his family, which was of noble extraction but poor, was quite dysfunctional. His father, a vicious man, made a living as a mercenary and disappeared when Johannes was five. His mother dealt in herbal cures and potions and was later put on trial for witchcraft. Johannes was born prematurely and was a sickly child from birth. Nevertheless, he was a prodigy and displayed his astounding mathematical ability from an early age.

Johannes's lifelong love for astronomy was apparently inspired by his observing at age six the Great Comet in 1577 and a full lunar eclipse in 1580, at age nine. His childhood smallpox caused him to have defective vision, which limited his efforts to perform astronomical observations. In 1589, at the age of 18, Johannes became a student of theology and philosophy at the University of Tübingen. There he demonstrated exceptional intellectual abilities. He studied astronomy under Michael Maestlin and learned about both the Ptolemaic, geocentric model of the solar system and the Copernican, heliocentric one. Kepler strongly supported the Copernican scheme, which he defended theoretically and theologically in a public debate. Upon completing his studies at Tübingen in 1594, Kepler took a position as professor of mathematics and astronomy at the Protestant School (later to become the University of Graz) in Graz, in what is now Austria. He remained in Graz for six years, until 1600.

During Kepler's tenure at Graz, he developed a scheme for explaining the ratios of the distances of the six planets that were known then—Mercury, Venus, Earth, Mars, Jupiter, and Saturn—from the Sun, the radii of their orbits—assumed circular—about the Sun, in the Copernican model. The explanation was very compelling for Kepler, as it was based on the existence of precisely five Platonic solids, which are the only regular three-dimensional polyhedra whose faces are all identical regular polygons. These bodies are the tetrahedron (whose faces consist of four identical equilateral triangles, with three meeting at each of its four vertices), the cube (six identical squares, three meeting at a vertex, eight vertices), the octahedron (eight identical equilateral triangles, four meeting at a vertex, six vertices), the dodecahedron (12 identical regular pentagons, three meeting at a vertex, 20 vertices), and the icosahedron (20 identical equilat-

Johannes Kepler was a 16th–17th-century astronomer and mathematician who is famous for his laws of planetary motion. *(Bildarchiv Preussischer Kulturbesitz/Art Resource, NY)*

eral triangles, five meeting at a vertex, 12 vertices). Kepler envisioned a model starting with a sphere, representing the location of the orbit of Mercury in space. One of the Platonic solids is then fitted snugly about the sphere. Then another sphere is fitted about the Platonic solid. This sphere represents the orbit of Venus. A second Platonic solid, different from the first, is fitted about the second sphere, with a third sphere, representing Earth's orbit, fitted about it. And so on, using the five different Platonic solids to obtain a nested structure containing six spheres representing the orbits of the six planets. By choosing the sequence of Platonic solids correctly, Kepler hoped to obtain ratios of radii of spheres that would match the known ratios of the radii of the planetary orbits.

Kepler indeed found such a sequence of Platonic solids that gave a rough approximation to the orbital radii. He was greatly encouraged and, assuming that the scheme was correct, blamed the model's discrepancies on errors in Copernicus's astronomical tables that he was using. The reason Kepler found this scheme to be so compelling was that he was convinced there must exist a fundamental harmony in the universe, so the various aspects of the universe must be strongly interrelated. This model, relating the six planets to the five Platonic solids, was so beautiful it could not be wrong! Viewed from the perspective of today's understanding and knowledge, Kepler's model was nonsense. As one example of this, the solar system contains more than six planets, as is now known. (The exact number depends on an arbitrary distinction between planets and dwarf planets. The current official specification has eight planets—the six that Kepler knew with the addition of Uranus and Neptune. Pluto, which for many years had been considered a planet, now belongs to the category of plutoids—named in Pluto's honor—which contains more than 40 other objects.) In addition, according to modern understanding of the mechanics of the solar system, the solar system could exist just as well with different ratios of orbital radii. On the other hand, Kepler's approach to nature, his conviction that a fundamental harmony must underlie the universe, is very recognizable in modern science.

Kepler published his ideas in a book on astronomy, *Mysterium Cosmographicum* (*Cosmographic Mystery,* or perhaps better, the *Mystery of the Universe*). The book appeared in 1596, and in 1597 Kepler distributed copies of it to important European astronomers. It drew Kepler to the attention of the astronomy community and established his reputation as an astronomer.

In 1597, Kepler married Barbara Müller, 25 years old and twice widowed, daughter of a local mill owner. Barbara entered the marriage with a young daughter, and the couple produced five additional children. The first two died while still infants. Of the other three, a daughter, Susanna, was born in 1602, and their sons, Friedrich and Ludwig, were born in 1604 and 1607, respectively.

After publication of his astronomy book, Kepler planned to continue, widen, and deepen his investigations. But he soon realized that the astronomical data from Copernicus that he had were not sufficiently precise. He could neither further his work nor answer objections to it that were based on his using Copernicus's data. Where were located the most accurate astronomical data that existed at that time? They were in Prague, jealously guarded by the person who obtained them through painstaking measurements, Tycho Brahe. Tycho, as he was known, was an extraordinary astronomer. He designed and constructed his own instruments and with very precise observational work managed to reach the highest level of precision for that time. Tycho had his own idea about the correct model of the solar system, one that was neither geocentric nor heliocentric. In his scheme, all the planets except Earth revolved around the Sun, while the Sun revolved around Earth. Tycho amassed his accurate observational data in order to use them eventually to prove his model correct. But in the meanwhile he kept the data to himself, unpublished, so that nobody else could use them to claim priority for his scheme. In order to gain access to Tycho's data, Kepler applied to work for him as a data analyst, was offered a position, accepted it, and moved with his family to Prague in 1600.

Kepler spent about a year analyzing Tycho's planetary data for him, especially Mars data, in which Kepler himself was interested in order to bolster his own ideas. Tycho grudged Kepler the necessary data for those analyses. But it soon became clear to Kepler that it would be very difficult, if not impossible, to access the masses of data that the secretive Tycho had hidden away, which Kepler badly needed. Fate intervened, however, and in late 1601, Tycho died of a bladder infection brought on, it is told, by overstretching his bladder at a banquet where it was considered extremely rude to leave, or even step out, before the conclusion. Recent investigations suggest that Tycho died instead of mercury poisoning. Some have claimed that circumstantial evidence exists that Kepler was his poisoner, as Kepler had the means, the motivation, and the opportunity. As for motivation, immediately upon Tycho's death, Kepler and his wife ransacked Tycho's home and stole all the astronomical data before the heirs had the opportunity to take hold of the material. Kepler immediately became court mathematician to Rudolph, king of Bohemia, in which position he served for the next 11 years in Prague. That good fortune together with Tycho's data gave Kepler 11 very fruitful years.

In 1611, Kepler's wife and one of his sons both died. And in 1612, at the death of his employer and patron Rudolph, Kepler moved to Linz, in what is now Austria, where he taught, studied, and wrote. In 1613, at age 42, Kepler married Susanna Reuttinger, who was 24 years old. They had six children, of whom the first three died in childhood. The surviving three—Cordula, Fridmar, and Hildebert—were born in 1621, 1623, and 1625, respectively. Because of the accusations of someone in a financial quarrel with Kepler's brother, their mother, Katherine, was tried for witchcraft, starting in 1617, and was imprisoned in 1620. Kepler hired lawyers and devoted a great deal of time and labor to her defense, including much travel back and forth between Linz and Weil. Katherine was released in 1621 as a result of Kepler's determined efforts on her behalf. The years 1626–30 were years of turmoil for Kepler and his family. In 1826, Linz was besieged in the Thirty Years' War and Kepler moved to Ulm, in what is now Germany. His jobs required much travel among Prague, Linz, and Ulm. He moved to Regensburg, Germany, in 1630, where he died the same year.

WORK AND PUBLICATIONS

Kepler's most significant accomplishment for modern science were his three laws of planetary motion, which he discovered while analyzing Tycho's astronomical data during his 11 years in Prague after Tycho's death in 1601. The first law made a revolutionary departure from the preconceived idea, held by all since ancient times, that the motions of the heavenly bodies must be circular. Kepler was surprised to discover that the planets revolved around the Sun not in circular orbits, but in elliptical ones. An ellipse is an oval curve, a "stretched" circle. If one views a circle at some angle to its perpendicular, it appears as an ellipse. Or, if one cuts through a cylindrical surface, such as a paper towel core or toilet paper core, at some angle to the perpendicular to its axis, the edge has the form of an ellipse. The usual definition of an ellipse is the curve comprising all the points in a plane for which the sum of the distances of each point from two fixed points, the foci, is the same for all the points. Kepler described this motion in a precise law.

> *Kepler's first law.* Each planet moves in an elliptical orbit with the Sun at one of the foci.

Kepler's data analysis showed that as a planet orbits the Sun in its elliptical path, its speed changes. It moves faster the closer it is to the Sun and more slowly as its path takes it farther away. Now, imagine a straight line running from the planet to the Sun. As the planet moves, this line, called the radius vector, sweeps out an area. Kepler discovered that for each planet the radius vector sweeps out equal areas in equal time intervals. Take a short time interval, say, one second. In each second, a planet sweeps out an area that is approximately an isosceles triangle, with the radius vector forming the equal sides and an almost-straight segment of the orbit serving as the triangle's base. Compare the situation when the planet is closer to the Sun with that when the planet is farther away. In the closer position, the radius vector, and thus the height of the triangle, is relatively short compared to the radius vector and height at the farther location. The areas of the two triangles are equal. The area equals one-half the height times the base. So with a shorter height, the closer triangle must have a longer base for it to possess the same area as the farther triangle, which has a longer height and thus a shorter base. Since the bases of the triangles are orbit segments, it follows that in one second the planet covers a longer distance closer to the Sun than when it is farther. Stated in terms of speed, the planet moves faster when it is closer to the Sun than when it is farther. That is indeed what Kepler observed.

> *Kepler's second law.* Each planet sweeps out equal areas in equal time intervals.

Kepler also compared the "year" of each planet, the time it took to make a complete revolution around the Sun, with the size of the planet's orbit. Clearly, the planets that were more distant from the Sun took longer to complete their orbit, and Kepler found a precise relationship.

> *Kepler's third law.* For any pair of planets the ratio of the squares of the times to complete one orbit (their "years") equals the ratio of the cubes of their respective average distances from the Sun.

An equivalent expression of this is: for each planet the square of the time to complete one orbit is proportional to the cube of its average distance from the Sun, with the same proportionality constant for all the planets.

Kepler's laws of planetary motion served as a challenge to Isaac Newton, who succeeded in explaining them with his own laws of motion and law of gravitation.

Kepler's most important publications on astronomy were his book *Astronomia Nova* (*The New Astronomy*), published in 1609, and the three-volume work *Epitome Astronomia Copernicanae* (*Epitome of Copernican Astronomy*), completed in 1615 and published in 1617, 1620, and 1621. In the first book, Kepler expounded his views at that time and included his first two laws as they apply to Mars. His second work, a major undertaking, was intended as an astronomy textbook. In it, he presented his version (with elliptical orbits) of the heliocentric system, while extending his first two laws to all the planets and even to the Moon and the moons of Jupiter.

Kepler also published an interdisciplinary work, titled *Harmonices Mundi* (*Harmonies of the World*), in 1619. In this book he attempted to find unifying relationships among various aspects of the universe, such as geometry, planets, astrology, meteorology, and souls. In this book, he presented his third law of planetary motion.

In addition to astronomy, Kepler studied optics and the application of optics to telescope design. His investigations resulted in two more books: *Astronomiae Pars Optica* (*The Optical Part of Astronomy*), finished in 1604, and *Dioptrice,* which he published in 1611. The former book covered theoretical optics extensively. The latter book reported on Kepler's investigations, both theoretical and practical, of telescope optics.

LEGACY

Kepler's most significant direct contributions to human understanding were his heliocentric model of the solar system and his three laws of planetary motion. The laws' significance lies only secondarily in their description of planetary motion. More importantly, they, together with Galileo's experimental results, served as the foundation upon which Newton constructed the magnificent edifice of his laws of motion and law of gravitation.

More generally, though, Kepler furthered the idea of nature's unity, an idea that serves as a guiding light in modern science and especially in physics. It is true that he included spiritual and religious considerations in his quest for unification, while they have no place in science, as Galileo insisted and as modern science concurs. Nevertheless, echoes of Kepler's ideas still reverberate loudly.

See also GALILEI, GALILEO; MOTION; NEWTON, SIR ISAAC; OPTICS; TELESCOPES.

FURTHER READING

Caspar, Max. *Kepler.* Mineola, N.Y.: Dover, 1993.

Connor, James A. *Kepler's Witch: An Astronomer's Discovery of Cosmic Order amid Religious War, Political Intrigue, and the Heresy Trial of His Mother.* New York: HarperCollins, 2004.

Gilder, Joshua, and Anne-Lee Gilder. *Heavenly Intrigue: Johannes Kepler, Tycho Brahe, and the Murder behind One of History's Greatest Scientific Discoveries.* New York: Doubleday, 2004.

Kepler, Johannes. *Epitome of Copernican Astronomy and Harmonies of the World.* Amherst N.Y.: Prometheus, 1995.

———. *The Harmony of the World.* Philadelphia: American Philosophical Society, 1997.

Kepler, Johannes, and Stephen W. Hawking, commentator. *Harmonies of the World.* Philadelphia: Running Press, 2005.

Koestler, Arthur. *The Sleepwalkers: A History of Man's Changing Vision of the Universe.* New York: Penguin, 1990.

Kornberg, Roger (1947–) American *Biochemist* Roger Kornberg is an American biochemist whose research helped elucidate the mechanisms and machinery of the process of copying deoxyribonucleic acid (DNA) into a ribonucleic acid (RNA) message, known as transcription. Kornberg began his scientific career working on the transport of particles across cellular membranes and developed the concept of membrane flip-flop. He later began work-

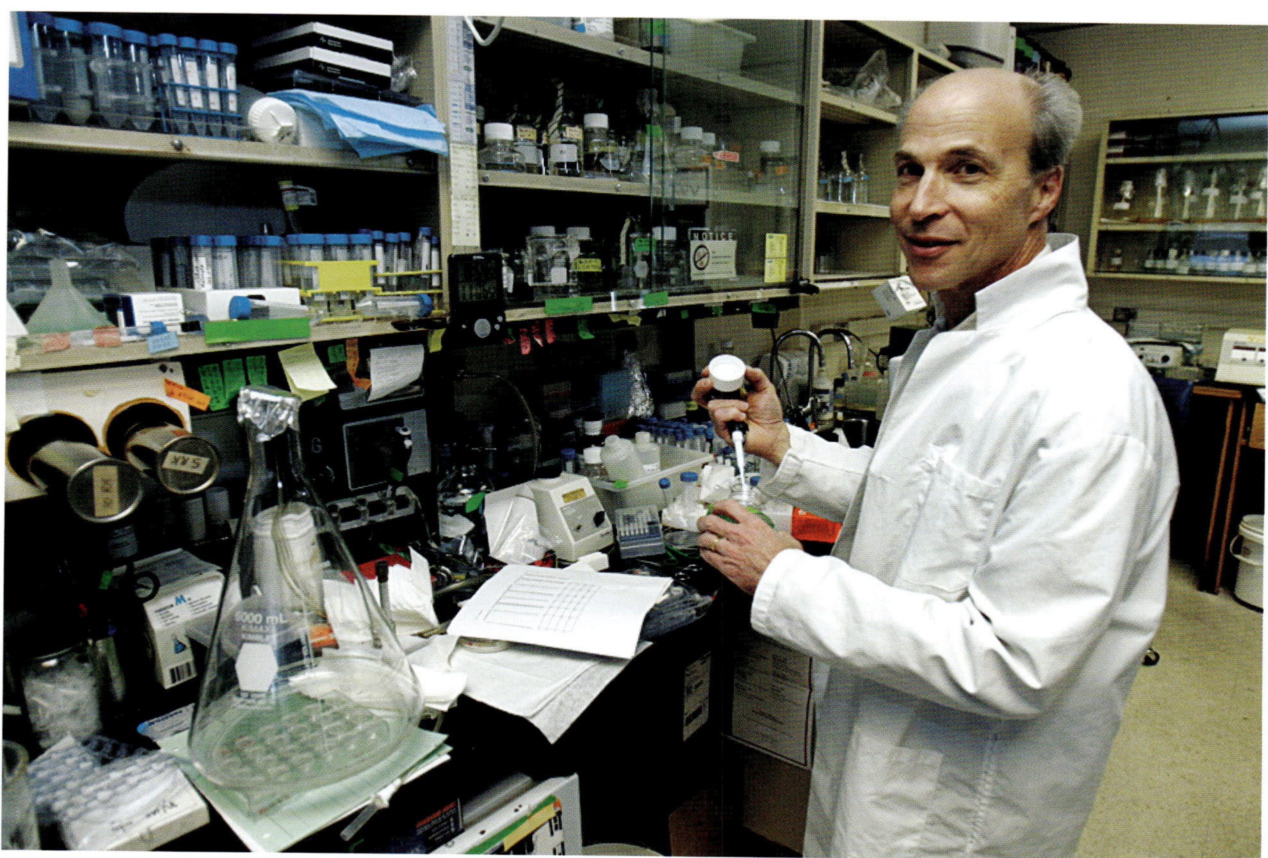

Roger D. Kornberg is an American biochemist whose work contributed to the current understanding of DNA packaging and transcriptional regulation. Kornberg is shown here in his lab at Stanford University. *(AP Images)*

ing on the identification of the histone complexes that make up the nucleosome, the basic unit of packaging of DNA into chromatin. Kornberg received the Nobel Prize in chemistry in 2006 for his work on the eukaryotic transcriptional machinery.

EARLY YEARS AND EDUCATION

Roger Kornberg was born on April 24, 1947, to Arthur and Sylvy Ruth Kornberg in St. Louis, Missouri. Roger came by his interest in biochemistry naturally. His father received the Nobel Prize in physiology and medicine in 1959 for his work on DNA replication, the process by which the genetic material is duplicated before a single cell divides to form two new cells. The younger Kornberg received his bachelor's degree in chemistry from Harvard University in 1967 and a Ph.D. in chemistry from Stanford University in 1972. Kornberg then accepted a postdoctoral fellowship at the Laboratory of Molecular Biology of the Medical Research Council (MRC) in Cambridge, England. After his fellowship, he took a teaching position at the MRC. Kornberg eventually returned to the United States in 1976, when he was offered a position at Harvard University Medical School in the biological chemistry department. Completing his travels, Kornberg moved in 1978 to the Stanford University School of Medicine in California, where he became an assistant professor of structural biology. Kornberg eventually became a professor of structural biology and remains at Stanford today.

MEMBRANE FLIP-FLOP

The membranes within plant and animal cells are phospholipid bilayers, meaning they consist of two layers of phospholipids. The hydrophilic (water-loving) phosphate head groups of the phospholipids surround the outer and inner surfaces of the membrane, while the hydrophobic (water-fearing) hydrocarbon tails make up the interior of the membrane. When two of these layers join to exclude water, the fatty acid tails become sandwiched in the middle and the structure is known as a phospholipid bilayer. The phospholipid heads interact with the exterior and interior of the cell, and the fatty acid tails make up the central interior portion of the double-layered membrane. Control of molecule movement through the cell membrane is one important focus of biochemical research.

Cell membranes are selectively permeable and give the cell tight control of the movement of materials into and out of the cell. The chemistry of membrane potential plays an important role in many cellular processes. The generation and alteration of chemical concentration gradients and electrochemical gradients across membranes drive processes such as nerve impulse transmission, muscle contraction, photosynthesis, and cellular respiration.

Movement within one side of a phospholipid bilayer was considered possible by simple lateral diffusion, the movement of substances from areas of high concentration to areas of low concentration. The lipids and proteins found in the cell membrane can move laterally within the same side of the membrane relatively easily and quickly. Movement from one side of the lipid bilayer to the other was not thought possible until the early 1970s. The thermodynamics of such a process, the movement of a hydrophilic phosphate head group through the hydrophobic membrane core, was considered unfeasible. As a graduate student at Stanford, Kornberg was studying membrane movement in 1971, when he determined that the motion of phospholipids from one side of the membrane to the other did indeed occur, although it was a slow process. Kornberg helped coin the term *flip-flop* for the movement of those phospholipids that moved from one side of the membrane to the other. This concept aided transport studies in biochemistry for years.

X-RAY CRYSTALLOGRAPHY AND THE DISCOVERY OF THE NUCLEOSOME

While studying at the MRC Laboratory of Molecular Biology in Cambridge, Kornberg began learning X-ray crystallography, a technique used to understand the molecular structure of a substance. Crystallography depends on the analysis of diffraction patterns (the way rays are scattered by the molecules they hit) that emerge from a closely packed, highly organized lattice, or crystal, targeted by a beam of short waves. Scientists most frequently use X-rays, a form of electromagnetic radiation, as the beam source; neutrons and electrons (two kinds of subatomic particles possessing wavelike properties) are alternative sources.

X-ray crystallography is useful for studying the structure of a microscopic entity or molecule at the atomic level. Light microscopy permits the examination of gross features of microscopic objects by focusing light waves, or rays, of the visible light spectrum through a series of lenses to permit visualization of the object. At the atomic level, wavelengths of visible light are longer than not only the atoms but also the bonds linking the atoms together to form a structure. X-rays have shorter wavelengths than the atoms being studied, making it possible to differentiate atomic positions relative to each other in a molecule. Because no lens exists that focuses X-rays, researchers must examine the cumulative effect of many atoms in similar configuration to detect the position and relationship of one atom to another. The atoms within a crystal possess the same relative position and orientation throughout, giving rise to a detectable signal.

Using these techniques, Kornberg began studying the structure of packaged DNA. His X-ray crystallography work led to the visualization of the histone complex containing a protein core with a short stretch of DNA wrapped around it. In 1977, Kornberg obtained electron microscopy results that showed a histone octamer with DNA wrapped around it for the first time. This structure became known later as the nucleosome, the fundamental unit in the packaged form of DNA known as chromatin. The relationship between the packaging of DNA and the activity of the genes is an area that Kornberg continues to study.

TRANSCRIPTION STUDIES

Kornberg's work at Stanford University in the 1970s and 1980s revolved primarily around the function of the nucleosome rather than its structure. He began working in earnest on the process of eukaryotic transcription at the molecular level. The organized series of events that lead to the formation of a messenger RNA molecule from a DNA template is known as transcription. In eukaryotic organisms transcription takes place within the nucleus. Messenger RNA (mRNA) serves as a temporary copy of genetic instructions that can be used by the protein synthesizing machinery in the cytoplasm. The composition of an mRNA molecule corresponds to the DNA template, except mRNA contains uracil residues in place of the thymine residues that exist in DNAs.

RNA is a long, unbranched molecule consisting of ribonucleotides joined by $3'\rightarrow5'$ phosphodiester bonds. As the name implies, the ribonucleotides consist of the pentose sugar ribose (in contrast to deoxyribose sugars in DNA), a phosphate group, and a nitrogenous base. RNA and DNA consist of three of the four same bases: adenine, cytosine, and guanine. RNA molecules replace the fourth nitrogenous base found in DNA, thymine, with uracil (U). The number of nucleotides in a molecule of RNA ranges from as few as 75 to many thousands. RNA molecules are usually single-stranded, except in some viruses; consequently, an RNA molecule need not have complementary base pair ratios (equal amounts of A and U and equal amounts of G and C).

Although the fundamental concepts regarding transcription were known to Kornberg and others, the mechanism for the process was still quite unclear.

The enzyme responsible for transcription of protein-coding genes, RNA polymerase II, adds nucleotides to the growing mRNA chain and serves a proofreading function. Kornberg set out to determine how this process functioned at the atomic level.

Using simple baker's yeast, *Saccaromyces cerevisiae,* as a model eukaryotic system Kornberg began studying eukaryotic transcription. Before yeast could be used as the system, it was necessary to see how similar the process of transcription was in yeast and in animals. Research performed by Kornberg's wife, Yahli Lorch, and others determined that RNA polymerase II functions similarly in both yeast and other animals. This simpler eukaryotic system became the basis for Kornberg's transcriptional work.

Roger Kornberg's research returned to structural studies to produce detailed X-ray crystallography images of RNA polymerase II as well as the rest of the transcription machinery. The impact of these images led to huge advances in the understanding of eukaryotic transcription. His images revealed the growing mRNA strand during its synthesis, still attached through the RNA polymerase complexed to the DNA. From these images, Kornberg discovered that the presence of nucleosomes surrounding the DNA did not inhibit RNA polymerase II from reading the DNA sequence. The presence of histones at the promoter, however, could inhibit the start of transcription. The structural determination of the transcription machinery led to the understanding of each of the steps of eukaryotic transcription including the discovery of a 20-protein complex termed *mediator* that is involved in transcriptional control. Kornberg's structural studies also revealed a nine-protein complex known as TFIIH that is thought to be involved in cell cycle control and DNA repair.

PRESENT RESEARCH STUDIES

The primary focus of Kornberg's present research is the characterization of the transcriptional regulatory complex known as mediator. His group is interested in determining how mediator functions in the regulation of genes. His work involves both functional and structural studies of the transcriptional machinery.

Roger Kornberg is a gifted biochemist whose contributions to the field of biochemistry have advanced the understanding of membrane structure and transport, DNA packaging, and transcriptional regulation. He earned the Nobel Prize in chemistry in 2006 for his work on eukaryotic gene transcription in yeast.

See also BIOCHEMISTRY; CRICK, FRANCIS; FRANKLIN, ROSALIND; HODGKIN, DOROTHY CROWFOOT; INORGANIC CHEMISTRY; NUCLEIC ACIDS; ORGANIC CHEMISTRY; PAULING, LINUS; PERUTZ, MAX; PROTEINS; WATSON, JAMES; WILKINS, MAURICE.

FURTHER READING

Berg, Jeremy M., John L. Tymoczko, and Lubert Stryer. *Biochemistry,* 6th ed. New York: W. H. Freeman, 2007.

The Nobel Foundation. "The Nobel Prize in Chemistry 2006." Available online. URL: http://nobelprize.org/nobel_prizes/chemistry/laureates/2006/index.html. Accessed March 30, 2008.

Roger D. Kornberg Laboratory Web site. Available online. URL: kornberg.stanford.edu. Accessed March 30, 2008.

laboratory safety Federal right-to-know laws and federal Hazardous Communications Standards require every employee in any workplace to be aware of potential chemical and industrial hazards in the workplace. The safety requirements as they apply to educational laboratory safety are known as the Lab Standard, which requires public schools and universities to develop a Chemical Hygiene Plan. According to the Occupational Safety and Health Administration, the U.S. government agency known as OSHA, the Chemical Hygiene Plan must address each of these six components: material safety data sheets (MSDSs), hazardous materials list, chemical inventory, notification, training, and labeling.

MSDSs summarize vital information relating to different chemicals. The MSDSs for companies in the United States must meet the following criteria:

- be written in English
- include the chemical name
- list all hazardous components
- describe the physical characteristics of the chemical (density, melting point, boiling point, and flash point)
- list all physical hazards including fire, reactivity, and explosion
- list all chronic and acute health hazards including signs and symptoms of exposure and any carcinogenic behavior
- list all primary routes of entry and target organs
- include the exposure limits
- instruct as to applicable precautions (such as gloves or goggles)
- outline first aid and emergency procedures
- state the date prepared
- display the name and address of manufacturer or MSDS preparer and phone number

Many MSDSs also include the disposal information for that chemical in order to ensure that it is properly contained. The hazardous materials list and the inventory need to be updated regularly and include the chemical names, quantities on hand, and location of all chemicals present in the workplace. The institution must inform all employees who work in the lab of the location of the inventory and the MSDS books. Independent contractors who do work on the school building or university are also required to carry copies of the MSDSs that apply to their chemicals. If a hired company refinished the gym floor in a school, and suddenly many students felt ill from the fumes, proper treatment for the students would require knowledge of the composition of the floor cleaner, information included in the MSDS. The personnel responsible for ordering need to understand that updating the chemical inventory is a priority as a safety concern. Knowing what chemicals are present in the school or laboratory is necessary for the safety of the teachers, workers, and students. The notification component of the Chemical Hygiene Plan involves notifying all employees who have been exposed to any potentially dangerous chemical while at the workplace of their rights. The training aspect of the Chemical Hygiene Plan of the Lab Standard requires that the institution hold training sessions for all employees so they understand the risks of the chemicals to which they may be exposed at the workplace. The employer must also train the employees and lab workers on the proper way to read a MSDS and inform them of their rights according to the right-to-know laws. The final piece of the Lab Standard is the labeling component. Every chemical that enters the lab must contain the original manufacturer's label that indicates the contents of the containers. When a chemical is removed from the original container, the secondary container must also be appropriately labeled. That labeling often includes the chemical name, safety information, concentration, and date of preparation.

Because of the importance of laboratory safety, in addition to the federal government standards on training, organization, and labeling, teaching laboratories have their own safety requirements that cover everything from behavior and dress to proper ways of handing chemicals and equipment. Before entering any laboratory to visit or perform an experiment, one must be familiar with the laboratory safety techniques specific to that lab and safety procedures of labs in general. Safety in the lab environment is everyone's concern. The students, visitors, scientists, and even custodians who enter a lab need to be aware of the safety procedures in place in a science laboratory.

The first required safety precaution is personal protective equipment (PPE), including items such as: safety goggles, lab aprons or lab coats, gloves, and respirators or masks. Wearing the proper personal protection equipment can prevent many laboratory accidents. Clothing and apparel are also important in science laboratories because improper clothing can lead to personal injuries. Open-toed shoes leave feet exposed to potential harm from hot, corrosive, or sharp objects. Dangling jewelry or clothing is a safety threat when it gets caught in equipment and traps the wearer, or pulls or knocks over laboratory equipment, leading to spills, breakage, and potential personal injury. Requiring safety goggles protects the eyes, which can be easily damaged by splashed or spilled chemicals. Lab aprons protect clothing, and most are chemically resistant to protect the body against the chemicals being spilled. Gloves protect hands during the handling of toxic or corrosive substances, and heat-resistant gloves protect against burns when handling hot glassware. In some cases, masks or respirators are required for the use of certain chemicals that are highly toxic or carcinogenic. Regulations for these would be determined and enforced for the specific lab situation.

Appropriate behavior when visiting or working in a lab is also crucial to maintaining a safe lab environment. All motions and movements should be controlled and calm. Erratic behavior can lead to spills, breakage, or injury. No eating or drinking is permitted inside a science laboratory. The potential transfer of chemicals to the food or drink creates a hazard. Gum chewing or application of cosmetics in the laboratory can also cause chemicals to be transferred from the laboratory surfaces to the person.

Despite taking all precautions, laboratory accidents still occur. Spilled chemicals are one of the most common types of lab accidents. Controlling spills in an educational science lab requires care by the students and planning by the lab instructor. Chemicals should be dispensed in plastic (unbreakable) bottles whenever possible. Chemicals should be ordered and stored in the smallest container possible. For example, five milliliters of hydrochloric acid should not be dispensed from a 2.5-liter bottle. This sets students up for spills and maximizes the damage done if a spill does occur. Providing students with only the amount of the chemical they will be using decreases the chance for waste and spills. Many chemicals have secondary containers that are able to contain the spill if the internal container were to break, minimizing the spill. Many acid storage cabinets have a containment shelf on the top of the cabinet with sides that allows for dispensing of the acid from a large container (where there is a greater chance for a spill), preventing spills from spreading past the containment tray. The chemical containers should not be opened until needed for use.

A chemical spill kit should be available in case a spill does occur. This kit could be as simple as a pair of gloves, sand, and kitty litter. When a liquid spill occurs, it is necessary to ventilate the area, contain the spill, and then clean up the spill. Opening windows and turning on exhaust fans and hoods will help remove fumes from volatile liquids. Pouring sand and/or kitty litter around the spill will contain the spill and facilitate cleanup. Spills of solid chemicals should be cleaned up immediately to prevent spreading and contamination of people from contact with the lab surfaces.

Glassware presents special safety concerns because it can shatter and break. Many people forget that hot glassware looks just like cold glassware, and one should know whether the glassware was heated before touching it to prevent burns. All glassware should be inspected prior to use to ensure that there are no cracks in the glass. Cracked or chipped glassware can shatter when heated or put under pressure during the course of an experiment, creating dangerous glass fragments that may be coated in the chemicals involved in the experiment. All broken glassware should be immediately disposed of in a designated broken glass container. Broken glass should not ever be put into the regular trash. Removal of the trash bag for disposal could cause the glass to puncture the bag or cut the person carrying the bag.

Electrical equipment used in laboratories can also present safety concerns. Equipment should be inspected regularly in order to determine whether it is in good working order and to be sure that the cords and plugs are not damaged. Any equipment that appears to be broken should not be used.

Effective and organized chemical storage is worth the effort when setting up a lab. Proper storage of chemicals simplifies experimental design, reduces double ordering when one item cannot be found, and aids in safety by having compatible chemicals stored together. When working in a lab, the first rule of hazardous chemical storage is to store the minimal amounts that are required. Keeping and storing a quantity of a hazardous chemical larger than necessary only increase the risk for accidents or injury. Some of the most dangerous chemicals in a lab are corrosives such as acids, bases, and flammable substances. All chemical labs should undergo periodic cleaning to remove any older chemicals that are no longer needed or have passed their useful lifetime. Acids and flammable substances require special storage containers in order to prevent spills and keep the corrosive and flammable chemicals away from other chemicals.

In order to ensure the safety of those who work in and visit the lab, one must safely secure and store the chemicals in that laboratory according to reactivity. Incompatible chemicals must be stored separately. Storing chemicals by class is the safest way to accomplish this. Organic chemicals (chemicals containing carbon) should be separated from inorganic chemicals (chemicals that do not contain carbon). All other chemicals should be stored by family, as indicated in the following list; failure to store chemicals properly can lead to disastrous results. According to the U.S. Environmental Protection Agency (EPA), chemicals should be separated into the following groups for storage: inorganic acids, organic acids, caustics, amines, halogenated compounds, alcohols, glycols and glycol ether, aldehydes, ketones, saturated hydrocarbons, aromatic hydrocarbons, petroleum oils, esters, monomers, phenols, nitriles, ammonia, and halogens.

FURTHER READING

U.S. Department of Labor. Occupational Safety and Health Administration Web site. Available online. URL: http://www.osha.gov. Accessed July 25, 2008.

lasers

Laser stands for *l*ight *a*mplification by *s*timulated *e*mission of *r*adiation. Lasers utilize a monochromatic (single-color) light source that emits all its light in the same direction, at the same wavelength, and in phase, that is, lined up crest to crest and trough to trough. Such light is termed coherent. Lasers can produce exceptionally concentrated beams that are useful in many fields including medicine, the military, building, and household use.

Electrons are negatively charged subatomic particles that are located in electron clouds surrounding the nucleus of an atom. The electrons reside in the lowest energy levels available to them unless they absorb energy from an outside source. The energy boosts them to a higher energy level, and the electrons, as well as the atoms that contain them, are said to be in an excited state. As an atom deexcites, its excited electron falls to a lower energy level, and the atom releases energy in the form of a photon of electromagnetic radiation. Photons are the bundles, or quanta, of energy that form electromagnetic waves, according to quantum mechanics.

The wavelength of a wave is the distance between two adjacent crests of the wave. The wavelength is inversely proportional to the energy of each of the individual photons making up the wave. Thus, each photon of a wave with a long wavelength has less energy than one of a shorter-wavelength wave. The frequency of a wave is a measure of how many cycles, that is, wavelengths, of the wave pass a given point per second. Frequency and wavelength are inversely proportional. Consequently, the energy of an individual photon is proportional to the frequency of its wave. Thus, the energy of each photon of a wave with a high frequency is greater than that of a photon of a lower-frequency wave.

The form of electromagnetic radiation with the shortest wavelength and the highest frequency is gamma rays, followed by X-rays and ultraviolet radiation with longer wavelengths and lower frequencies. Visible light includes all of the wavelengths that can be seen with the naked eye. Visible light has lower frequencies and longer wavelengths than ultraviolet light, with violet at the high-frequency (short-wavelength) end and red at the low-frequency (long-wavelength) end of the range of the visible portion of the electromagnetic spectrum. The types of electromagnetic radiation with lower frequencies and longer wavelengths than visible light include, in order of descending frequency, infrared radiation, microwaves, and radio waves. Since photon energy is proportional to frequency, the preceding ranking of forms of electromagnetic radiation from high frequency (short wavelength) to low frequency (long wavelenth) also ranks the associated photons from high energy to low. For example, an ultraviolet photon is more energetic than an infrared one. Or a photon of blue light (about 460 nm) is more energetic than a photon of yellow light (about 565 nm).

Once an atom becomes excited, it can be used to produce a laser beam. Lasers are made of light rays that all travel in the same direction and that have perfectly aligned crests and troughs. Their operation depends crucially on an effect called stimulated emission, which works in this way. Assume that an atom is in an excited state. After some time, it will normally spontaneously deexcite, that is, make a spontaneous transition to some lower-energy state, and at the same time emit a photon whose energy equals the difference between the energies of the initial and final states. However, if the photon energy of a passing electromagnetic wave happens to match the energy the atom would emit in such a transition, then the atom will tend to make the transition sooner than it would spontaneously. It is then "stimulated," so to speak, to deexcite. Moreover, the emitted photon will join the stimulating wave and add to it in a coherent manner, meaning with the same direction, wavelength, and phase.

Knowledge of photons and stimulated emission led to the development of functional lasers. In order to operate, a laser needs three main components: an active medium, a source of energy known as pumping energy, and an optical cavity. An active medium must contain appropriate energy levels in its atoms in order to give off the desired wavelength of light; this active medium can be solid or gas.

The pumping energy excites the atoms in the laser medium and creates what is known as a population inversion. In this situation, there are more atoms excited to a higher-energy state than there are excited to the next-lower-energy level. When atoms spontaneously deexcite, their emitted radiation stimulates further deexcitations, and a strong, coherent electromagnetic wave builds up. The process would eventually stop when the high-energy states all deexcite. But the pumping source maintains the population inversion and allows continuous laser action.

Laser light is stronger than similar electromagnetic radiation from other sources. Waves (including electromagnetic waves) that are lined up crest to crest and trough to trough are said to be in phase, and they reinforce each other. When waves are not aligned this way, they are said to be out of phase, and some, or much, cancellation and weakening can occur. In light from a source such as an incandescent or fluorescent lamp, the photons all move in different directions and the emitted waves are out of phase. In lasers, all of the waves are in phase, so intensity is maximized.

In addition, a laser is made up of an optical cavity that is designed to contain the signal, causing all of the photons to exit in the same direction and in phase, thus amplifying the signal. Finely adjusted mirrors located at either end of the optical cavity contain the laser beam by reflecting the light back and forth between them. The mirror at the exit point is generally 10 to 90 percent nonreflecting to allow some of the laser beam to leave the optical cavity and perform a useful function. As the photons bounce around in the optical cavity, they stimulate further deexcitations and, therefore, produce even more photons, creating a stronger beam. The nonreflecting mirror allows the photons to leave the laser at only one point and to move in only one direction.

The laser was originally developed from its predecessor, the maser (*m*icrowave *a*mplification by *s*timulated *e*mission of *r*adiation). Similar technology to that of the laser was used in the maser, which produced microwaves, rather than visible light, in a beam. The first maser was developed in 1953 by the American physicist Charles H. Townes at Bell Laboratories, and it was functional although it was only able to pulse a beam, not emit a continuous beam. After Townes's work, Nikolay G. Basov and Aleksander Prokhorov of the USSR overcame the previous problems with the maser and population inversion, leading to a continuous-output maser. The collective work of Townes, Basov, and Prokhorov earned them the Nobel Prize in physics in 1964. In 1957, Gordon Gould of Columbia University coined the term *laser*. Theodore H. Maiman of the Hughes Research Laboratory manufactured the first work-

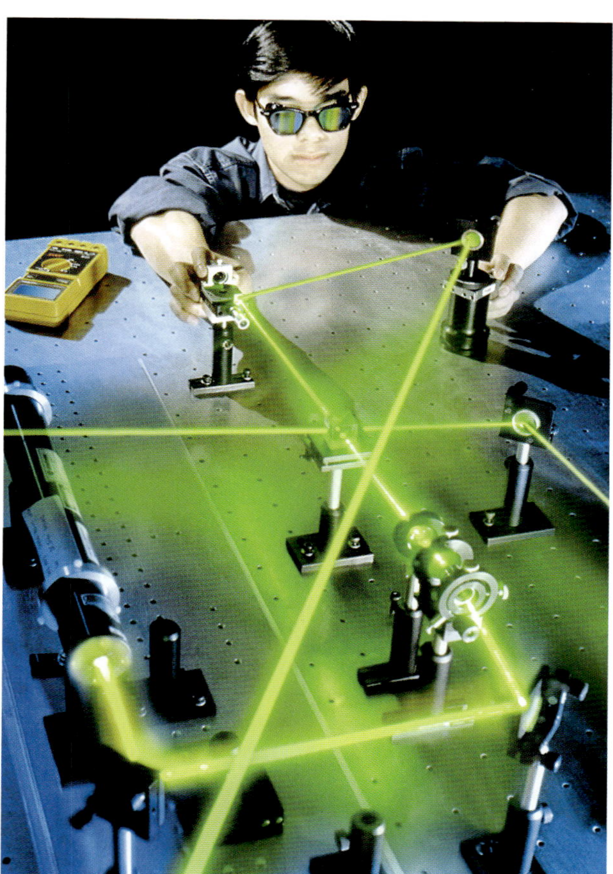

The laser (on the left) is supplying the high-intensity light beam required for this experiment in optical computing research. *(© Firefly Productions/CORBIS)*

ing pulsed laser using ruby as a medium, a success he published in the journal *Nature* in 1960. Some controversy surrounds who deserves the credit for this invention. Townes technically invented the laser, Gould technically named it, but Maiman was the first to make it work.

TYPES AND USES OF LASERS

There are many types of laser media including solid-state lasers, gas lasers, and semiconductor lasers. Each of these laser types excites a different medium, allowing for the production of light in a multitude of wavelengths and strengths. Solid-state lasers contain solid pieces of insulators that serve as the excitable medium. Reflecting mirrors contain this medium in a chamber. An example of this type of laser is the holmium YAG solid-state laser used in dentistry and for destroying kidney stones.

Gas lasers contain a gas, sealed in a tube with mirrors at its ends. Examples of this type of laser include the helium-neon gas laser that is used in supermarket barcode scanning, holographs, and spectroscopy (the study of the interaction between light and matter). Argon-ion gas lasers are used in general and ophthalmologic surgeries and in treatment of certain medical disorders, and the xenon-krypton gas lasers are used mainly for scientific research. The Excimer chemical laser is used in laser surgery and LASIK eye surgery. Military weaponry utilizes the hydrogen fluoride laser, and the carbon monoxide gas laser is used for cutting and welding.

Semiconductor lasers excite a medium made of semiconductors. The light generated is bounced back and forth to give the stimulated emission. The most widely used semiconductor laser type, the aluminum-gallium-arsenide laser, emits a light source at 780 nm and is found in compact disk players and in laser pointers that have a small laser run by battery in a compartment that contains mirrors.

The uses of lasers are as varied as the types of media available to produce them. Lasers are used in medicine, carpentry, weapons, household goods, and other technologies. In medicine, lasers are used as tools to remove damaged tissues and in the removal of unwanted imperfections of the skin including hair, tattoos, birthmarks, and scars. Lasers are useful in dentistry for surgery as well as for stain removal. Laser eye surgery, or refractive eye surgery, one of the fastest-growing surgeries in the country, includes photorefractive keratectomy (PRK) and *l*aser-*a*ssisted *i*n *s*itu *k*eratomileusis (LASIK). Defects of the cornea and shape of the eye cause many vision problems: myopia, or nearsightedness; hyperopia, or farsightedness; and astigmatism, a condition characterized by an irregularly shaped cornea (most common) or lens. Refractive eye surgery consists of the temporary removal of the front of the cornea, subsequent removal of tissue from the middle of the cornea using a laser to correct the shape, then replacement of the front of the cornea. Successful laser eye surgery can eliminate a patient's need for wearing glasses or contacts by correcting vision problems.

Lasers are also used in scientific research, especially in spectroscopy. They are used in laser levels and tools such as saws in order to ensure that lines are straight in the building industry. With the use of lasers, a builder can determine whether two points across a room are at equal heights without measuring. This has simplified the installation of pictures, flooring, and even major structural components in homes today. Many industrial processes also depend on laser technology for cutting through metal and etching glass and metals.

Military and hunting weapons have lasers installed as sighting mechanisms. Many guns include lasers that improve aim and by beaming light at a specific point on the target, greatly increasing accuracy. Even the entertainment industry exploits laser technology. The game of laser tag makes millions of dollars and has surfaced in nearly every city as a form of group entertainment. Vests are worn that have sensors that detect laser light. The "guns" give off a laser beam that activates the sensor and can be recorded as a score.

Lasers are encountered in one's everyday life in a variety of ways. Lasers are critical to the functioning of the telecommunication system. In 1974, the development of the laser bar code reader changed the way that we shop, because checkout lines do not require keying in the price of every item. Americans can scan their own purchases on a laser bar code reader at self-checkout lanes in grocery stores. The laser disk player was developed in 1978 and although large and cumbersome by today's standards was able to store a large amount of information accessibly at the time. Compact disk (CD) players developed in 1982 are now found in home stereo systems, cars, and computers. They revolutionized the way music, movies, and information were played and stored. Today digital technology is once again changing the way we store information and use entertainment.

See also ATOMIC STRUCTURE; ELECTROMAGNETIC WAVES; ELECTROMAGNETISM; ELECTRON CONFIGURATIONS; QUANTUM MECHANICS; SPECTROPHOTOMETRY.

FURTHER READING

Chang, William. *Principles of Lasers and Optics.* Cambridge: Cambridge University Press, 2006.

Silfvast, William T. *Laser Fundamentals.* Cambridge: Cambridge University Press, 2004.

Lauterbur, Paul (1929–2007) American *Chemist*

Paul Lauterbur was an American-born chemist who spent most of his life studying the uses of nuclear magnetic resonance (NMR) technology. His greatest contributions were his discoveries in magnetic resonance imaging (MRI) technology. Lauterbur studied the NMR properties of water and applied them to studying living tissues. He discovered that using magnetic gradients rather than one continuous magnetic field would create variations in the image that would allow for the creation of two-dimensional images from a NMR scan. His work in this field led to a Nobel Prize in physiology or medicine, which he shared with the Scottish physicist Sir Peter Mansfield in 2003.

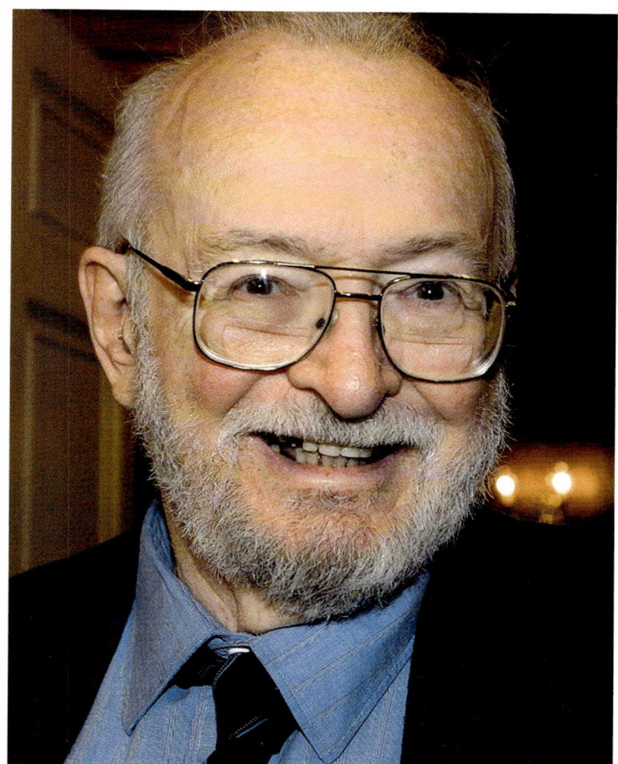

Paul Lauterbur was an American chemist who received the Nobel Prize in physiology or medicine in 2003 for the development of magnetic resonance imaging (MRI). Lauterbur shared this prize with the Scottish physicist Sir Peter Mansfield. *(AP Images)*

EARLY YEARS AND EDUCATION AND TRAINING

Paul Lauterbur was born to Edward Joseph Lauterbur and Gertrude Wagner in Sidney, Ohio. The family spent much time on their family farm, and Paul was avidly interested in geology and animals. His interest in science continued to grow through high school. In 1951, Lauterbur completed his undergraduate work at Case Institute of Technology, which is now part of Case Western Reserve University in Cleveland, Ohio. He majored in chemistry and was especially interested in the processes of organometallic substances. Lauterbur did not have any immediate desire to continue his education after his undergraduate work, so he took a position at Mellon Institute of Dow Corning in Pittsburgh, Pennsylvania.

While working at the Mellon Institute, Lauterbur was called up into military service. Because of his background he was assigned to a science department of the army. Rather than hinder his career, the time he spent away from school in the army actually encouraged his career in the field of NMR. He had the unique opportunity to work on a NMR machine, and he even published several papers from his work during that time. After his service in the army, Lauterbur returned to the Mellon Institute, where they had purchased a new NMR machine for his use.

Lauterbur completed his doctorate in chemistry 1962 at the University of Pittsburgh. While working and taking courses there, Lauterbur studied the process of strengthening rubber, and he based his research on the addition of phthalocyanine dyes to strengthen rubber. While he was studying at the University of Pittsburgh, Lauterbur's interest in NMR was cemented. His personal life also changed as he married Rose Mary Caputo in 1962. The couple had two children. Lauterbur passed on postdoctoral fellowship offers to take an assistant professor position at the State University of New York (SUNY), where he worked from 1969 to 1985. Here Lauterbur spent the majority of his research career studying the possible applications of NMR technology to imaging living tissues.

NUCLEAR MAGNETIC RESONANCE (NMR)

The use of nuclear magnetic resonance in condensed matter had been introduced in the 1940s. When a magnetic field is applied to atoms, they will rotate in the magnetic field according to the strength of the magnetic field. When these nuclei are subsequently subjected to radio waves, the amount of energy given off by the individual atoms is received as a signal that represents the location of the atoms and therefore the structure of the matter being studied. Different atomic nuclei (e.g., hydrogen, carbon, phosphorus, and sodium) possess signature patterns that produce specific NMR spectra. Within different compounds, the properties depend upon the location of the atoms as well as the effect of the neighboring atoms. This relationship became known as the chemical shift of NMR.

Lauterbur studied NMR for nearly his entire career. From early on, he sought methods to apply NMR technology to biological imaging systems. Such applications became highly important. The high concentration of water (which was easily studied in NMR) made living organisms perfect candidates for the application of this technology. Many NMR scientists prior to Lauterbur and Mansfield investigated the possibility of using NMR for human studies, but none with any seriously designed research. Stories circulated that both Edward Purcell and Felix Bloch (innovators in the discovery of NMR) used themselves as test subjects. Purcell was thought to have put his head in the NMR device, and Felix Bloch had put his thumb in the device to create a scan. Lauterbur wanted to present well-designed and well-controlled experiments that could lead to biological applications of NMR.

DISCOVERIES IN MRI

The development of MRI from NMR technology required a shift in thinking of those who were studying the NMR processes. Many believed that the process of utilizing NMR in living tissues was not possible. Lauterbur observed studies done on excised rat tissues that led him to believe it was possible. The rat tissue experiments appeared to be accurate and reproducible, convincing Lauterbur that it was possible to differentiate between the NMR spectra of normal tissues, cancerous tissue, and damaged tissue. He set about developing a method to distinguish these tissue differences without sacrificing the animal. He wanted to demonstrate that living tissue within organisms could be examined the same way as tissues excised from animals. The differences in water content between the normal and diseased/damaged tissues formed the basis for his study of the NMR patterns of tissues.

Lauterbur had to overcome several problems in order to develop this technology. These problems included the necessary mathematical calculations, the ability to create a strong enough signal-to-noise ratio, and the practicality of creating a large enough magnet to produce the signal. At the time, Lauterbur was unaware of any calculations available to determine a method in which a one-dimensional NMR signal could create a visual image that was useful to physicians and technicians. Lauterbur utilized differentially oriented magnetic fields (known as magnetic field gradients) and hand-calculated simple tests of individual one-dimensional NMR scans. Although many around him said that a multidimensional image could not be calculated from the one-dimensional scans, he continued his efforts until he determined that it could be done. He continually searched for mathematical and computer science solutions to this problem. The methods that he finally came upon were repeat methods that had already been published in other fields such as radioastronomy.

The second problem was determining whether there could be enough signal intensity over the noise from the machine. Since the machine would have to be large enough to hold an entire human body, the background noise was a significant issue. Lauterbur performed his own calculations and established enough data to convince himself that the signal would be strong enough to measure. The third problem that Lauterbur needed to solve was the practicality of large enough magnets. His engineering background greatly benefitted Lauterbur, and with the help of engineers and mathematicians he determined that such a magnet could be built.

When Lauterbur believed that he had reasonable solutions to each of these problems, he began his journey into the development of MRI. Although most scientists at the time argued that the process simply could not work, Lauterbur continued his diligent research and ultimately successfully imaged living tissues using NMR. Lauterbur was once told that the MRI "would not repay the cost of getting the patent." His initial work involved all types of tissues from plants and animals, and all of the MRI imaging processes showed accurate and reproducible results.

APPLICATIONS OF MRI

The best-known application of NMR is magnetic resonance imaging (MRI). This technique applies the principles of NMR to diagnosing biological injuries, diseases, and conditions. MRI has multiple advantages over procedures such as X-rays. MRI uses radio waves, which cause no known side effects, whereas X-rays can cause damage to living cells. In contrast to X-rays, which allow the observation of bones, MRI allows for the visualization of soft tissues. The hydrogen atoms in water are ideal subjects for NMR technology; thus, the large amount of water in living tissues makes possible their imaging by NMR techniques. Determining the location and quantity of water in tissues makes it possible to locate and identify tissue damage, tumors, and tissue degeneration. This makes MRI irreplaceable in diagnosing brain and spinal cord diseases and injuries, cardiovascular issues, aneurysms, strokes, and multiple sclerosis, as well as sports injuries and other soft tissue damage.

MRI scans take approximately 20–90 minutes, depending on the extent of the tissue to be scanned. The patient is placed into a large hollow cylinder that contains the magnet and encircles the patient. When the magnetic field is applied, the patient hears pounding noises that indicate radio waves are being pulsed through the chamber. Prior to having an MRI scan, patients must remove all metal objects. The intense magnetic field used during the test could be very dangerous. In 2001, a child was killed when an unsecured oxygen tank was inadvertently pulled into the MRI machine. Many conditions exist that would exclude a patient from having a MRI done. These include the following:

- pacemaker
- metal implants
- insulin pump
- ear implant
- bullets
- tattoos
- metal fragments
- prosthesis
- hearing aids
- claustrophobia
- brain surgery
- pregnancy
- epilepsy
- history as a machinist or welder

During the MRI, the technician maintains communication with the patient via headphones. The subject must remain perfectly still during the course of the test to prevent blurred images.

RESEARCH INTERESTS AFTER THE NOBEL PRIZE

In 1985, Paul Lauterbur left SUNY to accept a professorship at the University of Illinois, in Chicago; he retired in 2004. He spent much of his time there as director of the Biomedical Medical Resonance Laboratory, where he continued his research on magnetic resonance imaging. The contributions to medical imaging made by Paul Lauterbur and Peter Mansfield have benefited millions of people worldwide, as approximately 60 million MRI scans are performed each year. The technology that led to the development of this common, useful, noninvasive procedure is due to the work of scientists such as Paul Lauterbur. He died on March 27, 2007.

See also MANSFIELD, SIR PETER; NUCLEAR MAGNETIC RESONANCE (NMR).

FURTHER READING

Liang, Zhi-Pei, and Paul C. Lauterbur. *Principles of Magnetic Resonance Imaging: A Signal Processing Perspective.* New York: Institute of Electrical and Electronics Engineers, 2000.

The Nobel Foundation. "The Nobel Prize in Physiology or Medicine 2003." Available online. URL: http://nobelprize.org/nobel_prizes/medicine/laureates/2003/. Accessed March 26, 2008.

Lavoisier, Antoine-Laurent (1743–1794)

French *Chemist* Antoine-Laurent Lavoisier was a French-born scientist who is best known for experimentally determining the law of conservation of mass, stating that mass can be neither created nor destroyed. He also disproved the phlogiston theory of matter, a strongly held concept of alchemists. Lavoisier received credit for naming two elements, oxygen and hydrogen, and he wrote the first chemistry textbook. Called the father of modern chemistry, Lavoisier revolutionized the thinking and nomenclature of the field of chemistry.

BIRTH AND EDUCATION

Antoine-Laurent Lavoisier was born on August 26, 1743, into a wealthy and well-educated family in Paris, France: Jean-Antoine Lavoisier, his father, was a respected lawyer, and his mother, Émilie Punctis Lavoisier, was a member of a wealthy family. Lavoisier inherited a vast amount of money upon his mother's death, when he was only five years old. His family expected him to follow the family pattern and pursue a career in law. He went to Collège des Quatre-Nations, also called Collège Mazarin, with

Antoine-Laurent Lavoisier was an 18th-century French chemist known as the father of modern chemistry. His contributions include the law of conservation of mass and an organized system of chemical nomenclature. *(National Library of Medicine)*

the original plan to enter law, but he also studied chemistry, geology, mathematics, astronomy, and botany. Lavoisier received his bachelor of law degree in 1763. By this time, Lavoisier knew his true love was science, and he never practiced law. The French Academy of Science recognized and admitted him after he wrote an essay on potential improvements to street lighting in Paris.

In 1771, Lavoisier married Marie-Anne Pirrette Paulze, who was only 13 years old. She later became his assistant, and translated many of his works into English, and drew sketches and pictures to accompany his work. The couple did not have any children.

TRANSMUTATION AND PHLOGISTON THEORY

The careful experimental work of Lavoisier advanced the field of chemistry by disproving some long-held bogus theories about matter and combustion. Until Lavoisier's time, most chemistry was performed by alchemists, who believed that all matter consisted of various amounts of four basic elements—earth, water, air, and fire—that could be combined to make different types of matter. While working at the French Academy of Sciences, Lavoisier examined the purity of Paris's drinking water. Boiling water in sealed vessels for months left behind particulate matter. Alchemists believed that the water had transmuted into earth since the vessel and the water together retained the same mass as was originally tested. Lavoisier took the experiment one step further. He removed the water from the vessel and discovered the mass of the water was identical to the original mass before boiling. The solids found in the water sample were shown to have emerged from the glass after prolonged boiling. This experiment demonstrated that water was not transmuted into another substance.

The scientific community also believed that matter underwent a chemical reaction by gaining or losing phlogiston, a mysterious substance. The phlogiston theory stated that all materials released a substance known as phlogiston during burning, and the process of burning and releasing phlogiston left behind a dephlogisticated substance. The more phlogiston a type of matter had, the more it would burn. Phlogiston was considered to have mass in some cases and negative mass in others (as suited their purposes). When a substance lost mass upon combustion, it was said to lose phlogiston; when a substance gained mass, such as the combustion of magnesium (forming magnesium oxide), alchemists claimed the cause to be the loss of phlogiston that had negative mass. Strong support for the phlogiston theory and a fundamental misunderstanding of how chemical reactions occurred had prevented any significant advances in chemistry for many years. Lavoisier disproved the phlogiston theory by performing repeated, controlled experiments in which he burned multiple types of metals. He showed that the reason that substances gained weight during burning was not the loss of phlogiston carrying a negative mass, but the combination of the metal with oxygen from the air.

LAW OF CONSERVATION OF MASS

Some of Lavoisier's greatest contributions to modern chemistry relate to the stoichiometric relationships of elements in a chemical reaction. Lavoisier's careful, methodical quantitative experiments went a long way in demonstrating that the amount of substances that enter a chemical reaction is always equal to the amount of substances that leave the reaction. No matter can be lost along the way without being gained by the surroundings, and no mass can be gained unless it is subtracted from the surroundings. This led to his establishment of the law of conservation of mass (or matter), which states that mass (or matter) can be neither created nor destroyed. This relatively simple concept revolutionized 18th-century chemistry and is fundamental to chemistry today.

NAMING OF OXYGEN

Lavoisier is also credited with naming the element oxygen. Although most sources credit Joseph Priestley with the discovery of oxygen, Lavoisier took credit for much of his work. Throughout his career, Lavoisier demonstrated a pattern of taking credit for the work of others. Some science historians claim that it was Lavoisier's synthesis of other scientists' work that enabled him to impact the field of chemistry. Lavoisier recognized the existence of oxygen and its relationship to combustion, naming oxygen as the element that was required for combustion to occur. The phlogiston theory stated that combustion occurred because the air contained phlogiston that was required for burning. When a fire burning inside a closed container died, supporters of the theory believed the depletion of phlogiston was responsible for ending the combustion and said the air was then dephlogisticated. Lavoisier demonstrated that oxygen was the substance that supported a burning flame and that the lack of oxygen was responsible for the extinguished flame. Lavoisier demonstrated that processes such as cellular respiration and rusting required oxygen as well. The name *oxygen* means "acid-former," as Lavoisier incorrectly proposed that all acids contained oxygen.

CHEMICAL NOMENCLATURE

Lavoisier recognized that a stumbling block of chemistry at the time was the inability of scientists to communicate and share information in an organized fashion. Many scientists purposely named compounds cryptically to prevent people from stealing their ideas. Lavoisier recognized that in order for chemistry to advance, chemists worldwide needed to adopt a systematic method for naming compounds. In 1787, Lavoisier published a nomenclature system that he developed in *Méthode de nomenclature chimique* (Methods of chemical nomenclature), with the help of three other French chemists, Louis-Bernard Guyton de Morveau, Claude Berthollet, and Antoine Fourcroy. The first chemistry textbook, *Traité élémentair de chimie* (Elementary treatise on chemistry), was published by Lavoisier in 1789 and included such concepts as the definition of an element as the smallest substance that cannot be broken down further, a list of what he considered to be the known elements at the time, the law of conservation of mass, methods of chemical nomenclature, states of matter, and a definition of compounds.

WORK ON CELLULAR RESPIRATION

Lavoisier expanded his work on combustion to include the process in animals and plants known as cellular respiration. The study of the amount of heat gained or lost during a chemical reaction is known as calorimetry. Experiments designed to measure heat exchange use an instrument called a calorimeter. The outer portion of the calorimeter is insulated and filled with water, and the sample is placed in an interior vessel. When the reaction inside the reaction vessel gives off heat (exothermic), the water temperature in the outer chamber rises, and when the reaction requires heat (endothermic), the temperature of the water decreases. Calculations based on this change in temperature reveal the amount of energy given off or taken in by the chemical reaction.

Lavoisier utilized a calorimeter to measure the amount of energy given off by a respiring guinea pig in order to study the heat transfer of a living organism. Lavoisier first determined the amount of heat given off by the guinea pig by measuring the amount of ice melted. He took the experiment one step further to determine the amount of carbon dioxide produced while the temperature changed. Lavoisier then burned the amount of charcoal required to produce the equivalent amount of carbon dioxide and showed that it produced the same amount of heat. This led Lavoisier to propose that cellular respiration using oxygen in animals and plants was a combustion reaction similar to burning charcoal.

LAVOISIER'S HISTORY AS A TAX COLLECTOR

In addition to his fascination with science, Lavoisier had a strong interest in business. In 1768, Lavoisier invested in a business, Ferme Générale, which was responsible for collecting taxes for the French government. Lavoisier's father-in-law, Jacques Paulze, was one of his partners in this business venture. Many of these tax collectors were not ethical and took more than they needed to collect. Lavoisier's and Paulze's association with the Ferme Générale eventually led to their deaths. When the French Revolution broke out in 1789, Lavoisier was targeted for his association with the Ferme as a governmental organization. He was arrested, tried, and sentenced to death on November 1793 and beheaded on the guillotine on May 8, 1794.

Lavoisier had been a strong supporter of the revolution from its early days, so some believe it was for his scientific findings as much as his political beliefs that he was executed. Chemistry lost a valuable asset that day. Lavoisier did not necessarily discover anything new, but he was a master experimenter and could understand the larger picture. He was able to take a large amount of work done by other scientists, synthesize it, repeat it, and show the significance of it to the greater body of work in chemistry at the time. Without him, much of the work of individuals would never have been synthesized into a coherent field of chemistry. Lavoisier contributed much to scientific experimentation by his reliance on experimental evidence to guide his findings. His belief in this is found in his quote "I have tried . . . to arrive at the truth by linking up facts; to suppress as much as possible the use of reasoning, which is often an unreliable instrument which deceives us, in order to follow as much as possible the torch of observation and of experiment." Lavoisier's evidence-based approach helped found the field of modern chemistry.

See also CALORIMETRY; CHEMICAL REACTIONS; PRIESTLEY, JOSEPH; STATES OF MATTER.

FURTHER READING

Donovan, Arthur. *Antoine Lavoisier: Science, Administration, and Revolution.* New York: Cambridge University Press, 1996.

Lavoisier, Antoine. *Elements of Chemistry.* With introduction by Douglas McKie. Mineola, N.Y.: Dover, 1984.

lipids

Lipids are one of four types of biomolecules found within all living things. This group of biomolecules includes triglycerides, steroids, phospholipids, and waxes. Both eukaryotic and prokaryotic cells have lipids. Within the eukaryotic cell, lipids are made inside a cellular organelle called the smooth endoplasmic reticulum. Lipids dissolve easily in organic solvents, such as ether and chloroform, but not in water. The hydrophobic (water-hating) nature of lipids distinguishes them from the other three types of biomolecules: carbohydrates, proteins, and nucleic acids. The hydrophobicity of lipids makes them excellent cellular barriers. Lipids also efficiently store energy in cells and organisms.

TRIGLYCERIDES

Triglycerides, commonly referred to as fats and oils, consist of one molecule of glycerol and three attached long-chain fatty, or carboxylic, acids. An ester bond is formed between each of the three hydroxyl groups of glycerol and the carboxyl end of a fatty acid that ranges between 12 and 24 carbons in length.

If the carbons in the fatty acids are linked together via single bonds, the fatty acid is said to be saturated. Saturated fatty acids can fit tightly together because the chains have a tendency to extend straight out from the glycerol backbone. Because the saturated triglycerides, or fats, can make so many intermolecular contacts, fats exist as solids at room temperature. Most fats, such as lard, are from animals; however, coconuts and palm kernels also contain fats.

Oils are the second type of triglycerides, and they exist as liquids at room temperature. Oils are triglycerides composed of unsaturated fatty acids: that is, some of the carbons in these chains are linked via double bonds. The double bonds create bends in the long carbon chains, hindering or preventing intermolecular interactions, and the dispersion forces are less effective in these triglycerides. Most oils are plant products, such as olive and canola oils.

As are other esters, fats and oils are easily hydrolyzed by acids and bases. Saponification, the process of soap making, involves the hydrolysis of oils or fats in a boiling aqueous solution containing alkali metal hydroxide. A fat or oil is heated with an excess of an alkali metal hydroxide, such as sodium hydroxide. Sodium chloride is then added to this mixture, causing the sodium salts of the fatty acids to rise in a crude curd to the top of the mixture. The crude soap is then further purified. A by-product of this reaction is glycerol, which can be obtained by evaporating the aqueous solution.

As more Americans struggle with health problems such as heart disease and obesity, many members of the health community have made it a priority to educate the general public about good and bad fats. A recent target for public awareness has been trans fat, a synthetic hydrogenated triglyceride. Trans fats are made from partially hydrogenated oils that are exposed to additional hydrogen under high pressure. The result of this process is a solid fat with the consistency of shortening. Trans fat can be found in store-bought cookies, crackers, potato chips, cake frosting, and margarine. Many fast food restaurants use trans fat to prepare french fries, fried pies, biscuits, and doughnuts. Twenty years ago, food manufacturers started using trans fat to increase the shelf life of their products. Research indicates that trans fat in the diet increases the risk of obesity and heart disease. Trans fat has been shown to clog arteries, a condition that can lead to stroke or heart attack and that increases the overall level of cholesterol in the body. Naturally derived saturated fats such as butter also increase cholesterol levels; however, only trans fat actually reduces the level of high-density lipoprotein (HDL) cholesterol, the good cholesterol, while increasing the levels of low-density lipoprotein (LDL) cholesterol, the bad cholesterol. To address the problem, the U.S. Food and Drug Administration now requires that food manufacturers show on food labels the amount of trans fat in each serving.

STEROIDS

Steroids are nonester lipids, composed of three six-carbon and one five-carbon aromatic ring linked together to form a chicken wire structure common to all steroids. Functional groups attached at different points around this structure give rise to many different types of steroids. Seven major groups of steroids exist. They are cholesterol, progestagens, androgens, estrogens, glucocorticoids, mineralocorticoids, and vitamin D.

Cholesterol is the precursor to all steroids. In eukaryotes, it is essential for growth and viability in higher organisms because it modulates the fluidity of the cell membrane; however, high serum levels of cholesterol give rise to atherosclerotic plaques, which can lead to disease and death. Progestagens such as progesterone aid in initial implantation and later maintenance of pregnancy. Synthesis of these steroids occurs in the corpus luteum. Androgens, the class of steroids that includes testosterone, are responsible for the development of male secondary sexual characteristics and are made in the testis. Estrogens, such as estrone, are responsible for the development of female secondary sexual characteristics and are synthesized by the ovary. Glucocorticoids such as cortisol regulate gluconeogenesis, the formation of glycogen, and the metabolism of fats and proteins. Mineralocorticoids, mainly aldosterone, maintain sufficient blood volume and pressure by positively affecting sodium reabsorption by the kidneys. The adrenal cortex synthesizes both glucocorticoids and mineralocorticoids. Another steroid derivative, vitamin D, plays an essential role in calcium and phosphorus metabolism. Vitamin D deficiency results in the debilitating disorders of rickets in children and osteomalacia in adults. Bones become so soft that they twist, bend, and readily break from the weight of the individual.

PHOSPHOLIPIDS

Phospholipids are a third class of lipids. As the name implies, these lipids contain a phosphate group. Phospholipids consist of two distinct parts—a hydrophilic (water-loving) head and two hydrophobic tails. Phosphoglycerides contain a glycerol backbone with two long fatty acid chains and a phosphorylated alcohol. The fatty acid chains usually have an even number of carbon atoms ranging from 14 to 24 carbons. The most common phosphoglycerides are phosphatidyl choline, phosphatidyl serine, phosphatidyl ethanolamine, phosphatidyl inositol, and diphosphatidyl glycerol. Sphingomyelin is a phospholipid found in biological membranes with a sphingosine backbone. In this phospholipid, a fatty acid is linked to the backbone via an amide bond. Esterification of a hydroxyl group to phosphoryl choline causes sphingomyelin to behave as phosphatidyl choline.

Phospholipids are the main component of biological membranes. In an aqueous solution, the hydrophobic portions of these lipids join to exclude water while the hydrophilic portions of these molecules form polar interactions with the water around them. One arrangement that these amphipathic (both hydrophilic and hydrophobic) molecules can spontaneously form in an aqueous medium is a micelle, a spherical structure in which the hydrophilic parts of the lipid face out toward the medium and the hydrophobic tails are inside the sphere itself. A second, more prevalent arrangement formed by these molecules is the formation of a double layer of phospholipids called a lipid bilayer. The hydrophobic tails join in the middle of this entity, and the hydrophilic moieties face out toward the aqueous environment. Proteins are also embedded in this biomolecular sheet in biological membranes. The bilayer is not rigid in nature. Both phospholipids and proteins continue to diffuse laterally along the bilayer, but they rarely flip from one side to the other. The result is that each side of the bilayer has its own unique components and properties. In other words, a cell membrane is asymmetric with a different makeup of lipids and proteins within each layer.

A cell membrane is selectively permeable, allowing the entry and exit of specific ions and molecules. Proteins found within the lipid bilayer regulate this selective permeability. Some proteins serve as gates, or channels, permitting recognized molecules in or out of the cell. Other proteins join to form pumps that regulate ion concentration in and out of the cell. Still others act as receptors and enzymes. Often a carbohydrate group attached to a protein that faces the exterior of the cell forms hydrogen bonds with water and serves as a means of cell recognition. All of these proteins act in concert with the phospholipid bilayer to serve as a barrier between the inside and outside of the cell.

A method called electroporation circumvents the selective permeability of the cell membrane by creating temporary holes in the membrane using an electrical current. These holes permit molecules, including drugs, to enter a cell that would normally be inaccessible. Because components of the cell membrane are made up of charged molecules, the structure of the membrane is sensitive to external electrical sources. Exposure of a cell to short bursts of electrical pulses induces structural rearrangements of the lipid bilayer by changing the voltage gradient across the cell membrane. Transient holes, or electropores, that cross through the bilayer form at various points and allow for the passive exchange of material inside the cell with material outside the cell. The pores appear to be lined with hydrophilic, polar lipid heads that have temporarily turned inward. The hydrophilic nature of these electropores attracts hydrophilic, polar molecules on either side of the disrupted membrane. Results in laboratory experiments suggest that preexisting channels and pumps temporarily malfunction so that nonselective material exchange occurs at these sites as well. If the structural rearrangement is not extensive, the cell recovers and the electropores disappear as the lipid bilayer reanneals together and the channels and pumps regain their selectivity. If the current is too strong, the contents of the cell leak into the environment as cell lysis and then death occurs. To date, electroporation has been used to introduce genetic material into mouse cells and fungi. Currently this method is being used in clinical studies to deliver in vivo chemotherapeutic drugs to tumor cells.

Cell-to-cell communication is critical for the survival of multicellular organisms. A cell must work in concert with other cells to know when it is time to divide, to differentiate, or to die. The cell membrane plays an integral part in this communication. In some cases, adjacent cells form gap junctions characterized by pores that permit molecules to move freely from one cytoplasm to another. Cardiac cells in humans have gap junctions that allow for the coordinated contraction of the heart muscle. Some cells release signal molecules that travel quite a distance to exert an effect on a target cell. In order for the cell signal to be detected, a cell must have a membrane-bound protein called a receptor that recognizes the specific signal. These receptors are on the surface of the bilayer facing outward. When the signal molecule, called a ligand, binds to its specific receptor, the receptor transfers the message via a conformational change, leading to a domino effect within the cell itself as it responds to the signal. Neurotransmitters and hormones exert their effects on cells in this type of cell signaling.

WAXES

A fourth type of lipid is waxes, stable solids that have a low melting temperature. These lipids are made

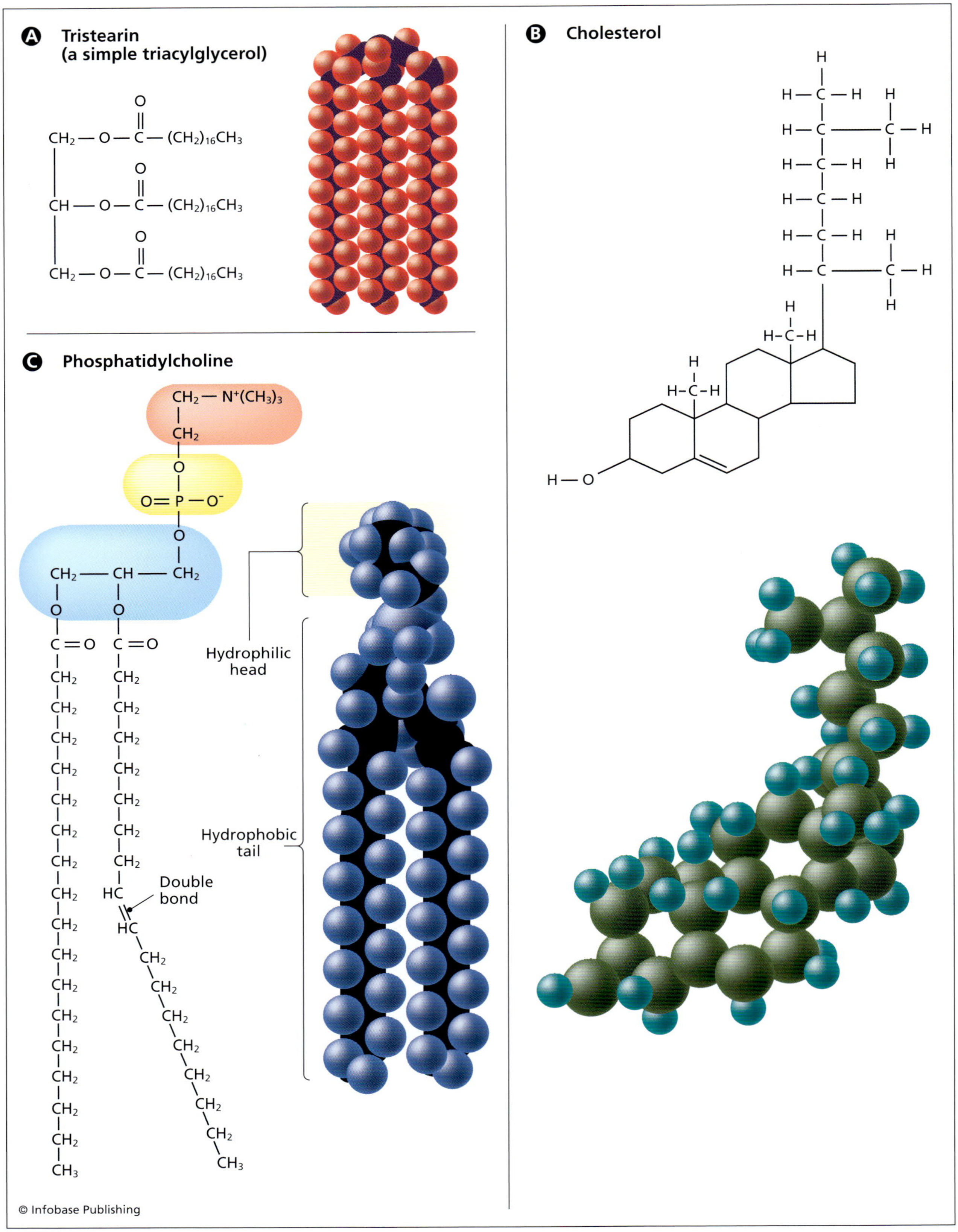

A Tristearin
(a simple triacylglycerol)

B Cholesterol

C Phosphatidylcholine

Hydrophilic head

Hydrophobic tail

Double bond

Lipids are biomolecules that include triglycerides, steroids, and phospholipids. (a) Tristearin is a triglyceride consisting of a glycerol backbone and three long-chain fatty acids. (b) Cholesterol is the precursor molecule of all steroids. (c) As all phospholipids do, phosphatidyl choline consists of a hydrophilic region that contains a phosphate group and a hydrophobic region that contains two long-chain fatty acids.

from long-chain fatty acids and long-chain alcohols. The hydrocarbon chains in both the fatty acids and the alcohols that combine to form waxes are 10 to 30 carbons. Found in both plants and animals, waxes protect an organism from water loss and prohibit the attack of microorganisms in many plants by providing a barrier between the plant and its environment. Beeswax, the building material of honeycombs, is a wax consisting of myricyl alcohol and palmitic acid, which combine to form myricyl palmitate. Carnauba wax, taken from the leaves of the South American palm tree, is used to make car wax and floor polish.

See also BIOCHEMISTRY; CARBOHYDRATES; FATTY ACID METABOLISM; NUCLEIC ACIDS; ORGANIC CHEMISTRY; PROTEINS.

FURTHER READING

Neumann, Eberhard. "The Relaxation Hysteresis of Membrane Electroporation." In *Electroporation and Electrofusion in Cell Biology,* edited by E. Neumann, A. E. Sowers, and C. A. Jordan, 61–82. New York: Plenum Press, 1989.

Watters, Chris. "Lipids." In *Biology*. Vol. 3, edited by Richard Robinson. New York: Macmillan Reference USA, 2002.

———. "Plasma Membrane." In *Biology*. Vol. 3, edited by Richard Robinson. New York: Macmillan Reference USA, 2002.

liquid crystals Liquid crystals are compounds that have an ordered crystalline array to their liquid form rather than a nonordered form that results when most solids melt into liquids. Friedrich Reinitzer, an Austrian chemist, discovered liquid crystals in 1888, when he was studying the properties of cholesterol benzoate, and realized that, upon heating, the cholesterol appeared to have two melting points. At one temperature, 293°F (145°C), the cholesterol transitioned from a solid to a cloudy liquidlike phase, and then upon further heating, at 354°F (179°C), the cloudy phase turned into a clear liquid phase. Reinitzer conferred with the German physicist Otto Lehmann, who was a leading expert on phase changes. They concluded this cloudy phase represented a new phase of matter that has since been observed in many substances and termed this state liquid crystal. The apparent orderliness of the particles in what should be a disordered liquid is the fundamental characteristic of liquid crystals. Changes in temperature, pressure, or electric or magnetic fields strongly affect the properties of compounds in this state. Their unique properties allow liquid crystals to be used in liquid crystal displays (LCDs) in many electronic devices including thermometers, watches, and even laptop computer screens.

A mesogen is the fundamental part of the liquid crystal molecule that confers the liquid crystal properties on it. Liquid crystals can be divided into thermotropic and lyotropic liquid crystals. Thermotropic liquid crystals, such as those used in thermometers with color readouts, change phase upon a change in temperature. Lyotropic liquid crystals change phase on the basis of concentration as well as temperature. The uses for the different types of liquid crystals are varied. After the discovery of liquid crystals, scientists continued to characterize the substances and to discover all of their properties, but the applications were still elusive. In the 1960s, Pierre-Gilles de Gennes, a French physicist, received the Nobel Prize in physics for his work on the applications of liquid crystals. De Gennes described the relationship between the structure and orientation of the liquid crystal substance and the rotation of polarized light that passes through them, thus paving the way for the use of liquid crystals in computer screens, calculators, watches, and more, in the form of (LCDs).

The molecules in a liquid crystal, known as mesogens, are lined up relative to perfect order in a solid (known as the director) and measured in terms of their deviation from the director, known as an order parameter (S). Mathematically defined,

$$S = \tfrac{1}{2}\,(3\cos^2\theta - 1)$$

where θ is the average angle value from the director. The value for S is close to 1 if the substance is a crystalline solid, and the value of S is close to 0 if it is a true liquid. Liquid crystals have S values between 0.3 and 0.9.

A convenient method of characterizing liquid crystals depends on their orientational order and their positional order. Orientational order describes how likely it is that the mesogen will be aligned along the director. Positional order determines the likelihood of showing translational symmetry that is characteristic of crystalline structures, such as being in layers or rows.

The positional order divides liquid crystals into two different phases dependent on their structures and arrangements known as nematic and smectic. Nematic phase is closer to the liquid phase. Nematic crystals are still ordered relative to the director, yet they float around as in a liquid and do not have any layering order. The smectic phase is more like the solid phase with molecules ordered in layers and with some variability within the layers.

Within the nematic and smectic phases, orientational order creates different classifications. There are multiple classifications of the smectic phase. In the smectic A phase, the particles lie perpendicular to the layer. The smectic C phase has the particles

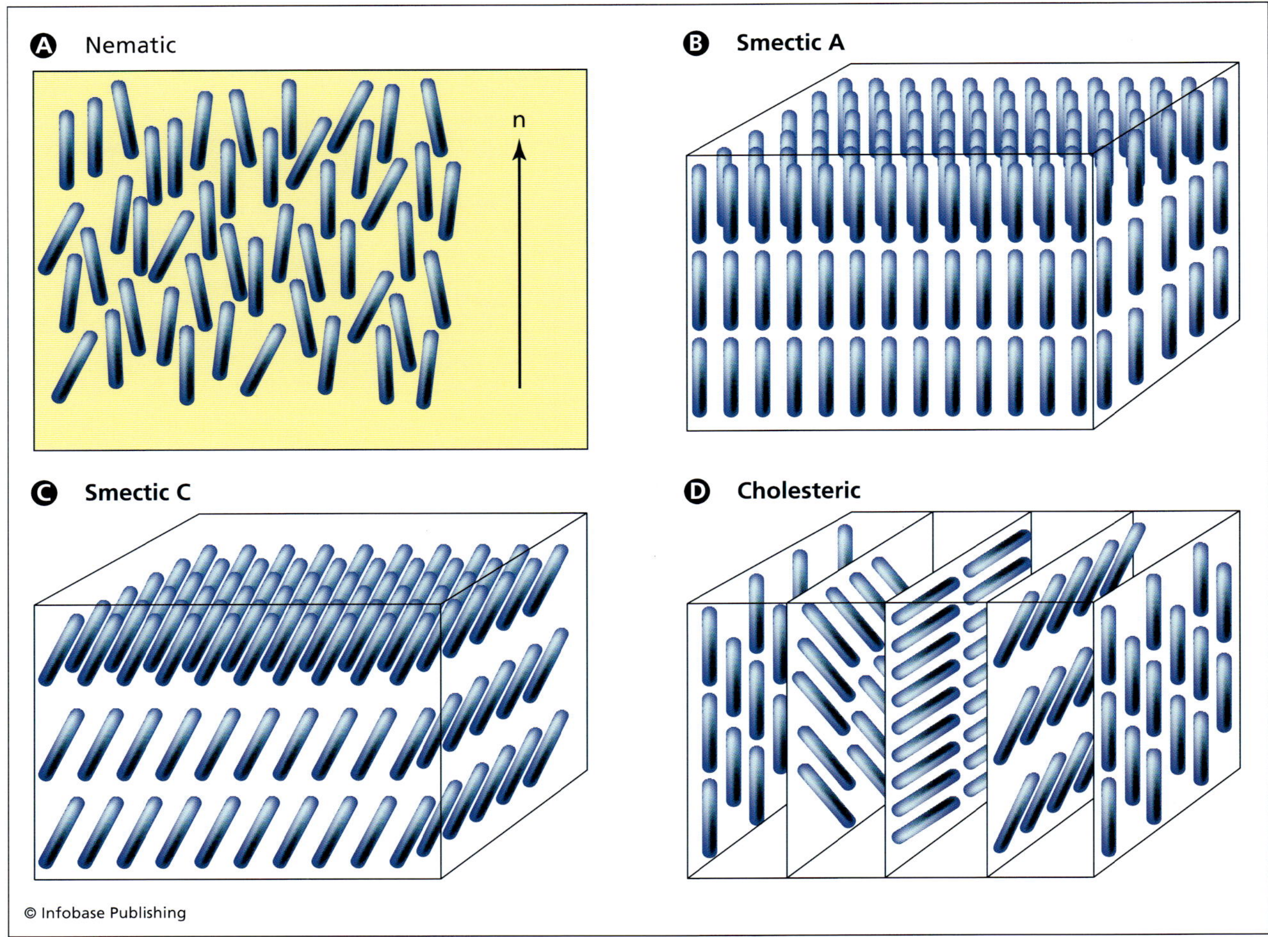

Liquid crystals are substances that are not quite solid or liquid and possess properties characteristic of both phases. (a) The nematic phase of liquid crystals shows parallel order between the crystals. (b), (c) The smectic phases contain a further type of order as they demonstrate organization into layers of ordered planes. (d) In the cholesteric form of the nematic liquid crystal phase, the ordered planes are rotated with respect to each other, forming a twisted structure.

at an angle to the layer, yet they are all in the same direction. In the smectic phase of antiferroelectric compounds (compounds that act just as magnets but have electric fields instead of magnetic fields), the layers have particles that are tilted relative to the direction of the layer, and each layer has an opposite tilt from the adjacent layers. Particles in a liquid crystal of strongly chiral molecules, non-superimposable mirror images, are classified as smectic C. The chiral nature of the particles causes them to change orientation continuously, resulting in a twisted structure. This type of nematic phase is known as the cholesteric phase. The chiral nature of the molecule causes the orientation to change as earlier, leading to a twisted orientation. Chiral molecules float around as in a liquid with an ordered orientation. This twisted orientation of liquid crystal molecules gives them different colors, and since changing the temperature changes the orientation, they can be easily adapted for use in thermometers. Instant-read forehead thermometers for children contain this type of liquid crystal. After placement on the skin, the temperature of the skin determines the orientation of the liquid crystal molecules and therefore the color of the temperature indicator on the thermometer.

LIQUID CRYSTAL DISPLAYS (LCDs)

Liquid crystal displays are based on the ability of liquid crystals to rotate light, especially when in the twisted nematic or cholesteric phase. The liquid crystal components are placed between two flat pieces of glass, allowing LCDs to be much lighter and thinner than displays using a traditional cathode-ray tube. LCDs also run on much less power than traditional displays. Most LCDs utilize the twisted nematic

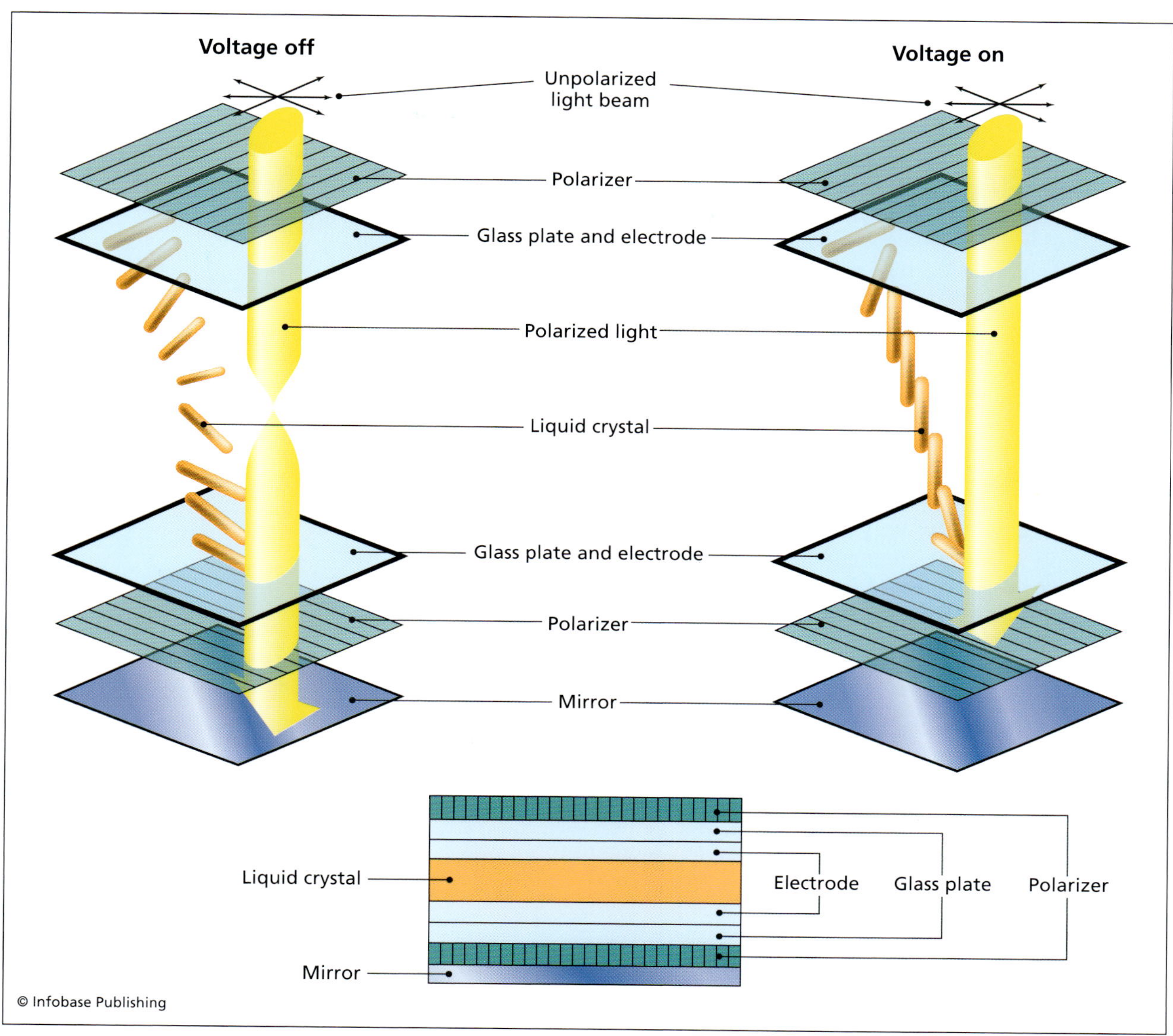

A liquid crystal display (LCD) contains a liquid crystal between two parallel glass plates flanked by crossed polarizers (i.e., polarizers whose axes are perpendicular to each other). When unpolarized light passes through the first polarizer, the light becomes plane-polarized, and, unless the polarization is rotated by the liquid crystal, the beam is blocked by the second polarizer. When the liquid crystal is subjected to a voltage, it does not rotate the plane-polarized light and the screen remains dark. In the absence of a voltage, the liquid crystal rotates the light beam's polarization, the beam then passes through the second polarizer, and the screen appears light.

phase of liquid crystals between two polarizers with perpendicular orientation.

When ordinary light is sent through a vertical polarizer, only the light that is in the same direction as the slits in the polarizer will pass through. Therefore, the light leaving is all in the same direction. When this polarized light passes through the second (horizontal) polarizer, it will not be able to pass through. In an LCD, a layer of liquid crystal is between these polarizers and, when in its twisted phase, rotates the polarized light 90 degrees, directing it through the second polarizer, which is horizontal. Only the rotated light can be reflected back and appear bright on the screen of the watch, thermometer, or computer screen. When the electric field on the liquid crystal changes, the liquid crystal substance changes its orientation and can be made to align with the incoming light, so it will not rotate the vertically polarized light and the light will not be able to pass through the second polarizer; thus that screen or area of the screen will appear dark. Changing which circuits receive electricity controls which portions of the display are light and dark, creating an image.

Compounds in the liquid state that have an ordered arrangement, rather than the disordered arrangement of standard liquids, are known as liquid crystals. They are utilized in many types of liquid crystal displays, as shown here in the LCD monitor. *(jamalludin, 2008, used under license from Shutterstock, Inc.)*

The proper function of an LCD depends on the reaction of the liquid crystal material to the input of current. *Addressing* is the term that describes how the electrical current is supplied to the LCD. *Direct addressing* means that each individual pixel requires its own circuit so that it is either on or off. Most watches and calculators utilize direct addressing. The display of each number depends on which area is lit. For example, a number 3 on a calculator is simply a number 8 without two of the pixels lit. Direct addressing works fine for simpler systems, but when the screens get larger and the number of pixels involved increases, it is difficult for each pixel to have its own circuit. Multiplex addressing prevents this problem: each pixel is referenced by its row and column rather than run by its own circuit.

Multiplex addressing can be either passive or active. Passive LCDs utilize two perpendicular electrodes that can turn on and off any one pixel depending on its location by row and column with a faster response time than direct addressing. Multiple pixels can be on and off at the same time. Active matrix displays have the electrodes behind the LCD and are viewable at multiple angles, in contrast to passive matrix displays. All of these types of displays require some form of backlighting in order to utilize the displays fully. Backlighting requires much energy and is usually minimized on battery-operated devices in order to save on power. Cell phones and watches have backlights that can be turned on and off when needed. Future areas of research on LCDs include angle-of-view problems and increasing the response time in order to maximize the usefulness of liquid crystal materials.

See also OPTICS; STATES OF MATTER.

FURTHER READING

Collings, P. J., and M. Hird. *Introduction to Liquid Crystals: Chemistry and Physics.* Philadelphia: Taylor & Francis, 1997.

Strobl, Gert. *Condensed Matter Physics: Crystals, Liquids, Liquid Crystals, and Polymers.* New York: Springer, 2004.

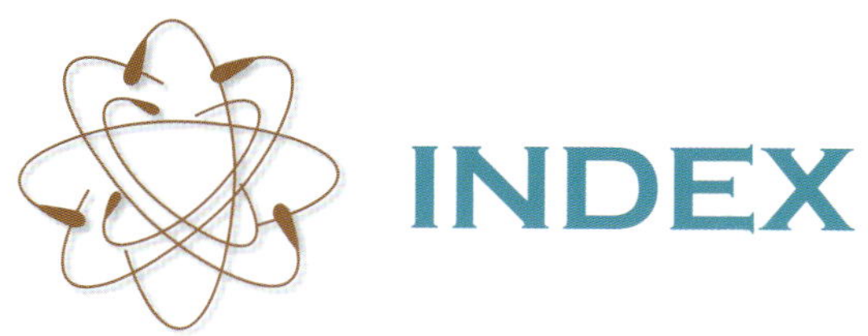

INDEX

Note: Page numbers in **boldface** indicate main entries; *italic* page numbers indicate photographs and illustrations.

C

D

E

N

O

Q

T

U

V